Urban Watershed Microbiology, Volume 1

Sonia M. Tiquia-Arashiro • Gertjan Medema
Eunice Ubomba-Jaswa • Hidetoshi Urakawa
Victor Ella • Nelson Libardi
Editors

Urban Watershed Microbiology, Volume 1

Metagenomic Insights and Resistance Patterns, and Monitoring Approaches

Springer

Editors
Sonia M. Tiquia-Arashiro
Department of Natural Sciences
University of Michigan-Dearborn
Dearborn, MI, USA

Gertjan Medema
Civil Engineering and Geosciences
Delft University of Technology
Delft, The Netherlands

Eunice Ubomba-Jaswa
Water Research Commission
Pretoria, South Africa

Hidetoshi Urakawa
Department of Marine and Ecological Sciences
Florida Gulf Coast University
Fort Myers, FL, USA

Victor Ella
Land and Water Resources Engineering Division, Institute of Agricultural and Biosystems Engineering
University of the Philippines Los Baños
Laguna, Philippines

Nelson Libardi
Department of Sanitary and Environmental Engineering
Universidade Federal de Santa Catarina
Santa Catarina, Brazil

ISBN 978-3-031-89070-3 ISBN 978-3-031-89071-0 (eBook)
https://doi.org/10.1007/978-3-031-89071-0

This Springer imprint is published by the registered company Springer Nature Switzerland AG
The registered company address is: Gewerbestrasse 11, 6330 Cham, Switzerland

Preface

The rapid pace of urbanization is transforming natural environments worldwide, driving profound changes in biological communities and ecological functions. These transformations arise both from direct alterations to local environmental conditions and from the reduced capacity of species to thrive within increasingly fragmented habitats. Urban environments, with their distinctive atmospheric chemistry, geochemistry, climate, hydrology, and vegetation cover, depart dramatically from their natural counterparts, creating unique ecosystems whose dynamics and intricacies are still not fully understood.

Recognizing the urgency of understanding how urbanization influences biodiversity and ecosystem functioning, this work focuses on an often-overlooked aspect: microbial responses to urbanization across diverse spatial scales. Urban watersheds are central to this discussion. They provide essential ecosystem services, such as water purification, degradation of xenobiotics, and nutrient recycling, services largely mediated by microbial communities. Yet, these systems are increasingly stressed by pollution and urban sprawl, culminating in the condition often referred to as the "urban watershed syndrome."

Despite advances in watershed science, several knowledge gaps remain. Microbial communities in urban streams, crucial to global biogeochemical cycles, remain understudied. Urban runoff introduces microbes from diverse sources, posing risks that include the spread of pathogens, disruption of aquatic food webs, and genetic exchanges that may accelerate antimicrobial resistance (AMR). Human activities create pathways that enable urban- and human-derived microbes to infiltrate aquatic systems, historically leading to disease outbreaks. Moreover, microbes are significant contributors to the spread of multidrug resistance, underscoring the urgent need to understand the sources, mechanisms, and ecological contexts of antibiotic resistance genes (ARGs) in urban watersheds.

Watershed Microbiology: Biodiversity Composition, Ecological Function, and Technological Advances is the result of a truly global collaboration with 40 contributions from authors across five continents (Africa, Asia, Europe, North America, and South America) and 24 countries. The breadth of perspectives enriches the exploration of critical and unresolved questions in urban watershed microbiology.

To present the scope of work effectively, the book is published as a two-volume set, each with its own thematic focus and subtitle.

Volume 1: Metagenomic Insights and Resistance Patterns, and Monitoring Approaches

Parts I–III

- *Part I: Metagenomics and Molecular Biology in Watershed Microbial Community Studies*
  This Part synthesizes studies illustrating the application of advanced molecular tools to watershed microbiology. Santiago-Rodriguez et al. highlight viral metagenomics for tracing fecal contamination using crAssphage and Torque teno virus (TTV). Ifon et al. examine how dissolved and anthropogenic organic matter shape microbial communities in urban rivers. Ng et al. and Zemskaya et al. investigate microbial biodiversity in lakes, reservoirs, and sedimentary layers using amplicon sequencing and metabarcoding. Ma et al. assess land cover effects on dissolved organic matter and microbial communities, while Te et al. explore cyanobacterial harmful algal blooms (CHABs) in tropical urban catchments, emphasizing microbial interactions in bloom dynamics and nutrient cycling.
- *Part II: Urban Development and the Occurrence and Persistence of Antimicrobial Resistance and Microbial Source Tracking Technologies*
  This Part examines the link between urban development and AMR spread, as well as the use of microbial source tracking (MST). Henriquez and Marvasi show how urbanization and industrialization intensify AMR spread. Ghosh Roy and Mester highlight microbial bioremediation using *Bacillus* and *Pseudomonas*. Van Wyk and Ubomba-Jaswa discuss advanced treatment technologies for AMR in aquaculture and wastewater. Catalfano et al. explore AMR in *Escherichia coli* from urban watersheds. Book et al. review ARG detection advances, advocating integrated sequencing and biosensor approaches. Urakawa and Hancock demonstrate Quanti-tray-based amplicon sequencing (QT-AMP) for detecting fecal indicator bacteria after Hurricane Ian.
- *Part III: Advances in Biological Monitoring and Modeling in Urban Watersheds and Community Science Monitoring*
  Here, the focus is on monitoring innovations. Acuña-Alonso et al. use Partial Least Squares Path Modeling (PLS-PM) and satellite imagery to predict harmful algal blooms (HABs). Lobo et al. employ the Algae Bloom Monitoring application (AlgaeMAp) for Brazilian water quality assessments. Paulino et al. review

remote sensing in aquatic microbiology. Ella discusses simulation models such as SWAT and HSPF, as well as deep learning and IoT integration. Tolin presents a case study on citizen science mussel monitoring in the Huron River Watershed, illustrating public engagement in aquatic monitoring.

Volume 2: Environmental Indicators, Regional Case Studies, and Bioremediation Strategies

Parts IV–VI

- *Part IV: Microbial Response Indicators of Water Quality and Aquatic Ecosystems Change*
  This Part investigates microbial bioindicators for water quality assessment. Jiang et al. link climate change to intensified cyanobacterial blooms. Murrella et al. study enzymatic responses to fecal and chemical pollutants. Hancock and Urakawa evaluate hydrogen peroxide as a predictive tool for blooms. Gao et al. review benthic biofilm roles in nutrient cycling. Patel et al. use Biolog EcoPlate™ to assess carbon utilization shifts. De Mello et al. advocate integrating microbial metrics with biological indicators. Leareng et al. emphasize microbes as bioindicators of chemical pollution. Otim and Otim analyze climate indicator–algal growth correlations in California inland waters, revealing habitat-specific variability.
- *Part V: Regional and Country-Specific Perspectives on Urban Watershed Health Worldwide*
  This Part offers global case studies. Zhao and Ni (China) discuss nitrogen cycle disruptions. Gorshkova et al. (Russia) examine Lake Baikal tributary pollution. Ogola and Odhiambo (Kenya) address Lake Victoria challenges. Rodríguez-López et al. (Chile) apply remote sensing to study turbidity in Lake Maihue. Krishna et al. (India) assess extreme weather impacts on Vembanad Lake contamination. Cortez and Siringan (Philippines) call for updated microbial water criteria. Karmakar et al. (Bangladesh) emphasize conservation in the Chittagong Hill Tracts.
- *Part VI: Emerging Contaminants and Bioremediation Strategies*
  This Part addresses microbial–pollutant interactions. Bai et al. review steroid estrogen biodegradation and NOM interactions. Garcia et al. focus on hydrogen peroxide's role in contaminant breakdown. Kuroda et al. explore anaerobic treatment for purified terephthalic acid (PTA). Sanchez et al. identify bacterial candidates for organic contaminant biodegradation. Opara et al. examine metal biosorption mechanisms. Chamlee et al. study siderophore production for heavy metal mitigation.

Together, these two volumes present a comprehensive, state-of-the-art synthesis of research, practical applications, and global perspectives in urban watershed microbiology. They are intended as essential references for microbiologists, ecologists, hydrologists, environmental engineers, and public health professionals, as well as for policymakers seeking to safeguard water resources in an era of rapid urban change.

Dearborn, MI, USA — Sonia M. Tiquia-Arashiro
Delft, The Netherlands — Gertjan Medema
Pretoria, South Africa — Eunice Ubomba-Jaswa
Fort Myers, FL, USA — Hidetoshi Urakawa
Laguna, Philippines — Victor Ella
Santa Catarina, Brazil — Nelson Libardi

Contents of Volume 1

Contents of Volume 2

Editors and Contributors

About the Editors

Sonia M. Tiquia-Arashiro is a Professor of Microbiology at the University of Michigan-Dearborn, with primary research interests in biotechnology for clean water, renewable energy, and healthy ecosystems. Her broader interests include the analysis and management of complex adaptive systems, as well as microbiology education and science literacy. She serves as Associate Editor for *Environmental Technology* and *Frontiers in Microbiology*, and continues to contribute to global discussions on water sustainability, environmental microbiology, and biotechnology.

Gertjan Medema is a part-time Chair on Water & Health at Delft University of Technology and Principal Microbiologist at Kiwa Watercycle Research (KWR) Water Research Institute. He has expertise in waterborne pathogen detection and epidemiology, leading research on SARS-CoV-2 sewage surveillance and investigating a *Legionella* outbreak in Flint, Michigan. Since 1991, he has advised the WHO on water microbiological guidelines and contributed to responses during outbreaks like SARS and Ebola. He also advises the European Commission on water guidelines and directs KWR's WHO Collaborating Centre on Water Quality.

Hidetoshi Urakawa is a Professor of Environmental Science at Florida Gulf Coast University and a Courtesy Professor at the University of South Florida. His research covers aquatic ecology, environmental microbiology, and ecosystem processes, with a focus on the nitrogen cycle, cyanobacteria dynamics, and marine species feeding ecology. He also works in aquaculture, biotechnology, and nutrient reduction technologies. Dr. Urakawa is the Editor-in-Chief of the *International Journal of Microbiology*.

Nelson Libardi is a faculty member at the Federal University of Santa Catarina (UFSC) in Brazil, where he focuses on research and academic activities related to bioprocesses. He holds a PhD in Bioprocess Engineering and Biotechnology from the Federal University of Paraná. His expertise is in environmental and industrial biotechnology, specializing in biological processes for wastewater treatment and resource recovery, with a strong emphasis on developing sustainable biotechnological solutions to improve environmental practices.

Eunice Ubomba-Jaswa is a Research Manager in Water Resources Quality & Management at the Water Research Commission in Pretoria, South Africa. She specializes in water quality and pollution control, focusing on sustainable environmental remediation technologies. Her research covers wastewater treatment, waterborne pathogens, and the broader impacts of pollution on health and ecosystems. Dr. Ubomba-Jaswa is an advocate for water sustainability and has contributed to peer-reviewed journals and international conferences on environmental science and policy.

Victor Ella is a Professor of Land and Water Resources Engineering at the University of the Philippines Los Baños. His research focuses on hydrologic modeling, groundwater hydrology, water quality, irrigation engineering, and soil conservation, with broader interests in contaminant transport, erosion, soil physics, water resources planning, climate change, IoT-based irrigation, and watershed-scale hydrologic modeling.

Contributors

Carolina Acuña-Alonso Universidade de Vigo, Hydro-Forestry Geomodeling Research Group, School of Forestry Engineering, Pontevedra, Spain

Centre for the Research and Technology of Agro-Environmental and Biological Sciences – CITAB, University of Trás-os-Montes and Alto Douro (UTAD), Vila Real, Portugal

Bob Adyari Department of Environmental Engineering, Universitas Pertamina, Jakarta, Indonesia

Beatriz Romeu Alvarez Department of Microbiology and Virology, Biology Faculty, University of Havana, Havana, Cuba

Xana Álvarez Universidade de Vigo, Hydro-Forestry Geomodeling Research Group, School of Forestry Engineering, Pontevedra, Spain

Abdulaziz Anas CSIR-National Institute of Oceanography, Regional Centre Kochi, Kochi, India

Academy of Scientific and Innovative Research (AcSIR), Ghaziabad, India

Keerthi Sri Senarathna Atapaththu Department of Limnology and Water Technology, Faculty of Fisheries and Marine Sciences & Technology, University of Ruhuna, Matara, Sri Lanka

Marcia María Rojas Badía Department of Microbiology and Virology, Biology Faculty, University of Havana, Havana, Cuba

Leilei Bai State Key Laboratory of Lake and Watershed Science for Water Security, Nanjing Institute of Geography and Limnology, Chinese Academy of Sciences, Nanjing, China

Sakcham Bairoliya School of Civil and Environmental Engineering, Nanyang Technological University, Singapore, Singapore

Singapore Centre for Environmental Life Sciences Engineering, Nanyang Technological University, Singapore, Singapore

Misha Bakhsh Department of Natural Sciences, University of Michigan-Dearborn, Dearborn, MI, USA

Denisse Balbina Faculty of Sciences, University Santo Tomas, Concepción, Chile

Tarit Kumar Baul Institute of Forestry and Environmental Sciences, University of Chittagong, Chattogram, Bangladesh, 4331

Karina Boltes Universidad de Alcalá, Departamento de Química Analítica, Química Física e Ingeniería Química, Alcalá de Henares, Madrid, Spain

IMDEA Water, Alcalá de Henares, Madrid, Spain

Reeghan Book Department of Natural Sciences, University of Michigan-Dearborn, Dearborn, MI, USA

Lisandra Bravo Alvarez Department of Electrical Engineering, Universidad de Concepción, Concepción, Chile

Laura Martins Bueno Center for Technological Development (CDTec), Universidade Federal de Pelotas (UFPel), Pelotas, Brazil

Sergei V. Bukin Limnological Institute, Siberian Branch of the Russian Academy of Sciences, Irkutsk, Russia

Bin Cao School of Civil and Environmental Engineering, Nanyang Technological University, Singapore, Singapore

Singapore Centre for Environmental Life Sciences Engineering, Nanyang Technological University, Singapore, Singapore

Bryan Catalfano Department of Natural Sciences, University of Michigan-Dearborn, Dearborn, MI, USA

Nikhil Chakma Department of Forestry and Environmental Sciences, Rangamati Science and Technology University, Rangamati, Bangladesh 4500

Yanesa Chamlee Department of Natural Sciences, University of Michigan-Dearborn, Dearborn, MI, USA

Svetlana M. Chernitsyna Limnological Institute, Siberian Branch of the Russian Academy of Sciences, Irkutsk, Russia

Luis Angelo A. Cortez Department of Agriculture—Bureau of Soils and Water Management, Soil and Water Resources Research Division, Quezon City, Philippines

Oliver Coutinho Department of Natural Sciences, University of Michigan-Dearborn, Dearborn, MI, USA

Iongel Duran-Llacer Hémera Centro de Observación de la Tierra, Facultad de Ciencias, Ingeniería y Tecnología, Universidad Mayor, Huechuraba, Santiago, Chile

Danielle de Almeida Bressiani Center for Technological Development (CDTec), Universidade Federal de Pelotas (UFPel), Pelotas, Brazil

Lino Augusto Sander de Carvalho Department of Meteorology, Federal University of Rio de Janeiro (UFRJ), Rio de Janeiro, Brazil

Gilberto Queiroz de Lima Filho Agência Pernambucana de Águas e Clima (APAC), Recife, Brazil

Patricio de los Rios Universidad Católica de Temuco, Temuco, Araucania, Chile

Felipe de Lucia Lobo Center for Technological Development (CDTec), Universidade Federal de Pelotas (UFPel), Pelotas, Brazil

Kaline de Mello Research Institute for the Environment and Livelihoods, Charles Darwin University, Casuarina, NT, Australia

CSIRO Environment, Berrimah, NT, Australia

Victor Ella Land and Water Resources Engineering Division, Institute of Agricultural and Biosystems Engineering, College of Engineering and Agro-industrial Technology, University of the Philippines Los Baños, College, Laguna, Philippines

Nathalie Fagel UR Argile, Geochimie et Environment sedimentary (AGEs), Geology Department University of Liege, Liege, Belgium

James Farinas Department of Natural Sciences, University of Michigan-Dearborn, Dearborn, MI, USA

Mitchell Fenner Department of Natural Sciences, University of Michigan-Dearborn, Dearborn, MI, USA

Luís Filipe Sanches Fernandes Centre for the Research and Technology of Agro-Environmental and Biological Sciences – CITAB, University of Trás-os-Montes and Alto Douro (UTAD), Vila Real, Portugal

Aleksa Fortuna Department of Natural Sciences, University of Michigan-Dearborn, Dearborn, MI, USA

Huimin Gao Department of Geography and Spatial Information Techniques, Ningbo University, Ningbo, China

Erica A. Garcia Research Institute for the Environment and Livelihoods, Charles Darwin University, Casuarina, NT, Australia

Patricia E. Garcia Grupo de Ecología de Sistemas Acuáticos a escala de Paisaje (GESAP) INIBIOMA, Universidad Nacional del Comahue, CONICET, San Carlos de Bariloche, Argentina

Grinson George ICAR-Central Marine Fisheries Research Institute, Kochi, India

Subhomita Ghosh Roy Department of Biological Sciences, Northern Kentucky University, Highland Heights, KY, USA

Karina Yew-Hoong Gin Department of Civil and Environmental Engineering, National University of Singapore, Singapore, Singapore

Adrian Salcedo Gómez Department of Microbiology and Virology, Biology Faculty, University of Havana, Havana, Cuba

Anna S. Gorshkova Limnological Institute, Siberian Branch of the Russian Academy of Sciences, Irkutsk, Russian Federation

Zeinab Hammoud Department of Natural Sciences, University of Michigan-Dearborn, Dearborn, MI, USA

Taylor L. Hancock Department of Ecology and Environmental Studies, Florida Gulf Coast University, Fort Myers, FL, USA

School of Geosciences, University of South Florida, Tampa, FL, USA

Department of Biology, Doane University, Crete, NE, USA

Tania Henriquez Department of Medical Biotechnologies, University of Siena, Siena, Italy

Emily B. Hollister Sandia National Laboratories, Albuquerque, NM, USA

Zhicheng Hong College of Environment and Ecology, Key Laboratory of the Three Gorges Reservoir Region's Eco-Environment, Ministry of Education, Chongqing University, Chongqing, China

Anyi Hu Carbon Neutral Innovation Research Center, Xiamen University, Xiamen, China

Binessi Edouard Ifon College of Engineering, Shantou University, Shantou, China

CAS Key Laboratory of Urban Pollutant Conversion, Institute of Urban Environment, Chinese Academy of Sciences, Xiamen, China

Laboratory of Physical-Chemistry, Materials and Molecular Modeling (LCP3M), University of Abomey-Calavi, Republic of Benin, Cotonou, Benin

Rino Isshiki Biomanufacturing Process Research Center, National Institute of Advanced Industrial Science and Technology (AIST), Sapporo, Hokkaido, Japan

Integrated Research Center for Circular Technology, National Institute of Advanced Industrial Science and Technology (AIST), Sapporo, Hokkaido, Japan

Mohammed Jashimuddin Institute of Forestry and Environmental Sciences, University of Chittagong, Chattogram, Bangladesh, 4331

Helong Jiang State Key Laboratory of Lake and Watershed Science for Water Security, Nanjing Institute of Geography and Limnology, Chinese Academy of Sciences, Nanjing, China

Sunny Jiang Department of Civil and Environmental Engineering, University of California-Irvine, Irvine, CA, USA

Vaughn Karaim Department of Natural Sciences, University of Michigan-Dearborn, Dearborn, MI, USA

Shyamal Karmakar Institute of Forestry and Environmental Sciences, University of Chittagong, Chattogram, Bangladesh, 4331

Diana Kasperek Department of Natural Sciences, University of Michigan-Dearborn, Dearborn, MI, USA

Amin Kassem Department of Natural Sciences, University of Michigan-Dearborn, Dearborn, MI, USA

Victoria Khamis Department of Natural Sciences, University of Michigan-Dearborn, Dearborn, MI, USA

Sangar Khan Department of Geography and Spatial Information Techniques, Ningbo University, Ningbo, China

Jerome Wai Kit Kok NUS College, National University of Singapore, Singapore, Singapore

R. Keller Kopf Research Institute for the Environment and Livelihoods, Charles Darwin University, Casuarina, NT, Australia

Kiran Krishna CSIR-National Institute of Oceanography, Regional Centre Kochi, Kochi, India

Academy of Scientific and Innovative Research (AcSIR), Ghaziabad, India

Widana Gamage Shirani Manel Kumari Department of Botany, Faculty of Science, University of Ruhuna, Matara, Sri Lanka

Kyohei Kuroda Biomanufacturing Process Research Center, National Institute of Advanced Industrial Science and Technology (AIST), Sapporo, Hokkaido, Japan

Division of Environmental Engineering, Faculty of Engineering, Hokkaido University, Sapporo, Hokkaido, Japan

Integrated Research Center for Circular Technology, National Institute of Advanced Industrial Science and Technology (AIST), Sapporo, Hokkaido, Japan

Samuel K. Leareng Department of Microbiology, Stellenbosch University, Stellenbosch, South Africa

School of Data Science and Computational Thinking, Stellenbosch University, Stellenbosch, South Africa

Sheng-Ao Li College of Environment and Ecology, Key Laboratory of the Three Gorges Reservoir Region's Eco-Environment, Ministry of Education, Chongqing University, Chongqing, China

Xin Liu College of Biology and Environment, Nanjing Forestry University, Nanjing, China

Anna V. Lomakina Limnological Institute, Siberian Branch of the Russian Academy of Sciences, Irkutsk, Russia

Hua Ma College of Environment and Ecology, Key Laboratory of the Three Gorges Reservoir Region's Eco-Environment, Ministry of Education, Chongqing University, Chongqing, China

Daniel Andrade Maciel Instrumentation Lab for Aquatic Systems (LabISA), Earth Observation Coordination of National Institute for Space Research (INPE), São José dos Campos, Brazil

Thulani P. Makhalanyane Department of Microbiology, Stellenbosch University, Stellenbosch, South Africa

School of Data Science and Computational Thinking, Stellenbosch University, Stellenbosch, South Africa

John P. Makumbi DSI/NRF SARChI Chair in Marine Microbiomics, Department of Biochemistry, Genetics and Microbiology, University of Pretoria, Pretoria, South Africa

Valerii V. Malnik Limnological Institute, Siberian Branch of the Russian Academy of Sciences, Irkutsk, Russian Federation

Derek C. Manheim Civil and Environmental Engineering Department, California Polytechnic State University, San Luis Obispo, CA, USA

Vitor S. Martins Department of Agricultural & Biological Engineering, Mississippi State University (MSU), Starkville, MS, USA

Massimiliano Marvasi Department of Biology, University of Florence, Sesto Fiorentino, Florence, Italy

Kanaji Masakorala Department of Botany, Faculty of Science, University of Ruhuna, Matara, Sri Lanka

Nandini Menon Nansen Environmental Research Centre India, KUFOS Amenity Centre, Kochi, India

Joseph C. Mester Department of Biological Sciences, Northern Kentucky University, Highland Heights, KY, USA

Sohag Miah Institute of Forestry and Environmental Sciences, University of Chittagong, Chattogram, Bangladesh, 4331

Suzana Gico Montenegro Centro de Tecnologia, Engenharia Civil, Universidade Federal de Pernambuco, Recife, Brazil

Agência Pernambucana de Águas e Clima (APAC), Recife, Brazil

Daysi Lugo Moya Department of Microbiology and Virology, Biology Faculty, University of Havana, Havana, Cuba

Hongli Mu Department of Geography and Spatial Information Techniques, Ningbo University, Ningbo, China

Jeny Adina Larrea Murrell Department of Microbiology and Virology, Biology Faculty, University of Havana, Havana, Cuba

Gustavo Willy Nagel Ocean and Earth Science, University of Southampton, Southampton, UK

Meri Nakajima Biomanufacturing Process Research Center, National Institute of Advanced Industrial Science and Technology (AIST), Sapporo, Hokkaido, Japan

Division of Environmental Engineering, Faculty of Engineering, Hokkaido University, Sapporo, Hokkaido, Japan

Takashi Narihiro Biomanufacturing Process Research Center, National Institute of Advanced Industrial Science and Technology (AIST), Sapporo, Hokkaido, Japan

Division of Environmental Engineering, Faculty of Engineering, Hokkaido University, Sapporo, Hokkaido, Japan

Integrated Research Center for Circular Technology, National Institute of Advanced Industrial Science and Technology (AIST), Sapporo, Hokkaido, Japan

Zainab Nassereddine Department of Natural Sciences, University of Michigan-Dearborn, Dearborn, MI, USA

Daphne H.P. Ng School of Civil and Environmental Engineering, Nanyang Technological University, Singapore, Singapore

Singapore Centre for Environmental Life Sciences Engineering, Nanyang Technological University, Singapore, Singapore

Shou-Qing Ni School of Environmental Science and Engineering, Shandong University, Qingdao, Shandong, China

Marisa C. Nielsen Department of Pathology and Laboratory Medicine, Boston University Chobanian and Avedisian School of Medicine, Boston, MA, USA

Evlyn M.L.M. Novo Earth Observation and Geoinformatics Division (DIOTG), National Institute for Space Research (INPE), São José dos Campos, SP, Brazil

Kennedy Achieng' Odhiambo School of Biological, Physical and Mathematics Sciences, Jaramogi Oginga Odinga University of Science and Technology, Bondo, Kenya

Adam Oest Department of Natural Sciences, University of Michigan-Dearborn, Dearborn, MI, USA

Henry Joseph Oduor Ogola Department of Environmental Science, University of South Africa, Florida Science Campus, Roodepoort, South Africa

Okugbe Ebiotubo Ohore Ministry of Education Key Laboratory of Tropical Translational Medicine, School of Tropical Medicine, Hainan Medical University, Haikou, People's Republic of China

Natalya A. Onishchuk Limnological Institute, Siberian Branch of the Russian Academy of Sciences, Irkutsk, Russian Federation

Somie Opara Department of Natural Sciences, University of Michigan-Dearborn, Dearborn, MI, USA

Francesca N. Otim Department of Anthropology, University of California, Davis, Davis, CA, USA

Ochan Otim Department of Health and Sciences, University of California – Los Angeles, Los Angeles, CA, USA

Dustin Overfield Department of Natural Sciences, University of Michigan-Dearborn, Dearborn, MI, USA

Fernando António Leal Pacheco Chemistry Centre of Vila Real – CQVR, University of Trás-os-Montes and Alto Douro (UTAD), Vila Real, Portugal

Grace Pagnucco Department of Natural Sciences, University of Michigan-Dearborn, Dearborn, MI, USA

Erica Parmenter Department of Natural Sciences, University of Michigan-Dearborn, Dearborn, MI, USA

Himanshu Patel Department of Natural Sciences, University of Michigan-Dearborn, Dearborn, MI, USA

Rejane S. Paulino Department of Agricultural & Biological Engineering, Mississippi State University (MSU), Starkville, MS, USA

Olga N. Pavlova Limnological Institute, Siberian Branch of the Russian Academy of Sciences, Irkutsk, Russia

Alice L. Petre Universidad de Alcalá, Departamento de Química Analítica, Química Física e Ingeniería Química, Alcalá de Henares, Madrid, Spain

IMDEA Water, Alcalá de Henares, Madrid, Spain

Florent Qyteti Department of Natural Sciences, University of Michigan-Dearborn, Dearborn, MI, USA

Ranith Rajamohananpillai Nansen Environmental Research Centre India, KUFOS Amenity Centre, Kochi, India

Anna E. Richards CSIRO Environment, Berrimah, NT, Australia

Lien Rodríguez-López Facultad de Ingeniería, Lientur 1457, Universidad San Sebastián, Concepción, Chile

Erin Rubley Department of Natural Sciences, University of Michigan-Dearborn, Dearborn, MI, USA

Rasmieh Saab Department of Natural Sciences, University of Michigan-Dearborn, Dearborn, MI, USA

Maria V. Sakirko Limnological Institute, Siberian Branch of the Russian Academy of Sciences, Irkutsk, Russian Federation

Alyssia Sanchez Department of Natural Sciences, University of Michigan-Dearborn, Dearborn, MI, USA

Kelum Sanjaya Department of Limnology and Water Technology, Faculty of Fisheries and Marine Sciences & Technology, University of Ruhuna, Matara, Sri Lanka

Tasha M. Santiago-Rodriguez Baylor College of Medicine, Houston, TX, USA

Shubha Sathyendranath National Centre for Earth Observation, Plymouth Marine Laboratory, Devon, UK

Trevor Platt Science Foundation, KochiIndia

Hisashi Satoh Division of Environmental Engineering, Faculty of Engineering, Hokkaido University, Sapporo, Hokkaido, Japan

Reis Schittenhelm Department of Natural Sciences, University of Michigan-Dearborn, Dearborn, MI, USA

Elena Seoane-Martínez Universidade de Vigo, Hydro-Forestry Geomodeling Research Group, School of Forestry Engineering, Pontevedra, Spain

Claudia Shuler Department of Natural Sciences, University of Michigan-Dearborn, Dearborn, MI, USA

Zhi Yang Sim Department of Civil and Environmental Engineering, National University of Singapore, Singapore, Singapore

Maria Auxilia T. Siringan University of the Philippines, Natural Science Research Institute-Microbiological Research and Services Laboratory, Quezon City, Philippines

Hannah Solanick Department of Natural Sciences, University of Michigan-Dearborn, Dearborn, MI, USA

Qian Sun CAS Key Laboratory of Urban Pollutant Conversion, Institute of Urban Environment, Chinese Academy of Sciences, Xiamen, China

Donald Sweetapple Department of Natural Sciences, University of Michigan-Dearborn, Dearborn, MI, USA

Shu Harn Te Department of Civil and Environmental Engineering, National University of Singapore, Singapore, Singapore

Oleg A. Timoshkin Limnological Institute, Siberian Branch of the Russian Academy of Sciences, Irkutsk, Russian Federation

Sonia M. Tiquia-Arashiro Department of Natural Sciences, University of Michigan-Dearborn, Dearborn, MI, USA

Irina V. Tomberg Limnological Institute, Siberian Branch of the Russian Academy of Sciences, Irkutsk, Russian Federation

Amani Tolin School for Environment and Sustainability, University of Michigan, Ann Arbor, MI, USAGary A. Toranzos Environmental Microbiology Laboratory, Biology Department, University of Puerto Rico, San Juan, PR, USA

Eunice Ubomba-Jaswa Water Research Commission, Pretoria, South Africa

Hidetoshi Urakawa Department of Marine and Ecological Sciences, Florida Gulf Coast University, Fort Myers, FL, USA

Roberto Urrutia Facultad de Ciencias Ambientales, Universidad de Concepci6n, Concepci6n, Chile

Yazeed van Wyk Water Research Commission, Pretoria, South Africa

Department of Geology, University of Pretoria, Pretoria, South Africa

Changhui Wang State Key Laboratory of Lake and Watershed Science for Water Security, Nanjing Institute of Geography and Limnology, Chinese Academy of Sciences, Nanjing, China

Naicheng Wu Department of Geography and Spatial Information Techniques, Ningbo University, Ningbo, China

Yapa Mudiyanselage Ajith Lalith Weerasinghe Yapa Department of Chemistry, Faculty of Science, University of Ruhuna, Matara, Sri Lanka

Chang-Ping Yu CAS Key Laboratory of Urban Pollutant Conversion, Institute of Urban Environment, Chinese Academy of Sciences, Xiamen, China

Adel Zeibo Department of Natural Sciences, University of Michigan-Dearborn, Dearborn, MI, USA

Tamara I. Zemskaya Limnological Institute, Siberian Branch of the Russian Academy of Sciences. Irkutsk, Russia

Dong Zhang Singapore Centre for Environmental Life Science Engineering. Nanyang Technological University, Singapore, Singapore

Yiyi Zhao School of Environmental Science and Engineering, Shandong University, Qingdao, Shandong, China

School of Public Health, Shandong First Medical University and Shandong Academy of Medical Sciences, Jinan, Shandong, China

Eric Zielinkski Department of Natural Sciences, University of Michigan-Dearborn, Dearborn, MI, USA

Part I
Metagenomics and Molecular Biology in Watershed Microbial Community Studies

Chapter 1
Viral Metagenomics: Assessing Potential Sources of Fecal Contamination in Watershed Systems

Tasha M. Santiago-Rodriguez, Emily B. Hollister, and Gary A. Toranzos

1 Introduction

Fecal material originating from point and non-point sources impacts watershed systems and thus affects human, animal, and environmental health (Holt, 2000; Santiago-Rodriguez & Hollister, 2023). Point sources of fecal contamination include those that are easily identifiable (e.g., sewage effluents) (Li et al., 2021a, b), while non-point pollution sources of fecal contamination are often more difficult to identify due to unknown and diffuse origin(s) that can be intermittent (e.g., leaking septic systems, and agricultural runoff) (Hayes et al., 1990; Calderon & Mood, 1991; Nguyen et al., 2013; Shrestha et al., 2020; Pagnucco et al., 2023). Close contact with water contaminated by point and non-point sources of fecal contamination affects millions of people every year through exposure to enteric pathogens such as bacteria, viruses, and protozoans. Collectively known as waterborne pathogens, these microorganisms can be transmitted through the fecal-oral route (Bain et al., 2014; Oon et al., 2023), and cause gastrointestinal (GI) and respiratory illnesses, as well as eye and skin infections (Magana-Arachchi & Wanigatunge, 2020), affecting over one billion people worldwide (UN-Water, 2021). In the USA alone, over seven million infectious waterborne illnesses have been reported, where over 70% of the cases have been attributed to exposure to contaminated recreational waters, and 16% have been attributed to exposure to contaminated drinking water. Annual costs

T. M. Santiago-Rodriguez (✉)
Baylor College of Medicine, Houston, TX, USA
e-mail: tasha.santiagorodriguez@bcm.edu

E. B. Hollister
Sandia National Laboratories, Albuquerque, NM, USA

G. A. Toranzos
Environmental Microbiology Laboratory, Biology Department, University of Puerto Rico, San Juan, PR, USA

S. M. Tiquia-Arashiro et al. (eds.), *Urban Watershed Microbiology, Volume 1*,
https://doi.org/10.1007/978-3-031-89071-0_1

of emergency doctor visits and hospitalizations from exposure to fecally contaminated drinking or recreational water in the USA are estimated to be over $2 billion (Gerdes et al., 2023). Transmission of waterborne pathogens may also result from aging wastewater treatment infrastructure (Damashek et al., 2022), increasing prevalence of treatment-resistant enteric pathogens (Zerva et al., 2021; Nassereddine et al., 2024), and, potentially, climate change (Levy et al., 2018). Reduced water quality due to fecal contamination may also result in economic losses related to increased frequency of food recalls, as well as reduced foodstuff/crop production (Geldreich & Bordner, 1971; Tiquia et al., 1998; Kim et al., 2007; Pachepsky et al., 2011; Garvey, 2019; Garcia et al., 2020), and impacts to fishing and shellfish harvesting (Ralston et al., 2011).

Among the potential enteric pathogens that can be introduced into watersheds and cause disease, viruses pose one of the greatest threats (Bouseettine et al., 2019). Enteric pathogenic viruses, in particular, are associated with gastroenteritis (Jain & Jain, 2014), as well as skin (Gall et al., 2015) and eye infections (Ahmad, 2018). In addition to affecting humans, enteric pathogenic viruses can have adverse effects on animal health, causing a range of diseases and mortality (Fong & Lipp, 2005). Enteric pathogenic viruses are also transmitted through contaminated food and fomites (Rzezutka & Cook, 2004; Alidjinou et al., 2019), and can cause disease at low infectious doses (Teunis et al., 2008; Li et al., 2023). Most enteric viruses are not eliminated by conventional water treatment barriers such as filtration and may be refractive to disinfection under certain conditions, such as the presence of biochemical oxygen demand (BOD)(Montazeri et al., 2015), and are stable over a wide pH range (e.g., pH 3–10, as in the case of most enteroviruses) (Wyn-Jones & Sellwood, 2001). Moreover, certain enteric pathogenic viruses are capable of extended survival periods under specific conditions (e.g., temperature and presence/absence of endogenous microbiota), increasing the risk of exposure (Ward et al., 1986; Moresco et al., 2016). Noroviruses, for instance, are the most common agents of non-bacterial gastroenteritis outbreaks, and their transmission occurs through contact with infected individuals, exposure to contaminated food and water, or to infectious aerosols (De Graaf et al., 2016). While noroviruses circulate all year round, infections peak during winter (Ahmed et al., 2013). In the USA, a total of 24,995 norovirus cases were reported from 2009 to 2019, with season durations ranging from 20 to 36 weeks depending on the US region (Kambhampati et al., 2023).

Detection of enteric pathogenic viruses may include the use of culture and/or molecular methods; however, culture assays can be time-consuming (Malik et al., 2019), and molecular methods are often developed to target specific viral pathogens (Hamza et al., 2011b). Thus, a degree of a priori knowledge of the potential source(s) of fecal contamination is often needed when applying viral culture and molecular identification methods. The need to find untargeted approaches to identify potential sources of fecal contamination has enabled the benchmarking of tools, such as viral metagenomics (described below), to circumvent the potential drawbacks of culture and molecular methods for virus detection. Therefore, this chapter will discuss: (i) historical aspects of viruses as indicators of fecal contamination; (ii) examples of current and proposed viral indicators; (iii) current culture and molecular detection

methods of viral indicators; and (iv) advancements and future directions of viral metagenomics as a tool to identify viruses in watershed systems impacted by fecal contamination.

2 Viruses as Indicators of Sourses of Fecal Contamination

Historically, non-pathogenic enteric microorganisms, particularly bacteria, have been used to infer fecal contamination and potential source(s), as well as to identify the potential presence of enteric pathogens (Toranzos, 1991; Saxena et al., 2015; Holcomb & Stewart, 2020). Also known as microbial indicators of fecal contamination, these microorganisms should satisfy several criteria including the following: (i) detection using simple and reproducible laboratory methods across various geographical regions, (ii) presence in fecally polluted waters and absence in pristine waters, (iii) survival and inactivation rate(s) similar to those of the pathogen of concern under similar environmental and growth conditions, and (iv) inability to reproduce/replicate outside their human/animal host (Yates, 2007). While no microbial indicator of fecal contamination usually fulfills all the mentioned characteristics, a toolbox of indicators and detection methods is usually applied when tracking specific sources of fecal contamination. Certain enteric bacteria, however, do not satisfy several of the mentioned requirements and thus may not be considered the most suitable proxies of enteric pathogenic viruses (Toribio-Avedillo et al., 2021). Certain enteric viruses, on the other hand, exhibit host-specificity, and exclusivity to the gastrointestinal tract of warm-blooded animals, and thus may be considered more suitable indicators of specific sources of fecal contamination and the potential presence of pathogenic enteric viruses. Other viruses are acquired through diet and are shed through fecal material or are residents of the human gut and do not cause disease; thus, these viruses have also been proposed and tested as indicators of fecal contamination. In the following sections, the suitability of specific enteric viruses as indicators of fecal contamination is discussed. The section will include bacterial-, diet-, and host-associated (non-pathogenic) viruses.

2.1 Phages

Phages are ubiquitous and present where suitable bacterial host(s) are available, including the human gut. Enteric phages or those originating from the gastrointestinal (GI) tract, particularly, can exhibit source-specificity (e.g., human) (Dutilh et al., 2014; Jofre et al., 2014), making them suitable proxies of specific sources of fecal contamination. Phages can be strictly lytic, where replication and cell lysis occur immediately after attachment and introduction of nucleic acids, or lysogenic, where viral nucleic acids remain inserted into the host bacterial genome until induced. Most phages utilized for the purpose of identifying sources of fecal contamination

are strictly lytic phages. The following section will discuss *Escherichia coli* (coliphages) phages, *Bacteroides fragilis* phages, *Enterococcus* phages, crAssphages, and crAss-like phages as examples of viral indicators of fecal contamination.

2.1.1 *Escherichia coli* Phages or Coliphages

Among the group of phages applied to infer fecal contamination are those infecting *E. coli* or coliphages. The term coliphage is broad and refers to bacteriophages that infect coliform bacteria; it is primarily used to refer to *E. coli* phages (Jofre et al., 2016). Coliphages are classified mainly based on the mode of infection, namely, somatic and F-male specific if infection occurs through the cell wall, or sex pili, respectively (Singh et al., 2022). Somatic coliphages are a diverse group of DNA viruses that include siphoviruses (long non-contractile tails), myoviruses (long contractile tails), podoviruses (short contractile tails), all with double-stranded (ds) DNA genomes, and microviruses, which possess single-stranded (ss) DNA genomes (Jofre et al., 2016). F-male-specific coliphages, on the other hand, include both DNA and RNA phages from the *Inoviridae* and *Leviviridae* families, respectively (Jofre et al., 2016). Coliphages are inhabitants of the gastrointestinal tract of humans and other warm-blooded animals, and their presence in raw domestic and animal sewage makes them reliable indicators of fecal contamination (Cole et al., 2003; Jones et al., 2022). In addition, coliphages often exhibit survival and inactivation rates similar to those of certain human enteric viruses (Geldenhuys & Pretorius, 1989; Vergara et al., 2015), and correlate with enteric pathogens (Fitzmorris-Brisolara et al., 2022).

The feasibility of coliphages as indicators of fecal contamination in water sources, however, has been challenged. For instance, it has been suggested that the ability of *E. coli* to reproduce in the water environment may facilitate replication of somatic coliphages, particularly under higher temperatures (e.g., summer or in tropical or subtropical geographic areas) (Vaughn & Metcalf, 1975; Seeley & Primrose, 1980). However, other studies have shown that coliphage replication in the environment may not significantly impact coliphage numbers in water sources impacted by fecal contamination (Muniesa & Jofre, 2004). For this reason, F-male-specific coliphages have been proposed as the most suited indicators of specific sources of fecal contamination as they require specific temperatures to replicate (i.e., 32 °C, required for the sex pilli to form) (Jofre, 2009). Studies have also shown that neither somatic nor F-male-specific coliphages replicate under environmental conditions in tropical waters (Hernández-Delgado & Toranzos, 1995), supporting the use of these viruses as indicators of fecal contamination.

2.1.2 *Bacteroides fragilis* Phages

Another group of phages widely tested as indicators of specific sources of fecal contamination are those infecting *B. fragilis*. The suitability of *B. fragilis* phages as indicators of fecal contamination, particularly that of a human origin, has been evaluated for approximately four decades (Jofre et al., 1986; Mcminn et al., 2014; Tariq et al., 2020). The rationale of using *B. fragilis* phages as indicators of fecal contamination resulted from the biology and ecology of the bacterial host itself. *Bacteroides fragilis* is a common inhabitant of the human gut, and its obligate anaerobic nature inhibits its replication in the water environment (Elsaghir & Reddivari, 2022), and thus, it is also expected that phages infecting these bacteria are restricted to the human gut and do not replicate in water systems. Indeed, studies have shown that *B. fragilis* phages are highly prevalent in raw domestic sewage, as well as freshwaters contaminated with human fecal material, and are absent from water sources impacted with wildlife fecal pollution (Tartera et al., 1989; da Silva et al., 2022). Certain *B. fragilis* phages also exhibit survival rates and resistance to water treatment that resemble that of certain pathogenic enteric viruses (Tartera et al., 1988). One potential limitation of *B. fragilis* phages as indicators of human fecal contamination is their low concentrations relative to coliphages and fecal indicator bacteria in raw sewage and waters impacted by fecal pollution (Tartera et al., 1989; McLaughlin & Rose, 2006). In addition, studies have shown that certain *B. fragilis* phages are more prevalent in certain European countries, and thus their applicability as fecal contamination indicators may be limited to certain geographical regions (McLaughlin & Rose, 2006).

2.1.3 *Enterococcus* Phages (Enterophages)

The need to find alternative indicators of human fecal contamination that could circumvent the drawbacks of coliphages and *B. fragilis* phages, as well as indicator bacteria, resulted in the evaluation and proposal of phages infecting the genus *Enterococcus*, also known as enterophages. Early studies identified *Enterococcus faecalis* ATCC 49372 as a suitable strain for the identification of phages originating from domestic wastewater and developed culture methods for their detection with as few as 4 hours of incubation (Bonilla et al., 2010). Enterophages possess several characteristics of a suitable indicator of fecal contamination. For instance, enterophages are highly prevalent in raw domestic sewage in Puerto Rico and Portugal (Santiago-Rodriguez et al., 2013) and possess survival and inactivation rates similar to pathogenic enteric viruses in freshwater, marine waters, and sand microcosms (Santiago-Rodríguez et al., 2010; Santiago-Rodriguez et al., 2013). Additional studies have demonstrated that enterophages have the potential to be part of a toolbox of methods that would enable the identification of specific sources of fecal contamination in tropical watersheds impacted by point and non-point sources during rainfall and dry events (Santiago-Rodriguez et al., 2012, 2016).

While the original studies focused on the characterization of enterophages infecting *E. faecalis* strain ATCC 49372 as indicators of human fecal contamination, other *Enterococcus* species have also been evaluated as potential hosts for phages originating from various fecal sources. For instance, a study found that newly isolated *Enterococcus* strains were suitable for the detection of phages originating from human fecal material (Vijayavel et al., 2014). Other *Enterococcus* spp., such as *Enterococcus casseliflavus* and *Enterococcus gallinarum*, have also been evaluated as potential hosts for phages originating from other animal sources, such as cattle and poultry (Purnell et al., 2011). While these studies have shown that enterophages possess several characteristics of a suitable indicator of fecal contamination, the low concentrations of enterophages in raw domestic sewage have been considered a limitation for their use. However, it has also been argued that the relatively low concentrations of this indicator phage in raw sewage are similar to the concentrations of enteric viruses found in this matrix. This limitation associated with culture-based method of detection, however, opens the opportunity to explore the potential of molecular methods for the detection of enterophages in sewage and various water sources. In addition, more studies are needed to assess the reliability of enterophages as indicators of fecal pollution in additional geographical regions using the same bacterial strain.

2.1.4 CrAssphages and crAss-Like Phages

Unlike certain phages infecting *Bacteroides* and *Enterococcus*, where data have shown geographical specificity, crAssphages are ubiquitous in humans across various geographical regions (Edwards et al., 2019). CrAssphages were discovered as part of the viral dark matter of the human gut, and its characterization required assembly and binning of human gut metagenomes, identifying potential bacterial host(s), and determining genome size and gene function (Dutilh et al., 2014). While the original study identified a crAssphage possessing a genome size of approximately 97 kbp, predicted *Bacteroides* sp. as a presumptive bacterial host, and implied that this virus is ubiquitous to the human gut, crAssphages are now clustered into various groups including alpha-gamma, beta, delta, epsilon, and zeta groups (Yutin et al., 2021). Additional metagenomic, microscopy, and culture-based studies have since aided in the classification of crAssphage and crAss-like phages into several families and subfamilies according to their morphology, genome structure, and putative bacterial host (Guerin et al., 2018; Shkoporov et al., 2018).

Sequences homologous to crAssphage have been identified in sewage metagenomes from the USA, Europe, Africa, and Asia, with a lower prevalence in the latter two continents (Stachler & Bibby, 2014). The discovery and characterization of crAssphage from metagenomic data enabled the design of new technologies for its rapid identification in sewage and water sources impacted by fecal contamination. This study also demonstrated the feasibility of using metagenomics as a potential tool to identify enteric viruses originating from specific fecal sources. Quantitative PCR (qPCR) assays, in particular, have since been developed and successfully

tested in several countries and demonstrated that crAssphage are highly prevalent in raw domestic sewage and possess several characteristics of a suitable indicator of human fecal contamination (García-Aljaro et al., 2017; Stachler et al., 2017; Ahmed et al., 2018; Farkas et al., 2019; Kongprajug et al., 2019). In addition, crAssphage concentrations in sewage and river samples have been found to be comparable to the human-specific indicators HF183 and HMBif, but higher than culturable *E. coli* (Ballesté et al., 2019). This same study showed that the survival of crAssphage was longer compared to other bacterial markers identified using qPCR, but lower than culturable markers in a mesocosm composed of diluted sewage fed with well water (Ballesté et al., 2019). As evidence that crAssphage may be a suitable indicator of human fecal contamination continues to grow, additional studies may open the opportunity to understand their prevalence in other countries, their seasonal variations/stability, and the potential effect of environmental variables such as rainfall and temperature on their prevalence, survival, and inactivation rates. Moreover, determining the prevalence of crAssphage in additional species of warm-blooded animals may open the opportunity to understand the potential transmission of crAssphage to/from various animal species, which can aid to reconstruct their phylogenetic diversity (Li et al., 2021b).

2.2 *Diet-Associated*

While certain viruses are tightly coupled with gut bacteria and can be considered innate members of the gut microbiome (and human viromes), others are acquired through diet and are present as transient members of the gut virome (Gyawali et al., 2019). The following section will discuss Pepper mild mottle virus (PMMoV) as an example of a diet-associated virus utilized as an indicator of human fecal contamination.

2.2.1 Pepper Mild Mottle Virus

PMMoV is a pathogenic RNA plant virus that causes white mottles and chlorosis in infected plants (Jarret et al., 2008), resulting in significant global economic loss (Martínez-Ochoa et al., 2003; Wang et al., 2006). In some cases, however, PMMoV can replicate without signs of the disease or may cause mild foliar damage (Tsuda et al., 2007). Since many plants harbor PMMoV without showing signs of the disease, human consumption of PMMoV-infected plants may result in shedding of the virus through fecal material, ending up in sewage and/or water systems. Indeed, RNA metagenomic datasets have demonstrated that PMMoV is highly prevalent in the human gut, with a large proportion of viral RNA sequences being attributed to this virus (Zhang et al., 2006). As a result of its frequent shedding, common prevalence, and high relative abundance in the human gut, PMMoV has been proposed as an indicator of human fecal contamination (Rosario et al., 2009).

Original PMMoV qPCR detection assays targeted a replication-associated protein (Rosario et al., 2009). The assay successfully detected PMMoV in raw and treated domestic sewage in the USA and demonstrated that the virus is resistant to various removal strategies as well as to tertiary treatment methods (Rosario et al., 2009). Notably, this study did not detect PMMoV in fecal material originating from cows, horses, sheep, coyotes, raccoons, pigs, and dogs (Rosario et al., 2009). This, in addition to the co-occurrence of PMMoV with indicators of fecal contamination such as *Enterococci*, and human-specific *Bacteroides* spp., as well as viral pathogens, suggests that the plant pathogen is promising in identifying and tracking sources of human fecal contamination. PMMoV, however, has been detected in the feces of chicken and seagulls (Rosario et al., 2009; Gyawali et al., 2019), which are known to contribute to fecal pollution in various water sources. PMMoV often exhibits extended survival rates and, thus, may not be suited to indicate fresh fecal pollution (Kitajima et al., 2018). Finally, the fact that this is a plant virus also opens the possibility that sequences may be detected in water matrices in the absence of human fecal contamination.

2.3 *Host-Associated (Non-pathogenic)*

In addition to phages and diet-associated viruses, non-pathogenic viruses have also been tested as indicators of fecal contamination. While human enteric viruses are usually associated with disease, there is also a number of non-pathogenic enteric viruses (in addition to phages) that have also been proposed and tested as indicators of fecal pollution (Voroshilova, 1989). The following section will discuss Torque teno virus (TTV) as an example of a human-associated non-pathogenic virus proposed and tested as an indicator of fecal pollution.

2.3.1 Torque Teno Virus (TTV)

TTV is a human virus estimated to be present in 80% of the population and can be transmitted through the fecal-oral route (Griffin et al., 2008). Although it has also been identified in samples such as plasma and serum, TTV does not seem to cause disease (Gore et al., 2023). TTV has also demonstrated resistance to dry heat at 65 °C for 96 hours (Takayama et al., 1999). Since TTV is highly ubiquitous and resistant to high temperatures, it has been proposed and tested as a potential indicator of human fecal contamination in the water environment. Early studies showed the prevalence of TTV in 25% of river samples known to be contaminated with human fecal material, as confirmed with the presence of human enteric viruses (Verani et al., 2006). Follow-up studies demonstrated that TTV was detected in approximately 50% of sewage samples, in concentrations ranging from 10^5 to 10^2 PFU/L in raw and treated sewage, respectively (Hamza et al., 2011a). Interestingly, this study also challenged TTV as an indicator of fecal contamination due to its

relatively low prevalence compared to other indicators tested and its variable prevalence across geographical regions (Hamza et al., 2011a; Nick et al., 2019).

3 Overview of Current Methods to Identify Viral Indicators of Fecal Contamination

The cost and time-efficient identification of viral indicators is essential when tracking, identifying, and remediating sources of fecal contamination. Viral indicators of fecal contamination, such as those described in the previous section, are commonly identified using culture- and/or molecular-based methods, which may depend on the viral indicator of interest as well as the reasons for monitoring waters. In the following sections, a general overview of culture- and molecular-based methods for the detection of viral indicators of fecal pollution is discussed.

3.1 Culture-Based Methods

While molecular-based methods are widely preferred for the detection of viral indicators of fecal contamination, mainly due to their short turn-around times, culture methods continue to be essential to characterize, classify, and propagate viruses, including enteric viruses (Gholipour et al., 2022). Viral indicators of fecal contamination may require specific host cells, culture media, and growth conditions for their enumeration, isolation, and propagation. One of the main aspects to consider when developing culture assays for the detection of viral indicators of fecal contamination is the identification of a suitable host cell—human, animal, or bacterial—that would support efficient viral replication. An ideal host cell should enable the maximum/optimal enumeration and isolation of the virus of interest. In addition to the identification of an ideal host strain, culture media should support optimal host cell growth, which in turn is needed for the successful enumeration of the virus of interest. Finally, growth conditions, which include, but are not limited to, incubation temperature, would also need to be optimized depending on the host strain used (Hyman, 2019). For instance, assays targeting the detection of enterophages require the use of Tryptic Soy (TS) Broth (TSB) and TS Agar (TSA) media for plaque enumeration, and complemented with $CaCl_2$ and NaN_3 for better phage adsorption and to minimize the background growth of other bacteria, respectively (Bonilla et al., 2010; Santiago-Rodríguez et al., 2010).

Notably, one of the drawbacks of culture methods for the detection of phages as indicators of the presence of enteric viruses or as indicators of fecal contamination is that these may underestimate the true abundance of enteric viruses in the environmental waters being tested (Fong & Lipp, 2005; Worley-Morse et al., 2019). This may be due to the aggregation of virions in aqueous conditions and thus, cell culture

methods may detect "clumps" rather than individual virions, leading to the appearance of lower concentrations than when using molecular methods. Similarly, the limitations of culturing viruses are impacted by many variables, including physicochemical variables, which may also influence the intrinsic electrochemical charge of the viral proteins and their adsorbent capabilities to compounds present in the inoculum (Ibrahim et al., 2021). Notably, novel culture detection methods continue to be developed and tested for the detection of enteric viruses that may circumvent some of the drawbacks of traditional culture assays. For instance, colorimetric thiazolyl blue (MTT) assays have been tested to determine the infectiousness of enteric viruses. This assay utilizes 3-(4,5-dimethylthiazol-2-yl)-2,5-diphenyl tetrazolium bromide to generate percentages based on colorimetric results (Lee et al., 2013). Culture methods and assays, such as this one, are expected to be developed and benchmarked for specific viruses under various conditions.

3.2 *Molecular-Based Methods*

Detection of viral indicators of fecal contamination using molecular methods has been facilitated, in part, by advancements in genome sequencing. Genome sequencing has enabled the development of molecular methods targeting shared and conserved regions in viral indicators of fecal contamination, such as F-male-specific coliphages (Yong et al., 2006) and *B. fragilis* phages (Puig & Gironés, 1999). Molecular-based methods targeting viral indicators are often applied in water sources suspected to be contaminated with fecal material due to their fast turnaround time compared to most culture-based methods (Donaldson et al., 2002; Fernández-Baca et al., 2021). Short turn-around times enable prompter decision-making, which aids in the process of remediating potential fecal sources of contamination and avoiding potential pathogen transmission. In general, molecular-based assays targeting viruses can be qualitative and/or quantitative and are applied depending on the aim(s) of the study (Buesa & Rodriguez-Díaz, 2016). For instance, PCR is a qualitative assay that may be used to quickly assess the presence/absence of a virus in a sensitive and specific manner. As with the many variations of PCR-based assays, PCR primers can be designed to target a specific viral strain/variant (Buesa & Rodriguez-Díaz, 2016), or larger groups of viruses (e.g., at the order and family levels) (Nanda et al., 2008). Other PCR-based assays, such as real-time quantitative PCR (qPCR) and reverse-transcription quantitative PCR (RT-qPCR), provide quantitative results in shorter periods compared to PCR, as an agarose gel is not needed to visualize results (Fong & Lipp, 2005), and are often preferred for virus detection in various water systems, including watersheds (Stobnicka-Kupiec et al., 2022).

Modifications of molecular-based methods for the detection of enteric viruses continue to be developed and tested. For instance, high-throughput molecular techniques applying qPCR and microfluidic chips have shown to be suitable for the detection of multiple pathogens in water samples (Ishii et al., 2014). One drawback

of such techniques is that the limit of detection is not amenable for public health monitoring since small volumes of amplification reaction and samples are used (Ishii et al., 2014). Another technique, known as digital PCR (dPCR), has been applied to infer fecal pollution and identify potential sources. In dPCR, the reaction mix is dispersed in thousands to millions of wells on a chip and results are presented as presence/absence (Jahne et al., 2020). Quantitative variations of dPCR are also highly sensitive and can detect low genome copy numbers in a few hours (Jahne et al., 2020). In addition, dPCR and its variants possess the advantage of not needing standards and is not as influenced by inhibitors, as with qPCR. Modifications to dPCR protocols have proven to be effective for the absolute quantification of norovirus gene copies with a sensitivity identical to that of RT-qPCR (Monteiro & Santos, 2017).

Molecular-based methods, however, are not without limitations. For instance, some drawbacks originate from the nature of the method as it is highly sensitive to cross-contamination, contamination originating from reagents, and inhibitors (Borst et al., 2004; Lauri & Mariani, 2009; Mancini et al., 2010; McKindles & Tiquia-Arashiro, 2012; Saxena et al., 2015). As such, PCR-based methods need to be performed by highly trained personnel who can troubleshoot with the proper/relevant controls when needed. In addition, molecular-based methods can overestimate viral abundance as naked nucleic acids (if no nuclease treatment is used as part of the procedures), and non-infectious viral particles may also be detected (Lodder et al., 2010). In addition, nucleic acid purification steps can be time-consuming, adding to the problem of immediacy when it comes to determining contamination vis-a-vis risks to public health. PCR-based methods are also restricted to the viruses whose primers are available, requiring a degree of a priori knowledge of the virus that may be present. For this reason, an increasing number of studies have explored the use of non-targeted, high-throughput sequencing approaches to identify enteric viruses that could indicate fecal pollution and potential origin(s), a field known as viral metagenomics (Bibby & Peccia, 2013).

4 Viral Metagenomics

4.1 What Is Viral Metagenomics?

Viral metagenomics refers to the high-throughput sequencing and bioinformatic analysis of the DNA and/or RNA viral genomes in a sample, also known as the virome. Viral metagenomics was first applied over 20 years ago to characterize marine viromes (Breitbart et al., 2002), and it is increasingly being utilized as a tool to characterize virus community composition in clinical (Batool & Galloway-Peña, 2023) and environmental samples (Dávila-Ramos et al., 2019; Sommers et al., 2021). Viral metagenomics, unlike culture and molecular methods targeting enteric viruses, usually does not require a priori knowledge of the identity of a virus (unless

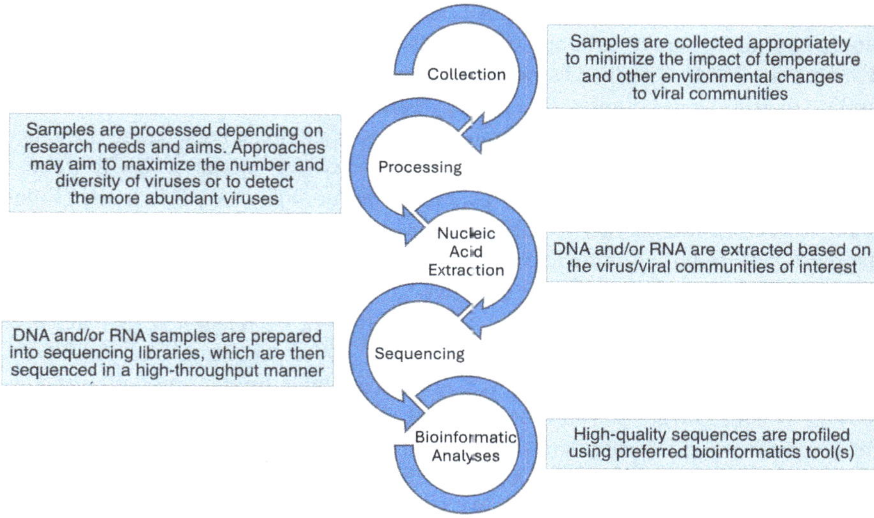

Fig. 1.1 General overview of a viral metagenomics pipeline. A viral metagenomics pipeline usually includes sample collection and processing, nucleic acid extraction, sequencing, and bioinformatic analyses

a targeted approach is considered, not discussed in this chapter), and thus multiple viruses can be identified simultaneously. Viral metagenomics involves a series of steps prior to achieving the sequencing and profiling of high-quality viral genomes. Common steps in a viral metagenomic workflow include sample collection using appropriate sterile containers or collection devices (clinical and biological) (Ilett et al., 2019; Maghini et al., 2024), and nucleic acids are extracted following the protocol/method most suitable for the sample type and/or targeted nucleic acid (Sabatier et al., 2020). Nucleic acids are then prepared into DNA and/or RNA libraries that are sequenced, and analyzed using bioinformatic tools to profile viruses in a community. Selection of a bioinformatic tool will depend on the aims of the study (Santiago-Rodriguez & Hollister, 2020) (Fig. 1.1). The following sections will briefly describe the mentioned steps in a viral metagenomics pipeline.

4.2 Viral Metagenomics as a Tool to Identify Indicators of Fecal Contamination

Since viral metagenomics as a tool to track sources of fecal contamination is still in its infancy, studies applying viral metagenomics in sewage and fecal samples, as well as various water sources, including watershed systems, are discussed below.

Viral metagenomics has been applied to identify viral pathogens and indicators of fecal contamination in water, sewage, and fecal samples (Stachler & Bibby, 2014;

Wang et al., 2024). For instance, a study found sequences homologous to crAssphage, PMMoV, adenoviruses, TTV, and noroviruses in sewage originating from the USA, Europe, Africa, and Asia (Stachler & Bibby, 2014). Interestingly, sewage samples from the USA and Europe, in particular, showed the highest number of reads homologous to crAssphage compared to sewage samples originating from Africa and Asia, opening the opportunity to understand the global distribution and transmission of this virus. The higher prevalence of crAssphage in sewage samples originating from the USA and Europe may be in agreement with studies demonstrating the higher prevalence of phages infecting other *Bacteroides* spp., such as those infecting *B. fragilis*, in the USA and Europe (McLaughlin & Rose, 2006). The number of crAssphage reads was also higher compared to the other viruses considered. Intriguingly, this study also identified crAssphage homologous sequences in bat guano metagenomes (Stachler & Bibby, 2014), opening the opportunity to investigate the potential transmission or spillover of viruses across animal species. These results also merit further investigation to determine the prevalence of crAssphage and crAss-like phages in the gut metagenomes of additional warm-blooded animals, and not only domestic animals.

Due to the ability of viral metagenomics to identify viruses in various sample types, it can also potentially be used as a tool to track sources of fecal contamination in watershed systems affected by point and non-point sources of fecal contamination. While this has not been extensively evaluated in watersheds specifically, several studies have identified viruses in environmental water metagenomes. For instance, a study investigating viruses in a watershed contaminated with microplastics (MPs) identified viral communities dominated by bacteriophages from the former *Siphoviridae*, *Myoviridae*, and *Podoviridae* families, and hundreds of viral species were found along the river (Li et al., 2022). Intriguingly, human herpesviruses were found in polypropylene (PP), which merits further investigation. While the study did not find significant sources of enteric viruses, probably due to the need of adding a virus concentration step after sample collection, and/or a deeper sequencing depth (discussed later in this chapter), similar studies could be performed in watersheds impacted by various sources of fecal contamination. Notably, this study also highlights that the viral dark matter may account for most viruses in watershed systems. This, in turn, opens the opportunity to further explore the identity and role of the viral dark matter in watersheds contaminated with MPs and fecal material originating from various sources. Similarly, another study analyzing metagenomic datasets identified crAssphage and crAss-like phage sequences in a river system at both pristine and contaminated sites (Mafumo et al., 2023). This study produced metagenome-assembled genomes (MAGs) from six different metagenomes representing different sites of the river system. The study found higher relative abundances and greater diversity of crAss-like phages in contaminated sites compared to pristine sites, which supports the fecal origin of crAssphages and crAss-like phages. Intriguingly, results also showed a site-specific prevalence of certain crAssphages and crAss-like phages, which merits further investigation as it could suggest a higher association of certain phylogenetic groups with the degree of fecal pollution. The study also reported a phylogenetically novel

crAss-like phage group, which opens the opportunity for further characterization. Similar approaches can be applied to identify other gut-specific viruses, and, potentially, enable the discovery of novel viral groups from metagenomic datasets.

While viral metagenomics has been mainly applied to characterize the DNA virome, this approach has also been utilized to investigate the RNA virome. RNA viruses are often associated with disease, and thus their monitoring is essential from a source tracking and surveillance perspective. A study evaluating wastewater samples in a large city in Latin America applied viral metagenomics to characterize the RNA virome and identified RNA phages and RNA viruses infecting invertebrates and humans (Guajardo-Leiva et al., 2020). While this study focused on the applications of viral metagenomics as a tool for public health surveillance, similar approaches may provide information of the specific sources of enteric RNA viruses in wastewater samples and fecally contaminated waters (Guajardo-Leiva et al., 2020). In addition to gut-, insect-, and diet-associated RNA viruses, those inhabiting the respiratory tract have also been identified in wastewater using metagenomics since some are known to be shed through feces (Vicente et al., 2007; Chan et al., 2011; Xiao et al., 2020). Indeed, a similar approach was evaluated during the COVID-19 pandemic as monitoring of SARS-CoV-2 in sewage samples provided insights into potential spreading and transmission mechanisms of the virus in specific areas (Medema et al., 2020; Nemudryi et al., 2020; Agrawal et al., 2021; Galani et al., 2022). The SARS-CoV-2 pandemic pushed researchers and public health workers to utilize (and develop) methods for the rapid detection of these viruses in the environment, and especially as part of what is called Wastewater-Based Epidemiology (Novoa et al., 2022). While efforts have been monumental, many studies were too focused on SARS-CoV-2 and few were based on other viral pathogens as well. Some studies did focus on different OMICS and have shown very promising results (Ginebreda et al., 2023).

5 Considerations and Limitations of Viral Metagenomics

A key limitation of using viral metagenomics for the identification and tracking of fecal contamination is that little is still known about the resiliency of nucleic acids in the environment and how this may affect analyses and results. Nucleic acids may be at least (partially) protected from the actions of nucleases if they are within the capsid or if viruses are aggregated. Other considerations for the use of viral metagenomics in MST and water quality usually include similar limitations encountered in other virome studies. Limitations of the use of viral metagenomics usually include those associated with sample pre-processing, nucleic acid extraction, sequencing depth, and bioinformatic analysis (Fig. 1.2), which are discussed below.

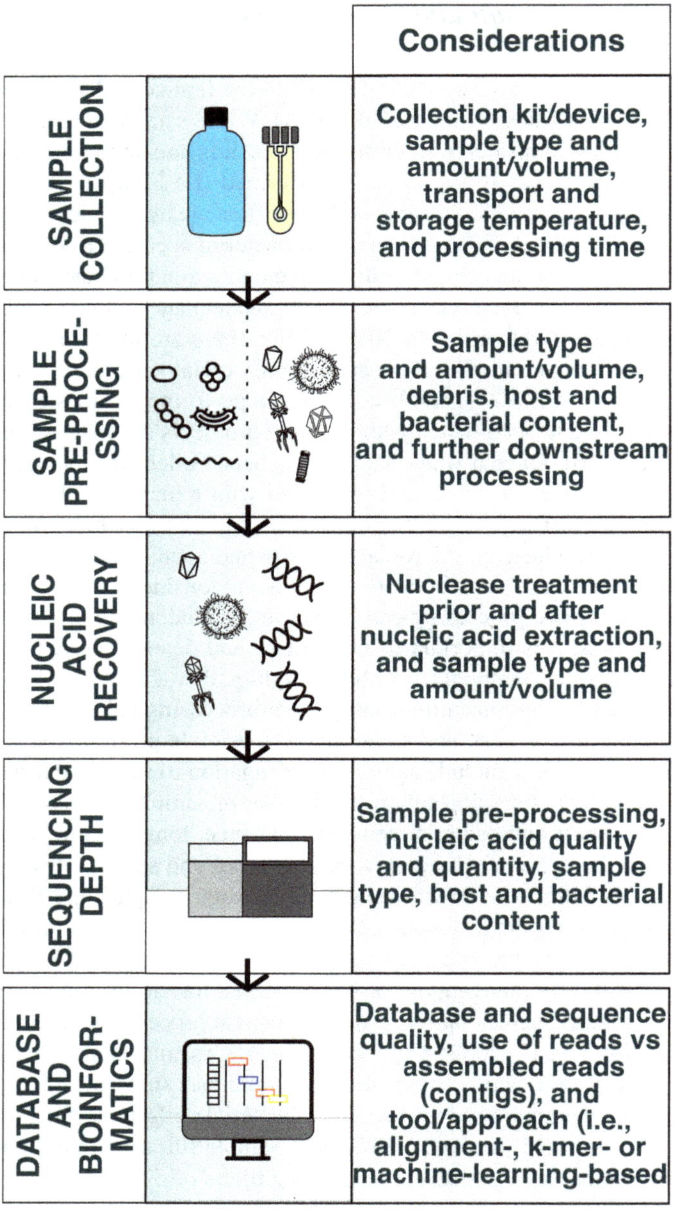

Fig. 1.2 Considerations in viral metagenomics studies include, but are not limited to, sample collection and pre-processing, nucleic acid recovery, sequencing depth, as well as database and bioinformatics tool selection (see text)

5.1 Sample Collection and Pre-processing

As mentioned above, sample collection is one of the first considerations in designing a metagenomics study (Kumar et al., 2014). While sample collection is not an aspect that is heavily considered in virome studies, it is known to affect the concentration of specific viruses if not properly performed (Le François & Doukhanine, 2019). Improper sample collection can add biases to a metagenomics study and may skew the diversity and abundance of specific bacterial species (Kool et al., 2023). These biases have been associated with sample contamination, and inactivation or proliferation of specific bacterial species due to inadequate sample handling, transport, and/or storage (Watson et al., 2019). While there are microbiome collection kits intended for human samples, there are no commercially available collection kits for water samples; thus, water sample collection is performed using sterile plastic or glass containers and in accordance with standard protocols for water sample collection (USEPA, 2016). Once a water sample has been collected, it should be transported to the laboratory at <4 °C and processed with a maximum holding time of 48 hours (USEPA, 2016). This may minimize the impact that sample transport time and temperature may have on microbial diversity and abundance.

Once a sample has been collected, the next consideration is sample pre-processing. Sample pre-processing can influence the outcome of any viral metagenomics study; thus, it is important to benchmark and determine the most suitable sample pre-processing method(s) to obtain high-quality viromes. Although more labor-intensive and time-consuming, sample pre-processing enhances the viral signal by removing debris, host, and bacterial cells. Sample pre-processing in a viral metagenomics study may include sample centrifugation to remove debris, and centrifugation parameters may depend on the amount of sample, debris, and expected concentration of host and bacterial cells. For instance, longer centrifugation times may be needed for debris-rich samples, such as stool and saliva. For large-volume samples, such as freshwater, however, techniques such as tangential flow filtration (TFF) may be better suited. Notably, these methods were mostly developed and tested for virus concentration decades before the first virome study (Watanabe et al., 1988); thus, these techniques should be benchmarked for virome studies taking into consideration factors such as sample type, subsequent processing steps, and aims of the study (Breitbart et al., 2002; Angly et al., 2006; Cassman et al., 2012).

After centrifugation, and/or application of techniques such as TFF, further sample filtration may be applied to trap host and bacterial cells using filters of specific sizes (e.g., 0.8 μm, 0.45 μm, and/or 0.2 μm). Sample filtration may be performed using one of the mentioned filter sizes, or using filters of different sizes (i.e., serial filtration), and this will depend on if/which downstream sample processing will be further applied. For instance, studies applying filtration or serial filtration may further apply cesium chloride (CsCl) gradient ultracentrifugation to concentrate viral particles (Santiago-Rodriguez et al., 2015a, b). Notably, CsCl gradient ultracentrifugation has been widely benchmarked for the isolation of the phageome, or the viral fraction corresponding to phages. CsCl gradient ultracentrifugation, like other

techniques applied in sample pre-processing, are not without limitations. Particularly, a study showed that, while CsCl gradient ultracentrifugation was the most efficient in removing host DNA, the technique also showed a strong bias against the isolation of specific phages, and may also result in lower DNA yields (Kleiner et al., 2015). Notably, while CsCl gradient ultracentrifugation is a common technique used to concentrate and purify viruses, there are several considerations associated with selecting the most suitable technique for these purposes. The reason is that the various adsorptive properties and electrostatic interactions of viruses can affect the outcome and performance of the concentration method. For instance, chemical precipitation methods that require low pH may impact infectivity of certain viruses (Lewis & Metcalf, 1988). Another approach known as PEG (polyethylene glycol) precipitation is widely accepted to concentrate viruses from large volume-samples, as it is known to preferentially crystallize biological molecules such as viruses (Colombet et al., 2007). Virus concentration using PEG, however, has shown to promote the formation of a viscous, high molecular weight compound when added to the sample filtrate (Kleiner et al., 2015). These results show that virus concentration for viral metagenomic studies should be carefully benchmarked and tested in different sample types.

The characterization of viromes without pre-processing steps, such as those mentioned above, is also widely accepted. Since not all laboratories have access to the equipment and reagents necessary to perform sample pre-processing for viral metagenomics as described above, total DNA and/or messenger RNA (mRNA) sequencing have also been used to characterize the virome of various sample types. While this approach requires little to no sample pre-processing, it also results in most of the reads being attributed to host and/or bacteria. The small proportion of viral reads resulting from this approach usually corresponds to the most highly abundant and common viruses, which may be phages or diet-associated viruses (Bajaj et al., 2020); thus, this approach may be suitable for certain studies.

5.2 Nucleic Acid Recovery

Following sample pre-processing and prior nucleic acid extraction, a nuclease treatment step is strongly recommended for the recovery of high-quality viral sequences. Nuclease treatment may also be applied following nucleic acid extraction for the selection of the viral community of interest (DNA or RNA) if total nucleic acids were extracted. Nuclease treatment selection may also be dependent on the sample type of interest. For instance, water samples are usually dominated by DNA viruses (e.g., bacteriophages), and thus, an RNAse treatment may be appropriate. It should be noted, however, that this does not indicate that RNA viruses are not present in such sample types. Nuclease treatment, therefore, will depend on the target viral group and aims of the study.

Viral nucleic acids are then extracted following nuclease treatment. A successful viral nucleic acid isolation protocol generally includes disruption of the viral

capsid, protein denaturation, separation of proteins and other components from the nucleic acid, followed by nucleic acid purification (El-Ashram et al., 2016). Nucleic acid extraction can be classified into organic, where proteins and lipids are extracted from a nucleic acid solution using organic substances such as phenol-chloroform (Steward & Culley, 2010), inorganic, where inorganic solvents are utilized (e.g., proteinase K and cetyl trimethylammonium bromide [CTAB]), and solid phase extractions, where nucleic acids bind to silica under conditions of high salt concentration and low pH, and are then eluted (Steward & Culley, 2010). A study testing three different viral DNA recovery methods from soil samples, including CTAB, Wizard mini columns, and a modification of the DNeasy PowerSoil DNA extraction kit with and without further purification using AMPure beads, showed that PowerSoil recovers the highest DNA yield. Interestingly, each tested nucleic acid extraction method captured unique viral communities (Trubl et al., 2019), demonstrating the intrinsic biases associated with nucleic acid extraction kits and protocols. As with nuclease treatment, selection of the nucleic acid extraction method will also depend on the sample type and aims of the study. In addition, as with sample pre-processing, nucleic acid extraction and recovery should be benchmarked using appropriate controls (e.g., viral mock community), if possible.

5.3 Sequencing Depth

High-quality viral nucleic acids will then go through library preparation and sequencing. While library preparation and sequencing protocols are relatively standardized, an important consideration during the sequencing process is determining the number of sequences needed to obtain an unbiased representation of viral abundance and diversity; yet, few studies have investigated this. The sequencing depth required to obtain an unbiased viral community representation will depend on several factors including sample type, amount of host content, and virus concentration/purification step(s), if applied. For instance, a study looking at the human urinary virome showed that 10,000–20,000 sequences were sufficient to reach a plateau in terms of viral diversity (Santiago-Rodriguez et al., 2015a). This study applied centrifugation, followed by sample concentration using CsCl gradient ultracentrifugation. Another study looking into the virome of colonic mucosal aspirate samples investigated the effect of sequencing depth on viral contig recovery (Yan et al., 2023). The study showed that the yield for new contigs per million bases decreased from approximately 38 viral contigs at 1 million reads to approximately 9 viral contigs at 10 million reads to approximately 2 viral contigs at 100 million reads. This indicates that a deeper sequencing depth may not necessarily result in a significant recovery of new contigs in low biomass samples.

Sequencing depth analyses may also translate to water samples, but further studies are needed to establish the sequencing depth needed to accomplish an unbiased representation of viral diversity and abundance in such sample types. In the case of utilizing viral metagenomics for tracking sources of fecal contamination in water

sources and/or surveillance, it is essential to estimate the sequencing depth needed for the detection of viral indicators of fecal contamination and/or enteric pathogenic viruses as these may be present in low abundances compared to autochthonous viral communities. This may be accomplished by, for example, spiking a known virus(es) or their respective nucleic acids at a specific concentration, sequencing at a relatively deep depth, and subsampling the sequences in silico to mimic different sequencing depths. The spiked virus(es) can then be detected using direct mapping to the reference genome or performing a viral community analysis to determine the presence and/or abundance of the virus(es) of interest.

5.4 Database Selection and Bioinformatic Analysis

Other considerations in viral metagenomic study design and execution are the database(s) and bioinformatic tools used to assign taxonomy. The use of viral metagenomics as a tool to track sources of fecal contamination may require sequences of the target viral indicator and/or enteric pathogenic virus to be available in the selected database and to share a high degree of homology to available viral reference genomes; thus, it is important to use an updated and curated database. One of the most common databases used to identify viruses from virome datasets is the NCBI RefSeq database, which can be curated as needed. In addition to the database, the bioinformatic tool of choice should be able to assign viral taxonomy in a precise manner, providing the lowest number of false-positives and false-negatives; thus, this may require analysis tool benchmarking using a control, such as a viral mock community, for example. While several bioinformatic tools utilized in viral metagenomics are database dependent (e.g., alignment- and k-mer-based algorithms), others apply machine learning algorithms.

Alignment- and k-mer-based tools have been summarized in a previous review (Santiago-Rodriguez & Hollister, 2019), and both have shown advantages and disadvantages in viral metagenomic studies. Since 2019, additional tools have been developed for the analysis of virome datasets, mostly being based on machine learning or hybrid methods. Several of these tools are summarized in Table 1.1. Some of these tools include, but are not limited to, machine learning-based such as VirionFinder (Fang & Zhou, 2021), DeePVP (Fang et al., 2022), DeepVirFinder (Ren et al., 2020), VirSorter 2 (Guo et al., 2021), and VirHunter (Sukhorukov et al., 2022). Hybrid methods that incorporate machine learning, such as VIBRANT (Kieft et al., 2020) and MetaPhage (Pandolfo et al., 2022), are also available for the analysis of virome datasets. While the mentioned tools have been benchmarked, there is still the need to test these tools in water virome datasets to determine if they offer an additional advantage(s) over other methods including those that are alignment- and k-mer based.

Regardless of the bioinformatic tool and approach used, a challenge associated with virome datasets is the lack of similarity of sequences to reference viral genomes. In certain virome datasets, up to 99% of the viral sequences cannot be

Table 1.1 Examples of machine learning and hybrid bioinformatic tools for virus taxonomy from metagenomic datasets

Bioinformatic tool	Description	Source	Reference
Machine learning			
DeepVirFinder	Predicts viral sequences using deep learning method. The method has good prediction accuracy for short viral sequences, so it can be used to predict sequences from the metagenomic data	https://github.com/jessieren/DeepVirFinder	Ren et al. (2020)
VirionFinder	Applies deep learning to identify whether a given protein is a prokaryote virus virion protein	https://github.com/zhenchengfang/VirionFinder	Fang and Zhou (2021)
VirSorter2	A DNA and RNA virus identification machine learning tool that leverages genome-informed database advances across a collection of customized automatic classifiers to improve the accuracy and range of virus sequence detection	https://bitbucket.org/MAVERICLab/virsorter2/src/master/	Guo et al. (2021)
DeePVP	Uses machine learning to classify prokaryote virus virion proteins into ten major classes in a highly precise manner	https://github.com/fangzcbio/DeePVP/	Fang et al. (2022)
VirHunter	A deep learning convolutional neural network approach, to detect novel and known viruses in assemblies of sequencing datasets	https://github.com/cbib/virhunter	Sukhorukov et al. (2022)
Hybrids			
VIBRANT	Uses neural networks of protein annotation signatures from non-reference-based similarity searches with Hidden Markov Models, as well as a unique "v-score" metric to maximize identification of diverse and novel viruses	https://github.com/AnantharamanLab/VIBRANT	Kieft et al. (2020)
MetaPhage	A fully automated computational pipeline that applies mapping and machine learning for quality control, assembly, and phage detection as well as classification and quantification of these phages in metagenomics data	https://github.com/MattiaPandolfoVR/MetaPhage	Pandolfo et al. (2022)

attributed to known viruses (Shkoporov & Hill, 2019). Also known as the viral dark matter, these sequences often need to be processed in numerous ways (e.g., comprehensive removal of host sequences, possible assembly, and searches in both nucleic acid and protein space) to obtain a potential novel viral genomic sequence (Krishnamurthy & Wang, 2017; Santiago-Rodriguez & Hollister, 2022). This was the case of crAssphage (Dutilh et al., 2014), which represents one of the most evaluated indicators of human fecal contamination to have emerged from the viral dark matter. While discovering and characterizing novel viral indicators of specific sources of fecal contamination is not necessarily an easy task given the variety of bioinformatic techniques needed to accomplish this, the information contained in sewage, fecal, and water viral dark matter represents an opportunity to discover potential novel viral indicators of fecal contamination.

6 Conclusions and Future Perspectives

Viral metagenomics has been proposed and is increasingly being applied as a tool to track sources of fecal contamination by identifying specific viral indicators and/or enteric pathogenic viruses. Although viral metagenomics as a tool for tracking sources of fecal contamination is still in its infancy, studies thus far have been successful in identifying enteric viruses in fecal, sewage, and water samples, including watershed systems. Viral metagenomics has successfully identified crAssphage, PMMoV, adenoviruses, TTV, and noroviruses in sewage and other water sample types. The application of viral metagenomics to track sources of fecal contamination has also demonstrated that a significant proportion of the sequences in a sample are attributed to the viral dark matter, or viruses not sharing homology to those available in sequence databases. The viral dark matter, however, also represents a venue for virus discovery and their characterization as potential indicators of fecal pollution, as in the case of crAssphage. The RNA virome has also been characterized using metagenomics and has helped in surveillance efforts of viruses such as SARS-CoV-2; thus, viral metagenomics is actively being explored as a surveillance tool to detect emerging and re-emerging pathogens that are often found in sewage and water sources.

As with culture and molecular methods utilized for the detection of viruses, considerations around viral metagenomics are important to obtain quality viromes and viral genomes, which in turn will enable to successfully track sources of fecal contamination. Several of these considerations for viral metagenomics studies in the field of microbial source tracking include the need to standardize sample collection, storing and processing methods, as well as viral nucleic acid extraction, and bioinformatic analyses across academic laboratories and research facilities. The addition of a sample with known composition, such as commercially available viral mock communities from sources including the American Type Culture Collection (ATCC), in a viral metagenomics study may aid in benchmarking the mentioned steps of a viral metagenomics pipeline.

It is anticipated that viral metagenomics will continue to develop as an enabling technology to support the identification and tracking of specific sources of fecal contamination in watershed systems impacted by fecal contamination from a variety of sources. As highlighted in this chapter, viral metagenomics offers tremendous potential in its ability to provide potentially host-specific targets that mirror the dynamics of fecal bacteria and other pathogens of concern in water and the environment. However, in order to move the field forward, additional work is needed to confirm the reliability of viral markers of fecal contamination. Specifically, additional study is needed to verify our understanding of host- and source-specificity of viral marker candidates, their biogeographical distribution and applicability, interactions with environmental factors (e.g., temperature, plastics), and limits of detection in systems of interest. It is anticipated that studies of viral dark matter will continue to reveal new viruses with the potential to serve as host-specific indicators of fecal contamination, and the considerations mentioned above with respect to reliability will also need to be evaluated. We also expect that the development of novel culture- and molecular-based methods will also move the field forward as we improve our ability to culture, detect, and sequence previously unknown viruses and understand their biology and ecology. In addition, coupling collections of novel host-specific markers with rapid, reliable, and potentially fieldable testing systems will allow for timely, accurate detection at the point of need to support decision-making and stewardship of water, agricultural resources, and public health. The SARS-CoV-2 pandemic has pushed many researchers and public health workers to revisit the concept of sewage epidemiology; a concept that is several decades-old, but had not been fully implemented mainly because of cost and the lack of proper methods to deal with the complexity of environmental samples, and especially sewage. It is expected that as more methods and OMICS approaches are developed for use in the analyses of waters and sewage, we will get a better picture of total population health at speeds approaching real time. This has the potential to completely change the current paradigm on how we detect and deal with fecal contamination of waters.

Disclosure Statements Sandia National Laboratories is a multi-mission laboratory managed and operated by National Technology & Engineering Solutions of Sandia, LLC (NTESS), a wholly owned subsidiary of Honeywell International Inc., for the U.S. Department of Energy's National Nuclear Security Administration (DOE/NNSA) under contract DE-NA0003525. This written work is authored by an employee of NTESS. The employee, not NTESS, owns the right, title, and interest in and to the written work and is responsible for its contents. Any subjective views or opinions that might be expressed in the written work do not necessarily represent the views of the U.S. Government. The publisher acknowledges that the U.S. Government retains a non-exclusive, paid-up, irrevocable, worldwide license to publish or reproduce the published form of this written work or allow others to do so, for U.S. Government purposes. The DOE will provide public access to results of federally sponsored research in accordance with the DOE Public Access Plan.

References

Agrawal, S., Orschler, L., & Lackner, S. (2021). Long-term monitoring of SARS-CoV-2 RNA in wastewater of the Frankfurt metropolitan area in Southern Germany. *Scientific Reports, 11*(1), 5372. https://doi.org/10.1038/s41598-021-84914-2

Ahmad, S. S. (2018). Water related ocular diseases. *Saudi Journal of Ophthalmology., 32*, 227. https://doi.org/10.1016/j.sjopt.2017.10.009

Ahmed, S. M., Lopman, B. A., & Levy, K. (2013). A systematic review and meta-analysis of the global seasonality of norovirus. *PLoS One, 8*(10), e75922. https://doi.org/10.1371/journal.pone.0075922

Ahmed, W., et al. (2018). Precipitation influences pathogenic bacteria and antibiotic resistance gene abundance in storm drain outfalls in coastal sub-tropical waters. *Environment International, 116*, 308. https://doi.org/10.1016/j.envint.2018.04.005

Alidjinou, E. K., et al. (2019). Resistance of Enteric Viruses on Fomites. *Intervirology, 61*, 205. https://doi.org/10.1159/000448807

Angly, F. E., et al. (2006). The marine viromes of four oceanic regions. *PLoS Biology, 4*(11), e368. https://doi.org/10.1371/journal.pbio.0040368

Bain, R., et al. (2014). Global assessment of exposure to faecal contamination through drinking water based on a systematic review. *Tropical Medicine and International Health, 19*(8), 917. https://doi.org/10.1111/tmi.12334

Bajaj, J. S., et al. (2020). Interaction of bacterial metagenome and virome in patients with cirrhosis and hepatic encephalopathy. *Gut, 70*, 1162. https://doi.org/10.1136/gutjnl-2020-322470

Ballesté, E., et al. (2019). Dynamics of crAssphage as a human source tracking marker in potentially faecally polluted environments. *Water Research, 155*, 233. https://doi.org/10.1016/j.watres.2019.02.042

Batool, M., & Galloway-Peña, J. (2023). Clinical metagenomics—challenges and future prospects. *Frontiers in Microbiology, 14*. https://doi.org/10.3389/fmicb.2023.1186424

Bibby, K., & Peccia, J. (2013). Identification of viral pathogen diversity in sewage sludge by metagenome analysis. *Environmental Science and Technology, 47*(4), 1945. https://doi.org/10.1021/es305181x

Bonilla, N., et al. (2010). Enterophages, a group of phages infecting Enterococcus faecalis, and their potential as alternate indicators of human faecal contamination. *Water Science and Technology (IWA Publishing), 61*(2), 293–300.

Borst, A., Box, A. T. A., & Fluit, A. C. (2004). False-positive results and contamination in nucleic acid amplification assays: Suggestions for a prevent and destroy strategy. *European Journal of Clinical Microbiology & Infectious Diseases, 23*, 289–299. https://doi.org/10.1007/s10096-004-1100-1

Bouseettine, R., et al. (2019). Waterborne transmission of enteric viruses and their impact on public health. In *Emerging and reemerging viral pathogens: Volume 1: Fundamental and basic virology aspects of human, animal and plant pathogens*. https://doi.org/10.1016/B978-0-12-819400-3.00040-5

Breitbart, M., et al. (2002). Genomic analysis of uncultured marine viral communities. *Proceedings of the National Academy of Sciences of the United States of America., 99*, 14250. https://doi.org/10.1073/pnas.202488399

Buesa, J., & Rodriguez-Díaz, J. (2016). The molecular virology of enteric viruses', *Viruses in Foods*. https://doi.org/10.1007/978-3-319-30723-7_3

Calderon, R. L., & Mood, E. W. (1991). Health effects of swimmers and nonpoint sources of contaminated water. *International Journal of Environmental Health Research, 1*(1), 21. https://doi.org/10.1080/09603129109356701

Cassman, N., et al. (2012). Oxygen minimum zones harbour novel viral communities with low diversity. *Environmental Microbiology, 14*(11), 3043. https://doi.org/10.1111/j.1462-2920.2012.02891.x

Chan, M. C. W., et al. (2011). Seasonal influenza a virus in feces of hospitalized adults. *Emerging Infectious Diseases, 17*(11), 2038. https://doi.org/10.3201/eid1711.110205

Cole, D., Long, S. C., & Sobsey, M. D. (2003). Evaluation of F+ RNA and DNA coliphages as source-specific indicators of fecal contamination in surface waters. *Applied and Environmental Microbiology, 69*(11), 6507. https://doi.org/10.1128/AEM.69.11.6507-6514.2003

Colombet, J., et al. (2007). Virioplankton "pegylation": Use of PEG (polyethylene glycol) to concentrate and purify viruses in pelagic ecosystems. *Journal of Microbiological Methods, 71*(3), 212. https://doi.org/10.1016/j.mimet.2007.08.012

da Silva, D. T. G., et al. (2022). A Longitudinal study of Bacteriophages as indicators of norovirus contamination of mussels (Mytilus edulis) and their overlying waters. *Pollutants, 2*(1), 66. https://doi.org/10.3390/pollutants2010008

Damashek, J., et al. (2022). Non-point source fecal contamination from aging wastewater infrastructure is a primary driver of antibiotic resistance in surface waters. *Water Research, 222*, 118853. https://doi.org/10.1016/j.watres.2022.118853

Dávila-Ramos, S., et al. (2019). A review on viral metagenomics in extreme environments. *Frontiers in Microbiology, 10*(OCT). https://doi.org/10.3389/fmicb.2019.02403

De Graaf, M., Van Beek, J., & Koopmans, M. P. G. (2016). Human norovirus transmission and evolution in a changing world. *Nature Reviews Microbiology, 14*, 421. https://doi.org/10.1038/nrmicro.2016.48

Donaldson, K. A., Griffin, D. W., & Paul, J. H. (2002). Detection, quantitation and identification of enteroviruses from surface waters and sponge tissue from the Florida Keys using real-time RT-PCR. *Water Research, 36*(10), 2505. https://doi.org/10.1016/S0043-1354(01)00479-1

Dutilh, B. E., et al. (2014). A highly abundant bacteriophage discovered in the unknown sequences of human faecal metagenomes. *Nature Communications, 5*, 4498. https://doi.org/10.1038/ncomms5498

Edwards, R. A., et al. (2019). Global phylogeography and ancient evolution of the widespread human gut virus crAssphage. *Nature Microbiology, 4*, 1727. https://doi.org/10.1038/s41564-019-0494-6

El-Ashram, S., Al Nasr, I., & Suo, X. (2016). Nucleic acid protocols: Extraction and optimization. *Biotechnology Reports, 12*, 33. https://doi.org/10.1016/j.btre.2016.10.001

Elsaghir, H., & Reddivari, A. K. R. (2022). Bacteroides fragilis. In *StatPearls [Internet]*. StatPearls Publishing.

Fang, Z., & Zhou, H. (2021). VirionFinder: Identification of complete and partial prokaryote virus virion protein from virome data using the sequence and biochemical properties of amino acids. *Frontiers in Microbiology, 12*. https://doi.org/10.3389/fmicb.2021.615711

Fang, Z., et al. (2022). DeePVP: Identification and classification of phage virion proteins using deep learning. *GigaScience, 11*. https://doi.org/10.1093/gigascience/giac076

Farkas, K., et al. (2019). Critical evaluation of CrAssphage as a molecular marker for human-derived wastewater contamination in the aquatic environment. *Food and Environmental Virology, 11*(2), 113. https://doi.org/10.1007/s12560-019-09369-1

Fernández-Baca, C. P., et al. (2021). Rapid qPCR-based water quality monitoring in New York State recreational waters. *Frontiers in Water, 3*. https://doi.org/10.3389/frwa.2021.711477

Fitzmorris-Brisolara, K., et al. (2022). Monitoring coliphages to reduce waterborne infectious disease transmission in the One Water framework. *International Journal of Hygiene and Environmental Health., 240*, 113921. https://doi.org/10.1016/j.ijheh.2022.113921

Fong, T.-T., & Lipp, E. K. (2005). Enteric viruses of humans and animals in aquatic environments: Health risks, detection, and potential water quality assessment tools. *Microbiology and Molecular Biology Reviews, 69*(2), 357. https://doi.org/10.1128/mmbr.69.2.357-371.2005

Galani, A., et al. (2022). SARS-CoV-2 wastewater surveillance data can predict hospitalizations and ICU admissions. *Science of the Total Environment, 804*, 150151. https://doi.org/10.1016/j.scitotenv.2021.150151

Gall, A. M., et al. (2015). Waterborne viruses: A Barrier to safe drinking water. *PLoS Pathogens, 11*, e1004867. https://doi.org/10.1371/journal.ppat.1004867

Garcia, S. N., Osburn, B. I., & Jay-Russell, M. T. (2020). One health for food safety, food security, and sustainable food production. *Frontiers in Sustainable Food Systems, 4*, 1. https://doi.org/10.3389/fsufs.2020.00001

García-Aljaro, C., et al. (2017). Determination of crAssphage in water samples and applicability for tracking human faecal pollution. *Microbial Biotechnology, 10*(6), 1775. https://doi.org/10.1111/1751-7915.12841

Garvey, M. (2019). Food pollution: a comprehensive review of chemical and biological sources of food contamination and impact on human health. *Nutrire, 44*, 1. https://doi.org/10.1186/s41110-019-0096-3

Geldenhuys, J. C., & Pretorius, P. D. (1989). The occurrence of enteric viruses in polluted water, correlation to indicator organisms and factors influencing their numbers. *Water Science and Technology, 21*(3), 105. https://doi.org/10.2166/wst.1989.0086

Geldreich, E. E., & Bordner, R. H. (1971). Fecal contamination of fruits and vegetables during cultivation and processing for market. A review. *Journal of Food Protection, 34*(4), 184–195. https://doi.org/10.4315/0022-2747-34.4.184

Gerdes, M. E., et al. (2023). Estimating waterborne infectious disease burden by exposure route, United States, 2014. *Emerging Infectious Diseases, 29*(7), 1357. https://doi.org/10.3201/eid2907.230231

Gholipour, S., et al. (2022). Occurrence of viruses in sewage sludge: A systematic review. *Science of the Total Environment, 824*, 153886. https://doi.org/10.1016/j.scitotenv.2022.153886

Ginebreda, A., et al. (2023). Environmental proteomics: A potential tool in wastewater-based epidemiology. *Water Emerging Contaminants & Nanoplastics, 2*. https://doi.org/10.20517/wecn.2023.19

Gore, E. J., et al. (2023). Understanding torquetenovirus (TTV) as an immune marker. *Frontiers in Medicine, 10*. https://doi.org/10.3389/fmed.2023.1168400

Griffin, J. S., Plummer, J. D., & Long, S. C. (2008). Torque teno virus: An improved indicator for viral pathogens in drinking waters. *Virology Journal, 5*, 112. https://doi.org/10.1186/1743-422X-5-112

Guajardo-Leiva, S., et al. (2020). Metagenomic insights into the sewage RNA virosphere of a large city. *Viruses, 12*(9). https://doi.org/10.3390/v12091050

Guerin, E., et al. (2018). Biology and taxonomy of crAss-like bacteriophages, the most abundant virus in the human gut. *Cell Host and Microbe, 24*, 653. https://doi.org/10.1016/j.chom.2018.10.002

Guo, J., et al. (2021). VirSorter2: A multi-classifier, expert-guided approach to detect diverse DNA and RNA viruses. *Microbiome, 9*(1), 37. https://doi.org/10.1186/s40168-020-00990-y

Gyawali, P., et al. (2019). Evaluation of pepper mild mottle virus as an indicator of human faecal pollution in shellfish and growing waters. *Water Research, 154*, 370. https://doi.org/10.1016/j.watres.2019.02.003

Hamza, I. A., et al. (2011a). Evaluation of pepper mild mottle virus, human picobirnavirus and Torque teno virus as indicators of fecal contamination in river water. *Water Research, 45*(3), 1358. https://doi.org/10.1016/j.watres.2010.10.021

Hamza, I. A., et al. (2011b). Methods to detect infectious human enteric viruses in environmental water samples. *International Journal of Hygiene and Environmental Health, 214*(6), 424. https://doi.org/10.1016/j.ijheh.2011.07.014

Hayes, S., et al. (1990). Septic tank and agricultural non-point source pollution within a rural watershed. *Toxicological & Environmental Chemistry, 26*(1–4), 137. https://doi.org/10.1080/02772249009357541

Hernández-Delgado, E. A., & Toranzos, G. A. (1995). In situ replication studies of somatic and male-specific coliphages in a tropical pristine river. *Water Science and Technology, 31*(5–6). https://doi.org/10.1016/0273-1223(95)00274-Q

Holcomb, D. A., & Stewart, J. R. (2020). Microbial indicators of fecal pollution: recent progress and challenges in assessing water quality. *Current Environmental Health Reports, 7*, 311. https://doi.org/10.1007/s40572-020-00278-1

Holt, M. S. (2000). Sources of chemical contaminants and routes into the freshwater environment. *Food and Chemical Toxicology, 38*(SUPPL.1), S21. https://doi.org/10.1016/s0278-6915(99)00136-2

Hyman, P. (2019). Phages for phage therapy: Isolation, characterization, and host range breadth. *Pharmaceuticals, 12*. https://doi.org/10.3390/ph12010035

Ibrahim, Y., et al. (2021). Detection and removal of waterborne enteric viruses from wastewater: A comprehensive review. *Journal of Environmental Chemical Engineering, 9*, 105613. https://doi.org/10.1016/j.jece.2021.105613

Ilett, E. E., et al. (2019). Gut microbiome comparability of fresh-frozen versus stabilized-frozen samples from hospitalized patients using 16S rRNA gene and shotgun metagenomic sequencing. *Scientific Reports, 9*(1), 13351. https://doi.org/10.1038/s41598-019-49956-7

Ishii, S., et al. (2014). Microfluidic quantitative PCR for simultaneous quantification of multiple viruses in environmental water samples. *Applied and Environmental Microbiology, 80*(24), 7505. https://doi.org/10.1128/AEM.02578-14

Jahne, M. A., et al. (2020). Droplet digital PCR quantification of norovirus and adenovirus in decentralized wastewater and graywater collections: Implications for onsite reuse. *Water Research, 169*. https://doi.org/10.1016/j.watres.2019.115213

Jain, P., & Jain, A. (2014). Waterborne viral gastroenteritis: An introduction to common agents. *Water and Health*. https://doi.org/10.1007/978-81-322-1029-0_4

Jarret, R. L., et al. (2008). The occurrence and control of pepper mild mottle virus (PMMoV) in the USDA/ARS Capsicum germplasm collection. *Seed Technology, 30*(1).

Jofre, J. (2009). Is the replication of somatic coliphages in water environments significant? *Journal of Applied Microbiology, 106*, 1059. https://doi.org/10.1111/j.1365-2672.2008.03957.x

Jofre, J. *et al.* (1986) 'Evaluation of Bacteroides fragilis bacteriophages as indicators of the virological quality of water.', Water Science and Technology, 18(10). doi: https://doi.org/10.2166/wst.1986.0126.

Jofre, J., et al. (2014). Bacteriophages infecting Bacteroides as a marker for microbial source tracking. *Water Research, 55*, 1. https://doi.org/10.1016/j.watres.2014.02.006

Jofre, J., et al. (2016). Coliphages as model organisms in the characterization and management of water resources. *Water (Switzerland), 8*. https://doi.org/10.3390/w8050199

Jones, K. R., et al. (2022). Occurrence of coliphage in effluent: A systematic literature review and meta-analysis. *Hygiene and Environmental Health Advances, 3*, 100014. https://doi.org/10.1016/j.heha.2022.100014

Kambhampati, A. K., et al. (2023). Spatiotemporal trends in norovirus outbreaks in the United States, 2009–2019. *Clinical Infectious Diseases, 76*(4), 667. https://doi.org/10.1093/cid/ciac627

Kieft, K., Zhou, Z., & Anantharaman, K. (2020). VIBRANT: Automated recovery, annotation and curation of microbial viruses, and evaluation of viral community function from genomic sequences. *Microbiome (BioMed Central), 8*(1), 1–23.

Kim, M. S., Chen, Y. R., Cho, B. K., et al. (2007). Hyperspectral reflectance and fluorescence line-scan imaging for online defect and fecal contamination inspection of apples. *Sensing and Instrumentation for Food Quality and Safety, 1*, 151–159. https://doi.org/10.1007/s11694-007-9017-x

Kitajima, M., Sassi, H. P., & Torrey, J. R. (2018). 'Pepper mild mottle virus as a water quality indicator', *npj Clean. Water, 1*. https://doi.org/10.1038/s41545-018-0019-5

Kleiner, M., Hooper, L. V., & Duerkop, B. A. (2015). Evaluation of methods to purify virus-like particles for metagenomic sequencing of intestinal viromes. *BMC Genomics, 16*. https://doi.org/10.1186/s12864-014-1207-4

Kongprajug, A., Mongkolsuk, S., & Sirikanchana, K. (2019). CrAssphage as a potential human sewage marker for microbial source tracking in Southeast Asia. *Environmental Science and Technology Letters, 6*(3), 159. https://doi.org/10.1021/acs.estlett.9b00041

Kool, J., et al. (2023). Reducing bias in microbiome research: Comparing methods from sample collection to sequencing. *Frontiers in Microbiology, 14*. https://doi.org/10.3389/fmicb.2023.1094800

Krishnamurthy, S. R., & Wang, D. (2017). Origins and challenges of viral dark matter. *Virus Research, 239*, 136–142. https://doi.org/10.1016/j.virusres.2017.02.002

Kumar, R., et al. (2014). Getting started with microbiome analysis: Sample acquisition to bioinformatics. *Current Protocols in Human Genetics, 2014*. https://doi.org/10.1002/0471142905.hg1808s82

Lauri, A., & Mariani, P. O. (2009 Mar). Potentials and limitations of molecular diagnostic methods in food safety. *Genes & Nutrition, 4*(1), 1–12. https://doi.org/10.1007/s12263-008-0106-1

Le François, B., & Doukhanine, E. (2019) OMNIgene®• GUT accurately captures and stabilizes the human fecal dsDNA virome.

Lee, J. H., et al. (2013). Development of a new cell culture-based method and optimized protocol for the detection of enteric viruses. *Journal of Virological Methods, 191*(1), 16. https://doi.org/10.1016/j.jviromet.2013.03.019

Levy, K., Smith, S. M., & Carlton, E. J. (2018). Climate change impacts on waterborne diseases: Moving toward designing interventions. *Current Environmental Health Reports, 5*, 272. https://doi.org/10.1007/s40572-018-0199-7

Lewis, G. D., & Metcalf, T. G. (1988). Polyethylene glycol precipitation for recovery of pathogenic viruses, including hepatitis A virus and human rotavirus, from oyster, water, and sediment samples. *Applied and Environmental Microbiology, 54*(8), 1983. https://doi.org/10.1128/aem.54.8.1983-1988.1988

Li, E., et al. (2021a). Biological indicators for fecal pollution detection and source tracking: A review. *PRO, 9*. https://doi.org/10.3390/pr9112058

Li, Y., et al. (2021b). CrAssphage and its bacterial host in cat feces. *Scientific Reports, 11*(1), 815. https://doi.org/10.1038/s41598-020-80076-9

Li, R., et al. (2022). Viral diversity and potential environmental risk in microplastic at watershed scale: Evidence from metagenomic analysis of plastisphere. *Environment International, 161*, 107146. https://doi.org/10.1016/j.envint.2022.107146

Li, C., et al. (2023). Waterborne virus transport and the associated risks in a large lake. *Water Research (Elsevier), 229*, 119437.

Lodder, W. J., et al. (2010). Presence of enteric viruses in source waters for drinking water production in The Netherlands. *Applied and Environmental Microbiology, 76*(17), 5965. https://doi.org/10.1128/AEM.00245-10

Mafumo, N., et al. (2023). CrAssphage may be viable markers of contamination in pristine and contaminated river water. *mSystems, 8*(1), e0128222. https://doi.org/10.1128/msystems.01282-22

Magana-Arachchi, D. N., & Wanigatunge, R. P. (2020). Ubiquitous waterborne pathogens. *Waterborne Pathogens: Detection and Treatment*. https://doi.org/10.1016/B978-0-12-818783-8.00002-5

Maghini, D. G., et al. (2024). Quantifying bias introduced by sample collection in relative and absolute microbiome measurements. *Nature Biotechnology, 42*(2), 328. https://doi.org/10.1038/s41587-023-01754-3

Malik, Y. S., et al. (2019). Advances in diagnostic approaches for viral etiologies of diarrhea: From the lab to the field. *Frontiers in Microbiology, 10*. https://doi.org/10.3389/fmicb.2019.01957

Mancini, N., Carletti, S., Ghidoli, N., Cichero, P., Burioni, R., & Clementi, M. (2010). The era of molecular and other non-culture-based methods in diagnosis of sepsis. *Clinical Microbiology Reviews, 23*(1), 235–251. https://doi.org/10.1128/CMR.00043-09

Martínez-Ochoa, N., et al. (2003). First Report of Pepper mild mottle virus in Jalapeno Pepper in Georgia. *Plant Health Progress, 4*(1). https://doi.org/10.1094/php-2003-1223-01-hn

McKindles, K. M., & Tiquia-Arashiro, S. M. (2012). Functional gene arrays for analysis of microbial communities on ocean platform. In S. Tiquia-Arashiro (Ed.), *Molecular biological technologies for ocean sensing* (Springer protocols handbooks). Humana Press. https://doi.org/10.1007/978-1-61779-915-0_9

McLaughlin, M. R., & Rose, J. B. (2006). Application of Bacteroides fragilis phage as an alternative indicator of sewage pollution in Tampa Bay, Florida. *Estuaries and Coasts (Springer), 29*, 246–256.

Mcminn, B. R., Korajkic, A., & Ashbolt, N. J. (2014). Evaluation of Bacteroides fragilis GB-124 bacteriophages as novel human-associated faecal indicators in the United States. *Letters in Applied Microbiology, 59*(1), 115. https://doi.org/10.1111/lam.12252

Medema, G., et al. (2020). Presence of SARS-Coronavirus-2 RNA in sewage and correlation with reported COVID-19 prevalence in the early stage of the Epidemic in The Netherlands. *Environmental Science and Technology Letters, 7*, 511. https://doi.org/10.1021/acs.estlett.0c00357

Montazeri, N., et al. (2015). Pathogenic enteric viruses and microbial indicators during secondary treatment of municipal wastewater. *Applied and Environmental Microbiology, 81*(18), 6436. https://doi.org/10.1128/AEM.01218-15

Monteiro, S., & Santos, R. (2017). Nanofluidic digital PCR for the quantification of Norovirus for water quality assessment. *PLoS One, 12*(7), e0179985. https://doi.org/10.1371/journal.pone.0179985

Moresco, V., Damazo, N. A., & Barardi, C. R. M. (2016). Thermal and temporal stability on the enteric viruses infectivity in surface freshwater. *Water Science and Technology: Water Supply, 16*(3), 620. https://doi.org/10.2166/ws.2015.171

Muniesa, M., & Jofre, J. (2004). Factors influencing the replication of somatic coliphages in the water environment, Antonie van Leeuwenhoek. *International Journal of General and Molecular Microbiology, 86*(1), 65. https://doi.org/10.1023/B:ANTO.0000024909.75523.be

Nanda, S., et al. (2008). Universal virus detection by degenerate-oligonucleotide primed polymerase chain reaction of purified viral nucleic acids. *Journal of Virological Methods, 152*(1–2), 18. https://doi.org/10.1016/j.jviromet.2008.06.007

Nassereddine, Z., Opara, S., Coutinho, O., Qyteti, F., Book, R., Heinicke, M. P., Napieralski, J., & Tiquia-Arashiro, S. M. (2024). Critical perspectives on advancing antibiotic resistant gene (ARG) detection technologies in aquatic ecosystems. *Science of the Total Environment, 957*, 177775. https://doi.org/10.1016/j.scitotenv.2024.177775

Nemudryi, A., et al. (2020). Temporal detection and phylogenetic assessment of SARS-CoV-2 in municipal wastewater. *Cell Reports Medicine, 1*(6), 100098. https://doi.org/10.1016/j.xcrm.2020.100098

Nguyen, S., Ala, F., Cardwell, C., Cai, D., McKindles, K. M., Lotvola, A., Hodges, S., Deng, Y., & Tiquia-Arashiro, S. M. (2013). Isolation and screening of carboxydotrophs isolated from composts and their potential for butanol synthesis. *Environmental Technology, 34*, 1995–2007. https://doi.org/10.1080/09593330.2013.795987

Nick, S. T., et al. (2019). Occurrence and molecular characterization of Torque teno virus (TTV) in a wastewater treatment plant in Tehran. *Journal of Water and Health, 17*(6), 971. https://doi.org/10.2166/wh.2019.137

Novoa, B., et al. (2022). Wastewater and marine bioindicators surveillance to anticipate COVID-19 prevalence and to explore SARS-CoV-2 diversity by next generation sequencing: One-year study. *Science of the Total Environment, 833*, 155140. https://doi.org/10.1016/j.scitotenv.2022.155140

Oon, Y.-L., et al. (2023). Waterborne pathogens detection technologies: advances, challenges, and future perspectives. *Frontiers in Microbiology, 14*, 1286923.

Pachepsky, Y., et al. (2011). Irrigation waters as a source of pathogenic microorganisms in produce. A review. *Advances in Agronomy.* https://doi.org/10.1016/B978-0-12-386473-4.00007-5

Pagnucco, G., Overfield, D., Chamlee, Y., Fortuna, A., Sulaiman, F., Farinas, J., Schittenhelm, R., Shuler, C., Kassem, A., Opara, S., Najaf, H., Abbas, L., Coutinho, O., Catalfano, B., Li, X., & Tiquia-Arashiro, S. M. (2023). Metal Tolerance and Biosorption Capacities of Bacterial Strains Isolated from an Urban Watershed. *Frontiers in Microbiology, 14*, 1278886. https://doi.org/10.3389/fmicb.2023.1278886

Pandolfo, M., et al. (2022). MetaPhage: An automated pipeline for analyzing, annotating, and classifying bacteriophages in metagenomics sequencing data. *mSystems, 7*(5), e0074122. https://doi.org/10.1128/msystems.00741-22
Puig, M., & Gironés, R. (1999). Genomic structure of phage B40-8 of Bacteroides fragilis. *Microbiology, 145*(7), 1661. https://doi.org/10.1099/13500872-145-7-1661
Purnell, S. E., Ebdon, J. E., & Taylor, H. D. (2011). Bacteriophage lysis of enterococcus host strains: A tool for microbial source tracking? *Environmental Science and Technology, 45*(24), 10699. https://doi.org/10.1021/es202141x
Ralston, E. P., Kite-Powell, H., & Beet, A. (2011). An estimate of the cost of acute health effects from food- and water-borne marine pathogens and toxins in the USA. *Journal of Water and Health, 9*(4), 680. https://doi.org/10.2166/wh.2011.157
Ren, J., et al. (2020). Identifying viruses from metagenomic data using deep learning. *Quantitative Biology, 8*, 64. https://doi.org/10.1007/s40484-019-0187-4
Rosario, K., et al. (2009). Pepper mild mottle virus as an indicator of fecal pollution. *Applied and Environmental Microbiology, 75*(22), 7261. https://doi.org/10.1128/AEM.00410-09
Rzezutka, A., & Cook, N. (2004). Survival of human enteric viruses in the environment and food. *FEMS Microbiology Reviews, 28*, 441. https://doi.org/10.1016/j.femsre.2004.02.001
Sabatier, M., et al. (2020). Comparison of nucleic acid extraction methods for a viral metagenomics analysis of respiratory viruses. *Microorganisms, 8*(10). https://doi.org/10.3390/microorganisms8101539
Santiago-Rodriguez, T. M., & Hollister, E. B. (2019). Human virome and disease: High-throughput sequencing for virus discovery, identification of phage-bacteria dysbiosis and development of therapeutic approaches with emphasis on the human gut. *Viruses, 11*. https://doi.org/10.3390/v11070656
Santiago-Rodriguez, T. M., & Hollister, E. B. (2020). Potential applications of human viral metagenomics and reference materials: Considerations for current and future viruses. *Applied and Environmental Microbiology, 86*. https://doi.org/10.1128/AEM.01794-20
Santiago-Rodriguez, T. M., & Hollister, E. B. (2022). Unraveling the viral dark matter through viral metagenomics. *Frontiers in Immunology (Frontiers Media SA), 13*, 1005107.
Santiago-Rodriguez, T. M., & Hollister, E. B. (2023). Viral metagenomics as a tool to track sources of fecal contamination: A one health approach. *Viruses, 15*. https://doi.org/10.3390/v15010236
Santiago-Rodríguez, T. M., et al. (2010). Characterization of Enterococcus faecalis-infecting phages (enterophages) as markers of human fecal pollution in recreational waters. *Water Research, 44*, 4716. https://doi.org/10.1016/j.watres.2010.07.078
Santiago-Rodriguez, T. M., et al. (2012). Microbial quality of tropical inland waters and effects of rainfall events. *Applied and Environmental Microbiology, 78*, 5160. https://doi.org/10.1128/AEM.07773-11
Santiago-Rodriguez, T. M., et al. (2013). Evaluation of Enterococcus-infecting phages as indices of fecal pollution. *Journal of Water and Health, 11*(1), 51. https://doi.org/10.2166/wh.2012.100
Santiago-Rodriguez, T. M., Ly, M., Bonilla, N., et al. (2015a). The human urine virome in association with urinary tract infections. *Frontiers in Microbiology, 6*. https://doi.org/10.3389/fmicb.2015.00014
Santiago-Rodriguez, T. M., Ly, M., Daigneault, M. C., et al. (2015b). Chemostat culture systems support diverse bacteriophage communities from human feces. *Microbiome, 3*, 58. https://doi.org/10.1186/s40168-015-0124-3
Santiago-Rodriguez, T. M., Toranzos, G. A., & Arce-Nazario, J. A. (2016). Assessing the microbial quality of a tropical watershed with an urbanization gradient using traditional and alternate fecal indicators. *Journal of Water and Health, 14*, 796. https://doi.org/10.2166/wh.2016.041
Saxena, G., Bharagava, R. N., Kaithwas, G., & Raj, A. (2015). Microbial indicators, pathogens and methods for their monitoring in water environment. *Journal of Water and Health, 13*(2), 319–339. https://doi.org/10.2166/wh.2014.275
Seeley, N. D., & Primrose, S. B. (1980). The effect of temperature on the ecology of aquatic bacteriophages. *Journal of General Virology, 46*(1), 87. https://doi.org/10.1099/0022-1317-46-1-87

Shkoporov, A. N., & Hill, C. (2019). Bacteriophages of the Human Gut: The "Known Unknown" of the microbiome. *Cell Host and Microbe, 25*, 195. https://doi.org/10.1016/j.chom.2019.01.017
Shkoporov, A. N., et al. (2018). ΦCrAss001 represents the most abundant bacteriophage family in the human gut and infects Bacteroides intestinalis. *Nature Communications, 9*, 4781. https://doi.org/10.1038/s41467-018-07225-7
Shrestha, A., et al. (2020). Fecal pollution source characterization at non-point source impacted beaches under dry and wet weather conditions. *Water Research, 182*, 116014. https://doi.org/10.1016/j.watres.2020.116014
Singh, S., Pitchers, R., & Hassard, F. (2022). Coliphages as viral indicators of sanitary significance for drinking water. *Frontiers in Microbiology, 13*. https://doi.org/10.3389/fmicb.2022.941532
Sommers, P., et al. (2021). Integrating viral metagenomics into an ecological framework. *Annual Review of Virology., 8*, 133. https://doi.org/10.1146/annurev-virology-010421-053015
Stachler, E., & Bibby, K. (2014). Metagenomic evaluation of the highly abundant human gut bacteriophage CrAssphage for source tracking of human fecal pollution. *Environmental Science and Technology Letters., 1*, 405. https://doi.org/10.1021/ez500266s
Stachler, E., et al. (2017). Quantitative CrAssphage PCR assays for human fecal pollution measurement. *Environmental Science and Technology, 51*(16), 9146. https://doi.org/10.1021/acs.est.7b02703
Steward, G. F., & Culley, A. I. (2010). Extraction and purification of nucleic acids from viruses. *Manual of Aquatic Viral Ecology*. https://doi.org/10.4319/mave.2010.978-0-9845591-0-7.154
Stobnicka-Kupiec, A., et al. (2022). Detection and identification of potentially infectious gastrointestinal and respiratory viruses at workplaces of wastewater treatment plants with viability qPCR/RT-qPCR. *Scientific Reports, 12*(1), 4517. https://doi.org/10.1038/s41598-022-08452-1
Sukhorukov, G., et al. (2022). VirHunter: A deep learning-based method for detection of novel RNA viruses in plant sequencing data. *Frontiers in Bioinformatics, 2*. https://doi.org/10.3389/fbinf.2022.867111
Takayama, S., et al. (1999). Prevalence and persistence of a novel DNA TT virus (TTV) infection in Japanese haemophiliacs. *British Journal of Haematology, 104*(3), 626. https://doi.org/10.1046/j.1365-2141.1999.01207.x
Tariq, M. A., et al. (2020). Genome characterization of a novel wastewater Bacteroides fragilis Bacteriophage (vB_BfrS_23) and its host GB124. *Frontiers in Microbiology, 11*. https://doi.org/10.3389/fmicb.2020.583378
Tartera, C., Bosch, A., & Jofre, J. (1988). The inactivation of bacteriophages infecting Bacteroides fragilis by chlorine treatment and UV-irradiation. *FEMS Microbiology Letters, 56*(3), 313. https://doi.org/10.1111/j.1574-6968.1988.tb03198.x
Tartera, C., Lucena, F., & Jofre, J. (1989). Human origin of Bacteroides fragilis bacteriophages present in the environment. *Applied and Environmental Microbiology., 55*, 2696. https://doi.org/10.1128/aem.55.10.2696-2701.1989
Teunis, P. F. M., et al. (2008). Norwalk virus: How infectious is it? *Journal of Medical Virology, 80*(8), 1468. https://doi.org/10.1002/jmv.21237
Tiquia, S. M., Tam, N. F. Y., & Hodgkiss, I. J. (1998). Salmonella elimination during composting of spent pig litter. *Bioresource Technology, 63*(2), 193–196. https://doi.org/10.1016/S0960-8524(97)00113-2
Toranzos, G. A. (1991). Current and possible alternate indicators of fecal contamination in tropical waters: A short review. *Environmental Toxicology and Water Quality, 6*(2), 121. https://doi.org/10.1002/tox.2530060202
Toribio-Avedillo, D., et al. (2021). Bacteriophages as fecal pollution indicators. *Viruses, 13*. https://doi.org/10.3390/v13061089
Trubl, G., et al. (2019). Towards optimized viral metagenomes for double-stranded and single-stranded DNA viruses from challenging soils. *PeerJ, 2019*(7), e7265. https://doi.org/10.7717/peerj.7265

Tsuda, S., et al. (2007). Pathogenicity of Pepper mild mottle virus is controlled by the RNA silencing suppression activity of its replication protein but not the viral accumulation. *Phytopathology, 97*(4), 412. https://doi.org/10.1094/PHYTO-97-4-0412

UN-Water. (2021). *Summary progress update 2021: SDG 6—water and sanitation for all.*

USEPA. (2016). *Quick guide to drinking water sample collection.* U.S. Environmental Protection Agency. Available at: https://www.epa.gov/sites/production/files/2015-11/documents/drinking_water_sample_collection.pdf

Vaughn, J. M., & Metcalf, T. G. (1975). Coliphages as indicators of enteric viruses in shellfish and shellfish raising estuarine waters. *Water Research, 9*(7), 613. https://doi.org/10.1016/0043-1354(75)90165-7

Verani, M., et al. (2006). One-year monthly monitoring of Torque teno virus (TTV) in river water in Italy. *Water Science and Technology, 54*(3), 191. https://doi.org/10.2166/wst.2006.468

Vergara, G. G. R. V., et al. (2015). Evaluation of FRNA coliphages as indicators of human enteric viruses in a tropical urban freshwater catchment. *Water Research, 79*, 39. https://doi.org/10.1016/j.watres.2015.04.022

Vicente, D., et al. (2007). Human bocavirus, a respiratory and enteric virus. *Emerging Infectious Diseases, 13*(4), 636. https://doi.org/10.3201/eid1304.061501

Vijayavel, K., et al. (2014). Enterococcus phages as potential tool for identifying sewage inputs in the Great Lakes region. *Journal of Great Lakes Research, 40*(4), 989. https://doi.org/10.1016/j.jglr.2014.09.011

Voroshilova, M. K. (1989). Potential use of nonpathogenic enteroviruses for control of human disease. *Progress in Medical Virology, 36, 191–202.*

Wang, X., et al. (2006). Detection and molecular characterization of Pepper mild mottle virus in China. *Journal of Phytopathology, 154*(11–12). https://doi.org/10.1111/j.1439-0434.2006.01186.x

Wang, Y., et al. (2024). Metagenomic analysis of antibiotic-resistance genes and viruses released from glaciers into downstream habitats. *Science of the Total Environment, 908*, 168310. https://doi.org/10.1016/j.scitotenv.2023.168310

Ward, R. L., Knowlton, D. R., & Winston, P. E. (1986). Mechanism of inactivation of enteric viruses in fresh water. *Applied and Environmental Microbiology, 52*(3), 450. https://doi.org/10.1128/aem.52.3.450-459.1986

Watanabe, R. A., Fryer, J. L., & Rohovec, J. S. (1988). Molecular filtration for recovery of waterborne viruses of fish. *Applied and Environmental Microbiology, 54*(6), 1606. https://doi.org/10.1128/aem.54.6.1606-1609.1988

Watson, E. J., et al. (2019). Human faecal collection methods demonstrate a bias in microbiome composition by cell wall structure. *Scientific Reports, 9*(1), 16831. https://doi.org/10.1038/s41598-019-53183-5

Worley-Morse, T., et al. (2019). Evaluating the fate of bacterial indicators, viral indicators, and viruses in water resource recovery facilities. *Water Environment Research, 91*(9), 830. https://doi.org/10.1002/wer.1096

Wyn-Jones, A. P., & Sellwood, J. (2001). Enteric viruses in the aquatic environment. *Journal of Applied Microbiology, 91*(6), 945–962.

Xiao, F., et al. (2020). Infectious SARS-CoV-2 in feces of patient with severe COVID-19. *Emerging Infectious Diseases, 26*(8), 1920. https://doi.org/10.3201/eid2608.200681

Yan, A., et al. (2023). Multiomic spatial analysis reveals a distinct mucosa-associated virome. *Gut Microbes, 15*(1). https://doi.org/10.1080/19490976.2023.2177488

Yates, M. V. (2007). Classical Indicators in the 21st Century-Far and Beyond the Coliform. *Water Environment Research, 79*(3), 279. https://doi.org/10.2175/106143006x123085

Yong, S. F., et al. (2006). Real-time PCR detection of male-specific coliphages. *The Malaysian Journal of Pathology, 28*(2), 79.

Yutin, N., et al. (2021). Analysis of metagenome-assembled viral genomes from the human gut reveals diverse putative CrAss-like phages with unique genomic features. *Nature Communications, 12*(1). https://doi.org/10.1038/s41467-021-21350-w

Zerva, I., et al. (2021). Effect of chlorination on microbiological quality of effluent of a full-scale wastewater treatment plant. *Life, 11*(1). https://doi.org/10.3390/life11010068

Zhang, T., et al. (2006). RNA viral community in human feces: Prevalence of plant pathogenic viruses. *PLoS Biology, 4*(1), e3. https://doi.org/10.1371/journal.pbio.0040003

Chapter 2
Evaluation of Microbial Community Assembly and Their Interactions with Riverine and Anthropogenic DOM Under Different Urbanization Levels

Binessi Edouard Ifon, Bob Adyari, Qian Sun, Okugbe Ebiotubo Ohore, Chang-Ping Yu, and Anyi Hu

1 Introduction

Chemical engineering has significantly influenced the extensive production, transformation, and utilization of chemicals, materials, pharmaceuticals, and energy, thereby advancing various industries and sectors (e.g., cosmetics, hospitals, and veterinary centers) and benefiting humanity. While chemical engineering also prioritizes environmental considerations by developing sustainable and eco-friendly solutions like waste treatment and pollution control technologies, the spread and effects of anthropogenic organic matters (AOMs) in aquatic ecosystems remain a

B. E. Ifon
College of Engineering, Shantou University, Shantou, China

CAS Key Laboratory of Urban Pollutant Conversion, Institute of Urban Environment, Chinese Academy of Sciences, Xiamen, China

Laboratory of Physical-Chemistry, Materials and Molecular Modeling (LCP3M), University of Abomey-Calavi, Republic of Benin, Cotonou, Benin

B. Adyari
Department of Environmental Engineering, Universitas Pertamina, Jakarta, Indonesia

Q. Sun · C.-P. Yu
CAS Key Laboratory of Urban Pollutant Conversion, Institute of Urban Environment, Chinese Academy of Sciences, Xiamen, China

O. E. Ohore
Ministry of Education Key Laboratory of Tropical Translational Medicine, School of Tropical Medicine, Hainan Medical University, Haikou, People's Republic of China

A. Hu (✉)
Carbon Neutral Innovation Research Center, Xiamen University, Xiamen, China
e-mail: ayhu@xmu.edu.cn

S. M. Tiquia-Arashiro et al. (eds.), *Urban Watershed Microbiology, Volume 1*,
https://doi.org/10.1007/978-3-031-89071-0_2

significant concern. The spread of AOMs such as bisphenols, antibiotics pharmaceuticals, and personal care products and metals in aquatic environments is increasing due to increasing urbanization and economic growth (Cizmas et al., 2015; Oest et al., 2018; Patel et al., 2019; Ávila & García, 2014; Anand et al., 2022; Zhang et al., 2022). Waste management of AOMs is a global issue faced by the modern world due to the increase in the variety of anthropogenic activities such as agriculture, industries, and personal care, which generate large waste of AOMs (Ávila & García, 2014; Nassereddine et al., 2024). These AOMs are among a growing group of micropollutants known as contaminants of emerging concerns (Ávila & García, 2014), which can persist in the environment as anthropogenic dissolved organic matters (ADOMs). Recently, it was revealed that aside from drug manufacturing plants, the primary sources of ADOMs' introduction into the environment are hospitals, veterinary centers, animal husbandry, and aquaculture (Dey et al., 2019). The discharge of effluents from wastewater treatment plants (WWTPs) and untreated sewages, where ADOMs removal is usually incomplete, is one of their main continuous sources into the natural aquatic environment (Zhao et al., 2010; Tiquia, 2011; Cho et al., 2012; Ávila & García, 2014; Gan et al., 2022; Rizzuto et al., 2021). Therefore, it is paramount to continuously monitor their impacts on aquatic ecosystems.

In recent years, there has been a growing awareness of the increasing release of ADOMs into various compartments of the aquatic environment (Jennifer et al., 2017; Rizzuto et al., 2021). Although some studies have been conducted on ADOMs in natural aquatic environments, these were mainly focused on their sources' occurrences, characterization, and distribution. For instance, downstream areas were said to be the major contributors (55.6–100%) to the spatial variability of most ADOMs, and the local contribution of ADOMs predominated over neighbor influence with an average value of 75.5% in a typical urbanizing Houxi river (Zhang et al., 2022). The pool of dissolved organic matters (DOMs) encompasses a diverse array of substances, including natural DOMs (herein referred to as DOMs) and xenobiotics DOMs (herein referred to as ADOMs, which are typically included in the class of contaminants of emerging concern) (Khanzada et al., 2020; Wünsch & Murphy, 2021; Feng et al., 2023). ADOMs and DOMs significantly impact aquatic ecosystems and disrupt aquatic biodiversity by affecting organism growth and survival, leading to altered species composition (Sinsabaugh & Foreman, 2003; Tiquia et al., 2007; Adyari et al., 2020; Perveen et al., 2023). These components disrupt key processes like nutrient cycling and primary production, cascading through the ecosystem and causing trophic shifts (Paerl et al., 2003; Tiquia-Arashiro, 2012; Kim et al., 2020; Liang et al., 2021). They pose health risks through drinking water, recreational activities, and seafood consumption, with potential toxic effects and antibiotic resistance (Tak & Vellanki, 2018; Feng et al., 2023). Additionally, they reduce ecosystem resilience to stressors like climate change, weakening organisms' ability to adapt and recover, necessitating effective management and further research for mitigation. It was said that natural DOMs and water pH alter the possible risks posed by ADOMs and that the binding of ADOMs to DOMs was driven by the presence of carboxylic groups and shifting of the structural configuration of DOMs

(Rizzuto et al., 2021; Kassem et al., 2023). However, it is poorly known how ADOMs influence riverine DOM component composition and fluorescence.

Microorganisms as well as DOMs govern the majority of biogeochemical cycles and are key elements for the degradation and removal of chemical pollutants such as ADOMs from aquatic ecosystems (Paerl & Pinckney, 1996; Tiquia et al., 2006a; Tiquia, 2010a; Tiquia, 2010b; Subashchandrabose et al., 2013; Chen et al., 2015; Flood et al., 2015; Hu et al., 2017b; Wang et al., 2019). Bacterial communities have been known for degrading riverine DOMs, but they may also consume the ADOMs for their growth (Chen et al., 2015; Li et al., 2020b). In our recent publication, we have revealed the importance of DOM optical properties (e.g., fluorescence components) for bacterial composition and structure and further demonstrated that >85% of DOM-bacterial taxa interactions differed between urbanized rivers (Ifon et al., 2023). Additionally, it was said that ADOMs can cause resistance development in several pathogenic bacterial strains in the environment (Dey et al., 2019). However, the weight of the negative compared to the positive associations of ADOMs, DOMs, and bacterial community assemblies in rivers still needs exploration.

Microbial communities were shaped by deterministic and stochastic processes as determinants of their community assembly in an aquatic ecosystem (Leibold et al., 2004; Tiquia et al., 2006b; van der Gast et al., 2008; Caruso et al., 2011; Chen et al., 2019; Gad et al., 2020). Environmental factors and the biotic interactions that are responsible for shaping the microbial communities stand as deterministic processes that drive the non-neutral bacterial community (N_NBC) (Gad et al., 2020). However, stochastic processes refer to birth, death, speciation, and immigration as the mechanisms driving the neutral bacterial community (NBC) (Gad et al., 2020). Thus, NBC is assumed not to be influenced by environmental factors while N_NBC is mainly influenced by environmental factors. However, increasing evidence has shown that changes in environmental conditions can influence the composition and behaviors of bacteria and their ecosystem function (McKindles & Tiquia-Arashiro, 2012; Hu et al., 2017b; Qin et al., 2018; Chen et al., 2019). This can further alter the processes shaping the bacterial communities, especially in urbanized river ecosystems undergoing mounting anthropogenic influences. However, the response of NBC and N_NBC assembly associations to ADOMs and DOMs in urbanized river ecosystems is yet to be investigated. Moreover, bacteria can use both natural DOMs and some ADOMs (e.g., bisphenol A) as energy sources in urban rivers (Oni et al., 2015; Hu et al., 2017a; Oest et al., 2018). Compared to physicochemical, natural DOMs have been shown to have fewer effects on bacterial communities (Zhou et al., 2021). Although DOMs' effect on bacterial communities was revealed to increase with high urbanization (Ifon et al., 2023), ADOMs' effect on bacterial communities demands further investigation. Additionally, investigating the effects of ADOMs and DOMs as well as other environmental parameters on NBC and N_NBC will provide important ecological insights. We hypothesized that riverine DOMs and ADOMs components will enhance deterministic processes on bacterial community assembly in urbanized rivers.

The peculiarity of this study was to (i) unravel ADOMs' influence on riverine DOMs components' composition and fluorescence via a laboratory assay, (ii)

untangle the ecological interactions of the neutral and non-neutral bacterial communities with ADOMs and DOMs in the urbanized rivers, and (iii) highlight the extent to which ADOMs and DOMs affect the neutral and non-neutral community compositions compared to other environmental factors. To achieve these, we conducted our study in the less urbanized North River (NR) and a more urbanized West River (WR) of the Jiulong River (JR) watershed in Fujian province, China. Jiulong River is the largest river in southern Fujian province and the second largest in this province and stands as an important source of drinking and agricultural and industrial water. NR and WR are largely influenced by anthropogenic activities that result in high input of terrestrial and autochthonous DOMs as well as important discharges of ADOMs (Ifon et al., 2023; Sun et al., 2016; Tiquia et al., 2008) thus, making a suitable study area for our current investigation. The results of this study will provide important novel insights into the impacts of ADOMs on riverine DOM pools and introduce the ecological significance of ADOMs and DOMs to the dynamics of NBC and N_NBC assemblies in human-disturbed urbanized rivers.

2 Materials and Methods

2.1 *Description of the Study Area and Sample Collection*

JR, which consists of two major rivers (i.e., NR and WR), is an anthropogenically impacted urbanized watershed in the southeast coast of China with a catchment area of 14,741 km^2, a length of 258 km, and annual mean temperatures of 19.9–21.1 °C (Hu et al., 2017a; Ifon et al., 2023). The JR ecosystem is generally influenced by agricultural over-fertilization, livestock breeding, and industrial, domestic, and municipal sewage and effluent discharge (Qu et al., 2020). Noteworthy, WR is influenced by higher levels of urbanization and human disturbances (Ifon et al., 2023). However, NR has a larger drainage area of 9803 km^2 (which accounts for 66.5% of the watershed's total area) (Hong, 2011; Yang et al., 2013). Samples were collected from 6 sites (N1, N2, N3 for NR and W1, W2, and W3 for WR) (Fig. 2.1) for 11 continuous days from August 5 to 16 of the year 2020. Following a previous study (Erdogan & Schlautman, 2015), 66 samples for dissolved organic carbon (DOC) and DOMs were syringe-filtered through a 0.45 μm pore-size polyethersulfone filter (Gelman Supor) into clean 15 mL centrifuge tubes and 30 mL pretreated amber color bottles, respectively. Moreover, 66 samples for ADOMs were collected in 1000 mL pretreated amber color bottles. All DOMs samples were kept in the dark at 4 °C for 48–72 h maximum before measurements, while ADOMs samples were processed immediately upon arrival in the laboratory. For each sampling site, the land use ratio (i.e., the proportion of woodland, grassland, cropland, built-up land, shrubland, bare land, and water bodies from a 2 km buffer zone) was determined using ArcGIS 10.2 software and Landsat-7 ETM+ (2020) data with a resolution of 30 m.

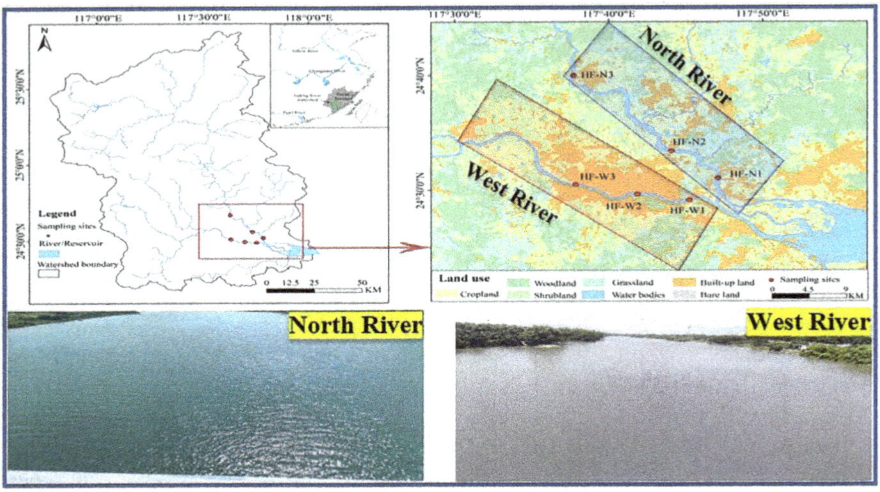

Fig. 2.1 Location of Jiulong River (JR) watershed and sampling sites at North River (NR) and West River (WR)

2.2 *Environmental Variables Measurement and Spectroscopy Analysis*

Environmental parameters including temperature (Temp), pH, electric conductivity (EC), and dissolved oxygen (DO) were measured in situ using a probe (WTW, Multi 3630 IDS SET G, Weilheim, Germany). Total phosphorous (TP) and total nitrogen (TN) were measured following a previous study (Miranda et al., 2001). However, concentrations of nitrite (NO_2—N), nitrate (NO_3^-N), ammonia (NH_4^+-N), and phosphate (PO_4^{3-}P) were determined via ion chromatography system (ICS 3000, Dionex, Sunnyvale, CA, USA); total inorganic nitrogen (TIN) was also calculated. DOC concentrations were measured using a TOC-VCPH analyzer (Shimadzu, Kyoto, Japan) (Santinelli, 2015). For DOM analysis, absorption spectra of samples were obtained from 200 to 800 nm at 1 nm intervals using an Evolution 300 PC UV-Vis spectrophotometer (Thermo Scientific, USA). However, the excitation-emission matrix (EEMs) spectra of DOM fluorescence were measured using a fluorescence spectrophotometer (Hitachi FL4600, Japan) with each series of measurements starting with Milli-Q water as the blank. EEMs wavelengths were set to 5 nm intervals with an excitation range tuned between 250 and 450 nm in 10 nm increments, while the emission range was set between 300 and 560 nm in 2 nm increments. Absorption data were used to correct a total of 66 EEMs before PARAFAC analysis via StaRdom package in R version 4.0.3 with RStudio, and the corrected factors include blank subtraction, instrument-specific influences on EEMs, absorbance baseline, and inner-filter effect (Kothawala et al., 2013; Li & Hur, 2017). Furthermore, the fluorescence intensities (Fmax; i.e., used to quantify DOM fluorescence) for each component were calculated, as was the biological index (BIX; i.e., used to track fresh autochthonous production) (Ifon et al., 2022a;

Table 2.1 Chemical properties of anthropogenic organic matters

Compounds	CAS number	Molecular formula	Molecular weight (g/mol)	Type
Carbamazepine	298-46-4	$C_{15}H_{12}N_2O$	236.269	Drugs
Diazepam	439-14-5	$C_{16}H_{13}ClN_2O$	284.7	Drugs
Losartan	11479826-4	$C_{22}H_{23}ClN_6O$	422.91	Drugs
Metoprolol	37350-58-6	$C_{15}H_{25}NO_3$	267.364	Drugs
Metronidazole	443-48-1	$C_6H_9N_3O_3$	171.16	Drugs
Miconazole	22916-47-8	$C_{18}H_{14}Cl_4N_2O$	416.127	Drugs
Propyphenazone	479-92-5	$C_{14}H_{18}N_2O$	230.31	Drugs
Sildenafil citrate	17159983-0	$C_{28}H_{38}N_6O_{11}S$	666.70	Drugs
Thiabendazole	148-79-8	$C_{10}H_7N_3S$	201.249	Drugs
Metoprolol acid	56392-14-4	$C_{14}H_{21}NO_4$	267.32	Drugs
Paraxanthine	611-59-6	$C_7H_8N_4O_2$	180.16	Drugs
Clarithromycin	81103-11-9	$C_{38}H_{69}NO_{13}$	747.953	Antibiotics
Flumequine	42835-25-6	$C_{14}H_{12}FNO_3$	261.25	Antibiotics
Lomefloxacin	98079-51-7	$C_{17}H_{19}F_2N_3O_3$	351.348	Antibiotics
Levofloxacin	10098685-4	$C_{18}H_{20}FN_3O_4$	361.368	Antibiotics
Lincomycin	154-21-2	$C_{18}H_{34}N_2O_6S$	406.5	Antibiotics
Ofloxacin	8241936-1	$C_{18}H_{20}FN_3O_4$	361.368	Antibiotics
Trimethoprim	738-70-5	$C_{14}H_{18}N_4O_3$	290.32	Antibiotics
Bisphenol A	80-05-7	$C_{15}H_{16}O_2$	228.29	Bisphenols
Bisphenol AF	1478-61-1	$C_{15}H_{10}F_6O_2$	336.23	Bisphenols
Bisphenol E	2081-08-5	C_{14} H1$_4$ O_2	214.26	Bisphenols

Ma et al., 2021). The concentrations of 21 ADOMs including carbamazepine, diazepam, losartan, metoprolol, metronidazole, miconazole, propyphenazone, sildenafil citrate, thiabendazole, metoprolol acid, paraxanthine, clarithromycin, flumequine, lomefloxacin, levofloxacin, lincomycin, ofloxacin, trimethoprim, bisphenol A, bisphenol AF, and bisphenol E were determined following a well-described previous methods (Zhang et al., 2022). Detailed information related to ADOMs is provided in Table 2.1.

2.3 Laboratory Assay

A three-litter pretreated amber bottle was used to collect riverine water, which was kept at 4 °C and used the following day for treatments (riverine water was filtered through a 0.45 μm pore-size polyethersulfone filter [Gelman Supor] before each treatment). High purity (>97%) of seven selected ADOMs including Trimethoprim, Metoprolol, Paraxanthine, Lomefloxacin, Lincomycin, Bisphenol A, and Bisphenol AF were purchased from Sigma Aldrich and Aladdin companies. Stock solutions of 1 ppm (*stock1*) and 10 ppm (*stock2*) that were filtered through a 0.45 μm pore-size polyethersulfone filter (Gelman Supor) into clean 30 mL pretreated amber color bottles were prepared for each ADOM. Two concentrations (0.1 and 1 g/L) were

used for each ADOM to treat riverine water and the reason for these two concentrations was to override the environmental concentrations in aim to forecast ADOMs' effects on DOMs' properties if there were any. Two different treatments were conducted as follows:

- *Treatment1: 2 mL of stock1 + 18 mL of riverine water.*
- *Treatment2: 2 mL of stock2 + 18 mL of riverine water.*

The pH of all treated samples was adjusted to 7.4 ± 0.2 using 1 mL of 500 ppm of sodium hydroxide (NaOH) or 1 M HCl solution for the solution with pH below 6.5 or above 8.5, respectively. All samples were shaken for 20 min and incubated in the dark for 48 h before they were taken for measurements. Before absorption and EEM spectra measurements, all samples were ten times diluted to minimize inner filter effects. To identify how ADOMs affect the riverine DOMs, absorption and EEMs spectra determination, and EEMs-PARAFAC analysis were conducted as described above. The attenuated total reflection Fourier-transform infrared (ATR-FTIR) spectroscopy (Nicolet 6700 FTIR, Thermo Fisher Scientific Inc., Waltham, MA, USA) was used to analyze the interferences of ADOMs with functional groups of riverine DOMs. We acquired the ATR-FTIR spectra within a wave number ranging from 4000 to 400 cm^{-1} with a resolution of 4 cm^{-1} (Kassem et al., 2023) and 32 scans per spectrum. The derivation of a chemometric linear system was used to solve the water interference issues by first smoothening and conducting baseline corrections before spectra plotting (Gordon et al., 2010).

2.4 Extraction, PCR Amplification, and Bioinformatics Analysis

At each site, a jug (4 L) of water was collected to determine environmental DNA samples. For environmental DNA extraction, water samples (500–1000 mL) were collected from each site, pre-filtered through 200 μm mesh (Millipore, Bedford, MA, USA), and then filtered onto 0.2 μm pore-size polycarbonate filters (47 mm, Millipore, Bedford, MA, USA). DNA samples were stored at −80 °C until extraction. The extraction of DNA was performed using a fastDNA SPIN kit for soil (Qbiogene-MP Biomedicals, Irvine, CA, USA) according to the manufacturer's protocol. The hypervariable V4-V5 region of the prokaryotic 16S rRNA genes was amplified using the universal primers 515F (5′-GTG YCA GCM GCC GCG GTA-3′) and 907R (5′-CCG YCA ATT YMT TTR AGT TT-3′). PCR reaction was conducted under the following conditions: 95 °C for 5 min, 25 cycles of 95 °C for 30 s, 55 °C for 30 s, 72 °C for 90 s, and lastly, 72 °C for 10 min (Adyari et al., 2020). Both purification and quantification of the 16S rDNA PCR products were performed according to a previous study (Hu et al., 2017b). The sequencing of the purified PCR products was done at Majorbio Bio-Pharm Technology Co., Ltd. (Shanghai, China) on an Illumina HiSeq platform using a paired-end sequencing approach (2 × 250 bp). BioProject number PRJNA861934 could be used to access the raw sequences deposited in the NCBI short reads archive database. The processing of the amplicon sequencing data was done using the LotuS pipeline (Hildebrand et al.,

2014) as reported previously (Adyari et al., 2020). The quality controls were performed to eliminate low-quality reads (Adyari et al., 2020). The reads were subsequently checked for chimera sequences and clustered into operational taxonomic units (OTUs) at 97% similarity threshold using UPARSE (Edgar, 2013). Taxonomic classification was done using an RDP classifier with SILVA database v132 (Quast et al., 2013) at 80% bootstrap cut-off. All samples were rarefied to the smallest library size (90,000 reads for the bacterial group and 50,000 reads for the micro-eukaryotic group).

2.5 Neutral Community Model for Bacterial Communities

We assess the proportion of stochastic processes on bacterial community assembly by employing a neutral community model (NCM, i.e., an adaptation of the neutral theory for large microbial populations) in R (version 4.1.0). NCM predicts the relationship between the detection frequency and relative abundance across the larger metacommunity (Chen et al., 2019; Mo et al., 2021; Wang et al., 2023b). Nm (number of metacommunities) is an estimate of community dispersal in this model. Nm describes the relationship between occurrence frequency and regional relative abundance, with N describing the size of the metacommunity and m describing the immigration rate, while R^2 represents the overall fit to the neutral model (Chen et al., 2019; Wang et al., 2023b). Bootstrapping with 1000 bootstrap replication was used to compute 95% confidence intervals around all fitting statistics. In this study, we used operational taxonomic unit (OTU) datasets from NR, WR, and both rivers combined as JR watershed. NCM divides OTUs from each dataset into three partitions based on their occurrence frequency. Some OTUs occurred more frequently or less than (above or below partition, respectively), or within (neutral partition) the 95% confidence interval of the NCM predictions. OTUs above and below prediction are classified as the non-neutral bacterial community (N_NBC) assembly likely shaped by the deterministic forces, while those within prediction are classified as the neutral bacterial community (NBC) assembly likely shaped by the stochastic forces. To investigate deviations from the NCM predictions, we compared the composition, and diversity, and calculated the estimated migration rate (m) of neutral and non-neutral bacterial partitions.

2.6 Ecological Network Analysis

To investigate the neutral and non-neutral bacterial OTUs' co-occurrence and their association with ADOMs and DOMs, networks for NR, WR, and JR were constructed using Sparse InversE Covariance estimation for Ecological Association Inference (SPIEC-EASI) (Kurtz et al., 2015). SPIEC-EASI is an interesting tool that includes functions for suitable data transformation before building the network. Each dataset (for NR, WR, and JR) contains a total of 23,110 OTUs. We set a

threshold using occurrence frequency (≥50%) to simplify each OTU dataset, which resulted in 1256, 1300, and 1210 OTUs obtained for NR, WR, and JR, respectively. We set the same threshold of occurrence frequency (≥50%) to selected ADOMs and DOMs variables. In NR 18 ADOMs were selected while 19 were selected for WR as well as for JR. However, four DOMs' variables were selected for both NR, WR, and JR. To avoid bias that may be generated by unit dissimilarity and huge differential abundance between OTUs, we converted each simplified OTU dataset into relative abundance, and the same thing for each ADOMs and DOMs variables dataset. The relative abundance of each simplified dataset was combined with its related selected ADOMs and DOMs variables dataset to run the network analysis. The model selection was performed with a variability threshold of 0.05% using the Stability Approach to Regularization Selection (StARS) and networks were visualized using Gephi 0.9.2. Neutral and non-neutral community taxa were identified in the network of NR, WR, and JR based on the classification obtained from NCM (see Sect. 2.5).

2.7 Statistical Analyses

Principal component analysis (PCA) for environmental factors distribution and Principal Coordinate Analysis (PCoA) based on the Bray–Curtis distance matrix for exploring the β-diversity pattern of the compositions of bacteria in NR and WR were performed. Permutational multivariate analysis of variance (ANOVA; using the Adonis function) was performed based on the Bray–Curtis distance to examine the significance of bacterial communities between NR and WR. Two-way analysis of variance (ANOVA) was also used to examine other significant differences. Alpha-diversity indexes including Chao1 and ACE (non-parametric species richness estimator), Pielou's evenness, and the Shannon index (a mixture of richness and evenness) were computed using the "vegan" package (Oksanen et al., 2019). Redundancy analysis (RDA) (Tang et al., 2019b) was applied to evaluate the effects of ADOMs, DOMs, and other environmental variables on the whole bacterial community alpha-diversities. Mantel's test (Mo et al., 2021) was used to identify the correlations of environmental parameters with the richness and diversity of N_NBC and NBC.

Moreover, we employ the partial least square path model (PLS-PM) (Sanchez, 2013) to specifically examine the effects of ADOMs and DOMs on the alpha- and beta-diversities of N_NBC and NBC in urbanized rivers. Eight-factor blocks (i.e., Land use, Physicochemical, Antibiotics, Bisphenols, Drugs, Humics, Proteins, and nutrient) were built against four variables' blocks (i.e., Alpha-diversity of NBC, Beta-diversity of NBC, Alpha-diversity of N_NBC, and Beta-diversity of N_NBC,) to run the PLS-PM, and only parameters with loading ≥0.7 (Benitez et al., 2020; Sanchez, 2013) were kept within each block (Table 2.2) for both rivers. To evaluate the stability of estimates by PLS-PM and the significance of the total effects of each factor on the diversities of NBC and N_NBC, the resampling method regarding the bootstrapping validation (1000 resampling) (Chin, 2010; Sanchez, 2013) was used.

Table 2.2 Loading of variables within each block of the constructed PLS-PM models

MODEL NR	Blocks	Variables	Weight	Loading	Communality	Redundancy	MODEL WR	Weight	Loading	Communality	Redundancy
	Landcover										
		Cultivated land	0.125	0.779	0.607	0		0.278	0.998	0.997	0
		Residence	0.371	0.923	0.851	0		0.206	0.883	0.78	0
		Grassland	0.247	0.956	0.913	0		0.271	0.999	0.998	0
		Woodland	0.324	0.998	0.999	0		0.285	0.946	0.895	0
	Physicochemical										
		Temp	0.303	0.725	0.525	0.035		0.275	0.729	0.395	0.024
		pH	0.485	0.913	0.834	0.057		0.5	0.918	0.843	0.051
		DO	0.364	0.927	0.859	0.058		0.417	0.885	0.783	0.048
	Antibiotics										
		Flumequine	0.373	0.957	0.916	0.067		–	–	–	–
		Trimethoprim	0.417	0.963	0.928	0.068		–	–	–	–
		Clarithromycin	0.319	0.755	0.57	0.042		0.265	0.995	0.991	0.013
		Lomeofloxacin	–	–	–	–		0.252	0.996	0.992	0.013
		Levofloxacin	–	–	–	–		0.241	0.998	0.996	0.013
		Ofloxacin	–	–	–	–		0.245	0.999	0.997	0.013
	Bisphenols										
		Bisphenol.E	–	–	–	–		0.344	0.701	0.372	0.039
		Bisphenol.A	0.737	0.904	0.818	0.092		0.836	0.945	0.894	0.095
		Bisphenol.AF	0.458	0.728	0.529	0.059		–	–	–	–

	Drugs										
		Losartan	0.441	0.778	0.605	0.147		0.195	0.831	0.691	0.161
		Carbamazepine	0.485	0.773	0.598	0.145		0.179	0.721	0.521	0.121
		Sildenafil. Citrate	0.386	0.731	0.534	0.129		0.166	0.759	0.576	0.134
		Paraxanthine	–	–	–	–		0.211	0.852	0.726	0.169
		Metoprolol	–	–	–	–		0.223	0.954	0.91	0.212
		Metoprolol. Acid	–	–	–	–		0.211	0.902	0.814	0.191
	Humics										
		C1	0.601	0.943	0.889	0.195		0.338	0.78	0.608	0.298
		C2	0.478	0.909	0.826	0.181		0.309	0.82	0.673	0.331
		C4						0.602	0.802	0.644	0.316
	Proteins										
		C3	1	1	1	0.299		1	1	1	0.487
	Nutrient										
		NH4	0.305	0.758	0.574	0.255		0.247	0.826	0.683	0.512
		SO4	0.247	0.732	0.536	0.238		0.186	0.793	0.628	0.471
		Mg	0.333	0.954	0.909	0.405		0.166	0.751	0.563	0.423
		Na	0.296	0.915	0.838	0.373		–	–	–	–
		TIN.N	–	–	–	–		0.191	0.925	0.855	0.641
		NO3	–	–	–	–		0.182	0.903	0.816	0.613
		Ca	–	–	–	–		0.203	0.908	0.825	0.618
	NBC_beta										
		PCoA1	0.747	0.908	0.824	0.537		0.506	0.803	0.645	0.219
		PCoA2	0.449	0.716	0.513	0.331		0.666	0.892	0.795	0.271
	N_NBC_beta										
		PCoA1	0.736	0.824	0.681	0.372		0.505	0.743	0.295	0.127

(continued)

Table 2.2 (continued)

MODEL NR	Blocks	Variables	Weight	Loading	Communality	Redundancy	MODEL WR	Weight	Loading	Communality	Redundancy
		PCoA2	0.573	0.706	0.471	0.258		0.841	0.863	0.745	0.327
	NBC_alpha										
		Chao1	0.716	0.889	0.791	0.344		0.618	0.701	0.492	0.171
		Shannon 'H'	0.491	0.742	0.551	0.241		0.718	0.791	0.624	0.217
	N_NBC_alpha										
		Chao1	0.862	0.849	0.721	0.367		0.381	0.766	0.321	0.256
		Shannon 'H'	0.529	0.707	0.257	0.131		0.845	0.929	0.863	0.679

NBC neutral bacterial communities, **N_NBC* non-neutral bacterial communities, *beta* beta-diversity, *alpha* alpha-diversity

We also determined the path coefficients and coefficients of determination (R^2) via PLS-PM in RStudio and the goodness of fit index (GoF) was used to evaluate the overall performance of the conducted PLS-PM models.

3 Results and Discussion

3.1 *Spatial Dynamics of ADOMs, DOMs, and Other Environmental Variables*

The identified ADOMs were grouped into antibiotics, drugs, and bisphenols according to previous reports (Chen et al., 2016; Sun et al., 2016). We observed that the concentrations (i.e., expressed in ngL^{-1}) of antibiotics (mean values: 480.87 and 319.66), drugs (mean values: 284.65 and 91.91), and bisphenols (mean values: 2690.6 and 2370.28) were significantly higher in the river with high urbanization compared to that with low urbanization (Table 2.3). The proportion of the total ADOMs concentration (60.27 against 39.73%) was higher in the river with high urbanization. These results are also in agreement with previous studies that found higher concentrations of ADOMs in rivers with higher anthropogenic impacts (Chen et al., 2016; Sun et al., 2016). Moreover, three components (C1 [Ex/Em = 250/390 nm], C2 [Ex/Em = (260, 355)/442 nm], and C4 [Ex/Em = (250, 390)/516 nm]) of DOM related to humic-like (HLS) while one (C3 [Ex/Em = 280/332 nm]) related to protein-like (PLS) substances were identified (Ifon et al., 2022a). Both HLS (mean values of 0.63 in NR and 1.09 in WR), PLS (mean values: 0.20 in NR and 0.38 in WR), and the relative abundance of the total DOM fluorescence (38.95 in NR and 61.05% in WR) were significantly higher in the river with high urbanization ($p < 0.01$) (Table 2.3). Further, the statistical summary (Table 2.3) revealed that the majority of the environmental parameters were higher in rivers with higher urbanization. Together, these results demonstrate that high urbanization and human disturbances drive ADOMs, DOM, and water quality in aquatic ecosystems thus, supporting previous studies (Choi, 2014; Li et al., 2018; Tang et al., 2019b). Further, PCA explained a total variance of 79.2% between the rivers (Fig. 2.2a) and indicated that both DOMs and ADOMs majorly contribute to the spatial difference (Fig. 2.2b). We observed opposite loading between different ADOMs (e.g., Bisphenol A and Bisphenol AF compared to Propyphenazone and Sildenafil citrate in PC1) thus, indicating that their sources occurrence to the rivers may be dissimilar. Moreover, DOMs' fluorescence components (C1, C2, C3, and C4) show positive and negative relationships with ADOMs as demonstrated by PCA (Fig. 2.2a); this may signify that ADOMs could significantly alter DOMs components' fluorescence in those rivers. A previous study has demonstrated that bisphenol A decreases soil and plants' DOM components fluorescence (Xiao et al., 2011). In our previous study, we also found that bisphenol A, AF, F, and S decrease DOM fluorescence of commercial humic substances while they increase that of amino acids (which are often classified as protein-like in river samples) (Ifon et al., 2022b;

Table 2.3 Summary of environmental variables

	Temp	EC	pH	DO	Antibiotics	Bisphenols	Drugs	HLS	PLS	TIN	TN	TP
Samples	33	33	33	33	33	33	33	33	33	33	33	33
NR												
Min	29.5	192.4	7.09	5.66	9.22	64	56.63	0.47	0.13	1.23	2.56	0.06
Max	33.1	234	8.54	9.01	319.66	2370.28	91.91	0.76	0.28	4.68	29.76	0.34
Sum	1026.4	6755.1	244.03	231.88	1518.91	14,936.6	2458.5	20.78	6.32	69.67	464.75	4.21
Mean	31.11	204.7	7.4	7.03	46.03	452.63	74.5	0.63	0.2	2.12	14.09	0.13
Stand. dev	0.84	9.71	0.28	0.83	58.97	686.43	9.57	0.06	0.04	0.77	7.38	0.07
WR												
Min	21.9	191	6.88	4.91	27.24	86.88	46.54	0.82	0.29	1.75	2.45	0.06
Max	32.7	483	7.69	8.79	480.87	2690.6	284.65	1.41	0.5	11.61	29.05	0.29
Sum	999.9	7889.4	238.3	209.73	2341.32	12,549.2	3899.19	35.87	12.26	170.43	438.23	4.4
Mean	30.3	239.08	7.23	6.36	70.95	380.28	118.16	1.09	0.38	5.17	13.28	0.14
Stand. dev	1.9	53.77	0.18	0.85	76.51	462.26	48.52	0.15	0.06	2.75	7.03	0.06

DO dissolved oxygen, *EC* electric conductivity, *HLS* humic-like substances, *NR* North River, *PLS* protein-like substances, *Temp* temperature, *TIN* total inorganic nitrogen, *TN* total organic nitrogen, *TP* total phosphorous, *WR* West River

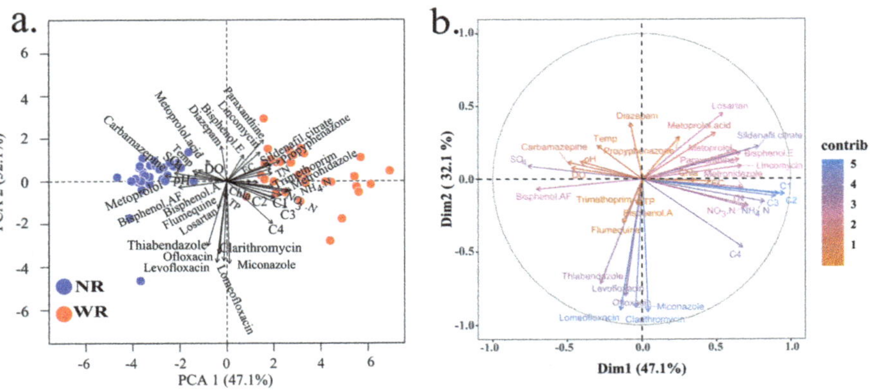

Fig. 2.2 Principal component analysis (**a**) and contribution of environmental variables to the spatial differences between NR and WR (**b**)

Ma et al., 2021). Therefore, it is paramount to evaluate how ADOMs influence DOM components' fluorescence within rivers.

3.2 *Effects of ADOMs on DOMs in River Samples: A Laboratory Assay*

To evaluate ADOMs' impacts on the riverine DOMs components fluorescence, seven ADOMs including two bisphenols (bisphenol A and AF), two antibiotics (Lomefloxacin and Lincomycin), and three drugs (Metoprolol, Paraxanthine, and Trimethoprim) were used to treat river samples for 48 h. The ATR-FTIR spectroscopy demonstrated that there were similar peaks between the control and the treated river samples; however, there was attenuation of some peaks and apparition of others in the treated river samples (Fig. 2.3e). Noticeably, peaks at 603.95, 603.98, 607.15, 584.21, 588.47, 605.3, 605.25, and 584.52 cm^{-1} all, which could be assigned to C–O/aromatic or C–C stretching (Lingegowda et al., 2013; Ifon et al., 2022b), were present in both the control and all treatments. However, peaks at 3752.65, 3332.55, and 2258.33 Cm^{-1}, which represent O–H stretching vibrations, N–H stretching secondary amine, and C=C stretching (Lingegowda et al., 2013; Ifon et al., 2022b), respectively, were attenuated in the treatments. Moreover, new peaks at 2610.23, 2670.89, 2790.35, 2855.25, and 3190.55 cm^{-1}, which represent carboxylic acid groups (Lingegowda et al., 2013; Kassem et al., 2023; Pagnucco et al., 2023), appeared in treatment with bisphenol A, Lincomycin, Metoprolol, Paraxanthine, and Trimethoprim, respectively. New peaks at 2353.22 and 2352.75 cm^{-1} corresponding to N-H/C-O stretching (Ifon et al., 2022b) also appear in treatment with bisphenol AF and Lomefloxacin. These results indicated a complex interaction between ADOMs and riverine DOMs pool, which may alter the structure and conformation of DOMs fluorescence components in rivers and further

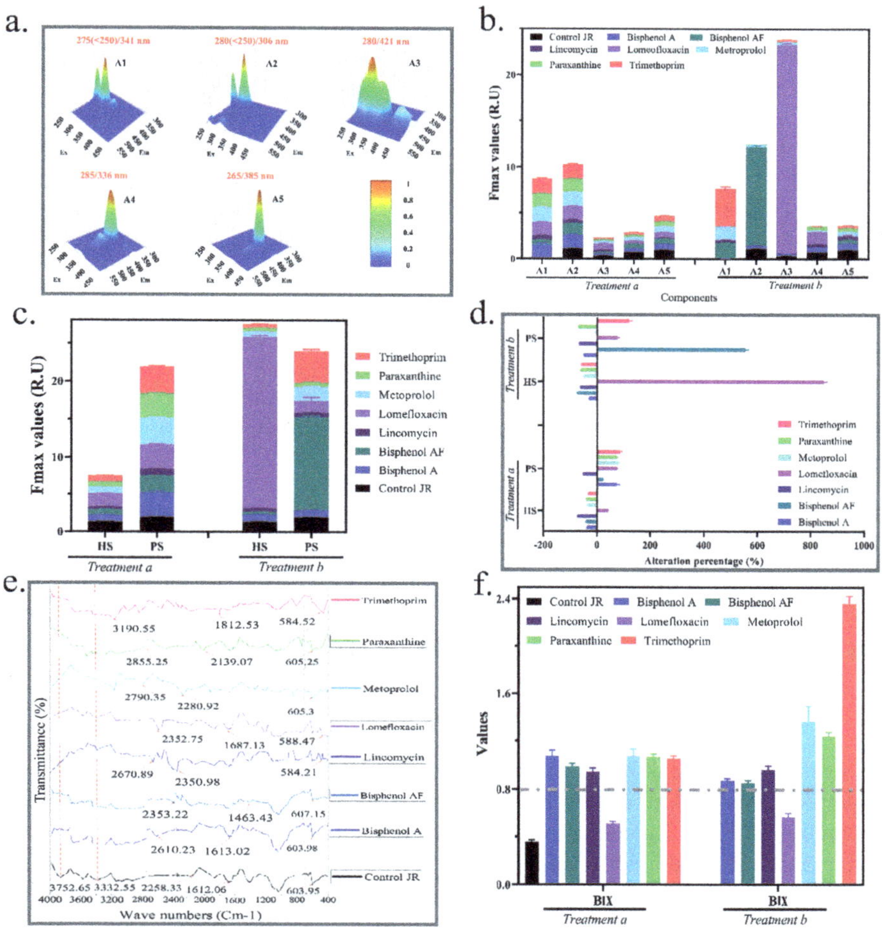

Fig. 2.3 Results of anthropogenic dissolved organic matter (ADOM) effects on riverine DOMs after treatments: contour plot of fluorescence components (**a**); changes in each component's fluorescence (**b**); change in humic-like substance (HLS) and protein-like substance (PLS) fluorescence (**c**); alteration proportions of HLS and PLS (**d**); attenuated total reflection Fourier-transform infrared (ATR-FTIR) spectra (**e**); changes in the biological index (BIX) (**f**). (**Treatment a = treatment with 0.1 mg/L; *Treatment b = treatment with 1 mg/L of each ADOMs*)

lead to changes in components fluorescence and even generate new fluorescence components as also revealed in our previous study (Ifon et al., 2022b).

EEM-PARAFAC analysis validly identified five components (A1, A2, A3, A4, and A5) (Fig. 2.3a). Compared to previous studies, components A3 and A5 were HLS while A1, A2, and A4 were PLS (Ifon et al., 2022b; Madonia et al., 2020; Murphy et al., 2014). Although their peaks were slightly altered after treatments, components A3, A4, and A5 were, respectively, similar to C2, C3, and C1

previously identified (see Sect. 3.1 above). Interestingly, we found that component A1 (275[250]/341 nm, i.e., similar to a tryptophan-like) was not found in the control river samples but only found in the treated river samples (Fig. 2.3a, b). This suggests that ADOMs could cause the generation of new fluorescence components in riverine DOM pools. Additionally, regardless of low or high concentration (0.1 and 1 mg/L), we found that Metoprolol (minimum of 88.43 and −67.74%), Trimethoprim (minimum of 90.38 and −53.75%), and Bisphenol AF (minimum of 51.62 and −65.95%) significantly increase the fluorescence intensity of PLS and reduce that of HLS, respectively ($p < 0.01$) (Fig. 2.3c, d).

However, Lomefloxacin (minimum of 60.35%) significantly increases the fluorescence intensity of both HLS and PLS while Lincomycin (minimum of 73.22%) decreases them ($p < 0.01$) (Fig. 2.3d). Although, generally, the effects of ADOMs on DOMs fluorescence was not concentration dependent, we found that the highest increase of PLS (up to 600%) and HLS (up to 900%) were observed in the treatment with Bisphenol AF and Lomefloxacin, respectively, at high concentration (Fig. 2.3d). However, the highest decrease of PLS (up to 80%) was observed in the treatment with Lincomycin while that of HLS (up to 92%) was in the treatment with Bisphenol AF at high concentration (Fig. 2.3d). Although no standard mechanism has been assigned to the complex interactions of DOMs and ADOMs, we suggest that the alteration of DOMs component fluorescence by ADOMs could be due to their significant interferences with DOMs functional groups and conformational changes as revealed by ATR-FTIR spectroscopy (Fig. 2.3e). The reduction of HLS and increase in PLS fluorescence by Bisphenol AF was in agreement with our previous study (Ifon et al., 2022b). We also found that Bisphenol A decreases HLS fluorescence (Fig. 2.3e), which is in line with previous studies although not in river samples (Gan et al., 2019; Ifon et al., 2022b; Xiao et al., 2011). However, the effects of antibiotics (Lomefloxacin and Lincomycin) and drugs (Metoprolol and Trimethoprim) on DOMs component fluorescence observed in this study have not been previously reported. The effects of ADOMs on DOM components fluorescence could directly or indirectly influence the ecological functions of bacterial communities, especially the heterotrophic bacteria that strongly rely on DOMs in rivers (Kirchman et al., 2004; Zhou et al., 2019). Noteworthy, we found that within 48 h the biological index (BIX; i.e., used to track fresh biological release of DOMs) was 0.36 in the control river sample; however, this was increased to above 0.8 in all treated river samples at both low and high concentrations of ADOMs (except for Lomefloxacin). This result demonstrates that ADOMs stimulated the rise of BIX. Since BIX > 0.8 indicates fresh generation of DOM components fluorescence (Huguet et al., 2009; Kida et al., 2018), our results demonstrate that bacterial communities generated fresh DOMs probably by degrading ADOMs in those treated samples. Therefore, investigating the ecological associations and the combined effects of both ADOMs and DOMs on bacterial communities will provide important ecological insights.

3.3 Beta- and Alpha-Diversity Analysis, Redundancy Analysis

A clear and significant spatial pattern of bacterial communities' structure between urban rivers ($R^2 = 0.303$, $p < 0.001$, Permanova) was observed by PCoA, which explained about 55.36% of the total variance (Fig. 2.4a).

A total of 23,109 operational taxonomic units (OTUs) were identified from 10,456,958 high-quality sequences at a 97% similarity level for the 66 samples. We found that the relative abundance of phyla Acidobacteria, Actinobacteria, Armatimonadetes, Bacteroidetes, Chloroflexi, Cyanobacteria, Firmicutes, Planctomycetes, Proteobacteria, and Verrucomicrobia accounted for 82.8% and 85.5% of the total bacterial communities in NR and WR, respectively. However, phyla Actinobacteria, Armatimonadetes, Cyanobacteria, and Firmicutes were significantly higher in NR than WR, while phyla Acidobacteria and Proteobacteria do the same in WR than NR ($p < 0.05$) (Fig. 2.4b). Additionally, we observed that in NR, phyla Proteobacteria (30.02%), Cyanobacteria (21.35%), Actinobacteria (17.54%), and Bacteroidetes (17.26%) while in WR, Proteobacteria (39.91%), Cyanobacteria (18.50%), Bacteroidetes (15.70%), and Actinobacteria (13.51%) were the most dominant (Fig. 2.4c, d). Rarefaction curves for individual and combined sets of 66 samples showed that most samples from both rivers tended to approach saturation, and bacterial communities in rivers with high urbanization showed higher species richness with the same sequencing effort (Fig. 2.5). In general, the bacterial metacommunity in rivers with high urbanization showed higher

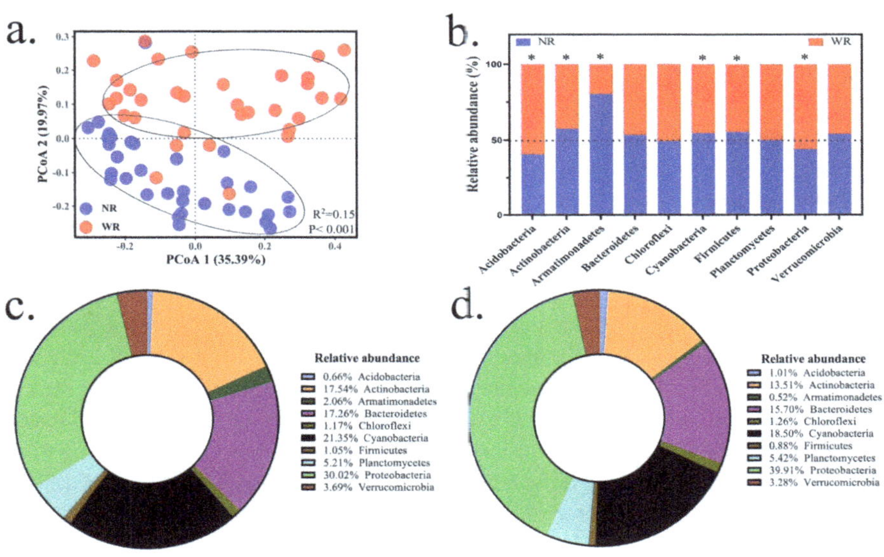

Fig. 2.4 Principal coordinate analysis for north (NR) and west (WR) rivers (**a**); proportion of major phyla between NR and WR (**b**); proportion of major phyla within NR (**c**); proportion of major phyla within WR (d)

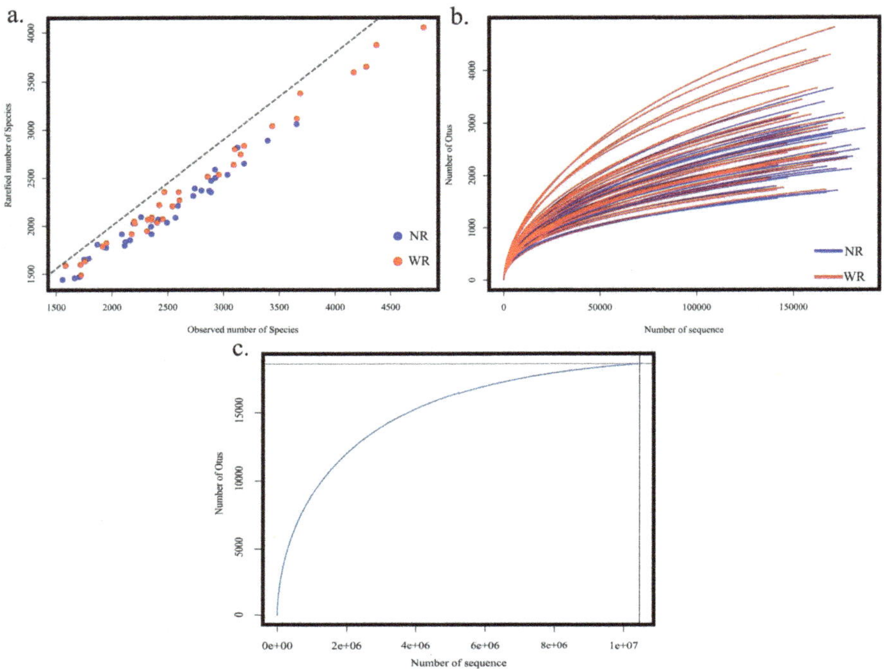

Fig. 2.5 Rarefaction of similarity-based observed number of species for all samples (**a**); rarefaction curves of similarity-based operational taxonomic units (OTUs) at 97% sequence similarity level for individual samples (**b**); combined set of 66 samples (**c**)

alpha-diversity compared to those in river with low urbanization. This shows that river with high urbanization demonstrates higher species richness, heterogeneity, and distribution thus, suggesting that higher human disturbances could lead to more introduction of non-river native microbial communities (such as from effluent or sewage disposal) along with more energy and bioavailable nutrient loading that could accelerate bacteria growth, proliferation, and communities' changes (Nguyen et al., 2013; Plecha et al., 2013; Mittelbach et al., 2016). RDA (Fig. 2.6a–c). As indicated by RDA1 (31.44%), Losartan, Diazepam, all DOMs components, TN, NO_3^--N, and NH_4^+-N display positive effects while Bisphenol A, Metoprolol, and Clarithromycin display negative effects on bacteria species richness, diversities, and distribution in river with low urbanization. As for the highly urbanized river, RDA1 (68.56%) demonstrated that Clarithromycin, Bisphenol AF, DOMs components (C1, C2, and C4), TN, NO_3^--N, and NH_4^+-N show positive effects while Bisphenol E, Trimethoprim, and Diazepam show negative effects on bacterial alpha-diversities.

However, for the whole watershed, RDA1 (23.47%) indicates that Clarithromycin, Metoprolol, DOMs components (C1, C2, and C4), TN, NO_3^--N, and NH_4^+-N show positive effects while Bisphenol AF, pH, and Carbamazepine show negative effects

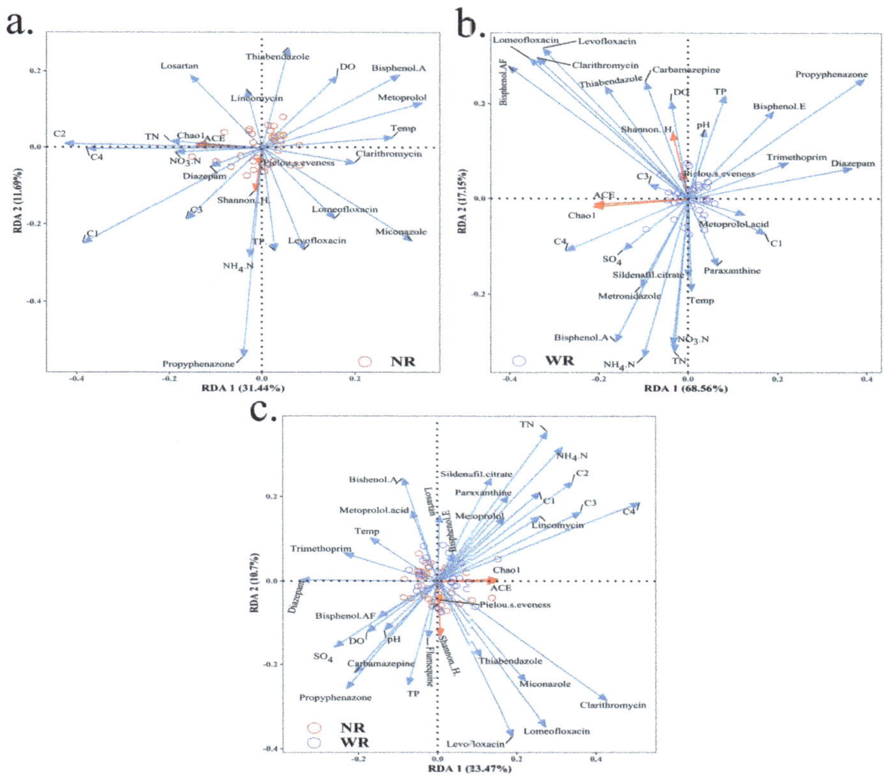

Fig. 2.6 Redundancy analysis (RDA) showing relationships between environmental variables and bacterial alpha-diversity in NR (**a**), WR (**b**), and both rivers (**c**)

on bacterial alpha-diversities. These results demonstrate the complex ecological impacts of physicochemical, nutrients, DOMs, and ADOMs on the bacterial communities' compositions across gradients of urbanization and human disturbances. It has been reported that environmental parameters such as water temperature, nutrient concentrations, and heavy metal can display significant influence on the microbial community composition thus, affecting the relative abundance and occurrence frequency of microorganisms (Smith, 2002; Curtis & Sloan, 2004; Tiquia, 2008; Zeglin, 2015; Bowman et al., 2018; Chen et al., 2019; Tiquia-Arashiro & Pant, 2020; Wang et al., 2023b), thereby resulting in the neutral and non-neutral partition (Liu et al., 2013; Chen et al., 2019). A need for unraveling other factors that also can separate the taxa into neutral and non-neutral partitions has been reported. Therefore, we judge paramount to examine the neutral and non-neutral microbial communities' associations with DOMs and ADOMs, which are important environmental factors.

3.4 *Neutral Community Model Fit and Stochastic Forces on Bacterial Communities*

The NCM successfully predicted a large portion of the correspondence between OTUs' occurrence frequency and relative abundance variations (Fig. 2.7a), with 88.75%, 87.98%, and 88.63% of explained community variance for rivers with low and high urbanization (NR and WR, respectively), and both rivers (JR), respectively. Additionally, bacterial communities within prediction (Neutral) were largely

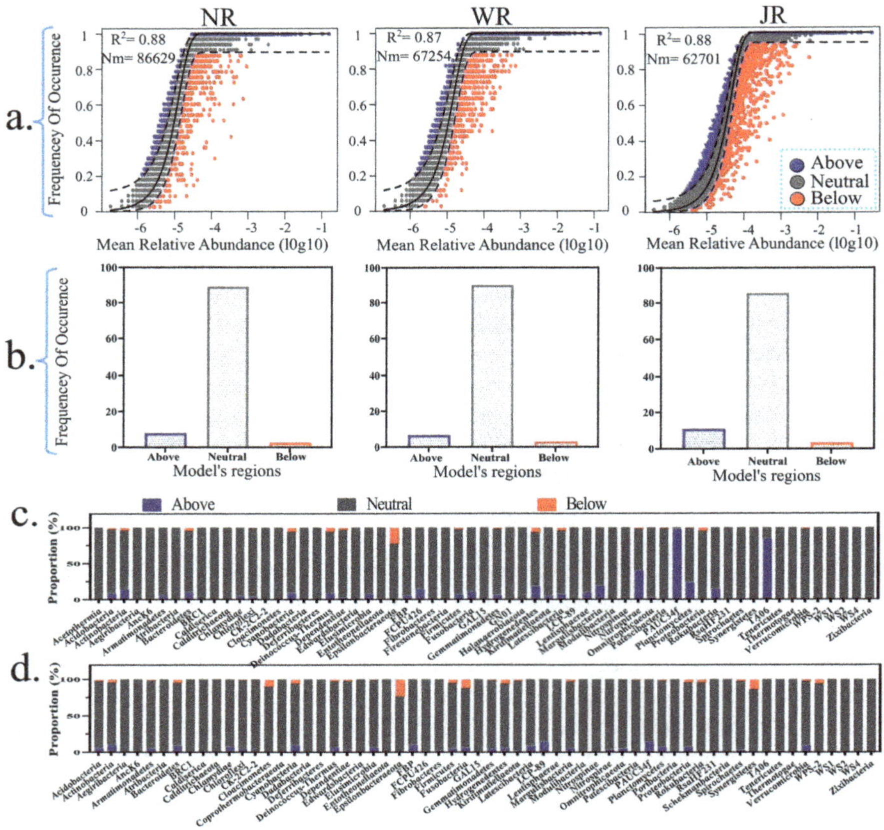

Fig. 2.7 Fit of the neutral community model (NCM) of community assembly: the predicted occurrence frequencies for NR, WR, and both rivers as JR watershed showing partitions of bacterial communities from NR, WR, and JR, respectively (**a**). The solid black lines indicate the best fit to the NCM, and the dashed black lines represent 95% confidence intervals around the model prediction. OTUs that occur more (*Above*) or less (*Below*) frequently than predicted (Neutral) by the NCM are shown in different colors. Nm and R^2 indicate the metacommunity size times immigration and the fit to each model, respectively. Model regions: Above, Below, and Neutral occurrence proportions predicted bacterial communities (**b**). The distinct fitting proportions of richness (OTU number) of bacterial communities' phyla for NR (**c**) and WR (**d**)

higher than those above and below prediction (Above and Below, respectively) (Fig. 2.7b). Further, the explained variation of NCM revealed that the member of each phylum has large neutral members in river with low urbanization (ranged from 75.67% to 100% [except for phyla PAUC34f, Nitrospirae, and TA06], with mean value 87.84%) (Fig. 2.7c) and in rivers with high urbanization (ranged from 80.55% to 100%, with mean value 90.28%) (Fig. 2.7c). Generally, this indicates that taxa within the communities (phylum) show similar responses to stochastic processes.

The above observations clearly indicate that stochastic processes were important forces driving bacterial communities' assembly in both rivers. Additionally, the Normalized Stochastic Ratio (NST) (Ning et al., 2019) reveals that the numerical distribution of NST was higher than 50% for the bacterial communities of the river with low urbanization (86.240%) and that with high urbanization (72.25%), as well as that of the whole watershed (79.75%), indicating that the stochastic processes were dominant in both rivers and the whole watershed thus, confirming the results of NCM. The NCM (i.e., a valid method for the inference of stochastic forces influencing community assembly) is a valid neutral-based process model widely used to estimate the significance of processes that are complex to observe directly but can have large and significant impacts on microbial communities (i.e., dispersal and ecological drift) (Burns et al., 2016; Chen et al., 2019; Tong et al., 2019).

The NCM highly predicted the bacterial OTU changes as also revealed at the phylum level (Fig. 2.7a–d), implying that stochastic stability between the loss and gain of bacterial communities (e.g., stochastic births, deaths, and immigration) was a crucial process driving their community assembly in urban rivers. Previous studies that used NCM to investigate the bacterial community in lakes of Paris, France ($R^2 = 0.76$) (Roguet et al., 2015), in freshwater/brackish habitats in southern and central Sweden ($R^2 = 0.85$) (Östman et al., 2010), and urban Tingjiang River, Fujian province, China ($R^2 = 0.89$) (Chen et al., 2019) also revealed that bacterial community structure was mainly driven by stochastic forces. The power of stochastic forces is able to significantly cause large community diversity within and between different rivers with different urbanization levels and human disturbances (Chen et al., 2019).

Both values of $R^2 = 0.88$ and the number of metacommunity (Nm = 86,629) for bacterial communities' taxa in the river with low urbanization were higher than in river with high urbanization ($R^2 = 0.87$ and Nm = 67,254) (Fig. 2.7a). Moreover, NST results demonstrated a significant difference in the stochastic ratio between rivers with low and high urbanization (i.e., 86.24% and 72.25%, respectively; two Sample t-test, $p < 0.001$). This first suggests that the influence of stochastic forces on bacterial communities was stronger in the river with low urbanization. Second, because the number of sequences in both river samples was the same, the higher Nm observed in the river with low urbanization indicates that species dispersal of bacterial communities was higher in that river. The higher stochastic forces and species dispersal of bacterial communities in rivers with low urbanization could be attributed to the fact that it has a higher drainage area (9803 km^2) and mean annual discharge (82.3×10^8 m^3) (Yang et al., 2013). The higher drainage area and discharge as well as high river connectivity can enhance the possibility of the movement and successful establishment of microorganisms across space, leading to higher

bacterial communities' immigration rate in this river. Although only in part, a high dispersal rate can have a significant impact on both environmental selection and ecological drift. The wide distribution range of bacterial communities, combined with the fact that communities in rivers with low urbanization are likely to receive lower spatial influences (Wang et al., 2023b), suggests that dispersal limitation has only a minor effect on spatial turnover in bacterial communities in such rivers. Although NCM revealed that stochastic processes are the driving forces of riverine bacterial communities, it is difficult to ignore the influence of environmental factors on bacterial communities' composition and structure, especially with the impacts of urbanization and other human disturbances on the rivers. Therefore, it is paramount to evaluate the ecological significance of environmental factors to the bacterial community assemblies.

3.5 Neutral and Non-neutral Bacterial Communities' Assembly and Their Association with Environmental Variables: Mantel Test

The NCM indicates less than 100% stochastic forces on the bacterial communities' changes, demonstrating that other community assembly mechanisms might be existing simultaneously, which caused the N_NBC distribution. These other mechanisms include the influence of environmental factors and ecological interactions (Chen et al., 2019; Hu et al., 2017a), which make it difficult to quantify their relative contribution to bacterial communities. The NBC and N_NBC, as classified by NCM, are two fractions probably formed of different richness, diversities, and abundance thus, constituting different communities' compositions. We first evaluated the composition of NBC or N_NBC between rivers and found no significant difference as revealed by non-metric multidimensional scaling ordination (NMDS) based on the Bray–Curtis similarity among partitions and the Permanova test (Fig. 2.8a, b). The proportion of NBC and N_NBC at the phylum level between rivers also shows no strong dissimilarity (Fig. 2.8c, d). These results indicate that NCM likely predicted the same bacterial taxa as NBC or N_NBC in both rivers. However, it is important to note that taxa above prediction and those below prediction (Above vs. Below) display significant distinct partitions within each river as demonstrated by NMDS ($p < 0.05$, *Permanova*) (Fig. 2.8a). Taxa above predictions are those found more frequently than expected, indicating their higher migration and dispersal ability to more locations, while those below predictions are considered less frequent than expected thus, having lower dispersal ability (Chen et al., 2019). Therefore, we suggest that the significant difference revealed by NMDS between the taxa above and those below neutral prediction was due to their different migration or dispersal abilities and the possibility of their strong response to environmental influences. Second, as indicated by PCoA, we found a significant spatial pattern (beta-diversity) of NBC ($R^2 = 0.10$, $p < 0.05$ and $R^2 = 0.21$, $p < 0.001$) and

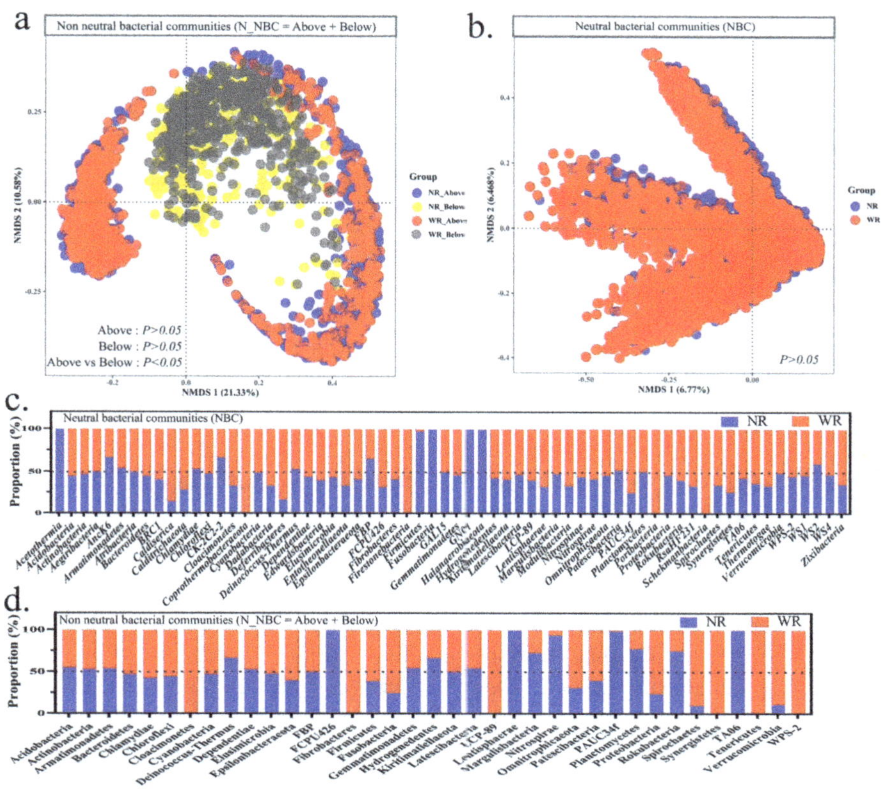

Fig. 2.8 Non-metric multidimensional scaling ordination (NMDS) based on the Bray-Curtis similarity for non-neutral bacterial communities (N_NBC) (**a**) and neutral bacterial communities (NBC) (**b**) between NR and WR. Occurrence proportion of NBC (**c**) and N_NBC (**d**) at phylum level between NR and WR as revealed by the neutral community model

N_NBC ($R^2 = 0.11$, $p < 0.05$ and $R^2 = 0.28$, $p < 0.001$) for both rivers (Fig. 2.9), suggesting their dependence on a more widespread and heterogeneous effect of human disturbance (Chase, 2005; Maria, 2017).

Moreover, we observed that the alpha-diversity (species richness and diversity) of N_NBC is significantly higher in the river with high urbanization compared to that with low urbanization ($p < 0.05$) (Fig. 2.10a, d). As shown by Mantel test, N_NBC species richness and diversity significantly associated with several environmental variables such as physicochemical and nutrient (pH, TP, NO_3-N, NH_4-N), DOMs components (C1, C2, and C3) as well as ADOMs (Bisphenol AF, Paraxanthine, Miconazole, Flumequine) in both urban rivers ($p < 0.01$) (Fig. 2.10b, c). This indicates that N_NBC composition is associated with those environmental variables, and this is in line with NCM results which indicated that N_NBC was not driven by stochastic forces. Thus, N_NBC is driven by deterministic forces that include both physicochemical and nutrient, and DOM components and ADOMs for

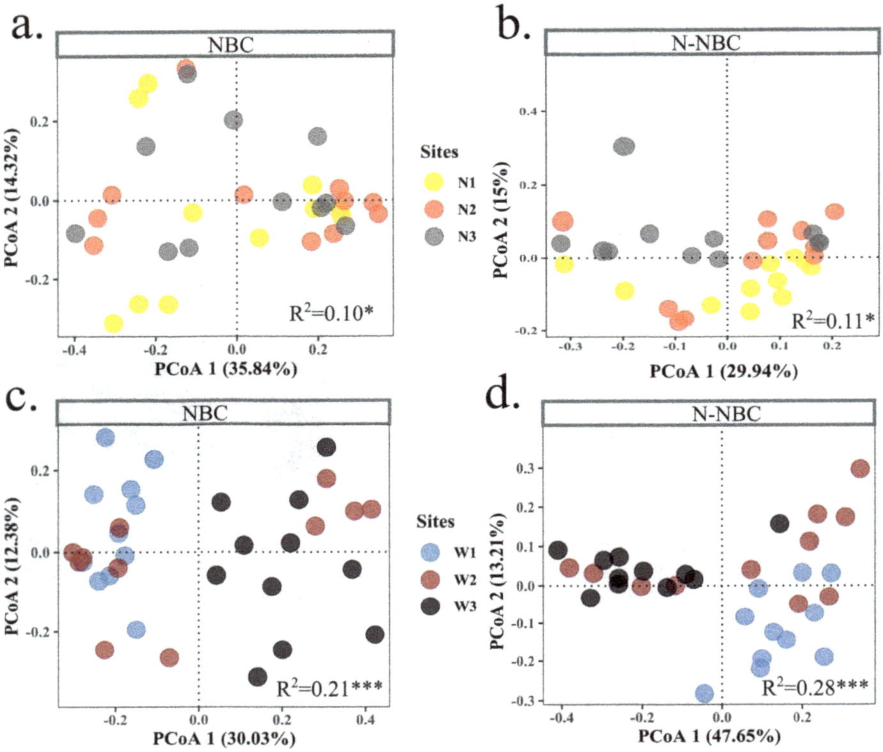

Fig. 2.9 Principal coordinate analysis for neutral bacterial communities (NBC) in NR (**a**) and WR (**c**), and for non-neutral bacterial communities (N_NBC) in NR (**b**) and WR (**d**)

both urban rivers. Additionally, we found that the species richness of NBC was significantly higher in rivers with high urbanization ($p < 0.05$) (Fig. 2.11a); however, there was no significant difference between their diversity ($p > 0.05$) (Fig. 2.11d). Mantel test reveals that NBC alpha-diversity was significantly associated with NO_3-N and Miconazole but with DOMs components (C1 and C3) and Flumequine in rivers with high and low urbanization, respectively ($p < 0.01$) (Fig. 2.11b, c). The Mantel results in this study are comparable to a previous study which indicated that both NBC and N_NBC were significantly associated with environmental variables in urban Tingjiang River, Fujian province, China (Chen et al., 2019). However, our findings further revealed higher associations of species richness and diversity of both NBC and N_NBC with DOMs, ADOMs, and other environmental variables in the river with higher than with low urbanization. Several previous studies have demonstrated that environmental factors (i.e., water temperature, nutrient concentrations, and heavy metal) display significant impacts on the microbial communities' composition by influencing their relative abundance and occurrence frequency, thereby resulting in the neutral and non-neutral distribution (Chen et al., 2019; Liu et al., 2013; Zhang et al., 2018). Herein, our findings

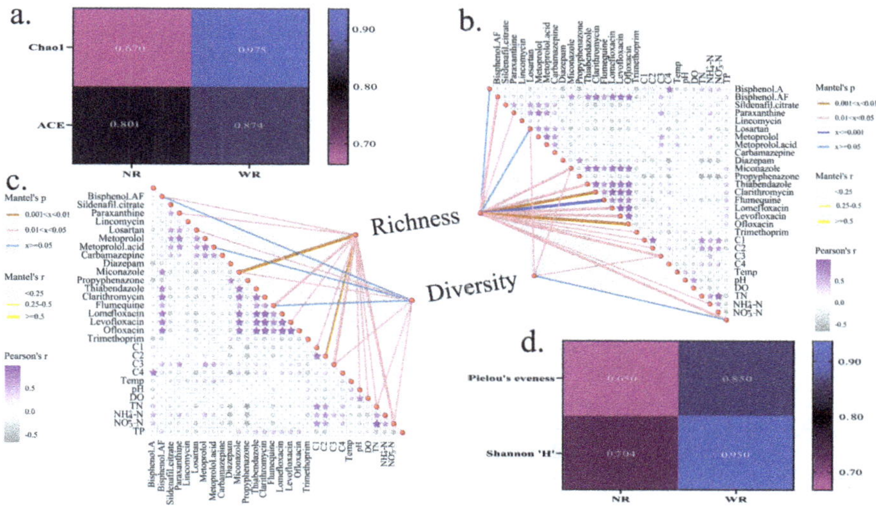

Fig. 2.10 Species richness of non-neutral bacterial communities (N_NBC) between NR and WR (**a**). Mantel test for N_NBC alpha-diversity associations with physicochemical, nutrients, dissolved organic matter (DOM) components, and anthropogenic DOM variables in WR (**b**) and NR (**c**). Species diversity and distribution of N_NBC between NR and WR (**d**)

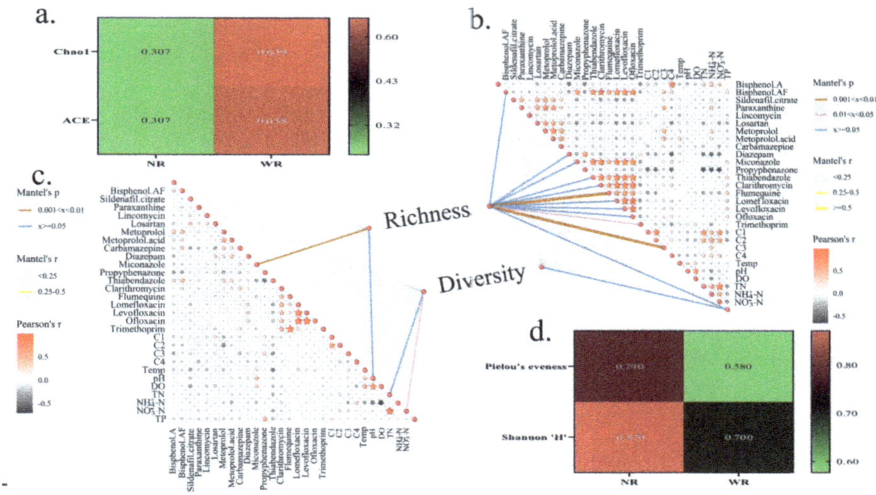

Fig. 2.11 Species richness of neutral bacterial communities (NBC) between NR and WR (**a**). Mantel test for N_NBC alpha-diversity associations with physicochemical, nutrients, dissolved organic matter (DOM) components, and anthropogenic DOM variables in WR (**b**) and NR (**c**). Species diversity and distribution of N_NBC between NR and WR (**d**)

demonstrate that both DOM components and ADOMs may be other important environmental factors that can separate taxa into neutral and non-neutral bacterial communities.

3.6 Untangling the Ecological Association Between Neutral and Non-neutral Bacterial Communities and ADOMs and DOMs in Urban Rivers

Details on the topological features of each network are displayed in Fig. 2.12. Modularity, which measures how well a network separates into modules were 0.504, 0.541, and 0.505 for rivers with low and high urbanization (NR and WR, respectively), and whole watershed (JR), respectively, indicating a good separation of community composition and structure (Deng et al., 2012).

We observed that DOMs and ADOMs variables were significantly associated with various OTUs within and between modules in each network (Fig. 2.13a). This indicates that DOM components and ADOMs variables may play significant roles in determining the bacterial communities' structure in urbanized rivers. These findings are in part similar to previous studies which also revealed a significant role of DOM components in bacterial community structure in other urbanized rivers (Li et al., 2020b; Zhou et al., 2019). Together with the results of the Mantel test, the significant roles of DOM components and ADOM variables in determining the bacterial communities' structure could demonstrate how ADOMs and DOMs may affect bacterial relative abundance and occurrence frequency and further indicate their importance in separating bacterial taxa into neutral and non-neutral communities. Based on its higher degree or betweenness centrality, a node is assumed to have a greater influence on the bacterial structures and thus, can be considered a key element or keystones or dominant species in a network (Banerjee et al., 2016; Gao et al., 2021). Hypothetically, it is generally viewed that dominant species and their traits contribute most to ecosystem functions because energies primarily pass through them, hence this concept was applied to unravel the microbial interactions. The top five primary determinants of ecosystem functions are these keystone genera *Xanthobacter, Candidatus Megaira, Conexibacter, Dinghuibacter,* and *hgcI clade* while these were *Oligoflexus, Bradyrhizobium, LD29, Bacteroides,* and *Nordella* in rivers with low and high urbanization, respectively (Table 2.4). Noteworthy, we observed that the keystone species were of both NBC and N_NBC, indicating that both neutral and non-neutral taxa control urban river ecosystem functions. Moreover, we examined the key molecules that may potentially drive the ecological associations and found that the top five were C3, C4, Bisphenol.A, Miconazole, and Thiabendazole while these were Paraxanthine, Trimethoprim, Metoprolol.acid, Bisphenol.A, and Bishenol.AF in river with low and high urbanization, respectively (Fig. 2.12b, c; Table 2.4). These results suggest that compared to ADOMs, DOM molecules may contribute more to bacterial communities' composition and structure in rivers with low urbanization. However, in the rivers with high urbanization,

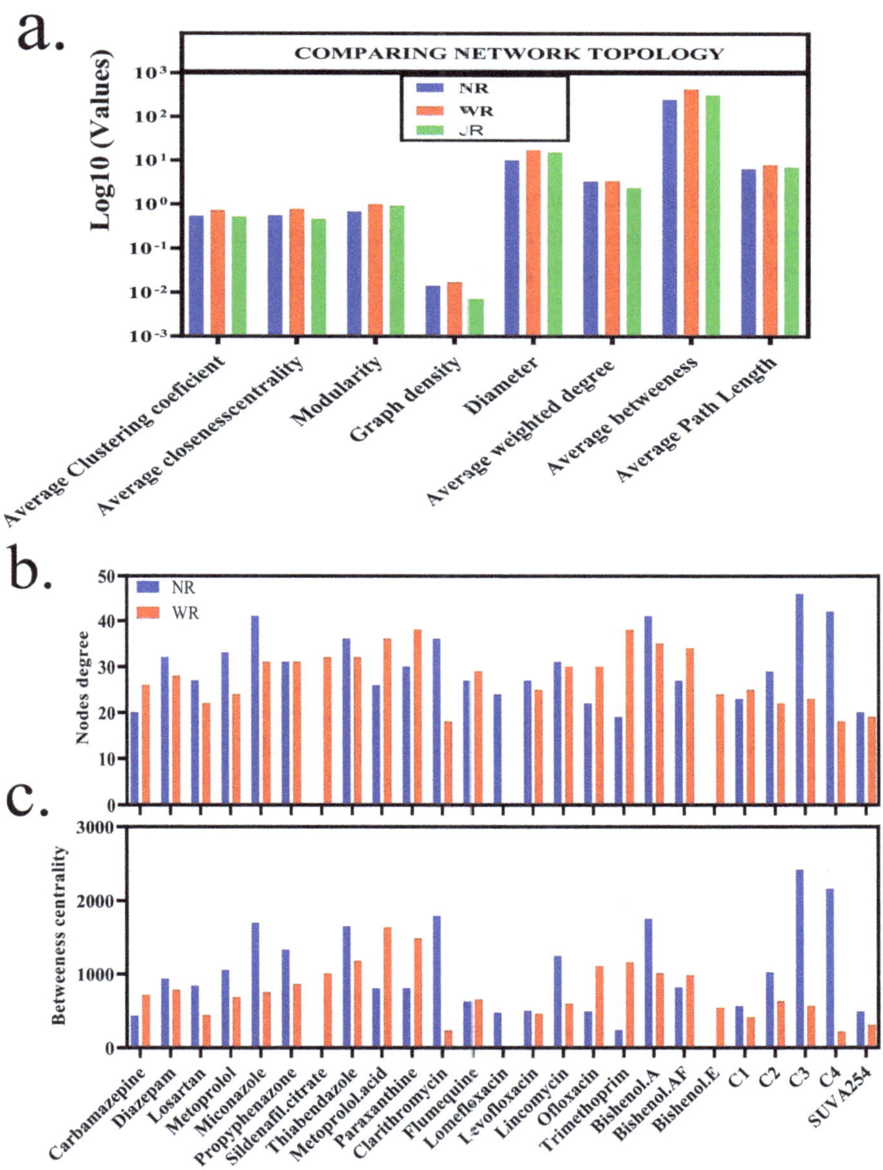

Fig. 2.12 Summary of network topology for the north (NR) and west (WR) rivers, and whole Jiulong River (JR) (**a**); summary of nodes degree for each variable of dissolved organic matter (DOM) and anthropogenic DOMs (**b**); summary of betweenness centrality for each variable of DOMs and anthropogenic DOMs (**c**)

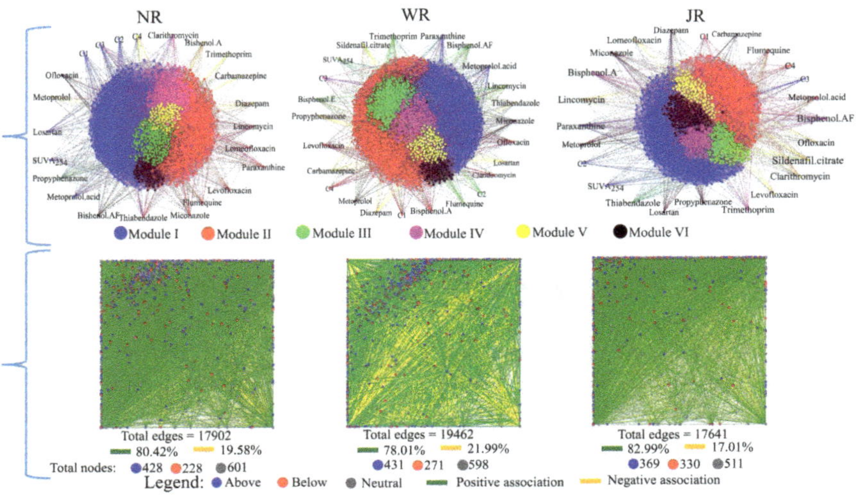

Fig. 2.13 Ecological network between bacterial communities' assembly, dissolved organic matter (DOM) components, and anthropogenic DOM variables (**a**), and between only neutral (NBC) and non-neutral (N_NBC) bacterial communities' assembly (**b**) in NR, WR, and JR. (**N_NBC is represented by Above and Below; *NBC is represented by Neutral*)

ADOM molecules may be contributing more. This could be because rivers with high urbanization have higher ADOM concentration thus, leading to their higher impacts on the bacterial communities' composition and structure in this river. This finding also supports our results of the Mantel test which demonstrated that the species richness and diversity of both NBC and N_NBC associated more with ADOMs in the river with high urbanization (see Sect. 3.5 above).

3.7 Insight into the Positive Versus Negative Associations of Neutral and Non-neutral Bacterial Communities with ADOMs and DOMs in Urbanized Rivers

To examine the degree of positive and negative associations between NBC and N_NBC, and estimate the type of associations between them, a network that only maintained the links between microbial OTUs of NBC (Neutral) and N_NBC (Above and Below) was constructed (Fig. 2.13b). Out of 17,902 total edges in the river with low urbanization, the results indicated 80.42% positive and 19.58% negative associations (Fig. 2.13b, NR). However, of 19,462 total edges in rivers with high urbanization 78.01% positive and 21.99% negative associations were observed (Fig. 2.13b, WR) while there were 82.99% positive and 17.01% negative associations out of 17,641 total edges in the whole watershed (Fig. 2.13b, JR). These results demonstrate that the ecological associations between neutral and non-neutral bacterial

Table 2.4 Key elements and keystone species in the ecological networks

Molecules	Category	Class	Bet-centrality	Compounds	Category	Class	Bet-centrality
C3	DOMs	PLS	2411	Paraxanthine	ADOMs	Drugs	1485
C4	DOMs	HLS	2159	Trimethoprim	ADOMs	Drugs	1158
Bisphenol.A	ADOMs	Bisphenols	1752	Metoprolol. acid	ADOMs	Drugs	1636
Miconazole	ADOMs	Drugs	1698	Bisphenol.A	ADOMs	Bisphenols	1008
Thiabendazole	ADOMs	Drugs	1652	Bisphenol.AF	ADOMs	Bisphenols	982
Bacteria taxa							
Genera	**Phylum**	**Assembly**	**Bet-centrality**	**Genera**	**Phylum**	**Assembly**	**Bet-centrality**
Xanthobacter	Proteobacteria	NBC	37378	*Oligoflexus*	Proteobacteria	NBC	7914
Candidatus Megaira	Proteobacteria	N_NBC	3234	*Bradyrhizobium*	Proteobacteria	NBC	7606
Conexibacter	Actinobacteria	N_NBC	3223	*LD29*	Verrucomicrobia	NBC	4927
Dinghuibacter	Bacteroidetes	N_NBC	3174	*Bacteroides*	Bacteroidetes	N_NBC	4528
hgcI clade	Actinobacteria	N_NBC	2987	*Nordella*	Proteobacteria	N_NBC	3338

ADOMs anthropogenic dissolved organic matters, *DOMs* dissolved organic matters, *HLS* humic-like substances, *NR* river with decreased urbanization, *PLS* protein-like substances, *WR* river with increased urbanization, *Bet-centrality* betweenness centrality

communities exhibited more positive relationships. Negative relationships in the ecological networks are frequently attributed to competition or amensalism, while positive associations are frequently attributed to cross-feeding, co-aggregation, co-colonization, niche overlap, or other factors (Deng et al., 2012; Zhou et al., 2021). Although the proportion of positive links was three times higher than that of negative links in all networks, we noticed a slight decrease in the proportion of positive connections in the network of the river with high urbanization (Fig. 2.13b, WR). This may indicate that a higher urbanization level decreases positive interactions such as cross-feeding, syntrophic, and mutualistic interactions between neutral and non-neutral bacterial communities. This could be attributed to higher environmental disturbances thus increasing the network stability because previous reports have indicated that an increase in negative associations within ecological communities increases the stability of networks under disturbed conditions (Gentile et al., 2006; Olesen et al., 2007; Zhou et al., 2021). Although previous studies have also reported a higher proportion of positive relationships in the networks of bacterial communities in 33 lakes located in the middle and lower reaches of the Yangtze and Huaihe River (Zhou et al., 2021), in the Sea in Sweden (Deutschmann et al., 2021), our study is potentially the first to evaluate the proportion of positive and negative relationships between neutral and non-neutral bacterial communities in urbanized rivers.

Further, as displayed in Fig. 2.14, we constructed networks that only maintained the links between bacterial OTUs of NBC (Neutral) or N_NBC (Above and Below) and ADOMs or DOMs. Thus, the links OTUs–OTUs, ADOMs–ADOMs, DOMs–DOMs, and ADOMs–DOMs were excluded. The goal was to examine the number of OTUs of NBC or N_NBC that have positive or negative associations with ADOMs or DOMs. We found that for N_NBC, there were 225 OTUs positively associated and 46 OTUs negatively associated with ADOMs and DOMs in the river with low urbanization, 191 OTUs positively associated and 72 OTUs negatively associated in the river with high urbanization, and 255 OTUs positively associated and 61 OTUs negatively associated in the whole watershed. For NBC, there were 207 OTUs positively associated and 40 OTUs negatively associated with ADOMs and DOMs in the river with low urbanization, 179 OTUs positively associated and 63 OTUs negatively associated in the river with high urbanization, and 179 OTUs positively associated and 58 OTUs negatively associated in the whole watershed (Fig. 2.14a, b). The results demonstrate that both NBC and N_NBC could degrade or consume ADOMs and that the positive impacts of ADOMs and DOMs are higher than their negative impacts on NBC and N_NBC. However, regardless of river, we found that both the positive and negative associations of ADOMs and DOMs with N_NBC were higher than that of NBC, indicating that N_NBC responds more to ADOMs and DOMs, as also demonstrated by our Mantel test results (Figs. 2.10 and 2.11). Although bacterial communities might be able to consume both ADOMs and DOMs for their growth and other ecological functions, they are likely not producers of ADOMs. Therefore, we assume that the positive associations may signify that bacteria use ADOMs and DOMs as energy and nutrient sources as also suggested in previous studies that bisphenol A, and humic- and protein-like substances of DOMs could be used by bacteria as carbon and nutrient sources (Hu et al., 2017a; Zhou

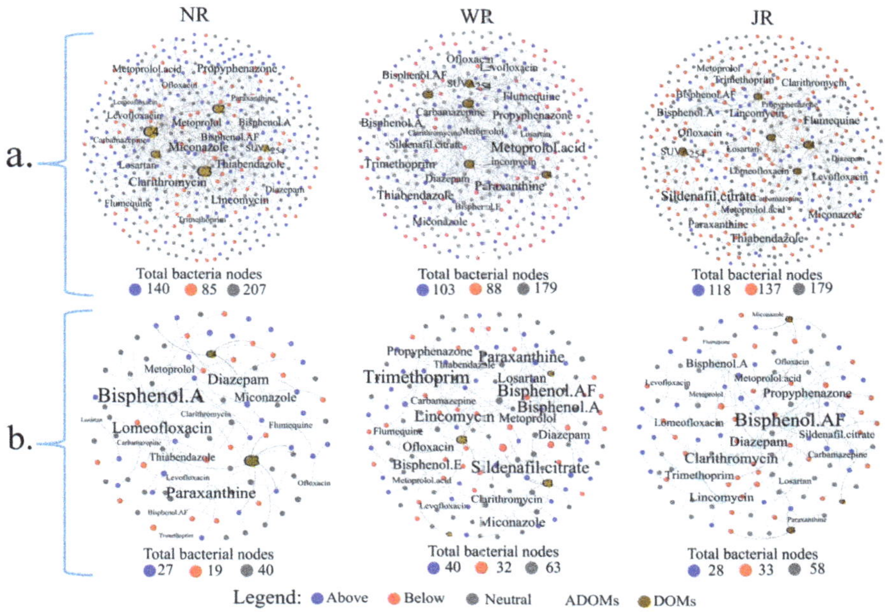

Fig. 2.14 Positive (**a**) and negative (**b**) associations of neutral (NBC) and non-neutral (N_NBC) bacterial communities' assembly with dissolved organic matter (DOM) components and anthropogenic DOM (ADOM) variables in NR, WR, and both rivers as JR. (**N_NBC is represented by Above and Below; *NBC is represented by Neutral. Note: Only links Above-DOMs, Above-ADOMs, Below-DOMs, Below-ADOMs, Neutral-DOMs, and Neutral-ADOMs are maintained in each network*)

et al., 2019). However, we assign the negative associations to detrimental effects (including endocrine, developmental, epigenetic systems, and other ecological functions) of ADOMs and DOMs on the bacterial communities (Dey et al., 2019; Touliabah et al., 2022). Our findings thus demonstrate that ADOMs and DOMs are more beneficial than detrimental to bacterial communities in urban rivers. This may be explained by resistance development in several bacterial communities in the environments (Dey et al., 2019), thus potentially enabling them to inhibit negative ADOMs or DOMs effects while degrading them.

Thus, the results of network analysis combined with that of the Mantel test demonstrate that physicochemical, nutrients, biotic interactions, DOMs, and ADOMs are important factors shaping the bacterial communities' assembly in urban rivers. It is noteworthy that although NBC or N_NBC taxa that are associated with ADOMs and DOMs majorly belong to similar phyla for both rivers (Fig. 2.15a), these were largely different at genus level (Fig. 2.15b). This may be due to differences in environmental conditions and bacterial ecological adaptation. However, regardless of NBC or N_NBC, 50 genera (*Acidibacter, Acinetobacter, ADurb.Bin063-1, Algoriphagus, Agrococcus, alphaI cluster, Anaeromyxobacter, Armatimonas,*

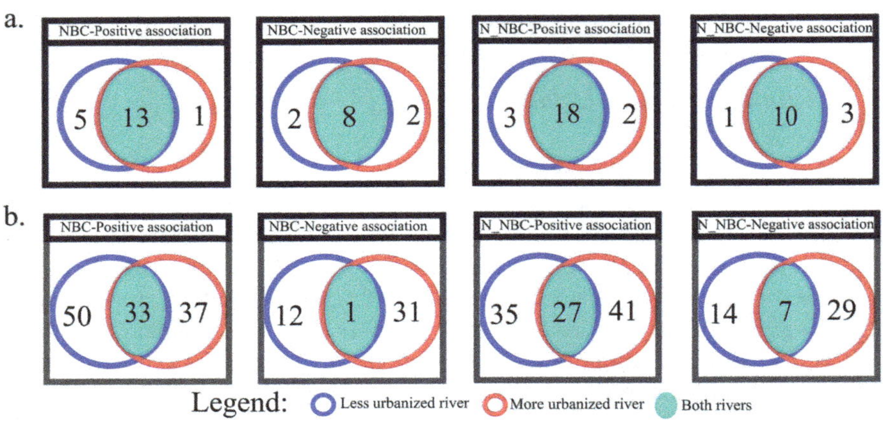

Fig. 2.15 Taxonomical dissimilarity and similarity between rivers for neutral (NBC) or non-neutral (N_NBC) bacterial OTUs that show positive or negative associations with dissolved organic matter (DOM) and anthropogenic DOMs at phylum level (**a**) and genus level (**b**). (**The numbers in the Wayne diagram (in the white space of the blue and red circles) indicate the number of dissimilar taxa at phylum and genus levels. The numbers for both rivers (in space colored with cyan) indicate the number of overlap (similar) taxa for each phylum and genus levels*)

bacterium SH3-11, BD1-7 clade, Blyi10, Candidatus planktophila, Candidimonas, Caulobacter, Chthoniobacter, CL500–29 marine group, Cyanobium PCC-6307, Dechloromonas, Deinococcus, Dinghuibacter, Dunaliella tertiolecta, Ellin6067, Epipyxis sp. *PR26KG, Escherichia-Shigella, Fibrella, Flavobacterium, Fluviicola, Gemmata, Haliscomenobacter, hgcI clade, IMCC26134, Legionella, Limnohabitans, Novosphingobium, Marmoricola, Monoraphidium neglectum, Nocardioides, oc32, Oligoflexus, OM27 clade, Opitutus, Paenibacillus, Phaselicystis, Planctonema lauterbornii, Polynucleobacter, Pseudarcicella, Roseomonas, Reyranella, Sphingomonas, Xanthobacter*) show positive associations, while only 8 genera (*Anaeromyxobacter, Cyanobium PCC-6307, Deinococcus, Dinghuibacter, hgcI clade, Paenibacillus, Prevotella 9, Pseudarcicella*) show negative associations with ADOMs and DOMs for both river. We suggest that those genera could be used to explore ADOMs and DOMs biodegradation and environmental stress on bacteria (e.g., endocrine, developmental, and epigenetic systems) in urbanized rivers. Our findings also support previous studies which indicated that genera like *Anaeromyxobacter, Flavobacterium, Limnohabitans, Lacihabitans, Novosphingobium,* and *Sphingomonas* were vital for biodegradation processes of organic materials (Kujawinski, 2011; Kulikova, 2021; Li et al., 2020b). Although our findings together with those of a previous study (Chen et al., 2019) demonstrated that both NBC and N_NBC taxa associate with environmental variables in urban rivers, quantifying the effects of land use, physicochemical, nutrients, DOMs, and ADOMs on NBC and N_NBC is important to evaluate the extent to which they influence each bacterial community assembly in urban rivers.

3.8 Effects of Land Use, Physicochemical, Nutrients, ADOMs, and DOMs on Neutral and Non-neutral Bacterial Communities in Urban Rivers

Evaluating the effect of different factors, namely, abiotic, on NBC and N_NBC will provide insights into how each of these influenced microbial assembly distribution and diversity. In this section, we used PLS-PM to estimate the effect of land use, physicochemical, nutrients, ADOMs, and DOMs on NBC and N_NBC alpha- and beta-diversities. Overall, the goodness of fit was 0.50 for NR and 0.52 for WR, indicating a good PLS-PM model fit. In NR, the river with low urbanization, DOMs (i.e., HLS) had significant negative effects on NBC beta-diversity, and DOMs (i.e., PLS) had significant positive effects on N_NBC alpha-diversity ($p < 0.05$) (Fig. 2.16a, c). However, the total effect of land use was significant on both NBC and N_NBC beta diversity while physicochemical and ADOMs (i.e., bisphenol) also had a significant total effect on N_NBC beta-diversity (Fig. 2.16a, c). Additionally, only DOMs (i.e., PLS) followed by ADOMs (i.e., drugs) displayed significant total effects on N_NBC alpha-diversity.

Similarly, to NR, DOMs (i.e., HLS) had the strongest negative effects on NBC and N_NBC beta diversity in WR (i.e., river with high urbanization) although not significant; however, only nutrients had the highest significant negative effects on N_NBC alpha-diversity in WR ($p < 0.05$) (Fig. 2.16a, b). Contrary to NR, the total effect of land use on both NBC and N_NBC beta-diversity, that of physicochemical and ADOMs (i.e., bisphenol) on N_NBC beta-diversity, as well as that of DOMs (i.e., PLS) on N_NBC alpha-diversity was insignificant in WR (Fig. 2.16c, d). However, the total effect of physicochemical on N_NBC beta-diversity and that of ADOMs (i.e., drugs) on N_NBC alpha-diversity was significant but opposite while that of DOMs (i.e., HLS) on N_NBC alpha-diversity was similar for both rivers (Fig. 2.16c, d). Additionally, the total effect of ADOMs (i.e., bisphenol) on both NBC and N_NBC alpha-diversity was only significant in WR. More importantly, the total effect of ADOMs on both NBC and N_NBC alpha- and beta-diversity was generally higher in WR than NR while that of DOMs was generally higher in NR than WR (Fig. 2.16c, d). To some extent, these results demonstrate that land use, physicochemicals, nutrients, ADOMs (Antibiotics, Bisphenols, and Drugs), and DOMs (HLS and PLS) play different roles in influencing the NBC and N_NBC between those rivers. This could be partially attributed to differences in anthropogenic activities and biogeochemical processes affecting each river, as well as differences in the composition and abundance of native bacterial communities.

Overall, our findings demonstrated that both NBC and N_NBC were significantly influenced by both ADOMs and DOMs in aquatic ecosystems. Interestingly, ADOMs and nutrients had a higher significant influence on community assembly in the river with high urbanization, while the influence of DOMs, land use, and physicochemical was significantly higher in rivers with low urbanization (Fig. 2.16c, d).

In agreement with our Mantel test (see Sect. 3.5) and network analysis (see Sects. 3.6 and 3.7) results, PLS-PM also shows the importance of both ADOMs and DOMs

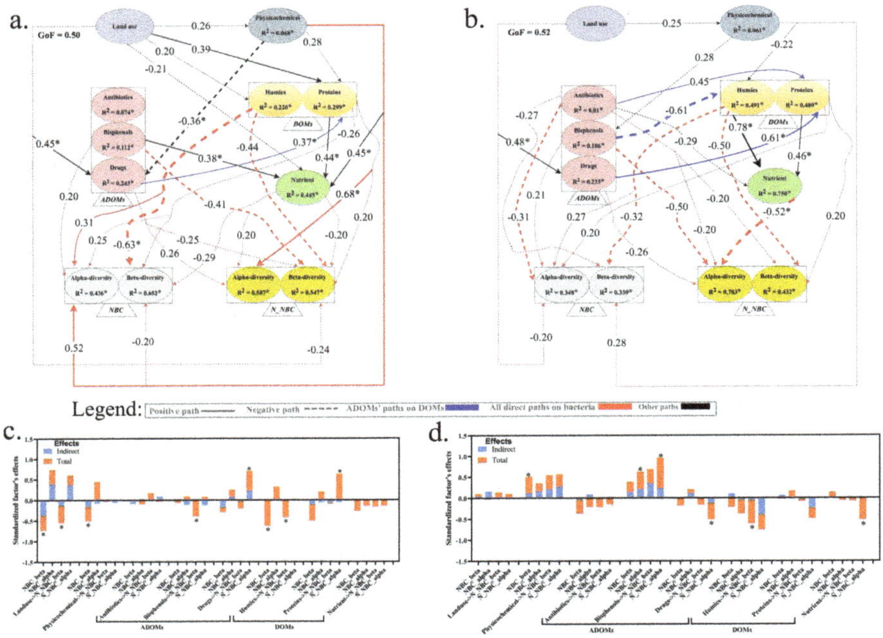

Fig. 2.16 PLS-PM showing the effects of land use, physicochemical, nutrients, DOMs (*Humics and Proteins*), and ADOMs (*Antibiotics, Bisphenols, and Drugs*) on alpha-diversity and beta-diversity of neutral (NBC) and non-neutral (N_NBC) bacterial communities' assembly in NR (**a**) and WR (**b**). Legend**:** solid and dashed lines indicate positive and negative path coefficients, respectively; only path coefficients with an absolute value of coefficient ≥ 0.2; line thicknesses represent the absolute value of path coefficient and are categorized into three (low = 0.20–0. 25, medium = 0.31–0.5, and high = 0.51–0.8); (R^2) represents the coefficient of determination; the Asterix indicates significance ($p < 0.05$). (**Red lines indicate the direct paths to alpha-diversity and Beta-diversity of NBC and N_NBC. *Blue lines indicate the direct paths to Humics and proteins*). The bar chart displays the standardized indirect, and total effects of land use, physicochemical, nutrients, DOMs (*Humics and Proteins*), and ADOMs (*Antibiotics, Bisphenols, and Drugs*) on alpha-diversity and Beta-diversity of neutral and non-neutral bacterial communities' assembly in NR (**c**) and WR (**d**). The Asterix represents the significance ($p < 0.05$). (*NBC_beta, NBC_aplha, N_NBC_beta, and N_NBC_alpha represent the beta- and alpha-diversity, respectively*)

impacts on the neutral and non-neutral bacterial assembly in urbanized rivers. These results first indicate significant impacts of those environmental factors on the relative abundance, occurrence frequency, and distribution of the neutral and non-neutral bacterial communities in urbanized rivers (Chen et al., 2019; Liu et al., 2013; Zhang et al., 2018). Second, the higher significant effects of ADOMs on neutral and non-neutral bacterial assembly in the river with high urbanization could be assigned to their higher concentrations in this river thus, increasing their importance to the community assembly. Similarly, the higher influence of DOMs on neutral and non-neutral bacterial assembly in the river with low urbanization could be assigned to lower ADOMs and nutrients concentrations thus, enhancing the significance of

DOMs to bacterial communities' assembly in this river (Chen et al., 2019; Zhou et al., 2021).

So far, relatively no study has quantified the influence of land use, physicochemical, nutrients, ADOMs, and DOMs on neutral and non-neutral bacterial communities in urbanized rivers, especially using PLS-PM. Although stochastic processes were the primary drivers of bacterial communities, PLS-PM results revealed that ADOMs and DOMs were important factors controlling the compositional variations of bacterial community assembly. Thus, we suggest that both ADOMs and DOMs could enhance the deterministic processes in urbanized rivers and significantly influence the assembly of the so-called neutral communities.

4 Conclusions

This study sheds light on the intricate mechanisms governing assembly processes and interactions of microbial communities with ADOMs and DOMs in urbanized river ecosystems undergoing different anthropogenic impacts. By employing the neutral community model and normalized stochastic ratio, we found that stochastic forces predominated in bacterial community assembly, particularly in rivers with low urbanization levels. Co-occurrence network analysis revealed strong positive associations of non-neutral and nearly neutral bacterial clusters with ADOMs and DOMs, indicating their significant ecological roles in shaping community structure. Furthermore, our PLS-PM results highlighted the influence of land use, physicochemical parameters, nutrients, ADOMs, and DOMs on both alpha- and beta-diversities of bacterial communities. Notably, humic substances exhibited substantial direct effects on beta diversity, suggesting their pivotal role in shaping microbial community composition. Moreover, the total effect of ADOMs on alpha- and beta-diversities was more pronounced in highly urbanized rivers, underscoring the heightened impact of urbanization on microbial community dynamics. These findings imply that ADOMs and DOMs act as key environmental drivers, enhancing deterministic processes in river ecosystems and potentially influencing the ecological functions and growth of bacterial taxa. We suggest that increased human disturbances can lower stochastic forces in urbanized rivers and thus, several NBC taxa may also be significantly influenced by deterministic processes. Our study provides new and useful insight for understanding the ecological impact of ADOMs on DOMs pools and the dynamics of NBC and N_NBC in relationship with ADOMs and DOMs components in urbanized river ecosystems. However, it is essential to acknowledge the limitations of this study including sample size and short-term monitoring, potential biases or uncertainties inherent in the sampling, and analytical methods employed especially in determining ADOMs. Future research should aim to elucidate the specific biochemical and ecological mechanisms underlying microbial community responses to ADOMs and DOMs, as well as their implications for ecosystem functioning and human health. Additionally, exploring the long-term effects of urbanization on microbial community and DOM dynamics and

identifying novel avenues for mitigating anthropogenic impacts on aquatic ecosystems would further enrich our understanding of the complex interactions between microbial communities and their environment.

Acknowledgments This work was supported by the National Natural Science Foundation of China (31870475) and the Ocean Negative Carbon Emissions (ONCE) Program.

Author Contributions **Binessi Edouard Ifon:** Investigation, Methodology, Software, Data curation, Formal analysis, Visualization, Writing original draft, and Review & editing. **Bob Adyari:** Investigation, Methodology, Software, Data curation, Writing, and Review & editing. **Qian Sun:** Investigation and Methodology. **Okugbe Ebiotubo Ohore:** Validation, Writing, and Review & editing. **Chang-Ping Yu:** Conceptualization, Resources, Supervision, and Funding acquisition. **Anyi Hu:** Conceptualization, Data curation, Software, Validation, Writing, Review & editing, Supervision, Project administration, and Funding acquisition.

Conflicts of Interest The authors declare no conflict of interest.

Data Availability Upon request, data will be made available by the corresponding author.

References

Adyari, B., Shen, D., Li, S., Zhang, L., Rashid, A., Sun, Q., Hu, A., Chen, N., & Yu, C. P. (2020). Strong impact of micropollutants on prokaryotic communities at the horizontal but not vertical scales in a subtropical reservoir, China. *Science of the Total Environment, 721*, 137767. https://doi.org/10.1016/j.scitotenv.2020.137767

Anand, U., Adelodun, B., Cabreros, C., et al. (2022). Occurrence, transformation, bioaccumulation, risk and analysis of pharmaceutical and personal care products from wastewater: A review. *Environmental Chemistry Letters, 20*, 3883–3904. https://doi.org/10.1007/s10311-022-01498-7

Ávila, C., & García, J. (2014). Chapter 7 . Pharmaceuticals and personal care products (PPCPs) in the environment and their removal from wastewater through constructed wetlands 1–37.

Banerjee, S., Kirkby, C. A., Schmutter, D., Bissett, A., Kirkegaard, J. A., & Richardson, A. E. (2016). Network analysis reveals functional redundancy and keystone taxa amongst bacterial and fungal communities during organic matter decomposition in an arable soil. *Soil Biology and Biochemistry, 97*, 188–198. https://doi.org/10.1016/j.soilbio.2016.03.017

Benitez, J., Henseler, J., Castillo, A., & Schuberth, F. (2020). How to perform and report an impactful analysis using partial least squares: Guidelines for confirmatory and explanatory IS research. *Information Management, 57*, 103168. https://doi.org/10.1016/j.im.2019.05.003

Bowman, N., Patel, P., Xu, W., Alsaffar, A., & Tiquia-Arashiro, S. M. (2018). Enrichment and isolation of Pb-resistant bacteria from Saint Clair River sediments and their potential for Pb removal in aqueous solutions. *Applied Microbiology and Biotechnology, 102*, 2391–2398. https://doi.org/10.1007/s00253-018-8772-4

Burns, A. R., Stephens, W. Z., Stagaman, K., Wong, S., Rawls, J. F., Guillemin, K., & Bohannan, B. J. M. (2016). Contribution of neutral processes to the assembly of gut microbial communities in the zebrafish over host development. *The ISME Journal, 10*, 655–664. https://doi.org/10.1038/ismej.2015.142

Caruso, T., Chan, Y., Lacap, D. C., Lau, M. C., McKay, C. P., & Pointing, S. B. (2011). Stochastic and deterministic processes interact in the assembly of desert microbial communities on a global scale. *The ISME Journal, 5*(9), 1406–1413. https://doi.org/10.1038/ismej.2011.21

Chase, J. M. (2005). Towards a really unified theory for metacommunities. *Functional Ecology, 19*, 182–186.

Chen, X., Vollertsen, J., & Lund, J. (2015). Degradation of PPCPs in activated sludge from different WWTPs in Denmark. *Ecotoxicology, 24*, 2073. https://doi.org/10.1007/s10646-015-1548-z

Chen, D., Kannan, K., Tan, H., Zheng, Z., Feng, Y. L., Wu, Y., & Widelka, M. (2016). Bisphenol analogues other than BPA: Environmental occurrence, human exposure, and toxicity – A review. *Environmental Science & Technology, 50*, 5438–5453. https://doi.org/10.1021/acs.est.5b05387

Chen, W., Ren, K., Isabwe, A., Chen, H., Liu, M., & Yang, J. (2019). Stochastic processes shape microeukaryotic community assembly in a subtropical river across wet and dry seasons. *Microbiome., 7*(1), 138.

Chin, W. W. (2010). How to write up and report PLS analyses. In V. Esposito Vinzi, W. Chin, J. Henseler, & H. Wang (Eds.), *Handbook of partial least squares. Springer handbooks of computational statistics*. Springer. https://doi.org/10.1007/978-3-540-32827-8

Cho, K., Zholi, A., Frabutt, D., Flood, M., Floyd, D., & Tiquia, S. M. (2012). Linking bacterial diversity and geochemistry of uranium-contaminated groundwater. *Environmental Technology, 33*, 1629–1640. https://doi.org/10.1080/09593330.2011.641036

Choi, Y. (2014). Disturbtion and degradation of Bisphenol A (BPA) substitutes BPAF and BPS compared to BPA in aerobic soil and anaerobic sludge. Dr. Thesis.

Cizmas, L., Sharma, V. K., Gray, C. M., & McDonald, T. J. (2015). Pharmaceuticals and personal care products in waters: Occurrence, toxicity, and risk. *Environmental Chemistry Letters, 13*(4), 381–394. https://doi.org/10.1007/s10311-015-0524-4

Curtis, T. P., & Sloan, W. T. (2004). Prokaryotic diversity and its limits: Microbial community structure in nature and implications for microbial ecology. *Current Opinion in Microbiology, 7*(3), 221–226. https://doi.org/10.1016/j.mib.2004.04.010

Deng, Y., Jiang, Y. H., Yang, Y., He, Z., Luo, F., & Zhou, J. (2012). Molecular ecological network analyses. *BMC Bioinformatics, 13*. https://doi.org/10.1186/1471-2105-13-113

Deutschmann, I. M., Lima-Mendez, G., Krabberød, A. K., Raes, J., Vallina, S. M., Faust, K., & Logares, R. (2021). Disentangling environmental effects in microbial association networks. *Microbiome, 9*, 1–18. https://doi.org/10.1186/s40168-021-01141-7

Dey, S., Bano, F., & Malik, A. (2019). Pharmaceuticals and personal care product (PPCP) contamination—A global discharge inventory. In *Pharmaceuticals and personal care products: Waste management and treatment technology*. https://doi.org/10.1016/B978-0-12-816189-0.00001-9

Edgar, R.C., 2013. UPARSE: highly accurate OTU sequences from microbial amplicon reads. Nat. Methods 10, 996–998. https://doi.org/10.1038/nmeth.2604.

Erdogan, I., & Schlautman, M. A. (2015). Selecting filter membranes for measuring DOC and UV_{254}. *Journal AWWA, 95*, 86–100.

Feng, W., Deng, Y., Yang, F., Miao, Q., & Ngien, S. K. (2023). Systematic review of contaminants of emerging concern (CECs): Distribution, risks, and implications for water quality and health. *Water (Switzerland), 15*. https://doi.org/10.3390/w15223922

Flood, M., Frabutt, D., Floyd, D., Powers, A., Ezegwe, U., Devol, A., & Tiquia-Arashiro, S. M. (2015). Ammonia-oxidizing bacteria and archaea in sediments of the Gulf of Mexico. *Environmental Technology, 36*, 124–135. https://doi.org/10.1080/09593330.2014.942385

Gad, M., Hou, L., Li, J., Wu, Y., Rashid, A., Chen, N., & Hu, A. (2020). Distinct mechanisms underlying the assembly of microeukaryotic generalists and specialists in an anthropogenically impacted river. *Science of the Total Environment, 748*, 141434. https://doi.org/10.1016/j.scitotenv.2020.141434

Gan, L. H., Yan, Z. R., Ma, Y. F., Zhu, Y. Y., Li, X. Y., Xu, J., & Zhang, W. (2019). pH dependence of the binding interactions between humic acids and bisphenol A – A thermodynamic perspective. *Environmental Pollution, 255*, 113292. https://doi.org/10.1016/j.envpol.2019.113292

Gan, S., Wu, P., Song, Y., Guo, P., Cai, N., Yuan, F., Yang, Q., Wu, Y., Liu, N., & Pan, J. (2022). Non-targeted characterization of dissolved organic matter from a wastewater treatment plant

by FT-ICR-MS: A case study of hospital sewage. *Journal of Water Process Engineering, 48*, 102834. https://doi.org/10.1016/j.jwpe.2022.102834

Gao, Y., Zhang, W., Li, Y., Wu, H., Yang, N., & Hui, C. (2021). Dams shift microbial community assembly and imprint nitrogen transformation along the Yangtze River. *Water Research, 189*, 116979. https://doi.org/10.1016/j.watres.2020.116579

Gentile, M., Yan, T., Tiquia, S. M., et al. (2006). Stability in a denitrifying fluidized bed reactor. *Microbial Ecology, 52*, 311–321. https://doi.org/10.1007/s00248-006-9024-1

Gordon, S. H., Mohamed, A., Kuru, R. E. H., & Imam, S. H. (2010). A chemometric method for correcting Fourier transform infrared spectra of biomaterials for interference from water in KBr Discs. *Applied Spectroscopy, 64*(4), 448–457. https://doi.org/10.1366/000370210791114301

Hildebrand, F., Tito, R.Y., Voigt, A.Y., Bork, P., Raes, J., 2014. LotuS: An efficient and user-friendly OTU processing pipeline. Microbiome 2, 1–7. https://doi.org/10.1186/2049-2618-2-37.

Hong, H. (2011). Characterization of dissolved organic matter under contrasting hydrologie regimes in a subtropical watershed using PARAFAC model. *Production, 7*, 2008–2010.

Hu, A., Ju, F., Hou, L., Li, J., Yang, X., Wang, H., Mulla, S. I., Sun, Q., Bürgmann, H., & Yu, C. P. (2017a). Strong impact of anthropogenic contamination on the co-occurrence patterns of a riverine microbial community. *Environmental Microbiology, 19*, 4993–5009. https://doi.org/10.1111/1462-2920.13942

Hu, A., Wang, H., Yang, X., Hou, L., Li, J., Li, S., & Yu, C. (2017b). Seasonal and spatial variations of prokaryoplankton communities in a salinity-influenced watershed, China. *FEMS Microbiology Ecology, 93*(8). https://doi.org/10.1093/femsec/fix093

Huguet, A., Vacher, L., Relexans, S., Saubusse, S., Froidefond, J. M., & Parlanti, E. (2009). Properties of fluorescent dissolved organic matter in the Gironde estuary. *Organic Geochemistry, 40*, 706–719. https://doi.org/10.1016/j.orggeochem.2009.03.002

Ifon, B. E., Adyari, B., Hou, L., Zhang, L., & Liao, X. (2022a). Insight into variation and controlling factors of dissolved organic matter between urban rivers undergoing different anthropogenic influences. *Journal of Environmental Management, 326*, 116737. https://doi.org/10.1016/j.jenvman.2022.116737

Ifon, B. E., Kiki, C., & Hassan, K. (2022b). Effects of bisphenols and perfluoroalkylated substances on fluorescence properties of humic and amino acids substances of dissolved organic matter: EEM-PARAFAC and ATR-FTIR analysis. *Journal of Environmental Chemical Engineering, 10*, 108186. https://doi.org/10.1016/j.jece.2022.108186

Ifon, B. E., Adyari, B., Hou, L., Ebiotubo, O., Rashid, A., Yu, C., & Anyi, H. (2023). Urbanization influenced the interactions between dissolved organic matter and bacterial communities in rivers. *Journal of Environmental Management, 341*, 117986. https://doi.org/10.1016/j.jenvman.2023.117986

Jennifer, A., Abdallah, M. A., & Harrad, S. (2017). Pharmaceuticals and personal care products (PPCPs) in the freshwater aquatic environment. *Emerging Contaminants, 3*, 1–16. https://doi.org/10.1016/j.emcon.2016.12.004

Kassem, A., Abbas, L., Coutinho, O., Opara, S., Najaf, H., Kasperek, D., Li, X., & Tiquia-Arashiro, S. M. (2023). Applications of Fourier transform-infrared spectroscopy in microbial cell biology and environmental microbiology: Advances, challenges, and future perspectives. *Frontiers in Microbiology, 14*, 1304081. https://doi.org/10.3389/fmicb.2023.1304081

Khanzada, N. K., Farid, M. U., Kharraz, J. A., Choi, J., Tang, C. Y., Nghiem, L. D., Jang, A., & An, A. K. (2020). Removal of organic micropollutants using advanced membrane-based water and wastewater treatment: A review. *Journal of Membrane Science, 598*, 117672. https://doi.org/10.1016/j.memsci.2019.117672

Kida, M., Fujitake, N., Suchewaboripont, V., Poungparn, S., Tomotsune, M., Kondo, M., Yoshitake, S., Iimura, Y., Kinjo, K., Maknual, C., & Ohtsuka, T. (2018). Contribution of humic substances to dissolved organic matter optical properties and iron mobilization. *Aquatic Sciences, 80*. https://doi.org/10.1007/s00027-018-0578-z

Kim, S. H., Kim, J. H., Baek, S. H., Kim, J. H., Ajani, P. A., Park, B. S., & Han, M. S. (2020). The influence of dissolved organic carbon on the microbial community associated with tetraselmis striata for bio-diesel production. *Applied Sciences, 10*. https://doi.org/10.3390/app10103601

Kirchman, D. L., Dittel, A. I., Findlay, S. E. G., & Fischer, D. (2004). Changes in bacterial activity and community structure in response to dissolved organic matter in the Hudson River, New York. *Aquatic Microbial Ecology, 35*, 243–257. https://doi.org/10.3354/ame035243

Kothawala, D. N., Murphy, K. R., Stedmon, C. A., Weyhenmeyer, G. A., & Tranvik, L. J. (2013). Inner filter correction of dissolved organic matter fluorescence. *Limnology and Oceanography: Methods, 11*, 616–630. https://doi.org/10.4319/lcm.2013.11.616

Kujawinski, E. B. (2011). The impact of microbial metabolism on marine dissolved organic matter. *Annual Review of Marine Science, 3*, 567–599. https://doi.org/10.1146/annurev-marine-120308-081003

Kulikova, N. A. (2021). Interactions between humic substances and microorganisms and their implications for nature-like bioremediation technologies. *Molecules, 26*, 1–32.

Kurtz, Z. D., Müller, C. L., Miraldi, E. R., Littman, D. R., Blaser, M. J., & Bonneau, R. A. (2015). Sparse and compositionally robust inference of microbial ecological networks. *PLoS Computational Biology, 11*, 1–26. https://doi.org/10.1371/journal.pcbi.1004226

Leibold, M. A., Holyoak, M., Mouquet, N., Amarasekare, P., Chase, J. M., Hoopes, M. F., Holt, R. D., Shurin, J. B., Law, R., Tilman, D., Loreau, M., & Gonzalez, A. (2004). The metacommunity concept: A framework for multi-scale community ecology. *Ecology Letters, 7*(7), 601–613. https://doi.org/10.1111/j.1461-0248.2004.00608.x

Li, P., & Hur, J. (2017). Utilization of UV-vis spectroscopy and related data analyses for dissolved organic matter (DOM) studies: A review. *Critical Reviews in Environmental Science and Technology, 47*, 131–154. https://doi.org/10.1080/10643389.2017.1309186

Li, S., Ju, H., Ji, M., Zhang, J., Song, K., Chen, P., & Mu, G. (2018). Terrestrial humic-like fluorescence peak of chromophoric dissolved organic matter as a new potential indicator tracing the antibiotics in typical polluted watershed. *Journal of Environmental Management, 228*, 65–76. https://doi.org/10.1016/j.jenvman.2018.09.013

Li, J., Wang, L., Geng, J., Li, S., Yu, Q., Xu, K., & Ren, H. (2020a). Distribution and removal of fluorescent dissolved organic matter in 15 municipal wastewater treatment plants in China. *Chemosphere, 251*, 126375. https://doi.org/10.1016/j.chemosphere.2020.126375

Li, Y., Xu, C., Zhang, W., Lin, L., Wang, L., Niu, L., Zhang, H., Wang, P., & Wang, C. (2020b). Response of bacterial community in composition and function to the various DOM at river confluences in the urban area. *Water Research, 169*, 115293. https://doi.org/10.1016/j.watres.2019.115293

Liang, J., Liu, J., Zhan, Y., Zhou, S., Xue, C. X., Sun, C., Lin, Y., Luo, C., Wang, X., & Zhang, X. H. (2021). Succession of marine bacteria in response to Ulva prolifera-derived dissolved organic matter. *Environment International, 155*, 106687. https://doi.org/10.1016/j.envint.2021.106687

Lingegowda, D. C., Komal Kumar, J., Devi Prasad, A. G., & Mahsa Zarei, S. G. (2013). FTIR spectroscopic studies on cleome gynandra—comparative analysis of functional group. *Romanian Journal of Biophysics, 22*, 137–143.

Liu, L., Yang, J., Yu, X., Chen, G., & Yu, Z. (2013). Patterns in the composition of microbial communities from a subtropical river: Effects of environmental, spatial and temporal factors. *PLoS One, 8*, 1–10. https://doi.org/10.1371/journal.pone.0081232

Ma, Y., Mao, R., & Li, S. (2021). Hydrological seasonality largely contributes to riverine dissolved organic matter chemical composition: Insights from EEM-PARAFAC and optical indicators. *Journal of Hydrology, 595*, 125993. https://doi.org/10.1016/j.jhydrol.2021.125993

Madonia, A., Caruso, G., Piazzolla, D., Bonamano, S., Piermattei, V., Zappalà, G., & Marcelli, M. (2020). Chromophoric dissolved organic matter as a tracer of fecal contamination for bathing water quality monitoring in the northern Tyrrhenian Sea (Latium, Italy). *Journal of Marine Science and Engineering, 8*, 1–17. https://doi.org/10.3390/JMSE8060430

Maria, F. (2017). Beta-diversity of central European forests decreases along an elevational gradient due to the variation in local community assembly processes. *Ecography (Cop.)., 41*, 1038. https://doi.org/10.1111/ecog.02809

McKindles, K. M., & Tiquia-Arashiro, S. M. (2012). Functional gene arrays for analysis of microbial communities on ocean platform. In S. Tiquia-Arashiro (Ed.), *Molecular biological technologies for ocean sensing* (Springer protocols handbooks). Humana Press. https://doi.org/10.1007/978-1-61779-915-0_9

Miranda, K. M., Espey, M. G., & Wink, D. A. (2001). A rapid, simple spectrophotometric method for simultaneous detection of nitrate and nitrite. *Nitric Oxide, 5*, 62–71. https://doi.org/10.1006/niox.2000.0319

Mittelbach, G. G., Steiner, C. F., Scheiner, S. M., Katherine, L., Reynolds, H. L., Waide, R. B., Willig, M. R., & Dodson, S. I. (2016). *What is the observed relationship between species richness and productivity?* Laura Gough Published by: Wiley Stable. URL: http://www.jstor.org/stable/2679922. Accessed: 27-04-2016 22: 38 UTC Your use of the JSTOR archive indicates your acceptance of th 9, 2381–2396

Mo, Y., Peng, F., Gao, X., Xiao, P., Logares, R., Jeppesen, E., Ren, K., Xue, Y., & Yang, J. (2021). Low shifts in salinity determined assembly processes and network stability of microeukaryotic plankton communities in a subtropical urban reservoir. *Microbiome, 9*, 1–18. https://doi.org/10.1186/s40168-021-01079-w

Murphy, K. R., Stedmon, C. A., Wenig, P., & Bro, R. (2014). OpenFluor- An online spectral library of auto-fluorescence by organic compounds in the environment. *Analytical Methods, 6*, 658–661. https://doi.org/10.1039/c3ay41935e

Nassereddine, Z., Opara, S., Coutinho, O., Qyteti, F., Book, R., Heinicke, M. P., Napieralski, J., & Tiquia-Arashiro, S. M. (2024). Critical perspectives on advancing antibiotic resistant gene (ARG) detection technologies in aquatic ecosystems. *Science of the Total Environment, 957*, 177775. https://doi.org/10.1016/j.scitotenv.2024.177775

Nguyen, S., Ala, F., Cardwell, C., Cai, D., McKindles, K. M., Lotvola, A., Hodges, S., Deng, Y., & Tiquia-Arashiro, S. M. (2013). Isolation and screening of carboxydotrophs isolated from composts and their potential for butanol synthesis. *Environmental Technology, 34*, 1995–2007. https://doi.org/10.1080/09593330.2013.795987

Ning, D., Deng, Y., Tiedje, J. M., & Zhou, J. (2019). A general framework for quantitatively assessing ecological stochasticity. *Proceedings of the National Academy of Sciences of the United States of America, 116*, 16892–16898. https://doi.org/10.1073/pnas.1904623116

Oest, A., Alsaffar, A., Fenner, M., Azzopardi, D., & Tiquia-Arashiro, S. M. (2018). Patterns of change in metabolic capabilities of sediment microbial communities in river and lake ecosystems. *International Journal of Microbiology, 2018*, 1. https://doi.org/10.1155/2018/6234931

Oksanen, A. J., Blanchet, F. G., Friendly, M., Kindt, R., Legendre, P., Mcglinn, D., Minchin, P. R., Hara, R. B. O., Simpson, G. L., Solymos, P., Stevens, M. H. H., & Szoecs, E. (2019). Vegan. In *Encyclopedia of food and agricultural ethics* (pp. 2395–2396). Springer. https://doi.org/10.1007/978-94-024-1179-9_301576

Olesen, J. M., Bascompte, J., Dupont, Y. L., & Jordano, P. (2007). The modularity of pollination networks. *Proceedings of the National Academy of Sciences of the United States of America, 104*, 19891–19896. https://doi.org/10.1073/pnas.0706375104

Oni, O. E., Schmidt, F., Miyatake, T., Kasten, S., Witt, M., Hinrichs, K. U., & Friedrich, M. W. (2015). Microbial communities and organic matter composition in surface and subsurface sediments of the Helgoland mud area, North Sea. *Frontiers in Microbiology, 6*, 1–16. https://doi.org/10.3389/fmicb.2015.01290

Östman, Ö., Drakare, S., Kritzberg, E. S., Langenheder, S., Logue, J. B., & Lindström, E. S. (2010). Regional invariance among microbial communities. *Ecology Letters, 13*, 118–127. https://doi.org/10.1111/j.1461-0248.2009.01413.x

Paerl, H. W., & Pinckney, J. L. (1996). A mini-review of microbial consortia: Their roles in aquatic production and biogeochemical cycling. *Microbial Ecology, 31*, 225–247. https://doi.org/10.1007/BF00171569

Paerl, H. W., Dyble, J., Moisander, P. H., Noble, R. T., Piehler, M. F., Pinckney, J. L., Steppe, T. F., Twomey, L., & Valdes, L. M. (2003). Microbial indicators of aquatic ecosystem change: Current applications to eutrophication studies. *FEMS Microbiology Ecology, 46*(3), 233–246. https://doi.org/10.1016/S0168-6496(03)00200-9

Pagnucco, G., Overfield, D., Chamlee, Y., Fortuna, A., Sulaiman, F., Farinas, J., Schittenhelm, R., Shuler, C., Kassem, A., Opara, S., Najaf, H., Abbas, L., Coutinho, O., Catalfano, B., Li, X., & Tiquia-Arashiro, S. M. (2023). Metal tolerance and biosorption capacities of bacterial strains isolated from an urban watershed. *Frontiers in Microbiology, 14*, 1278886. https://doi.org/10.3389/fmicb.2023.1278886

Patel, D., Gismondi, R., Alsaffar, A., & Tiquia-Arashiro, S. M. (2019). Applicability of API ZYM to capture seasonal and spatial variabilities in lake and river sediments. *Environmental Technology, 40*(24), 3227–3239. https://doi.org/10.1080/09593330.2018.1468492

Perveen, S., Pablos, C., Reynolds, K., Stanley, S., & Marugán, J. (2023). Growth and prevalence of antibiotic-resistant bacteria in microplastic biofilm from wastewater treatment plant effluents. *Science of the Total Environment, 856*, 159024. https://doi.org/10.1016/j.scitotenv.2022.159024

Plecha, S., Hall, D., & Tiquia-Arashiro, S. M. (2013). Screening for novel bacteria from the bioenergy feedstock switchgrass (*Panicum virgatum* L.). *Environmental Technology, 34*, 1895–1904. https://doi.org/10.1080/09593330.2013.818701

Qin, W., Amin, S. A., Lundeen, R. A., Heal, K. R., Martens-Habbena, W., Turkarslan, S., Urakawa, H., Costa, K. C., Hendrickson, E. L., Wang, T., Beck, D. A., Tiquia-Arashiro, S. M., Taub, F., Holmes, A. D., Vajrala, N., Berube, P. M., Lowe, T. M., Moffett, J. W., Devol, A. H., Baliga, N. S., Arp, D. J., Sayavedra-Soto, L. A., Hackett, M., Armbrust, E. V., Ingalls, A. E., & Stahl, D. A. (2018). Stress response of a marine ammonia-oxidizing archaeon informs physiological status of environmental populations. *The ISME Journal, 12*(2), 508–519. https://doi.org/10.1038/ismej.2017.186

Quast, C., Pruesse, E., Yilmaz, P., Gerken, J., Schweer, T., Yarza, P., Peplies, J., Gl¨ockner, F.O., 2013. The SILVA ribosomal RNA gene database project: improved data processing and web-based tools. Nucleic Acids Res. 41, 590–596. https://doi.org/10.1093/nar/gks1219.

Qu, L., Wu, Y., Li, Y., Stubbins, A., Dahlgren, R. A., Chen, N., & Guo, W. (2020). El Niño-driven dry season Flushing enhances dissolved organic matter export from a subtropical watershed. *Geophysical Research Letters, 47*. https://doi.org/10.1029/2020GL089877

Rizzuto, S., Baho, D. L., Jones, K. C., Zhang, H., Leu, E., & Nizzetto, L. (2021). Science of the total environment binding of waterborne pharmaceutical and personal care products to natural dissolved organic matter. *Science of the Total Environment, 784*, 147208. https://doi.org/10.1016/j.scitotenv.2021.147208

Roguet, A., Laigle, G. S., Therial, C., Bressy, A., & Soulignac, F. (2015). Neutral community model explains the bacterial community assembly in freshwater lakes. *FEMS Microbiology Ecology, 91*(11), fiv125.

Sanchez, G. (2013). PLS path modeling with R. R Packag. Notes 235.

Santinelli, C. (2015). DOC in the Mediterranean Sea. In D. A. Hansell & C. A. Carlson (Eds.), *Biogeochemistry of marine dissolved organic matter* (2nd ed., pp. 579–608). Elsevier Inc.. https://doi.org/10.1016/B978-0-12-405940-5.00013-3

Sinsabaugh, R. L., & Foreman, C. M. (2003). Integrating dissolved organic matter metabolism and microbial diversity: An overview of conceptual models. In S. E. G. Findlay & R. L. Sinsabaugh (Eds.), *Aquatic ecology, aquatic ecosystems* (pp. 425–454). Academic Press. https://doi.org/10.1016/B978-012256371-3/50019-6

Smith, V. H. (2002). Effects of resource supplies on the structure and function of microbial communities. *Antonie Van Leeuwenhoek, 81*, 99–106. https://doi.org/10.1023/A:1020533727307

Subashchandrabose, S. R., Ramakrishnan, B., Megharaj, M., Venkateswarlu, K., & Naidu, R. (2013 Jan). Mixotrophic cyanobacteria and microalgae as distinctive biological agents for organic pollutant degradation. *Environment International, 51*, 59–72. https://doi.org/10.1016/j.envint.2012.10.007

Sun, Q., Li, Y., Li, M., Ashfaq, M., Lv, M., Wang, H., Hu, A., & Yu, C. (2016). Chemosphere PPCPs in Jiulong River estuary (China): Spatiotemporal distributions, fate , and their use as chemical markers of wastewater. *Chemosphere, 150*, 596–604. https://doi.org/10.1016/j.chemosphere.2016.02.036

Tak, S., & Vellanki, B. P. (2018). Natural organic matter as precursor to disinfection byproducts and its removal using conventional and advanced processes: State of the art review. *Journal of Water and Health, 16*, 681–703. https://doi.org/10.2166/wh.2018.032

Tang, J., Li, X., Cao, C., Lin, M., Qiu, Q., Xu, Y., & Ren, Y. (2019a). Compositional variety of dissolved organic matter and its correlation with water quality in peri-urban and urban river watersheds. *Ecological Indicators, 104*, 459–469. https://doi.org/10.1016/j.ecolind.2019.05.025

Tang, J., Wang, W., Yang, L., Cao, C., & Li, X. (2019b). Variation in quantity and chemical composition of soil dissolved organic matter in a peri-urban critical zone observatory watershed in Eastern China. *Science of the Total Environment, 688*, 622–631. https://doi.org/10.1016/j.scitotenv.2019.06.270

Tiquia, S. M. (2008). Diversity of sulfate-reducing genes (dsrAB) in sediments from Puget Sound. *Environmental Technology, 29*(10), 1095–1108. https://doi.org/10.1080/09593330802190608

Tiquia, S. M. (2010a). Salt-adapted bacteria isolated from the Rouge River and potential for degradation of contaminants and biotechnological applications. *Environmental Technology, 31*, 967–978. https://doi.org/10.1080/09593331003706226

Tiquia, S. M. (2010b). Metabolic diversity of the heterotrophic microorganisms and potential link to pollution of the Rouge River. *Environmental Pollution, 158*, 1435–1443. https://doi.org/10.1016/j.envpol.2009.12.035

Tiquia, S. M. (2011). Extracellular hydrolytic enzyme activities of the heterotrophic microbial communities of the Rouge River: An approach to evaluate ecosystem response to urbanization. *Microbial Ecology, 62*, 679–689. https://doi.org/10.1007/s00248-011-9871-2

Tiquia, S. M., Gurczynski, S., Zholi, A., & Devol, A. (2006a). Diversity of biogeochemical cycling genes from Puget Sound sediments using DNA microarrays. *Environmental Technology, 27*(12), 1377–1389. https://doi.org/10.1080/09593332708618756

Tiquia, S. M., Masson, S. A., & Devol, A. (2006b). Vertical distribution of nitrite reductase genes (nirS) in continental margin sediments of the Gulf of Mexico. *FEMS Microbiology Ecology, 58*(3), 464–475. https://doi.org/10.1111/j.1574-6941.2006.00173.x

Tiquia, S. M., Davis, D., Hadid, H., Kasparian, S., Ismail, M., Sahly, R., Shim, J., Singh, S., & Murray, K. S. (2007). Halophilic and halotolerant bacteria from river waters and shallow groundwater along the Rouge River of southeastern Michigan. *Environmental Technology, 28*(3), 297–307. https://doi.org/10.1080/09593332808618789

Tiquia, S. M., Schleibak, M., Schlaff, J., Floyd, J. C., Benipal, B., Zakhem, E., & Murray, K. S. (2008). Microbial community profiling and characterization of some heterotrophic bacterial isolates from river waters and shallow groundwater wells along the Rouge River, Southeast Michigan. *Environmental Technology, 29*, 651–663. https://doi.org/10.1080/09593330801986998

Tiquia-Arashiro, S. M. (2012). *Molecular biological technologies for ocean sensing. Springer protocols handbooks* (p. 307). Humana. https://doi.org/10.1007/978-1-61779-915-0

Tiquia-Arashiro, S. M., & Pant, D. (2020). *Microbial electrochemical technologies* (p. 518). CRC Press. https://doi.org/10.1201/9780429487118

Tong, X., Leung, M. H. Y., Wilkins, D., Cheung, H. H. L., & Lee, P. K. H. (2019). Neutral processes drive seasonal assembly of the skin mycobiome. *mSystems*, 4. https://doi.org/10.1128/msystems.00004-19

Touliabah, H. E. S., El-Sheekh, M. M., Ismail, M. M., & El-Kassas, H. (2022). A review of microalgae-and cyanobacteria-based biodegradation of organic pollutants. *Molecules, 27*. https://doi.org/10.3390/molecules27031141

van der Gast, C. J., Ager, D., & Lilley, A. K. (2008 Jun). Temporal scaling of bacterial taxa is influenced by both stochastic and deterministic ecological factors. *Environmental Microbiology, 10*(6), 1411–1418. https://doi.org/10.1111/j.1462-2920.2007.01550.x

Wang, K., Pang, Y., He, C., Li, P., Xiao, S., Sun, Y., Pan, Q., Zhang, Y., Shi, Q., & He, D. (2019). Optical and molecular signatures of dissolved organic matter in Xiangxi Bay and mainstream of three gorges reservoir, China: Spatial variations and environmental implications. *Science of the Total Environment, 657*, 1274–1284. https://doi.org/10.1016/j.scitotenv.2018.12.117

Wang, D., Luo, Q., Huang, K., & Zhang, X. X. (2023b). Distinct mechanisms underlying assembly processes and interactions of microbial communities in two single-stage bioreactors coupling anammox with denitrification. *Chemical Engineering Journal, 452*, 139319. https://doi.org/10.1016/j.cej.2022.139319

Wünsch, U. J., & Murphy, K. (2021). A simple method to isolate fluorescence spectra from small dissolved organic matter datasets. *Water Research, 190*, 116730. https://doi.org/10.1016/j.watres.2020.116730

Xiao, Y., Huang, Q., Chen, L., Wang, F., Li, P. (2011). Effects of bisphenol A on dissolved organic matter fluorescence characteristics using parallel factor analysis (PARAFAC) modeling. 2011 2nd Int. Conf. Mech. Autom. Control Eng. MACE 2011 – Proc. (pp. 2124–2127). https://doi.org/10.1109/MACE.2011.5987395.

Yang, L., Guo, W., Chen, N., Hong, H., Huang, J., Xu, J., & Huang, S. (2013). Influence of a summer storm event on the flux and composition of dissolved organic matter in a subtropical river, China. *Applied Geochemistry, 28*, 164–171. https://doi.org/10.1016/j.apgeochem.2012.10.004

Zeglin, L. H. (2015). Stream microbial diversity in response to environmental changes: Review and synthesis of existing research. *Frontiers in Microbiology, 6*, 454. https://doi.org/10.3389/fmicb.2015.00454

Zhang, W., Pan, Y., Yang, J., Chen, H., Holohan, B., Vaudrey, J., Lin, S., & McManus, G. B. (2018). The diversity and biogeography of abundant and rare intertidal marine microeukaryotes explained by environment and dispersal limitation. *Environmental Microbiology, 20*, 462–476. https://doi.org/10.1111/1462-2920.13916

Zhang, Y., Rashid, A., Guo, S., Jing, Y., Zeng, Q., & Li, Y. (2022). Spatial autocorrelation and temporal variation of contaminants of emerging concern in a typical urbanizing river. *Water Research, 212*, 118120. https://doi.org/10.1016/j.watres.2022.118120

Zhao, J. L., Ying, G. G., Liu, Y. S., Chen, F., Yang, J. F., & Wang, L. (2010). Occurrence and risks of triclosan and triclocarban in the Pearl River system, South China: From source to the receiving environment. *Journal of Hazardous Materials, 179*(1–3), 215–222. https://doi.org/10.1016/j.jhazmat.2010.02.082

Zhou, L., Zhou, Y., Hu, Y., Cai, J., Liu, X., Bai, C., Tang, X., Zhang, Y., Jang, K. S., Spencer, R. G. M., & Jeppesen, E. (2019). Microbial production and consumption of dissolved organic matter in glacial ecosystems on the Tibetan Plateau. *Water Research, 160*, 18–28. https://doi.org/10.1016/j.watres.2019.05.048

Zhou, L., Zhou, Y., Tang, X., Zhang, Y., Zhu, G., Székely, A. J., & Jeppesen, E. (2021). Eutrophication alters bacterial co-occurrence networks and increases the importance of chromophoric dissolved organic matter composition. *Limnology and Oceanography, 66*, 2319–2332. https://doi.org/10.1002/lno.11756

Chapter 3
Microbial Biodiversity of Man-Made Water Features

Daphne H. P. Ng, Sakcham Bairoliya, and Bin Cao

1 Introduction

Microorganisms such as bacteria, fungi, algae, and viruses are ubiquitous in natural and man-made habitats (Gilbert & Stephens, 2018; Gupta et al., 2017; Grossart et al., 2019; Hedrich & Schippers, 2021; Lee et al., 2021; Reineke & Schlömann, 2023; Sharma et al., 2023). Essential to maintaining ecosystem health and function, microorganisms play a leading role in several natural processes, such as recycling nutrients and breaking down waste and dead matter as well as degrading harmful contaminants in the environment (Gupta et al., 2017; Oest et al., 2018; Patel et al., 2019; Grossart et al., 2019; Yang et al., 2021; Li et al., 2024). Water features created by human activities are important habitats for wildlife and plants. In man-made water bodies, microorganisms influence water quality, cause disease outbreaks, and are indicators of pollution.

Several culture-independent and culture-dependent methods have been developed to study microorganisms from habitats such as man-made water bodies (Fig. 3.1). However, conventional culture-based methods greatly underestimate the total microbial diversity found in the environment, as a very small fraction (sometimes even less than 1%) of microorganisms can be cultured in laboratory conditions, a phenomenon known as "the great plate count anomaly" (Staley & Konopka, 1985). On the other hand, culture-independent methods such as quantitative polymerase chain reaction (qPCR), droplet digital PCR (ddPCR), and high-throughput DNA sequencing technologies are now commonly used to profile microbial

D. H. P. Ng (✉) · S. Bairoliya · B. Cao (✉)
School of Civil and Environmental Engineering, Nanyang Technological University, Singapore, Singapore

Singapore Centre for Environmental Life Sciences Engineering, Nanyang Technological University, Singapore, Singapore
e-mail: daphneng@ntu.edu.sg; bincao@ntu.edu.sg

S. M. Tiquia-Arashiro et al. (eds.), *Urban Watershed Microbiology, Volume 1*,
https://doi.org/10.1007/978-3-031-89071-0_3

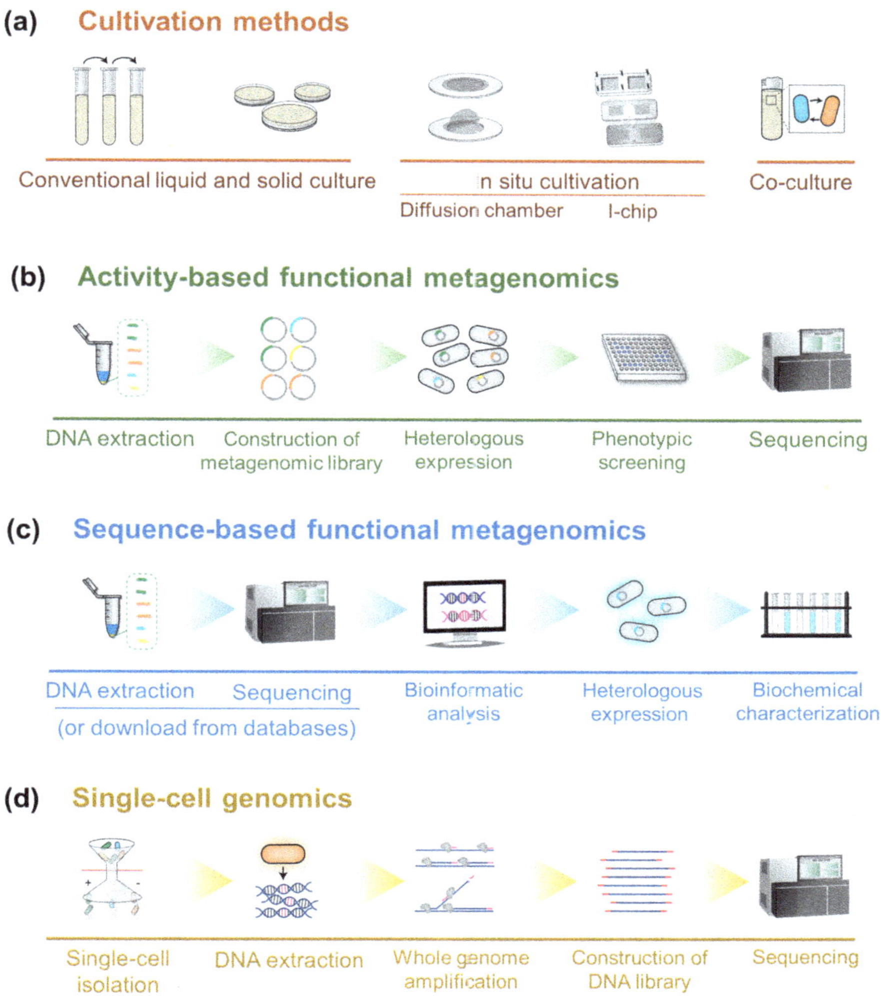

Fig. 3.1 Various culture-independent and culture-dependent methods that can be used to investigate microbial biodiversity. (Reprinted from Liu et al., 2023, with permission from Elsevier)

communities in man-made water features as they do not require microorganisms to be isolated (Abellan-Schneyder et al., 2021; Jantharadej et al., 2021; Jiang et al., 2021a, b; Kumar et al., 2019; She et al., 2021). Advances in single-cell genomics and transcriptomics have also opened the door to investigating important and valuable characteristics of uncultured microorganisms in various water bodies. However, new developments in microbial culture techniques have enabled microorganisms to be isolated and cultured from urban waters and other aquatic environments (Berdy et al., 2017; Polrot et al., 2022).

2 Methods to Study Microbial Biodiversity

2.1 Quantitative Polymerase Chain Reaction (qPCR) and Droplet Digital PCR (ddPCR)

Polymerase chain reaction (PCR) is widely used for various applications in environmental microbiology. PCR enables targeted amplification of genes of interest from DNA samples. The DNA is first denatured at high temperature followed by annealing of primers that are designed to target a specific gene. Once the primers anneal to the target region, the complementary DNA (cDNA) strand is synthesized by a thermostable DNA polymerase enzyme. Multiple cycles of this process produce high copy numbers of the target region. PCR can be used to confirm the presence or absence of target genes in particular samples and combined with sequencing techniques to study the target region amplified from single (Sanger sequencing) or multiple organisms (High-Throughput Sequencing, discussed in following sections) (Botes et al., 2013).

Quantitative PCR (qPCR) is commonly used to quantify microorganisms in the environment. qPCR enables DNA concentration to be monitored in real time during the amplification cycles. A standard curve is created by amplification of samples containing known copy numbers of the target gene, facilitating the absolute quantification of the target in unknown samples. qPCR has been used to quantify the total number of bacterial cells (Sakcham et al., 2019), the number of antibiotic-resistant genes (Keenum et al., 2022; Nassereddine et al., 2024; Srathongneam et al., 2024), the presence of pathogens like *Legionella* (Donohue, 2021), *Mycobacterium* (Radomski et al., 2010), or *Vibrio* (Kim et al., 2015), and potential functions like nitrification, denitrification (Gentile et al., 2006; Geets et al., 2007; Qin et al., 2018), methanogenesis (Steinberg & Regan, 2009), and sulfate reduction (Ben-Dov et al., 2007; McKindles & Tiquia-Arashiro, 2012). qPCR can also be used to measure expression of a target gene and changes in its relative expression under different conditions relative to a housekeeping gene.

In reverse transcriptase qPCR (RT-qPCR), RNA is converted into cDNA through reverse transcription, followed by qPCR to determine the expression level of target transcripts. RT-qPCR has been used to study the differential expression of target genes under varying environmental conditions (Pinto et al., 2012; Tiquia-Arashiro, 2012; Gómez-Silván et al., 2014; Wan et al., 2017; Kumar et al., 2022) and stresses (Goswami & Narayana Rao, 2018; Qin et al., 2018) to elucidate the mechanism of different interactions and observations (Augsburger et al., 2019; Mantilla-Calderon et al., 2019).

An advantage of qPCR is that it can be used to study microorganisms without the need for culture. However, qPCR is limited by our knowledge of the target genes responsible for specific functions and RT-qPCR is dependent on the choice of housekeeping genes (Alqarni et al., 2016). qPCR is also dependent on the initial copy number of the target gene in the samples, yielding spurious results with low target copy numbers.

Droplet digital PCR (ddPCR) was recently introduced to analyze microorganisms that are in low abundance in a sample. In ddPCR, PCR take place independently in individual droplets instead of a larger reaction volume, resulting in higher sensitivity (Abellan-Schneyder et al., 2021). ddPCR has significantly improved the detection of low abundance target genes in samples while providing absolute quantification without the need of a calibration curve (Taylor et al., 2017). However, it is more expensive than qPCR and provides variable improvement depending on the target gene (Nshimyimana et al., 2019). PCR amplification also cannot distinguish between viable and non-viable microorganisms (Botes et al., 2013). To quantify viable microorganisms, samples are pre-treated with viability dyes such as propidium monoazide (PMA) before undergoing qPCR to prevent the amplification of DNA from compromised cells (Acosta et al., 2018; Desneux et al., 2015). qPCR may also be affected by inhibitors in environmental samples, which need to be removed for accurate results (Hamza & Leifels, 2023; Jones et al., 2015).

2.2 *High-Throughput Amplicon Sequencing*

High-throughput amplicon sequencing is a culture-independent method that is commonly used to study the microbial ecology in various ecosystems. The method involves sequencing of a gene-specific region to elucidate the diversity and composition of the taxonomic groups of microorganisms in a sample. The main genetic markers used in microbial ecology are the 16S rRNA gene for archaea and bacteria, the internal transcribed spacer region for fungi, and the 18S rRNA gene for eukaryotes such as microalgae and protists (Lawley et al., 2004; Tiquia, 2005; Tiquia et al., 2005; Cho et al., 2012; Nguyen et al., 2013; Plecha et al., 2013; Hugerth et al., 2014; Kumar et al., 2021; Lee et al., 2020; Taylor et al., 2016). These genetic sequences are highly conserved but contain hypervariable regions that vary considerably between different microorganisms, making them effective molecular markers for identifying microbial species (Bukin et al., 2019; Schoch et al., 2012).

In marker gene analysis, PCR is performed to amplify the marker genes before they are sequenced. The Illumina sequencing technology is an example of a high-throughput sequencing approach that is used in marker gene analysis. First, environmental samples are collected, and genetic material (DNA or RNA) is extracted from them. If RNA is extracted, it is converted to DNA by reverse transcription. Universal primers are usually used to amplify the marker genes from DNA extracts. Primers also include overhang adapter sequences that are compatible with the indexing and sequencing adapters of the sequencing platform. After purification, the PCR products are sequenced on a high-throughput sequencing platform, which generates millions of short reads, each 100–300 bp long. Paired-end sequencing technology, in which one read is generated from each end of the DNA fragment, is usually used as it offers greater coverage and accuracy. The reads are then trimmed to remove low-quality regions and merged to obtain high-quality representative sequences before they are compared against databases to identify microorganisms (Kumar et al., 2019).

As short amplicons (300 bp) are used for Illumina technology, few sequences are accurately classified at the species level. Introduction and improvement of third-generation long-read sequencing technologies like Nanopore (Zhang et al., 2023) and PacBio (Schloss et al., 2016) enable high-throughput sequencing of the whole marker gene, which allows species-level identification (Bertolo et al., 2024; Matsuo et al., 2021). This is essential for identifying pathogens and other organisms of interest with high specificity and confidence.

Although marker gene analysis is a cost-effective and efficient way to analyze microbial communities, it is prone to biases during nucleic acid extraction and PCR. For example, Gram-positive bacteria and other microorganisms with thick cell walls are more difficult to lyse than Gram-negative bacteria, thus potentially skewing the DNA and amplicon profiles. To address this, researchers have developed various chemical and mechanical disruption methods to facilitate DNA extraction from resistant species, such as bead-beating or the use of chemical or enzymatic methods (Hoorzook & Barnard, 2022; Krsek & Wellington, 1999; Roose-Amsaleg et al., 2001). PCR-based techniques are also sensitive to biases due to the presence of extracellular DNA in environmental samples (eDNA), which may obscure community profiling results (Bairoliya et al., 2022). Treatment with PMA to inhibit PCR amplification of eDNA (Lennon et al., 2018) or DNase to remove eDNA (Sakcham et al., 2019) before DNA extraction can reduce these biases.

Due to differing GC content and secondary structures, DNA from different species of bacteria may be copied unevenly during amplicon generation, which affects the resulting taxonomic profile (Browne et al., 2020). Low DNA concentrations can also affect the accuracy of amplicon sequencing. Various library amplification procedures optimized for biased genomes and low biomass samples have been developed (Abellan-Schneyder et al., 2021; Oyola et al., 2012). As only the 16S rRNA or 18S rRNA amplicons are sequenced during marker gene analysis, the functions of the microbial community in the ecosystem cannot be directly elucidated using this method. However, tools such as FAPROTAX (Louca et al., 2016), PICRUSt2 (Douglas et al., 2020), and Tax4Fun2 (Wemheuer et al., 2020) have been developed to predict functions based on 16S rRNA amplicon data. These predictions are highly reliant on the quality and accuracy of data annotation.

2.3 *Shotgun Metagenomic Sequencing and Metatranscriptomics*

Unlike marker gene analysis, shotgun metagenomic sequencing and metatranscriptomics are methods that sequence all the DNA or RNA transcripts respectively in an environmental sample to identify and characterize the microorganisms present. Metagenomics and metatranscriptomics involve the random fragmentation and sequencing of all DNA or RNA (converted to cDNA) in a sample, otherwise known as shotgun sequencing (Almeida & De Martinis, 2019; Bashiardes et al., 2016;

Shakya et al., 2019; Wooley et al., 2010). Shotgun metagenomic sequencing and metatranscriptomics approaches are usually not subject to the biases encountered during amplicon generation by PCR, if there is sufficient input nucleic acid concentration. However, like amplicon sequencing, these methods are still prone to biases during nucleic acid extraction.

A major advantage of shotgun metagenomic sequencing as compared to marker gene analysis is that both taxonomy and functional genes of microbial communities can be characterized, that is, who is there and what can they do. As all the genetic information from the sample is recovered, the complete diversity of a habitat, including archaea, bacteria, eukaryotes, and viruses, can be captured (Hempel et al., 2022). Metagenomics approaches have been used to profile antibiotic-resistant genes in urban canals, water distribution systems, and lakes (Chen et al., 2019; Garner et al., 2018; Jantharadej et al., 2022; Nassereddine et al., 2024). Similarly, metatranscriptomics offers a snapshot of the microbial genes that are expressed in the sample. By analyzing total RNA, metatranscriptomics can provide a deeper insight into how a microbial community functions and responds to changes in environmental conditions.

Another advantage of shotgun metagenomic sequencing is that full genomes can be assembled for an in-depth insight into the genomic diversity of the microbial community to uncover novel functions. Shotgun sequencing is also a boon for the study of novel microbes as draft genomes of uncultured microorganisms can also be obtained (Garza & Dutilh, 2015). However, due to the challenges of assembling multiple genomes and the need for greater sequencing depth to identify low-abundance microorganisms, shotgun metagenomic sequencing is potentially more expensive than marker gene analysis. This makes it less cost-effective for long-term studies and studies with large numbers of samples (Pérez-Cobas et al., 2020). Long read sequencers like Nanopore and PacBio are addressing this issue with single DNA molecule sequencing, eliminating the need for library preparation and reducing the complexity of computationally intensive steps such as genome assembly (Xie et al., 2020; Yang et al., 2022, 2023).

2.4 Single-Cell Genomics and Transcriptomics

Unlike metagenomics and metatranscriptomics techniques that profile a wide range of microorganisms, single-cell genomics and transcriptomics enable the detailed investigation of individual microbes. The approach is especially useful for decoding individual genomes from highly diverse microbial samples and studying rare microorganisms in the environment that may be underrepresented by bulk analysis metagenomics and metatranscriptomics methods (Hosokawa & Nishikawa, 2024).

In single-cell genomics and transcriptomics, single microbial cells are recovered from the environment and DNA or RNA is extracted from them respectively. DNA undergoes whole-genome amplification and is sequenced before being assembled into draft genomes. Similarly, in single-cell transcriptomics, RNA is converted to

cDNA, amplified, and sequenced to construct DNA libraries. Single-cell genomics has been used to identify microorganisms with genes that can degrade organic pollutants in aquatic environments, facilitating the mining of microbial resources (Liu et al., 2023). It can also be used to track mutations in microorganisms that enable them to resist antibiotics, which can shed light on how microorganisms in urban water bodies acquire antibiotic resistance (Bawn et al., 2022; Nassereddine et al., 2024).

2.5 Culture-Based Methods

Nevertheless, culturing and isolating more "microbial dark matter" is key to gaining a detailed understanding of their physiology. Microorganisms in the environment are a treasure trove of metabolites and the ability to culture these microorganisms paves the way for the discovery of pharmaceuticals such as new antibiotics (Hamamoto et al., 2021; Liu et al., 2023; Mohseni et al., 2013; Nassereddine et al., 2024; Zhou et al., 2021). Culture-based methods have conventionally been used to detect specific types of microorganisms in the environment. For example, selective and differential media are commonly used to detect and enumerate fecal indicator bacteria and pathogens in microbiological quality assessments of urban water bodies and systems (Robert-Pillot et al., 2002; Tiquia, 2010; Boczek et al., 2021; Hassanein et al., 2023; Mishra et al., 2018). Culture-based methods have also been used in the enrichment and isolation of hydrocarbon-degrading bacteria (Spini et al., 2018; Tiquia-Arashiro & Pant, 2020), metal-resistant bacteria (Wu et al., 2005; Bowman et al., 2018; Kassem et al., 2023; Pagnucco et al., 2023), and extremophilic microorganisms (Tiquia & Mormile, 2010; Tiquia-Arashiro & Rodrigues, 2016a, b; Tiquia-Arashiro & Grube, 2019).

Unlike culture-independent methods, culture-based methods can distinguish between viable and dead cells as only viable cells will grow in the culture media. However, competition and the presence of inhibitory factors in the environment prevent the growth of slow-growing bacteria during isolation, resulting in an underestimation of biodiversity (Vartoukian et al., 2010). To survive environmental stresses and unfavorable conditions, bacteria can also enter the viable but not culturable (VBNC) state, where they are metabolically active but do not grow on culture media (Alvear-Daza et al., 2021; Fu et al., 2021; Guo et al., 2021; Ye et al., 2020). Several strategies have been developed to improve the cultivation efficiency of unculturable microorganisms from the environment, including mimicking the natural environment and physically separating cells to decrease competition or reduce growth inhibitors. In addition to growth factors, another factor that may affect the culturability of environmental microorganisms is the preparation of culture media. Tanaka et al. (2014) have reported that separately autoclaving the agar and phosphate, instead of together as is usually done, improved the recovery of environmental bacteria in pure culture. This is because when autoclaved together, the agar and

phosphate react together to form hydrogen peroxide, a disinfectant that kills bacteria (Tanaka et al., 2014).

To make the entire range of microbial diversity more accessible, microbial cultivation platforms such as the Isolation Chip (iChip) that mimic the natural environment of the microorganisms have also been devised. The iChip uses naturally occurring growth factors to promote the growth of bacteria that cannot be grown under standard cultivation conditions, by simulating relevant environmental conditions. The innovation has been shown to increase microbial recovery from 5- to 300-fold, depending on the study (Berdy et al., 2017). The iChip and its variants employ the concept of a diffusion chamber in which bacterial cells from environmental samples are placed between two semi-permeable membranes. The chamber is returned to the cells' environment, where naturally occurring nutrients and growth factors diffuse into the chamber. As the pores of the membrane are too small for the bacteria to move in or out of the chamber, they are retained and divide to form colonies. The isolated bacteria can then be identified and characterized. Using the iChip, bacteria have been isolated from soil and estuarine sediment. By enabling more bacteria to be isolated from the environment, the iChip increases culturable bacteria diversity and reduces cultivation bias (Liu et al., 2021; Polrot et al., 2022) (Table 3.1).

3 Measuring Microbial Biodiversity

As microorganisms play essential roles in the environment, understanding microbial biodiversity is central to studying ecosystem structure and function. In microbial community DNA sequencing, microbial diversity is commonly represented as operational taxonomic units (OTUs). OTUs are used to group sequences together based on their similarity (97%), especially when analyzing 16S or 18S rRNA sequence datasets. This approach minimizes the impact of sequencing errors as it clusters similar sequences into a consensus sequence based on a similarity threshold (Bolyen et al., 2019). The Amplicon Sequence Variant (ASV) approach has been proposed as a more accurate alternative to the OTU approach. The ASV approach differentiates actual sequences from spurious sequences using error models and also takes into account abundance and sequence similarity (Callahan et al., 2017; Joos et al., 2020).

Microbial diversity is usually described as alpha diversity (diversity within samples), beta diversity (diversity between samples), and gamma diversity (total diversity of the landscape) (Fig. 3.2) (Andermann et al., 2022). Mathematical functions known as diversity indices are quantitative measurements of microbial diversity, allowing biodiversity to be compared between different samples.

Richness and evenness are the most commonly measured parameters of diversity. Richness refers to the number of species or units of diversity present in a population of interest while evenness reflects the distribution of the abundance of the different species or units of diversity. Richness is commonly estimated using the Chao1 and Abundance Coverage-based Estimator (ACE) indices (Chao, 1984; Chao

Table 3.1 Summary of the advantages and disadvantages of each method at investigating microbial diversity

Method	Advantages	Disadvantages
qPCR and ddPCR	Fast, specific, and sensitive Can be used to study microorganisms without the need for culture Requires low amounts of template	Sample pre-treatment required to differentiate between viable and non-viable microorganisms Requires optimization Requires prior knowledge of target genes Inhibitors in environmental samples can affect PCR ddPCR is expensive
High-throughput amplicon sequencing	Microbial diversity can be studied without requiring microorganisms to be cultured	Cannot be used to directly study the function of the microbial community as only the marker gene is sequenced Sample pre-treatment required to differentiate between viable and non-viable microorganisms
Metagenomics and metatranscriptomics	Microbial diversity can be studied without requiring microorganisms to be cultured Usually not subject to PCR biases (depends on input nucleic acid concentration and library preparation methods) Able to profile microbial community function	Expensive compared to amplicon sequencing Computationally intensive Affected by extracellular nucleic acids in the environment (eDNA)
Single-cell genomics	Able to investigate rare microorganisms in the environment Can identify uncultured microorganisms with useful properties	Challenging to isolate single cells Bias during DNA amplification Expensive compared to bulk sequencing methods
Culture-based	Able to distinguish between viable and non-viable microorganisms Enable the in-depth study of the physiology of isolated microorganisms	Underestimate microbial biodiversity as most microorganisms in the environment cannot be cultured under standard laboratory culture conditions

et al., 2000). Other widely used biodiversity metrics are Faith Phylogenetic Diversity, the Hill diversity index, Shannon index, and Simpson index that quantify both richness and evenness (Faith, 1992; Roswell et al., 2021; Shannon, 1948; Simpson, 1949). As a result of the different mathematical formulation of each index, it is important to understand how each index estimates biodiversity to choose the appropriate index for the objective of the study and avoid bias (Cazzolla Gatti et al., 2020).

When measuring diversity, it is also important to ensure an unbiased estimation of the true diversity of a community from a random sample. Several methods have

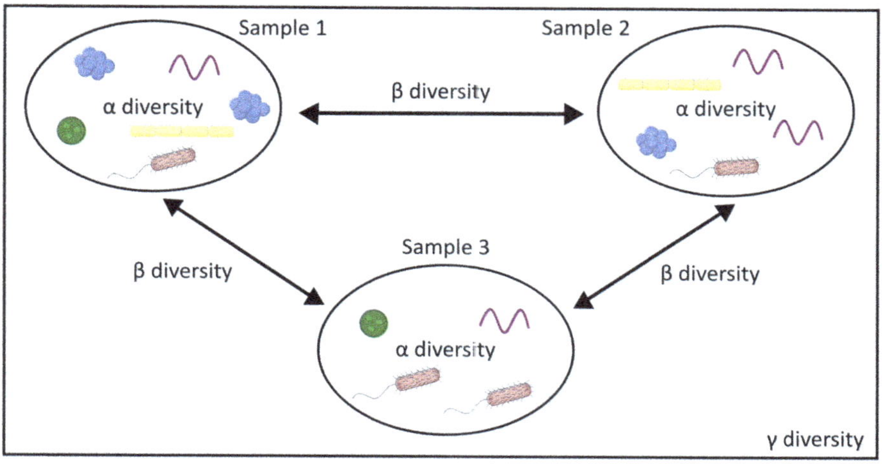

Fig. 3.2 Schematic diagram showing alpha, beta, and gamma diversity. Alpha diversity is the measure of diversity within samples, beta diversity quantifies the difference in diversity between samples, while gamma diversity represents the total diversity of the landscape

been developed to address bias between different samples. For example, rarefaction (random subsampling with or without replacement) is conventionally used to reduce samples of different sizes to a single standard library size so that the diversity of the communities can be comparable. However, rarefaction does not take into account the distribution of relative abundances in the whole community (Cho et al., 2012; Flood et al., 2015). This results in a larger underestimate of diversity for more diverse communities with rarer species. On the other hand, a newer method called coverage measures how completely a community is sampled and takes into account that more diverse communities need to be sampled to a greater extent to be equally well characterized. Coverage describes how accurately a sample captures the true diversity of the community, including species that have not yet been detected (Roswell et al., 2021).

4 Microbial Diversity of Man-Made Water Features

Created to collect, treat, and discharge water or as the result of activities such as quarrying and mining, man-made water features are a common sight in urban and suburban areas (Fig. 3.3). To meet the needs of the community, a network of man-made canals, reservoirs, and water treatment facilities capture, treat, and channel water to where it is needed. On the other hand, quarrying and mining lakes, water fountains, and cooling towers serve recreational and decorative purposes.

As interfaces between human activity and the natural aquatic environment, man-made water features represent new ecological niches for microbial communities.

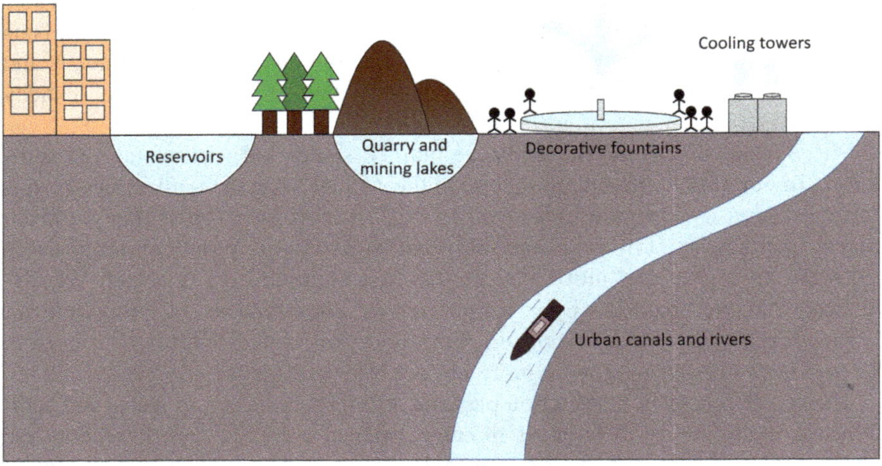

Fig. 3.3 Some examples of man-made water features that are used for recreation, transport, cooling, and water collection. Reservoirs are artificial lakes created primarily to store water for various purposes. Similarly, quarry and mining lakes may be used as a source of agricultural, potable, or industrial water. The benefits of decorative fountains include cooling and beautifying the surrounding environment. Waterways such as urban canals and rivers play important roles in sustaining trade and commerce

Microorganisms from the environment can be transported into man-made water bodies via sewage discharge and stormwater runoff. Consequently, man-made water bodies are home to a diverse array of microorganisms (McLellan et al., 2015).

Out of all the microorganisms present in man-made water bodies, bacteria are among the most studied and well understood. A multitude of studies using culture-independent and culture-dependent methods have investigated the microbial community composition and function of urban water infrastructure, including quarry and mining lakes, reservoirs, fountains, cooling towers, as well as urban canals and rivers (Kumar et al., 2019; Jantharadej et al., 2021; Jiang et al., 2021a, b; Pinel et al., 2020; She et al., 2021; Carpio Vallejo et al., 2024). Such findings shed light on the water quality and the likelihood of disease outbreaks associated with these urban water bodies, with important implications for water and public health policies.

5 Lakes Created by Quarrying and Mining Activities

Quarry lakes are man-made lakes formed after mining activities cease and the disused sites fill with rainwater and/or groundwater. Re-naturalized quarry lakes are important ecosystems that support communities of flora and fauna. They are also popular scenic spots for various water activities. Microorganisms play important roles in the ecological restoration of polluted water bodies such as quarry lakes as

they break down toxic compounds left behind from anthropogenic activities (Guezennec et al., 2015; Aguinaga et al., 2021; Skinder et al., 2020). Microorganisms in aquatic environments such as quarry lakes also interact with solid minerals and contribute to the biogeochemical cycling of metals (Ng et al., 2016; Pandey et al., 2021).

Microorganisms in quarry lakes are mainly associated with water and sediments. In our study where we sequenced the 16S rRNA genes of bacterial communities from two granite quarry lakes in Singapore, we found that different microorganisms exist in water and sediment, and that microbial communities in sediment are richer and more diverse than the communities in water. We discovered that the bacteria phyla *Actinobacteria*, *Bacteroides*, *Cyanobacteria*, *Chloroflexi*, and *Verrucomicrobia* dominate in water while *Nitrospirae*, *Firmicutes* and *Aquificae* were more abundant in sediments.

These differences in distribution patterns of microorganisms in water and sediment are likely due to differences in environmental conditions. Sediments at the bottom of the lake create a low-oxygen environment for anaerobic bacteria to thrive while photosynthetic bacteria like cyanobacteria are present in the water column that is exposed to sunlight.

The key microbial players in water and sediment shed light on the potential functions that are taking place in the habitat. For example, *Archaea* that are predominant in sediments play important roles in the oxidation of ammonia to nitrite and methanogenesis. Other functions driven by bacteria in the sediment include sulfur transformation and degradation of aromatic compounds. On the other hand, microorganisms that perform aerobic chemoheterotrophy and phototrophy are abundant in water. Taken together, our findings provide valuable insight into the putative microbial processes in water and sediment that contribute to the biogeochemical processes of granite quarry lakes (Kumar et al., 2019).

Acidification is a common problem of mining operations due to the oxidation of exposed sulfide minerals, and acidophilic and metal-tolerant microorganisms are commonly found in water bodies contaminated by the outflow of acidic water from mines (Sánchez-España et al., 2020; Tiquia-Arashiro & Rodrigues, 2016a, b; Bernardelli et al., 2021; Bonilla et al., 2021). Through high-throughput 16S rRNA sequencing, a study by She et al. (2021) found that the prokaryotic community compositions varied with the chemical gradients in an acid mine drainage lake, with taxonomic and phylogenetic diversity increasing with depth.

In this study, pH was the key environmental driver for the vertical distribution pattern of microbial communities in the lake, with stronger acid stresses driving more prominent environmental selection in the upper layers of the water column, reducing microbial diversity. The study also revealed that at lower depths, random processes instead of environmental factors were responsible for changes in relative population abundance (She et al., 2021).

Metagenomics and metatranscriptomics approaches have also identified several metal-resistant genes in mine drainage. These enable microorganisms to survive high concentrations of heavy metals by transforming metals into less toxic forms, exporting metals via efflux pumps and bioabsorption (Bowman et al., 2018;

Giddings et al., 2020; Jiang et al., 2021a; Pagnucco et al., 2023). Microorganisms from these extreme environments could have important applications in bioremediation (Tiquia-Arashiro & Grube, 2019; Bonilla et al., 2021; Shylla et al., 2021).

6 Reservoirs

Like quarry and acid mine drainage lakes, reservoirs are artificial lakes that collect and store rainwater. The stored water is treated and supplied to homes and industries for consumption and other purposes. Reservoirs are also commonly used for recreational activities such as boating and fishing. As such, maintaining the quality of the water in reservoirs is of utmost importance. As drivers of biogeochemical processes, the microbial communities present in reservoir water can affect the quality of the water, depending on how environmental conditions influence microbial community function.

A variety of hydrological, physicochemical, and biological processes influence the composition of microbial communities in reservoirs. For example, sunlight promotes the growth of photosynthetic microorganisms such as algae present in surface reservoirs. In tropical regions where temperatures are relatively uniform throughout the year, microbial communities in reservoirs were found to be relatively constant (Nshimyimana et al., 2017). In contrast, the composition of microbial communities of reservoirs in regions that experience distinct seasons is influenced by the different environmental conditions in each season (Patel et al., 2019; Jiang et al., 2021b; Zhong et al., 2022).

The use of underground reservoirs to store raw water is currently being explored as a solution to meet the rising demand for clean water in densely populated cities with limited land to build surface reservoirs. Surface runoff can also be channeled into underground reservoirs to reduce flooding. Unlike surface reservoirs, subsurface reservoirs are not exposed to light. How the absence of sunlight in these reservoirs influences microbial community function and thus water quality is not well known. This knowledge has important implications on the necessary water treatment processes required to ensure that the stored water is safe for use.

In our study, we simulated the dark conditions of a subsurface reservoir and stored untreated surface water obtained from an urban surface reservoir in the absence of light. We then monitored changes in the physicochemical and microbiological quality of the stored water over time. Notably, we observed an increase in nitrate levels as time passed. The nitrifying bacteria *Nitrospira* and *Nitrosomonas* were also more abundant in the dark (Kumar et al., 2022). Based on our observations, we proposed that algae in the water decayed rapidly in the dark due to the inhibition of photosynthesis. This degradation of biomass released free nitrogen into the water, which was converted by nitrifying bacteria to nitrate. When consumed, nitrate in drinking water is converted to nitrite, which binds to hemoglobin in the blood and prevents it from transporting oxygen efficiently. Levels of nitrate higher than the United States Environmental Protection Agency's maximum

contaminant level of 10 mg/L can result in a condition known as "blue baby syndrome" in infants, where poorly oxygenated blood causes a baby's skin to turn blue (United States Environmental Protection Agency, 2023). Consumption of water contaminated with nitrates has also been linked to an increased risk of developing thyroid disease and cancer (García Torres et al., 2022; Temkin et al., 2019). Our findings indicate that it is important to monitor nitrate closely when untreated reservoir water is stored in the dark and the potential need of downstream treatment to remove nitrate (Kumar et al., 2022).

It is also important to understand the dynamics of eukaryotic algae and prokaryotic cyanobacteria communities in reservoirs, as they can produce toxins and cause algal blooms (Rastogi et al., 2015). Microscopic analysis is commonly used to identify species of microalgae, due to their distinctive morphology (Zheng et al., 2021). More recently, 18S rRNA sequencing has been applied to sequence dinoflagellates to detect potentially harmful algae (Hatfield et al., 2020). Chloroplast 23S rRNA gene analysis using domain V in the 23S rRNA gene can also be used to study the diversity of photosynthetic eukaryotic algae and cyanobacteria communities. For cyanobacteria, this marker also provides better coverage of community diversity than 16S rRNA sequencing (Djemiel et al., 2020).

Water temperature, nutrient availability, pH, and sunlight are some important parameters that influence microalgal communities. For optimal light and nutrient assimilation, different microalgae have unique adaptations to live at different depths of the water column (Edwards et al., 2013). For example, some algae such as *Dinophyta* and *Euglenophyta* have flagella that enables them to move to suitable water layers. Others like *Cyanophyta* have special structures that allow them to control their buoyancy to suspend in different water layers (Zhong et al., 2022). Microalgae also respond to the variations in the environment by adjusting their physiology, such as increasing the number of photosynthetic reaction centers or increasing antennae size to capture more light under low light conditions (Michel-Rodriguez et al., 2021; Young & Schmidt, 2020; Zhong et al., 2022). Understanding how microalgal communities in reservoirs vary with environmental conditions has important applications in the management of water quality in reservoirs. For instance, nutrient runoff from phosphate fertilizers should not be discharged into reservoirs, to avoid algal blooms.

Viruses are the most abundant biological entity on Earth and are present in both the sediment and water column of aquatic environments, where they play an important role in regulating bacteria populations (He et al., 2019). There is evidence that they direct carbon cycling in aquatic environments. Viruses infect prokaryotic cells, causing them to lyse and release carbon. By introducing and expressing metabolic genes in host cells, viruses influence metabolic processes such as carbon fixation (Braga et al., 2022; Howard-Varona et al., 2020). Pathogenic viruses in water bodies contaminated by sewage pose a significant health risk as they cause several diseases in humans (Burrell et al., 2017; Hayes et al., 2023). For example, rotavirus is one of the main waterborne causes of gastroenteritis while hepatitis A is transmitted via the fecal-oral route when a person ingests food or water contaminated with the feces of an infected person (Shamsollahi et al., 2019; Takuissu et al., 2023). Hence,

metagenomic analyses of viruses can aid in the development of tools to monitor water quality (Acharya et al., 2020; Bibby et al., 2019; Wang et al., 2020).

7 Decorative Fountains and Cooling Towers

Urban water infrastructure such as ornamental water fountains and cooling towers help cities cope with high temperatures, by generating cooling effects through the evaporation of water mists (Carpio Vallejo et al., 2024; Girolamini et al., 2023). Ornamental water fountains are also popular indoor decorations and outdoor attractions. However, pathogens may lurk in these water features. An example of a waterborne pathogen is the *Legionella* bacterium. The bacterium grows in the warm and stagnant water of decorative water fountains and cooling towers and causes Legionnaires Disease, a lung infection that can be fatal. As outbreaks of the disease have been linked to the inhalation of aerosolized water generated by decorative fountains and cooling towers, the presence of *Legionella* in these water features is a public health concern (Fitzhenry et al., 2017; Girolamini et al., 2023; Haupt et al., 2012; Smith et al., 2015).

qPCR is commonly used to detect and quantify *Legionella* in environmental samples, in addition to culture methods (Donohue, 2021). Unlike culture methods, qPCR can detect *Legionella* when it is in the VBNC state (Whiley & Taylor, 2016). This is important as *Legionella* in the VBNC state is still metabolically active and able to cause disease (Dietersdorfer et al., 2018). Although qPCR is faster and more sensitive than culture methods, one disadvantage of the technique is that it does not directly differentiate between viable and non-viable cells, resulting in a tendency to overestimate cell numbers (Whiley & Taylor, 2016). 16S rRNA gene amplicon and shotgun metagenomic sequencing have also been used to detect the presence of *Legionella* (Borthong et al., 2018; Kanatani et al., 2021). Hence, detecting *Legionella* using qPCR and next-generation sequencing methods could help prevent disease outbreaks.

Next-generation and whole-genome sequencing can also be used to accurately characterize parasites such as *Cryptosporidium* and *Giardia* that contaminate water and cause gastroenteritis (Xiao & Feng, 2017). The presence of pathogens in ornamental water fountains and cooling towers indicates a need for such urban water features to be routinely maintained and disinfected so they can be used safely. Next-generation sequencing can be used to evaluate and monitor if disinfection has been successful. In a study, 16S rRNA gene amplicon sequencing analyses of water samples from an industrial cooling water system revealed that chlorination prevented the growth of *Legionella*. However, some species of bacteria were resistant to the treatment and persisted (Pinel et al., 2020). Likewise, 16S rRNA gene amplicon sequencing can be used to characterize chlorine-resistant bacteria in swimming pools that survive chlorination (Peters et al., 2018), and antibiotic-resistant bacteria (Shuai et al., 2021; Nassereddine et al., 2024). A better understanding of how

different pathogens respond to disinfection could be used to predict and develop strategies for disease outbreak prevention.

8 Urban Canals and Rivers

Urban waterways such as canals and rivers are important transport routes for people and goods and a source of water. Man-made canals are constructed to divert water to where it is needed, such as to supply water to areas experiencing drought (Zhang et al., 2021). Anthropogenic activities can affect the profile of microbial communities in urban canals and rivers, resulting in adverse outcomes such as disease transmission, heavy metal contamination, and foul odors (Feng et al., 2022; Yao et al., 2022). Often located in densely populated human settlements, canals and rivers can be polluted by urban runoff containing domestic sewage and industrial waste. These wastewater discharges introduce pathogens and toxic heavy metals into the waterways. Anaerobic microbial degradation of organic compounds in the wastewater discharge also contributes to unpleasant odors. Exposure to nuisance odors can negatively affect human health and well-being. Therefore, it is important to ensure that urban canals and rivers are safe and pleasant for use.

Analysis of microbial communities in canal sediments can provide a better understanding of the microorganisms that cause malodors, to aid in the restoration of these impaired waterways. 16S rRNA amplicon sequencing studies have revealed that fermentative bacteria, methanogens, and sulfate-reducing bacteria are present in the sediments of polluted canals. Sulfate-reducing bacteria such as *Desulfobulbus* were the main producers of hydrogen sulfide, which causes an odor that is reminiscent of rotten eggs (Jantharadej et al., 2021). Treatments to control the populations of sulfate-reducing bacteria include biocides and chlorination.

Waterborne pathogens from untreated sewage discharged into urban waterways also pose a serious threat to human health. By sequencing the 16S rRNA gene amplicon, several pathogenic bacteria have been detected in urban rivers. The presence of human-associated enteric pathogens such as *Escherichia-Shigella*, *Streptococcus*, and *Enterococcus* suggests that human waste commonly contaminates urban rivers (Gu et al., 2021).

Sewage pollution can also introduce antibiotics from urine and feces into urban waterways. The presence of antibiotics in the environment acts as a selection pressure that promotes the persistence of antibiotic-resistant bacteria. Pathogenic bacteria can become resistant themselves when they acquire antibiotic-resistant genes through horizontal transmission (Pepi & Focardi, 2021).

Through metagenomics, antibiotic-resistant genes have been detected in a variety of man-made water features. As such, the transmission and enrichment of antibiotic-resistant genes in pathogens in these water features may pose a major health risk. Detecting and monitoring the presence of antibiotic-resistant genes in urban waterways using bioinformatics methods can help authorities monitor antibiotic-resistant genes in urban water bodies and mitigate the risk of pathogens

becoming resistant to antimicrobials (Jantharadej et al., 2022; Malayil et al., 2022; Nassereddine et al., 2024).

9 Conclusions

Man-made water features play important roles in sustainable development as sources of water, scenic spots, and transportation routes. Due to their intimate association with human society, they are also associated with infectious disease outbreaks and can harbor antibiotic-resistant microorganisms. A detailed understanding of the microorganisms and their roles in man-made water infrastructure is key to ensuring the functionality and safety of urban water resources. PCR methods, high-throughput sequencing technologies such as 16S rRNA gene amplicon, and shotgun metagenomic sequencing address the shortcomings of conventional culture-based methods used to elucidate the microorganisms in ecosystems. Advances in sequencing and culture techniques will continue to reveal more of these unseen inhabitants and what they do.

10 Future Perspectives

Future challenges in urban water management include the development of new solutions to address water-related issues for a sustainable future. Several new infectious diseases, including Coronavirus Disease-2019 (COVID-19) and Severe Acute Respiratory Syndrome (SARS), have emerged in the twenty-first century (Baker et al., 2022). In the face of emerging infectious threats, there is a need for advanced microbial monitoring and detection technologies that can rapidly identify and quantify pathogenic microorganisms in real time. Artificial intelligence (AI) models that identify pathogens or biosensors that use cutting-edge molecular techniques to continuously monitor microbial indicators in water bodies could enable timely responses to infectious disease outbreaks (Chen et al., 2020; Lin et al., 2020; Yi et al., 2023). Flow cytometry is another example of a rapid method for monitoring aquatic bacteria concentration to ensure water quality (Buysschaert et al., 2018).

AI techniques such as machine learning have also revolutionized the assessment and prediction of microbial water quality, providing valuable insights for proactive water management strategies of man-made water bodies (Chhipi-Shrestha et al., 2023; Pérez-Beltrán et al., 2024). In the future, smart water monitoring infrastructure utilizing AI could be developed to manage urban water resources (Xiang et al., 2021).

At the same time, to prevent the emergence of antibiotic-resistant strains in man-made water bodies, there is an urgent need to devise novel antimicrobial strategies that target specific pathogens while minimizing the impact on non-pathogenic microorganisms. Research into eco-friendly antimicrobial agents or the exploration

of phage therapy as an alternative to traditional antibiotics represents a frontier in mitigating microbial risks in urban water systems (Plotniece et al., 2023; Strathdee et al., 2023).

These innovative approaches collectively contribute to addressing the microbial dimension of urban water challenges, fostering a sustainable and resilient water management framework for the future.

Author Contributions Daphne H.P. Ng had the idea for the article, performed the literature search and data analysis, and drafted the article. Sakcham Bairoliya performed the literature search, analyzed the data, and drafted the article. Cao Bin critically revised the work.

Declaration The authors have no relevant financial or non-financial interests to disclose.

References

Abellan-Schneyder, I., Schusser, A. J., & Neuhaus, K. (2021). ddPCR allows 16S rRNA gene amplicon sequencing of very small DNA amounts from low-biomass samples. *BMC Microbiology, 21*(1), 349.

Acharya, K., Blackburn, A., Mohammed, J., Haile, A. T., Hiruy, A. M., & Werner, D. (2020). Metagenomic water quality monitoring with a portable laboratory. *Water Research, 184*, 116112.

Acosta, C. P., Codony, F., Fittipaldi, M., Sierra-Torres, C. H., & Morató, J. (2018). Monitoring levels of viable *Helicobacter pylori* in surface water by qPCR in Northeast Spain. *Journal of Water and Health, 16*(5), 839–845.

Aguinaga, O. E., White, K. N., Dean, A. P., & Pittman, J. K. (2021). Addition of organic acids to acid mine drainage polluted wetland sediment leads to microbial community structure and functional changes and improved water quality. *Environmental Pollution, 290*, 118064.

Almeida, O. G. G., & De Martinis, E. C. P. (2019). Bioinformatics tools to assess metagenomic data for applied microbiology. *Applied Microbiology and Biotechnology, 103*(1), 69–82.

Alqarni, B., Colley, B., Klebensberger, J., McDougald, D., & Rice, S. A. (2016). Expression stability of 13 housekeeping genes during carbon starvation of *Pseudomonas aeruginosa*. *Journal of Microbiological Methods, 127*, 182–187.

Alvear-Daza, J. J., García-Barco, A., Osorio-Vargas, P., Gutiérrez-Zapata, H. M., Sanabria, J., & Rengifo-Herrera, J. A. (2021). Resistance and induction of viable but non culturable states (VBNC) during inactivation of *E.coli* and *Klebsiella pneumoniae* by addition of H_2O_2 to natural well water under simulated solar irradiation. *Water Research, 188*, 116499.

Andermann, T., Antonelli, A., Barrett, R. L., & Silvestro, D. (2022). Estimating alpha, beta, and gamma diversity through deep learning. *Frontiers in Plant Science, 13*, 839407.

Augsburger, N., Mantilla-Calderon, D., Daffonchio, D., & Hong, P.-Y. (2019). Acquisition of Extracellular DNA by *Acinetobacter baylyi* ADP1 in response to solar and UV-C $_{254nm}$ disinfection. *Environmental Science & Technology, 53*(17), 10312–10319.

Bairoliya, S., Xiang, K. Z., & J., & Cao, B. (2022). Extracellular DNA in environmental samples: Occurrence, extraction, quantification, and impact on microbial biodiversity assessment. *Applied and Environmental Microbiology, 88*(3), e01845–e01821.

Baker, R. E., Mahmud, A. S., Miller, I. F., Rajeev, M., Rasambainarivo, F., Rice, B. L., Takahashi, S., Tatem, A. J., Wagner, C. E., Wang, L. F., Wesolowski, A., & Metcalf, C. J. E. (2022). Infectious disease in an era of global change. *Nature Reviews. Microbiology, 20*(4), 193–205.

Bashiardes, S., Zilberman-Schapira, G., & Elinav, E. (2016). Use of metatranscriptomics in microbiome research. *Bioinformatics and Biology Insights, 10*, 19–25.

Bawn, M., Hernandez, J., Trampari, E., Thilliez, G., Quince, C., Webber, M. A., Kingsley, R. A., Hall, N., & Macaulay, I. C. (2022). Single-cell genomics reveals population structures from in vitro evolutionary studies of *Salmonella*. *Microbial Genomics, 8*(9), mgen000871.

Ben-Dov, E., Brenner, A., & Kushmaro, A. (2007). Quantification of sulfate-reducing bacteria in industrial wastewater, by real-time polymerase chain reaction (PCR) using dsrA and apsA genes. *Microbial Ecology, 54*(3), 439–451.

Berdy, B., Spoering, A. L., Ling, L. L., & Epstein, S. S. (2017). In situ cultivation of previously uncultivable microorganisms using the ichip. *Nature Protocols, 12*(10), 2232–2242.

Bernardelli, C. E., Maza, S. N., Lecomte, K. L., Collo, G., Astini, R. A., & Donati, E. R. (2021). Acidophilic microorganisms enhancing geochemical dynamics in an acidic drainage system, Amarillo river in La Rioja, Argentina. *Chemosphere, 263*, 128098.

Bertolo, A., Valido, E., & Stoyanov, J. (2024). Optimized bacterial community characterization through full-length 16S rRNA gene sequencing utilizing MinION nanopore technology. *BMC Microbiology, 24*(1), 58.

Bibby, K., Crank, K., Greaves, J., Li, X., Wu, Z., Hamza, I. A., & Stachler, E. (2019). Metagenomics and the development of viral water quality tools. *npj Clean Water, 2*, 9.

Boczek, L. A., Tang, M., Formal, C., Lytle, D., & Ryu, H. (2021). Comparison of two culture methods for the enumeration of *Legionella pneumophila* from potable water samples. *Journal of Water and Health, 19*(3), 468–477.

Bolyen, E., Dillon, M., Bokulich, N., Abnet, C., Al-Ghalith, G., Alexander, H., Alm, E., Arumugam, M., Asnicar, F., Bai, Y., Bisanz, J., Bittinger, K., Brejnrod, A., Brislawn, C., Brown, T., Callahan, B., Chase, J., Cope, E., Dorrestein, P., et al. (2019). Reproducible, interactive, scalable and extensible microbiome data science using QIIME 2. *Nature Biotechnology, 37*(8), 852–857.

Bonilla, J. O., Callegari, E. A., Paez, M. D., Gil, R. A., & Villegas, L. B. (2021). Characterization of copper stress response in *Fusarium tricinctum* M6: A metal-resistant microorganism isolated from an acid mine drainage-affected environment. *Journal of Hazardous Materials, 412*, 125216.

Borthong, J., Omori, R., Sugimoto, C., Suthienkul, O., Nakao, R., & Ito, K. (2018). Comparison of database search methods for the detection of *Legionella pneumophila* in water samples using metagenomic analysis. *Frontiers in Microbiology, 9*, 1272.

Botes, M., de Kwaadsteniet, M., & Cloete, T. E. (2013). Application of quantitative PCR for the detection of microorganisms in water. *Analytical and Bioanalytical Chemistry, 405*(1), 91–108.

Bowman, N., Patel, P., Xu, W., Alsaffar, A., & Tiquia-Arashiro, S. M. (2018). Enrichment and isolation of Pb-resistant bacteria from Saint Clair River sediments and their potential for Pb removal in aqueous solutions. *Applied Microbiology and Biotechnology, 102*, 2391–2398. https://doi.org/10.1007/s00253-018-8772-4

Braga, L. P. P., Orland, C., Emilson, E. J. S., Fitch, A. A., Osterholz, H., Dittmar, T., Basiliko, N., Mykytczuk, N. C. S., & Tanentzap, A. J. (2022). Viruses direct carbon cycling in lake sediments under global change. *PNAS, 119*(41), e2202261119.

Browne, P. D., Nielsen, T. K., Kot, W., Aggerholm, A., Gilbert, M. T. P., Puetz, L., Rasmussen, M., Zervas, A., & Hansen, L. H. (2020). GC bias affects genomic and metagenomic reconstructions, underrepresenting GC-poor organisms. *GigaScience, 9*(2), giaa008.

Bukin, Y. S., Galachyants, Y. P., Morozov, I. V., Bukin, S. V., Zakharenko, A. S., & Zemskaya, T. I. (2019). The effect of 16s rRNA region choice on bacterial community metabarcoding results. *Scientific Data, 6*, 190007.

Burrell, C. J., Howard, C. R., & Murphy, F. A. (2017). Emerging virus diseases. In *Fenner and White's medical virology* (pp. 217–225). Academic Press.

Buysschaert, B., Vermijs, L., Naka, A., Boon, N., & De Gusseme, B. (2018). Online flow cytometric monitoring of microbial water quality in a full-scale water treatment plant. *npj Clean Water, 1*(1), 1–7.

Callahan, B. J., McMurdie, P. J., & Holmes, S. P. (2017). Exact sequence variants should replace operational taxonomic units in marker-gene data analysis. *The ISME Journal, 11*(12), 2639–2643.

Carpio Vallejo, E., Düker, U., Meyer, F., Berding, U., & Nogueira, R. (2024). Microbiological water quality and derived health risks from exposure to ornamental water fountains in the city of Hannover. *Risk Analysis, 44*(1), 24–39.

Cazzolla Gatti, R., Amoroso, N., & Monaco, A. (2020). Estimating and comparing biodiversity with a single universal metric. *Ecological Modelling, 424*, 109020.

Chao, A. (1984). Nonparametric estimation of the number of classes in a population. *Scandinavian Journal of Statistics, 11*(4), 265–270.

Chao, A., Hwang, W. H., Chen, Y. C., & Kuo, C. Y. (2000). Estimating the number of shared species in two communities. *Statistica Sinica, 10*(1), 227–246.

Chen, H., Jing, L., Yao, Z., Meng, F., & Teng, Y. (2019). Prevalence, source and risk of antibiotic resistance genes in the sediments of Lake Tai (China) deciphered by metagenomic assembly: A comparison with other global lakes. *Environment International, 127*, 267–275.

Chen, Y., Qian, C., Liu, C., Shen, H., Wang, Z., Ping, J., Wu, J., & Chen, H. (2020). Nucleic acid amplification free biosensors for pathogen detection. *Biosensors & Bioelectronics, 153*, 112049.

Chhipi-Shrestha, G., Mian, H. R., Mohammadiun, S., Rodriguez, M., Hewage, K., & Sadiq, R. (2023). Digital water: Artificial intelligence and soft computing applications for drinking water quality assessment. *Clean Technologies and Environmental Policy, 25*(5), 1409–1438.

Cho, K., Zholi, A., Frabutt, D., Flood, M., Floyd, D., & Tiquia, S. M. (2012). Linking bacterial diversity and geochemistry of uranium-contaminated groundwater. *Environmental Technology, 33*, 1629–1640. https://doi.org/10.1080/09593330.2011.641036

Desneux, J., Chemaly, M., & Pourcher, A. M. (2015). Experimental design for the optimization of propidium monoazide treatment to quantify viable and non-viable bacteria in piggery effluents. *BMC Microbiology, 15*(1), 1–9.

Dietersdorfer, E., Kirschner, A., Schrammel, B., Ohradanova-Repic, A., Stockinger, H., Sommer, R., Walochnik, J., & Cervero-Aragó, S. (2018) Starved viable but non-culturable (VBNC) *Legionella* strains can infect and replicate in amoebae and human macrophages. *Water Research, 141*, 428–438.

Djemiel, C., Plassard, D., Terrat, S., Crouzet, O., Sauze, J., Mondy, S., Nowak, V., Wingate, L., Ogée, J., & Maron, P. A. (2020). µgreen-db: A reference database for the 23S rRNA gene of eukaryotic plastids and cyanobacteria. *Scientific Reports, 10*(1), 5915.

Donohue, M. J. (2021). Quantification of *Legionella pneumophila* by qPCR and culture in tap water with different concentrations of residual disinfectants and heterotrophic bacteria. *Science of the Total Environment, 774*, 145142.

Douglas, G. M., Maffei, V. J., Zaneveld, J. R., Yurgel, S. N., Brown, J. R., Taylor, C. M., Huttenhower, C., & Langille, M. G. I. (2020). PICRUSt2 for prediction of metagenome functions. *Nature Biotechnology, 38*(6), 685–688.

Edwards, K. F., Litchman, E., & Klausmeier, C. A. (2013). Functional traits explain phytoplankton responses to environmental gradients across lakes of the United States. *Ecology, 94*(7), 1626–1635.

Faith, D. P. (1992). Conservation evaluation and phylogenetic diversity. *Biological Conservation, 61*(1), 1–10.

Feng, W., Gao, J., Wei, Y., Liu, D., Yang, F., Zhang, Q., & Bai, Y. (2022). Pattern changes of microbial communities in urban rivers affected by anthropogenic activities and their environmental driving mechanisms. *Environmental Sciences Europe, 34*, 93.

Fitzhenry, R., Weiss, D., Cimini, D., Balter, S., Boyd, C., Alleyne, L., Stewart, R., McIntosh, N., Econome, A., Lin, Y., Rubinstein, I., Passaretti, T., Kidney, A., Lapierre, P., Kass, D., & Varma, J. K. (2017). Legionnaires' disease outbreaks and cooling towers, New York City, New York, USA. *Emerging Infectious Diseases, 23*(11), 1769–1776.

Flood, M., Frabutt, D., Floyd, D., Powers, E. U., Devol, A., & Tiquia-Arashiro, S. M. (2015). Ammonia-oxidizing bacteria and archaea in sediments of the Gulf of Mexico. *Environmental Technology, 36*, 124–135. https://doi.org/10.1080/09593330.2014.942385

Fu, Y., Peng, H., Liu, J., Nguyen, T. H., Hashmi, M. Z., & Shen, C. (2021). Occurrence and quantification of culturable and viable but non-culturable (VBNC) pathogens in biofilm on different pipes from a metropolitan drinking water distribution system. *Science of the Total Environment, 764*, 142851.

García Torres, E., Pérez Morales, R., González Zamora, A., Ríos Sánchez, E., Olivas Calderón, E. H., Alba Romero, J. D. J., & Calleros Rincon, E. Y. (2022). Consumption of water contaminated by nitrate and its deleterious effects on the human thyroid gland: A review and update. *International Journal of Environmental Health Research, 32*(5), 984–1001.

Garner, E., Chen, C., Xia, K., Bowers, J., Engelthaler, D. M., McLain, J., Edwards, M. A., & Pruden, A. (2018). Metagenomic characterization of antibiotic resistance genes in full-scale reclaimed water distribution systems and corresponding potable systems. *Environmental Science & Technology, 52*(11), 6113–6125.

Garza, D. R., & Dutilh, B. E. (2015). From cultured to uncultured genome sequences: Metagenomics and modeling microbial ecosystems. *Cellular and Molecular Life Sciences, 72*(22), 4287–4308.

Geets, J., de Cooman, M., Wittebolle, L., Heylen, K., Vanparys, B., De Vos, P., Verstraete, W., & Boon, N. (2007). Real-time PCR assay for the simultaneous quantification of nitrifying and denitrifying bacteria in activated sludge. *Applied Microbiology and Biotechnology, 75*(1), 211–221.

Gentile, M., Yan, T., Tiquia, S. M., et al. (2006). Stability in a denitrifying fluidized bed reactor. *Microbial Ecology, 52*, 311–321. https://doi.org/10.1007/s00248-006-9024-1

Giddings, L. A., Chlipala, G., Kunstman, K., Green, S., Morillo, K., Bhave, K., Peterson, H., Driscoll, H., & Maienschein-Cline, M. (2020). Characterization of an acid rock drainage microbiome and transcriptome at the Ely Copper Mine Superfund site. *PLoS One, 15*(8), e0237599.

Gilbert, J. A., & Stephens, B. (2018). Microbiology of the built environment. *Nature Reviews Microbiology, 16*(11), 661–670.

Girolamini, L., Brattich, E., Marino, F., Pascale, M. R., Mazzotta, M., Spiteri, S., Derelitto, C., Tositti, L., & Cristino, S. (2023). Cooling towers influence in an urban environment: A predictive model to control and prevent *Legionella* risk and Legionellosis events. *Building and Environment, 228*, 109891.

Gómez-Silván, C., Vílchez-Vargas, R., Arévalo, J., Gómez, M. A., González-López, J., Pieper, D. H., & Rodelas, B. (2014). Quantitative response of nitrifying and denitrifying communities to environmental variables in a full-scale membrane bioreactor. *Bioresource Technology, 169*, 126–133.

Goswami, M., & Narayana Rao, A. V. S. S. (2018). Transcriptome profiling reveals interplay of multifaceted stress response in *Escherichia coli* on exposure to glutathione and ciprofloxacin. *MSystems, 3*(1), 10–1128. https://doi.org/10.1128/msystems.00001-18

Grossart, H. P., Van den Wyngaert, S., Kagami, M., et al. (2019). Fungi in aquatic ecosystems. *Nature Reviews. Microbiology, 17*, 339–354. https://doi.org/10.1038/s41579-019-0175-8

Gu, Q., Wang, W., Li, G., Liu, Y., Wong, P. K., & An, T. (2021). Pollution profile of waterborne bacterial and fungal community in urban rivers of Pearl River estuary: Microbial safety assessment. *Journal of Freshwater Ecology, 36*(1), 305–322.

Guezennec, A. G., Michel, C., Ozturk, S., Togola, A., Guzzo, J., & Desroche, N. (2015). Microbial aerobic and anaerobic degradation of acrylamide in sludge and water under environmental conditions—Case study in a sand and gravel quarry. *Environmental Science and Pollution Research, 22*(9), 6440–6451.

Guo, L., Wan, K., Zhu, J., Ye, C., Chabi, K., & Yu, X. (2021). Detection and distribution of vbnc/viable pathogenic bacteria in full-scale drinking water treatment plants. *Journal of Hazardous Materials, 406*, 124335.

Gupta, A., Gupta, R., & Singh, R. L. (2017). Microbes and environment. In *Principles and applications of environmental biotechnology for a sustainable future* (pp. 43–84). Springer.

Hamamoto, H., Panthee, S., Hashimoto, K., Tsuchida, T., & Sekimizu, K. (2021). An efficient method to screen for the soil bacteria producing therapeutically effective antibiotics. *The Journal of Antibiotics, 74*(12), 850–855.

Hamza, I. A., & Leifels, M. (2023). Assessment of PCR inhibitor removal methods to monitor viruses in environmental water samples: DAX-8 outperforms competitors. *Water, Air, & Soil Pollution, 235*(1), 20.

Hassanein, F., Masoud, I. M., Fekry, M. M., Abdel-Latif, M. S., Abdel-Salam, H., Salem, M., & Shehata, A. I. (2023). Environmental health aspects and microbial infections of the recreational water: Microbial infections and swimming pools. *BMC Public Health, 23*(1), 302.

Hatfield, R. G., Batista, F. M., Bean, T. P., Fonseca, V. G., Santos, A., Turner, A. D., Lewis, A., Dean, K. J., & Martinez-Urtaza, J. (2020). The application of Nanopore sequencing technology to the study of dinoflagellates: A proof of concept study for rapid sequence-based discrimination of potentially harmful algae. *Frontiers in Microbiology, 11*, 844.

Haupt, T. E., Heffernan, R. T., Kazmierczak, J. J., Nehls-Lowe, H., Rheineck, B., Powell, C., Leonhardt, K. K., Chitnis, A. S., & Davis, J. P. (2012). An outbreak of legionnaires disease associated with a decorative water wall fountain in a hospital. *Infection Control and Hospital Epidemiology, 33*(2), 185–191.

Hayes, E. K., Gouthro, M. T., Fuller, M., Redden, D. J., & Gagnon, G. A. (2023). Enhanced detection of viruses for improved water safety. *Scientific Reports, 13*(1), 17336.

He, L., Yin, K., & Yuan, X. (2019). Double maximum ratios of viruses to bacteria in the water column: Implications for different regulating mechanisms. *Frontiers in Microbiology, 10*, 455705.

Hedrich, S., & Schippers, A. (2021). Distribution of acidophilic microorganisms in natural and man-made acidic environments. *Current Issues in Molecular Biology, 40*(1), 25–48. https://doi.org/10.21775/cimb.040.025

Hempel, C. A., Wright, N., Harvie, J., Hleap, J. S., Adamowicz, S. J., & Steinke, D. (2022). Metagenomics versus total RNA sequencing: Most accurate data-processing tools, microbial identification accuracy and perspectives for ecological assessments. *Nucleic Acids Research, 50*(16), 9279–9293.

Hoorzook, K. B., & Barnard, T. G. (2022). Culture independent DNA extraction method for bacterial cells concentrated from water. *MethodsX, 9*, 101653.

Hosokawa, M., & Nishikawa, Y. (2024). Tools for microbial single-cell genomics for obtaining uncultured microbial genomes. *Biophysical Reviews, 16*, 69–77.

Howard-Varona, C., Lindback, M. M., Bastien, G. E., Solonenko, N., Zayed, A. A., Jang, H. B., Andreopoulos, B., Brewer, H. M., Glavina del Rio, T., Adkins, J. N., Paul, S., Sullivan, M. B., & Duhaime, M. B. (2020). Phage-specific metabolic reprogramming of virocells. *The ISME Journal, 14*(4), 881–895.

Hugerth, L. W., Muller, E. E., Hu, Y. O., Lebrun, L. A., Roume, H., Lundin, D., Wilmes, P., & Andersson, A. F. (2014). Systematic design of 18S rRNA gene primers for determining eukaryotic diversity in microbial consortia. *PLoS One, 9*(4), e95567. https://doi.org/10.1371/journal.pone.0095567

Jantharadej, K., Limpiyakorn, T., Kongprajug, A., Mongkolsuk, S., Sirikanchana, K., & Suwannasilp, B. B. (2021). Microbial community compositions and sulfate-reducing bacterial profiles in malodorous urban canal sediments. *Archives of Microbiology, 203*(5), 1981–1993.

Jantharadej, K., Kongprajug, A., Mhuantong, W., Limpiyakorn, T., Suwannasilp, B. B., Mongkolsuk, S., & Sirikanchana, K. (2022). Comparative genomic analyses of pathogenic bacteria and viruses and antimicrobial resistance genes in an urban transportation canal. *Science of the Total Environment, 848*, 157652.

Jiang, X., Liu, W., Xu, H., Cui, X., Li, J., Chen, J., & Zheng, B. (2021a). Characterizations of heavy metal contamination, microbial community, and resistance genes in a tailing of the largest copper mine in China. *Environmental Pollution, 280*, 116947.

Jiang, T., Sun, S., Chen, Y., Qian, Y., Guo, J., Dai, R., & An, D. (2021b). Microbial diversity characteristics and the influence of environmental factors in a large drinking-water source. *Science of the Total Environment, 769*, 144698.

Jones, L. R., Uyua, N. M., & Manrique, J. M. (2015). The peril of PCR inhibitors in environmental samples: The case of *Didymosphenia geminata*. *Biodiversity and Conservation, 24*(6), 1541–1548.

Joos, L., Beirinckx, S., Haegeman, A., Debode, J., Vandecasteele, B., Baeyen, S., Goormachtig, S., Clement, L., & De Tender, C. (2020). Daring to be differential: Metabarcoding analysis of soil and plant-related microbial communities using amplicon sequence variants and operational taxonomical units. *BMC Genomics, 21*(1), 1–17.

Kanatani, J. I., Watahiki, M., Kimata, K., Kato, T., Uchida, K., Kura, F., Amemura-Maekawa, J., & Isobe, J. (2021). Detection of Legionella species, the influence of precipitation on the amount of *Legionella* DNA, and bacterial microbiome in aerosols from outdoor sites near asphalt roads in Toyama Prefecture, Japan. *BMC Microbiology, 21*, 215.

Kassem, A., Abbas, L., Coutinho, O., Opara, S., Najaf, H., Kasperek, D., Li, X., & Tiquia-Arashiro, S. M. (2023). Applications of Fourier Transform-Infrared spectroscopy in microbial cell biology and environmental microbiology: Advances, challenges, and future perspectives. *Frontiers in Microbiology, 14*, 1304081. https://doi.org/10.3389/fmicb.2023.1304081

Keenum, I., Liguori, K., Calarco, J., Davis, B. C., Milligan, E., Harwood, V. J., & Pruden, A. (2022). A framework for standardized qPCR-targets and protocols for quantifying antibiotic resistance in surface water, recycled water and wastewater. *Critical Reviews in Environmental Science and Technology, 52*(24), 4395–4419.

Kim, H.-J., Ryu, J.-O., Lee, S.-Y., Kim, E.-S., & Kim, H.-Y. (2015). Multiplex PCR for detection of the *Vibrio genus* and five pathogenic *Vibrio* species with primer sets designed using comparative genomics. *BMC Microbiology, 15*(1), 239.

Krsek, M., & Wellington, E. M. H. (1999). Comparison of different methods for the isolation and purification of total community DNA from soil. *Journal of Microbiological Methods, 39*(1), 1–16.

Kumar, A., Ng, D. H. P., Wu, Y., & Cao, B. (2019). Microbial community composition and putative biogeochemical functions in the sediment and water of tropical granite quarry lakes. *Microbial Ecology, 77*(1), 1–11.

Kumar, V., Al Momin, S., Kumar, V. V., Ahmed, J., Al-Musallam, L., Shajan, A. B., Al-Aqeel, H., Al-Mansour, H., & Al-Zakri, W. M. (2021). Distribution and diversity of eukaryotic microalgae in Kuwait waters assessed using 18S rRNA gene sequencing. *PLoS One, 16*(4), e0250645.

Kumar, A., Ng, D. H. P., Bairoliya, S., & Cao, B. (2022). The dark side of microbial processes: Accumulation of nitrate during storage of surface water in the dark and the underlying mechanism. *Microbiology Spectrum, 10*(1), e0223221.

Lawley, B., Ripley, S., Bridge, P., & Convey, P. (2004). Molecular analysis of geographic patterns of eukaryotic diversity in Antarctic soils. *Applied and Environmental Microbiology, 70*(10), 5963–5972. https://doi.org/10.1128/AEM.70.10.5963-5972.2004

Lee, D. I., Park, S. H., Baek, J. H., Yoon, J. W., Jin, S. I., Han, K. E., & Yu, H. S. (2020). Identification of free-living amoebas in tap water of buildings with storage tanks in Korea. *The Korean Journal of Parasitology, 58*(2), 191–194.

Lee, Y., Lam, S. Q. Y., Tay, T. S., et al. (2021). Composition and structure of tropical intertidal hard coral communities on natural and man-made habitats. *Coral Reefs, 40*, 685–700. https://doi.org/10.1007/s00338-021-02059-0

Lennon, J. T., Muscarella, M. E., Placella, S. A., & Lehmkuhl, B. K. (2018). How, when, and where relic DNA affects microbial diversity. *MBio, 9*(3), 10–1128. https://doi.org/10.1128/mbio.00637-18

Li, X., Shen, X., Jiang, W., Xi, Y., & Li, S. (2024). Comprehensive review of emerging contaminants: Detection technologies, environmental impact, and management strategies. *Ecotoxicology and Environmental Safety*, 278, 116420.

Lin, L., Zheng, Q., Lin, J., Yuk, H. G., & Guo, L. (2020). Immuno- and nucleic acid-based current technique for *Salmonella* detection in food. *European Food Research and Technology, 246*(3), 373–395.

Liu, X., Wang, M., Nie, Y., & Wu, X. L. (2021). Isolation chip increases culturable bacterial diversity and reduces cultivation bias. *Current Microbiology, 78*(5), 2025–2032.

Liu, Y. N., Deng, S., Bairoliya, S., & Cao, B. (2023). Mining microbial resources from water. *Resources, Conservation and Recycling, 191*, 106883.

Louca, S., Parfrey, L. W., & Doebeli, M. (2016). Decoupling function and taxonomy in the global ocean microbiome. *Science, 353*(6305), 1272–1277.

Malayil, L., Ramachandran, P., Chattopadhyay, S., Allard, S. M., Bui, A., Butron, J., Callahan, M. T., Craddock, H. A., Murray, R., East, C., Sharma, M., Kniel, K., Micallef, S., Hashem, F., Gerba, C. P., Ravishankar, S., Parveen, S., May, E., Handy, E., et al. (2022). Variations in bacterial communities and antibiotic resistance genes across diverse recycled and surface water irrigation sources in the mid-Atlantic and Southwest United States: A CONSERVE two-year field study. *Environmental Science & Technology, 56*(21), 15019–15033.

Mantilla-Calderon, D., Plewa, M. J., Michoud, G., Fodelianakis, S., Daffonchio, D., & Hong, P.-Y. (2019). Water disinfection byproducts increase natural transformation rates of environmental DNA in *Acinetobacter baylyi* ADP1. *Environmental Science & Technology, 53*(11), 6520–6528.

Matsuo, Y., Komiya, S., Yasumizu, Y., Yasuoka, Y., Mizushima, K., Takagi, T., Kryukov, K., Fukuda, A., Morimoto, Y., Naito, Y., Okada, H., Bono, H., Nakagawa, S., & Hirota, K. (2021). Full-length 16S rRNA gene amplicon analysis of human gut microbiota using MinION™ nanopore sequencing confers species-level resolution. *BMC Microbiology, 21*(1), 35.

McKindles, K. M., & Tiquia-Arashiro, S. M. (2012). Functional gene arrays for analysis of microbial communities on ocean platform. In S. Tiquia-Arashiro (Ed.), *Molecular biological technologies for ocean sensing* (Springer protocols handbooks). Humana Press. https://doi.org/10.1007/978-1-61779-915-0_9

McLellan, S. L., Fisher, J. C., & Newton, R. J. (2015). The microbiome of urban waters. *International Microbiology, 18*(3), 141–149.

Michel-Rodriguez, M., Lefebvre, S., Crouvoisier, M., Mériaux, X., & Lizon, F. (2021). Underwater light climate and wavelength dependence of microalgae photosynthetic parameters in a temperate sea. *PeerJ, 9*, e12101.

Mishra, M., Arukha, A. P., Patel, A. K., Behera, N., Mohanta, T. K., & Yadav, D. (2018). Multi-drug resistant coliform: Water sanitary standards and health hazards. *Frontiers in Pharmacology, 9*, 311.

Mohseni, M., Norouzi, H., Hamedi, J., & Roohi, A. (2013). Screening of antibacterial producing actinomycetes from sediments of the Caspian Sea. *International Journal of Molecular and Cellular Medicine, 2*(2), 64–71.

Nassereddine, Z., Opara, S., Coutinho, O., Qyteti, F., Book, R., Heinicke, M. P., Napieralski, J., & Tiquia-Arashiro, S. M. (2024). Critical perspectives on advancing antibiotic resistant gene (ARG) detection technologies in aquatic ecosystems. *Science of the Total Environment, 957*(20), 177775. https://doi.org/10.1016/j.scitotenv.2024.177775

Ng, D. H. P., Kumar, A., & Cao, B. (2016). Microorganisms meet solid minerals: Interactions and biotechnological applications. *Applied Microbiology and Biotechnology, 100*(16), 6935–6946.

Nguyen, S., Ala, F., Cardwell, C., Cai, D., McKindles, K. M., Lotvola, A., Hodges, S., Deng, Y., & Tiquia-Arashiro, S. M. (2013). Isolation and screening of carboxydotrophs isolated from composts and their potential for butanol synthesis. *Environmental Technology, 34*, 1995–2007. https://doi.org/10.1080/09593330.2013.795987

Nshimyimana, J. P., Freedman, A. J. E., Shanahan, P., Chua, L. C. H., & Thompson, J. R. (2017). Variation of bacterial communities with water quality in an urban tropical catchment. *Environmental Science & Technology, 51*(10), 5591–5601.

Nshimyimana, J. P., Cruz, M. C., Wuertz, S., & Thompson, J. R. (2019). Variably improved microbial source tracking with digital droplet PCR. *Water Research, 159*, 192–202.

Oest, A., Alsaffar, A., Fenner, M., Azzopardi, D., & Tiquia-Arashiro, S. M. (2018). Patterns of change in metabolic capabilities of sediment microbial communities in river and

lake ecosystems. *International Journal of Microbiology, 2018*(1), 6234931. https://doi.org/10.1155/2018/6234931

Oyola, S. O., Otto, T. D., Gu, Y., Maslen, G., Manske, M., Campino, S., Turner, D. J., MacInnis, B., Kwiatkowski, D. P., Swerdlow, H. P., & Quail, M. A. (2012). Optimizing illumina next-generation sequencing library preparation for extremely at-biased genomes. *BMC Genomics, 13*(1), 1–12.

Pagnucco, G., Overfield, D., Chamlee, Y., Fortuna, A., Sulaiman, F., Farinas, J., Schittenhelm, R., Shuler, C., Kassem, A., Opara, S., Najaf, H., Abbas, L., Coutinho, O., Catalfano, B., Li, X., & Tiquia-Arashiro, S. M. (2023). Metal tolerance and biosorption capacities of bacterial strains isolated from an urban watershed. *Frontiers in Microbiology, 14*, 1278886. https://doi.org/10.3389/fmicb.2023.1278886

Pandey, S., Fosso-Kankeu, E., Redelinghuys, J., Kim, J., & Kang, M. (2021). Implication of biofilms in the sustainability of acid mine drainage and metal dispersion near coal tailings. *Science of the Total Environment, 788*, 147851.

Patel, D., Gismondi, R., Alsaffar, A., & Tiquia-Arashiro, S. M. (2019). Applicability of API ZYM to capture seasonal and spatial variabilities in lake and river sediments. *Environmental Technology, 40*(24), 3227–3239. https://doi.org/10.1080/09593330.2018.1468492

Pepi, M., & Focardi, S. (2021). Antibiotic-resistant bacteria in aquaculture and climate change: A challenge for health in the mediterranean area. *International Journal of Environmental Research and Public Health, 18*(11), 5723.

Pérez-Beltrán, C. H., Robles, A. D., Rodriguez, N. A., Ortega-Gavilán, F., & Jiménez-Carvelo, A. M. (2024). Artificial intelligence and water quality: From drinking water to wastewater. *TrAC Trends in Analytical Chemistry, 172*, 117597.

Pérez-Cobas, A. E., Gomez-Valero, L., & Buchrieser, C. (2020). Metagenomic approaches in microbial ecology: An update on whole-genome and marker gene sequencing analyses. *Microbial Genomics, 6*(8), e000409.

Peters, M. C. F. M., Keuten, M. G. A., Knezev, A., Van Loosdrecht, M. C. M., Vrouwenvelder, J. S., Rietveld, L. C., & de Kreuk, M. K. (2018). Characterization of the bacterial community in shower water before and after chlorination. *Journal of Water and Health, 16*(2), 233–243.

Pinel, I. S. M., Moed, D. H., Vrouwenvelder, J. S., & van Loosdrecht, M. C. M. (2020). Bacterial community dynamics and disinfection impact in cooling water systems. *Water Research, 172*, 115505.

Pinto, F., Pacheco, C. C., Ferreira, D., Moradas-Ferreira, P., & Tamagnini, P. (2012). Selection of suitable reference genes for RT-qPCR analyses in cyanobacteria. *PLoS One, 7*(4), e34983. https://doi.org/10.1371/journal.pone.0034983

Plecha, S., Hall, D., & Tiquia-Arashiro, S. M. (2013). Screening for novel bacteria from the bioenergy feedstock switchgrass (Panicum virgatum L.). *Environmental Technology, 34*, 1895–1904. https://doi.org/10.1080/09593330.2013.818701

Plotniece, A., Sobolev, A., Supuran, C. T., Carta, F., Björkling, F., Franzyk, H., Yli-Kauhaluoma, J., Augustyns, K., Cos, P., De Vooght, L., Govaerts, M., Aizawa, J., Tammela, P., & Žalubovskis, R. (2023). Selected strategies to fight pathogenic bacteria. *Journal of Enzyme Inhibition and Medicinal Chemistry, 38*(1), 2155816.

Polrot, A., Kirby, J. R., Olorunniji, F. J., Birkett, J. W., & Sharples, G. P. (2022). iChip increases the success of cultivation of TBT-resistant and TBT-degrading bacteria from estuarine sediment. *World Journal of Microbiology and Biotechnology, 38*(10), 180.

Qin, W., Amin, S. A., Lundeen, R. A., Heal, K. R., Martens-Habbena, W., Turkarslan, S., Urakawa, H., Costa, K. C., Hendrickson, E. L., Wang, T., Beck, D. A., Tiquia-Arashiro, S. M., Taub, F., Holmes, A. D., Vajrala, N., Berube, P. M., Lowe, T. M., Moffett, J. W., Devol, A. H., Baliga, N. S., Arp, D. J., Sayavedra-Soto, L. A., Hackett, M., Armbrust, E. V., Ingalls, A. E., & Stahl, D. A. (2018). Stress response of a marine ammonia-oxidizing archaeon informs physiological status of environmental populations. *The ISME Journal, 12*(2), 508–519. https://doi.org/10.1038/ismej.2017.186

Radomski, N., Lucas, F. S., Moilleron, R., Cambau, E., Haenn, S., & Moulin, L. (2010). Development of a real-time qPCR method for detection and enumeration of *mycobacterium* spp. in surface water. *Applied and Environmental Microbiology, 76*(21), 7348–7351.

Rastogi, R. P., Madamwar, D., & Incharoensakdi, A. (2015). Bloom dynamics of cyanobacteria and their toxins: Environmental health impacts and mitigation strategies. *Frontiers in Microbiology, 6*, 1254.

Reineke, W., & Schlömann, M. (2023). Microorganisms at different sites: Living conditions and adaptation strategies. In *Environmental microbiology*. Springer Spektrum. https://doi.org/10.1007/978-3-662-66547-3_10

Robert-Pillot, A., Baron, S., Lesne, J., Fournier, J. M., & Quilici, M. L. (2002). Improved specific detection of Vibrio cholerae in environmental water samples by culture on selective medium and colony hybridization assay with an oligonucleotide probe. *FEMS Microbiology Ecology, 40*(1), 39–46.

Roose-Amsaleg, C. L., Garnier-Sillam, E., & Harry, M. (2001). Extraction and purification of microbial DNA from soil and sediment samples. *Applied Soil Ecology, 18*(1), 47–60.

Roswell, M., Dushoff, J., & Winfree, R. (2021). A conceptual guide to measuring species diversity. *Oikos, 130*(3), 321–338.

Sakcham, B., Kumar, A., & Cao, B. (2019). Extracellular DNA in monochloraminated drinking water and its influence on DNA-based profiling of a microbial community. *Environmental Science & Technology Letters, 6*(5), 306–312.

Sánchez-España, J., Yusta, I., Ilin, A., van der Graaf, C., & Sánchez-Andrea, I. (2020). Microbial geochemistry of the acidic saline pit Lake of Brunita mine (La Unión, SE Spain). *Mine Water and the Environment, 39*(3), 535–555.

Schloss, P. D., Jenior, M. L., Koumpouras, C. C., Westcott, S. L., & Highlander, S. K. (2016). Sequencing 16S rRNA gene fragments using the PacBio SMRT DNA sequencing system. *PeerJ, 4*, e1869.

Schoch, C. L., Seifert, K. A., Huhndorf, S., Robert, V., Spouge, J. L., Levesque, C. A., Chen, W., Bolchacova, E., Voigt, K., Crous, P. W., Miller, A. N., Wingfield, M. J., Aime, M. C., An, K. D., Bai, F. Y., Barreto, R. W., Begerow, D., Bergeron, M. J., Blackwell, M., et al. (2012). Nuclear ribosomal internal transcribed spacer (ITS) region as a universal DNA barcode marker for Fungi. *PNAS, 109*(16), 6241–6246.

Shakya, M., Lo, C. C., & Chain, P. S. G. (2019). Advances and challenges in metatranscriptomic analysis. *Frontiers in Genetics, 10*, 904.

Shamsollahi, H. R., Ghoochani, M., Sadeghi, K., Jaafari, J., Masinaei, M., Sillanpää, M., Yousefi, M., Mirtalb, S. T., & Alimohammadi, M. (2019). Evaluation of the physical and chemical characteristics of water on the removal efficiency of rotavirus in drinking water treatment plants and change in induced health risk. *Process Safety and Environment Protection, 130*, 6–13.

Shannon, C. E. (1948). A mathematical theory of communication. *The Bell System Technical Journal, 27*(3), 3–55.

Sharma, H., Devi, Q., Borah, D., et al. (2023). Algal communities in man-made marshy wetlands near a coal-fired brick kiln industry in India. *Vegetos, 36*, 1037–1048. https://doi.org/10.1007/s42535-022-00462-8

She, Z., Pan, X., Wang, J., Shao, R., Wang, G., Wang, S., & Yue, Z. (2021). Vertical environmental gradient drives prokaryotic microbial community assembly and species coexistence in a stratified acid mine drainage lake. *Water Research, 206*, 117739.

Shuai, X., Sun, Y., Meng, L., Zhou, Z., Zhu, L., Lin, Z., & Chen, H. (2021). Dissemination of antibiotic resistance genes in swimming pools and implication for human skin. *Science of the Total Environment, 794*, 148693.

Shylla, L., Barik, S. K., & Joshi, S. R. (2021). Characterization and bioremediation potential of native heavy-metal tolerant bacteria isolated from rat-hole coal mine environment. *Archives of Microbiology, 203*(5), 2379–2392.

Simpson, E. H. (1949). Measurement of diversity. *Nature, 163*(4148), 688.

Skinder, B. M., Uqab, B., & Ganai, B. A. (2020). Bioremediation: A sustainable and emerging tool for restoration of polluted aquatic ecosystem. In H. Qadri, R. Bhat, & M. D. G. Mehmood (Eds.), *Fresh water pollution dynamics and remediation*. Springer Singapore.

Smith, S. S., Ritger, K., Samala, U., Black, S. R., Okodua, M., Miller, L., Kozak-Muiznieks, N. A., Hicks, L. A., Steinheimer, C., Ewaidah, S., Presser, L., & Siston, A. M. (2015). Legionellosis outbreak associated with a hotel fountain. *Open Forum Infectious Diseases, 2*(4), ofv164.

Spini, G., Spina, F., Poli, A., Blieux, A.-L., Regnier, T., Gramellini, C., Varese, G. C., & Puglisi, E. (2018). Molecular and microbiological insights on the enrichment procedures for the isolation of petroleum degrading bacteria and fungi. *Frontiers in Microbiology, 9*, 2543. https://doi.org/10.3389/fmicb.2018.02543

Srathongneam, T., Sresung, M., Paisantham, P., Ruksakul, P., Singer, A. C., Sukchawalit, R., Satayavivad, J., Mongkolsuk, S., & Sirikanchana, K. (2024). High throughput qPCR unveils shared antibiotic resistance genes in tropical wastewater and river water. *Science of the Total Environment, 908*, 167867.

Staley, J. T., & Konopka, A. (1985). Measurement of in situ activities of nonphotosynthetic microorganisms in aquatic and terrestrial habitats. *Annual Review of Microbiology, 39*(1), 321–346.

Steinberg, L. M., & Regan, J. M. (2009). mcrA-targeted real-time quantitative PCR method to examine methanogen communities. *Applied and Environmental Microbiology, 75*(13), 4435–4442.

Strathdee, S. A., Hatfull, G. F., Mutalik, V. K., & Schooley, R. T. (2023). Phage therapy: From biological mechanisms to future directions. *Cell, 186*(1), 17–31.

Takuissu, G. R., Kenmoe, S., Ebogo-Belobo, J. T., Kengne-Ndé, C., Mbaga, D. S., Bowo-Ngandji, A., Ndzie Ondigui, J. L., Kenfack-Momo, R., Tchatchouang, S., Kenfack-Zanguim, J., Lontuo Fogang, R., Zeuko'o Menkem, E., Kame-Ngasse, G. I., Magoudjou-Pekam, J. N., Veneri, C., Mancini, P., Bonanno Ferraro, G., Iaconelli, M., Orlandi, L., et al. (2023). Occurrence of hepatitis a virus in water matrices: A systematic review and meta-analysis. *International Journal of Environmental Research and Public Health, 20*(2), 1054.

Tanaka, T., Kawasaki, K., Daimon, S., Kitagawa, W., Yamamoto, K., Tamaki, H., Tanaka, M., Nakatsu, C. H., & Kamagata, Y. (2014). A hidden pitfall in the preparation of agar media undermines microorganism cultivability. *Applied and Environmental Microbiology, 80*(24), 7659–7666.

Taylor, D. L., Walters, W. A., Lennon, N. J., Bochicchio, J., Krohn, A., Caporaso, J. G., & Pennanen, T. (2016). Accurate estimation of fungal diversity and abundance through improved lineage-specific primers optimized for Illumina amplicon sequencing. *Applied and Environmental Microbiology, 82*(24), 7217–7226.

Taylor, S. C., Laperriere, G., & Germain, H. (2017). Droplet digital PCR versus qPCR for gene expression analysis with low abundant targets: From variable nonsense to publication quality data. *Scientific Reports, 7*(1), 2409.

Temkin, A., Evans, S., Manidis, T., Campbell, C., & Naidenko, O. V. (2019). Exposure-based assessment and economic valuation of adverse birth outcomes and cancer risk due to nitrate in United States drinking water. *Environmental Research, 176*, 108442.

Tiquia, S. M. (2005). Microbial community dynamics in manure composts based on 16S and 18S rDNA T-RFLP profiles. *Environmental Technology, 26*(10), 1101–1113. https://doi.org/10.1080/09593332608618482

Tiquia, S. M. (2010). Metabolic diversity of the heterotrophic microorganisms and potential link to pollution of the Rouge River. *Environmental Pollution, 158*, 1435–1443. https://doi.org/10.1016/j.envpol.2009.12.035

Tiquia, S. M., & Mormile, M. (2010). *Extremophiles: A source of innovation for industrial and environmental applications* (Vol. 31, pp. 1061–1181). Taylor and Francis. https://doi.org/10.1080/09593330.2010.489333

Tiquia, S. M., Ichida, J. M., Keener, H. M., et al. (2005). Bacterial community profiles on feathers during composting as determined by terminal restriction fragment length polymorphism

analysis of 16S rDNA genes. *Applied Microbiology and Biotechnology, 67*, 412–419. https://doi.org/10.1007/s00253-004-1788-y

Tiquia-Arashiro, S. M. (2012). *Molecular biological technologies for ocean sensing* (Springer protocols handbooks) (p. 307). Humana Press. https://doi.org/10.1007/978-1-61779-915-0

Tiquia-Arashiro, S. M., & Grube, M. (2019). *Fungi in extreme environments: Ecological role and biotechnological significance* (p. 626). Springer Nature Switzerland AG. https://doi.org/10.1007/978-3-030-19030-9

Tiquia-Arashiro, S. M., & Pant, D. (2020). *Microbial electrochemical technologies* (p. 518). CRC Press. https://doi.org/10.1201/9780429487118

Tiquia-Arashiro, S. M., & Rodrigues, D. (2016a). Alkaliphiles and acidophiles in nanotechnology. In *Extremophiles: Applications in nanotechnology* (pp. 129–162). Springer International Publishing.

Tiquia-Arashiro, S. M., & Rodrigues, D. (2016b). *Extremophiles: Applications in nanotechnology* (Springerbriefs in microbiology: Extremophilic microorganisms) (p. 193). Springer International Publishing. https://doi.org/10.1007/978-3-319-45215-9

United States Environmental Protection Agency. (2023, November 30). Estimated nitrate concentrations in groundwater used for drinking. https://www.epa.gov/nutrientpollution/estimated-nitrate-concentrations-groundwater-used-drinking

Vartoukian, S. R., Palmer, R. M., & Wade, W. G. (2010). Strategies for culture of "unculturable" bacteria. *FEMS Microbiology Letters, 309*(1), 1–7.

Wan, Q., Chen, S., Shan, Z., Yang, Z., Chen, L., Zhang, C., Yuan, S., Hao, Q., Zhang, X., Qiu, D., Chen, H., & Zhou, X. (2017). Stability evaluation of reference genes for gene expression analysis by RT-qPCR in soybean under different conditions. *PLoS One, 12*(12), e0189405. https://doi.org/10.1371/journal.pone.0189405

Wang, H., Neyvaldt, J., Enache, L., Sikora, P., Mattsson, A., Johansson, A., Lindh, M., Bergstedt, O., & Norder, H. (2020). Variations among viruses in influent water and effluent water at a wastewater plant over one year as assessed by quantitative PCR and metagenomics. *Applied and Environmental Microbiology, 86*(24), e02073-20.

Wemheuer, F., Taylor, J. A., Daniel, R., Johnston, E., Meinicke, P., Thomas, T., & Wemheuer, B. (2020). Tax4Fun2: Prediction of habitat-specific functional profiles and functional redundancy based on 16S rRNA gene sequences. *Environmental Microbiomes, 15*(1), 11.

Whiley, H., & Taylor, M. (2016). Legionella detection by culture and qPCR: Comparing apples and oranges. *Critical Reviews in Microbiology, 42*(1), 65–74.

Wooley, J. C., Godzik, A., & Friedberg, I. (2010). A primer on metagenomics. *PLoS Computational Biology, 6*(2), e1000667.

Wu, W., Gu, B., Fields, M. W., Gentile, M., Ku, Y.-K., Yan, H., Tiquia, S. M., Yan, T., Nyman, J., Zhou, J., Jardine, P. M., & Criddle, C. S. (2005). Uranium (VI) reduction by denitrifying biomass. *Bioremediation Journal, 9*, 1–13. https://doi.org/10.1080/10889860590929628

Xiang, X., Li, Q., Khan, S., & Khalaf, O. I. (2021). Urban water resource management for sustainable environment planning using artificial intelligence techniques. *Environmental Impact Assessment Review, 86*, 106515.

Xiao, L., & Feng, Y. (2017). Molecular epidemiologic tools for waterborne pathogens *Cryptosporidium* spp. and *Giardia duodenalis*. *Food and Waterborne Parasitology, 8*, 14–32.

Xie, H., Yang, C., Sun, Y., Igarashi, Y., Jin, T., & Luo, F. (2020). PacBio long reads improve metagenomic assemblies, gene catalogs, and genome binning. *Frontiers in Genetics, 11*, 516269.

Yang, N., Zhang, C., Wang, L., Li, Y., Zhang, W., Niu, L., Zhang, H., & Wang, L. (2021). Nitrogen cycling processes and the role of multi-trophic microbiota in dam-induced river-reservoir systems. *Water Research, 206*, 117730. https://doi.org/10.1016/j.watres.2021.117730

Yang, Y., Che, Y., Liu, L., Wang, C., Yin, X., Deng, Y., Yang, C., & Zhang, T. (2022). Rapid absolute quantification of pathogens and ARGs by nanopore sequencing. *Science of the Total Environment, 809*, 152190.

Yang, Y., Deng, Y., Shi, X., Liu, L., Yin, X., Zhao, W., Li, S., Yang, C., & Zhang, T. (2023). QMRA of beach water by Nanopore sequencing-based viability-metagenomics absolute quantification. *Water Research, 235*, 119858.

Yao, Y., Zhao, J., Miao, L., & Hou, J. (2022). Effects of sediment physicochemical factors and heavy metals on the diversity, structure, and functions of bacterial and fungal communities from a eutrophic river. *Environmental Pollution, 303*, 119129.

Ye, C., Lin, H., Zhang, M., Chen, S., & Yu, X. (2020). Characterization and potential mechanisms of highly antibiotic tolerant VBNC *Escherichia coli* induced by low level chlorination. *Scientific Reports, 10*(1), 1957.

Yi, J., Wisuthiphaet, N., Raja, P., Nitin, N., & Earles, J. M. (2023). AI-enabled biosensing for rapid pathogen detection: From liquid food to agricultural water. *Water Research, 242*, 120258.

Young, J. N., & Schmidt, K. (2020). It's what's inside that matters: Physiological adaptations of high-latitude marine microalgae to environmental change. *The New Phytologist, 227*(5), 1307–1318.

Zhang, L., Yin, W., Wang, C., Zhang, A., Zhang, H., Zhang, T., & Ju, F. (2021). Untangling microbiota diversity and assembly patterns in the world's largest water diversion canal. *Water Research, 204*, 117617.

Zhang, T., Li, H., Ma, S., Cao, J., Liao, H., Huang, Q., & Chen, W. (2023). The newest Oxford Nanopore R10.4.1 full-length 16S rRNA sequencing enables the accurate resolution of species-level microbial community profiling. *Applied and Environmental Microbiology, 89*(10), e00605–e00623.

Zheng, X., Duan, X., Tu, X., Jiang, S., & Song, C. (2021). The fusion of microfluidics and optics for on-chip detection and characterization of microalgae. *Micromachines, 12*(10), 1137.

Zhong, Y., Su, Y., Zhang, D., She, C., Chen, N., Chen, J., Yang, H., & Balaji-Prasath, B. (2022). The spatiotemporal variations in microalgae communities in vertical waters of a subtropical reservoir. *Journal of Environmental Management, 317*, 115379.

Zhou, L., Song, C., Li, Z., & Kuipers, O. P. (2021). Antimicrobial activity screening of rhizosphere soil bacteria from tomato and genome-based analysis of their antimicrobial biosynthetic potential. *BMC Genomics, 22*(1), 1–14.

Chapter 4
Microbial Communities in the Sediments of Rift Lake Baikal and Their Role in Diagenesis Processes

Tamara I. Zemskaya, Sergei V. Bukin, Svetlana M. Chernitsyna, Anna V. Lomakina, and Olga N. Pavlova

1 Introduction

Lake Baikal is located in Central Asia (East Siberia, Russia) where it straddles the boundary between the Siberian platform and the Central Asian fold belt. In terms of area, Lake Baikal ranks seventh among the world's lakes, behind the Caspian Sea, African Great Lakes (Victoria and Tanganyika), and North American Lakes (Superior, Huron, and Michigan). In terms of depth and age, it is the world's deepest (maximum depth 1637 m) and the oldest (over 40 million years, according to Logachev (1974) and the Atlas of Lake Baikal (1993). Baikal's surface is 456 m above the level of the Baltic Sea. The volume of water In Lake Baikal is 23,000 km^3, and the volume of sediments is ~75,000 km^3, the most ancient age dating back to the Oligocene (Florensov, 1966; Hutchinson et al., 1992; Golmshtok et al., 2000; Mats et al., 2001). Underwater elevations morphologically divide the lake into three deep basins. The maximum thickness of sediments in the southern (SB) and central basins (CB) does not exceed 7.5 km, and in the northeast of the northern basin (NB), 4.4 km (Hutchinson et al., 1992). The major tributaries (Selenga, Barguzin, and Turka rivers) contribute 80% of the suspended particles and about 57% of the particulate organic carbon (C_{org}) input, while the Upper Angara river in the north supplies about 15% of the total suspended sediments entering the lake (Müller et al., 2005).

Lake Baikal is also oligotrophic. Its water has low mineral content, with total dissolved solids of ~0.1 g/L (Votintsev et al., 1975; Falkner et al., 1991; Khodzher et al., 2018). Among deep lakes (> 800 m), a unique feature of Baikal is the presence of oxygen up to the maximum depth, which to some extent makes it similar to the ocean. Based on these parameters, Lake Baikal has many similarities with the

T. I. Zemskaya (✉) · S. V. Bukin · S. M. Chernitsyna · A. V. Lomakina · O. N. Pavlova
Limnological Institute, Siberian Branch of the Russian Academy of Sciences, Irkutsk, Russia
e-mail: tzema@lin.irk.ru

S. M. Tiquia-Arashiro et al. (eds.), *Urban Watershed Microbiology, Volume 1*,
https://doi.org/10.1007/978-3-031-89071-0_4

oxygenated, oligotrophic, and shallower (>1000 m) Crater Lake (USA), differing from deep-rift Lake Tanganyika (1470 m), which is anoxic in the hypolimnion (Tran et al., 2021; Stromecki et al., 2022).

The concentrations of biogenic elements, nitrogen, phosphorus, and silica in the surface and near-bottom waters of Lake Baikal are subject to seasonal changes due to their consumption and regeneration by biota in the upper water layers and the unique mechanism for renovation of the near-bottom waters (Weiss et al., 1991; Khodzher et al., 2016). In spring and early winter, the surface waters of the lake with low (below 3.1–3.3 °C) temperature and high (over 10–10.5 mg/l) O_2 concentration enter the deep and near-bottom areas where a cold layer forms with a thickness of up to 100–200 m, persisting for several months (Wuest et al., 2005; Shimaraev et al., 2011, 2015; Khodzher et al., 2018). The fast downwelling of water from the photic zone leads to the enrichment of deep layers of the water column with not only oxygen and nutrients but also prokaryotes and small eukaryotes (Likhoshway et al., 1996; Parfenova et al., 2000; Shimaraev et al., 2012).

The organic matter productivity in the water column is low and comparable to that of oligotrophic water masses in the ocean (Weiss et al., 1991; Yoshida et al., 2003; Yoshioka et al., 2002). Picoplankton prevails in the summer primary production. Its production is comparable with that of diatoms that develop en masse in spring and autumn (Votintsev et al., 1975; Nagata et al., 1994). However, only 10% of the picoplankton biomass reaches the bottom; the rest decomposes in the photic zone (Bondarenko et al., 1996).

Bottom sediments are a special zone of Lake Baikal, the structure of which is determined by tectonic and climatic factors (Fig. 4.1a–f). Thanks to the enormous dimensions and depth, Lake Baikal has never dried out during warm seasons and never froze to the bottom; therefore, lake sediments have never been disturbed (plowed) by glaciers during ice ages, unlike most lakes of the Northern Hemisphere (Karabanov, 1999). These characteristics allowed researchers to use Lake Baikal sediments for reconstructing paleoclimates in the continental part of the Northern Hemisphere (Colman et al., 1992, 1995; Baikal Drilling Project (BDP-93) 1997; Williams et al., 1997). The long sedimentation history and many kilometers of sediment with high concentrations of organic matter contributed to the formation of large quantities of liquid and gaseous hydrocarbons (Fig. 4.1d, e). Lake Baikal is the only fresh-water lake in the world containing gas (GH) in its sedimentary infill (Fig. 4.1g). Areas with fluid seepages to the lake's floor contain different types of geological structures (mud volcanoes, methane seeps, and natural oil seeps) (Khlystov et al., 2013). Moreover, their cold seep systems are comparable in size and frequency to those of oceanic, high-productivity continental margins (Aloisi et al., 2019). Various diagenetic, catagenetic, and circulation processes in the sediments of the Baikal Rift Zone substantiate their formation (Kuzmin et al., 2001; Klerkx et al., 2003; Krylov et al., 2008; Khlystov et al., 2013, 2022; Krylov et al., 2008; Morgunova et al., 2022). The appearance of new oil seeps recorded in the form of oil films on the surface of the water column or detected in the surface bottom sediments through geological and geophysical methods (Khlystov et al., 2007, 2022) confirms the activity of the Baikal rift.

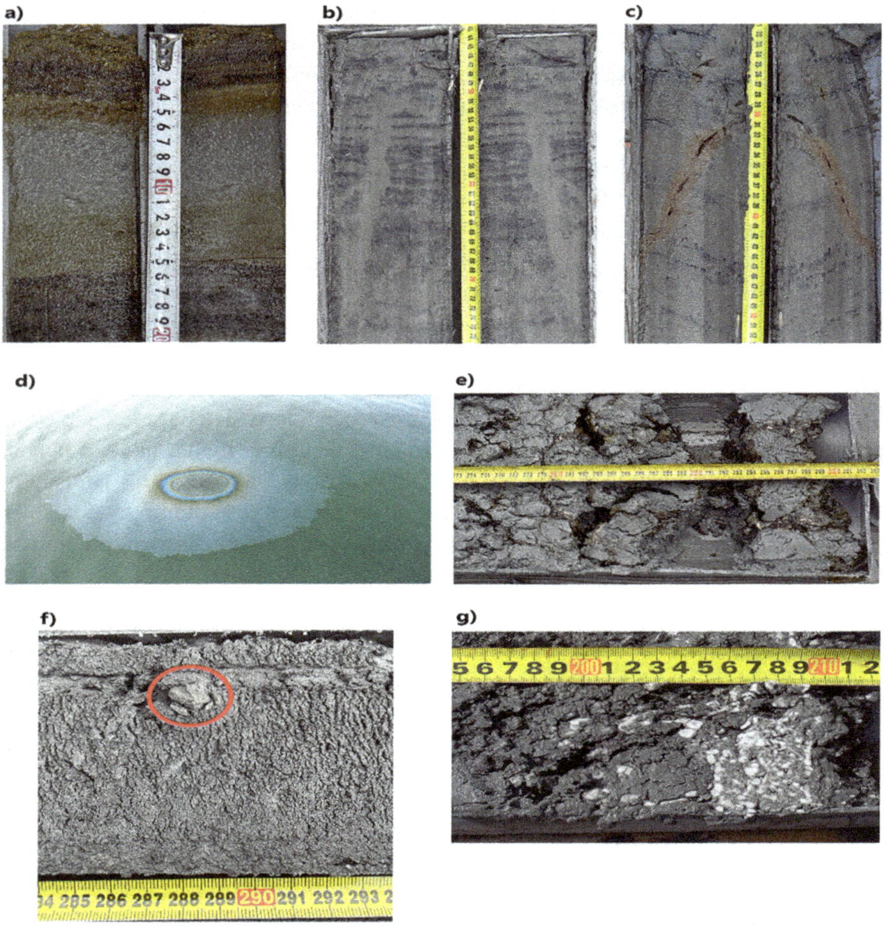

Fig. 4.1 (**a**) Structure of surface sediments in the area with quiet sedimentation in the southern basin of Lake Baikal; (**b, c**) fragments of sediments from the area with methane seep and channels caused by fluid flows; (**d**) oil film on the surface of the water column near the oil seep B. Zelenovskaya (BZ); (**e**) oil and gas hydrates in the sediments of oil seep Gorevoy Utes (GU); (**f**) sediments from the area of the Kedr mud volcanoes (MV) with authigenic carbonate inclusions; (**g**) gas hydrate layers in the sediments of the K-9 MV

Distinctive features of Lake Baikal hydrology and hydrochemistry have made this lake a very peculiar environment for its inhabitants (Kozhov, 1972; Grachev, 1994; Timoshkin, 2001), including microbial communities. The diversity of flora and fauna in Lake Baikal is much greater than in other ancient lakes (Kozhov, 1972), making this ecosystem an ideal tool to understand the evolution of life on the Earth. Eukaryotic and prokaryotic organisms inhabit the water column and bottom sediments. Some of these organisms are related to those from fresh waters or sea waters with different salinity, or are endemic, being adapted to low-temperature water of

the lake depleted of organic substrates (Timoshkin, 2001; Cabello-Yeves et al., 2020; Coutinho et al., 2020; Usoltseva et al., 2020; Bukin et al., 2023; Chernitsyna et al., 2024). Most likely, the constant supply of various hydrocarbons from deep Lake Baikal sediments maintains the high diversity in the bathypelagic zone of Lake Baikal and bottom sediments. Their involvement in food webs could ensure survival of various benthic organisms in periods of unfavorable climates in this ancient ecosystem (Grachev et al., 1995).

Thanks to the above features, Lake Baikal was included on the UNESCO World Heritage List in 1996. This imposes certain requirements for the conservation of this unique ecosystem. Currently, two main causes of the changes in the ecosystem of Lake Baikal are being discussed: increased eutrophication in the littoral zone of the lake and global climate change (Hampton et al., 2008; Moore et al., 2009; Timoshkin et al., 2015, 2018; Katz et al., 2015). Researchers primarily observed changes in the littoral zone of the lake in settlements and tourist recreations. They studied the composition and abundance of eukaryotic organisms, including higher aquatic plants, sponges, and phytoplankton (Khanaev et al., 2016; Kravtsova et al., 2014; Timoshkin et al., 2015, 2018; Bondarenko et al., 2019), and obtained the data on hygienically relevant bacteria, an elevated number of which was also recorded previously (Maksimova & Maksimov, 1989; Malnik et al., 2019). Obviously, intensive touristic development along the lake's shores will influence its ecosystem. Therefore, studying its various elements, understanding the functioning mechanisms, and ensuring the conservation of species are necessary to monitor the state of this ecosystem and develop the fundamental task to study speciation in ancient ecosystems.

2 Geological and Geochemical Characterization of Lake Baikal Sediments

The structure of the upper sedimentary layer of all deep-water basins and lake rises is uniform. The sediments contain alternating layers of silty-pelitic diatom ooze and layers of thin diatom-free glacial clays (Colman et al., 1991, 1999; Nelson et al., 1999) (Fig. 4.1a–c). Sediments from different basins often included layers of coarse-grained material sediments of suspension flows (turbidites) (Karabanov, 1999; Vologina et al., 2007). The lithology of sediments sampled from the cold seep area had features typical of normal background sedimentation in SB and evidence of the seepage of methane-charge fluids. Holocene sediments had another type of layering caused by the presence of buried surface iron-manganese layers in the sediment strata (Granina et al., 1993; Och et al., 2012; Pagnucco et al., 2023). These layers form at the boundary of oxidized and reduced sediments, and their depth depends on the depth of oxygen penetration into the lake sediments and its concentration, as well as on the porosity, organic matter (OM) content, and sedimentation rate.

Diatom ooze ubiquitously occupies the central part of the lake. The concentration of biogenic silica averaged 19% in the pelagic zone, and the 90%

SiO_2biog-containing sediments covered <10% of the lake floor (Granina et al., 1993). In the deep parts of the three basins, sediments have identical chemical composition due to homogenization of fine fractions of the incoming terrigenous material transported by steady currents (Karabanov, 1999).

Hemipelagic sediments form due to a significant input of aquatic vegetation, mainly phytoplankton and diatoms (Kalmychkov et al., 2006; Garankina et al., 2009; Vologina & Sturm, 2017; Patel et al., 2019; Morgunova et al., 2022). The highest rates of sedimentation are characteristic of areas near the estuaries of large rivers, it is less intense in the deep-water zone, and its lowest rates are characteristic of underwater elevations (Vologina & Sturm, 2009). In open Lake Baikal, sedimentation rates vary from 0.015 to <1 mm per year (Colman et al., 1993; Vologina et al., 2000). The shallow and oligotrophic lake Laguna Potrok Aike (represents 51,000 years of sediment deposition) has similar sedimentation rates (up to 1 mm per year) (Zolitschka et al., 2013). The average concentration of organic carbon (C_{org}) in Lake Baikal sediments is 1.9%: 1.5–2.5% in the southern and central basins and 1.0–1.3% in the northern basin (Vykhistryuk, 1980), which is comparable with oceanic sediments in highly productive continental margins (Qiu et al., 1993; Prokopenko et al., 2007). In sediments with different types of geological structures, the C_{org} concentrations range from 0.2 to 4.8%, and organic matter (OM) from 0.3 to 8.7%. The highest concentrations are naturally associated with oil seeps and sediments from the hydrothermal discharge zone in Frolikha Bay (NB) (Vykhistryuk, 1980; Afonina & Karabanov, 2001; Morgunova et al., 2022). According to these authors, the composition of the buried organic matter (OM) includes humic acids (HA) and cellulose, highly decomposed remains of terrestrial higher plants and phytoplankton that are highly resistant to further degradation. Morgunova et al. (2022) recorded relatively high HA concentrations in the surface sediments of MV containing mud volcanic breccia that formed during the rise of older sedimentary material enriched in humins to the bottom surface. Oil and natural gas are present in lake sediments in free, dissolved, and solid (GH) states (Kontorovich et al., 2007; Khlystov & Kalmychkov, 2020). Hydrocarbon gases in Lake Baikal sediments (mainly methane and ethane) are both of microbial and thermogenic origin (Kalmychkov et al., 2006, 2017, 2019; Hachikubo et al., 2020, 2023). As shown in some works (Klerkx et al., 2003; Chernitsyna, Mamaeva, et al., 2016a; Pogodaeva et al., 2020), valves of small-celled diatoms and microorganisms migrate together with oil- and gas-containing pore waters along the faults to the bottom surface.

We have studied sedimentary layers in different basins of Lake Baikal. All of them consist mainly of Holocene deposits and belong to the diatomaceous ooze lithotypes characteristic of modern sedimentation. The surface sediments are mostly oxidized at the lake bottom; the thickness of the oxidized layer varies from 0.2 to >30 cm (Granina, 2008; Och et al., 2012; Pogodaeva & Khodzher, 2023). The oxygen concentration in the near-bottom water is >0.36 mmol L^{-1}, and the low C_{org} rate ensures high O_2 penetration depth into sediments (>2 to 50 mm) (Mizandrontsev, 1978; Och et al., 2012). Thus, from 60% to 75% of the total sediments in C_{org} are aerobically mineralized (Maerki et al., 2006). Oxic conditions allow aerobic heterotrophs to colonize the sediment-water interface, while low salinity leads to rapid

depletion of electron acceptors in the pore water, promoting fermentation and methylotrophic methanogenesis as the key processes in this sedimentary zone. The sediments near methane seeps are reduced from the surface down; the oxygen penetration depth here is minimal (Zemskaya et al., 2010; Lomakina et al., 2023a). The pH values of sediments are mainly within 6.9 and 7.4, ranging from 6 to >8 (Mizandrontsev, 1975; Granina, 2008; Och et al., 2012; Pogodaeva et al., 2017).

The chemical composition of pore water in the areas of quiet sedimentation is similar to that of lake water and is rather homogeneous in salinity (Mizandrontsev, 1975; Pogodaeva et al., 2007, 2017; 2023; Granina, 2008). HCO_3^- and Ca^{2+} are the most abundant ions in Lake Baikal. Other ions (Cl^-, Na^+, Mg^{2+}, and K^+) are present in pore waters of the lake, and their concentrations are very low (≤0.15 mM) (Pogodaeva et al., 2007, 2017). According to studies (Borrel et al., 2012; Wurzbacher et al. 2017a; Rissanen et al., 2019; Von Eggers et al., 2024), sedimentary compounds are metabolized by microbes following the typical electron acceptor sequence ($O_2 > NO^{3-} > NO^{2-} > Mn^{4+} > Fe^{3+} > SO_4^{2-} > CO_2$). As in other oligotrophic lakes, water and sediment layers of Lake Baikal have low SO_4^{2-} concentrations (Garankina & Dagurova, 2008; Pimenov et al., 2014; Aloisi et al., 2019). Their concentrations near bottom waters and upper layers remained in a narrow range between 55.4 and 77.5μmol l^{-1}. A renewed increase in SO_4^{2-} concentrations was observed deeper in the sediment around the buried Fe/Mn layer. In this regard, the zone of sulfate reduction in Lake Baikal sediments is very limited, and the zone of methane generation almost immediately replaces the zone of aerobic oxidation of methane (Namsaraev & Zemskaya, 2000; Zemskaya et al., 2008; Pimenov et al., 2014). This is consistent with the observed methane concentration profiles and with the presence of methanogenic archaea in this layer despite the oxic conditions (Och et al., 2012; Bukin et al., 2023). In the vicinity of MVs and hydrothermal vents (in 2008–2010 data), sulfate ion concentrations reached 2.9 to 15 mM (Granina, 2008; Zemskaya et al., 2008). Currently, there is another distribution pattern of sulfate profiles. Their concentrations along the sediment depth of different sediment layers correspond to the background values and do not exceed 0.08 mM (Krylov et al., 2008, 2023; Zemskaya et al., 2010; Lomakina et al., 2023b; Pogodaeva & Khodzher, 2023). NO_3^- concentrations did not exceed 10μmol l^{-1} in the pore waters of reference sites (Och et al., 2012) and tended to increase slightly within the oxygenated part of the sediment, decreasing below the O_2 penetration depth. The flux of nitrate ranges from 2.8 to 5.9 mmol m^2 d^{-1} at these sites (Och et al., 2012). The concentrations of nitrate and nitrite ions were >0.002 mM in the pore water of the hydrocarbon discharge areas (Zemskaya et al., 2015; Lomakina et al., 2014, 2018a, 2020). Ammonia concentrations (up to 0.2 mM) are more significant for Lake Baikal sediments because the ammonia oxidation produces nitrate and nitrite ions (van Kessel et al., 2015), which support a wide variety of nitrogen-cycling microorganisms, including those that perform anaerobic oxidation of methane (AOM) (Chernitsyna et al., 2016a, Lomakina et al., 2018a, 2020).

In Baikal sediments, the oxidation capacities of NO_3^- and SO_4^{2-} are low (due to low fluxes); therefore, Fe(III) and Mn(IV) are the main electron acceptors for the anaerobic mineralization of C_{org} (Och et al., 2012). The high concentrations of Mn

and Fe in Lake Baikal sediments may be associated with the Paleoproterozoic rocks in the watershed, which were formed during an enhanced Mn- and Fe- ore genesis at the Earth's surface (Maynard, 2010). These conditions, combined with a low sedimentation rate, ensure the reduction of Mn(IV) and Fe(III), a major anaerobic OM mineralization pathway. In terms of concentration gradients, reduced Mn^{2+} and Fe^{2+} ions in the pore water diffuse up and down the sediment, with upward diffusion being stopped due to reoxidation involving either O_2 or nitrate ions (Och et al., 2012). The maximum concentrations of iron (up to 0.49 and 0.65 mmol g^{-1}) and manganese (up to 0.26 and 2.14 mmol g^{-1}) were recorded at sediment depths ranging from 13 to 25 cm (Granina, 2008; Och et al., 2012; Lomakina et al., 2023a). In the deeper reduced sediments of hydrocarbon discharge areas, Fe and Mn concentrations were small (up to 0.018 and 0.007 mM, respectively) (Lomakina et al., 2018b, 2023b).

Two decades of our studies of pore water chemistry in the areas where mineralized gas-bearing fluids are discharged testify to variability in the concentrations of major ions and changes in the composition of their components both throughout the lake and at the same coordinates (Pogodaeva et al., 2007; Pogodaeva & Khodzher, 2023; Krylov et al., 2008, 2023; Zemskaya et al., 2010). It is most likely that these changes are caused by changes in the intensity of the fluid flows and the overall activity of the Baikal Rift.

3 Microbial Communities of Lake Baikal Sediments

3.1 Microbiological Studies of Lake Baikal Sediments

Studies of microbial communities from sediments of Lake Baikal have been carried out since the beginning of the past century using the approaches and methods characteristic of certain periods. Some publications provide information about the number of microorganisms in the lake sediments with different geochemical conditions, analyze the distribution of physiological groups of microorganisms in different areas, and assess the metabolic activity of microbial communities, including areas with increased anthropogenic impact (Nechaeva & Salimovskaya-Rodina, 1935; Rodina, 1954; Romanova, 1958; Kuznetsov, 1970; Goman, 1975; Mladova, 1975; Maksimova et al., 1982, 1995; Nassereddine et al., 2024). A study of the abundance of microorganisms in estuarine evidenced a correlation with the content of OM and nutrients in sediments, and the differences in the number of microorganisms in different sedimentary layers (Mladova, 1975). Long-term studies were carried out to determine the number of bacteria involved in the phosphorus cycle, cellulose-degrading and phenol-degrading bacteria, and the activity of bacterial enzymes (cellulase, dehydrogenase, and carbonate assimilation) in the zone of influence of the Baikal Pulp and Paper Mill (BPPM) located on the shore of the lake (now liquidated) (Goman, 1975; Maksimova et al., 1995; Namsaraev et al., 1995; Vinogradova

et al., 2004). These works indicated the dependence of a number of the investigated groups of microorganisms on the type of bottom sediments, trophicity degree in the study area, and anthropogenic impact, particularly near BPPM. A new stage of research into the role of microbial communities in the sediments began during the large-scale geological and geophysical work carried out by research teams at Lake Baikal since the late 1990s. At that time, of special interest was the study of microorganisms involved in different stages of OM destruction and the assessment of the rates of OM destruction in the sedimentary strata of Lake Baikal (Maksimova & Maksimov, 1989; Namsaraev et al., 1994, 1995; Namsaraev & Zemskaya, 2000). These studies revealed that the availability of substrates, oxidized or reduced environmental conditions at different depths of the sedimentary strata, and the supply of mineralized gas-bearing fluids from deep sediments determined the development of physiological groups of microorganisms in sediments (Namsaraev et al., 1994; Zemskaya et al., 2001; Klerkx et al., 2003; Pimenov et al., 2014). The annual supply of OM to sediments and low sulfate-ion concentrations ensures the dominance of the methane generation at the final stages of destruction (Namsaraev & Zemskaya, 2000; Namsaraev et al., 2006; Pimenov et al., 2014), which experimental and field studies with radiolabeled substrates confirm. Traditional ($^{13}C/^{12}C$ and deuterium-to-hydrogen (D/H)) measurements of methane isotopes are illustrative of a significant proportion of microbial methane generated in the lake sedimentary strata (Kalmychkov et al., 2006, 2017; Hachikubo et al., 2020, 2023). The decrease in the relative contribution of biogenic methane with depth validates its supply from the deep zone. In particular, the study of the diversity of microorganisms in sediments focused mainly on aerobic bacteria, whose taxonomic position was determined by morphological, physiological, and biochemical characteristics (Maksimova et al., 1995; Vinogradova et al., 2004).

3.2 *Taxonomic Diversity of Microbial Communities in Lake Baikal Sediments*

Molecular biology techniques and approaches have provided the achievement of progress in assessing the diversity of microbial communities. 16S rRNA gene sequencing has become a standard step in assessing the composition of microbial communities in different environments (Woese, 1987; Amann et al., 1995; Dunbar et al., 1999; Tiquia et al., 2005, 2007; Bernhard et al., 2005; Petrosino et al., 2009; Claesson et al., 2010; Cho et al., 2012; Kim et al., 2013; Nguyen et al., 2013; Plecha et al., 2013; Pomaranski & Tiquia-Arashiro, 2016; Myer et al., 2016). Different platforms that allow for a parallel analysis of a large number of samples, which were used in the past decade, have made this approach relevant for statistical testing of the influence of natural or experimental factors and variables. Identifying uncultured microbial communities is one of the advantages of NGS-based

metabarcoding, but readlength limitations only allow analysis of certain gene regions, resulting in reduced species identification accuracy (Myer et al., 2016).

In the analysis of different 16S rRNA gene regions, we have statistically substantiated the sequencing coverage and the relative number of the metabarcoding data on the samples from Lake Baikal. Examination of the V2-V3 region yielded the highest resolution for low-ranked bacterial taxa (species and genera) (Bukin et al., 2019). At the same time, the diversity of communities at the phylum level depended little on the examined 16S rRNA gene region and the platform used (Zemskaya et al., 2010, 2015a; Kadnikov et al., 2012, 2013; Lomakina et al., 2014, 2018b; Bukin et al., 2016; Bukin et al., 2019; Chernitsyna et al., 2016a). Lake Baikal sediments, like those in marine and freshwater ecosystems, contain various heterotrophic communities that carry out hydrolysis of organic carbon, fermentation, and respiration (Phillips et al., 2023). The diversity of microbial communities in sediments from different areas of Lake Baikal overlaps at the phylum level and demonstrates increasing dissimilarity of microorganisms below the class and family level (Zemskaya et al., 2010, 2015a; Kadnikov et al., 2012, 2013; Lomakina et al., 2018b). The structure of microbial communities at the phylum level clearly differs between oxidized and reduced layers of the sedimentary strata (Kuznetsov, 1970; Zavarzin, 2003; Huang et al., 2019; Han et al., 2020; Von Eggers et al., 2024). Studies of microbial communities from the sediments of Lake Stechlin, five lakes in Central Switzerland, Lake Superior, and 36 lakes in four disconnected mountain ranges in Wyoming (USA) revealed changes in the structure of the microbial community when oxidized conditions changed to reduced ones (Wurzbacher et al. 2017a, b; Han et al., 2020; Phillips et al., 2023; Von Eggers et al., 2024). In Lake Baikal sediments, both the analysis of clonal libraries and high-throughput sequencing data (Rosh 454 platform) recorded this fact. Organotrophic bacteria typical of freshwater lakes such as *Actinomycetota*, *Pseudomonadota*, and *Verrucomicrobiota* dominated the communities of oxidized sediments (Kadnikov et al., 2012; Zemskaya et al., 2015a, 2018; Zakharova et al., 2018). These bacteria are involved in the first stages of organic compound destruction (Newton et al., 2011).

Analysis of metabarcoding data (Illumina platform) on 78 bacterial (V2-V3 16S rRNA gene region) and 75 archaeal (V5-V6 region) samples obtained between 2018 and 2023 from different ecotopes (Fig. 4.2) of Lake Baikal expanded the taxonomic composition of microbial communities (Figs. 4.3, 4.4, and 4.5).

Calculation of the Bray-Curtis distance matrix with subsequent statistical analysis and plotting of the non-metric multidimensional scaling (NMDS) graph confirmed the difference in composition and beta diversity of the bacterial (Fig. 4.3a) and archaeal (Fig. 4.3b) communities between oxidized and reduced sediment layers. As previously demonstrated (Zemskaya et al., 2021), communities in the oxidized and reduced sediments differed the most at the level of dominant phyla, with increasing dissimilarities in communities at the level of lower-ranked taxa in different sediment types (reference sites, MVs, and oil seeps) (Figs. 4.4 and 4.5).

The statistical significance of the difference between communities of "oxidized" and "reduced" clusters (reference sites, MVs and oil seeps) was confirmed by non-parametric statistical tests (ANOSIM $R = 0.792–0.805$, $p_{\text{value}} < 0.001$; PERMANOVA

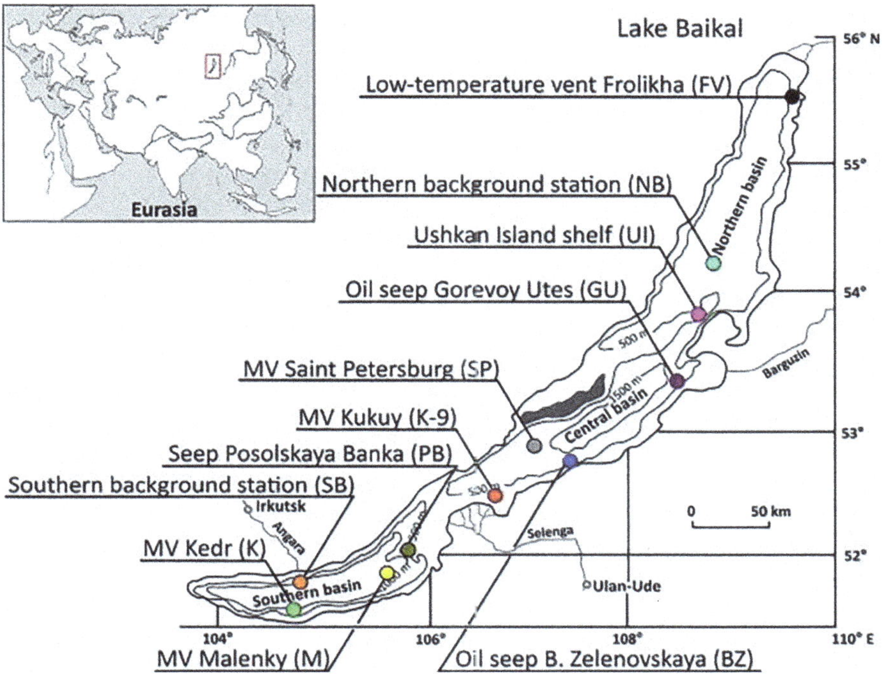

Fig. 4.2 Geographical distribution of sampling sites in the southern, central and northern basins of Lake Baikal

$R = 0.183–0.228$, $F = 17.182–21.617$, $p_{\text{value}} < 0.001$). Samples from the reference sites of all basins indicated obvious differences in the composition of bacterial and archaeal communities, which were most likely due to the different composition of the dominant complex of diatoms, a key source of OM in Lake Baikal (Votintsev et al., 1975). In oxidized sediments, the relative abundance of individual bacterial taxa showed a positive correlation with the concentration of nitrate and sulfate ions. Communities from surface sediments of MVs and oil seeps had a more similar composition in comparison with the reference sites, despite the differences in the oxygen regime (e.g., the Kedr MV) and the component composition of discharged mineralized gas-bearing fluids. The community structure in reduced sediments for the bulk of the communities was correlated with their occurrence depth and the concentration of bicarbonate ions, a major ion of pore waters in these sediments.

A higher relative abundance of metabolically diverse taxa (*Gammaproteobacteria*, *Desulfobacterota*, *Methylomirabilota* (formerly NC10), *Verrucomicrobiota*, and *Planctomycetota)* was characteristic of the communities from oxidized sediments. It is known that these bacteria mineralize CM and are common in lake sediments (Newton et al., 2011). The reduced sediments contained phyla *Atribacterota* and *Caldisericota*, with their greatest contribution to the communities of sediments from MVs and oil seeps. Communities from all sediment types showed a significant

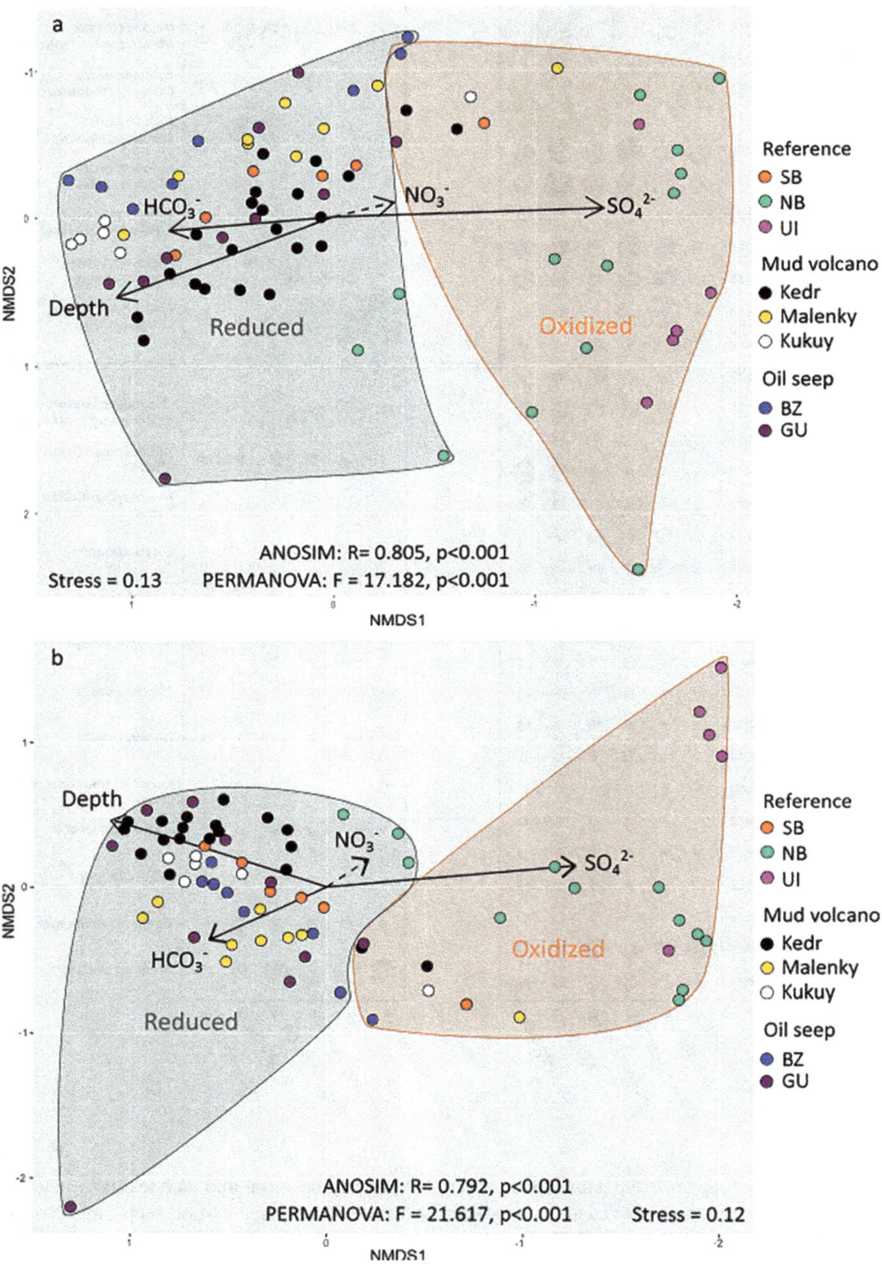

Fig. 4.3 NMDS plot showing the differences in composition and abundance of (**a**) bacterial and (**b**) archaeal communities from oxidized and reduced sediment layers of oil seeps, mud volcanoes and reference areas of Lake Baikal ($OTU_{0.03}$ level based on 16S rRNA). Beta diversity was estimated according to Bray–Curtis dissimilarities. Continuous arrows show environmental factors reliably correlated ($p_{value} < 0.05$) with community structure ordination

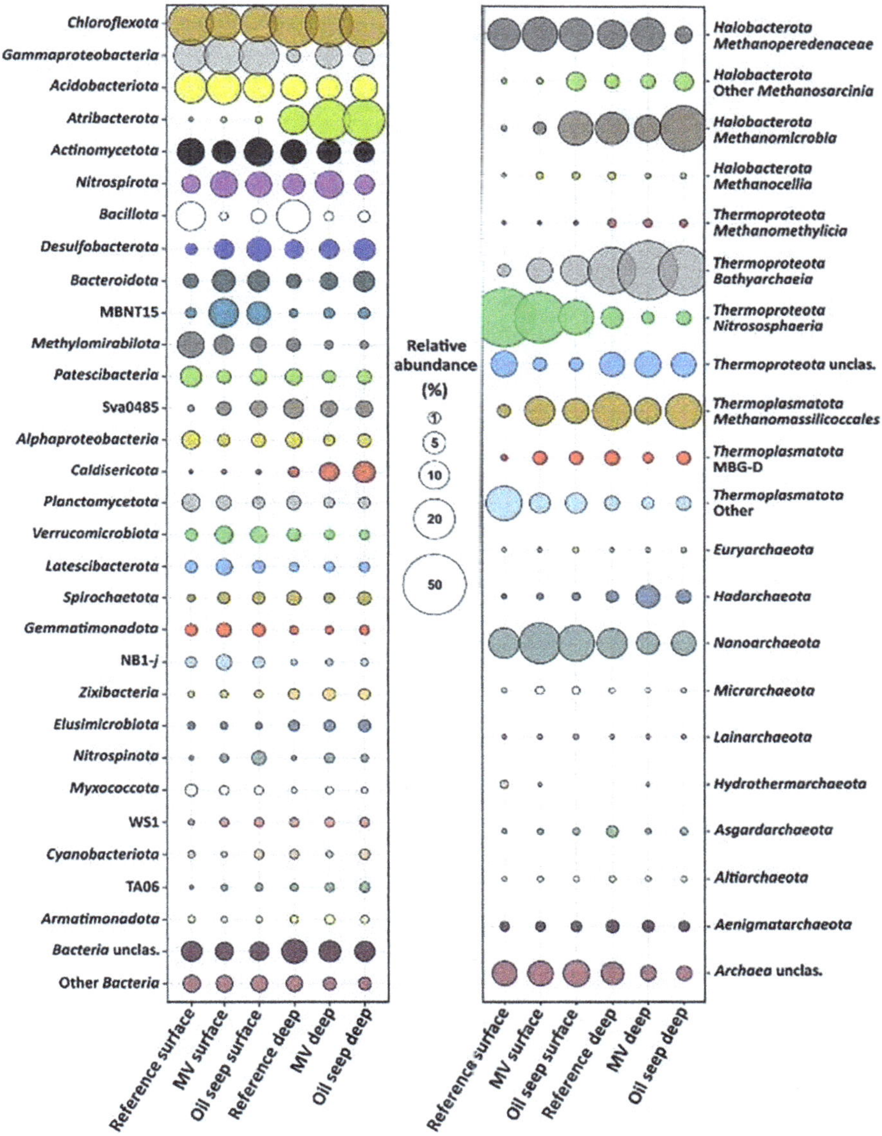

Fig. 4.4 The average relative abundance (16S rRNA) of key bacterial and archaeal taxa in communities from surface oxidized and deep reduced bottom sediment layers of oil seeps, mud volcanoes and reference sites of Lake Baikal

contribution of the *Chloroflexota, Actinomycetota, Acidobacteriota, Nitrospirota, Bacteroidota,* and *Desulfobacterota* members. Members of the organoheterotrophic *Chloroflexota* involved in different stages of fermentation of complex compounds

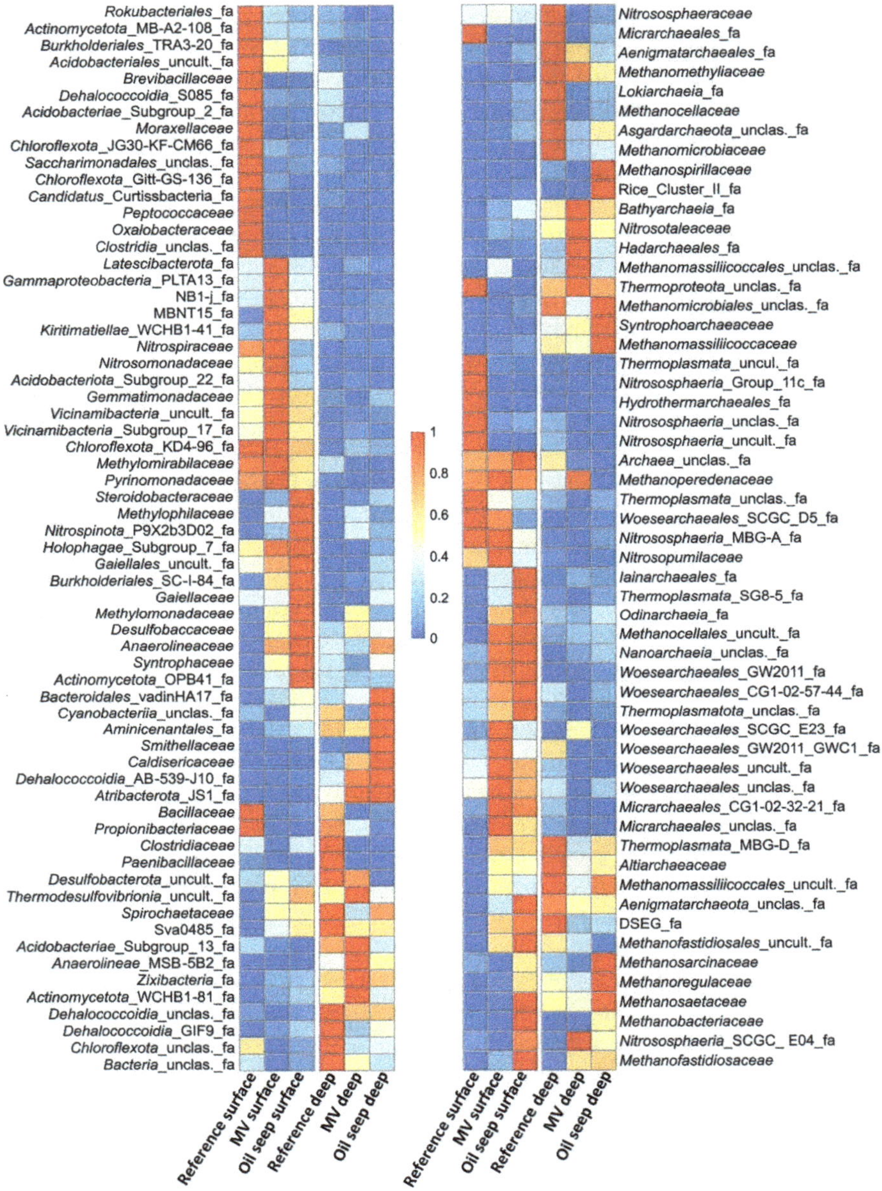

Fig. 4.5 Heatmap showing the normalized relative abundance (16S rRNA) of bacterial and archaeal families (>1%) in sediment communities of various ecotopes in Lake Baikal

were the most representative in the communities of oxidized/reduced sediments, MVs, oil seeps, and reference sites (Eiler, 2006; Suominen et al., 2021; Phillips et al., 2023). Among the members of this phylum, the sequences of uncultured

members of the families *Dehalococcoidaceae* and *Anaerolineaceae*, as well as uncultured *Chloroflexota* from different clusters (KD4–96, JG30-KF-CM66, and GIF-9), predominated in the investigated ecotopes. Members of this phylum are frequently found in various pristine and contaminated environments, and those from the uncultured clusters—in soil or river sediments, freshwater lakes (Liang et al., 2015; Mehrshad et al., 2018; Vuillemin et al., 2020), and in taberal sediments, including recently thawed ones (Winkel et al., 2019). Most of the class *Gammaproteobacteria* (phylum *Pseudomonadota*) were affiliated with the taxa represented by aerobic, microaerobic, and facultatively anaerobic organotrophs of the orders *Burkholderiales* and *Pseudomonadales.* Noteworthy is the higher contribution of the sequences of the phylum *Methylomirabilota* to the communities of the oxidized layer from the reference sites. Among them, the proportion of the order *Methylomirabilales* using nitrite as electron acceptor during AOM) did not exceed 5%.

Members of another order, mixotrophic *Rokubacteriales,* accounted for 15% of all analyzed bacterial sequences. Bacteria of the families *Methylophilaceae* and *Methylomonadaceae* involved in different stages of methane oxidation made the largest contribution to the communities of oxidized oil seep sediments. Methanotrophs of the family *Methylomonadaceae* prevailed in the water column of different lakes (Constance, Zug, La Cruz, etc.) with high methane concentrations (Milucka et al., 2015; Bornemann et al., 2016; Oswald et al. 2016a, b). Additionally, *Methylomonadaceae* played a crucial role in the communities from the near-bottom zone in Lake Baikal areas where mineralized gas-bearing fluids are discharged (Zakharenko et al., 2019). Members of *Acidobacteriota* were also numerous (12% to 18%) in the oxidized layers of the sediments. Their relative abundance decreased in deep sediments, albeit remaining rather significant (0.5% to 10%). Different taxa represented this phylum at different depths: *Acidobacteriota* unclassified predominated in the oxidized sediments from reference sites, and chemoorganoheterotrophic *Vicinamibacteria* fermenting carbohydrates and protein substrates under anaerobic and aerobic conditions (Dedysh & Yilmaz, 2018; Kadnikov et al., 2019) predominated in the oxidized sediments of MVs. The order *Aminicenantales* was more characteristic of the reduced sediments of oil seeps (Fig. 4.4). These bacteria predominantly occurred in environments exposed to hydrocarbons (Farag et al., 2014). Phylum *Actinomycetota* was more common in the communities of oxidized sediments from reference sites and oil seeps. Among them, chemoorganoheterotrophic *Gaiellaceae* and sequences of the family OPB_{41} (*Coriobacteriia*) were the most numerous. A metabolic analysis of cultured members of OPB41 indicated that these microorganisms were anaerobic lithotrophs using thiosulfate, elemental sulfur, or Fe(III) as electron acceptors (Khomyakova et al., 2022). Members of the phyla *Nitrospirota*, *Bacteroidota*, and *Desulfobacterota* were detected in different proportions in all investigated ecotopes (Fig. 4.5). The family vadinHA17, proteolytic bacteria fermenting carbohydrates to produce volatile fatty acids, represented phylum *Bacteroidota* (Kampmann et al., 2012). The *Desulfobacterota* sequences belonged to the families of anaerobic bacteria oxidizing either short-chain fatty acids (UBA5619 (*Syntrophaceae*) and *Smithellaceae*) or acetate (*Desulfobaccaceae*)

in synthrophic interaction with methanogenic archaea (Galushko & Kuever, 2019). Members of the chemoorganoheterotrophic *Atribacterota* and *Caldisericota* found only in reduced sediments participated in the oxidation of peptides, amino acids, fatty acids, and simple sugars through aerobic and anaerobic respiration (Mori et al., 2009; Carr et al., 2015; Nobu et al., 2016). *Atribacterota* is the most common in geothermal systems, petroleum reservoirs, anaerobic marine sediments, and methane seeps (Carr et al., 2015; Nobu et al., 2016). Their proportion reached 45% of all analyzed bacterial sequences for the communities from reduced sediments of MVs and oil seeps in Lake Baikal, while it was below 15% in deep sediments from the reference sites. Currently, metabolic reconstruction of *Atribacterota* genomes has revealed that bacteria can not only ferment carbohydrates to produce hydrogen, acetate, and ethanol resulted from glycolysis but also ferment short-chain *n*-alkanes into fatty acids that secondary destructors then consume (Liu et al., 2019). Thermophilic obligate anaerobic heterotrophs reducing sulfur compounds and assigned to the phylum *Caldisericota* were among the dominant taxa in deep sediments of oil seeps (up to 15%) and MVs (up to 9%) and were minor (1.8%) or not detected at reference sites. Recently, sequences belonging to this phylum have been identified in hydrothermal systems, hydrocarbon contaminated aquifers, soda lakes, bioreactors, etc. (Mori et al., 2009). Sequences of the phyla *Zixibacteria* and *Elusimicrobiota* (< 1%) classified as representatives of the "rare biosphere" or "microbial dark matter" were minor components in deep-sediment communities (Pascoal et al., 2021). Reconstruction of genomes proves their fermentative metabolism, magnetotaxis, and involvement in the iron and sulfur cycles (Castelle et al., 2013; Uzun et al., 2023).

The structure of archaeal components in the communities also varied significantly between oxidized and reduced layers of the sedimentary strata (Figs. 4.4 and 4.5). One of the principal differences is the composition of the phylum *Thermoproteota*, which on average comprised 25–55% of all identified archaeal sequences from different ecotopes. Aerobic microorganisms of the class *Nitrososphaeria*, among which ammonia-oxidizing archaea of the family *Nitrosopumiliaceae* had the highest relative abundance (12–31%), mostly represented *Thermoproteota* in the surface oozes. Members of the families *Nitrososphaeraceae* and *Nitrosotaleaceae* had significantly lower percentages (<0.5%) in the communities. The dominance of *Nitrosopumiliaceae*, which also had a substantial proportion in the microbiome of Lake Baikal water column (Cabello-Yeves et al., 2018, 2020), likely resulted from the highest affinity for ammonia among all ammonia-oxidizing microorganisms, allowing them to successfully adapt to oligotrophic ecosystems (Flood et al., 2015; Jung et al., 2022). *Archaea* of the class *Bathyarchaeia*, which had the largest proportion in the MV sediments, were the dominant members (27–46%) of the phylum *Thermoproteota* in the reduced layers. *Bathyarchaeia* plays an important role in the global carbon cycle and is the most abundant group of microorganisms in the anaerobic ecotopes. Genome reconstruction of the members of various phylogenetic *Bathyarchaeia* lineages indicated the presence of a similar set of genes for carbon metabolism in them, including glycolysis, gluconeogenesis, and H_4-MPT Wood-Ljungdahl pathway, as well as

carbohydrate-active enzymes and peptidases (Hou et al., 2023). This enables them to utilize a wide range of organic compounds, including detrital proteins, carbohydrates, lignin, short chain alkanes, and multiple C_1 compounds such as CO_2, CO, formate, formaldehyde, methanol, and methane (Tiquia et al., 2008; Evans et al., 2015; Yu et al., 2018; Zhou et al., 2018; De Anda et al., 2021). Microorganisms from at least four recently proposed orders of *Bathyarchaeia* (Hou et al., 2023)—*Xuanwuarculales, Baizomonadales, Houtuarculales, Wuzhiqibi*ales—have been found in sediment communities of Lake Baikal. Other members of the phylum *Thermoproteota* detected mainly in the reduced layers of sediments were methyl-reducing hydrogenotrophic methanogens of the class *Methanomethylicia*. However, their proportion in various ecotopes did not exceed 0.5%. *Archaea* of the order *Methanomassiliicoccales* (*Thermoplasmatota*) found both in the communities of reduced (7–17%) and oxidized layers (1–9%) were the main group of microorganisms capable of producing methane from C_1-methylated substrates. *Archaea* of the phyla *Halobacteriota* (*Methanocellaceae*, *Methanomicrobiaceae*, *Methanoregulaceae*, *Methanospirillaceae*, *Methanosarcinaceae*, *Methanotrichaceae,* and *Candidatus* Methanoflorentales (now Bog-38) and *Methanobacteriota* (*Methanobacteriaceae* and *Methanofastidiosaceae*) also represented acetoclastic, hydrogenotrophic, and methylotrophic methanogens. Communities in the sediments of oil and methane seeps characterized the highest relative abundance of methanogenic archaea (up to 50%). In contrast to the high diversity of methanogens, microorganisms of the family *Methanoperedenaceae* (formerly ANME-2d) were the only participants of anaerobic oxidation of methane among archaea. They were evenly represented in all ecotopes (9–14%), except for reduced sediments of oil and methane seeps (2.5%). Symbiotic and host-independent fermentative heterotrophs of the order *Woesearchaeales* (*Nanoarchaeota*) (Castelle et al., 2018; Liu et al., 2018; Huang et al., 2021) represented another group identified both on the surface (10–20%) and in deep layers (5–10%). Among other taxa widespread in subsurface ecosystems, archaea of Marine Benthic Group D (now *Thermoprofundales*)—mixotrophic H_2 producers in global sedimentary environments—were detected in the sedimentary deposits of Lake Baikal (Zhou et al., 2019; Zhang et al., 2022). *Hadarchaeota* members capable of autotrophy using CO_2 during acetogenesis and hydrogenogenic oxidation of CO to CO_2 (Tiquia-Arashiro, 2014a, b; Baker et al., 2016; Wang et al., 2019) were numerous in the sediments of MVs (up to 6%). Microorganisms of the phyla *Aenigmarchaeota*, *Altiarchaeota*, *Micrarchaeota*, and *Lainarchaeota* comprised a minor part of the archaeal component and were represented by both free-living and symbiont forms with a small genome size and limited metabolic abilities (Youssef et al., 2015; Probst et al., 2018; Qin et al., 2018; Kadnikov et al., 2020; Li et al., 2021).

3.3 Diversity and Structure of Communities in Different Layers of Reduced Sediments

The diversity and structure of communities in different layers of reduced sediments were most evident at the level of lower-ranked taxa (orders, classes, and families). In almost all areas, the surface sedimentary layers contain oxygen, the most energetically favorable and available electron acceptor in the respiratory chain (Granina, 2008; Och et al., 2012). The presence of labile OM sources, namely vegetated diatoms that develop en masse in Lake Baikal in spring and fall (Votintsev et al., 1975; Vologina et al., 2007), facilitates the development of organotrophic microorganisms. Their active destruction leads to an increase in thermodynamically favorable electron acceptors—sulfate and nitrate ions (Granina, 2008; Pogodaeva et al., 2017; Pogodaeva & Khodzher, 2023)—in the subsurface layers of sediments (up to a depth of 50 cm). This zone demonstrated a higher relative abundance of metabolically diverse bacterial taxa (Figs. 4.4 and 4.5) that mineralize OM and are widespread in lake sediments (Newton et al., 2011). As sediments deepened and substrates became depleted, the taxonomic composition of the communities changed. Organotrophic members of the phylum *Chloroflexota* predominated in all sedimentary layers studied, with varying taxonomic status in different layers of sediment. Sequences of the KD4–96 group often found in metal-contaminated soils (Wegner & Liesack, 2017; Kassem et al., 2023) prevailed in the communities of the surface sediments from the reference site. Uncultured sequences from different clusters of this phylum (KD4-96, JG30-KF-CM66, and Gitt-GS-136) contributed significantly at sediment depths of 10–11 cm, 12–13 cm, and 24–25 cm, whereas at a depth of 30–31 cm, organotrophic homoacetogens of the class *Dehalococcoidia* predominated (Fincker et al., 2020; Vuillemin et al., 2020). Additionally, at the 12–13 cm sediment depth (Fig. 4.6a), the contribution of the *Patescibacteria* members (to 23%) and of *Acidobacteriota* (up to 14%) increased. Also, the relative abundance of ANAMMOX bacteria of the class *Brocadiia* (*Planctomycetota*) and members of the phylum *Methylomirabilota* (to 14% and 16%, respectively) increased with depth.

In the communities of reduced surface sediments of the Kedr MV, obligate anaerobes of the class *Anaerolineae* (Fig. 4.6a) comprised the largest proportion (10%) among *Chloroflexota*. A characteristic feature of the structure of the bacterial community of the surface layer of the MB sediment was a more significant contribution of anaerobic representatives MBNT15 (11%) and a less of nitrite-oxidizing bacteria of the phylum *Nitrospirota* in comparison with the oxidized layers of the SB (see Fig. 4.6a). The proportion of the members of the class *Thermodesulfovibrionia (Nitrospirota)*, which are involved in nitrate- and sulfate-reduction was high in the communities from the MV surface sediments (Umezawa et al., 2022). Organotrophic bacteria of the phylum *Atribacterota* (up to 30.5%) and *Dehalococcoidia* (up to 28.5%) predominated in the microbial community of the deep reduced layers of the Kedr MV. Unlike communities from the reference sites, the MV communities lacked ANAMMOX bacteria, *Patescibacteria* (present only in the surface sediment,

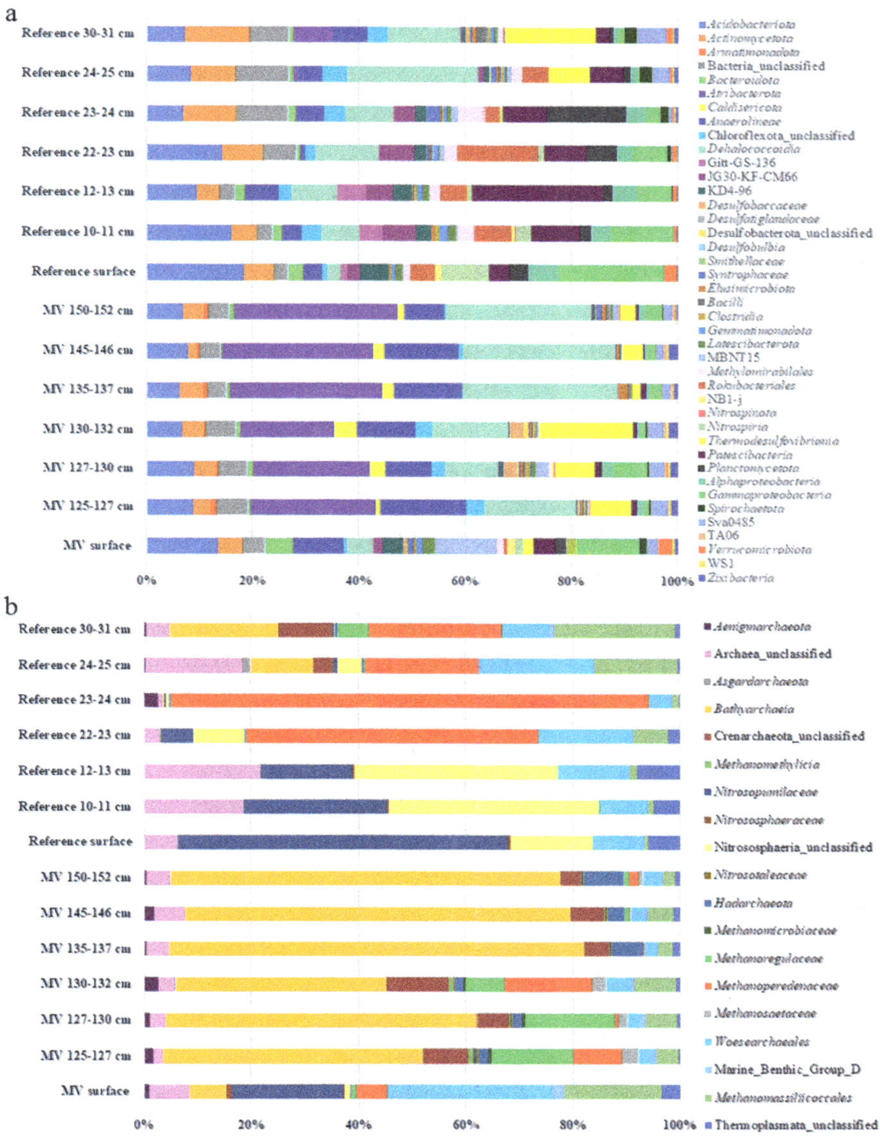

Fig. 4.6 Taxonomic composition microbial communities in different layers of the sediments of the Kedr MV and reference site in the northern basin of Lake Baikal (Lomakina et al. 2023a, b): (**a**) bacterial communities and (**b**) archaeal communities; more than 97% of reads

up to 3%), and uncultured *Chloroflexota* of different clusters (KD4-96, JG30-KF-CM66, and Gitt-GS-136).

The dominance of the members of aerobic ammonia-oxidizing *Nitrososphaeria* and their decrease with sediment depth was a characteristic pattern in the archaeal

distribution (Fig. 4.6b). All samples from the reference site showed a high contribution of the members of the order *Woesearchaeales* (up to 16%), while in MV communities, these archaea dominated (up to 30%) in the surface and were minor in deep sediments (less than 4%). Members of this order can function in the consortium with methanogens and participate in the carbon cycle of anaerobic environments, especially in soils or sediments (Liu et al., 2021). Methylotrophic methanogens (*Methanomassiliicoccales*) found at all sediment depths of these two areas (2–24%) were the most diverse among the members of the methane cycle. Hydrogenotrophic methanogens (*Methanoregulaceae*) and denitrifying anaerobic methane oxidizers (*Methanoperedenaceae*) were more representative in the communities from the MV sediments. A distinctive feature of the communities from this ecotope was the presence of various organoheterotrophic *Bathyarchaeia* often found in freshwater sediments (Vuillemin et al., 2018). Genomic data confirmed the Wood-Ljungdahl (WL) pathway in them, which was enabling them to participate in acetogenesis (He et al., 2016) or in the formation of H_2 and CO_2 from acetate (Vuillemin et al., 2018), substrates for methanogenic archaea.

In deep-lake sediments, the structure of microbial communities is based on the principle of energy-saving fermentative metabolism, with gradual exhaustion of electron acceptors in pore water (van den Berg et al., 2022) and labile OM (Glombitza et al., 2013). In the sediments of deep-water Lake Baikal, the structuring of microbial communities has specific characteristics. For example, methylotrophic methanogens and microorganisms involved at different stages of OM fermentation were identified throughout the investigated depth of the sedimentary strata. The mosaic structure of microbial communities in one coordinate indicated the variability of environmental conditions at different depths of the sedimentary strata and the wide plasticity of taxa belonging to one phylum, allowing them to colonize different layers of the sedimentary strata with limited energy sources.

4 Influence of Flow Migration on the Diversity of Microbial Communities

Active fault zones in marine and oceanic ecosystems demonstrate local heat flux anomalies. Moreover, along the active fault segments, researchers recorded upward flows of geothermal mineralized fluids containing high concentrations of sulfide, methane, and hydrogen (Suess, 2014; Joye, 2020; Teske & Carvalho, 2020). At the cold seep sites, the escaping fluid or gas is most often found to have temperatures close to that of the surrounding seawater (Joye, 2020). Fluid discharge rates at seeps are lower than fluid discharge rates at vents (Suess, 2014). As a rule, specific biological communities inhabit fluid discharge sites, the vital activity of which is based on the oxidation energy of the reduced sulfur and carbon compounds (Suess, 2014). The fauna of various reducing biotopes at great depths in seas and oceans is basically similar; the diversity is rather higher in hydrothermal springs (Gebruk, 2002).

Hydrothermal manifestations have long been known for freshwater bodies, in particular for Lake Tanganyika along the East African Rift (Tiercelin et al., 1993; Tran et al., 2021) and geothermal lakes in North America (Yellowstone Lake and Crater Lake, USA) (Klump et al., 1988; Yang et al., 2011). Cold seeps with elevated methane fluxes are more typical of Arctic thermokarst lakes (He et al., 2015; Bogoyavlensky, 2023). The formation of chemosynthesis-based fauna characteristic of deep thermal and gas seeps in seas and oceans was not observed in lakes. Nevertheless, in lacustrine ecosystems, similar to marine ecosystems, the chemical composition of discharged thermal waters and cold fluids influences the structure and diversity of microbial communities, promoting the formation of microbial mats (Yang et al., 2011; Tran et al., 2021). In the Crater Lake, where ferrous thermal water emerges, Dymond et al. (1989) detected bacterial fouling formed by bacteria of the genera *Gallionella* and *Leptothrix*, which derive energy from the oxidation of ferrous iron, white fouling consisting of colorless sulfur bacteria of the genera *Thiothrix* and *Beggiatoa* was found at the outlets of thermal hydrogen sulfide vents in Yellowstone Lake and Tanganyika, respectively (Konkol et al., 2010; Tiercelin et al., 1993). In areas with hydrothermal vents, temperature determines the structure of microbial communities, with high temperatures leading to archaeal dominance (Pagé et al., 2008). At the same time, the presence of electron donors and acceptors influences the diversity at the phylum level and below (Lagostina et al., 2021).

Modern tectonic activity of the Baikal Rift basin leads to extension and active subsidence of all three water-saturated sedimentary basins (Logatchev, 1993; Zonenshain, 1992). In this case, active fluid flows in the basins lead to local heat flux anomalies (Golubev et al., 1993; Poort & Klerkx, 2004) and the release of mineralized gas-saturated fluids to the bottom surface of the lake (Granin & Granina, 2002; De Batist et al., 2002; Rensbergen et al., 2002, 2003; Pogodaeva et al., 2007; Kalmychkov et al., 2006, 2017). Numerous seismic data and field measurements of the chemical composition of mineralized gas-bearing fluids confirm this, indicating the migration of methane with different genesis from the gas hydrate stability zone (GHSZ) to the bottom surface of the lake along large fault zones (De Batist et al., 2002; Rensbergen et al., 2002; Klerkx et al., 2006). Fluid flow migration is one of the mechanisms that affects the composition and structure of microbial communities near MVs and methane seeps (Aloisi et al., 2019). Studies near the Malenky MV (SB) revealed ancient diatoms of the genus *Tertiarius* in surface sediments, which developed in the ecosystem of Lake Baikal during the Pliocene (Klerkx et al., 2003). In CB, the oxygenated surface sediments and bottom water were supplied to more than the one-meter depth (Pogodaeva et al., 2020). In the sediments of the seep on the Posolsk Bank elevation, the deep microflora were supplied to the sediment surface through special migration pathways of fluid flows circulating in the GH stability zone (400 m) (Naudts et al., 2012; Chernitsyna et al., 2016a). This explains the presence of a significant relative abundance of sequences of *Aminicenantales* and *Atribacterota*, as well as methanotrophs of the genus *Methylobacter*, in reduced sediments at a depth of ~100 cm (Chernitsyna et al., 2016a).

In zones with particularly intense flows of mineralized gas-bearing fluids, there were two types of microbial mats on the surface of sediments: with dominance of

colorless sulfur bacteria and a significant contribution of the methane cycle microorganisms in Frolikha Bay (Kuznetsov et al., 1991; Namsaraev et al., 1994; Zemskaya et al., 2001) and with dominance of methanotrophic/methanogenic prokaryotes near the St. Petersburg MV (CB) above massive GH layers (Zemskaya et al., 2015b). Methanotrophy-based functioning of microbial communities is a known fact. Such communities are typical of forest soils, meadows, rice fields, wetlands, and sediments in fresh, saline, or alkaline lakes (Ding & Valentine, 2008; Wang et al., 2022; Zhang et al., 2023).

Underwater low-temperature hydrothermal spring in the Frolihka Bay located at the intersection of tectonic faults had the lake's maximum heat flux from the bottom (37,000 mW/m^2), elevated temperature (4.8–6.2°C), and electrical conductivity typical of bottom waters (Golubev et al., 1993; Golubev, 2007). pH values varied from 6.7 to 8.8 and Eh from −80 to +70 mV, which characterizes the conditions in the surface sedimentary layer of this area as microaerobic and anaerobic (Namsaraev et al., 1994). The maximum mineralization of pore waters in the vent sediments reached 1134 mg/L, which is higher than the maximum values recorded in the thermal springs of the Baikal region (up to 700 mg/L) (Lomonosov, 1974). Pore waters in this zone were enriched in HCO_3^- (up to 0.2 mM), chloride (up to 0.87 mM), Na^+ (up to 11 mM), Ca^{2+} (up to 2.3 mM), Mg^{2+} (up to 0.83 mM), and K^+ (up to 2.3 mM) (Namsaraev et al., 2002; Granina, 2008). High SO_4^{2-} concentrations (up to 3 mM) resulted from the sulfide oxidation and were recorded only in mineralized fluid seepage areas (Granina, 2008). In terms of the isotopic composition, pore waters in deep sedimentary layers were close to the water from a terrestrial thermal spring (Shanks & Callender, 1992). The influx of mineralized fluids led to the enrichment of pore waters in the vent zone with such trace elements as Sr, Sc, Zn, Mo, Cr, W, Be, As, Nb, and Ba (Granina et al., 1999).

During different years at the end of the twentieth century and the beginning of the twenty-first century, researchers used autonomous manned submarines and standard samplers to carry out comprehensive studies in this area (Crane et al., 1991; Kuznetsov et al., 1991; Namsaraev et al., 1994; Zemskaya et al., 2001; Chernitsyna et al., 2016b). The isotopic composition of carbon in sediments, benthic animals, and microbial mats indicated an important role of methane carbon in the functioning of this biotope (Namsaraev et al., 1994; Zemskaya et al., 2012). Fields of microbial mats alternating with sponge fields had a mosaic distribution on the sediment surface, covering a considerably large bottom area (Fig. 4.7a). Colorless sulfur bacteria of the genus *Thioploca* dominated microbial mats (Crane et al., 1991; Kuznetsov et al., 1991; Namsaraev et al., 1994). Like in marine areas with methane seeps, chemosynthesis and methanotrophy ensured their functioning in freshwater Lake Baikal (Namsaraev et al., 1994; Namsaraev & Zemskaya, 2000; Ding & Valentine, 2008). Similar to the marine members of this genus, sulfur bacteria inhabiting Lake Baikal are chemoorganoheterotrophs oxidizing sulfides via nitrate reduction (Namsaraev et al., 1994; Zemskaya et al., 2001). *Thioploca* from Lake Baikal can accumulate nitrates inside cells, receiving them from the bottom area and surface sediments (Zemskaya et al., 2001) and from various bacteria and archaea involved in the nitrogen cycle, which inhabit their sheaths (Chernitsyna

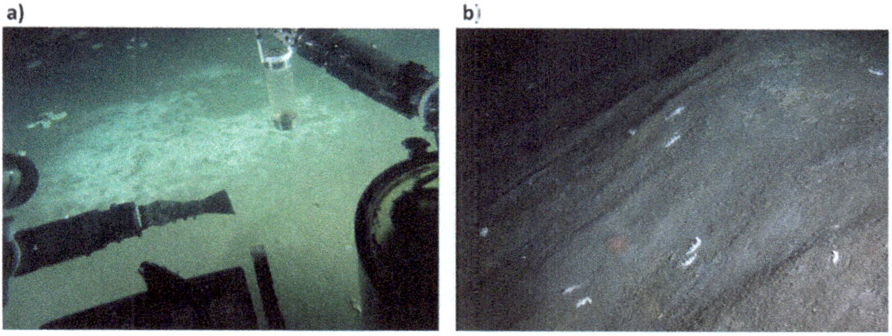

Fig. 4.7 (**a**) Bacterial mats near the underwater vent in Frolikha Bay; (**b**) microbial mats near the St. Petersburg MV. The photos were taken by the manned deep-sea submersibles of the MIR

et al., 2016b). This allows colorless sulfur bacteria to colonize a very wide zone of sediments (down to the 19 cm depth) in Lake Baikal and survive with a deficiency of nitrogen and sulfide sources in the surface sedimentary layers throughout practically the entire lake, in areas with high OM concentrations (Namsaraev et al., 1994; Zemskaya et al., 2001, 2009).

Freshwater bacteria of the genus *Thioploca*, including those from Lake Baikal, have smaller individual filaments and a common sheath compared to marine forms of this candidate genus *Marithioploca* (Salman et al., 2013). Based on morphological sizes, two species were distinguished among freshwater bacteria: *T. schmidlei* and *T. ingrica*. The former was described for Lake Constance (Southern Germany) and Lake Baikal (Lauterborn, 1907; Zemskaya et al., 2009). Based on the 16S rRNA gene structure, the bacteria from both lakes belonged to the species *T. ingrica* with 99% identity. The species T. *ingrica* (Wislouch, 1912) was also described for lakes Biwa (Japan), Constance (Germany), and Tonle Sap (Cambodia). Although members of the species *T. ingrica* were geographically distant from each other, they did not differ in phenotypic and genetic characteristics (Kojima et al., 2003; Nishino et al., 1998; Zemskaya et al., 2009). In the Great Lakes (USA), the *Thioploca* population also exhibited a wide range of morphological characters (McKay et al., 2023). An analysis of the microbial community associated with *Thioploca* from Lake Baikal revealed the members of the phyla *Acidobacteriota*, *Pseudomonadota*, *Bacteroidota*, *Actinomycetota*, *Nitrospirota*, *Aminicenantales*, *Verrucomicrobiota*, *Chloroflexota*, and *Methylomirabilota*, as well as archaea such as *Thermoproteota*, *Nitrososphaeria*, and *Methanobacteriota* (Chernitsyna et al., 2016b). Notably, the 16S rRNA gene structure of the obtained *Chloroflexota* sequences was close to those from the microbial community of the *Thioploca* bacteria from lakes Ogawara (Japan) and Constance (Germany) (Kojima et al., 2006). The *Thioploca*-associated microbial community from lakes Biwa (Japan) and Constance (Germany) also contained bacteria of the phylum *Bacteroidota* (Kojima et al., 2006). Members of the phylum *Actinomycetota* (~50%), *Pseudomonadota*, *Bacteroidota*, and *Cyanobacteriota* (13 to 17%) dominated the microbial communities of the

sediments with *Thioploca* filaments (Zemskaya et al., 2015b), with a minor contribution of the members of the phyla *Acidobacteriota, Chloroflexota, Nitrospirota,* and *Verrucomicrobiota*. Archaeal communities from the underwater vent were less diverse compared to other Lake Baikal sediments. Ammonia-oxidizing archaea—*Nitrosopumilaceae* (phylum *Thermoproteota*)—capable of chemoautotrophic growth in the presence of ammonia comprised the largest proportion (Lomakina et al., 2018b). Additionally, there were sequences of acetoclastic methanogens of the order *Methanosarcinales* (*Halobacteriota*) and members of the denitrifying anaerobic methane oxidation *Methanoperedenaceae* (28%).

Deep-sea manned submersibles "MIR" recorded another type of microbial mats above massive layers of methane hydrates in the bathypelagic zone of CB (Fig. 4.7b). Intense methane flows formed both due to methane hydrate dissolution and due to the activity of hydrogenotrophic methanogens ensured the mass development of methanotrophic and methylotrophic bacteria (Kadnikov et al., 2012; Zemskaya et al., 2015b), as well as primary organoheterotrophs forming the three-centimeter-thick layer of microbial mats, in the near-bottom aerobic layer. These mats ensure the vital activity of a wide range of benthic animals (Zemskaya et al., 2012; Sitnikova et al., 2017), which is confirmed by measurements of carbon isotope composition in the tissues of various benthic animals inhabiting this zone. Microscopy of microbial mats revealed morphologically diverse bacteria, diatoms, and picocyanobacteria in them. The phyla *Pseudomonadota, Verrucomicrobiota, Bacteroidota,* and *Planctomycetota* represented bacteria. There were also archaea of the phylum *Halobacteriota* of the genus *Methanothrix* and of the order *Methanoperedenaceae*, and the phylum *Thermoplasmatota* (order *Thermoplasmatales*). Members of the genera *Methylobacter, Methylophilus,* and *Methylotenera*, inhabiting various cold-water ecosystems (Dedysh et al., 2001; Pacheco-Oliver et al., 2002; Chen et al., 2007) and possessing unique enzymes providing various steps of methane oxidation via formaldehyde to CO_2 or biomass accumulation, were also abundant in the microbial mat community. Methane produced by methanogenic archaea directly under the layer of the microbial mat, as well as during the dissolution of GH layers, was the substrate for methanotrophic bacteria. Evidence for this was the high relative abundance of hydrogenotrophic methanogens of the genus *Methanosphaerula* and acetate-consuming methanogens (*Methanosarcinales*) in the surface sediment layers, with archaea making up about two-thirds of the microbial community (Kadnikov et al., 2012).

5 Diversity of Microorganisms Involved in Methane Generation

The sediments of Lake Baikal are a comfortable niche for methanogenic microorganisms. The low concentrations of electron acceptors (as SO_4^{2-} and NO_3^-), the depth of O_2 penetration, and the oxidation rate largely determine the upper limit of

the methane generation zone in Lake Baikal sediments (Namsaraev & Zemskaya, 2000; Dagurova et al., 2004; Pimenov et al., 2014). The activity of methanogenic archaea reliably extends hundreds of meters into the sedimentary layers, which is confirmed by the biogenic origin of methane sampled from deep-drilling cores (Kuzmin et al., 2001).

Sediments from reference sites in the pelagic zone of the lake showed an almost linear increase in methane concentrations with depth, which correlates with an increase in the concentration of bicarbonate ions released during diagenesis (Zemskaya et al., 2008; Pogodaeva et al., 2017). At the same time, methane generation rates were relatively low, amounting to 0.01–32.29μL CH_4/kg per day, which is comparable with the intensity of methane generation in sediments of highly productive areas of the World Ocean and Lake Michigan (Capone & Kiene, 1988). In hydrocarbon discharge zones, sediments of low-temperature vent, and sediments of coastal areas with elevated OM concentrations, methane generation rates increased to 5.0–534.7μL CH_4/kg per day (Namsaraev et al., 1995; Namsaraev & Zemskaya, 2000; Dagurova et al., 2004). This corresponds to the methane generation rates in deep lakes such as Lake Stechlin and Lake Geneva (Fuchs et al., 2016). The vertical distribution pattern of methane generation in these areas varied, but high methane generation rates were often recorded in the surface oozes to depths of 20–30 cm, followed by bursts of activity in some deep layers (Dagurova et al., 2004; Zemskaya et al., 2010). The study of the methanogenesis intensity using different isotopically labeled substrates determined that, in sediments of the deep zone, 75 to 100% of methane was generated from H_2/CO_2, whereas acetoclastic methanogenesis was dominant only in some layers of the sedimentary strata at shallow sites (Namsaraev et al., 1995; Namsaraev & Zemskaya, 2000; Zemskaya et al., 2010). Experiments on the cultivation of sediment methanogenic communities demonstrated a similar pattern (Bukin et al., 2018).

The first data on the taxonomic composition of methanogenic archaea in the sediments of Lake Baikal were obtained using immunofluorescent staining, which indicated microorganisms of the genera *Methanobacterium* and *Methanosarcina* in the communities (Namsaraev et al., 1995; Namsaraev & Zemskaya, 2000). Subsequently, molecular cloning techniques discovered markers of other main participants in the methane generation, including the members of the genera *Methanothrix* and *Methanoregula*, as well as archaea of the family *Methanomassiliicoccaceae* that did not have cultured homologues at that time (Chernitsina et al., 2007; Zemskaya et al., 2010; Lomakina et al., 2014). Thus, at this stage, researchers identified microorganisms capable of generating methane in Lake Baikal from three main types of substrates: H_2/CO_2, acetate, and C_1-methylated compounds. Moreover, cloning techniques detected methanogenic archaea in jelly-like microbial mats covering the surface of massive GH layers. This expanded our understanding of the niches occupied by them (Zemskaya et al., 2015b).

The development of methods for high-throughput sequencing of the 16S rRNA gene fragments provided the assessment of variability in the composition and structure of methanogenic communities with a higher resolution. The results of previous metabarcoding studies confirmed that the proportion of methanogens in oxidized

layers of surface oozes at the reference sites and in some hydrocarbon discharge zones did not exceed 1%, and ammonia-oxidizing microorganisms of the class *Nitrososphaeria* mainly represented archaea (Lomakina et al., 2018a; Zakharova et al., 2018; Zemskaya et al., 2018). At the same time, in oozes of low-temperature vent, reduced layers of sediments from methane seeps, and the St. Petersburg MV, as well as in bituminous rocks on the lake's floor, methanogens comprised 20–100% of all archaea detected in individual samples (Kadnikov et al., 2012, 2013; Lomakina et al., 2018a). Furthermore, their taxonomic composition was diverse and included, in various proportions, members of the orders *Methanomicrobiales, Methanosarcinales, Methanomassiliicoccales, Methanobacteriales*, and Ca. Methanoflorentales. However, in reduced sediments of other hydrocarbon discharge areas (the Posolsk Bank methane seep and the Malenky MV), only acetoclastic archaea of the genus *Methanothrix* (order *Methanosarcinales*) represented methanogenic microorganisms, with their proportion not exceeding 10%, or they were completely absent (Chernitsyna et al., 2016a; Lomakina et al., 2018a). These differences may be due to the influence of deep fluids with different chemical compositions, whose active discharges were recorded in the studied areas at that time (Klerkx et al., 2003; Zemskaya et al., 2020). This may be also due to the different primers, target regions of the 16 s rRNA gene, and molecular biology approaches used in different studies.

The standardized methods used in metabarcoding studies that have been conducted in the past decade systematically assessed the diversity of methanogenic archaeal communities and determined the features of their vertical distribution and the relationship with the chemical composition of pore waters. An analysis of microbial communities from 75 samples of Lake Baikal sediments collected both at the reference sites and in discharge zones of oil- and gas-bearing fluids revealed a high homogeneity in the composition of the methanogenic component. The total proportion of the known taxa of methanogenic archaea in the communities varied from 0.02% to 95.44% of all detected *Archaea*, averaging 21.3% (Fig. 4.8).

All detected methanogens were assigned to eight orders: *Methanomethylicales, Methanobacteriales, Methanofastidiosales, Methanocellales, Methanomassiliicoccales*, and Ca. Methanoflorentales represented by type families, and *Methanomicrobiales* (families *Methanomicrobiaceae, Methanoregulaceae*, and *Methanospirillaceae*) and *Methanosarcinales* (families *Methanotrichaceae* and *Methanosarcinaceae*). However, most of the detected sequences belonged to archaea of the order *Methanomassiliicoccales* and families *Methanoregulaceae* and *Methanotrichaceae* that can carry out three types of methanogenesis: hydrogenotrophic, acetoclastic, and methyl-dependent hydrogenotrophic. Their relative abundance averaged 8.9%, 8.5%, and 1.5%, respectively (Fig. 4.8). Microorganisms of these taxa form the basis of methanogenic communities in freshwater ecosystems worldwide (Borrel et al., 2011; Wen et al., 2017), which is associated with their high competitiveness to live under low concentrations of substrates and with the ability to create stable syntrophic associations during the degradation of volatile fatty acids (Borrel et al., 2011). Despite the wide distribution of the members of the three main taxa, all communities, including those from the surface oxidized layers, contained

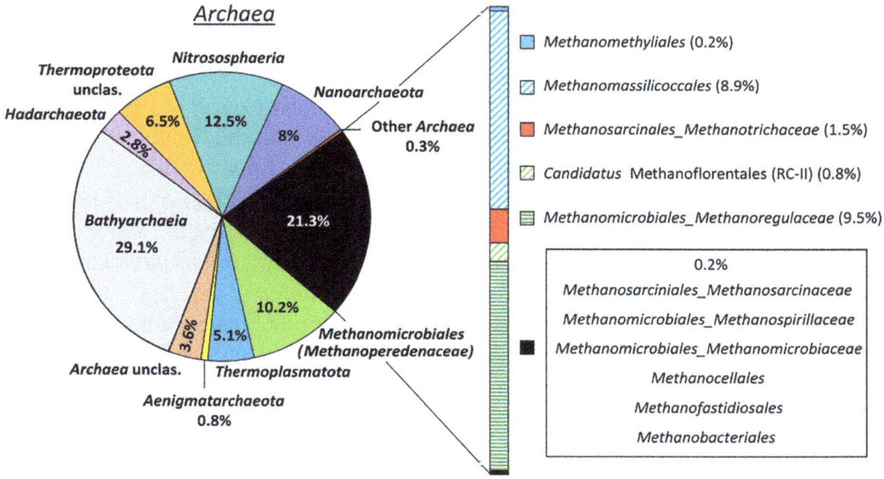

Fig. 4.8 The average relative abundance and taxonomic composition of methanogenic archaea in microbial communities from 75 samples of Lake Baikal sediments

only archaea of the order *Methanomassiliicoccales* consuming C_1-methylated compounds. Perhaps, the use of non-competitive methylated substrates allows *Methanomassiliicoccales* from Lake Baikal to occupy anaerobic microniches more actively in oxidized oozes where acetoclastic and hydrogenotrophic methanogens experience competitive pressure from sulfate-reducing and nitrate-reducing bacteria. In sediments of freshwater ecosystems, methanol formed during the destruction of pectin and lignocellulose, of which cell walls of algae and plants consist, can be the main substrate for this group of methanogens (Schink & Zeikus, 1980; Sollinger & Urich, 2019). Nevertheless, the metagenome-assembled genome (MAG) of *Methanomassiliicoccales* from Lake Baikal contained a set of methyltransferases, allowing for the use of both methanol and a full set of methylamines. This also expands their abilities to occupy various ecological niches.

A detailed analysis of microbial communities in sediments from different geological structures indicated that the proportion of methanogenic microorganisms increased from the reference sites to sediments of oil and methane seeps. In the communities from oxidized sediments of the reference sites, the relative abundance of methanogenic archaea averaged slightly over 1% of all detected *Archaea* (Fig. 4.4). At the same time, microorganisms of the order *Methanomassiliicoccales* almost entirely represented the methanogenic component. However, the relative abundance of methanogenic groups increased to 37% already in the subsurface reduced layers, involving the members of hydrogenotrophic and acetoclastic archaea of the families *Methanoregulaceae* and *Methanotrichaceae* (Figs. 4.4 and 4.5).

The proportion of methanogenic archaea varied from 12 to 39% in the sediments of MVs. Like in the reference sedimentary strata, members of *Methanomassiliicoccale*s predominated in the surface layers. With depth, the contribution of methanogens using H_2/CO_2 and acetate increased. The small

percentage of methylotrophic microorganisms of the order *Methanomethylicales* rarely found in other areas was a distinctive feature of the communities from the sediments of MVs (Figs. 4.4 and 4.5).

The sediments of oil seeps (GU and BZ) showed the highest relative abundance of methanogenic archaea (Fig. 4.4). Here, their proportion exceeded 20% even in the surface layers. Moreover, the community composition was similar to that from the reference sites and MVs, except for the community associated with oil-saturated GHs where the proportion of methanogenic archaea reached 95.6%, and the contribution of hydrogenotrophic members of the order *Ca.* Methanoflorentales (Rice Cluster II) (60.3%) distinguished this community. Previous studies also indicated the high relative abundance of archaea of the order *Ca.* Methanoflorentales in the sediments from GU containing GH interlayers (Lomakina et al., 2018b). Methanogenic archaea of this group were widespread in various ecotopes but often dominated the communities of permafrost Arctic soils subjected to short-term thawing (Mondav et al., 2014). We can assume that their confinement to Lake Baikal GHs characterized by periodical decomposition and crystallization processes is also associated with their adaptation to these conditions and the rapid restoration of activity and biomass in layers that were frozen until recently.

Results of assessing the composition and diversity of methanogenic communities, based on Bray–Curtis distance matrix followed by UPGMA clustering (unweighted pair group method with arithmetic mean) and plotting an NMDS graph, allowed us to reliably divide them into four groups (Fig. 4.9). The communities of the oxidized layers from the reference sites of NB formed the first group, which, on average, exhibited higher concentrations of sulfate ions and nitrate ions and an extremely low proportion of methanogenic archaea. The second group included the communities of the surface layers from the Malenky and K-9 MVs, as well as from the reference site of SB, and the communities of the reduced layers from the reference sites of NB. Microorganisms of the order *Methanomassiliicoccales* predominantly represented methanogenic archaea in these samples, though their relative abundance did not exceed 2%. The third group combined the bulk of the remaining samples. Among them, a subcluster of communities from MVs was distinguished by a high contribution of members of the order *Methanomethylicales* producing methane from C_1-methylted compounds. The fourth group included the methanogenic community associated with the oil-saturated GHs, which had a unique composition.

An analysis of the dependence of the composition and diversity of methanogenic communities on the type of the geological structure, the sediments of which they inhabit, revealed significant differences between reference sites and zones where hydrocarbon-containing fluids were discharged (ANOSIM $R = 0.290$, $p_{\text{value}} < 0.001$; PERMANOVA $R = 0.184$, $F = 8.117$, $p_{\text{value}} < 0.001$). A linear discriminant analysis (LDA) of effect size (LDA threshold value >2, false discovery rate adjusted $p_{\text{value}} < 0.05$) indicated that sediments from the reference sites had a high relative abundance of CO_2-reducing methanogenic archaea of the genera *Methanocella* and *Methanosphaera*. At the same time, methylotrophic microorganisms of the order *Methanomethylicales* were often detected in the MV sediments. Methanogens of

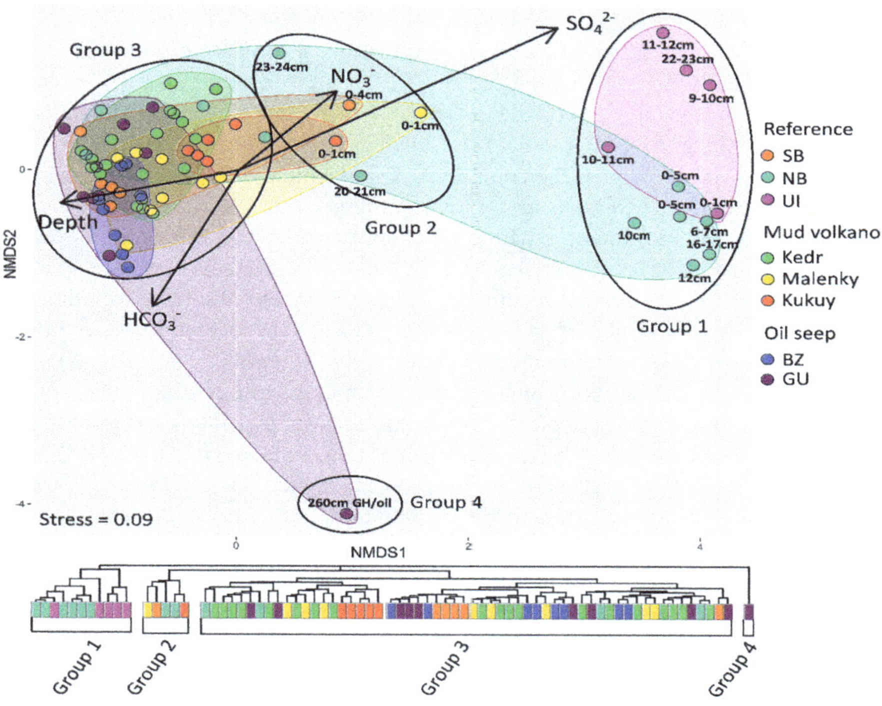

Fig. 4.9 NMDS plot and UPGMA clustering based on the Bray–Curtis distance matrix, reflecting the differences in the composition and abundance of methanogenic archaea in sediment communities from oil seeps, mud volcanoes and reference sites ofLake Baikal (OTU level with a genetic distance of 0.03)

the genera *Methanoregula, Methanothrix, Methanobacterium, Methanolinea, Methanosarcina*, and *Methanospirillum* had a large proportion in the communities of sediments from oil and methane seeps enriched in OM.

An analysis of the relationship between the relative abundance of individual methanogen groups and physicochemical parameters of the environment (Spearman correlation, $p_{\text{value}} < 0.05$) revealed that an increase in the proportion of all taxa has a positive correlation with the concentration of bicarbonate ion, a direct product of OM decomposition. Furthermore, a significant increase in the proportion of methanogenic archaea in the sediments from oil and methane seeps indicated that not only the destruction of buried organic remains but also fluid-brought hydrocarbons can be a source of methanogenesis substrates. Additionally, members of one of the most widespread families (*Methanoregulaceae* and *Methanotrichaceae*) demonstrated a negative relationship with the concentrations of sulfates and nitrates in pore waters. This may evidence the competition with sulfate- and nitrate-reducing microorganisms for common substrates, especially in the surface oxidized layers with the maximum concentrations of these ions.

6 Microorganisms Involved in AOM in Lake Baikal Sediments

The profiles of methane concentrations indicating its consumption evidenced the AOM processes in the sediments of Lake Baikal (Zemskaya et al., 2008, 2010). In sediment layers where methane concentrations decreased with high concentrations of sulfate ions (from 0.3 to 15 mM with a background value of 0.05 mM), we did not detect members of the ANME clusters typical of marine sediments (Zemskaya et al., 2008, 2010; Lomakina et al., 2014; 2018b; Chernitsyna et al., 2016a). Lake Baikal sequences from sediments with high concentrations of sulfate ions occupied an intermediate position between the AOM participants in the phylogenetic tree (Shubenkova et al., 2005; Zemskaya et al., 2010). Therefore, we suggested that not sulfate ions but nitrate and/or nitrite ions or Fe(III) ions function as electron acceptors in AOM in Lake Baikal sediments enriched in methane (Zemskaya et al., 2008, 2010; Krylov et al., 2010; Pimenov et al., 2014). Researchers previously described archaea of the family *Methanoperedenaceae* (Raghoebarsing et al., 2006) and bacteria of the phylum *Methylomirabilota* (Ettwig et al., 2010) using these oxidizers during AOM in freshwater ecosystems. Laboratory experiments confirmed the presence of these microorganisms in the microbial communities from the MV sediments of SB. On the 16th day of cultivating the microbial community under anaerobic conditions, methane concentration in the gaseous phase decreased, and consortia of microorganisms formed. Among them, fluorescent in situ hybridization (FISH) detected bacteria of the phylum *Methylomirabilota* and archaea of the family *Methanoperedenaceae* (Lomakina et al., 2018a). High-throughput sequencing of the 16S rRNA gene fragments confirmed that the *Methanoperedenaceae* and *Methylomirabilota* members were present in microbial communities from different sedimentary strata of freshwater Lake Baikal (Lomakina et al., 2018b, 2020). Sequences of the phylum *Methylomirabilota* and archaea of the family *Methanoperedenaceae* (Lomakina et al., 2020) clustered in the phylogenetic tree with nucleotide sequences from sediments of freshwater lakes Chaohu and Taihu (PRC), Constance (Alps), Biwa (Japan), mesophilic Lake Ørn (Denmark), the Coeur d'Alene River (USA), and various bioreactors where AOM processes were recorded (Rastogi et al., 2009; Deutzmann & Schink, 2011; Kojima et al., 2012; Ding et al., 2014; Weber et al., 2017). The detection in the total DNA of marker anaerobic methanotroph (*mcrA*) genes that ensure this process via reverse methanogenesis, and the *pmoA* gene that catalyzes methane oxidation similar to the aerobic pathway proved the ability of Lake Baikal archaea to carry out AOM (Lomakina et al., 2020). Subsequently (Lomakina et al., 2023a), we have confirmed the presence of the *mcrA* gene in the *Methanoperedenaceae* MAGs. Notably, despite the competition of these taxa for methane, in sediment communities from different areas, we detected bacteria and archaea that carry out AOM through nitrate- and nitrite-dependent pathways (Lomakina et al., 2018b, 2020, 2023a, b). As shown by Welte et al. (2016), the *Methanoperedenaceae* archaea can ensure their coexistence in the same biotope, supplying nitrite to the *Methylomirabilota* bacteria. It should

also be noted that the ability to carry out AOM due to nitrate and nitrite ions is consistent with a wide range of nitrogen-cycle microorganisms inhabiting the ecosystem of Lake Baikal (Cabello-Yeves et al., 2020). For instance, members of the class *Nitrososphaeria* were one of the dominant archaeal taxa in the near-bottom area of the water column and in the surface sediments of the reference sites (ca. 20% and up to 86% of 16S rRNA reads, respectively), and in the sediments of MVs and the PB, methane seep slightly exceeded 22% and 40% of 16S rRNA reads, respectively. Moreover, nitrite-oxidizing bacteria of the phylum *Nitrospirota* had a large percentage in the communities from the surface and deep sediments (to 33% near MVs, to 9% at the reference sites, and less than 2% near the PB methane seep), as well as anaerobic chemolithotrophs that carry out anaerobic oxidation of ammonia—ANAMMOX bacteria (Lomakina et al., 2014, 2018b; Chernitsyna et al., 2016a). The most common genes in their genomes are associated with the nitrogen cycle (Cabello-Yeves et al., 2020; Bukin et al., 2023; Lomakina et al., 2023a), which confirms a significant role of nitrogen compounds in the functioning of the Lake Baikal ecosystem. Noteworthy is the simultaneous presence of the *Methanoperedenaceae* members and bacteria of the phylum *Desulfobacterota* in the communities. They provide AOM in marine sedimentary strata via sulfate-dependent pathways. These microorganisms were present in the sediments of meromictic Lake Cadagno in Switzerland (Su et al., 2019). This allowed the authors to substantiate the participants of the *Methanoperedenaceae* members in AOM via sulfate-dependent pathway. Despite the high concentrations of sulfate ions in areas where mineralized gas-bearing fluids are discharged, microorganisms related to the marine participants in this process were not found in Lake Baikal sediments. Most likely, the absence of sulfate-dependent AOM was due to the different intensity of migration flows and the variable component composition of discharged fluids that did not ensure the development of microorganisms involved in this process (Pogodaeva et al., 2007, 2020). It should be noted that the *Methylomirabilota* MAGs did not contain the key methane oxidation gene (*pmoA*), which was most likely due to the incompleteness of their assembly (60 to 69%). However, nitrite reductase (NirS), which ensures the nitrite reduction during AOM, was present in all assembled genomes of *Methylomirabilota* from Lake Baikal (Lomakina et al., 2023a).

In the communities containing *Methanoperedenaceae* and *Methylomirabilota*, members of the class *Bathyarchaeia* (up to 80% in areas with MVs and up to 50% at the reference sites) and the *Methanomassiliicoccales* and *Methanoregulaceae* methanogens were also present. Members of this class predominated in many natural ecosystems, including sediments, and are among the most abundant on the planet (Lloyd et al., 2013). Genomic data predicted their ability to decompose different organic compounds such as detrital proteins, carbohydrates of plant origin, lignin, and aromatic compounds. These archaea possess genomic markers for acetogenesis and methane metabolism, the most fundamental and ancient microbial biochemical energy-saving processes (Loh et al., 2021). Recently, Hou et al. (2023) reclassified the 25 *Bathyarchaeia* subgroups into eight orders based on functional and ecological characteristics. Members of the class *Bathyarchaeia* were considered likely

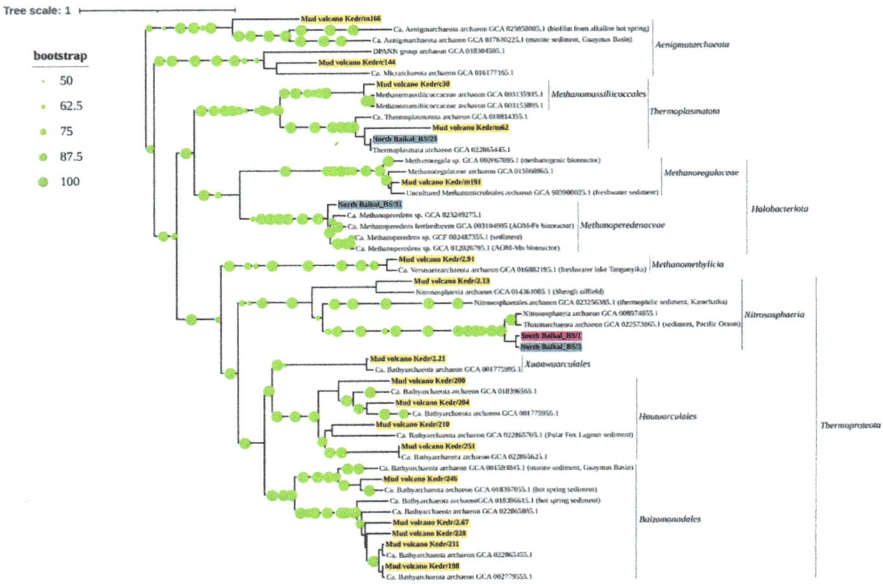

Fig. 4.10 Phylogenomic tree of archaeal MAGs from sediments of the Kedr MV (highlighted in yellow) and reconstructed genome of two reference sites in the northern (highlighted in blue) and southern (highlighted in purple) basins of Lake Baikal. Analysis was performed by Interactive Tree Of Life (iTOL). Taxonomy is shown according to the GTDB

participants in the methane cycle because the *mcrA* genes had been previously detected in some genomes (Evans et al., 2015). Nucleotide sequences of the 16S rRNA gene of *Bathyarchaeia* did not belong to the cluster with the known BA1 (subgroup-3, now *Baizomonadales*) and BA2 (subgroup-8, *Baizomonadales*) groups (Lomakina et al., 2023b), which indicated the presence of methyl coenzyme M reductase. However, some assembled *Bathyarchaeia* genomes contained *mcrA* genes. Phylogenomic analysis of the assembled genomes of *Bathyarchaeia* from the sediments of the MV area revealed (Fig. 4.10) their relationship with members of four orders: *Xuanwuarculales* (from groundwater sediments, formerly subgroup 17), *Baizomonadales* (the group with a different distribution but mostly in cold marine sediments), *Houtuarculales* (soil group, formerly 18), and *Wuzhiqibiales* (found in marine ecosystems and hydrothermal vent, formerly subgroup 15) (Hou et al., 2023). Some of the *Bathyarchaeia* MAGs formed separate lineages within the clusters of the known members of this class, suggesting metabolic differences in them. Along with the *mcrA* gene, there were marker genes such as *pta*, *ack*, and *ACSS* that ensure the ability of the *Bathyarchaeia* MAGs to acetogenesis. These members were also predicted to have the ability to break down aromatic hydrocarbons with the participation of the *bcrA* gene encoding the enzyme of the benzoyl-CoA reductase complex.

7 Microorganisms of the Fe and Mn Cycles

Fe and Mn are widespread in freshwater bodies, and, at low concentrations of sulfate and nitrate ions, they can be electron sources for AOM (Kallistova et al., 2017; Savvichev et al., 2017). This process, coupled with the reduction of iron, was observed in the sediments of lakes Kinneret, Israel (Sivan et al., 2011), Danish, Denmark (Norði et al., 2013), Ørn, Denmark (Weber et al., 2017), and Cadango, Switzerland (Su et al., 2019). The formation of ferromanganese layers is typical of most of the sediments in different basins of Lake Baikal (Granina, 2008; Baturin et al., 2009; Och et al., 2012). Like sediments in other freshwater lakes (Edgington & Callender, 1970; Hauck, 2001; Sommers et al., 2002; Dittrich et al., 2015), Lake Baikal sediments have obvious Fe and Mn accumulation zones in the form of interlayers (Fig. 4.1a) or crusts (Granina, 2008; Baturin et al., 2009; Och et al., 2012; Torres et al., 2014; Strakhovenko et al., 2020). Studies by Dubinina (1976), Granina (2008), and Zakharova et al. (2010) confirmed the hypothesis of diagenetic formation of Fe-Mn layers resulted from biogenic activity in freshwater sediments of Lake Baikal. They indicated the presence of the cultured Fe- and Mn-oxidizing bacteria in Lake Baikal sediments, the dependence of distribution and abundance on the redox environmental conditions, the OM concentration, and the presence of dissolved forms of Fe and Mn in pore waters. Members of six genera (*Metallogenium*, *Leptothrix*, *Siderocapsa*, *Naumanniella*, *Bacillus*, and *Pseudomonas*) participated in the oxidation of Fe (II) and Mn(II) in the sediments of Lake Baikal. In Lake Baikal sediments completely lacking dissolved SO_4^{2-} and having insignificant NO_3^- concentrations (Pogodaeva et al., 2007; Pogodaeva et al., 2017; Granina, 2008), the OM diagenesis can proceed predominantly through the oxidation of Mn(IV) and Fe(III) oxides (Baturin & Granina, 2009; Och et al., 2012; Aloisi et al., 2019). Microbial communities from iron-enriched sediments of Lake Superior were dominated by members of the phylum *Methylomirabilota* known as anaerobic methane oxidizers and by the *Nitrospira* species that were the most identical to bacteria from Fe-enriched seeps (Dittrich et al., 2015). In sediments of Lake Constance, Fe-enriched layers were formed through nitrate-dependent anaerobic oxidation (Hauck, 2001). *Methanoperedens ferrireducens* showed the participation in Fe-dependent AOM (Cai et al., 2018), and *Methanoperedens manganicus* and *Methanoperedens manganireducens* in Mn-dependent AOM (Leu et al., 2020).

Some researchers discussed the likely participation of Fe and/or Mn in AOM processes in Lake Baikal sediments (Tiquia et al., 2006a; Zemskaya et al., 2008; Och et al., 2012; Torres et al., 2014). According to Och et al. (2012), the dynamic growth of Fe and Mn begins immediately below the layers of maximum O_2 penetration, and the dissolution of Fe and Mn solid particles is associated with anaerobic CH_4 oxidation (AOM) either through the sulfate reduction (SO_4^{2-}) and subsequent Fe(II) formation through S(II) oxidation or directly related to the Fe reduction. Torres et al. (2014) proposed an alternative pathway of Fe(II) dissolution via anaerobic NH_4^+ nitrification. In Lake Baikal communities confined to the Fe-Mn layers, a large contribution was from the members of the nitrogen cycle, the

ammonia-oxidizing *Nitrososphaeria* archaea, and the *Nitrosomonadaceae* bacteria that can carry out denitrification and dissimilatory reduction of nitrates to ammonia and be involved in a chain of trophic interactions with anaerobic methanotrophs (Wu et al., 2005; Gentile et al., 2006; Zemskaya et al., 2022; Lomakina et al., 2023a). We analyzed bacterial and archaeal MAGs in enrichment culture (NB) from the Fe-Mn layer. Three MAGs belonged to the order *Methylomirabilales*, and one MAG to the genus *Methanoperedens*. A phylogenetic analysis of the assembled *Methanoperedens* genome based on 53 sequences of the conservative marker genes indicated the formation of an individual branch within the clade with *Methanoperedens ferrireducens* (Fig. 4.10). The assembled genome contained the genes (*mcrA, mtr, mer, mtd, mch*, and *ftr*) encoding the central methanogenesis pathway and responsible for steps 1 to 6 of this process, which suggests its ability to oxidize methane through the mechanism of reverse methanogenesis. The presence of Nar genes required for electron transport confirmed the ability of *Methanoperedens* to carry out AOM through a nitrate-dependent pathway. In this case, the syntrophic interaction with the ANAMMOX bacteria (in the presence of ammonia in the environment) or with members of the *Methylomirabilales* bacteria allows for the reduction of the formed nitrite to gaseous nitrogen. Nitrite reductase (NrfAH) in the *Methanoperedens* genome suggests its subsequent participation in the nitrite conversion, and terminal iron reductases (MtrC) in iron-dependent AOM. Wang et al. (2015) indicated that reduction of Fe(III) compounds, in particular ferrihydrite, occurs extracellularly, with the participation of Mtr genes. The presence of genes in the assembled genome of *Methanoperedens* from Lake Baikal, which provide both nitrate-dependent and iron-dependent AOM, may evidence the broad adaptive capacity of these archaea. They can be considered universal anaerobic methanotrophs playing an important role in methane emission.

The presence in the assembled genomes of functional genes associated with Fe and Mn cycles also confirmed the involvement of microorganisms in reduction/oxidation of these elements and the formation of Fe-Mn layers. The genomes of different taxa had type IV pili mediating microbial motility and cell attachment to solid surfaces, including insoluble substrates such as Fe and Mn (Lovley, 2012). MAGs from Lake Baikal showed multiheme c-type cytochromes that ensure electron transport to extracellular soluble and insoluble substrates. In the genomes of *Gammaproteobacteria, Methylomirabilota*, and *Deinococcota*, we detected multiheme c1-type cytochromes (Lomakina et al., 2023a). They act as terminal reductases and pass electrons to the inner bc1 membrane complex (Ciancio Casalini et al., 2022). The *Halobacteriota* and *Planctomycetota* MAGs contained the *mtrABC* genes that ensure the participation of these microorganisms in the iron reduction on the cell surface (Liu et al., 2012). Notably, the electron transport chain for metal transfer (cytochrome cbb3 oxidase) typical of Fe(II)-oxidizing microorganisms was present in a wide range of the bacterial genomes. It was detected in the MAGs of *Methylomirabilota, Nitrospirota, Chloroflexota, Bacteroidota, Actinomycetota,* and *Planctomycetota,* as well as of *Desulfobacterota* and some *Gammaproteobacteria.* Nitrite reductase (NrfA) was identified in some reconstructed MAGs, including *Methylomirabilota, Planctomycetota, Desulfobacterota, Acidomycetota,* and

Halobacteriota. It is responsible for electron transport functions in the NrfAB nitrite-reductase complex, predicting the interaction of the nitrogen metal cycles and nitrate-dependent aerobic oxidation of Fe and Mn through denitrification (Liu et al., 2003; Tiquia et al., 2006b; McKindles & Tiquia-Arashiro, 2012). The bulk of Lake Baikal genomes contained genes that encode ABC transporters of Fe and Mn and the *afuABC* genes that ensure ion metal transfer into the cytoplasm. These data support the hypothesis about the active role of microorganisms in the Fe and Mn cycles associated with AOM in sediments of freshwater Lake Baikal.

8 Microorganisms Associated with Authigenic Carbonates

Authigenic carbonate minerals are widespread in marine and terrestrial sediments (Kelts & Hsü, 1978; Greinert et al., 2001; Swart, 2015). Studies of the processes of carbonate nodule formation have been carried out mainly in marine environments, and these processes are explained by supersaturation for carbonate minerals archived by the shift of chemical equilibrium among the dissolved inorganic carbon species (Yanagawa et al., 2019). In marine environments where methane seeps have been detected, AOM is the main process for the formation of authigenic carbonate minerals (Aloisi et al., 2002; Gieskes et al., 2005; Greinert et al., 2001; Naehr et al., 2007). Elevated HCO_3^- fluxes is one of the important responses of this process, resulting in the formation of carbonates in marine ecosystems (e.g., Loyd & Smirnoff, 2022). Some studies (Aloisi et al., 2002; Lein, 2004; Reeburgh, 2007) suggested that, in the absence of HCO_3^- formation processes and increased alkalinity, authigenic carbonate minerals precipitated during methane production rather than oxidation (e.g., Soetaert et al., 2007; Teichert et al., 2014). In this case, CO_2 produced in the CH_4 methanogenic zone decreases pH and promotes the dissolution/weathering of silicate minerals by carbonic acid (Aloisi et al., 2004). Fermentation and marine silicate weathering result in favorable conditions for the precipitation of carbonate minerals and increased dissolved inorganic carbon (DIC) and cation fluxes (Torres et al., 2020). This contribution of DIC to methanogenesis enriches ^{13}C authigenic carbonates, a characteristic byproduct of methanogenesis, in which methane predominantly includes ^{12}C, leaving enriched ^{13}C DIC behind (Meister & Reyes, 2019).

Authigenic carbonates were also found in the water column and bottom sediments of lakes with different ages and diverse chemical regimes: in the water column of alkaline Great Salt Lake, UT (USA) (Ingalls et al., 2020), ferruginous sediments of stratified Lake Towuti, Indonesia (Vuillemin et al., 2019), maar lake Laguna Potrok Aike, Argentina (Hahn et al., 2014), ancient Lake Ohrid, North Macedonia, Albania (Francke et al., 2016), deep-drilling cores from lakes Khuvsgul, Mongolia (Solotchina et al., 2003) and Baikal, Russia (Sapota et al., 2006), and sediments of small saline endorheic lakes of the Holocene age located in the Olkhon District, Russia (Solotchina et al., 2008, 2011). Plee et al. (2010) indicated that photosynthetic activity of microorganisms is the main factor causing calcite

precipitation in the water column of Lake Geneva, increasing pH in the epilimnion. In oligotrophic lakes, calcite precipitation is associated with picocyanobacterial blooms (Dittrich & Obst, 2004). In hypersaline alkaline lake on the Eleuthera island, the Bahamas (Dupraz et al., 2004), precipitation of carbonate minerals was associated with the activity of microorganisms, changing alkalinity in microdomains and precipitating carbonate minerals. Stanton et al. (2022) described another mechanism for the formation of carbonate minerals in Fayetteville Green Lake, New York (USA), where fine-grained suspended calcium carbonate (belings) precipitates into the water column annually. In this case, particle formation occurs in spring through the deposition of fine amorphous calcium carbonate on microbial cells as well as on abundant extracellular polymeric substances often associated with centric diatoms. AOM promoting the formation of carbonate minerals in lacustrine ecosystems has not been widely studied.

In Lake Baikal sediments, unlike in marine ecosystems, authigenic carbonates have a local distribution, being confined to the areas of gas-bearing fluid discharge. Most of the dissolved inorganic matter in the sediments of different areas, which results from the OM decomposition and subsequent redox reactions, had a CO_2 form rather than HCO_3^-, leading to lower pH in the diagenetic environment (Mizandrontsev, 1975). One of the causes of rare formation of authigenic carbonates is the lack of electron acceptors resulting in low methane oxidation rates in the sediments of this freshwater ecosystem. This does not change pH, which would be sufficient for the equilibrium shift in the ecosystem and the precipitation of carbonate minerals (Namsaraev & Zemskaya, 2000; Pimenov et al., 2014; Aloisi et al., 2019). Low cation concentrations also limit the precipitation of carbonate minerals in Lake Baikal sediments. Nevertheless, all three basins of the lake had authigenic carbonates of rhodochrosite and siderite composition (Callender & Granina, 1992; Sapota et al., 2006; Krylov et al., 2008, 2010; Krylov et al., 2018). Authigenic carbonates with heavy (up to +21.9‰ VPDB) and extremely heavy (up to +36.3‰ VPDB) isotopic $\delta^{13}C$ values were found in some Lake Baikal areas, in particular near the Malenky and the Kedr MVs (Krylov et al., 2008, 2020). A study of carbon isotopy of authigenic carbonates in this area suggests different mechanisms of their formation (Krylov et al., 2008, 2020; Hachikubo et al., 2020): via OM destruction and methane generation. Hydrogenotrophic and acetoclastic methanogenesis could be a likely mechanism for the enrichment of dissolved carbon pools with a heavy isotope near the Kedr MV. The carbon isotope ratio of dissolved inorganic carbon (DIC) and carbonates, as well as the presence of taxonomically diverse microbial communities in the sediments of the Kedr MV, suggested two generations of carbonate minerals (Lomakina et al., 2023b) resulted from differently directed processes (AOM and methane generation). Data on the composition and metabolism of microorganisms inhabiting different sediment layers in this area confirmed the existence of microbial populations with different compositions (Fig. 4.6a, b). Organotrophic and fermenting taxa with a wide metabolic type predominated in deep sediments containing authigenic carbonates. Members of the classes *Anaerolineae* and *Dehalococcoidia* (*Chloroflexota*) capable of participating in the destruction of a wide range of substrates, including proteins, lipids, and

carbohydrates, made the largest contribution among bacteria. Analysis of the assembled *Chloroflexota* genomes from Lake Baikal confirmed the presence of individual genes for reductive acetyl-coenzyme A (acetyl-CoA) pathway (WL pathway) and genes for homoacetogenesis (*pta* and *ack*), during which acetate is formed from acetyl-CoA (Lomakina et al., 2023a). The *Atribacterota* members (Fig. 4.6a, b) were another dominant bacterial taxon in the sediment communities from this area. They showed the participation in the metabolism of sugars and organic acids followed by the formation of acetate/ethanol (Liu et al., 2019). Being homoacetogenes (Vuillemin et al., 2020), members of this phylum were observed in the analysis of carbonate minerals from different ecosystems (Marlow et al., 2014; Fussmann et al., 2019; Yanagawa et al., 2019). These bacteria fermenting OM could provide substrates for methanogens. Their percentage was significant in the microbial community associated with carbonates. Interestingly, among methanogens, there were hydrogenotrophic archaea of the family *Methanoregulaceae* (8–18% of the total number of reads), order *Methanomassiliicoccales* methylotrophic methanogens (up to 7%), and acetoclastic archaea of the family *Methanosarcinaceae* (up to 3%). Metabolic reconstruction of the *Methanoregulaceae* and *Methanomassiliicoccales* MAGs proved a complete set of genes responsible for hydrogenotrophic and methylotrophic methanogenesis. In addition to typical methanogens, the genomes of archaea belonging to other candidate taxa contained genes encoding key enzymes of methyl-reducing methanogenesis. These were members of the class *Bathyarchaeia* and *Methanomethylicia* (formerly *Verstraetearchaeota*) of the phylum *Thermoproteota* (Evans et al., 2015; Vanwonterghem et al., 2016). In the carbonate-associated community, *Bathyarchaeia* contributed up to 81% of all analyzed sequences, which was much higher than in other sediment layers from this area (Fig. 4.6b). The genomes of these archaea contained genes confirming their ability to use some complex organic compounds as substrates for growth, including aromatic compounds. The assembled genomes of *Bathyarchaeia* from Lake Baikal had genes responsible for benzoyl degradation, as well as genes partially encoding glycolysis/gluconeogenesis and beta-oxidation pathways. Only one *Bathyarchaeia* MAG had genes encoding methanogenesis. The capacity for methanogenesis was also predicted for the *Methanomethylicia* MAG and unclassified *Thermoplasmatota*, in the genomes of which a key gene for this process (*mcrA*) was identified. Thus, the microbial communities of sediments containing carbonates enriched with heavy carbon isotope ($\delta\ C^{13} \approx +30‰$ VPDB) were dominated by groups of bacteria and archaea involved in the degradation of complex organic and inorganic compounds, whose activity could lead to enrichment of the DIC pool ($\delta^{13}\ C \approx +10‰$ VPDB) with light carbon isotope in comparison with carbonates. The substrates formed during the activity of these groups of organotrophic microorganisms, in turn, stimulated the activity of methanogens with different metabolism, which could lead to the enrichment of carbonates with the heavy carbon isotope ^{13}C. Changes in the $\delta^{13}C$-DIC values and an increase in the contribution of anaerobic methanotrophs to the communities from other sediment layers in this area explained the shift in the methane-formation zone to the anaerobic methanotrophy, promoting the entry of ^{12}C in DIC of the pore fluid and formation of carbonates with lighter $\delta^{13}C$. The

contribution of the anaerobic methanotrophic *Methanoperedenaceae* archaea and the *Methylomirabilota* bacteria increased in the communities up to 16% and 1%, respectively. Taking into account the insignificant contribution of the latter and their presence only in the surface sediments (Figs. 4.5 and 4.6), we can assume that the *Methanoperedenaceae* archaea mainly ensured AOM in carbonate-containing layers through the nitrate-dependent pathway. The ionic concentrations acting as electron acceptors in AOM would also limit their activity. The overall decrease in the δ^{13}C-DIC values, which occurred between the crystallization of siderites with different isotopic compositions, could be associated with the evolution of the mud volcano and certain variability in situ of microbial processes. The presence of methanogenic and methanotrophic archaea in the sediments of this area confirms the possibility of simultaneous diagenetic methane generation and AOM. Li et al. (2018) also indicated the involvement of methane generation and AOM in sediments containing authigenic carbonate minerals.

9 Phylogenomic Analysis of Reconstructed MAGs of Some Bacterial and Archaeal Taxa from Lake Baikal Sediments

A phylogenetic analysis of 16S rRNA gene sequences and some functional genes revealed Lake Baikal clusters or individual lineages within clusters when compared with the closest relatives deposited in NCBI (Shubenkova et al., 2005; Kadnikov et al., 2012; Zemskaya et al., 2018). A subsequent analysis of water-column microbiomes evidenced that Lake Baikal is a source of new microbes (Cabello-Yeves et al., 2020; Zemskaya et al., 2020; Haro-Moreno et al., 2023). Reconstruction of the genomes of bacteria and archaea from Lake Baikal sediments and their phylogenetic analysis (based on 120 and 53 conserved marker genes, respectively) confirmed the presence of taxa different from those known in the GDTB database and involved in carbon, sulfur, and nitrogen cycles. The average nucleotide identity (ANI) indices for MAGs varied from 76 to 96% in NB, from 74 to 98% in the Kedr MV, and from 79 to 96% in SB, suggesting the presence of the new bacterial and archaeal taxa in Lake Baikal sediments. MAGs from different ecotopes within one phylum belonged to different classes and were phylogenetically distant from each other and compared to those from the GDTB database. For example, MAGs of dominant organotrophic members of the phylum *Chloroflexota* from oxidized and reduced sediments formed different clusters with genomes from freshwater sediments and hydrocarbon-contaminated marine waters (Fig. 4.11).

MAGs identified in the oxidized sediments of SB were phylogenetically close to bacteria of the class *Limnocylindria* that had not been previously detected in the metabarcoding datasets (Zemskaya et al., 2010, 2015a, 2022; Chernitsyna et al., 2016a; Lomakina et al., 2023a). Noteworthy is that members of this class are often found in freshwater ecosystems, including deep-water ones (Mehrshad et al., 2018). MAGs of anaerobic *Anaerolineae* from the reduced sediments of the reference site

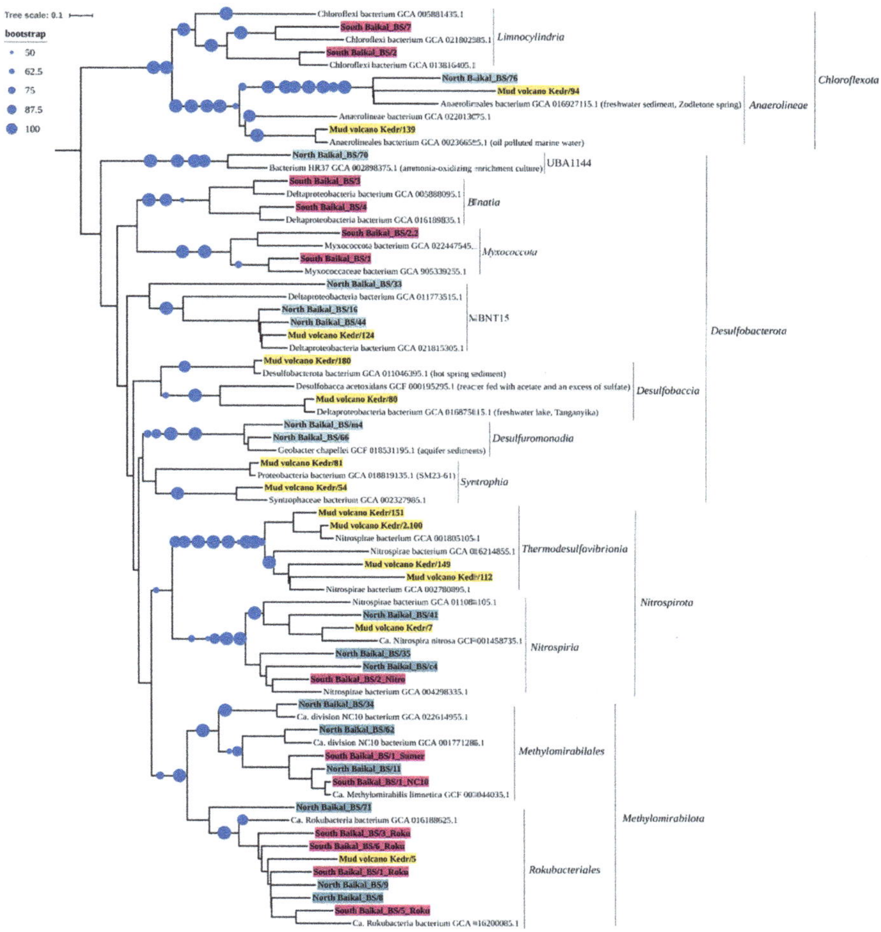

Fig. 4.11 Phylogenomic tree of bacterial MAGs from sediments of the Kedr MV (highlighted in yellow) and reconstructed genome of two reference sites in the northern (highlighted in blue) and southern (highlighted in purple) basins of Lake Baikal. Analysis was performed by Interactive Tree Of Life (iTOL). Taxonomy is shown according to the GTDB

in NB and MV also differed phylogenetically. Phylogenetically diverse MAGs were also recorded for the phylum *Nitrospirota*, which belonged to the classes *Nitrospiria* and *Thermodesulfovibrionia*. The latter were phylogenetically close to the genomes from groundwater. MAGs of nitrite-oxidizing *Nitrospiria* from the reference site of NB formed individual lineages within the cluster of these bacteria from groundwater and marine sponges. Phylogenetically diverse MAGs of the phylum *Desulfobacterota* involved in the sulfur cycle were widespread in the investigated

Lake Baikal sediments (Fig. 4.11). MAGs closely related to the class UBA1144 (*Desulfobacterota*) (Hug et al., 2015) and to the Fe(III)-reducing species *Geobacter chapellei* (class *Desulfuromonadia*) from aquifer sediments were identified only in the sediments from the reference site of NB. The MAG phylogenetically close to MBNT15 (Desulfobacterota) from the pelagic sediments of Khambhat Gulf was also detected in the sediments of the Kedr MV.

MAGs from the surface layer of sediments in SB formed two clusters in the phylogenetic tree: one with members of anaerobic *Myxococcota* and another with the uncultured class *Binatia* from groundwater, which showed the decomposition of aromatic compounds of plant biomass and xenobiotics (Rodríguez-Ramos et al., 2022). Moreover, MAGs from the sediments of the Kedr MV were closely related to sulfate-reducing bacteria of the class *Desulfobaccia* from Lake Tanganyika and hot spring sediments. MAGs formed an individual cluster with the members of anaerobic organoheterotrophs of the class *Syntrophia*, which can participate in syntrophic interactions, supplying fermentation products to methanogens or anaerobic methane-oxidizing archaea (Langwig et al., 2022). Bacteria of the phylum *Methylomirabilota* involved in the AOM were also phylogenetically diverse. In the phylogenetic tree, they formed two clusters with the members of the orders *Methylomirabilales* from Lake Zug (Switzerland) and *Rokubacteriales* from groundwater. Furthermore, MAGs assigned to the order *Rokubacteriales* formed individual lineages in this clade.

MAGs of archaea from different ecotopes were also phylogenetically diverse. Some of them formed individual clades or individual branches within clades with genomes of the known taxa from different ecosystems (Fig. 4.10). For instance, MAGs of ammonia-oxidizing *Nitrososphaeria* from oxidized sediments of reference sites in SB and NB formed an individual cluster related to genomes from the Pacific Ocean sediments. MAG from reduced sediments of the Kedr MV formed a cluster with genomes from the Shengli oil field. Within the phylum *Aenigmatarchaeota*, MAGs from the sediments of the Kedr MV formed an individual lineage with the genomes from the saltwater hot spring of the Guaymas basin. A genome phylogenetically different from the known participants of nitrate-dependent AOM (*Methanoperedens nitroreducens*) was assembled in the microbiome of this ecotope. It formed an individual branch with *Methanoperedens ferrireducens* from the bioreactor, which participates in AOM with Fe(III) reduction (Cai et al., 2018). Another MAG from this area related to the *Methanomassiliicoccales* methylotrophic methanogen was removed from the known species of this family in the phylogenetic tree. An analysis of ten MAGs phylogenetically related to organotrophs of the class *Bathyarchaeia* revealed the formation of individual lineages within the clusters of the known members of this class (Fig. 4.10). Thus, a phylogenomic analysis of the reconstructed MAGs confirmed the presence of phylogenetically diverse bacterial and archaeal groups, including those different from the known members, in the communities of Lake Baikal sediments. Their closest relatives inhabited freshwater lakes, marine sediments, and hot springs.

10 The Metabolic Potential of Microbiomes from Lake Baikal Sediments

We assessed the metabolic potential of microbiomes for some of the investigated ecotopes based on the ratio of the number of the reconstructed genes for the main metabolic pathways to the number of DNA-dependent ATPase genes (*rec*A/*rad*A) (Fig. 4.12). A distinctive feature of the metagenomes from the surface layers of Lake Baikal sediments is a rather large proportion of nitrogen metabolism genes that ensure nitrification and denitrification, which is also characteristic of metagenomes from the near-bottom region of the bathypelagic zone in Lake Baikal (Cabello-Yeves et al., 2020; Bukin et al., 2023). Additionally, in benthic genomes, we recorded high values of the *rec*A/*rad*A ratio for the methanol oxidation genes (*mxaABCDEFG*), methanesulfonates (*msmAB*), and alkanesulfonates (*ssuD* and *msuD*), as well as for genes for the destruction of some aromatic compounds. In the surface sediment layers from the reference site in SB and the Kedr MV, microbial communities differed significantly in the set of genes for carbon fixation and sulfur metabolism pathways. Genes of enzymes responsible for different stages of carbon oxidation, as well as genes of enzymes of the SOX system oxidizing thiosulfate and sulfate oxidase (*fccAB*), prevailed in the metagenome from the oxidized oozes of the reference site. At the same time, the relative proportion of key genes for autotrophic CO_2 fixation pathways was small. On the contrary, the metagenome of the reduced gas-saturated sediments from the Kedr MV had a significant number of genes for the WL pathway and reverse tricarboxylic acid (rTCA) cycle, as well as genes for assimilatory and dissimilatory sulfate reduction and methane oxidation. These data largely reflected differences in the physicochemical parameters of the environment. Taxa capable of combining the heterotrophic type of metabolism with the possibility of obtaining energy through oxidation or reduction of nitrogen compounds and thiosulfate and/or sulfur oxidation dominated the microbiome from the oxidized sediments. Autotrophic types of carbon fixation, nitrification, nitrate and sulfate reduction, and methane oxidation were more typical of the microbiome from the Kedr MV (sediments were recovered from the surface).

An analysis of the metagenomes from the sediments of the Kedr MV associated with authigenic carbonates revealed among all samples the largest proportion of genes providing both CO_2 fixation in the WL pathways, the 3-hydroxypropionate cycle, and the rTCA cycle and reverse reactions. This was consistent with an increase in the proportion of methanogenic/methanotrophic archaea and an increase in the *rec*A/*rad*A ratio for methyl-CoM reductase and CoM methyltransferases. This nature of the metabolic profile indicated that, in addition to methanogenesis and anaerobic oxidation of methane, fluctuations in the C_{org} isotopic composition recorded in the zone where extremely ^{13}C-enriched carbonates occur may be associated with the fixation or release of carbon dioxide in reactions catalyzed by the above enzymes. Moreover, microbiomes of this area contained a relatively large proportion of dissimilatory sulfate reduction genes, which may result from the SO_4^{2-} influx with deep fluid.

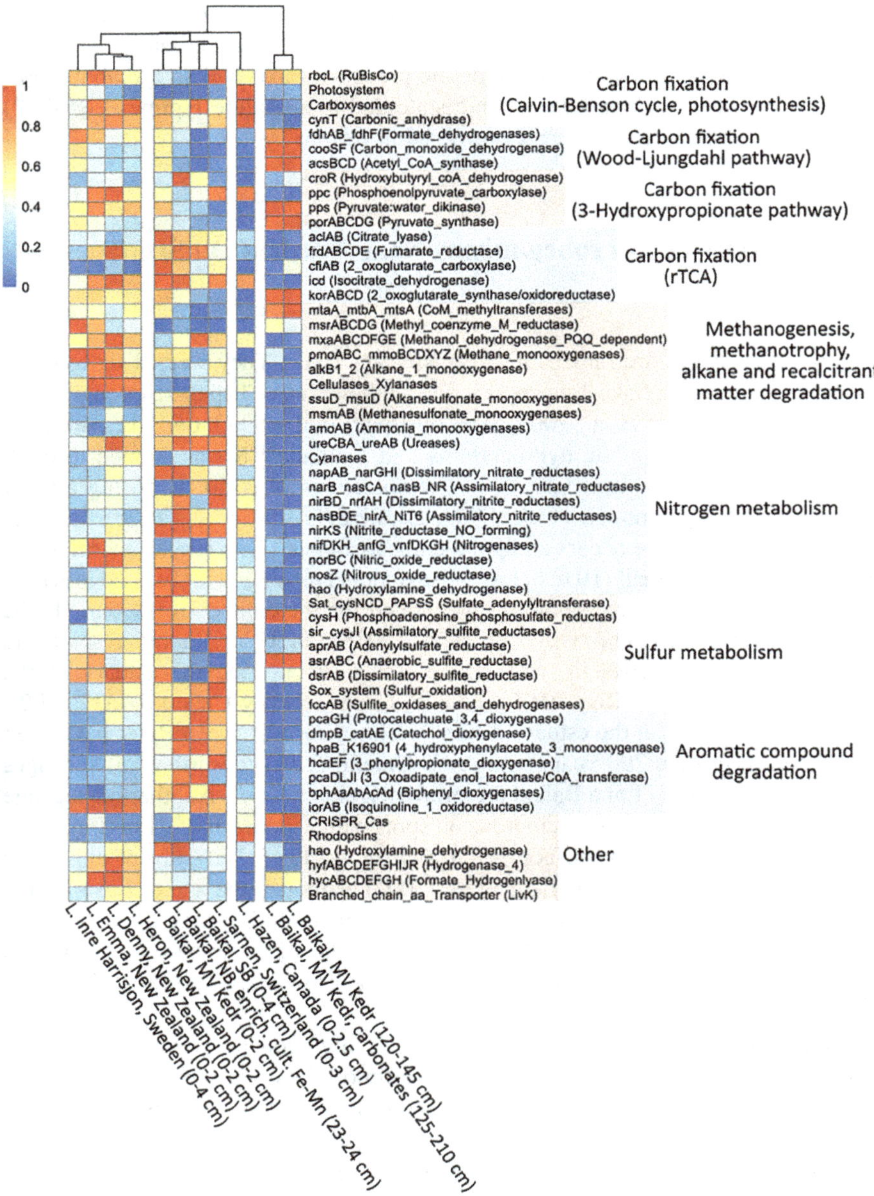

Fig. 4.12 The metabolic potential of microbiomes from Lake Baikal sediments and other freshwater lakes based on the normalized ratio of the number of individual genes of the main metabolic pathways to the *recA*/*radA* genes identified from the results of metagenomic analysis

The microbiome of the sediments from Lake Sarnen (Switzerland) with a significant contribution of genes for the metabolism of nitrogen compounds (Baumann et al., 2022) had the most similar metabolic profile to Lake Baikal. Metabolic profiles of other lakes with similar trophic state and temperature regime (Fig. 4.12) differed in the contribution of genes involved in various cycles.

11 Diversity of Microorganisms Near Natural Oil Seepages in Lake Baikal

Natural oil seepages are rather rare in freshwater lakes. They were discovered in Lake Baikal, deep rift Lake Tanganyika, and shallow Lake Chapala (Simoneit et al., 2000; Zárate-del Valle et al., 2006). The oil found in the two latter lakes does not contain polycyclic aromatic hydrocarbons and *n*-alkanes; it was formed through hydrothermal pathway from lake OM at temperatures of ~250 °C during a short geological period (Simoneit et al., 2000; Zárate-del Valle et al., 2006). In Lake Tanganyika, oil seepage occurs along the faults associated with a hydrothermal system that forms very small (10 m below the lake floor) oil source rocks aged less than 25,000 years. Here, hydrocarbons are generated via the contact of the buried OM with hydrothermal waters (Oppo & Hurst, 2018). In Lake Baikal, researchers recorded natural seepages in two areas of the CB: opposite the estuary of the Bolshaya Zelenovskaya River (BZ) and near Cape Gorevoy Utes (GU) (Fig. 4.2). Oil and gas seepages at the estuary of the Bolshaya Zelenovskaya River have been known for more than 200 years when I.G. Gmelin (1833) described oil seepages along the east coast of Lake Baikal (Kontorovich et al., 2007). Another oil seepage area was discovered in 2005. Numerous oil spots with a diameter of ~1 m (Fig. 4.1d) were found on the water surface in an area of ~1 km^2 (Khlystov et al., 2007; Kontorovich et al., 2007). Khlystov et al. (2007) estimated the age of Lake Baikal oil to be the Late Cretaceous and associated its source to oil source rocks from the Oligocene-Miocene strata of the Baikal Rift. The OM of the freshwater body with high concentrations of terrestrial plant components is the main source of oil in Lake Baikal (Kontorovich et al., 2007).

Initially, the oil near GU exhibited extremely high concentrations of *n*-alkanes and was identified as non-biodegradable paraffin oil. Its composition had an entire spectrum of petroleum hydrocarbons: *n*-alkanes, alkylcyclohexanes, isoprenoids (pristane, phytane), polycyclic aromatic hydrocarbons (PAHs), hopanes, and a complex of unique biomarker molecules (Khlystov et al., 2007; Kontorovich et al., 2007). Based on the diverse set of chemofossil molecules, sesquiterpanes, steranes, secogopanes, and new structures of hopanes (oleanane, carotane, etc.), Lake Baikal oils are considered unique. Currently, Gorshkov et al. (2020) have indicated a narrowing of the homologous series of *n*-alkanes and a decrease in the total amount of normal hydrocarbons and PAHs. In a more ancient oil seepage near the BZ, the oil collected both from the lake surface and in the upper layer of sediments was

biodegradable aromatic naphthenic consisting of a high molecular weight bitumen fraction (Kontorovich et al., 2007).

The discovery of new natural hydrocarbon seepage in 2005 allowed for online monitoring of the phenomena occurring in this water area and in situ observation of the qualitative and quantitative changes in the oil composition, structure of microbial communities, and their adaptive capabilities. Similar to cases of man-made oil spills occurring in different areas of the world ocean (Atlas & Hazen, 2011; King et al., 2015), an increase in the number of hydrocarbon-oxidizing microorganisms (HCOMs) was observed both in the water column and in the sediments during the first years after the discovery. In the zone of oil slicks formed on the surface of the water column, the number of cultured HCOMs was 12 times higher than the values from the water of the reference sites. Fifteen years of observations of the number of cultured microorganisms oxidizing oil and *n*-alkanes in the GU sediments indicated its increase in the first two years (2005–2007) and the subsequent decrease from 2017 to the present (Pavlova et al., 2020).

Based on metabarcoding data, the composition of microbial communities from areas with natural oil seepages did not differ at the phylum level from those from MV sediments (Fig. 4.3). As noted above, the taxa level of lower rank demonstrated more significant differences in the taxonomic composition both compared to those observed in other sedimentary strata of the lake and between the oxidized and reduced sediment layers (Fig. 4.4). In the surface sediment layers (usually oxidized), the class *Aminicenantia*, families *Acidobacteriaceae*, *Holophagaceae*, and *Vicinamibacteraceae*, capable of anaerobically decomposing aromatic compounds mostly represented the dominant phylum *Acidobacteriota* (Tiquia-Arashiro & Rodrigues, 2016). Sequences recorded in oil-bearing environments also represented numerous chemoorganoheterotrophic *Actinomycetota* (*Thermoleophilia*, *Coriobacteriia*) and *Pseudomonadota* (*Alpha-* and *Gammaproteobacteria*) (Pannekens et al., 2019; Khomyakova et al., 2022; Gittins et al., 2023). The families *Anaerolineaceae* (*Chloroflexota*) and *Syntrophaceae* (*Desulfobacterota*) dominated the communities from the surface sediments. They could carry out syntrophic oxidation of hydrocarbons, leading to methane generation with participation of methanogenic archaea (Liang et al., 2015). This was consistent with the high relative abundance of methanogenic archaea (up to 20%) in the communities from the sediments of GU and BZ. Furthermore, members of the family *Desulfobaccaceae* that reduce sulfate, sulfite, and thiosulfate through acetate oxidation (Galushko & Kuever, 2019) and the sequences of the families *Thermodesulfovibrionaceae* (*Nitrospirota*) and *Nitrospinaceae* (*Nitrospinota*)—sister phylotypes involved in the nitrogen cycle (D'Angelo et al., 2023)—had a significant percentage in the communities of these areas. Additionally, changes in the taxonomic composition with sediment depth within phyla were observed in the sediment strata of these ecotopes, for example, for *Chloroflexota* and *Desulfobacterota*. Members of the families *Caldisericaceae* (3.5–12%), JS1 (16–42%), *Chitinispirillaceae* and 2-02-FULL-39-32(*Elusimicrobiota*, to 4.6%), and *Zixibacteria* (to 4.2%) dominated in the communities of the reduced sediments (Fig. 4.4).

The archaeal diversity in the communities from the sediments of areas with oil seepages was similar to that from MV sediments. *Bathyarchaeia* were dominant (up to 60% of all sequences). Like in other areas, aerobic archaea of the family *Nitrosopumilaceae* prevailed in the communities from surface and subsurface sediments. Members of the families GW2011, GWC1_47_15, and order *Woesearchaeales* (*Nanoarchaeota*), as well as hydrogenotrophic archaea (*Methanomicrobiaceae*, *Methanoregulaceae*, *Methanoperedenaceae*, and *Methanosarcinaceae*) identified in various environments, including oil reservoirs and sulfur springs (Liu et al., 2021; Köhler et al., 2023), made various contributions to reduced sediments. Only reduced sediments contained sequences of the thermophilic *Hadarchaeota* (up to 3.8%) that possess genes for carbon fixation through the WL pathway, methanogenesis, and alkane metabolism (Hua et al., 2019; Wang et al., 2019).

Our experiments have shown that natural microbial communities from the water column and sediments of areas with oil seepages are actively involved in the degradation of various oil fractions. The degradation of oil *n*-alkanes by the natural community of the water column under anaerobic conditions was up to 90% in 60 days (Pavlova et al., 2012). Under anaerobic conditions, conversion of *n*-alkanes and PAHs was more efficient in the experiment with sediments sampled near the BZ oil seepage compared to GU. In the experiment with sediments from BZ, *n*-alkane conversion ranged from 20% to 48%, and PAH conversion ranged from 55% to 65%. In the experiment with sediments from GU, the loss of *n*-alkanes varied from 6% to 20% and PAHs from 20% to 46% (Pavlova et al., 2022). Indicators of hydrocarbon-oxidizing activity were correlated with the composition of OM and individual biomarker compounds in the sediments of these areas with oil seepages (Morgunova et al., 2022). As shown by these authors, different layers of sediments near GU were enriched in OM with different biodegradation degrees. The group composition of bitumen and the elevated productivity index indicate that oil saturation of GU sediments occurs constantly (Kostyreva et al., 2022). The presence of a naphthenic hump, a high maturity level of *n*-alkanes (CPI ≤ 2), and a decrease in the PAH concentration and n-C_{35}–n-C_{39} with depth confirm the biodegradation processes in GU sediments (Morgunova et al., 2022). In the sediments of BZ, the OM concentration was much lower; *n*-alkanes and isoprenoids were absent, and asphalt resin components predominated in the composition of bitumen (up to 94%) (Kostyreva et al., 2022; Khlystov et al., 2022; Morgunova et al., 2022).

Geological and geochemical processes can influence the composition of oil in areas with oil seepages. A fraction enriched in *n*-alkanes enters the lake water (Khlystov et al., 2009; Gorshkov et al., 2020) because oil fractionation and the formation of asphalt structures on the bottom surface in this area of the lake occur in the upper layers of sediments. Microorganisms very efficiently oxidize this fraction in the water column under aerobic conditions. Polyaromatic compounds subjected to destruction by microbial communities inhabiting this zone are accumulated in sediments under anaerobic conditions. Our model experiments and a decrease in the concentrations of these compounds in the sediments of this area confirm this fact (Gorshkov et al., 2020; Pavlova et al., 2022). In the more ancient BZ oil seepage, oil seeps on the bottom surface at a more significant distance from its generation source.

Oil, gradually accumulating in the upper layer of sediments, undergoes anaerobic biodegradation. The presence of perylene in bitumen, a stable end product of OM transformation in sediments, the concentration of which was two orders of magnitude higher than in oil from the GU oil seepage (Khlystov et al., 2022), evidence this. Most likely, Lake Baikal sediments of the oil seepages undergo the processes similar to those observed in the Persian Gulf containing ~48% of the world's reserves of oil subjected to constant natural leakages (Rezaei Somee et al., 2021). According to the authors, with constant exposure to oil, community composition changes toward the dominance of oil-decomposing microorganisms, simultaneously promoting a "division of labor" to initiate and carry out various stages of hydrocarbon biodegradation.

Notably, the generation of hydrocarbon gases (methane and ethane) accompanied the oil degradation in our experiments. Generation of methane during oil biodegradation was observed at many sites of oil drilling (Bonch-Osmolovskaya et al., 2003; Nazina et al., 2006; Jones et al., 2008; Heider & Schühle, 2013). The experiments with the natural microbial community from the GU oil seepage recorded the most significant methane concentrations. Ethane generation, on the contrary, was more typical during the cultivation of the communities from the BZ sediments where the C_2H_6 concentration was 2.3 to 7 times higher after one year of cultivation. In the experiment with natural communities from the BZ and GU sediments, the maximum concentration of the generated methane was 2.4 mmol/L (54 ml/L) and 32.54 mmol/L (728 ml/L), respectively, with methane concentrations in Lake Baikal sediments ranging from several μL/L to 20 mL/L (Zemskaya et al., 2008).

As mentioned above, methane generation could occur through the syntrophic interaction of the *Desulfobacterota* bacteria and the methanogenic *Halobacteriota* and *Methanobacteriota* archaea, whose percentage in the subsurface sediments of these areas was higher than in deep sediments. In reduced sediments, members of the phyla *Bacillota*, *Atribacterota*, *Chloroflexota*, and *Bathyarchaeia* could participate in the hydrocarbon conversion to generate hydrocarbon gases. Currently, metabolic reconstruction of the *Atribacterota* and *Bathyarchaeia* genomes has revealed that microorganisms can ferment not only carbohydrates but also short-chain *n*-alkanes into fatty acids that secondary decomposers then consume. *N*-alkanes are oxidized through hydroxylation, addition of fumarate to hydrocarbons, and degradation of anaerobic aromatic compounds by class I benzoyl-CoA reductase (Dong et al., 2019; Liu et al., 2019; Zhang et al., 2021). The benzoyl-CoA reductase gene (*bcr*) responsible for anaerobic oxidation of benzoates was detected in the assembled genomes of various taxonomic groups of bacteria and archaea from Lake Baikal which inhabit sediments of the Kedr MV: *Caldisericota, Chloroflexota, Armatimonadota, Atribacteriota, Verrucomicrobiota, Planctomycetota, Zixibacteria, Bacteroidota, Rokubacteriales, Nitrospirota, Desulfobacterota, Latescibacterota*, and *Bathyarchaeia*. A phylogenetic analysis of the 16S rRNA gene fragments of *Bathyarchaeia* from the GU and BZ sediments and their comparison with those from the MV sediments indicated both the common and unique sequences belonging to this class. It is very likely that these microorganisms from

the sediments of oil seepage areas also contain these genes and can biodegrade PAHs in Lake Baikal sediments.

Previously, an analysis of nucleotide sequences of *alk* genes encoding alkane 1-monooxygenase indicated that most cultured hydrocarbon-oxidizing microorganisms (76%) isolated from the water column and sediments in Lake Baikal areas with natural oil seepages contained *alkB* genes that were, in most cases, identical to the *alkB* genes previously detected in *Rhodococcus erythropolis* (Lomakina et al., 2014). The genome of the *Rhodococcus erythropolis* strain isolated from the asphalt structure of Lake Baikal had four types of *alkB* genes encoding alkane-1-monooxygenase (Likhoshvay et al., 2013, 2014). Thereafter, a greater abundance of genes mediating ammonia and methane oxidation, decomposition of aromatic compounds or alkane/methanesulfonate monooxygenase was identified in the deep part of SB compared to their oceanic analogues. The genomes of *Chloroflexota, Ignavibacteria, Bacteroidota, Limnohabitans, Methyloglobus, Myxococcota, Acidobacteriota, Rhodospirillaceae*, and unclassified *Pseudomonadota* contained *alk* genes (Cabello-Yeves et al., 2020). The authors suggested that the presence of endemic microorganisms and genes responsible for the oxidation of various compounds may be due to their entry along with gas-containing fluids from sediments during earthquakes characteristic of the Baikal depression (Cabello-Yeves et al., 2020). At the same time, the detection of *alk* genes in microorganisms of various taxonomic groups in the bathypelagic zone of Lake Baikal, as well as in the MV sediments, may result from horizontal gene transfer (HGT). Over the past 20 years, researchers have provided evidence that HGT plays an important role in the horizontal distribution of the existing catabolic pathways and in the natural construction of new ones. This transfer allows bacteria to quickly adapt to new xenobiotic compounds entering their environment. In particular, researchers proved the transfer of plasmids that carry genes for the oxidation of alkanes, phenol, and BTEX compounds (benzol, toluene, ethylbenzene, and xylene), and naphthalene (Top & Springael, 2003; Boronin & Kosheleva, 2014; Shahi et al., 2017). Therefore, bacteria from areas with natural oil seepages of CB are not only active biodestructors of hydrocarbons but also donors of catabolic genes for horizontal transfer to microorganisms inhabiting other areas of Lake Baikal, which are not directly associated with oil discharge.

12 Future Issues

Lake Baikal is located in a tectonically active zone, thereby suggesting exploratory geological and geophysical works aimed at well-timed identifying new petroleum hydrocarbon discharge sites, assessing the risk and scale of water pollution in Lake Baikal by incoming petroleum hydrocarbons, and their impact on water quality and the lake's biota. In this regard, we plan to continue monitoring in the water area of the lake using traditional limnological approaches and satellite observations, allowing for the study of quantitative and qualitative composition of oil, assessment of the

pollution area, and the ability of microbial communities to decompose oil. Here, it is very important to take into account the impact of global warming on the ecosystem of Lake Baikal, in particular the possible destabilization of GH accumulations and likely risks of this phenomenon. Another relevant task is to assess the carbon contribution of Lake Baikal in the Siberian region and forecasts for different amounts of discharged oil.

Previous metabarcoding studies revealed the complex composition of the microbial community responsible for the degradation of a wide range of substrates (organic carbon and petroleum hydrocarbons), key enzymes of aerobic degradation of hydrocarbons, and genes encoding synthesis of biosurfactants. The question of genome structure of bacteria and archaea involved in these processes and interaction mechanisms of prokaryotes in syntrophic processes of oil oxidation in low temperature and poorly mineralized Lake Baikal remains open. In this context, it is relevant to search for strains of psychrophilic hydrocarbon-oxidizing microorganisms having different genetic systems for hydrocarbon catabolism (OCT, TOL, NDO, NAH, etc.), as well as those involved in the destruction/formation of methane homologues (ethane, propane, and butane). Participants in these processes have been documented in marine ecosystems (Singh et al., 2017; Borrel et al., 2019; Hahn et al., 2020; Wu et al., 2024), while their presence and role in freshwater ecosystems is virtually unknown. The taxonomic affiliation of microorganisms and biogenic ethane production pathways are currently unknown for either marine or freshwater ecosystems. Identification of the participants in the processes, their enzymatic systems that ensure the transformation of short-chain hydrocarbons in the ultra-freshwater body are the next tasks in the study of microbial communities from Lake Baikal.

13 Conclusions

Various heterotrophic communities that provide hydrolysis of organic carbon, fermentation, and respiration inhabit the sedimentary strata of Lake Baikal and other marine and freshwater systems. In the analysis of the 16S rRNA gene libraries, the microbiome of Lake Baikal sediments at the phylum level did not differ from other freshwater bodies (Newton et al., 2011; Winters et al., 2014; Zhang et al., 2015; Vuillemin et al., 2016, 2018; Han et al., 2020). The diversity of microbial communities in the sediments of different Lake Baikal basins overlapped at the phylum level and differed at the level of lower-ranked taxa, with the greatest difference between communities of oxidized and reduced sediments. High relative abundance of metabolically diverse taxa that mineralize OM and are widespread in the sediments of other lakes is characteristic of the communities from oxidized layers of the sedimentary strata (Newton et al., 2011; Han et al., 2020). Lake Baikal sediments contain diverse participants in the early diagenesis such as *Gammaproteobacteria, Chloroflexota, Actinomycetota, Acidobacteriota, Nitrospirota, Bacteroidota,* and *Desulfobacterota* that destruct complex compounds during fermentation and

respiration (Eiler, 2006; Suominen et al., 2021; Phillips et al., 2023). The development of aerobic microorganisms is supported by the supply of OM into sediments after blooming phytoplankton (mainly diatoms, Votintsev et al., 1975). A large contribution of microorganisms involved in the carbon and nitrogen cycles is a distinctive feature of microbiomes in this zone. Our assembled genomes of bacteria and archaea showed a relatively high proportion of nitrogen metabolism genes that ensure nitrification and denitrification, which was also characteristic of metagenomes from the near-bottom area of the bathypelagic zone in Lake Baikal (Cabello-Yeves et al., 2020). At the same time, microbial communities from the surface sedimentary layers of various areas differed significantly in the set of genes for carbon fixation and sulfur metabolism pathways. Genes for enzymes responsible for various stages of carbon oxidation, as well as enzymes of the SOX system oxidizing thiosulfate and sulfite oxidase (*fccAB*), prevailed in the metagenome from oxidized sediments of the reference site, whereas the relative proportion of the key genes for the autotrophic CO_2 fixation pathways was small. In the discharge zone, the presence of methane of different genres explained high values of the *recA/radA* ratio concerning the genes for autotrophic carbon fixation, nitrification, and sulfate reduction pathways and for methane oxidation. Similar to lakes with different trophic states (Sivan et al., 2011; Norði et al., 2013; Weber et al., 2017; Su et al., 2019), Lake Baikal areas with quiet sedimentation form Fe-Mn layers, crusts, and nodules, with participation of a wide range of microorganisms. The presence of functional genes (*mtrABC*, *bc1*, *cbb3*, Cyc1 multiheme cytochromes, *afuABC* Fe transporters, and *pilA* electrically conductive pili) confirms the participation of microorganisms in reduction/oxidation of Fe and Mn in Lake Baikal sediments, which is coupled with the methane cycle.

Like in other oligotrophic lakes (Friese et al., 2021), low mineralization of pore waters and availability of carbon sources and electron acceptors are the main factors limiting the development of microorganisms along the sediment depth in Lake Baikal. In areas with quiet sedimentation, they are exhausted already in the first tens of centimeters of the sedimentary strata, leading to the development of other taxonomic complexes in the underlying sediment layers. A characteristic feature of these complexes is fermentation, and the end product of decomposition is methane (Namsaraev & Zemskaya, 2000; Bukin et al., 2016). Microorganisms involved in anaerobic fermentation, hydrolysis, and final stages of OM destruction develop in reduced sediments of various Lake Baikal ecotopes. In particular, primary and secondary fermenters (*Aminicenantia, Chloroflexota,* and *Caldisericota*), sulphate-reducing bacteria (e.g., *Desulfobacterota* and *Nitrospirota*), methanogens (e.g., *Halobacteriota* and *Methanobacteriota*), and anaerobic methanotrophs of the family *Methanoperedenaceae*, together with members of *Bathyarchaeia, Hadarchaeota,* and *Elusimicrobiota*, whose metabolism revolves around H_2, CO, acetate, and CH_4 (Colman et al., 2017), are indispensable participants in the communities of this zone. Mineralized oil- and gas-bearing fluid flows are additional sources of carbon and energy for benthic communities. Oxygenated near-bottom water enriched in

various oxidizing agents and aerobic microorganisms enters reduced sediments to a depth of more than one meter, which data on the analysis of the chemical composition of pore waters and metabarcoding studies of the 16S rRNA gene confirm. In areas with intense fluid discharge, microbial mats formed, which contained methanogenic and methanotrophic microorganisms. The communities of such sediments demonstrated a high relative abundance of chemoorganotrophic *Atribacterota* and *Bathyarchaeia*, as well as changes in the taxonomic composition of members of the phylum *Chloroflexota*. Metagenome of reduced gas-saturated sediments contained a significant number of genes for the WL and rTCA cycle pathways, as well as genes for assimilatory and dissimilatory sulfate reduction and methane oxidation.

The presence of both methanogenic and methanotrophic archaea at different depths of sediments evidences the possibility of simultaneous occurrence of diagenetic methane generation and AOM. The entry of methane with deep fluid and active methane generation processes in low-temperature conditions and high pressure in Lake Baikal leads to the formation of GHs in the sub-surface sediments of MVs and oil seeps. Taking into account that hydrates in Lake Baikal sediments have different structures, it is very likely that various archaeal taxa participate in their formation or destruction. Moreover, different contributions of methane oxidation/generation processes at different times can also lead to precipitation of authigenic carbonates with different carbon isotopic compositions, as reported for marine sediments (Li et al., 2018).

A phylogenomic analysis of the reconstructed MAGs revealed a distant relationship with genomes from freshwater lakes, marine sediments, and hot springs, which ANI values (74–79%) also confirmed. Despite our small range of assembled genomes, we can assume the presence of endemic taxa at the class and order level, which possess unique metabolic capabilities, ensuring their functioning in oligotrophic and cold-water Lake Baikal.

Acknowledgements This study was carried out within the framework of the LIN SB RAS State Project No. 0279-2021-0006 and supported by the Russian Science Foundation (grant No. 22-14-00084). The survey with MIR deep-sea manned submersible was organized and funded by the Fund for Conservation of Lake Baikal and the Metropol Company. The authors are grateful to O. Khlystov and A. Khabuev for assistance in sampling and choosing the study areas, the crew of the RV "G.Yu. Vereshchagin", and Yu. Vitushenko for translation of the manuscript.

The equipment of the Core Centrum "Genomic Technologies, Proteomics and Cell Biology" in ARRIAM was used in this study. The authors thank the staff of the Irkutsk Supercomputer Center SB RAS for providing access to HPCcluster "Akademik V.M. Matrosov".

Author Contributions TIZ conceptualized the manuscript; TIZ, SVB, SMC, AVL, and ONP wrote the manuscript and prepared the material. All authors read and approved the final manuscript.

Conflict of Interest The authors declare no conflicts of interest.

Ethical Approval This article does not contain any studies with human participants or animals performed by any of the authors.

References

Afonina, T. E., & Karabanov, E. B. (2001). Sources and distribution of organic matter in the pleistocene-holocene sediments of Northern Baikal by the example of n-alkane biomarkers. *Russian Geology and Geophysics, 42*(1–2), 241–253.

Aloisi, G., Bouloubassi, I., Heijs, S. K., Pancost, R. D., Pierre, C., SinningheDamsté, J. S., Gottschal, J. C., Forney, L. J., & Rouchy, J.-M. (2002). CH4-consuming microorganisms and the formation of carbonate crusts at cold seeps. *Earth and Planetary Science Letters, 203*, 195–203. https://doi.org/10.1016/S0012-821X(02)00878-6

Aloisi, G., Wallmann, K., Drews, M., & Bohrmann, G. (2004). Evidence for the submarine weathering of silicate minerals in Black Sea sediments: Possible implications for the marine Li and B cycles. *Geochemistry, Geophysics, Geosystems, 5*(4). https://doi.org/10.1029/2003gc000639

Aloisi, G., Pogodaeva, T. V., Poort, J., Khabuev, A. V., Kazakov, A. V., Akhmanov, G. G., & Khlystov, O. M. (2019). Biogeochemical processes at the Krasniy Yar seepage area (Lake Baikal) and a comparison with oceanic seeps. *Geo-Marine Letters, 39*(1), 59–75. https://doi.org/10.1007/s00367-019-00560-8

Amann, R. I., Ludwig, W., & Schleifer, K. H. (1995). Phylogenetic identification and in situ detection of individual microbial cells without cultivation. *Microbiological Reviews, 59*(1), 143–169. https://doi.org/10.1128/mr.59.1.143-169.1995

Atlas, R. M., & Hazen, T. C. (2011). Oil biodegradation and bioremediation: A tale of the two worst spills in U.S. History. *Environmental Science Technology, 45*(16), 6709–6715. https://doi.org/10.1021/es2013227

Atlas of Lake Baikal. (1993). In G. I. Galaziy (Ed.). (160 p.). Roskartografiya.

Baikal Drilling Project Members (BDP-93). (1997). Preliminary results of the first scientific drilling on Lake Baikal, Buguldeika site, southeastern Siberia. *Quaternary International, 37*, 3–17.

Baker, B. J., Saw, J. H., Lind, A. E., Lazar, C. S., Hinrichs, K.-U., Teske, A. P., & Ettema, T. J. G. (2016). Genomic inference of the metabolism of cosmopolitan subsurface *Archaea, Hadesarchaea*. *Nature Microbiology, 1*, 16002. Advance online publication. https://doi.org/10.1038/nmicrobiol.2016.2

Baturin, G. N., & Granina, L. Z. (2009). Rare earth elements in ferromanganese formations of Lake Baikal. *Doklady Earth Sciences, 428*(4), 511–514. https://doi.org/10.1134/S1028334X0907037X

Baturin, G. N., Yushina, I. G., & Zolotykh, E. O. (2009). Variations in the elemental composition of ferromanganese structures from Lake Baikal. *Oceanology, 49*(4), 505–514. https://doi.org/10.1134/s0001437009040092

Baumann, K. B. L., Thoma, R., Callbeck, C. M., Niederdorfer, R., Schubert, C. J., Müller, B., Lever, M. A., & Bürgmann, H. (2022). Microbial nitrogen transformation potential in sediments of two contrasting lakes is spatially structured but seasonally stable. *MSphere, 7*(1) Advance online publication. https://doi.org/10.1128/msphere.01013-21

Bernhard, A. E., Colbert, D., McManus, J., & Field, K. G. (2005). Microbial community dynamics based on 16S rRNA gene profiles in a Pacific Northwest estuary and its tributaries. *FEMS Microbiology Ecology, 52*(1), 115–128. https://doi.org/10.1016/j.femsec.2004.10.016

Bogoyavlensky, V. I. (2023). New data on mud volcanism in the arctic on the Yamal Peninsula. *Doklady Earth Sciences, 512*, 847–853. https://doi.org/10.1134/S1028334X23601116

Bonch-Osmolovskaya, E. A., Miroshnichenko, M. L., Lebedinsky, A. V., Chernyh, N. A., Nazina, T. N., Ivoilov, V. S., Belyaev, S. S., Boulygina, E. S., Lysov, Y. P., Perov, A. N., Mirzabekov, A. D., Hippe, H., Stackebrandt, E., L'Haridon, S., & Jeanthon, C. (2003). Radioisotopic, culture-based, and oligonucleotide microchip analyses of thermophilic microbial communities in a continental high-temperature petroleum reservoir. *Applied Environmental Microbiology, 69*(10), 6143–6151. https://doi.org/10.1128/aem.69.10.6143-6151.2003

Bondarenko, N. A., Guselnikova, N. E., Logacheva, N. F., & Pomazkina, G. V. (1996). Spatial distribution of phytoplankton in Lake Baikal, spring 1991. *Freshwater Biology, 35*(3), 517–523. https://doi.org/10.1111/j.1365-2427.1996.tb01765.x

Bondarenko, N. A., Ozersky, T., Obolkina, L. A., Tikhonova, I. V., Sorokovikova, E. G., Sakirko, M. V., Potapov, S. A., Blinov, V. V., Zhdanov, A. A., & Belykh, O. I. (2019). Recent changes in the spring microplankton of Lake Baikal, Russia. *Limnologica, 75*, 19–29. https://doi.org/10.1016/j.limno.2019.01.002

Bornemann, M., Bussmann, I., Tichy, L., Deutzmann, J., Schink, B., & Pester, M. (2016). Methane release from sediment seeps to the atmosphere is counteracted by highly active *Methylococcaceae* in the water column of deep oligotrophic Lake Constance. *FEMS Microbiology Ecology, 92*(8), fiw123. Advance online publication. https://doi.org/10.1093/femsec/fiw123

Boronin, A. M., & Kosheleva, I. A. (2014). The role of catabolic plasmids in biodegradation of petroleum hydrocarbons. In G. Cao & R. Orrù (Eds.), *Current environmental issues and challenges*. Springer. https://doi.org/10.1007/978-94-017-8777-2_9

Borrel, G., Jézéquel, D., Biderre-Petit, C., Morel-Desrosiers, N., Morel, J.-P., Peyret, P., Fonty, G., & Lehours, A.-C. (2011). Production and consumption of methane in freshwater lake ecosystems. *Research in Microbiology, 162*(9), 832–847. https://doi.org/10.1016/j.resmic.2011.06.004

Borrel, G., Lehours, A.-C., Crouzet, O., Jézéquel, D., Rockne, K., Kulczak, A., Duffaud, E., Joblin, K., & Fonty, G. (2012). Stratification of archaea in the deep sediments of a freshwater meromictic lake: Vertical shift from methanogenic to uncultured archaeal lineages. *PLoS One, 7*(8), e43346. https://doi.org/10.1371/journal.pone.0043346

Borrel, G., Adam, P. S., McKay, L. J., Chen, L.-X., Sierra-García, I. N., Sieber, C. M. K., Letourneur, Q., Ghozlane, A., Andersen, G. L., Li, W.-J., Hallam, S. J., Muyzer, G., de Oliveira, V. M., Inskeep, W. P., Banfield, J. F., & Gribaldo, S. (2019). Wide diversity of methane and short-chain alkane metabolisms in uncultured archaea. *Nature Microbiology, 4*(4), 603–613. https://doi.org/10.1038/s41564-019-0363-3

Bukin, S. V., Pavlova, O. N., Manakov, A. Y., Kostyreva, E. A., Chernitsyna, S. M., Mamaeva, E. V., Pogodaeva, T. V., & Zemskaya, T. I. (2016). The ability of microbial community of Lake Baikal bottom sediments associated with gas discharge to carry out the transformation of organic matter under thermobaric conditions. *Frontiers in Microbiology, 7*. Advance online publication. https://doi.org/10.3389/fmicb.2016.00690

Bukin, S. V., Pavlova, O. N., Kalmychkov, G. V., Ivanov, V. G., Pogodaeva, T. V., Galach'yants, Y. P., Bukin, Y. S., Khabuev, A. V., & Zemskaya, T. I. (2018). Substrate specificity of methanogenic communities from Lake Baikal bottom sediments associated with hydrocarbon gas discharge. *Microbiology, 87*(4), 549–558. https://doi.org/10.1134/s0026261718040045

Bukin, Y. S., Galachyants, Y. P., Morozov, I. V., Bukin, S. V., Zakharenko, A. S., & Zemskaya, T. I. (2019). The effect of 16S rRNA region choice on bacterial community metabarcoding results. *Scientific Data, 6*(1) Advance online publication. https://doi.org/10.1038/sdata.2019.7

Bukin, S. V., Lomakina, A. V., Rusanov, I. I., Ivanov, V. G., Pogodaeva, T. V., Kolesnikov, P. M., Shubenkova, O. V., Zakharenko, A. S., Bukin, Y. S., Rodriguez-Valera, F., & Zemskaya, T. I. (2023). Structure and metabolic potential of the microbiome of Lake Baikal upper sediment layers determined by metagenome analysis. *Microbiology, 96*(Suppl. 1), 56–62.

Cabello-Yeves, P. J., Zemskaya, T. I., Rosselli, R., Coutinho, F. H., Zakharenko, A. S., Blinov, V. V., & Rodriguez-Valera, F. (2018). Genomes of novel microbial lineages assembled from the sub-ice waters of Lake Baikal. *Applied and Environmental Microbiology, 84*(1) Advance online publication. https://doi.org/10.1128/aem.02132-17

Cabello-Yeves, P. J., Zemskaya, T. I., Zakharenko, A. S., Sakirko, M. V., Ivanov, V. G., Ghai, R., & Rodriguez-Valera, F. (2020). Microbiome of the deep Lake Baikal, a unique oxic bathypelagic habitat. *Limnology and Oceanography, 65*(7), 1471–1488. https://doi.org/10.1002/lno.11401

Cai, C., Leu, A. O., Xie, G.-J., Guo, J., Feng, Y., Zhao, J.-X., Tyson, G. W., Yuan, Z., & Hu, S. (2018). A methanotrophic archaeon couples anaerobic oxidation of methane to Fe(III) reduction. *The ISME Journal, 12*(8), 1929–1939. https://doi.org/10.1038/s41396-018-0109-x

Callender, E., & Granina, L. (1992). Transition metal geochemistry of sedimentary pore fluids associated with hydrothermal activity in Lake Baikal. In Y. K. Kharaka & A. S. Maest (Eds.),

Water-Rock Interaction. Proceedings of the 7th International Symposium on Water-Rock Interaction (pp. 621–626). Park City, Utah.

Capone, D. G., & Kiene, R. P. (1988). Comparison of microbial dynamics in marine and freshwater sediments: Contrasts in anaerobic carbon catabolism. *Limnology and Oceanography, 33*(4), 725–749. https://doi.org/10.4319/lo.1988.33.4_part_2.0725

Carr, S. A., Orcutt, B. N., Mandernack, K. W., & Spear, J. R. (2015). Abundant Atribacteria in deep marine sediment from the Adélie Basin, Antarctica. *Frontiers in Microbiology, 6*, 872. Advance online publication. https://doi.org/10.3389/fmicb.2015.00872

Castelle, C. J., Hug, L. A., Wrighton, K. C., Thomas, B. C., Williams, K. H., Wu, D., Tringe, S. G., Singer, S. W., Eisen, J. A., & Banfield, J. F. (2013). Extraordinary phylogenetic diversity and metabolic versatility in aquifer sediment. *Nature Communications, 4*(1) Advance online publication. https://doi.org/10.1038/ncomms3120

Castelle, C. J., Brown, C. T., Anantharaman, K., Probst, A. J., Huang, R. H., & Banfirld, J. F. (2018). Biosynthetic capacity, metabolic variety and unusual biology in the CPR and DPANN radiations. *Nature Reviews Microbiology, 16*, 629–645. https://doi.org/10.1038/s41579-018-0076-2

Chen, Y., Dumont, M. G., Cébron, A., & Murrell, J. C. (2007). Identification of active methanotrophs in a landfill cover soil through detection of expression of 16S rRNA and functional genes. *Environmental Microbiology, 9*(11), 2855–2869. https://doi.org/10.1111/j.1462-2920.2007.01401.x

Chernitsina, S. M., Zemskaya, T. I., Vorob'eva, S. S., Shubenkova, O. V., Khlystov, O. M., & Kostornova, T. Y. (2007). Comparative molecular biological analysis of the microbial community of the Holocene and Pleistocene deposits of Posol'skaya Shoal, Lake Baikal. *Microbiology, 76*(1), 102–111. https://doi.org/10.1134/s0026261707010146

Chernitsyna, S. M., Mamaeva, E. V., Lomakina, A. V., Pogodaeva, T. V., Galach'yants, Y. P., Bukin, S. V., Pimenov, N. V., Khlystov, O. M., & Zemskaya, T. I. (2016a). Phylogenetic diversity of microbial communities of the Posolsk Bank bottom sediments, Lake Baikal. *Microbiology, 85*(6), 672–680. https://doi.org/10.1134/s0026261716060060

Chernitsyna, S. M., Khal'zov, I. A., Khanaeva, T. A., Morozov, I. V., Klimenkov, I. V., Pimenov, N. V., & Zemskaya, T. I. (2016b). Microbial community associated with *Thioploca* sp. sheaths in the area of the Posolsk Bank methane seep, Southern Baikal. *Microbiology, 85*(5), 562–569. https://doi.org/10.1134/s0026261716050027

Chernitsyna, S. M., Elovskaya, I. S., Bukin, S. V., Bukin, Y. S., Pogodaeva, T. V., Kwon, D. A., & Zemskaya, T. I. (2024). Genomic and morphological characterization of a new *Thiothrix* species from a sulfide hot spring of the Zmeinaya bay (Northern Baikal, Russia). *Antonie Van Leeuwenhoek, 117*, 23. Advance online publication.. https://doi.org/10.1007/s10482-023-01918-w

Cho, K., Zholi, A., Frabutt, D., Flood, M., Floyd, D., & Tiquia, S. M. (2012). Linking bacterial diversity and geochemistry of uranium-contaminated groundwater. *Environmental Technology, 33*, 1629–1640. https://doi.org/10.1080/09593330.2011.641036

Ciancio Casalini, L., Piazza, A., Masotti, F., Garavaglia, B. S., Ottado, J., & Gottig, N. (2022). Manganese oxidation counteracts the deleterious effect of low temperatures on biofilm formation in *pseudomonas* sp. MOB-449. *Frontiers in Molecular Biosciences, 9*. Advance online publication. https://doi.org/10.3389/fmolb.2022.1015582

Claesson, M. J., Wang, Q., O'Sullivan, O., Greene-Diniz, R., Cole, J. R., Ross, R. P., & O'Toole, P. W. (2010). Comparison of two next-generation sequencing technologies for resolving highly complex microbiota composition using tandem variable 16S rRNA gene regions. *Nucleic Acids Research, 38*(22), e200–e200. Advance online publication. https://doi.org/10.1093/nar/gkq873

Colman, S. M., Karabanov, E. B., & Badardinov, A. A. (1991). Preliminary results of high resolution seismic-reflection surveys of Lake Baikal, Siberia. *Eos, Transactions American Geophysical Union, 72*(42), 306–315.

Colman, S. M., Williams, D. F., Hearn, P. P. Y., Grachev, M., Kuzmin, M., & Khahkaev, B. N. (1992). Initial results of U.S.-Soviet paleoclimate study of Lake Baikal. *Eos, Transactions American Geophysical Union, 73*(43), 457–462. https://doi.org/10.1029/91eo00340

Colman, G., Tanna, A., Efstratiou, A., & Gaworzewska, E. T. (1993). The serotypes of Streptococcus pyogenes present in Britain during 1980–1990 and their association with disease. *Journal of Medical Microbiology, 39*(3), 165–178. https://doi.org/10.1099/00222615-39-3-165

Colman, S. M., Peck, J. A., Karabanov, E. B., Carter, S. J., Bradbury, J. P., King, J. W., & Williams, D. F. (1995). Continental climate response to orbital forcing from biogenic silica records in Lake Baikal. *Nature, 378*(6559), 769–771. https://doi.org/10.1038/378769a0

Colman, S. M., Peck, J. A., Hatton, J., Karabanov, E. B., & King, J. W. (1999). Biogenic silica records from the BDP93 drill site and adjacent areas of the Selenga Delta, Lake Baikal, Siberia. *Journal of Paleolimnology, 21*, 9–17.

Colman, D. R., Poudel, S., Stamps, B. W., Boyd, E. S., & Spear, J. R. (2017). The deep, hot biosphere: Twenty-five years of retrospection. *Proceedings of the National Academy of Sciences of The United States of America, 114*(27), 6895–6903. https://doi.org/10.1073/pnas.1701266114

Coutinho, F. H., Cabello-Yeves, P. J., Gonzalez-Serrano, R., Rosselli, R., López-Pérez, M., Zemskaya, T. I., Zakharenko, A. S., Ivanov, V. G., & Rodriguez-Valera, F. (2020). New viral biogeochemical roles revealed through metagenomic analysis of Lake Baikal. *Microbiome, 8*(1), 1–15. https://doi.org/10.1186/s40168-020-00936-4

Crane, K., Hecker, B., & Golubev, V. (1991). Hydrothermal vents in Lake Baikal. *Nature, 350*(6316), 281–281. https://doi.org/10.1038/350281a0

D'Angelo, T., Goordial, J., Lindsay, M. R., McGonigle, J., Booker, A., Moser, D., Stepanauskus, R., & Orcutt, B. N. (2023). Replicated life-history patterns and subsurface origins of the bacterial sister phyla *Nitrospirota* and *Nitrospinota*. *The ISME Journal, 17*(6), 891–902. https://doi.org/10.1038/s41396-023-01397-x

Dagurova, O. P., Namsaraev, B. B., Kozyreva, L. P., Zemskaya, T. I., & Dulov, L. E. (2004). Bacterial processes of the methane cycle in bottom sediments of Lake Baikal. *Microbiology, 74*(2), 202–210. https://doi.org/10.1023/B:MICI.0000023990.71983.c1

De Anda, V., Chen, L.-X., Dombrowski, N., Hua, Z.-S., Jiang, H.-C., Banfield, J. F., Li, W.-J., & Baker, B. J. (2021). *Brockarchaeota*, a novel archaeal phylum with unique and versatile carbon cycling pathways. *Nature Communications, 12*(1), 2404. https://doi.org/10.1038/s41467-021-22736-6

De Batist, M., Klerkx, J., Van Rensbergen, P., Vanneste, M., Poort, J., Golmshtok, A. Y., Kremlev, A. A., Khlystov, O. M., & Krinitsky, P. (2002). Active hydrate destabilization in Lake Baikal, Siberia? *Terra Nova, 14*(6), 436–442. https://doi.org/10.1046/j.1365-3121.2002.00449.x

Dedysh, S. N., & Yilmaz, P. (2018). Refining the taxonomic structure of the phylum *Acidobacteria*. *International Journal of Systematic and Evolutionary Microbiology, 68*(12), 3796–3806. https://doi.org/10.1099/ijsem.0.003062

Dedysh, S. N., Derakshani, M., & Liesack, W. (2001). Detection and enumeration of methanotrophs in acidic Sphagnum peat by 16S rRNA fluorescence in situ hybridization, including the use of newly developed oligonucleotide probes for *Methylocella palustris*. *Applied and Environmental Microbiology, 67*(10), 4850–4857. https://doi.org/10.1128/aem.67.10.4850-4857.2001

Deutzmann, J. S., & Schink, B. (2011). Anaerobic oxidation of methane in sediments of Lake Constance, an oligotrophic freshwater lake. *Applied and Environmental Microbiology, 77*(13), 4429–4436. https://doi.org/10.1128/aem.00340-11

Ding, H., & Valentine, D. L. (2008). Methanotrophic bacteria occupy benthic microbial mats in shallow marine hydrocarbon seeps, Coal Oil Point, California. *Journal of Geophysical Research: Biogeosciences, 113*(G1) Advance online publication. https://doi.org/10.1029/2007jg000537

Ding, Z.-W., Ding, J., Fu, L., Zhang, F., & Zeng, R. J. (2014). Simultaneous enrichment of denitrifying methanotrophs and anammox bacteria. *Applied Microbiology and Biotechnology, 98*(24), 10211–10221. https://doi.org/10.1007/s00253-014-5936-8

Dittrich, M., & Obst, M. (2004). Are picoplankton responsible for calcite precipitation in lakes? *AMBIO: A Journal of the Human Environment, 33*, 559–564. https://doi.org/10.1579/0044-7447-33.8.559

Dittrich, M., Moreau, L., Gordon, J., Quazi, S., Palermo, C., Fulthorpe, R., Katsev, S., Bollmann, J., & Chesnyuk, A. (2015). Geomicrobiology of iron layers in the sediment of Lake Superior. *Aquatic Geochemistry, 21*(2–4), 123–140. https://doi.org/10.1007/s10498-015-9258-y

Dong, X., Greening, C., Rattray, J. E., Chakraborty, A., Chuvochina, M., Mayumi, D., Dolfing, J., Li, C., Brooks, J. M., Bernard, B. B., Groves, R. A., Lewis, I. A., & Hubert, C. R. J. (2019). Metabolic potential of uncultured bacteria and archaea associated with petroleum seepage in deep-sea sediments. *Nature Communications, 10*(1), 1816. Advance online publication. https://doi.org/10.1038/s41467-019-09747-0

Dubinina, G. A. (1976). Study of the ecology of iron bacteria in fresh water basins. *Biology Bulletin, 4*, 575–592.

Dunbar, J., Takala, S., Barns, S. M., Davis, J. A., & Kuske, C. R. (1999). Levels of bacterial community diversity in four arid soils compared by cultivation and 16S rRNA gene cloning. *Applied and Environmental Microbiology, 65*(4), 1662–1669. https://doi.org/10.1128/AEM.65.4.1662-1669.1999

Dupraz, C., Visscher, P. T., Baumgartner, L. K., & Reid, R. P. (2004). Microbe-mineral interactions: Early carbonate precipitation in a hypersaline lake (Eleuthera Island, Bahamas). *Sedimentology, 51*, 745–765. https://doi.org/10.1111/j.1365-3091.2004.00649.x

Dymond, J., Collier, R. W., & Watwood, M. E. (1989). Bacterial mats from Crater Lake, Oregon and their relationship to possible deep-lake hydrothermal venting. *Nature, 342*(6250), 673–675. https://doi.org/10.1038/342673a0

Edgington, D. N., & Callender, E. (1970). Minor element geochemistry of Lake Michigan ferromanganese nodules. *Earth and Planetary Science Letters, 8*(2), 97–100. https://doi.org/10.1016/0012-821x(70)90157-3

Eiler, A. (2006). Evidence for the ubiquity of mixotrophic bacteria in the upper ocean: Implications and consequences. *Applied and Environmental Microbiology, 72*(12), 7431–7437. https://doi.org/10.1128/aem.01559-06

Ettwig, K. F., Butler, M. K., Le Paslier, D., Pelletier, E., Mangenot, S., Kuypers, M. M. M., Schreiber, F., Dutilh, B. E., Zedelius, J., de Beer, D., Gloerich, J., Wessels, H. J. C. T., van Alen, T., Luesken, F., Wu, M. L., van de Pas-Schoonen, K. T., Op den Camp, H. J. M., Janssen-Megens, E. M., Francoijs, K.-J., Stunnenberg, H., Weissenbach, J., Jetten, M. S. M., & Strous, M. (2010). Nitrite-driven anaerobic methane oxidation by oxygenic bacteria. *Nature, 464*(7288), 543–548. https://doi.org/10.1038/nature08883

Evans, P. N., Parks, D. H., Chadwick, G. L., Robbins, S. J., Orphan, V. J., Golding, S. D., & Tyson, G. W. (2015). Methane metabolism in the archaeal phylum *Bathyarchaeota* revealed by genome-centric metagenomics. *Science, 350*(6259), 434–438. https://doi.org/10.1126/science.aac7745

Falkner, K. K., Measures, C. I., Herbelin, S. E., Edmond, J. M., & Weiss, R. F. (1991). The major and minor element geochemistry of Lake Baikal. *Limnology and Oceanography, 36*, 413–423. https://doi.org/10.4319/lo.1991.36.3.0413

Farag, I. F., Davis, J. P., Youssef, N. H., & Elshahed, M. S. (2014). Global patterns of abundance, diversity and community structure of the *Aminicenantes* (Candidate Phylum OP8). *PLoS One, 9*(3), e92139. Advance online publication. 10.1371-/journal.pone.0092139

Fincker, M., Huber, J. A., Orphan, V. J., Rappé, M. S., Teske, A., & Spormann, A. M. (2020). Metabolic strategies of marine subseafloor Chloroflexi inferred from genome reconstructions. *Environmental Microbiology, 22*(8), 3188–3204. https://doi.org/10.1111/1462-2920.15061

Flood, M., Frabutt, D., Floyd, D., Powers, A., Ezegwe, U., Devol, A., & Tiquia-Arashiro, S. M. (2015). Ammonia-oxidizing bacteria and archaea in sediments of the Gulf of Mexico. *Environmental Technology, 36*, 124–135. https://doi.org/10.1080/09593330.2014.942385

Florensov, N. A. (1966). The Baikal Rift Zone. Natural Resources Canada/CMSS/Information Management. https://doi.org/10.4095/103367.

Francke, A., Wagner, B., Just, J., Leicher, N., Gromig, R., Baumgarten, H., Vogel, H., Lacey, J. H., Sadori, L., Wonik, T., Leng, M. J., Zanchetta, G., Sulpizio, R., & Giaccio, B. (2016). Sedimentological processes and environmental variability at Lake Ohrid (Macedonia, Albania)

between 637 ka and the present. *Biogeosciences, 13*, 1179–1196. https://doi.org/10.5194/bg-13-1179-2016

Friese, A., Bauer, K., Glombitza, C., Ordoñez, L., Ariztegui, D., Heuer, V. B., Vuillemin, A., Henny, C., Nomosatryo, S., Simister, R., Wagner, D., Bijaksana, S., Vogel, H., Melles, M., Russell, J. M., Crowe, S. A., & Kallmeyer, J. (2021). Organic matter mineralization in modern and ancient ferruginous sediments. *Nature Communications, 12*(1). https://doi.org/10.1038/s41467-021-22453-0

Fuchs, A., Lyautey, E., Montuelle, B., & Casper, P. (2016). Effects of increasing temperatures on methane concentrations and methanogenesis during experimental incubation of sediments from oligotrophic and mesotrophic lakes. *Journal of Geophysical Research: Biogeosciences, 121*(5), 1394–1406. https://doi.org/10.1002/2016jg003328

Fussmann, D., von Hoyningen-Huene, A., Reimer, A., Schneider, D., Babková, H., Peticzka, R., Maier, A., Arp, G., Daniel, R., & Meister, P. (2019). Authigenic formation of Ca-Mg carbonates in the shallow alkaline Lake Neusiedl, Austria. *Biogeosciences, 17*, 2085–2106. https://doi.org/10.5194/bg-2019-449

Galushko, A., & Kuever, J. (2019). Desulfobacca. In M. E. Trujillo, S. Dedysh, P. DeVos, B. Hedlund, P. Kämpfer, F. A. Rainey, & W. B. Whitman (Eds.), *Bergey's manual of systematics of archaea and bacteria*. Advance online publication. https://doi.org/10.1002/9781118960608.gbm01061.pub2

Garankina, V. P., & Dagurova, O. P. (2008). Distribution and abundance of microorganisms in the soils of the Posolsky Sor Bay of Lake Baikal. *Bulletin of the Buryat State University Biology. Geography, 4*, 71–72.

Garankina, V. P., Dagurova, O. P., Dambaev, V. B., & Buriukhaev, S. P. (2009). Distribution and activity of microorganisms in the Proval Bay of Lake Baikal. *Bulletin of Buryat State University Biology. Geography, 4*, 84–88.

Gebruk, A. V. (Ed.). (2002). *Biology of hydrothermal systems*. KMK Press.

Gentile, M., Yan, T., Tiquia, S. M., et al. (2006). Stability in a denitrifying fluidized bed reactor. *Microbial Ecology, 52*, 311–321. https://doi.org/10.1007/s00248-006-9024-1

Gieskes, J., Mahn, C., Day, S., Martin, J. B., Greinert, J., Rathburn, T., & McAdoo, B. (2005). A study of the chemistry of pore fluids and authigenic carbonates in methane seep environments: Kodiak Trench, Hydrate Ridge, Monterey Bay, and Eel River Basin. *Chemical Geology, 220*, 329–345. https://doi.org/10.1016/j.chemgeo.2005.04.002

Gittins, D. A., Bhatnagar, S., & Hubert, C. R. J. (2023). Environmental selection and biogeography shape the microbiome of subsurface petroleum reservoirs. *MSystems, 8*(2) Advance online publication. https://doi.org/10.1128/msystems.00884-22

Glombitza, C., Stockhecke, M., Schubert, C. J., Vetter, A., & Kallmeyer, J. (2013). Sulfate reduction controlled by organic matter availability in deep sediment cores from the saline, alkaline Lake Van (Eastern Anatolia, Turkey). *Frontiers in Microbiology, 4*. Advance online publication. https://doi.org/10.3389/fmicb.2013.00209

Golmshtok, A. Y., Duchkov, A. D., Hutchinson, D. R., & Khanukaev, S. B. (2000). Heat flow and gas hydrates of the Baikal Rift Zone. *International Journal of Earth Sciences, 89*(2), 193–211. https://doi.org/10.1007/s005319900071

Golubev, V. A. (2007). *Conductive and convective heat output in the Baikal rift zone*. Geo.

Golubev, V. A., Klerkx, J., & Kipfer, R. (1993). Heat flow, hydrothermal vents and static stability of discharging thermal water in Lake Baikal (South-Eastern Siberia). *Bulletin des Centres de Recherches Exploration-Production Elf-Aquitaine, 17*(1), 53–66.

Goman, G. A. (1975). Bacterial reduction of sulfates and oxidation of sulfides in Baikal sediments. *Hydrobiological Journal, 11*(5), 18–22.

Gorshkov, A., Pavlova, O., Khlystov, O., & Zemskaya, T. (2020). Fractioning of petroleum hydrocarbons from seeped oil as a factor of purity preservation of water in Lake Baikal (Russia). *Journal of Great Lakes Research, 46*(1), 115–122. https://doi.org/10.1016/j.jglr.2019.10.010

Grachev, M. A. (1994). Formation of the Baikal International Center for Ecological Research. *Ecology International Bulletin, 21*, 75–88.

Grachev, M., Fialkov, V., Nakamura, T., Ohta, T., & Kawai, T. (1995). Extant fauna of ancient carbon. *Nature, 374*, 123–124.

Granin, N. G., & Granina, L. Z. (2002). Gas hydrates and gas venting in lake Baikal. *Russian Geology and Geophysics, 43*(7), 629–637.

Granina, L. Z. (2008). *Early diagenesis of bottom sediments in Lake Baikal*. Academical Publish House "Geo".

Granina, L. Z., Grachev, M. A., Karabanov, E. B., Kuptsov, V. M., Shimaraeva, M. K., & Williams, D. F. (1993). Accumulation of biogenic silica in the bottom sediments of Baikal. *Geology and Geophysics, 34*(10–11), 149–159.

Granina, L. Z., Callender, E., Clerks, J., Golobokova, L. P., & Gorbunova, L. A. (1999). Features of biogeochemistry of bottom sediments in the area of hydrothermal appearance on Baikal. In *Geochemistry of landscapes, human paleoecology and ethnogenesis. Abstracts of the International Symposium 1999* (pp. 297–299). Buryat Scientific Centre of Siberian Department of Russian Academy of Sciences.

Greinert, J., Bohrmann, G., & Suess, E. (2001). Gas hydrate-associated carbonates and ethane-venting at hydrate ridge: classification, distribution, and origin of authigenic lithologies. In W. P. Dillon (Ed.), *Natural gas hydrates* (pp. 99–113). American Geophysical Union.

Hachikubo, A., Minami, H., Yamashita, S., Khabuev, A., Krylov, A., Kalmychkov, G., Poort, J., De Batist, M., Chenskiy, A., Manakov, A., & Khlystov, O. (2020). Characteristics of hydrate-bound gas retrieved at the Kedr mud volcano (southern Lake Baikal). *Scientific Reports, 10*(1) Advance online publication. https://doi.org/10.1038/s41598-020-71410-2

Hachikubo, A., Minami, H., Sakagami, H., Yamashita, S., Krylov, A., Kalmychkov, G., Poort, J., De Batist, M., Manakov, A., & Khlystov, O. (2023). Characteristics and varieties of gases enclathrated in natural gas hydrates retrieved at Lake Baikal. *Scientific Reports, 13*(1) Advance online publication. https://doi.org/10.1038/s41598-023-31669-7

Hahn, A., Kliem, P., Oehlerich, M., Ohlendorf, C., & Zolitschka, B. (2014). Elemental composition of the Laguna Potrok Aike sediment sequence reveals paleoclimatic changes over the past 51 ka in southern Patagonia, Argentina. *Journal of Paleolimnology, 52*, 349–366. https://doi.org/10.1007/s10933-014-9798-y

Hahn, C. J., Laso-Pérez, R., Vulcano, F., Vaziourakis, K.-M., Stokke, R., Steen, I. H., Teske, A., Boetius, A., Liebeke, M., Amann, R., Knittel, K., & Wegener, G. (2020). "CandidatusEthanoperedens," a thermophilic genus of *Archaea* mediating the anaerobic oxidation of ethane. *MBio, 11*(2) Advance online publication. https://doi.org/10.1128/mbio.00600-20

Hampton, S. E., Izmest'eva, L. R., Moore, M. V., Katz, S. L., Dennis, B., & Silow, E. A. (2008). Sixty years of environmental change in the world's largest freshwater Lake—Lake Baikal, Siberia. *Global Change Biology, 14*, 1947–1958. https://doi.org/10.1111/j.1365-2486.2008.01616.x

Han, X., Schubert, C. J., Fiskal, A., Dubois, N., & Lever, M. A. (2020). Eutrophication as a driver of microbial community structure in lake sediments. *Environmental Microbiology, 22*(8), 3446–3462. https://doi.org/10.1111/1462-2920.15115

Haro-Moreno, J. M., Cabello-Yeves, P. J., Garcillán-Barcia, M. P., Zakharenko, A., Zemskaya, T. I., & Rodriguez-Valera, F. (2023). A novel and diverse group of CandidatusPatescibacteria from bathypelagic Lake Baikal revealed through long-read metagenomics. *Environmental Microbiomes, 18*(1) Advance online publication. https://doi.org/10.1186/s40793-023-00473-1

Hauck, S. (2001). Ferrous iron oxidation by denitrifying bacteria in profundal sediments of a deep lake (Lake Constance). *FEMS Microbiology Ecology, 37*(2), 127–134. https://doi.org/10.1016/s0168-6496(01)00153-2

He, R., Wooller, M. J., Pohlman, J. W., Tiedje, J. M., & Leigh, M. B. (2015). Methane-derived carbon flow through microbial communities in arctic lake sediments. *Environmental Microbiology, 17*(9), 3233–3250. https://doi.org/10.1111/1462-2920.12773

He, Y., Li, M., Perumal, V., Feng, X., Fang, J., Xie, J., Sievert, S. M., & Wang, F. (2016). Genomic and enzymatic evidence for acetogenesis among multiple lineages of the archaeal phylum *Bathyarchaeota* widespread in marine sediments. *Nature Microbiology, 1*(6). https://doi.org/10.1038/nmicrobiol.2016.35

Heider, J., & Schühle, K. (2013). Anaerobic biodegradation of hydrocarbons including methane. In E. Rosenberg, E. DeLong, S. Lory, E. Stackebrandt, & F. Thompson (Eds.), *The prokaryotes* (pp. 605–634). Springer. https://doi.org/10.1007/978-3-642-30141-4_80

Hou, J., Wang, Y., Zhu, P., Yang, N., Liang, L., Yu, T., Niu, M., Konhauser, K., Woodcroft, B. J., & Wang, F. (2023). Taxonomic and carbon metabolic diversification of *Bathyarchaeia* during its coevolution history with early Earth surface environment. *Science Advances, 9*(27). https://doi.org/10.1126/sciadv.adf5069

Hua, Z.-S., Wang, Y.-L., Evans, P. N., Qu, Y.-N., Goh, K. M., Rao, Y.-Z., Qi, Y.-L., Li, Y.-X., Huang, M.-J., Jiao, J.-Y., Chen, Y.-T., Mao, Y.-P., Shu, W.-S., Hozzein, W., Hedlund, B. P., Tyson, G. W., Zhang, T., & Li, W.-J. (2019). Insights into the ecological roles and evolution of methyl-coenzyme M reductase-containing hot spring *archaea. nature. Communications, 10*(1) Advance online publication. https://doi.org/10.1038/s41467-019-12574-y

Huang, W., Chen, X., Wang, K., Chen, J., Zheng, B., & Jiang, X. (2019). Comparison among the microbial communities in the lake, lake wetland, and estuary sediments of a plain river network. *Microbiology Open, 8*(2) Advance online publication. https://doi.org/10.1002/mbo3.644

Huang, W.-C., Liu, Y., Zhang, X., Zhang, C.-J., Zou, D., Zheng, S., Xu, W., Luo, Z., Liu, F., & Li, M. (2021). Comparative genomic analysis reveals metabolic flexibility of *Woesearchaeota. Nature Communications, 12*(1), 5281. Advance online publication. https://doi.org/10.1038/s41467-021-25565-9

Hug, L. A., Thomas, B. C., Sharon, I., Brown, C. T., Sharma, R., Hettich, R. L., Wilkins, M. J., Williams, K. H., Singh, A., & Banfield, J. F. (2015). Critical biogeochemical functions in the subsurface are associated with bacteria from new phyla and little studied lineages. *Environmental Microbiology, 18*(1), 159–173. https://doi.org/10.1111/1462-2920.12930

Hutchinson, D. R., Golmshtok, A. J., Zonenshain, L. P., Moore, T. C., Scholz, C. A., & Klitgord, K. D. (1992). Depositional and tectonic framework of the rift basins of Lake Baikal from multichannel seismic data. *Geology, 20*(7) Advance online publication. https://doi.org/10.1130/0091-7613(1992)020<0589:datfot>2.3.co;2

Ingalls, M., Frantz, C. M., Snell, K. E., & Trower, E. J. (2020). Carbonate facies-specific stable isotope data record climate, hydrology, and microbial communities in Great Salt Lake, UT. *Geobiology, 18*, 566–593. https://doi.org/10.1111/gbi.12386

Jones, D. M., Head, I. M., Gray, N. D., Adams, J. J., Rowan, A. K., Aitken, C. M., Bennett, B., Huang, H., Brown, A., Bowler, B. F. J., Oldenburg, T., Erdmann, M., & Larter, S. R. (2008). Crude-oil biodegradation via methanogenesis in subsurface petroleum reservoirs. *Nature, 451*(7175), 176–180. https://doi.org/10.1038/nature06484

Joye, S. B. (2020). The geology and biogeochemistry of hydrocarbon seeps. *Annual Review of Earth and Planetary Sciences, 48*(1), 205–231. https://doi.org/10.1146/annurev-earth-063016-020052

Jung, M.-Y., Sedlacek, C. J., Kits, K. D., Mueller, A. J., Rhee, S.-K., Hink, L., Nicol, G. W., Bayer, B., Lehtovirta-Morley, L., Wright, C., de la Torre, J. R., Herbold, C. W., Pjevac, P., Daims, H., & Wagner, M. (2022). Ammonia-oxidizing archaea possess a wide range of cellular ammonia affinities. *The ISME Journal, 16*(1), 272–283. https://doi.org/10.1038/s41396-021-01064-z

Kadnikov, V. V., Mardanov, A. V., Beletsky, A. V., Shubenkova, O. V., Pogodaeva, T. V., Zemskaya, T. I., Ravin, N. V., & Skryabin, K. G. (2012). Microbial community structure in methane hydrate-bearing sediments of freshwater Lake Baikal. *FEMS Microbiology Ecology, 79*(2), 348–358. https://doi.org/10.1111/j.1574-6941.2011.01221.x

Kadnikov, V. V., Lomakina, A. V., Likhoshvai, A. V., Gorshkov, A. G., Pogodaeva, T. V., Beletsky, A. V., Mardanov, A. V., Zemskaya, T. I., & Ravin, N. V. (2013). Composition of the microbial communities of bituminous constructions at natural oil seeps at the bottom of Lake Baikal. *Microbiology, 82*(3), 373–382. https://doi.org/10.1134/s0026261713030168

Kadnikov, V. V., Mardanov, A. V., Beletsky, A. V., Karnachuk, O. V., & Ravin, N. V. (2019). Genome of the candidate phylum *Aminicenantes* bacterium from a deep subsurface thermal aquifer revealed its fermentative saccharolytic lifestyle. *Extremophiles, 23*(2), 189–200. https://doi.org/10.1007/s00792-018-01073-5

Kadnikov, V. V., Savvichev, A. S., Mardanov, A. V., Beletsky, A. V., Chupakov, A. V., Kokryatskaya, N. M., Pimenov, N. V., & Ravin, N. V. (2020). Metabolic diversity and evolutionary history of the archaeal phylum "*Candidatus*Micrarchaeota" uncovered from a freshwater lake metagenome. *Applied and Environmental Microbiology, 86*(23) Advance online publication. https://doi.org/10.1128/aem.02199-20

Kallistova, A. Y., Merkel, A. Y., Tarnovetskii, I. Y., & Pimenov, N. V. (2017). Methane formation and oxidation by prokaryotes. *Microbiology, 86*(6), 671–691. https://doi.org/10.1134/s0026261717060091

Kalmychkov, G. V., Egorov, A. V., Kuz'min, M. I., & Khlystov, O. M. (2006). Genetic types of methane from Lake Baikal. *Doklady Earth Sciences, 411*(2), 1462–1465. https://doi.org/10.1134/s1028334x06090285

Kalmychkov, G. V., Pokrovsky, B. G., Hachikubo, A., & Khlystov, O. M. (2017). Geochemical characteristics of methane from sediments of the underwater high Posolskaya Bank (Lake Baikal). *Lithology and Mineral Resources, 52*(2), 102–110. https://doi.org/10.1134/s0024490217020055

Kalmychkov, G. V., Egorov, A. V., Hachikubo, A., & Khlystov, O. M. (2019). Hydrocarbon gases of the Gorevoi Utes underwater oil-gas seep (Lake Baikal, Russia). *Russian Geology and Geophysics, 60*(10), 1188–1194. https://doi.org/10.15372/rgg2019110

Kampmann, K., Ratering, S., Kramer, I., Schmidt, M., Zerr, W., & Schnell, S. (2012). Unexpected stability of *Bacteroidetes* and *Firmicutes* communities in laboratory biogas reactors fed with different defined dubstrates. *Applied and Environmental Microbiology, 78*(7), 2106–2119. https://doi.org/10.1128/aem.06394-11

Karabanov, E. B. (1999). *Geological structure of the sedimentary strata of Lake Baikal and reconstruction of climate changes in Central Asia in the late Cenozoic [in Russian].* (Abstract, Doctor Thesis). Institute of Lithosphere of Marginal Seas, RAS, Moscow.

Kassem, A., Abbas, L., Coutinho, O., Opara, S., Najaf, H., Kasperek, D., Li, X., & Tiquia-Arashiro, S. M. (2023). Applications of Fourier transform-infrared spectroscopy in microbial cell biology and environmental microbiology: Advances, challenges, and future perspectives. *Frontiers in Microbiology, 14*, 1304081. https://doi.org/10.3389/fmicb.2023.1304081

Katz, S. L., Izmest'eva, L. R., Hampton, S. E., Ozersky, T., Shchapov, K., Moore, M. V., Shimaraeva, S. V., & Silow, E. A. (2015). The "Melosira years" of Lake Baikal: Winter environmental conditionsat ice onset predict under-ice algal blooms in spring. *Limnology and Oceanography, 60*, 1950–1964. https://doi.org/10.1002/lno.10143

Kelts, K., & Hsü, K. J. (1978). Freshwater carbonate sedimentation. In A. Lerman (Ed.), *Lakes* (pp. 295–323). Springer. https://doi.org/10.1007/978-1-4757-1152-3_9

Khanaev, I. V., Dzyuba, E. V., Kravtsova, L. S., & Grachev, M. A. (2016). The effect of bloomof filamentous green algae on the reproduction of yellowfin sculpin Cottocomephorusgrewingkii (Dybowski, 1874) (Cottoidae) during ecological crisis in Lake Baikal. *Doklady Biological Sciences, 467*, 63–64. https://doi.org/10.1134/S0012496616020022

Khlystov, O. M., & Kalmychkov, G. V. (2020). Baikal gas hydrates: History of discovery and main results. In M. I. Kuzmin & E. V. Bezrukova (Eds.), *30 years of the Baikal-drilling program* (pp. 278–294). Academic Publishing House "Geo".

Khlystov, O. M., Gorshkov, A. G., Egorov, A. V., Zemskaya, T. I., Granin, N. G., Kalmychkov, G. V., Vorob'eva, S. S., Pavlova, O. N., Yakup, M. A., Makarov, M. M., Moskvin, V. I., & Grachev, M. A. (2007). Oil in the lake of world heritage. *Doklady Earth Science, 415*, 682–685. https://doi.org/10.1134/S1028334X07050042

Khlystov, O. M., Zemskaya, T. I., Sitnikova, T. Y., Mekhanikova, I. V., Kaigorodova, I. A., Gorshkov, A. G., Timoshkin, O. A., Shubenkova, O. V., Chernitsyna, S. M., Lomakina, A. V., Likhoshvai, A. V., Sagalevich, A. M., Moskvin, V. I., Peresypkin, V. I., Belyaev, N. A., Slipenchuk, M. V., Tulokhonov, A. K., & Grachev, M. A. (2009). Bottom bituminous constructions and biota inhabiting them according to investigation of Lake Baikal with the Mir submersible. *Doklady Earth Sciences, 429*(1), 1333–1336. https://doi.org/10.1134/s1028334x09080200

Khlystov, O., Batist, M. D., Shoji, H., Hachikubo, A., Nishio, S., Naudts, L., Poort, J., Khabuev, A., Belousov, O., Manakov, A., & Kalmychkov, G. (2013). Gas hydrate of Lake Baikal: Discovery and varieties. *Journal of Asian Earth Sciences, 62*, 162–166. https://doi.org/10.1016/j.jseaes.2012.03.009

Khlystov, O. M., Izosimova, O. N., Hachikubo, A., Minami, H., Makarov, M. M., & Gorshkov, A. G. (2022). A new oil and gas seep in Lake Baikal. *Petroleum Chemistry, 62*(5), 475–481. https://doi.org/10.1134/s096554412205005x

Khodzher, T. V., Domysheva, V. M., Sorokovikova, L. M., & Golobokova, L. P. (2016). Methods for monitoring the chemical composition of Baikal water. In L. Mueller, A. K. Sheudshen, & F. Eulenstein (Eds.), *Novel methods for monitoring and managing land and water resource es in Siberia* (pp. 113–132). Springer Nature. https://doi.org/10.1007/978-3-319-24409-9_3

Khodzher, T. V., Domysheva, V. M., Sorokovikova, L. M., Tomberg, I. V., & Sakirko, M. V. (2018). Hydrochemical studies in Lake Baikal: History and nowadays. *Limnology and Freshwater Biology, 1*, 2–9. https://doi.org/10.31951/2658-3518-2018-A-1-2

Khomyakova, M. A., Zavarzina, D. G., Merkel, A. Y., Klyukina, A. A., Pikhtereva, V. A., Gavrilov, S. N., & Slobodkin, A. I. (2022). The first cultivated representatives of the actinobacterial lineage OPB41 isolated from subsurface environments constitute a novel order *Anaerosomatales*. *Frontiers in Microbiology, 13*. Advance online publication. https://doi.org/10.3389/fmicb.2022.1047580

Kim, M., Lee, K.-H., Yoon, S.-W., Kim, B.-S., Chun, J., & Yi, H. (2013). Analytical tools and databases for metagenomics in the next-generation sequencing era. *Genomics & Informatics, 11*(3), 102. Advance online publication. https://doi.org/10.5808/gi.2013.11.3.102

King, G. M., Kostka, J. E., Hazen, T. C., & Sobecky, P. A. (2015). Microbial responses to the Deepwater Horizon Oil Spill: From coastal wetlands to the deep sea. *Annual Review of Marine Science, 7*(1), 377–401. https://doi.org/10.1146/annurev-marine-010814-015543

Klerkx, J., Zemskaya, T. I., Matveeva, T. V., Khlystov, O. M., Namsaraev, B. B., Dagurova, O. P., Golobokova, L. P., Vorobyova, S. S., Pogodaeva, T. P., Granin, N. G., Kalmichkov, G. V., Ponomarchuk, V. A., Shodgi, H., Mazurenko, L. L., Kaulio, V. V., Soloviev, V. A., & Grachev, M. A. (2003). Methane hydrates in surface layer of deep water sediments of Lake Baikal. *Doklady Earth Science, 393*(6), 822–826.

Klerkx, J., De Batist, M., Poort, J., Hus, R., Van Rensbergen, P., Khlystov, O., & Granin, N. (2006). Tectonically controlled methane escape in Lake Baikal. In *Advances in the geological storage of carbon dioxide* (pp. 203–219). https://doi.org/10.1007/1-4020-4471-2_17

Klump, J. V., Remsen, C. C., & Kaster, J. L. (1988). *The presence and potential impact of geothermal activity on the chemistry and biology of Yellowstone Lake, Wyoming*. NOAA-National Undersea Research Program, V. 6.

Köhler, J. M., Ehrhardt, L., & Günther, P. M. (2023). Archaeal and extremophilic bacteria from different archaeological excavation sites. *International Journal of Molecular Sciences, 24*(6), 5519. Advance online publication. https://doi.org/10.3390/ijms24065519

Kojima, H., Teske, A., & Fukui, M. (2003). Morphological and phylogenetic characterizations of freshwater *Thioploca* species from Lake Biwa, Japan, and Lake Constance, Germany. *Applied and Environmental Microbiology, 69*(1), 390–398. https://doi.org/10.1128/aem.69.1.390-398.2003

Kojima, H., Koizumi, Y., & Fukui, M. (2006). Community structure of bacteria associated with sheaths of freshwater and brackish *Thioploca* species. *Microbial Ecology, 52*(4), 765–773. https://doi.org/10.1007/s00248-006-9127-8

Kojima, H., Tsutsumi, M., Ishikawa, K., Iwata, T., Mußmann, M., & Fukui, M. (2012). Distribution of putative denitrifying methane oxidizing bacteria in sediment of a freshwater lake, Lake Biwa. *Systematic and Applied Microbiology, 35*(4), 233–238. https://doi.org/10.1016/j.syapm.2012.03.005

Konkol, N. R., Bruckner, J. C., Aguilar, C., Lovalvo, D., & Maki, J. S. (2010). Dominance of epiphytic filamentous Thiothrix spp. on an aquatic macrophyte in a hydrothermal vent Flume

in sedge bay, Yellowstone Lake, Wyoming. *Microbial Ecology, 60*(3), 528–538. https://doi.org/10.1007/s00248-010-9656-z

Kontorovich, A. E., Kashirtsev, V. A., Moskvin, V. I., Burshtein, L. M., Zemskaya, T. I., Kostyreva, E. A., Kalmychkov, G. V., & Khlystov, O. M. (2007). Petroleum potential of Baikal deposits. *Russian Geology and Geophysics, 48*(12), 1046–1053. https://doi.org/10.1016/j.rgg.2007.11.004

Kostyreva, E. A., Kashircev, V. A., Pavlova, O. N , Sotnich, I. S., & Khabuev, A. V. (2022). Chloroform extracts of bottom sediments in the region Central Baikal. *InterExpoGeoSiberia, 2*(1), 200–203. In Russian.

Kozhov, M. M. (1972). *Essays on Baikal science*. Irk State Univ.

Kravtsova, L. S., Izhboldina, L. A., Khanaev, I. V., Pomazkina, G. V., Rodionova, E. V., Domysheva, V. M., Sakirko, M. V., Tomberg, I. V., Kostornova, T. Y., Kravchenko, O. S., & Kupchinsky, A. B. (2014). Nearshore benthic blooms of filamentous green algaein Lake Baikal. *Journal of Great Lakes Research, 40*, 441–448. https://doi.org/10.1016/j.jglr.2014.02.019

Krylov, A., Khlystov, O., Zemskaya, T., Minami, H., Hachikubo, A., Nunokawa, Y., Kida, M., Shoji, H., Naudts, L., Poort, J., & Pogodaeva, T. (2008). First discovery and formation process of authigenic siderite from gas hydrate–bearing mud volcanoes in fresh water: Lake Baikal, eastern Siberia. *Geophysical Research Letters, 35*(5) Advance online publication. https://doi.org/10.1029/2007gl032917

Krylov, A. A., Khlystov, O. M., Hachikubo, A., Minami, H., Nunokawa, Y., Shoji, H., Zemskaya, T. I., Naudts, L., Pogodaeva, T. V., Kida, M., Kalmychkov, G. V., & Poort, J. (2010). Isotopic composition of dissolved inorganic carbon in subsurface sediments of gas hydrate-bearing mud volcanoes, Lake Baikal: Implications for methane and carbonate origin. *Geo-Marine Letters, 30*(3–4), 427–437. https://doi.org/10.1007/s00367-010-0190-2

Krylov, A. A., Hachikubo, A., Minami, H., Pogodaeva, T. V., Zemskaya, T. I., Krzhizhanovskaya, M. G., Poort, J., & Khlystov, O. M. (2018). Authigenic rhodochrosite from a gas hydrate-bearing structure in Lake Baikal. *International Journal of Earth Sciences, 107*(6), 2011–2022. 1437–3254. https://doi.org/10.1007/s00531-018-1584-z

Krylov, A., Khlystov, O., Hachikubo, A., Minami, H., Zemskaya, T., Logvina, E., Lomakina, A., & Semenov, P. (2020). The reconstruction of the mechanisms of problematic authigenic carbonates formation in diagenetic and catagenetic environments associated with the generation/oxidation of hydrocarbons. *Limnology and Freshwater Biology, 4*, 928–930. https://doi.org/10.31951/2658-3518-2020-a-4-928

Krylov, A. A., Khlystov, O. M., Semenov, P. B., Sagidullin, A. K., Malyshev, S. A., Bukin, S. V., Vidischeva, O. N., Manakov, A. Y., & Ismagilov, Z. R. (2023). Sources of hydrocarbon gases in the Kedr Mud Volcano, Southern Basin of Lake Baikal: Results of experimental studies. *Lithology and Mineral Resources, 58*(6), 534–543. https://doi.org/10.1134/s0024490223700335

Kuzmin, M. I., Karabanov, E. B., Kawai, T., Williams, D., Bychinskii, V. A., Kerber, E. V., Kravchinskii, V. A., Bezrukova, E. V., Prokopenko, A. A., & Geletii, V. F. (2001). Deep drilling on Lake Baikal: Main results. *Russian Geology and Geophysics, 42*, 8–34.

Kuznetsov, S. I. (1970). *The Microflora of lakes and its geochemical activity*. Nauka.

Kuznetsov, A. P., Strizhov, V. P., Kuzin, V. S., Fialkov, V. A., & Yastrebov, V. S. (1991). New in the nature of Baikal. community based on bacterial chemosynthesis. *Proceedings of the USSR Academy of Sciences Biological Series, 5*, 766–772.

Lagostina, L., Frandsen, S., MacGregor, B. J., Glombitza, C., Deng, L., Fiskal, A., Li, J., Doll, M., Geilert, S., Schmidt, M., Scholz, F., Bernasconi, S. M., Jørgensen, B. B., Hensen, C., Teske, A., & Lever, M. A. (2021). Interactions between temperature and energy supply drive microbial communities in hydrothermal sediment. *Communications Biology, 4*(1), 1006. https://doi.org/10.1038/s42003-021-02507-1

Langwig, M. V., De Anda, V., Dombrowski, N., Seitz, K. W., Rambo, I. M., Greening, C., Teske, A. P., & Baker, B. J. (2022). Large-scale protein level comparison of *Deltaproteobacteria*

reveals cohesive metabolic groups. *The ISME Journal, 16*(1), 307–320. https://doi.org/10.1038/s41396-021-01057-y

Lauterborn, R. (1907). Eine neue gattung der schwefel-bakterien (*Thioploca schmidlei* nov. gen. nov. spec.). *Berichte Der Deutschen Botanischen Gesellschaft, 25*(5), 238–242. https://doi.org/10.1111/j.1438-8677.1907.tb06615.x

Lein, A. Y. (2004). Authigenic carbonate formation in the ocean. *Lithology and Mineral Resources, 39*, 1–30. https://doi.org/10.1023/B:LIMI.0000010767.52720.8f

Leu, A. O., Cai, C., McIlroy, S. J., Southam, G., Orphan, V. J., Yuan, Z., Hu, S., & Tyson, G. W. (2020). Anaerobic methane oxidation coupled to manganese reduction by members of the *Methanoperedenaceae*. *The ISME Journal, 14*(4), 1030–1041. https://doi.org/10.1038/s41396-020-0590-x

Li, J., Peng, X., Bai, S., Chen, Z., & Nostrand, J. D. V. (2018). Biogeochemical processes controlling authigenic carbonate formation within the sediment column from the Okinawa Trough. *Geochimica et Cosmochimica Acta, 222*, 363–382. https://doi.org/10.1016/j.gca.2017.10.029

Li, L., Liu, Z., Zhou, Z., Zhang, M., Meng, D., Liu, X., Huang, Y., Li, X., Jiang, Z., Zhong, S., Drewniak, L., Yang, Z., Li, Q., Liu, Y., Nan, X., Jiang, B., Jiang, C., & Yin, H. (2021). Comparative genomics provides insights into the genetic diversity and evolution of the DPANN superphylum. *MSystems, 6*(4), e00602–e00621. Advance online publication. https://doi.org/10.1128/msystems.00602-21

Liang, B., Wang, L.-Y., Mbadinga, S. M., Liu, J.-F., Yang, S.-Z., Gu, J.-D., & Mu, B.-Z. (2015). *Anaerolineaceae* and *Methanosaeta* turned to be the dominant microorganisms in alkanes-dependent methanogenic culture after long-term of incubation. *AMB Express, 5*(1) Advance online publication. https://doi.org/10.1186/s13568-015-0117-4

Likhoshvay, A., Khanaeva, T., Gorshkov, A., Zemskaya, T., & Grachev, M. (2013). Do oil-degrading rhodococci contribute to the genesis of deep-water bitumen mounds in Lake Baikal? *Geomicrobiology Journal, 30*(3), 209–213. https://doi.org/10.1080/01490451.2012.665149

Likhoshvay, A., Lomakina, A., & Grachev, M. (2014). The complete *alk* sequences of *Rhodococcuserythropolis* from Lake Baikal. *SpringerPlus, 3*(1) Advance online publication. https://doi.org/10.1186/2193-1801-3-621

Likhoshway, E. V., Kuzmina, A. E., Potyomkina, T. G., Potemkin, V. L., & Shimaraev, M. N. (1996). The distribution of diatoms near a thermal bar in Lake Baikal. *Journal of Great Lakes Research, 22*(1), 5–14. https://doi.org/10.1016/S0380-1330(96)70929-2

Liu, X., Tiquia, S. M., Holguin, G., Wu, L., Nold, S. C., Devol, A. H., Luo, K., Palumbo, A. V., Tiedje, J. M., & Zhou, J. (2003). Molecular diversity of denitrifying genes in continental margin sediments within the oxygen-deficient zone off the Pacific coast of Mexico. *Applied and Environmental Microbiology, 69*(6), 3549–3560. https://doi.org/10.1128/AEM.69.6.3549-3560.2003

Liu, J., Wang, Z., Belchik, S. M., Edwards, M. J., Liu, C., Kennedy, D. W., Merkley, E. D., Lipton, M. S., Butt, J. N., Richardson, D. J., Zachara, J. M., Fredrickson, J. K., Rosso, K. M., & Shi, L. (2012). Identification and characterization of MtoA: A decaheme c-type cytochrome of the neutrophilic Fe(II)-oxidizing bacterium sideroxydans lithotrophicus ES-1. *Frontiers in Microbiology, 3*. Advance online publication. https://doi.org/10.3389/fmicb.2012.00037

Liu, X., Li, M., Castelle, C. J., Probst, A. J., Zhou, Z., Pan, J., Liu, Y., Banfield, J. F., & Gu, J.-D. (2018). Insights into the ecology, evolution, and metabolism of the widespread Woesearchaeotal lineages. *Microbiome, 6*(1), 102. Advance online publication. https://doi.org/10.1186/s40168-018-0488-2

Liu, Y.-F., Qi, Z.-Z., Shou, L.-B., Liu, J.-F., Yang, S.-Z., Gu, J.-D., & Mu, B.-Z. (2019). Anaerobic hydrocarbon degradation in candidate phylum 'Atribacteria' (JS1) inferred from genomics. *The ISME Journal, 13*(9), 2377–2390. https://doi.org/10.1038/s41396-019-0448-2

Liu, X., Wang, Y., & Gu, J.-D. (2021). Ecological distribution and potential roles of *Woesearchaeota* in anaerobic biogeochemical cycling unveiled by genomic analysis. *Computational and Structural Biotechnology Journal, 19*, 794–800. https://doi.org/10.1016/j.csbj.2021.01.013

Lloyd, K. G., May, M. K., Kevorkian, R. T., & Steen, A. D. (2013). Meta-analysis of quantification methods shows that Archaea and Bacteria have similar abundances in the subseafloor. *Applied and Environmental Microbiology, 79*(24), 7790–7799. https://doi.org/10.1128/aem.02090-13

Logachev, N. A. (1974). Sayan-Baikal Stanovoi Uplift. In N. A. Florensov (Ed.), *Uplifts of pre-Baikalian and Transbaikalian regions* (pp. 16–162). Nauka.

Logatchev, N. A. (1993). History and geodynamics of the Lake Baikal rift in the context of the Eastern Siberia rift system: A review. *Bulletin des Centres de Recherches Exploration – Production Elf-Aquitaine, 17*(2), 353–370.

Loh, H. Q., Hervé, V., & Brune, A. (2021). Metabolic potential for reductive acetogenesis and a novel energy-converting [NiFe] hydrogenase in *Bathyarchaeia* from termite guts—A genome-centric analysis. *Frontiers in Microbiology, 11*. Advance online publication. https://doi.org/10.3389/fmicb.2020.635786

Lomakina, A. V., Pogodaeva, T. V., Morozov, I. V., & Zemskaya, T. I. (2014). Microbial communities of the discharge zone of oil- and gas-bearing fluids in low-mineral Lake Baikal. *Microbiology, 83*(3), 278–287. https://doi.org/10.1134/s0026261714030126

Lomakina, A. V., Mamaeva, E. V., Pogodaeva, T. V., Kalmychkov, G. V., Khal'zov, I. A., & Zemskaya, T. I. (2018a). Anaerobic methane oxidation in enrichment cultures from deep sediments of a mud volcano Peschanka (South Baikal). *Microbiology, 87*(3), 317–325. https://doi.org/10.1134/s0026261718030049

Lomakina, A. V., Mamaeva, E. V., Galachyants, Y. P., Petrova, D. P., Pogodaeva, T. V., Shubenkova, O. V., Khabuev, A. V., Morozov, I. V., & Zemskaya, T. I. (2018b). Diversity of archaea in bottom sediments of the discharge area. *Geomicrobiology Journal, 35*, 50–63. https://doi.org/10.1080/01490451.2017.1315195

Lomakina, A., Pogodaeva, T., Kalmychkov, G., Chernitsyna, S., & Zemskaya, T. (2020). Diversity of NC10 bacteria and ANME-2d archaea in sediments of fault zones at Lake Baikal. *Diversity, 12*. Advance online publication. https://doi.org/10.3390/d12010010

Lomakina, A., Bukin, S., Shubenkova, O., Pogodaeva, T., Ivanov, V., Bukin, Y., & Zemskaya, T. (2023a). Microbial communities in ferromanganese sediments from the northern basin of Lake Baikal (Russia). *Microorganisms, 11*(7), 1865. Advance online publication. https://doi.org/10.3390/microorganisms11071865

Lomakina, A. V., Bukin, S. V., Pogodaeva, T. V., Turchyn, A. V., Khlystov, O. M., Khabuev, A. V., Ivanov, V. G., Krylov, A. A., & Zemskaya, T. I. (2023b). Microbial diversity and authigenic siderite mediation in sediments surrounding the Kedr-1 mud volcano, Lake Baikal. *Geobiology, 21*(6), 770–790. https://doi.org/10.1111/gbi.12575

Lomonosov, I. S. (1974). *Geochemistry and the formation of modern hydrotherms in the Baikal rift zone*. Nauka.

Lovley, D. R. (2012). Long-range electron transport to Fe(III) oxide via pili with metallic-like conductivity. *Biochemical Society Transactions, 40*(6), 1186–1190. https://doi.org/10.1042/bst20120131

Loyd, S. J., & Smirnoff, M. N. (2022). Progressive formation of authigenic carbonate with depth in siliciclastic marine sediments including substantial formation in sediments experiencing methanogenesis. *Chemical Geology, 594*, 120775. https://doi.org/10.1016/j.chemgeo.2022.120775

Maerki, M., Müller, B., & Wehrli, B. (2006). Microscale mineralization pathways in surface sediments: A chemical sensor study in Lake Baikal. *Limnology and Oceanography, 51*(3), 1342–1354. https://doi.org/10.4319/lo.2006.51.3.1342

Maksimova, E. A., & Maksimov, V. N. (1989). *Microbiology of Baikal water*. Irkutsk University Publishing House. In Russian.

Maksimova, E. A., Erenetenko, I. A., & Grebenkina, L. A. (1982). Seasonal course of microbial processes in bottom sediments of Southern Baikal. *Ecological Problems of the Baikal Region, 2*, 43–44. In Russian.

Maksimova, E. A., Maksimov, V. N., & Kolesnitskaya, G. N. (1995). Microbiological indication and assessment of the state of bottom sediments of southern Baikal. *Microbiology, 64*(3), 399–404. In Russian.

Malnik, V. V., Timoshkin, O. A., Suturin, A. N., Onishchuk, N. A., Sakirko, M. V., Tomberg, I. V., Gorshkova, A. S., & Zabanova, N. S. (2019). Anthropogenic changes in the hydrochemical and sanitary–microbiological characteristics of water quality in southern Baikal tributaries: Listvennichnyi Bay. *Water Resources, 46*, 748–758. https://doi.org/10.1134/S0097807819050154

Marlow, J. J., Steele, J. A., Case, D. H., Connon, S. A., Levin, L. A., & Orphan, V. J. (2014). Microbial abundance and diversity patterns associated with sediments and carbonates from the methane seep environments of Hydrate Ridge, OR. *Frontiers in Marine Science, 1*, 44. Advance online publication. https://doi.org/10.3389/fmars.2014.00044

Mats, V. D., Ufimtsev, G. F., Mandel'baum, M. M., Alakshin, A. M., Pospeev, A. V., Shimaraev, M. N., & Khlystov, O. M. (2001). *Baikal rift cenozoic of the Baikal rift depression: Structure and geological history*. Geo.

Maynard, J. B. (2010). The chemistry of manganese ores through time: A signal of increasing diversity of earth-surface environments. *Economic Geology, 105*(3), 535–552.

McKay, E., Katsev, S., Malkin, S., & Ozersky, T. (2023). Widespread occurrence of filamentous Thioploca bacteria in low-sulfate Great Lakes sediments with implications for sulfur and nitrogen cycling. *Journal of Great Lakes Research, 49*(5), 1111–1122. https://doi.org/10.1016/j.jglr.2023.07.003

McKindles, K. M., & Tiquia-Arashiro, S. M. (2012). Functional gene arrays for analysis of microbial communities on ocean platform. In S. Tiquia-Arashiro (Ed.), *Molecular biological technologies for ocean sensing* (Springer protocols handbooks). Humana Press. https://doi.org/10.1007/978-1-61779-915-0_9

Mehrshad, M., Salcher, M. M., Okazaki, Y., Nakano, S., Šimek, K., Andrei, A.-S., & Ghai, R. (2018). Hidden in plain sight—Highly abundant and diverse planktonic freshwater *Chloroflexi*. *Microbiome, 6*(1) Advance online publication. https://doi.org/10.1186/s40168-018-0563-8

Meister, P., & Reyes, C. (2019). The carbon-isotope record of the sub-seafloor biosphere. *Geosciences, 9*(12), 507. https://doi.org/10.3390/geosciences9120507

Milucka, J., Kirf, M., Lu, L., Krupke, A., Lam, P., Littmann, S., Kuypers, M. M. M., & Schubert, C. J. (2015). Methane oxidation coupled to oxygenic photosynthesis in anoxic waters. *The ISME Journal, 9*(9), 1991–2002. https://doi.org/10.1038/ismej.2015.12

Mizandrontsev, I. B. (1975). On the geochemistry of soil solutions. *Proceedings of LIN SB AN USSR, 21*(41), 203–230.

Mizandrontsev, I. B. (1978). Sedimentation. *Proceedings of LIN SB AN USSR, 16*(36), 33–46.

Mladova, T. A. (1975). Microbiological characteristics of bottom sediments of Lake Baikal. In G. I. Galaziy & K. K. Votintsev (Eds.), *Cycle of matter and energy in lake reservoirs* (pp. 246–250). Academy of Sciences of the USSR.

Mondav, R., Woodcroft, B. J., Kim, E.-H., McCalley, C. K., Hodgkins, S. B., Crill, P. M., Chanton, J., Hurst, G. B., VerBerkmoes, N. C., Saleska, S. R., Hugenholtz, P., Rich, V. I., & Tyson, G. W. (2014). Discovery of a novel methanogen prevalent in thawing permafrost. *Nature Communications, 5*(1), 3212. Advance online publication. https://doi.org/10.1038/ncomms4212

Moore, M. V., Hampton, S. E., Izmest'eva, L. R., Silow, E. A., Peshkova, E. V., & Pavlov, B. K. (2009). Climate change and the world's "Sacred Sea"—Lake Baikal, Siberia. *Bioscience, 59*, 405–417. https://doi.org/10.1525/bio.2009.59.5.8

Morgunova, I., Semenov, P., Kursheva, A., Litvinenko, I., Malyshev, S., Bukin, S., Khlystov, O., Pavlova, O., Zemskaya, T., & Krylov, A. A. (2022). Molecular indicators of sources and biodegradation of organic matter in sediments of fluid discharge zones of Lake Baikal. *Geosciences, 12*(2), 72. Advance online publication. https://doi.org/10.3390/geosciences12020072

Mori, K., Yamaguchi, K., Sakiyama, Y., Urabe, T., & Suzuki, K.-I. (2009). *Caldisericum exile* gen. nov., sp. nov., an anaerobic, thermophilic, filamentous bacterium of a novel bacterial phylum, *Caldiserica* phyl. nov., originally called the candidate phylum OP5, and description of *Caldisericaceae* fam. nov., *Caldisericales* ord. nov. and *Caldisericia* classis nov. *International*

Journal of Systematic and Evolutionary Microbiology, 59(11), 2894–2898. https://doi.org/10.1099/ijs.0.010033-0

Müller, B., Maerki, M., Schmid, M., Vologina, E. G., Wehrli, B., Wüest, A., & Sturm, M. (2005). Internal carbon and nutrient cycling in Lake Baikal: Sedimentation, upwelling, and early diagenesis. *Global and Planetary Change, 46*(1–4), 101–124. https://doi.org/10.1016/j.gloplacha.2004.11.008

Myer, P. R., Kim, M., Freetly, H. C., & Smith, T. P. L. (2016). Evaluation of 16S rRNA amplicon sequencing using two next-generation sequencing technologies for phylogenetic analysis of the rumen bacterial community in steers. *Journal of Microbiological Methods, 127*, 132–140. https://doi.org/10.1016/j.mimet.2016.06.004

Naehr, T. H., Eichhubl, P., Orphan, V. J., Hovland, M., Paull, K. P., Usslerlll, W., Lorenson, T. D., & Greene, G. (2007). Authigenic carbonate formation at hydrocarbon seeps in continental margin sediments: A comparative study. *Deep Sea Research Part II: Topical Studies in Oceanography, 54*, 1268–1291. https://doi.org/10.1016/j.dsr2.2007.04.010

Nagata, T., Takai, K., Kawanobe, K., Kim, D. S., Nakazato, R., Guselnikova, N., Bondarenko, N., Mologawaya, O., Kostornova, T., Drucker, V., Satoh, Y. A., & Watanabe, Y. A. (1994). Autotrophic picoplankton in southern Lake Baikal: Abundance, growth and grazing mortality during summer. *Journal of Plankton Research, 16*(8), 945–959. https://doi.org/10.1093/plankt/16.8.945

Namsaraev, B. B., & Zemskaya, T. I. (2000). *Microbial processes of carbon circulation in bottom sediments of Lake Baikal*. Geo.

Namsaraev, B. B., Dulov, L. E., Dubinina, G. A., Zemskaya, T. I., Granina, L. Z., & Karabanov, E. B. (1994). The participation of bacteria in the processes of synthesis and destruction of organic matter in the microbial mats of Lake Baikal. *Microbiology, 63*(2), 345–351.

Namsaraev, B. B., Dulov, L. E., Sokolova, E. N., & Zemskaya, T. I. (1995). Bacterial formation of methane in bottom sediments of Lake Baikal. *Microbiology, 65*(3), 411–417.

Namsaraev, B. B., Zemskaya, T. I., Dagurova, O. P., Granina, L. Z., Golobokova, L. P., & Sitnikova, T. Y. (2002). Bacterial communities of the bottom sediments near a hydrothermal source in Frolikha bay (Northern Baikal). *Russian Geology and Geophysics, 43*(7), 644–647.

Namsaraev, B. B., Dagurova, O. P., Zemskaya, T. I., & Golobokova, L. P. (2006). The functioning of the microbial community in bottom sediments of Lake Baikal, with special regard to hydrothermal and gas hydrate regions. *Hydrobiologia, 568*(S1), 83–85. https://doi.org/10.1007/s10750-006-0329-3

Nassereddine, Z., Opara, S., Coutinho, O., Qyteti, F., Book, R., Heinicke, M. P., Napieralski, J., & Tiquia-Arashiro, S. M. (2024). Critical perspectives on advancing antibiotic resistant gene (ARG) detection technologies in aquatic ecosystems. *Science of the Total Environment, 957*, 177775. https://doi.org/10.1016/j.scitotenv.2024.177775

Naudts, L., Khlystov, O., Granin, N., Chensky, A., Poort, J., & De Batist, M. (2012). Stratigraphic and structural control on the distribution of gas hydrates and active gas seeps on the Posolsky Bank, Lake Baikal. *Geo-Marine Letters, 32*(5–6), 395–406. https://doi.org/10.1007/s00367-012-0286-y

Nazina, T. N., Shestakova, N. M., Grigor'yan, A. A., Mikhailova, E. M., Tourova, T. P., Poltaraus, A. B., Feng, C., Ni, F., & Belyaev, S. S. (2006). Phylogenetic diversity and activity of anaerobic microorganisms of high-temperature horizons of the Dagang oil field (P. R. China). *Microbiology, 75*(1), 55–65. https://doi.org/10.1134/s0026261706010115

Nechaeva, N. B., & Salimovskaya-Rodina, A. G. (1935). Microbiological analysis of Baikal bottom sediments. *Trudy Baikalskoilimnologicheskoistantsii AN SSSR, 6*, 5–14. In Russian.

Nelson, H. C., Karabanov, E. B., Colman, S. M., & Escutia, C. (1999). Tectonic and sediment supply control of deep rift Lake turbidite systems: Lake Baikal, Russia. *Geology, 27*(2), 163. https://doi.org/10.1130/0091-7613(1999)027<0163:tassco>2.3.co;2

Newton, R. J., Jones, S. E., Eiler, A., McMahon, K. D., & Bertilsson, S. (2011). A guide to the natural history of freshwater lake bacteria. *Microbiology and Molecular Biology Reviews, 75*(1), 14–49. https://doi.org/10.1128/mmbr.00028-10

Nguyen, S., Ala, F., Cardwell, C., Cai, D., McKindles, K. M., Lotvola, A., Hodges, S., Deng, Y., & Tiquia-Arashiro, S. M. (2013). Isolation and screening of carboxydotrophs isolated from composts and their potential for butanol synthesis. *Environmental Technology, 34*, 1995–2007. https://doi.org/10.1080/09593330.2013.795987

Nishino, M., Fukui, M., & Nakajima, T. (1998). Dense mats of *Thioploca*, gliding filamentous sulfur-oxidizing bacteria in Lake Biwa, Central Japan. *Water Research, 32*(3), 953–957. https://doi.org/10.1016/s0043-1354(97)00227-3

Nobu, M. K., Dodsworth, J. A., Murugapiran, S. K., Rinke, C., Gies, E. A., Webster, G., Schwientek, P., Kille, P., Parkes, R. J., Sass, H., Jørgensen, B. B., Weightman, A. J., Liu, W. T., Hallam, S. J., Tsiamis, G., Woyke, T., & Hedlund, B. P. (2016). Phylogeny and physiology of candidate phylum 'Atribacteria' (OP9/JS1) inferred from cultivation-independent genomics. *The ISME Journal, 10*(2), 273–286. https://doi.org/10.1038/ismej.2015.97

Norði, K., Thamdrup, B., & Schubert, C. J. (2013). Anaerobic oxidation of methane in an iron-rich Danish freshwater lake sediment. *Limnology and Oceanography, 58*(2), 546–554. https://doi.org/10.4319/lo.2013.58.2.0546

Och, L. M., Müller, B., Voegelin, A., Ulrich, A., Göttlicher, J., Steiniger, R., Mangold, S., Vologina, E. G., & Sturm, M. (2012). New insights into the formation and burial of Fe/Mn accumulations in Lake Baikal sediments. *Chemical Geology, 330–331*, 244–259. https://doi.org/10.1016/j.chemgeo.2012.09.011

Oppo, D., & Hurst, A. (2018). Seepage rate of hydrothermally generated petroleum in East African Rift lakes: An example from Lake Tanganyika. *Marine and Petroleum Geology, 92*, 149–159. https://doi.org/10.1016/j.marpetgeo.2018.02.031

Oswald, K., Jegge, C., Tischer, J., Berg, J., Brand, A., Miracle, M. R., Soria, X., Vicente, E., Lehmann, M. F., Zopfi, J., & Schubert, C. J. (2016a). Methanotrophy under versatile conditions in the water column of the ferruginous meromictic Lake La Cruz (Spain). *Frontiers in Microbiology, 7*. Advance online publication. https://doi.org/10.3389/fmicb.2016.01762

Oswald, K., Milucka, J., Brand, A., Hach, P., Littmann, S., Wehrli, B., Kuypers, M. M. M., & Schubert, C. J. (2016b). Aerobic gammaproteobacterial methanotrophs mitigate methane emissions from oxic and anoxic lake waters. *Limnology and Oceanography, 61*(S1) Advance online publication. https://doi.org/10.1002/lno.10312

Pacheco-Oliver, M., McDonald, I. R., Groleau, D., Murrell, J. C., & Miguez, C. B. (2002). Detection of methanotrophs with highly divergent *pmoA* genes from Arctic soils. *FEMS Microbiology Letters, 209*(2), 313–319. https://doi.org/10.1111/j.1574-6968.2002.tb11150.x

Pagé, A., Tivey, M. K., Stakes, D. S., & Reysenbach, A. L. (2008). Temporal and spatial archaeal colonization of hydrothermal vent deposits. *Environmental Microbiology, 10*(4), 874–884. https://doi.org/10.1111/j.1462-2920.2007.01505.x

Pagnucco, G., Overfield, D., Chamlee, Y., Fortuna, A., Sulaiman, F., Farinas, J., Schittenhelm, R., Shuler, C., Kassem, A., Opara, S., Najaf, H., Abbas, L., Coutinho, O., Catalfano, B., Li, X., & Tiquia-Arashiro, S. M. (2023). Metal tolerance and biosorption capacities of bacterial strains isolated from an urban watershed. *Frontiers in Microbiology, 14*, 1278886. https://doi.org/10.3389/fmicb.2023.1278886

Pannekens, M., Kroll, L., Müller, H., Mbow, F. T., & Meckenstock, R. U. (2019). Oil reservoirs, an exceptional habitat for microorganisms. *New Biotechnology, 49*, 1–9. Advance online publication. https://doi.org/10.1016/j.nbt.2018.11.006

Parfenova, V. V., Shimaraev, M. N., Kostornova, T. Y., Domysheva, V. M., Levin, L. A., Dryukker, V. V., Zhdanov, A. A., Gnatovskii, R. Y., Tsekhanovsky, V. V., & Logacheva, N. F. (2000). On the vertical distribution of microorganisms in Lake Baikal during spring deep-water renewal. *Microbiology, 69*(3), 433–440.

Pascoal, F., Costa, R., & Magalhães, C. (2021). The microbial rare biosphere: Current concepts, methods and ecological principles. *FEMS Microbiology Ecology, 97*(1) Advance online publication. https://doi.org/10.1093/femsec/fiaa227

Patel, D., Gismondi, R., Alsaffar, A., & Tiquia-Arashiro, S. M. (2019). Applicability of API ZYM to capture seasonal and spatial variabilities in lake and river sediments. *Environmental Technology, 40*(24), 3227–3239. https://doi.org/10.1080/09593330.2018.1468492

Pavlova, O. N., Lomakina, A. V., Gorshkov, A. G., Suslova, M. Y., Likhoshvai, A. V., & Zemskaya, T. I. (2012). Microbial communities and their ability to oxidize n-alkanes in the area of release of gas- and oil-containing fluids in Mid-Baikal (Cape Gorevoi Utes). *Biology Bulletin, 39*(5), 458–463. https://doi.org/10.1134/s1062359012050123

Pavlova, O. N., Izosimova, O. N., Gorshkov, A. G., Novikova, A. S., Bukin, S. V., Ivanov, V. G., Khlystov, O. M., & Zemskaya, T. I. (2020). Current state of deep oil seepage near Cape Gorevoi Utes (Central Baikal). *Russian Geology and Geophysics, 61*(9), 1007–1014. https://doi.org/10.15372/rgg2019180

Pavlova, O. N., Izosimova, O. N., Chernitsyna, S. M., Ivanov, V. G., Pogodaeva, T. V., Khabuev, A. V., Gorshkov, A. G., & Zemskaya, T. I. (2022). Anaerobic oxidation of petroleum hydrocarbons in enrichment cultures from sediments of the Gorevoy Utes natural oil seep under methanogenic and sulfate-reducing conditions. *Microbial Ecology, 83*(4), 899–915. https://doi.org/10.1007/s00248-021-01802-y

Petrosino, J. F., Highlander, S., Luna, R. A., Gibbs, R. A., & Versalovic, J. (2009). Metagenomic pyrosequencing and microbial identification. *Clinical Chemistry, 55*(5), 856–866. https://doi.org/10.1373/clinchem.2008.107565

Phillips, A. A., Ulloa, I., Hyde, E., Agnich, J., Sharpnack, L., O'Malley, K. G., Webb, S. M., Schreiner, K. M., Sheik, C. S., Katsev, S., & Raven, M. R. (2023). Organic sulfur from source to sink in low-sulfate Lake Superior. *Limnology and Oceanography, 68*(12), 2716–2732. https://doi.org/10.1002/lno.12454

Pimenov, N. V., Zakharova, E. E., Bryukhanov, A. L., Korneeva, V. A., Kuznetsov, B. B., Tourova, T. P., Pogodaeva, T. V., Kalmychkov, G. V., & Zemskaya, T. I. (2014). Activity and structure of the sulfate-reducing bacterial community in the sediments of the southern part of Lake Baikal. *Microbiology, 83*(1–2), 47–55. https://doi.org/10.1134/s0026261714020167

Plecha, S., Hall, D., & Tiquia-Arashiro, S. M. (2013). Screening for novel bacteria from the bioenergy feedstock switchgrass (Panicum virgatum L.). *Environmental Technology, 34*, 1895–1904. https://doi.org/10.1080/09593330.2013.818701

Plee, K., Pacton, M., & Ariztegui, D. (2010). Discriminating the role of photosynthetic and heterotrophic microbes triggering low-mg calcite precipitation in freshwater biofilms (Lake Geneva, Switzerland). *Geomicrobiology Journal, 27*, 391–399. https://doi.org/10.1080/01490450903451526

Pogodaeva, T., & Khodzher, T. (2023). Organic carbon in the bottom sediments of Lake Baikal: Geochemical processes of burial and balance values. *Water, 15*(16), 2941. Advance online publication.. https://doi.org/10.3390/w15162941

Pogodaeva, T. V., Zemskaya, T. I., Golobokova, L. P., Khlystov, O. M., Minami, H., & Sakagami, H. (2007). Chemical composition of pore waters of bottom sediments in different Baikal basins. *Russian Geology and Geophysics, 48*(11), 886–900. https://doi.org/10.1016/j.rgg.2007.02.012

Pogodaeva, T. V., Lopatina, I. N., Khlystov, O. M., Egorov, A. V., & Zemskaya, T. I. (2017). Background composition of pore waters in Lake Baikal bottom sediments. *Journal of Great Lakes Research, 43*(6), 1030–1043. https://doi.org/10.1016/j.jglr.2017.09.003

Pogodaeva, T. V., Poort, J., Aloisi, G., Bataillard, L., Makarov, M. M., Khabuev, A. V., Kazakov, A. V., Chensky, A. G., & Khlystov, O. M. (2020). Fluid migrations at the Krasny Yar methane seep of Lake Baikal according to geochemical data. *Journal of Great Lakes Research, 46*(1), 123–131. https://doi.org/10.1016/j.jglr.2019.08.003

Pomaranski, E., & Tiquia-Arashiro, S. M. (2016). Butanol tolerance of carboxydotrophic bacteria isolated from manure composts. *Environmental Technology, 37*(15), 1970–1982. https://doi.org/10.1080/09593330.2015.1137360

Poort, J., & Klerkx, J. (2004). Absence of a regional surface thermal high in the Baikal rift; new insights from detailed contouring of heat flow anomalies. *Tectonophysics, 383*(3–4), 217–241. https://doi.org/10.1016/j.tecto.2004.03.011

Probst, A. J., Ladd, B., Jarett, J. K., Geller-McGrath, D. E., Sieber, C. M. K., Emerson, J. B., Anantharaman, K., Thomas, B. C., Malmstrom, R. R., Stieglmeier, M., Klingl, A., Woyke, T., Ryan, M. C., & Banfield, J. F. (2018). Differential depth distribution of microbial function and putative symbionts through sediment-hosted aquifers in the deep terrestrial subsurface. *Nature Microbiology, 3*(3), 328–336. https://doi.org/10.1038/s41564-017-0098-y

Prokopenko, A. A., Khursevich, G. K., Bezrukova, E. V., Kuzmin, M. I., Boes, X., Williams, D. F., Fedenya, S. A., Kulagina, N. V., Letunova, P. P., & Abzaeva, A. A. (2007). Paleoenvironmental proxy records from Lake Hovsgol, Mongolia, and a synthesis of Holocene climate change in the Lake Baikal watershed. *Quaternary Research, 68*(1), 2–17. https://doi.org/10.1016/j.yqres.2007.03.008

Qin, W., Amin, S. A., Lundeen, R. A., Heal, K. R., Martens-Habbena, W., Turkarslan, S., Urakawa, H., Costa, K. C., Hendrickson, E. L., Wang, T., Beck, D. A., Tiquia-Arashiro, S. M., Taub, F., Holmes, A. D., Vajrala, N., Berube, P. M., Lowe, T. M., Moffett, J. W., Devol, A. H., Baliga, N. S., Arp, D. J., Sayavedra-Soto, L. A., Hackett, M., Armbrust, E. V., Ingalls, A. E., & Stahl, D. A. (2018). Stress response of a marine ammonia-oxidizing archaeon informs physiological status of environmental populations. *The ISME Journal, 12*(2), 508–519. https://doi.org/10.1038/ismej.2017.186

Qiu, L., Williams, D. F., Gvorzdkov, A., Karabanov, E., & Shimaraeva, M. (1993). Biogenic silica accumulation and paleoproductivity in the northern basin of Lake Baikal during the Holocene. *Geology, 21*(1), 25. Advance online publication. https://doi.org/10.1130/0091-7613(1993)021<0025:bsaapi>2.3.co;2

Raghoebarsing, A. A., Pol, A., van de Pas-Schoonen, K. T., Smolders, A. J., Ettwig, K. F., Rijpstra, W. I., Schouten, S., Damsté, J. S., Op den Camp, H. J., Jetten, M. S., & Strous, M. (2006). A microbial consortium couples anaerobic methane oxidation to denitrification. *Nature, 440*(7086), 918–921. https://doi.org/10.1038/nature04617

Rastogi, G., Sani, R. K., Peyton, B. M., Moberly, J. G., & Ginn, T. R. (2009). Molecular studies on the microbial diversity associated with mining-impacted Coeur d'Alene River sediments. *Microbial Ecology, 58*(1), 129–139. https://doi.org/10.1007/s00248-008-9445-0

Reeburgh, W. S. (2007). Oceanic methane biogeochemistry. *ChemInform, 38*(20). https://doi.org/10.1002/chin.200720267

Rensbergen, P. V., De Batist, M., Klerkx, J., Hus, R., Poort, J., Vanneste, M., Granin, N., Khlystov, O., & Krinitsky, P. (2002). Sublacustrine mud volcanoes and methane seeps caused by dissociation of gas hydrates in Lake Baikal. *Geology, 30*(7), 631–634. https://doi.org/10.1130/0091-7613(2002)030<0631:smvams>2.0.co;2

Rensbergen, P., Poort, J., Kipfer, R., de Batist, M., Vanneste, M., Klerkx, J., Granin, N., Khlystov, O., & Krinitsky, P. (2003). Near-surface sediment mobilization and methane venting in relation to hydrate destabilization in Southern Lake Baikal, Siberia. *Geological Society, London, Special Publications, 216*(1), 207–221. https://doi.org/10.1144/gsl.sp.2003.216.01.14

Rezaei Somee, M., Dastgheib, S. M. M., Shavandi, M., Ghanbari Maman, L., Kavousi, K., Amoozegar, M. A., & Mehrshad, M. (2021). Distinct microbial community along the chronic oil pollution continuum of the Persian Gulf converge with oil spill accidents. *Scientific Reports, 11*(1) Advance online publication. https://doi.org/10.1038/s41598-021-90735-0

Rissanen, A. J., Peura, S., Mpamah, P. A., Taipale, S., Tiirola, M., Biasi, C., Mäki, A., & Nykänen, H. (2019). Vertical stratification of bacteria and archaea in sediments of a small boreal humic lake. *FEMS Microbiology Letters, 366*(5). https://doi.org/10.1093/femsle/fnz044

Rodina, A. G. (1954). Bacteria in the productivity of the rocky littoral zone of Lake Baikal. In *Problems of inland waters hydrobiology in USSR* (pp. 172–201) In Russian.

Rodríguez-Ramos, J. A., Borton, M. A., McGivern, B. B., Smith, G. J., Solden, L. M., Shaffer, M., Daly, R. A., Purvine, S. O., Nicora, C. D., Eder, E. K., Lipton, M., Hoyt, D. W., Stegen, J. C., & Wrighton, K. C. (2022). Genome-resolved metaproteomics decodes the microbial and viral contributions to coupled carbon and nitrogen cycling in river sediments. *mSystems, 7*(4), e00516–e00522. Advance online publication. https://doi.org/10.1101/2022.03.11.484050

Romanova, A. P. (1958). The intensity of the development of bacterial flora in the littoral zone of Lake Baikal (by fouling plates). *Microbiology, 27*, 634–640. In Russian.

Salman, V., Bailey, J. V., & Teske, A. (2013). Phylogenetic and morphologic complexity of giant Sulphur bacteria. *Antonie Van Leeuwenhoek, 104*(2), 169–186. https://doi.org/10.1007/s10482-013-9952-y

Sapota, T., Aldahan, A., & Al-Aasm, I. S. (2006). Sedimentary facies and climate control on formation of vivianite and siderite microconcretions in sediments of Lake Baikal, Siberia. *Journal of Paleolimnology, 36*(3), 245–257. https://doi.org/10.1007/s10933-006-9005-x

Savvichev, A. S., Kokryatskaya, N. M., Zabelina, S. A., Rusanov, I. I., Zakharova, E. E., Veslopolova, E. F., Lunina, O. N., Patutina, E. O., Bumazhkin, B. K., Gruzdev, D. S., Sigalevich, P. A., Pimenov, N. V., Kuznetsov, B. B., & Gorlenko, V. M. (2017). Microbial processes of the carbon and sulfur cycles in an ice-covered, iron-rich meromictic lake Svetloe (Arkhangelsk region, Russia). *Environmental Microbiology, 19*(2), 659–672. https://doi.org/10.1111/1462-2920.13591

Schink, B., & Zeikus, J. G. (1980). Microbial methanol formation: A major end product of pectin metabolism. *Current Microbiology, 4*, 387–389. https://doi.org/10.1007/BF02605383

Shahi, A., Ince, B., Aydin, S., & Ince, O. (2017). Assessment of the horizontal transfer of functional genes as a suitable approach for evaluation of the bioremediation potential of petroleum-contaminated sites: A mini-review. *Applied Microbiology and Biotechnology, 101*(11), 4341–4348. https://doi.org/10.1007/s00253-017-8306-5

Shanks, W. C., & Callender, E. (1992). Thermal springs in Lake Baikal. *Geology, 20*(6), 495. Advance online publication. https://doi.org/10.1130/0091-7613(1992)020<0495:tsilb>2.3.co;2

Shimaraev, M. N., Gnatovsky, R. Y., Blinov, V. V., & Ivanov, V. G. (2011). On the issue of updating deep waters in Baikal. *Doklady Earth Sciences, 438*(1), 121–124.

Shimaraev, M. N., Troitskaya, E. S., Blinov, V. V., Ivanov, V. G., & Gnatovsky, R. Y. (2012). Upwellings in Lake Baikal. *Doklady Earth Sciences, 442*(2), 272–276. https://doi.org/10.1134/S1028334X12020183

Shimaraev, M. N., Granin, N. G., Gnatovsky, R. Y., & Blinov, V. V. (2015). The mechanism of oxygen aeration of bottom waters of Lake Baikal. *Doklady Earth Sciences, 461*(2), 379–383. https://doi.org/10.1134/S1028334X15040078

Shubenkova, O. V., Zemskaya, T. I., Chernitsyna, S. M., Khlystov, O. M., & Triboi, T. I. (2005). The first results of an investigation into the phylogenetic diversity of microorganisms in Southern Baikal sediments in the region of subsurface discharge of methane hydrates. *Microbiology, 74*(3), 314–320. https://doi.org/10.1007/s11021-005-0069-9

Simoneit, B. R. T., Aboul-Kassim, T. A. T., & Tiercelin, J. J. (2000). Hydrothermal petroleum from lacustrine sedimentary organic matter in the East African Rift. *Applied Geochemistry, 15*(3), 355–368. https://doi.org/10.1016/s0883-2927(99)00044-x

Singh, R., Guzman, M. S., & Bose, A. (2017). Anaerobic oxidation of ethane, propane, and butane by marine microbes: A mini review. *Frontiers in Microbiology, 8*. Advance online publication. https://doi.org/10.3389/fmicb.2017.02056

Sitnikova, T. Y., Mekhanikova, I. V., Sideleva, V. G., Kiyashko, S. I., Naumova, T. V., Zemskaya, T. I., & Khlystov, O. M. (2017). Trophic relationships between macroinvertebrates and fish in St. Petersburg methane seep community in abyssal zone of Lake Baikal. *Contemporary Problems of Ecology, 10*(2), 147–156. https://doi.org/10.1134/s1995425517020123

Sivan, O., Adler, M., Pearson, A., Gelman, F., Bar-Or, I., John, S. G., & Eckert, W. (2011). Geochemical evidence for iron-mediated anaerobic oxidation of methane. *Limnology and Oceanography, 56*(4), 1536–1544. https://doi.org/10.4319/lo.2011.56.4.1536

Soetaert, K., Hofmann, A. F., Middelburg, J. J., Meysman, F. J. R., & Greenwood, J. (2007). The effect of biogeochemical processes on pH. *Marine Chemistry, 105*(1–2), 30–51. https://doi.org/10.1016/j.marchem.2006.12.012

Sollinger, A., & Urich, T. (2019). Methylotrophic methanogens everywhere — Physiology and ecology of novel players in global methane cycling. *Biochemical Society Transactions, 47*(6), 1895–1907. https://doi.org/10.1042/bst20180565

Solotchina, E. P., Kuz'min, M. I., Stolpovskaya, V. N., Karabanov, E. B., Prokopenko, A. A., & Tkachenko, L. L. (2003). Mineralogical and crystallochemical indicators of environmental and climatic changes in Pleistocene–Holocene sediments of Lake Khubsugul (Mongolia). *Doklady Earth Sciences, 391*, 775–778.

Solotchina, E. P., Sklyarov, E. V., Vologina, E. G., Orlova, L. A., Sklyarova, O. A., Solotchin, P. A., Stolpovskaya, V. N., Fedorovskii, V. S., & Khlystov, O. M. (2008). Carbonates in the sedimentary record of saline Tsagan-Tyrm Lake, west Baikal region: New type of high-resolution paleoclimatic signals. *Doklady Earth Sciences, 421*(2), 926–933. https://doi.org/10.1134/s1028334x08060147

Solotchina, E. P., Solotchin, P. A., Stolpovskaya, V. N., Izokh, O. P., Sklyarov, E. V., Vologina, E. G., Ukhova, N. N., & Sklyarova, O. A. (2011). Climatic signal in the Holocene carbonate sedimentary record of Namshi-Nur Lake, West Baikal region. *Doklady Earth Sciences, 436*, 295–300. https://doi.org/10.1134/S1028334X11020267

Sommers, M. G., Dollhopf, M. E., & Douglas, S. (2002). Freshwater ferromanganese stromatolites from Lake Vermilion, Minnesota: Microbial culturing and environmental scanning electron microscopy investigations. *Geomicrobiology Journal, 19*(4), 407–427. https://doi.org/10.1080/01490450290098513

Stanton, C., Barnes, B. D., Kump, L. R., & Cosmidis, J. (2022). A re-examination of the mechanism of whiting events: A new role for diatoms in Fayetteville Green Lake (New York, USA). *Geobiology*. https://doi.org/10.1111/gbi.12534

Strakhovenko, V., Subetto, D., Ovdina, E., Belkina, N., & Efremenko, N. (2020). Distribution of elements in iron-manganese formations in bottom sediments of Lake Onego (NW Russia) and Small Lakes (Shotozero and Surgubskoe) of adjacent territories. *Minerals, 10*(5), 440. Advance online publication.. https://doi.org/10.3390/min10050440

Stromecki, A., Murray, L., Fullerton, H., & Moyer, C. L. (2022). Unexpected diversity found within benthic microbial mats at hydrothermal springs in Crater Lake, Oregon. *Frontiers in Microbiology, 13*(87), 6044. https://doi.org/10.3389/fmicb.2022.876044

Su, G., Zopfi, J., Yao, H., Steinle, L., Niemann, H., & Lehmann, M. F. (2019). Manganese/iron-supported sulfate-dependent anaerobic oxidation of methane by archaea in lake sediments. *Limnology and Oceanography, 65*(4), 863–875. https://doi.org/10.1002/lno.11354

Suess, E. (2014). Marine cold seeps and their manifestations: Geological control, biogeochemical criteria and environmental conditions. *International Journal of Earth Sciences, 103*(7), 1889–1916. https://doi.org/10.1007/s00531-014-1010-0

Suominen, S., van Vliet, D. M., Sánchez-Andrea, I., van der Meer, M. T. J., Sinninghe Damsté, J. S., & Villanueva, L. (2021). Organic matter type defines the composition of active microbial communities originating from anoxic Baltic Sea sediments. *Frontiers in Microbiology, 12*. Advance online publication. https://doi.org/10.3389/fmicb.2021.628301

Swart, P. K. (2015). The geochemistry of carbonate diagenesis: The past, present and future. *Sedimentology, 62*, 1233–1304. https://doi.org/10.1111/sed.12205

Teichert, B. M. A., Johnson, J. E., Solomon, E. A., Giosan, L., Rose, K., Kocherla, M., Connolly, E. C., & Torres, M. E. (2014). Composition and origin of authigenic carbonates in the Krishna–Godavari and Mahanadi Basins, eastern continental margin of India. Marine and Petroleum Geology, 58, 438–460. https://doi.org/10.1016/j.marpetgeo.2014.08.023

Teske, A., & Carvalho, V. (2020). *Marine hydrocarbon seeps*. Springer International Publishing. https://doi.org/10.1007/978-3-030-34827-4

Tiercelin, J.-J., Boulègue, J., & Simoneit, B. R. T. (1993). Hydrocarbons, sulphides, and carbonate deposits related to sublacustrine hydrothermal seeps in the North Tanganyika Trough, East African Rift. *Bitumens in Ore Deposits*, 96–113. https://doi.org/10.1007/978-3-642-85806-2_7

Timoshkin, O. A. (2001). Index of animal species inhabiting Lake Baikal and its catchment. In O. A. Timoshkin (Ed.), *Lake Baikal: Diversity of fauna, problems of its immiscibility and origin, ecology and "exotic"* (pp. 74–117). Nauka. In Russian.

Timoshkin, O. A., Bondarenko, N. A., Volkova, Y. A., Tomberg, I. V., Vishnyakov, V. S., & Malnik, V. V. (2015). Mass development of green filamentous algae of the genera Spirogyra

and Stigeoclonium (Chlorophyta) in the littoral zone of the southern part of Lake Baikal. *Hydrobiological Journal, 51*(1), 13–23. https://doi.org/10.1615/hydrobj.v51.i1.20

Timoshkin, O. A., Moore, M. V., Kulikova, N. N., Tomberg, I. V., Malnik, V. V., Shimaraev, M. N., Troitskaya, E. S., Shirokaya, A. A., Sinyukovich, V. N., Zaitseva, E. P., Domysheva, V. M., Yamamuro, M., Poberezhnaya, A. E., & Timoshkina, E. M. (2018). Groundwater contamination by sewage causes benthic algal outbreaks in the littoral zone of Lake Baikal (East Siberia). *Journal of Great Lakes Research, 44*(2), 230–244. https://doi.org/10.1016/j.jglr.2018.01.008

Tiquia, S. M., Ichida, J. M., Keener, H. M., et al. (2005). Bacterial community profiles on feathers during composting as determined by terminal restriction fragment length polymorphism analysis of 16S rDNA genes. *Applied Microbiology and Biotechnology, 67*, 412–419. https://doi.org/10.1007/s00253-004-1788-y

Tiquia, S. M., Gurczynski, S., Zholi, A., & Devol, A. (2006a). Diversity of biogeochemical cycling genes from Puget Sound sediments using DNA microarrays. *Environmental Technology, 27*(12), 1377–1389. https://doi.org/10.1080/09593332708618756

Tiquia, S. M., Masson, S. A., & Devol, A. (2006b). Vertical distribution of nitrite reductase genes (nirS) in continental margin sediments of the Gulf of Mexico. *FEMS Microbiology Ecology, 58*(3), 464–475. https://doi.org/10.1111/j.1574-6941.2006.00173.x

Tiquia, S. M., Davis, D., Hadid, H., Kasparian, S., Ismail, M., & Murray, K. S. (2007). Halophilic and halotolerant bacteria from river waters and shallow groundwater along the Rouge River of Southeastern Michigan. *Environmental Technology, 28*, 297–307. https://doi.org/10.1080/09593332808618789

Tiquia, S. M., Schleibak, M., Schlaff, J., Floyd, J. C., Benipal, B., Zakhem, E., & Murray, K. S. (2008). Microbial community profiling and characterization of some heterotrophic bacterial isolates from river waters and shallow groundwater wells along the Rouge River, Southeast Michigan. *Environmental Technology, 29*, 651–663. https://doi.org/10.1080/09593330801986998

Tiquia-Arashiro, S. M. (2014a). *Thermophilic Carboxydotrophs and their applications in biotechnology* (SpringerBriefs in microbiology) (p. 131). Springer. https://doi.org/10.1007/978-3-319-11873-4

Tiquia-Arashiro, S. M. (2014b). Microbial CO metabolism. Chapter 2. In *Thermophilic Carboxydotrophs and their applications in biotechnology* (SpringerBriefs in microbiology) (pp. 5–9). Springer. https://doi.org/10.1007/978-3-319-11873-4_2

Tiquia-Arashiro, S., & Rodrigues, D. (2016). Alkaliphiles and Acidophiles in nanotechnology. In *Extremophiles: Applications in nanotechnology*. Springer. https://doi.org/10.1007/978-3-319-45215-9_4

Top, E. M., & Springael, D. (2003). The role of mobile genetic elements in bacterial adaptation to xenobiotic organic compounds. *Current Opinion in Biotechnology, 14*(3), 262–269. https://doi.org/10.1016/s0958-1669(03)00066-1

Torres, N. T., Och, L. M., Hauser, P. C., Furrer, G., Brandl, H., Vologina, E., Sturm, M., Bürgmann, H., & Müller, B. (2014). Early diagenetic processes generate iron and manganese oxide layers in the sediments of Lake Baikal. *Environmental Science: Processes & Impacts, 16*(4), 879–889. https://doi.org/10.1039/c3em00676j

Torres, M. E., Hong, W.-L., Solomon, E. A., Milliken, K., Kim, J.-H., Sample, J. C., Teichert, B. M. A., & Wallmann, K. (2020). Silicate weathering in anoxic marine sediment as a requirement for authigenic carbonate burial. *Earth-Science Reviews, 200*, 102960. https://doi.org/10.1016/j.earscirev.2019.102960

Tran, P. Q., Bachand, S. C., McIntyre, P. B., Kraemer, B. M., Vadeboncoeur, Y., Kimirei, I. A., Tamatamah, R., McMahon, K. D., & Anantharaman, K. (2021). Depth-discrete metagenomics reveals the roles of microbes in biogeochemical cycling in the tropical freshwater Lake Tanganyika. *The ISME Journal, 15*(7), 1971–1986. https://doi.org/10.1038/s41396-021-00898-x

Umezawa, K., Kojima, H., Kato, Y., & Fukui, M. (2022). Corrigendum to "*Dissulfurispirathermophila* gen. nov., sp. nov., a thermophilic chemolithoautotroph growing by sulfur disproportionation, and proposal of novel taxa in the phylum *Nitrospirota* to reclas-

sify the genus *Thermodesulfovibrio*". *Systematic and Applied Microbiology, 45*(6), 126323. Advance online publication.. https://doi.org/10.1016/j.syapm.2022.126323

Usoltseva, M., Kopyrina, L., Titova, L., Morozov, A., Firsova, A., Zakharova, Y., Bashenkhaeva, M., Maslennikova, M., & Likhoshway, Y. (2020). Finding of a putative Lake Baikal endemic, *Lindavia minuta*, in distant lakes near the Arctic pole in Yakutia (Russia). *Diatom Research, 35*(2), 141–153. https://doi.org/10.1080/0269249x.2020.1747551

Uzun, M., Koziaeva, V., Dziuba, M., Alekseeva, L., Krutkina, M., Sukhacheva, M., Baslerov, R., & Grouzdev, D. (2023). Recovery and genome reconstruction of novel magnetotactic *Elusimicrobiota* from bog soil. *The ISME Journal, 17*(2), 204–214. https://doi.org/10.1038/s41396-022-01339-z

van den Berg, N. I., Machado, D., Santos, S., Rocha, I., Chacon, J., William, H., Mitri, S., Patil., K.R. (2022). Ecological modelling approaches for predicting emergent properties in microbial communities. Nature Ecology Evolution, 6, 855–865. https://doi.org/10.1038/s41559-022-01746-7

van Kessel, M. A. H. J., Speth, D. R., Albertsen, M., Nielsen, P. H., Op den Camp, H. J. M., Kartal, B., Jetten, M. S. M., & Lücker, S. (2015). Complete nitrification by a single microorganism. *Nature, 528*(7583), 555–559. https://doi.org/10.1038/nature16459

Vanwonterghem, I., Evans, P. N., Parks, D. H., Jensen, P. D., Woodcroft, B. J., Hugenholtz, P., & Tyson, G. W. (2016). Methylotrophic methanogenesis discovered in the archaeal phylum *Verstraetearchaeota*. *Nature Microbiology, 1*(12) Advance online publication. https://doi.org/10.1038/nmicrobiol.2016.170

Vinogradova, T. P., Kerber, E. V., Drukker, V. V., Zavezenova, T. V., & Maksimova, E. A. (2004). *Microbiological heritage of the XX century*. Publishing House of the Institute of Geography SB RAS.

Vologina, E. G., & Sturm, M. (2009). Types of Holocene deposits and regional pattern of sedimentation in Lake Baikal. *Russian Geology and Geophysics, 50*(8), 722–727. https://doi.org/10.1016/j.rgg.2008.12.012

Vologina, E. G., & Sturm, M. (2017). Particulate fluxes in South Baikal: Evidence from sediment trap experiments. *Russian Geology and Geophysics, 58*(9), 1045–1052. https://doi.org/10.1016/j.rgg.2016.12.006

Vologina, E. G., Sturm, M., Vorobyova, S. S., & Granina, L. Z. (2000). New results of high-resolution studies of surface sediments of Lake Baikal. *Terra Nostra, 9*, 115–131.

Vologina, E. G., Kashik, S. A., Sturm, M., Vorob'eva, S. S., Lomonosova, T. K., Kalashnikova, I. A., Khramtsova, T. I., & Toshchakov, S. Y. (2007). Results of research into Holocene sediments of the South and Central basins of Lake Baikal (BDP-97 and short cores). *Russian Geology and Geophysics, 48*(4), 312–322. https://doi.org/10.1016/j.rgg.2007.03.002

Von Eggers, J. M., Wisnoski, N. I., Calder, J. W., Capo, E., Groff, D. V., Krist, A. C., & Shuman, B. (2024). Environmental filtering governs consistent vertical zonation in sedimentary microbial communities across disconnected mountain lakes. *Environmental Microbiology, 26*(3). https://doi.org/10.1111/1462-2920.16607

Votintsev, K. K., Meshcheryakova, A. I., & Popovskaya, G. I. (1975). *Cycleof organic matterinLakeBaikal*. Nauka.

Vuillemin, A., Friese, A., Alawi, M., Henny, C., Nomosatryo, S., Wagner, D., Crowe, S. A., & Kallmeyer, J. (2016). Geomicrobiological features of ferruginous sediments from Lake Towuti, Indonesia. *Frontiers in Microbiology, 7*. Advance online publication. https://doi.org/10.3389/fmicb.2016.01007

Vuillemin, A., Ariztegui, D., Horn, F., Kallmeyer, J., & Orsi, W. D. (2018). Microbial community composition along a 50 000-year lacustrine sediment sequence. *FEMS Microbiology Ecology, 94*(4) Advance online publication. https://doi.org/10.1093/femsec/fiy029

Vuillemin, A., Wirth, R., Kemnitz, H., Schleicher, A. M., Friese, A., Bauer, K. W., Simister, R., Nomosatryo, S., Ordoñez, L., Ariztegui, D., Henny, C., Crowe, S. A., Benning, L. G., Kallmeyer, J., Russell, J. M., Bijaksana, S., Vogel, H., & Towuti Drilling Project Science Team,

t. (2019). Formation of diagenetic siderite in modern ferruginous sediments. *Geology, 47*, 540–544. https://doi.org/10.1130/G46100.1

Vuillemin, A., Kerrigan, Z., D'Hondt, S., & Orsi, W. (2020). Exploring the abundance, metabolic potential and gene expression of subseafloor *Chloroflexi* in million-year-old oxic and anoxic abyssal clay. *FEMS Microbiology Ecology, 96*(12). https://doi.org/10.1093/femsec/fiaa223

Vykhistryuk, L. A. (1980). *Organic matter of Lake Baikal bottom sediments*. Nauka. In Russian.

Wang, Z., Shi, Z., Shi, L., White, G. F., Richardson, D. J., Clarke, T. A., Fredrickson, J. K., & Zachara, J. M. (2015). Effects of soluble flavin on heterogeneous electron transfer between surface-exposed bacterial cytochromes and iron oxides. *Geochimica et Cosmochimica Acta, 163*, 299–310. https://doi.org/10.1016/j.gca.2015.03.039

Wang, Y., Wegener, G., Hou, J., Wang, F., & Xiao, X. (2019). Expanding anaerobic alkane metabolism in the domain of Archaea. *Nature Microbiology, 4*(4), 595–602. https://doi.org/10.1038/s41564-019-0364-2

Wang, S., Sun, P., Liu, J., Xu, Y., Dolfing, J., & Wu, Y. (2022). Distribution of methanogenic and methanotrophic consortia at soil-water interfaces in rice paddies across climate zones. *iScience., 26*(1), 105851. https://doi.org/10.1016/j.isci.2022.105851

Weber, H. S., Habicht, K. S., & Thamdrup, B. (2017). Anaerobic methanotrophic archaea of the ANME-2d cluster are active in a low-sulfate, iron-rich freshwater sediment. *Frontiers in Microbiology, 8*. Advance online publication. https://doi.org/10.3389/fmicb.2017.00619

Wegner, C.-E., & Liesack, W. (2017). Unexpected dominance of elusive *Acidobacteria* in early industrial soft coal slags. *Frontiers in Microbiology, 8*. Advance online publication. https://doi.org/10.3389/fmicb.2017.01023

Weiss, R. F., Carmack, E. C., & Koropalov, V. M. (1991). Deep-water renewal and biological production in Lake Baikal. *Nature, 349*(6311), 665–669. https://doi.org/10.1038/349665a0

Welte, C. U., Rasigraf, O., Vaksmaa, A., Versantvoort, W., Arshad, A., Op den Camp, H. J., Jetten, M. S., Lüke, C., & Reimann, J. (2016). Nitrate- and nitrite-dependent anaerobic oxidation of methane. *Environmental Microbiology Reports, 8*(6), 941–955. https://doi.org/10.1111/1758-2229.12487

Wen, X., Yang, S., Horn, F., Winkel, M., Wagner, D., & Liebner, S. (2017). Global biogeographic analysis of methanogenic archaea identifies community-shaping environmental factors of natural environments. *Frontiers in Microbiology, 8*, 1339. Advance online publication.. https://doi.org/10.3389/fmicb.2017.01339

Williams, D. F., Peck, J., Karabanov, E. B., Prokopenko, A. A., Kravchinsky, V., King, J., & Kuzmin, M. I. (1997). Lake Baikal record of continental climate response to orbital insolation during the past 5 million years. *Science, 278*(5340), 1114–1117. https://doi.org/10.1126/science.278.5340.1114

Winkel, M., Sepulveda-Jauregui, A., Martinez-Cruz, K., Heslop, J. K., Rijkers, R., Horn, F., Liebner, S., & Walter Anthony, K. M. (2019). First evidence for cold-adapted anaerobic oxidation of methane in deep sediments of thermokarst lakes. *Environmental Research Communications, 1*(2), 021002. Advance online publication. https://doi.org/10.1088/2515-7620/ab1042

Winters, A. D., Marsh, T. L., Brenden, T. O., & Faisal, M. (2014). Molecular characterization of bacterial communities associated with sediments in the Laurentian Great Lakes. *Journal of Great Lakes Research, 40*(3), 640–645. https://doi.org/10.1016/j.jglr.2014.04.008

Wislouch, S. M. (1912). *Thioploca ingrica*nov. sp. *Berichte der Deutschen Botanischen Gesellschaft, 30*(8), 470–474. https://doi.org/10.1111/j.1438-8677.1912.tb07514.x

Woese, C. R. (1987). Bacterial evolution. *FEMS Microbiology Reviews, 51*, 221–271.

Wu, W., Gu, B., Fields, M. W., Gentile, M., Ku, Y.-K., Yan, H., Tiquia, S. M., Yan, T., Nyman, J., Zhou, J., Jardine, P. M., & Criddle, C. S. (2005). Uranium (VI) reduction by denitrifying biomass. *Bioremediation Journal, 9*, 1–13. https://doi.org/10.1080/10889860590929628

Wu, M., Li, J., Lai, C.-Y., Leu, A. O., Sun, S., Gu, R., Erler, D. V., Liu, L., Li, L., Tyson, G. W., Yuan, Z., McIlroy, S. J., & Guo, J. (2024). Nitrate-driven anaerobic oxidation of ethane and butane by bacteria. *The ISME Journal, 18*(1). https://doi.org/10.1093/ismejo/wrad011

Wuest, A., Ravens, T. M., Granin, N. G., Kosis, O., Schurter, M., & Sturm, M. (2005). Cold intrusions in Lake Baikal: Direct observational evidence for deep-water renewal. *Limnology and Oceanography, 50*(1), 184–196. https://doi.org/10.4319/lo.2005.50.1.0184

Wurzbacher, C., Fuchs, A., Attermeyer, K., Frindte, K., Grossart, H.-P., Hupfer, M., Casper, P., & Monaghan, M. T. (2017a). Shifts among Eukaryota, Bacteria, and Archaea define the vertical organization of a lake sediment. *Microbiome, 5*(1). https://doi.org/10.1186/s40168-017-0255-9

Wurzbacher, C., Nilsson, R. H., Rautio, M., & Peura, S. (2017b). Poorly known microbial taxa dominate the microbiome of permafrost thaw ponds. *The ISME Journal, 11*(8), 1938–1941. https://doi.org/10.1038/ismej.2017.54

Yanagawa, K., Shiraishi, F., Tanigawa, Y., Maeda, T., Mustapha, N. A., Owari, S., Tomaru, H., Matsumoto, R., & Kano, A. (2019). Endolithic microbial habitats hosted in carbonate nodules currently forming within sediment at a high methane flux site in the sea of Japan. *Geosciences, 9*, 463. Advance online publication.. https://doi.org/10.3390/geosciences9110463

Yang, T., Lyons, S., Aguilar, C., Cuhel, R., & Teske, A. (2011). Microbial communities and chemosynthesis in Yellowstone Lake sublacustrine hydrothermal vent waters. *Frontiers in Microbiology, 2*. https://doi.org/10.3389/fmicb.2011.00130

Yoshida, T., Sekino, T., Genkai-Kato, M., Logacheva, N. P., Bondarenko, N. A., Kawabata, Z., Khodzher, T. V., Melnik, N. G., Hino, S., Nozaki, K., Nishimura, Y., Nagata, T., Higashi, M., & Nakanishi, M. (2003). Seasonal dynamics of primary production in the pelagic zone of southern Lake Baikal. *Limnology, 4*(1), 53–62. https://doi.org/10.1007/s10201-002-0089-3

Yoshioka, T., Ueda, S., Khodzher, T., Bashenkhaeva, N., Korovyakova, I., Sorokovikova, L., & Gorbunova, L. (2002). Distribution of dissolved organic carbon in Lake Baikal and its watershed. *Limnology, 3*, 0159–0168. https://doi.org/10.1007/s102010200019

Youssef, N. H., Rinke, C., Stepanauskas, R., Farag, I., Woyke, T., & Elshahed, M. S. (2015). Insights into the metabolism, lifestyle and putative evolutionary history of the novel archaeal phylum 'Diapherotrites'. *The ISME Journal, 9*(2), 447–460. https://doi.org/10.1038/ismej.2014.141

Yu, T., Wu, W., Liang, W., Lever, M. A., Hinrichs, K.-U., & Wang, F. (2018). Growth of sedimentary *Bathyarchaeota* on lignin as an energy source. *Proceedings of the National Academy of Sciences, 115*(23), 6022–6027. https://doi.org/10.1073/pnas.1718854115

Zakharenko, A. S., Galachyants, Y. P., Morozov, I. V., Shubenkova, O. V., Morozov, A. A., Ivanov, V. G., Pimenov, N. V., Krasnopeev, A. Y., & Zemskaya, T. I. (2019). Bacterial communities in areas of oil and methane seeps in pelagic of Lake Baikal. *Microbial Ecology, 78*(2), 269–285. https://doi.org/10.1007/s00248-018-1299-5

Zakharova, Y. R., Parfenova, V. V., Granina, L. Z., Kravchenko, O. S., & Zemskaya, T. I. (2010). Distribution of iron- and manganese-oxidizing bacteria in the bottom sediments of Lake Baikal. *Inland Water Biology, 3*(4), 313–321. https://doi.org/10.1134/s1995082910040036

Zakharova, Y. R., Petrova, D. P., Galachyants, Y. P., Bashenkhaeva, M. V., Kurilkina, M. I., & Likhoshway, Y. V. (2018). Bacterial and archaeal community structure in the surface diatom sediments of deep freshwater Lake Baikal (Eastern Siberia). *Geomicrobiology Journal, 35*(8), 635–647. https://doi.org/10.1080/01490451.2018.1440036

Zárate-del Valle, P. F., Rushdi, A. I., & Simoneit, B. R. T. (2006). Hydrothermal petroleum of Lake Chapala, Citala Rift, western Mexico: Bitumen compositions from source sediments and application of hydrous pyrolysis. *Applied Geochemistry, 21*(4), 701–712. https://doi.org/10.1016/j.apgeochem.2006.01.002

Zavarzin, G. A. (2003). Formation of the system of biogeochemical cycles. *Paleontology Journal, 37*(6), 576–583.

Zemskaya, T. I., Namsaraev, B. B., Dul'tseva, N. M., Khanaeva, T. A., Golobokova, L. P., Dubinina, G. A., Dulov, L. E., & Wada, E. (2001). Ecophysiological characteristics of the mat-forming bacterium *Thioploca* in bottom sediments of the Frolikha Bay, Northern Baikal. *Microbiology, 70*, 335–341. https://doi.org/10.1023/a:1010463613498

Zemskaya, T. I., Khlystov, O. M., Egorov, A. V., Pogodaeva, T. V., Kalmychkov, G. V., Shubenkova, O. V., Chernitcyna, S. M., Vorob'eva, S. S., & Grachev, M. A. (2008). Comprehensive studies of gas hydrate manifestations in sediments of Lake Baikal. In G. A. Zavarzin & V. N. Kydeyarov

(Eds.), *Changes of natural environment and climate. Processes in Biosphere: changes in soil and plant cover and of territorial waters of Russia, turnover of biophilic elements as affected by global climatic changes and catastrophes* (pp. 125–152). Moscow, ISSP PBC RAS. In Russian.

Zemskaya, T. I., Chernitsyna, S. M., Dul'tseva, N. M., Sergeeva, V. N., Pogodaeva, T. V., & Namsaraev, B. B. (2009). Colorless sulfur bacteria *Thioploca* from different sites in Lake Baikal. *Microbiology, 78*(1), 117–124. https://doi.org/10.1134/s0026261709010159

Zemskaya, T. I., Pogodaeva, T. V., Shubenkova, O. V., Chernitsina, S. M., Dagurova, O. P., Buryukhaev, S. P., Namsaraev, B. B., Khlystov, O. M., Egorov, A. V., Krylov, A. A., & Kalmychkov, G. V. (2010). Geochemical and microbiological characteristics of sediments near the Malenky mud volcano (Lake Baikal, Russia), with evidence of Archaea intermediate between the marine anaerobic methanotrophs ANME-2 and ANME-3. *Geo-Marine Letters, 30*(3–4), 411–425. https://doi.org/10.1007/s00367-010-0199-6

Zemskaya, T. I., Sitnikova, T. Y., Kiyashko, S. I., Kalmychkov, G. V., Pogodaeva, T. V., Mekhanikova, I. V., Naumova, T. V., Shubenkova, O. V., Chernitsina, S. M., Kotsar, O. V., Chernyaev, E. S., & Khlystov, O. M. (2012). Faunal communities at sites of gas- and oil-bearing fluids in Lake Baikal. *Geo-Marine Letters, 32*(5–6), 437–451. https://doi.org/10.1007/s00367-012-0297-8

Zemskaya, T., Lomakina, A., Mamaeva, E., Zakharenko, A., Pogodaeva, T., Petrova, D., & Galachyants, Y. (2015a). Bacterial communities in sediments of Lake Baikal from areas with oil and gas discharge. *Aquatic Microbial Ecology, 76*(2), 95–109. https://doi.org/10.3354/ame01773

Zemskaya, T. I., Lomakina, A. V., Shubenkova, O. V., Pogodaeva, T. V., Morozov, I. V., Chernitsina, S. M., Sitnikova, T. Y., Khlystov, O., & Egorov, A. V. (2015b). Jelly-like microbial mats over subsurface fields of gas hydrates at the St. Petersburg methane seep (Central Baikal). *Geomicrobiology Journal, 32*(1), 89–100. https://doi.org/10.1080/01490451.2014.910572

Zemskaya, T. I., Lomakina, A. V., Mamaeva, E. V., Zakharenko, A. S., Likhoshvai, A. V., Galach'yants, Y. P., & Müller, B. (2018). Composition of microbial communities in sediments from Southern Baikal containing Fe/Mn concretions. *Microbiology, 87*(3), 382–392. https://doi.org/10.1134/s0026261718030165

Zemskaya, T. I., Cabello-Yeves, P. J., Pavlova, O. N., & Rodriguez-Valera, F. (2020). Microorganisms of Lake Baikal—The deepest and most ancient lake on Earth. *Applied Microbiology and Biotechnology, 104*(14), 6079–6090. https://doi.org/10.1007/s00253-020-10660-6

Zemskaya, T. I., Bukin, S. V., Lomakina, A. V., & Pavlova, O. N. (2021). Microorganisms in the sediments of Lake Baikal, the deepest and oldest Lake in the world. *Microbiology, 90*(3), 298–313. https://doi.org/10.1134/s0026261721030140

Zemskaya, T., Konstantinova, N., Shubenkova, O., Pogodaeva, T., Ivanov, V., Bukin, S., Khabuev, A., Khlystov, O., Vilkin, G., & Lomakina, A. (2022). Microbial communities of ferromanganese sedimentary layers and nodules of Lake Baikal (Bolshoy Ushkany Island). *Diversity, 14*(10), 868. Advance online publication.. https://doi.org/10.3390/d14100868

Zhang, J., Yang, Y., Zhao, L., Li, Y., Xie, S., & Liu, Y. (2015). Distribution of sediment bacterial and archaeal communities in plateau freshwater lakes. *Applied Microbiology and Biotechnology, 99*(7), 3291–3302. https://doi.org/10.1007/s00253-014-6262-x

Zhang, C., Meckenstock, R. U., Weng, S., Wei, G., Hubert, C. R. J., Wang, J.-H., & Dong, X. (2021). Marine sediments harbor diverse archaea and bacteria with potentials for anaerobic hydrocarbon degradation via fumarate addition. *FEMS Microbiology Ecology, 97*(5) Advance online publication. https://doi.org/10.1093/femsec/fiab045

Zhang, X., Huang, Y., Liu, Y., Xu, W., Pan, J., Zheng, X., Du, H., Zhang, C., Lu, Z., Zou, D., Liu, Z., Cai, M., Xiong, J., Zhu, Y., Dong, Z., Jiang, H., Dong, H., Jiang, J., Luo, Z., Huang, L., & Li, M. (2022). An ancient respiratory system in the widespread sedimentary archaea *Thermoprofundales*. *Molecular Biology and Evolution, 39*(10), msac213. https://doi.org/10.1093/molbev/msac213

Zhang, S., Yan, L., Cao, J., Wang, K., Luo, Y., Hu, H., Wang, L., Yu, R., Pan, B., Yu, K., Zhao, J., & Bao, Z. (2023 Jan). Salinity significantly affects methane oxidation and methanotrophic community in Inner Mongolia lake sediments. *Frontiers in Microbiology, 6*(13), 1067017. https://doi.org/10.3389/fmicb.2022.1067017

Zhou, Z., Pan, J., Wang, F., Gu, J.-D., & Li, M. (2018). *Bathyarchaeota*: Globally distributed metabolic generalists in anoxic environments. *FEMS Microbiology Reviews, 42*(5), 639–655. https://doi.org/10.1093/femsre/fuy023

Zhou, Z., Liu, Y., Lloyd, K. G., Pan, J., Yang, Y., Gu, J.-D., & Li, M. (2019). Genomic and transcriptomic insights into the ecology and metabolism of benthic archaeal cosmopolitan, Thermoprofundales (MBG-D archaea). *The ISME Journal, 13*(4), 885–901. https://doi.org/10.1038/s41396-018-0321-8

Zolitschka, B., Anselmetti, F., Ariztegui, D., Corbella, H., Francus, P., Lücke, A., Maidana, N. I., Ohlendorf, C., Schäbitz, F., & Wastegård, S. (2013). Environment and climate of the last 51,000 years—New insights from the Potrok Aike maar lake Sediment Archive Drilling prOject (PASADO). *Quaternary Science Reviews, 71*, 1–12. https://doi.org/10.1016/j.quascirev.2012.11.024

Zonenshain, L. P. (1992). Baikal rift structure. *Geotectonics, 26*(5), 396–407.

Chapter 5
Water Quality and Bacterial Community Characteristics of Lakes and Reservoirs with Different Land-Use Types

Hua Ma, Sheng-Ao Li, and Zhicheng Hong

1 Introduction

Natural lakes and man-made reservoirs store approximately 90% of the Earth's liquid freshwater resources on its surface and play a crucial role in supporting human economic and social development, as well as maintaining the stability of ecosystems (Verpoorter et al., 2014; Cantonati et al. 2020; Kalra et al. 2023). As major organic carbon sinks in the global biogeochemical cycle, lakes and reservoirs can not only play an important buffering role in mitigating climate warming (Cole and Caraco, 2001; Gentile et al. 2006; Raymond et al. 2014; Malhi et al. 2020; Soomro et al. 2023) but also contribute to global warming through the emission of greenhouse gasses (GHGs) such as carbon dioxide (CO_2) and methane (CH_4) (Hayes et al., 2017). Dissolved organic matter (DOM), as the main form of organic carbon in lakes and reservoirs, plays a crucial role in water quality, freshwater ecosystem health, and the global biogeochemical cycle of carbon, which is mainly divided into two components: terrestrial input and endogenous production (heterogenous and autogenous DOM, respectively) (Sobek et al., 2007; Tiquia, 2010a; Mckindles and Tiquia-Arashiro, 2012; Findlay and Parr, 2017; Oest et al. 2018; Patel et al. 2019). Thirty to seventy percent of all biogenic organic carbon content in lakes and reservoirs is derived from terrestrial sources of DOM, the composition and diversity of which shape the biological communities and ecosystems in lakes and reservoirs (Jansson et al., 2007; Tiquia, 2008; Tiquia, 2011; Oest et al. 2018; Xu et al. 2023). Lake DOM releases large amounts of carbon dioxide and methane to the atmosphere in response to light and microbial activity, and participates in the global carbon cycle, and is transported through freshwater systems to further influence the

H. Ma (✉) · S.-A. Li · Z. Hong
College of Environment and Ecology, Key Laboratory of the Three Gorges Reservoir Region's Eco-Environment, Ministry of Education, Chongqing University, Chongqing, China
e-mail: hma@cqu.edu.cn

S. M. Tiquia-Arashiro et al. (eds.), *Urban Watershed Microbiology, Volume 1*,
https://doi.org/10.1007/978-3-031-89071-0_5

total amount and composition of DOM in the oceans (Liu et al. 2003; Tiquia et al. 2006a; Tiquia et al. 2006b; ; Harrison et al., 2021).

Due to the increasing scope and intensity of human activities, the land use and land cover of lake basins have changed dramatically, which not only affects the total amount and type of carbon input into the lakes and reservoirs, but also alters the structure of microbial populations and the composition of dissolved organic matter in lakes and reservoirs, and thus has a profound impact on the ecological stability of the aquatic environment, the use of water resources, and the global carbon cycle. Wei et al. (2020) analyzed data on land cover changes and water quality indicators in 12 lakes in China over a 10-year period and found that degradation of natural land cover types was positively correlated with the deterioration of water quality in lakes and reservoirs (Wei et al., 2020). Ultrahigh-resolution mass spectrometry analysis of DOM in 120 lakes found that the chemical diversity of DOM molecules increased with increasing DOM and nutrient concentrations (Kellerman et al., 2014). Lakes and reservoirs have longer residence times and more stable hydraulic conditions than rivers, favoring stable metabolism of organic matter by microbial communities (Tiquia 2011; Oest et al. 2018; Wu et al., 2021; Dai et al. 2023). However, the combined effects of different land use compositions and microbial treatments on DOM in lakes and reservoirs have been much less explored.

High-throughput sequencing of microorganisms and high-resolution mass spectrometry of dissolved organic matter have enabled the detection of microbial community structure and abundance at the level of the basic constituent units of dissolved organic matter systems (Dudley et al. 2010; Tiquia-Arashiro, 2012; Tiquia-Arashiro and Pant, 2020; Kuang et al. 2023; Nassereddine et al. 2024), and both techniques have been used by a number of researchers to investigate microbial interactions with organic matter molecules in different systems. For example, Meng Wu et al. found that microorganisms contributed more to dissolved organic matter than consumed it by constructing a co-occurrence network diagram of soil organic matter and bacteria in a rice field (Wu et al., 2021). Zhao et al. determined the molecular formulae of organic matter associated with key bacteria in seawater related to the conversion of organic matter of algal origin by laboratory culture (Zhao et al., 2019). The ecological function analysis of microbial communities showed that the microbial community structure of different types of lakes and reservoirs exhibited unique ecological functions, which were mostly related to metabolic activities such as the synthesis and degradation of organic matter by microorganisms (Ma et al. 2018; Kiersztyn et al., 2019; Hong et al., 2023). Previous study shows that some organic molecules contained in the dissolved organic system of lakes and reservoirs are closely related to the metabolic activities of microorganisms (Hong et al., 2023). Researchers have explored the molecular composition of a variety of organic systems, such as vegetation, algae, sediments, sewage treatment plant effluent, and soil (Imai et al. 2002; Ratnadass et al., 2012; Nebbioso and Piccolo, 2013; Cao et al. 2018; Kurek et al. 2020; Kassem et al. 2023; Pagnucco et al. 2023; Zhu et al. 2023). A deep understanding of the sources of dissolved organic matter provides conditions for us to explore the formation of complex systems such as dissolved organic matter in natural water bodies. It has been shown that DOM systems in lakes and

reservoirs waters in watersheds with different land cover conditions exhibit unique chemical diversity, indicating the critical role of watershed land cover in shaping the DOM systems of lakes and reservoirs (Hong et al., 2023; Zhu et al. 2023; Wen et al. 2024). At present, most of the studies mainly focus on the effects of eutrophication on microbial community structure and DOM characteristics of lakes and reservoirs. Moreover, due to technical limitations, there is a lack of in-depth research on the composition of DOM.

In this study, we selected several lakes and reservoirs in Chongqing, China, and used ultra-high resolution mass spectrometry (UHRMS) to characterize DOM at the molecular compositional level in three types of land-covered watersheds, namely, forest, farmland, and urban. The sources of DOM in the lakes and reservoirs were investigated using EEM-PARAFAC and multivariate statistical methods. The ecological functions and co-occurrence network characteristics of microbial communities in various types of lakes and reservoirs were also analyzed using high-throughput sequencing. We also used multivariate statistical analyses to explore the influence of watershed land cover on the composition of DOM sources in natural waters by collecting high-resolution mass spectrometry characterization data on various DOM sources provided by other researchers. The aim of this review is to explore potential relationships between bacteria and the molecular formula of dissolved organic matter in lakes and reservoirs at both the operational taxonomic unit (OTU) and molecular levels.

2 Effect of Land Cover on Molecular Composition of DOM and Bacterial Community Structure

Seven lakes from watersheds with three different land cover conditions (forest-dominated, cropland-dominated, and urban-dominated) in Chongqing, China, were selected, and watershed information and land cover conditions were collected, and water quality characteristics were determined (Table 5.1). The concentration of

Table 5.1 Water quality of three types of lakes

Landcover situation	Lake name	Water quality (mg/L)				
		DOC	TP	TN	SRP	NH_4^+-N
Forest-dominated	Xiao Jiagou	3.40 ± 0.44	0.07 ± 0.01	2.65 ± 0.10	0.05 ± 0.01	0.12 ± 0.03
	Shuang He	2.06 ± 0.07	0.06 ± 0.01	3.98 ± 0.19	0.04 ± 0.01	0.10 ± 0.01
Cropland-dominated	Tong Xin	9.80 ± 0.14	0.53 ± 0.02	4.59 ± 0.10	0.37 ± 0.01	0.86 ± 0.07
	Jin Tang	9.40 ± 0.23	0.27 ± 0.02	4.52 ± 0.12	0.20 ± 0.01	0.45 ± 0.03
	Fan Shen	12.12 ± 0.20	1.38 ± 0.01	6.22 ± 0.18	1.15 ± 0.04	3.78 ± 0.05
Urban-dominated	Jiu Long	7.15 ± 0.15	1.56 ± 0.02	4.62 ± 0.05	1.32 ± 0.01	3.13 ± 0.06
	Cai Yun	4.92 ± 0.15	0.78 ± 0.02	7.47 ± 0.10	0.66 ± 0.01	1.38 ± 0.06

nutrients and dissolved organic matter was significantly higher than that in forested lakes with predominantly natural land cover (Donohue et al., 2006; Nielsen et al., 2012). Previous studies have shown that towns contribute more TN, NH_4^+, and TP to lakes and reservoirs than agricultural land, thus contributing significantly to eutrophication (Ding et al., 2015; Qin et al., 2007). The dissolved organic carbon (DOC) in the three types of lakes showed very significant differences, with forest-type lakes having the lowest DOC within 5 mg/L, town-type lakes having 5–10 mg/L, and farmland-type lakes having the highest DOC. Researchers have demonstrated that the type of land cover is a significant indicator in determining the concentration of DOC in lakes and reservoirs (Larson et al., 2014; Raghul et al. 2024). Additionally, higher population densities and increased anthropogenic impacts have been found to result in higher levels of DOM in watershed waters (Aitkenhead-Peterson et al., 2009; Tiquia, 2010b; Dumitran et al. 2024).

The optical properties and fluorescence components of the lake samples were analyzed by optical spectroscopy and parallel factor analysis (PARAFAC), respectively (Fig. 5.1). The molecular composition of DOM was characterized through

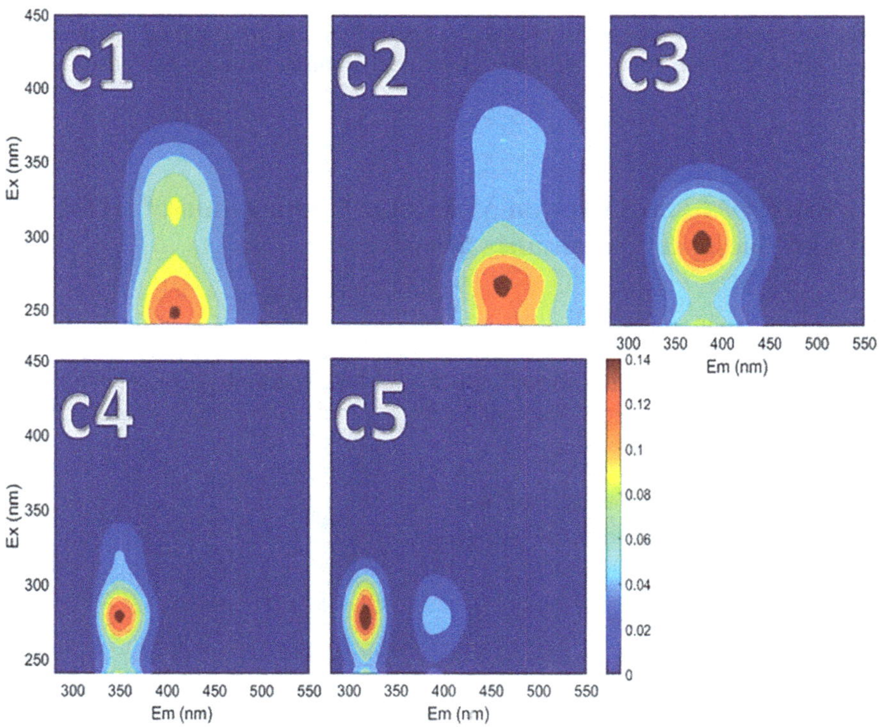

Fig. 5.1 Excitation and Emission spectra of 5 components. C1 was a readily biodegradable terrestrial humic substances; C2 represented a stable terrestrial humic substance; C3 represented a mixture of humic substances due to microbial production and tryptophan fraction; C4 was a newly generated autologous component; C5 represented a pure tryptophan protein fraction

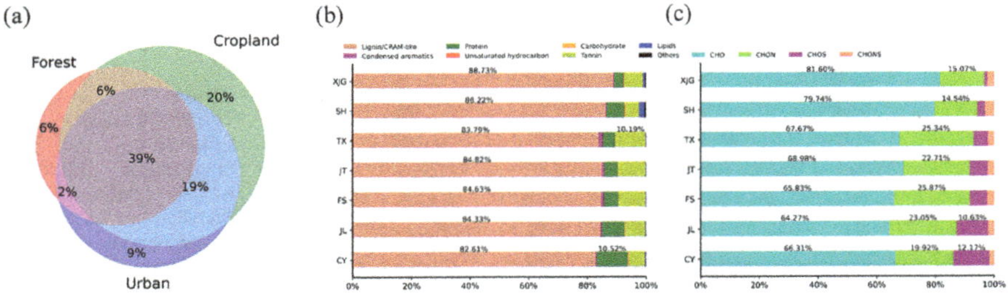

Fig. 5.2 Venn diagram of three kinds of lakes based on DOM molecular composition (**a**) and stack bar plot of two kinds of categories (**b-c**)

ultra-performance liquid chromatography plus Q-Exactive Orbitrap tandem mass spectrometry (UPLC-QE Plus Orbitrap/MS) (Fig. 5.2). Among the three types of lakes in this study, the most molecular formulas were detected in cropland-type lakes and the least in forest-type lakes (Fig. 5.2a). Cropland and urban lakes were located in watersheds where human daily life and agricultural activities led to a significant increase in the levels of organic matter concentration and molecular species diversity in both types of lakes and reservoirs. The greatest changes in total DOM and the number of molecular species were caused by agriculture in the watersheds. In addition, cropland lakes had the largest pool of shared organic matter molecules and the highest proportion of unique organic molecular weights, suggesting that lakes and reservoirs in cropland watersheds received a more diverse range of organic matter contamination compared to urban lakes. The higher density of vegetation cover in forested watersheds resulted in a higher proportion of lignin-like DOM compared to the other two types of lakes and reservoirs (Kaiser and Benner, 2012) (Fig. 5.2b). Urban lakes and reservoirs have a higher proportion of proteinaceous molecules than the other two types of lakes, and these organic matter molecules are readily available to microorganisms, making dissolved organic matter in urban lakes show a more pronounced endogenous character (Parr et al., 2015). Low nutrient content and favorable light conditions leading to the phenomenon of cyanobacterial dominance may be responsible for the higher proportion of lipids in forest-type lakes and reservoirs (Liu et al., 2020).

In order to investigate the characteristics of microbial population structure in lakes and reservoirs in watersheds dominated by the three land cover types, OTU taxonomic units were obtained for all sampling sites using 16S-rRNA-based sequencing, and a ternary map of the microbial composition of the three types of lakes was drawn based on the classification of OTUs at the class level (Fig. 5.3). *Cyanobacteria* and *Bacteroidia* dominated the forested lakes, while *Gammaproteobacteria*, *Alphaproteobacteria*, *Bacilli,* and *Bacteroidia* dominated the cropland lakes. *Acidimicrobiia*, *α-Aspergillus*, *Gammaproteobacteria,* and *Kapabacteria* were dominant in the urban lakes. Previous studies have shown that *Gammaproteobacteria* and *Alphaproteobacteria* are widely present in lakes that have received severe agricultural pollution as well as organic pollution (Wang et al., 2022). Meanwhile, the higher abundance of *Gammaproteobacteria* in cropland

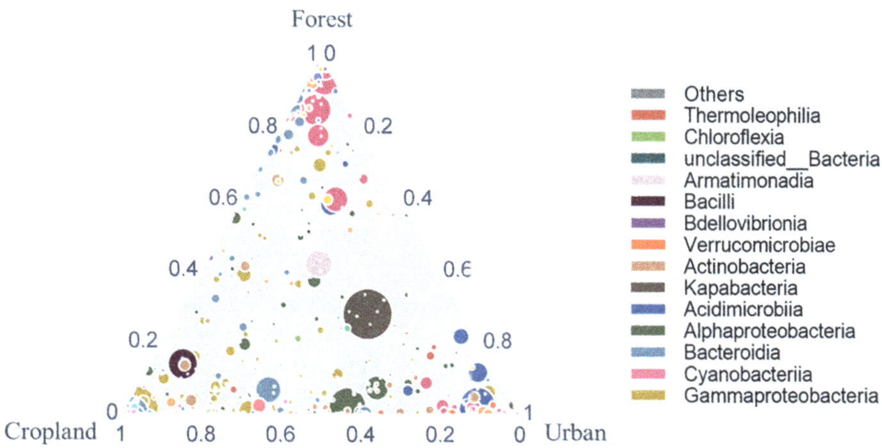

Fig. 5.3 Ternary plot of bacterial OTU abundance in three types of lakes and reservoirs

lakes may also originate from rainwater washing of agricultural soils in mountainous areas (Zhao et al., 2022). The higher abundance of *Cyanobacteria*, *Acidimicrobiia,* and *Bacteroidia* in forested lakes may be due to the competition of attenuated Aspergillus-like microorganisms (*Gammaproteobacteria* vs. *Alphaproteobacteria*) in low DOC concentrations on the one hand, and the sunlight transmittance ensured in the more transparent water of forested lakes on the other hand, which provides better conditions for *Cyanobacteria* to grow.

3 Correlation Between Molecular Composition of DOM and Environmental Factors

Mantel test is a statistical method used to test the correlation between two matrices with the same dimension. It is often used to find the potential correlation between different factors; for example, it is used to analyze the correlation between microbial community structure and soil fertilization method (Wu et al., 2021). The results showed that nutrient concentration indicators, microbial community structure, and dissolved organic matter composition in lakes and reservoirs were significantly different under different watershed land cover scenarios (Hong et al., 2023). Therefore, in this review, Mantel test was applied on the matrices of watershed land cover, water quality indicators, microbial community structure, and DOM composition to investigate the potential relationships among these indicators, and the results were presented in the form of heatmaps in Fig. 5.4. Meanwhile, in order to investigate the effects of single environmental indicators on the bacterial population structure and DOM molecular composition among them, as well as the effects of land cover on the fluorescence components and water quality indicators, this review conducted

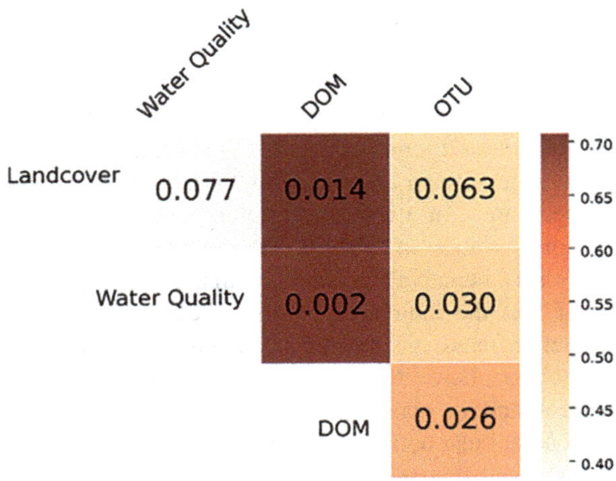

Fig. 5.4 Mantel test between matrix of environmental factors DOM molecular composition, and bacterial community composition. The color of the color blocks in the heatmaps represents the magnitude of the correlation r. The numbers in the color blocks are the significant p-values

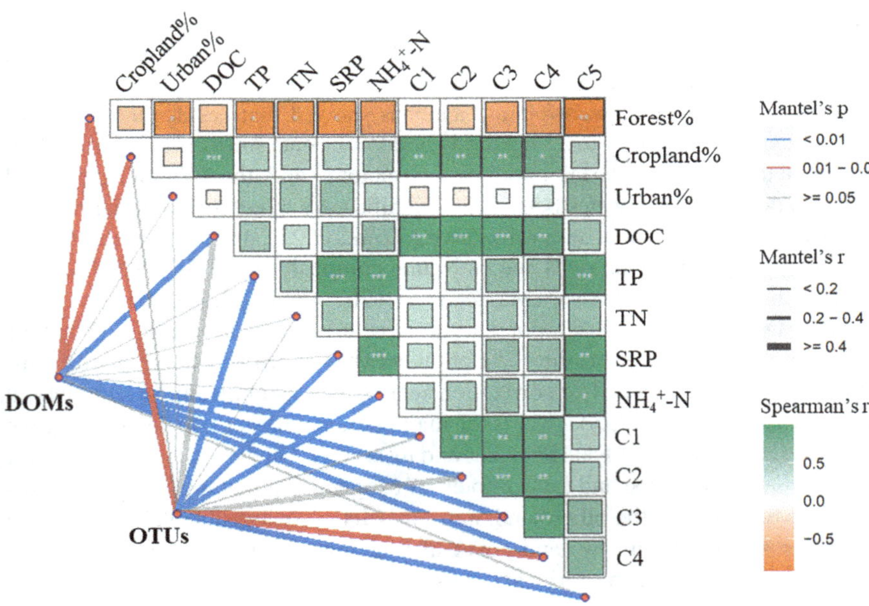

Fig. 5.5 Complex correlation analysis between matrix of environmental factors DOM molecular composition, and bacterial community composition. $*p < 0.05$, $**p < 0.01$, $***p < 0.001$

Spearman correlation analysis among all single watershed land cover percentage, fluorescence components, and water quality indicators, and these single indicators with DOM molecular composition matrices and bacterial population structure matrices. Mantel test and the results are presented in the form of composite plots in Fig. 5.5 (Mantel test was on the left line, and spearman correlation analysis was on the right heat map).

Significant differences in Spearman's correlations confirm that land use changes due to anthropogenic disturbances significantly increase nutrient inputs and influence the source and composition of total dissolved organic matter. For example, there is a strong positive correlation between the percentage of farmland cover and the fluorescence components C1 ~ C4, and a significant correlation between the percentage of town cover and the fluorescence component C5, and at the same time, the percentage of towns and farmland in the watershed is positively correlated with the concentration of DOM in lakes and reservoirs, whereas the percentage of forest cover has a negative correlation with the above mentioned fluorescence components as well as the indicators of the concentration of DOM. As concluded by high-resolution mass spectrometry analysis at the molecular level of DOM, watershed land cover altered the percentage of each type of organic matter, the heteroatom content, and the overall diversity of lakes and reservoirs DOM (Hong et al., 2023). The Mantel test between fluorescence components and bacterial community structure showed that microbial community (OTUs) was significantly correlated with three PARAFAC components related to microbial metabolism (C3, C4, and C5), but with terrigenous humus (C1 and C2, $p < 0.05$) was not significant, indicating that the metabolic activities of microorganisms had a significant interaction relationship with some DOM in lakes and reservoirs.

In addition, Spearman's correlation analysis showed that besides the composition and total amount of organic matter, N and P nutrients were also important factors influencing the bacterial population structure, and the total nitrogen and ammonia content in the water body could significantly affect the bacterial population structure in lakes and reservoirs (Hong et al., 2023). In terms of the overall correlation between the matrices, the molecular composition of DOM had the strongest correlation with land cover (Mantel's $r = 0.707$, $p < 0.05$), and the correlation between DOM and bacterial population structure was also strong (Mantel's $r = 0.525$, $p < 0.05$), whereas the correlation between DOM and water quality indicators was not significant. The correlation seen was not significant. In summary, it can be hypothesized that land cover is the main influencing factor in determining the molecular composition of DOM in lakes and reservoirs. On the one hand, land cover directly changes the type and total amount of exogenous pollutants input into lakes and reservoirs, and on the other hand, nutrient salts and exogenous DOM imported into lakes and reservoirs by land cover also change the population structure of microorganisms, such as bacteria, and thus indirectly change the molecular composition of DOM in lakes and reservoirs through the metabolism of microorganisms. Mantel test also showed significant correlations between the molecular composition of dissolved organic matter and the fluorescence fractions from C1 ~ C4 ($p < 0.01$), with only a non-significant correlation with the C5 fraction ($p > 0.05$), which indicates that although the fluorescent DOM is only a part of the whole DOM library, it still has a good representation of the molecular formula composition of DOM in lakes and reservoirs.

4 Influence of Watershed Land Cover on the Composition of Molecular Formula Sources of DOM

Simon et al. (2018) collected dissolved organic matter samples from soils, lakes, rivers, drinking water, and oceans, and used two instruments, the Orbit-trap and FT-ICR-MS, to compare and analyze these samples, and ultimately concluded that the sample data analyzed using the two instruments contained generally comparable information about the composition of the DOM molecular formulae and were comparable (Simon et al., 2018). Studies have shown that the composition of lake DOM is significantly different under different land cover conditions and that there is a significant correlation between the DOM molecular composition matrix and the watershed land cover matrix (Hong et al., 2023). Therefore, in this section of the study, we collected high-resolution mass spectrometry data from organic matter samples of various sources in previous studies (weighted values of mass spectrometry attributes of different sources of DOM are shown in Table 5.2), and attempted to analyze the effects of changing land cover conditions on the source of DOM in the watershed through a multidimensional statistical approach. The impact of changing land cover status on the source of DOM in watersheds was analyzed through a multidimensional statistical approach.

Among these DOM samples collected in previous studies, the Yanghe River is located in Hebei Province in northern China and is a tributary of the Yongding River, with a sampling time of June 2020 (Ge et al., 2022), and points YR6, YR8, and YR14 in the literature were selected based on the land-cover map as data points for the forested section of the Yanghe River, the agricultural section of the Yanghe River, and the urban section of the Yanghe River in this study. The sampling time of the DOM of the Yangtze River water samples was May 2017 and July 2018 (Pang et al., 2021), of which the Chongqing sampling sites were selected in this study as the data points of the Yangtze River during the flat-water and high-water periods, respectively. Suwannee and standard humic acid (SRFA) samples, and forest, town, and farmland-type lake DOM were sampled in winter (Hong et al., 2023). Dianchi and Taihu Lake samples were sampled in May 2018, and lake sediment, aquatic plant, and algae samples were collected from Lake Taihu (Liu et al., 2020). The DOM samples from Poyang Lake were collected in January 2022 (Xu et al., 2022), and in this study, the average of the four sample points collected was taken. Forest soil DOM was collected at the Heshan National Forest Ecosystem Field Research Station in August 2018 (Ye et al., 2020). Wastewater treatment plant effluent 1 sampled in Yiwu City (Gong et al., 2022). Urban Soil, Road Runoff, Sewage Plant 2, Marginal Plants, and Leaves DOM samples were sampled in Daegu, South Korea, during a heavy rainfall event between late July and August 2018 (Lee et al., 2019). Rice field soil DOM samples were collected in October 2014 from 16 farmlands distributed in China and were averaged for this study (Li et al., 2018).

Non-metric multidimensional scaling (NMDS) is a data analysis method that reduces research objects (samples or variables) in multidimensional space to low dimensions for localization, analysis, and categorization, while preserving the

Table 5.2 Weighted average molecular attributes for different sources of DOM

Number	Sample source	Sample type	$Mass_{wa}$	N_{wa}	S_{wa}	H/C_{wa}	O/C_{wa}	DBE_{wa}	$AImod_{wa}$
1	Standard humic acid of the Suwannee River (Hong et al., 2023)	River	444.26	0.10	0.18	1.07	0.52	10.75	0.35
2	Abundant water period in Chongqing section of the Yangtze River (Pang et al., 2021)	River	423.20	0.20	0.10	1.17	0.50	9.31	0.28
3	Flat water period in Chongqing section of the Yangtze River (Pang et al., 2021)	River	397.00	0.20	0.20	1.26	0.49	8.41	0.23
4	Forest section of Yanghe River (Ge et al., 2022)	River	402.99	0.39	0.10	1.20	0.42	9.30	0.31
5	Farmland section of Yanghe River (Ge et al., 2022)	River	371.43	0.44	0.16	1.23	0.43	8.01	0.28
6	Urban section of Yanghe River (Ge et al., 2022)	River	356.81	0.45	0.25	1.27	0.44	7.46	0.26
7	Urban lake (Hong et al., 2023)	Lake	365.23	0.37	0.15	1.25	0.51	7.74	0.16
8	Dianchi Lake (Liu et al., 2020)	Lake	341.55	0.15	0.04	1.54	0.33	5.27	0.15
9	Farmland lake (Hong et al., 2023)	Lake	386.03	0.43	0.12	1.17	0.52	8.90	0.22
10	Poyang Lake (Xu et al., 2022)	Lake	378.35	0.30	0.38	1.23	0.52	8.13	0.21
11	Forested lake (Hong et al., 2023)	Lake	401.00	0.33	0.10	1.19	0.49	9.13	0.28
12	Taihu Lake (Liu et al., 2020)	Lake	411.36	0.46	0.45	1.29	0.55	7.27	0.20
13	Summer Fanshen reservoir (Hong et al., 2023)	Reservoir	352.50	0.50	0.03	1.32	0.54	7.64	0.31
14	Lake sediments (Liu et al., 2020)	Endogenous pollution	366.99	0.58	0.12	1.41	0.43	6.61	0.18
15	Aquatic plants (Liu et al., 2020)	Endogenous pollution	380.60	0.24	0.04	1.29	0.55	7.27	0.20

(continued)

Table 5.2 (continued)

Number	Sample source	Sample type	$Mass_{wa}$	N_{wa}	S_{wa}	H/ C_{wa}	O/ C_{wa}	DBE_{wa}	$AImod_{wa}$
16	Algae (Liu et al., 2020)	Endogenous pollution	337.99	0.31	0.01	1.67	0.31	4.11	0.08
17	Marginal plants (Lee et al., 2019)	Exogenous pollution	318.00	0.28	0.25	1.61	0.32	4.40	0.13
18	Road run-off (Lee et al., 2019)	Exogenous pollution	432.00	1.15	0.34	1.13	0.20	14.30	0.42
19	Urban top soil (Ye et al., 2020)	Exogenous pollution	369.00	0.84	0.23	1.26	0.27	10.40	0.35
20	Urban deep soil (Lee et al., 2019)	Exogenous pollution	175.00	0.26	0.19	0.49	0.13	6.20	0.21
21	Paddy soil (Li et al., 2018)	Exogenous pollution	320.82	0.56	0.17	1.24	0.28	7.51	0.22
22	Forest top soil (Ye et al., 2020)	Exogenous pollution	440.00	0.25	0.10	1.06	0.52	10.90	0.36
23	Forest deep soil (Ye et al., 2020)	Exogenous pollution	456.00	0.30	0.10	1.17	0.49	9.65	0.34
24	Leaves (Lee et al., 2019)	Exogenous pollution	191.00	0.28	0.25	0.41	0.13	5.00	0.17
25	Sewage treatment plant 1 (Gong et al., 2022)	Exogenous pollution	411.39	0.16	0.42	1.29	0.42	7.15	0.20
26	Sewage treatment plant 2 (Lee et al., 2019)	Exogenous pollution	358.00	0.13	0.32	1.31	0.42	7.40	0.23

original relationships between the objects. NMDS can be applied to (1) data in which there are missing pairwise distances, (2) matrices generated by any distance algorithm, and (3) the analysis of quantitative, semiquantitative, qualitative, or mixed variables. NMDS can be used to analyze data in a variety of different ways, such as in the case of a large number of samples and species, and in the case of a large number of species. In the case of multiple samples and a large number of species, the NMDS model can more accurately reflect the numerical ordering information of the distance matrix. The NMDS model generated in this study is shown in Figs. 5.6 and 5.7, with a STRESS value <0.05 indicating that the NMDS model is well represented. Principal component analysis (PCA) is also a dimensionality reduction method for high-dimensional data, which removes redundancy by presentational transformation of similarity matrices between objects to discover hidden patterns in the data species. The PCA model was constructed based on the weighted values of high-resolution mass spectrometry attributes of the selected lake and river samples in Fig. 5.8, in which principal component 1 (PC1) explained 50.59% of the total variance; principal component 2 (PC2) explained 22.04% of the total variance; and principal component 3 (PC3) explained 15.65% of the total variance in this PCA model.

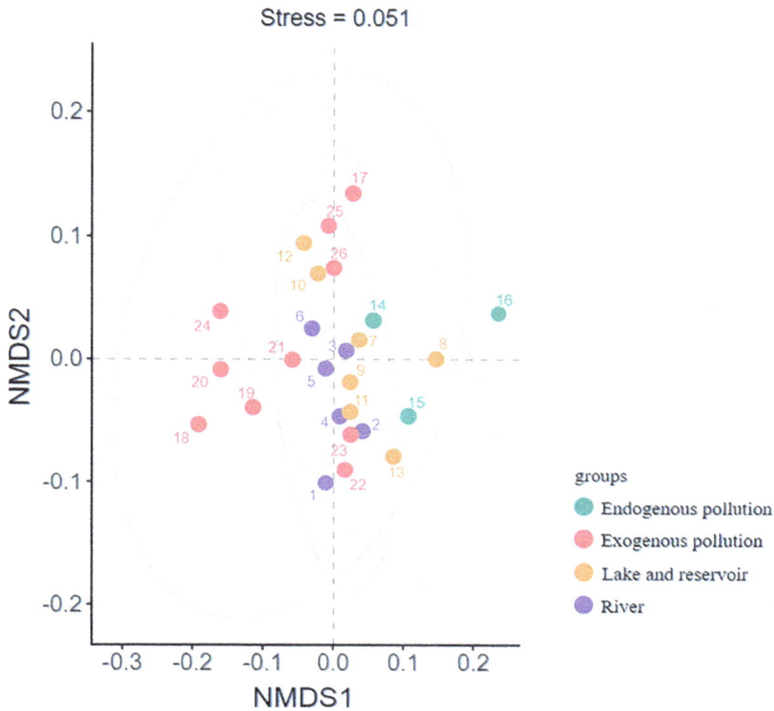

Fig. 5.6 NMDS analysis of DOM from different sources (2D-plot)

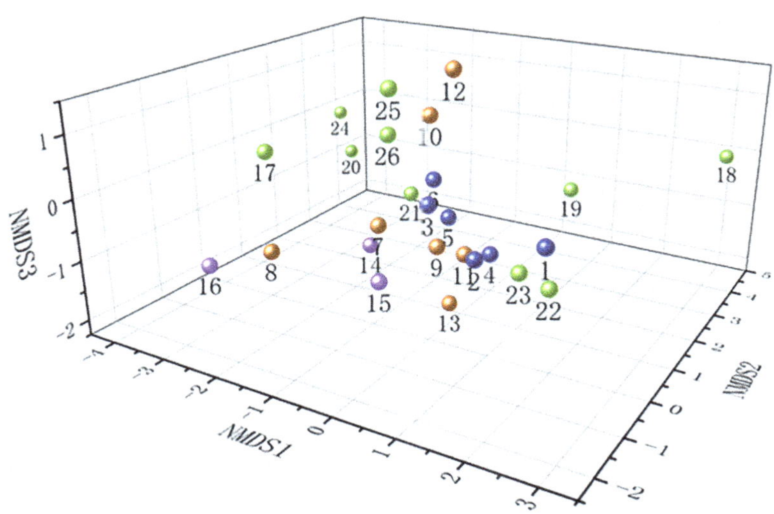

Fig. 5.7 NMDS analysis of DOM from different sources (3D-plot)

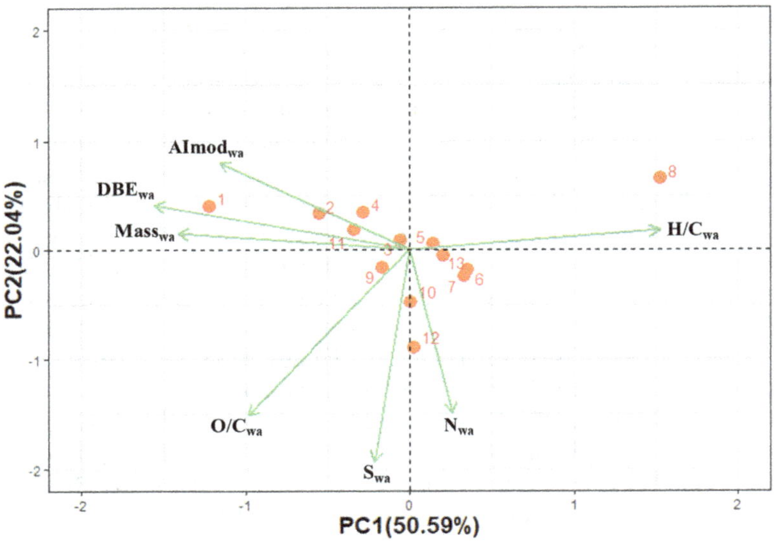

Fig. 5.8 PCA analysis of DOM from different sources

The results of the NMDS model showed that, except for the three sample points of standard humic acid in Poyang Lake, Taihu Lake and Suwannee River, the other 10 water sample points were located in the 13 DOM source samples, which indicated that these DOM sources had a good explanation for the composition of the water DOM, and that the out-of-range sample points were probably due to the absence of DOM sources with high N and S contents as well as those with high double-bond equivalent DOM sources. In addition, leaves, road runoff, and urban soils were distant from most points, presumably making these DOM sources less influential in the formation of DOM pools in water bodies.

According to the nature of the molecular formula represented by the weighted values of the high-resolution mass spectrometry attributes, if the PC1 value of a water sample in the PCA model is higher, the smaller the molecular mass of the DOM of the water sample is, the worse the stability in the aquatic environment, and the better the bioavailability is. On the contrary, a higher value of PC1 indicates a greater molecular mass of DOM, a higher abundance of double bonds and aromatic structures within its molecular formula, and consequently, enhanced stability in the aquatic environment. PC2, on the other hand, mainly reflects the content of N and S heteroatoms in the molecular formula. The farmland section of the Yanghe River, the summer Turning Reservoir, the urban section of the Yanghe River, the urban lakes, and the Dianchi Pool were located on the right side of the H/C$_{wa}$ axis in the PCA graphs, indicating that the DOM molecules in these samples were highly reactive, with the highest H/C$_{wa}$ in the Dianchi Pool of 1.54, but with fewer N and S heteroatoms. The NMDS model showed that the Dianchi Pool sample was located between four points, namely, the algal species, the lake sediment, aquatic plants,

and marginal plants, indicating that DOM in the lake in summer was mainly composed of proteins, lipids, and carbohydrates, and that carbohydrates dominated the organic matter, aquatic plants, and marginal plants, indicating that the DOM in the Dianchi pond in summer was mainly composed of proteins, lipids, and carbohydrates, with carbohydrates accounting for the majority of the DOM. The dominance of organic matter is due to the increased metabolic activity of algae, aquatic plant blooms, and the release of organic matter from lake sediments. Among all the samples, Tai Lake and Poyang Lake showed the highest N and S heteroatom contents, and the pollution sources closest to these two sample points in the NMDS model were the effluent from sewage treatment plant 1 and effluent from sewage treatment plant 2, according to which it can be assumed that the water quality pollution of Tai Lake and Poyang Lake is more affected by the discharge from the point source. Comparison of samples collected from the Yangtze River at two different time points revealed that the DOM of the Yangtze River during the abundance period has a larger average molecular mass than the DOM during the leveling period, as well as more double bonds and benzene ring structures. The two Yangtze River sample points in the NMDS model are closer to the forest soil DOM and are closer during the abundance period than during the leveling period; therefore, it can be assumed that the Yangtze River at the sample point is mainly affected by the forested watershed and that the input of humic acid substances to the forested soil is increased due to the rainfall flushing during the rainy season, it increases the input of humic acid substances from the forest soil.

Based on satellite maps of the sampling basins in the literature, of the three sampling sites selected for the Yanghe River, the forested section of the sampling site is located in the upper reaches of the river, and its upstream land cover consists of a large area of forests, grasslands, and a small amount of farmland. The farmland sampling site is located in the middle part of the river, and the land cover in the upper part of the river mainly consists of a large amount of farmland on both sides of the river. The town sampling site is located in the lowest reaches of the river, and the land cover upstream of this site consists mainly of impervious surfaces in the town and a small amount of agricultural land. The location of the Yanghe River sample point in the NMDS model changed accordingly due to the change in land cover status, gradually shifting from a location close to the forest soil to one close to the paddy soil before approaching the two pollution source points of the effluent from the sewage treatment plant 1 and 2. The changes reflected in the weighted attributes of the mass spectra of the Yanghe DOM were that the dominance of large molecule aromatic and unsaturated stable DOM from the source was changed to the dominance of low molecule unstable DOM, and the content of N and S heteroatoms in the molecular formula was also significantly elevated. Changes in the composition of lakes and reservoirs DOM resulting from the shift in land cover status from forested to agricultural and urban types in the lakes and reservoirs watershed generally showed a pattern consistent with that of the Ocean River. However, the three points in the NMDS model for Lakes and reservoirs were closer to algae, lake sediments, aquatic plants, and riparian plants than the Ocean River. Although the lakes and reservoirs DOM was sampled in winter, it still showed a stronger endogenous

signature than the river DOM sampled in summer, suggesting that endogenous DOM played a more influential role in shaping the molecular composition of lakes and reservoirs DOM than the river.

In summary, DOM in water bodies is mainly affected by algae, riparian plants, sediments, aquatic plants, forest soils, paddy soils, and effluent discharged from wastewater treatment plants. In watersheds dominated by forest land cover, DOM in water bodies mainly comes from inputs of organic matter from forest soils. A shift in watershed land cover from forest to farmland would result in paddy soil DOM replacing forest soil DOM as the dominant source of DOM, while a shift in watershed land cover from forest to towns would result in wastewater treatment plant output of organic matter as the dominant source. Land cover-induced changes in the composition of lakes and reservoirs DOM sources drive significant changes in the total amount, heteroatom content, and diversity of DOM in water.

5 Co-occurrence Network Relationships Between DOM Molecular Formulae and Bacterial Community Structure

The Mantel test showed significant correlations between bacterial population structure and the three biogenic fluorescence fractions (C1, C2, and C3), as well as significant correlations between the differences between the bacterial population structure and the molecular composition of the DOM, suggesting that there is a significant correlation between the structure of the bacterial community and the chemical diversity of the DOM (Figs. 5.4 and 5.5). In order to further explore the correlation between DOM chemical diversity and bacterial community composition, Spearman's correlation between all bacterial OTUs and DOM molecular formulas was calculated in this study and the results are presented as a co-occurrence/exclusion network diagram in Fig. 5.9. In the network graph, the number of other nodes linked to a node is called the degree of that node, and in the symbiotic network, the orders with higher degrees include *Gammaproteobacteria* (5.22%), *Bacteroidia* (3.60%), *Alphaproteobacteria* (2.44%), and *Cyanobacteria* (1.77%). In the co-exclusion network, the orders with higher degrees included *Gammaproteobacteria* (5.70%), *Bacteroidia* (3.97%), *Alphaproteobacteria* (2.67%), and *Cyanobacteria* (1.43%).

In order to better illustrate the symbiotic and co-exclusionary relationships between bacterial taxa and specific organic matter taxa, this study calculated the O/R value for each bacterial category and molecular category to form a composite network see Fig. 5.10. Briefly, the O/R value illustrates the ratio of the number of edges where an association exists between the two groups of nodes to the number of edges where an association was formed randomly. If O/R = 1, the formation of a relationship between two groups of nodes may be caused by random factors, and a larger value means that there is a higher level of confidence that an association exists between two groups of nodes.

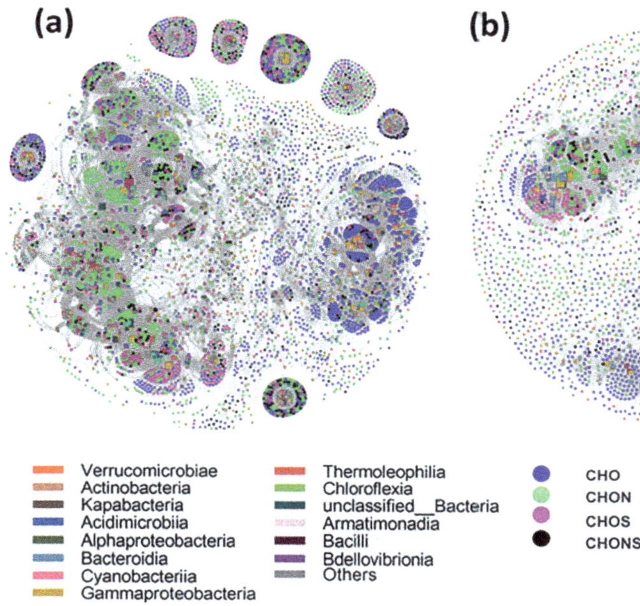

Fig. 5.9 Co-occurrence/exclusion network diagram between bacterial OTU and DOM formulas: (**a**) Co-occurrence network; (**b**) Co-exclusion network

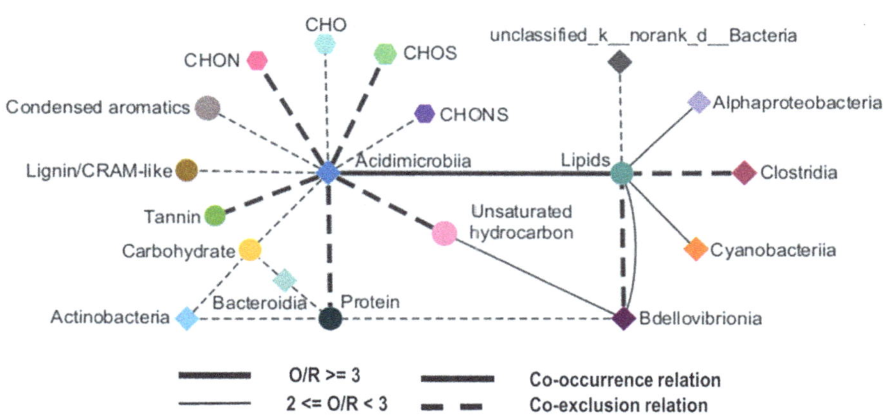

Fig. 5.10 O/R value-based composite network between bacterial OTU classification and DOM classification

The results of the O/R composite network analysis showed that among all the organic matter molecular classes, the Lipids class DOM was significantly associated with the largest number of bacteria. The co-associated bacteria included *Alphaproteobacteria*, *Acidimicrobiia*, *Cyanobacteria*, and part of *Bdellovibrionota*. Bacterial orders with co-excluded associations include *Clostridia*, some

Bdellovibrionota, and *unclassified_k_norank_d_Bacteria*. Protein showed co-exclusion associations with *Acidimicrobiia, Bacteroidia, Actinobacteria,* and *Bdellovibrionota.* Carbohydrates showed co-exclusive associations with *Acidimicrobiia, Bacteroidia,* and *Actinobacteria.* The complex correlation between these organisms and bacterial taxonomy may indicate their metabolic activity involved in bacterial activities. Other classes of DOM including Lignin, Tannin, and Condensed aromatics appeared to be associated only with *Acidimicrobiia*. Moreover, the organic matter classes presenting associations with more bacterial classes were all in the region of $H/C > 1.5$ in the paradigm diagram, while organic matter classes associated only with *Acidimicrobiia* were all in the region of $H/C < 1.5$ in the paradigm diagram (D'Andrilli et al., 2015), which is in line with the findings of D'Andrilli et al. by comparing 35 samples of DOM from natural water bodies tested by the same high-resolution mass spectrometry (HRMS) instrument for a decade, which suggests that it is possible to use $H/C = 1.5$ as a boundary to distinguish the stability of DOM in ecosystems (D'Andrilli et al., 2015).

In addition, *Acidimicrobiia* have strong co-occurrence and co-exclusion relationships with several molecular classes, including Lignin, Tannin, Condensed aromatics, CHOS, CHON, CHO, and CHONS, suggesting the sensitivity or metabolic activity of *Acidimicrobiia* to the composition of DOM in the water column. *Acidimicrobiia* are a widespread class of bacteria found in nature and, along with *γ-amoebae*, are considered to be a key metabolic class for dissolved organic sulfur (DOS) in the ocean (Landa et al., 2019). The strong co-occurrence exhibited with CHOS-like organisms in their lakes may indicate that this class of microorganisms is equally important in the transformation pathway of DOS in lake ecosystems. It also suggests that anthropogenic land cover inputs of heteroatom-containing compounds drive changes in the structure of bacterial populations in lakes and reservoirs (Hong et al., 2023). The symbiotic network suggests that there is indeed a strong link between specific bacterial community structures and DOM molecular classes, but how they interact remains difficult to determine only through the correlation analyses in this study, which may need to be further investigated by future in situ experiments or laboratory experiments conducted in a closed environment.

6 Conclusions

In this study, Mantel Test correlations were calculated for watershed land cover, water quality, bacterial community structure, and dissolved organic matter molecular composition matrices, and multivariate statistical analyses were used to downscale a wide range of dissolved organic matter systems collected, and co-occurrence/exclusion networks between bacterial OTUs and dissolved organic matter molecular formulas were also constructed, leading to the following conclusions:

1. Mantel Test showed that the molecular composition of dissolved organic matter was significantly correlated with land cover and bacterial community structure,

and the strongest correlation was with the land cover of the watershed $r = 0.707$; the bacterial population structure was significantly correlated with DOM, the land cover of the watershed, and the water quality of lakes and reservoirs; and the bacterial population structure was significantly correlated with the endogenous fluorescence fractions C1, C3, and C3.
2. NMDS and PCA downscaling analyses showed that the contribution of forest land cover to dissolved organic matter in lakes and reservoirs was mainly organic matter in forest soils, the contribution of agricultural land cover was mainly organic matter in paddy soils, and the contribution of urban land cover was mainly organic matter in the effluent from the wastewater treatment plant.
3. The co-occurrence/exclusion network between bacterial OTUs and dissolved organic matter molecules and O/R values showed that the organic matter classes presenting associations with a larger number of bacterial classes were all in the region of $H/C > 1.5$ in the paradigm diagram, corroborating the bio-metabolic instability of these classes of dissolved organic matter in lakes and reservoirs water body. *Acidimicrobiia* had strong co-occurrence and co-exclusion relationships with several molecular classes, suggesting the sensitivity of *Acidimicrobiia* to the composition of dissolved organic matter in the water column or its key role in the metabolism of dissolved organic matter in lakes and reservoirs.

In summary, this review investigated the effects of watershed land cover on the structure of bacterial communities and the molecular composition of dissolved organic matter DOM in lakes and reservoirs, and explored the interactions between DOM molecules and bacterial communities in lakes and reservoirs. This review will not only provide basic data to support the water resource utilization and ecological protection of lakes and reservoirs in the context of further anthropogenic land cover in the watershed in the future, but also contribute to the understanding of carbon cycling, photochemical transformations, and biogeochemical processes of dissolved organic matter in the natural water bodies to further address the potential threats to drinking water security and water resource use posed by the further agriculturalization and urbanization of the land cover in watershed in the future.

Acknowledgement HM is supported by the National Natural Science Foundation of China, China (No: 51979016).

Author Contributions Zhicheng Hong, Hua Ma: Experiment, Writing—Original draft preparation. Sheng-Ao Li: Instrumentation, Experiment, Methodology. Hua Ma: Conceptualization, Supervision, Writing—reviewing & editing, Funding acquisition.

Declaration of Competing Interest The authors declare that they have no known competing financial interests or personal relationships that could have appeared to influence the work reported in this paper.

Conflict of Interest The authors declare that they have no conflict of interest.

Data Availability Data will be made available on request.

References

Aitkenhead-Peterson, J. A., Steele, M. K., Nahar, N., & Santhy, K. (2009). Dissolved organic carbon and nitrogen in urban and rural watersheds of South-Central Texas: Land use and land management influences. *Biogeochemistry, 96*, 119–129. https://doi.org/10.1007/s10533-009-9348-2

Cantonati, M., Poikane, S., Pringle, C. M., Stevens, L. E., Turak, E., Heino, J., Richardson, J. S., Bolpagni, R., Borrini, A., Cid, N., et al. (2020). Characteristics, main impacts, and stewardship of natural and artificial freshwater environments: Consequences for biodiversity conservation. *Water., 12*(1), 260. https://doi.org/10.3390/w12010260

Cao, X., Aiken, G. R., Butler, K. D., Mao, J., & Schmidt-Rohr, K. (2018 Feb 20). Comparison of the chemical composition of dissolved organic matter in three lakes in Minnesota. *Environmental Science & Technology, 52*(4), 1747–1755. https://doi.org/10.1021/acs.est.7b04076

Cole, J. J., & Caraco, N. F. (2001). Carbon in catchments: Connecting terrestrial carbon losses with aquatic metabolism. *Marine and Freshwater Research., 52*(1), 101–110. https://doi.org/10.1071/MF00084

D'Andrilli, J., Cooper, W. T., Foreman, C. M., & Marshall, A. G. (2015). An ultrahigh-resolution mass spectrometry index to estimate natural organic matter lability. *Rapid Communications in Mass Spectrometry, 29*, 2385–2401. https://doi.org/10.1002/rcm.7400

Dai, T., Liu, R., Zhou, X., Zhang, J., Song, M., Zou, P., Bi, X., & Li, S. (2023). Role of lake aquatic–terrestrial ecotones in the ecological restoration of eutrophic water bodies. *Toxics., 11*(7), 560. https://doi.org/10.3390/toxics11070560

Ding, J., Jiang, Y., Fu, L., Liu, Q., Peng, Q., & Kang, M. (2015). Impacts of land use on surface water quality in a subtropical river basin: A case study of the dongjiang river basin, southeastern China. *Water (Switzerland), 7*, 4427–4445. https://doi.org/10.3390/w7084427

Donohue, I., McGarrigle, M. L., & Mills, P. (2006). Linking catchment characteristics and water chemistry with the ecological status of Irish rivers. *Water Research, 40*, 91–98. https://doi.org/10.1016/j.watres.2005.10.027

Dudley, E., Yousef, M., Wang, Y., & Griffiths, W. J. (2010). Chapter 2 – Targeted metabolomics and mass spectrometry, editor(s): Rossen Donev. In *Advances in Protein Chemistry and Structural Biology* (Vol. 80, pp. 45–83). Academic. https://doi.org/10.1016/B978-0-12-381264-3.00002-3

Dumitran, G. E., Vuta, L. I., & Popa, B. (2024). Overview of the Eutrophication in Romanian Lakes and reservoirs. *Limnological Review., 24*(1), 76–104. https://doi.org/10.3390/limnolrev24010005

Findlay, S. E. G., & Parr, T. B. (2017). Dissolved organic matter. In *Methods in stream ecology* (3rd ed., pp. 21–36). Elsevier. https://doi.org/10.1016/B978-0-12-813047-6.00002-4

Ge, J., Qi, Y., Li, C., Ma, J., Yi, Y., Hu, Q., Mostofa, K. M. G., Volmer, D. A., & Li, S. L. (2022). Fluorescence and molecular signatures of dissolved organic matter to monitor and assess its multiple sources from a polluted river in the farming-pastoral ecotone of northern China. *Science of the Total Environment, 837*, 154575. https://doi.org/10.1016/j.scitotenv.2022.154575

Gentile, M., Yan, T., Tiquia, S. M., et al. (2006). Stability in a denitrifying fluidized bed reactor. *Microbial Ecology, 52*, 311–321. https://doi.org/10.1007/s00248-006-9024-1

Gong, C., Jiao, R., Yan, W., Yu, Q., Li, Q., Zhang, P., Li, Y., & Wang, D. (2022). Enhanced chemodiversity, distinctive molecular signature and diurnal dynamics of dissolved organic matter in streams of two headwater catchments, Southeastern China. *Water Research, 211*, 118052. https://doi.org/10.1016/j.watres.2022.118052

Harrison, J. A., Prairie, Y. T., Mercier-Blais, S., & Soued, C. (2021). Year-2020 global distribution and pathways of reservoir methane and carbon dioxide emissions according to the greenhouse gas from reservoirs (G-res) model. *Global Biogeochemical Cycles, 35*. https://doi.org/10.1029/2020GB006888

Hayes, N. M., Deemer, B. R., Corman, J. R., Roxanna Razavi, N., & Strock, K. E. (2017). Current evidence key differences between lakes and reservoirs modify climate signals: A case for a new conceptual model scientific significance statement. *Limnology and Oceanography Letters, 2*, 47–62. https://doi.org/10.6073/pasta/17cb7958c74f8bfc135f3e7f04ee944e

Hong, Z., Ma, H., Zhang, T., Wang, Q., Chang, Y., Song, Y., Li, Z., & Cui, F. (2023). Joint role of land cover types and microbial processing on molecular composition of dissolved organic matter in inland lakes. *Science of the Total Environment, 857*. https://doi.org/10.1016/j.scitotenv.2022.159522

Imai, A., Fukushima, T., Matsushige, K., Kim, Y. H., & Choi, K. (2002 Feb). Characterization of dissolved organic matter in effluents from wastewater treatment plants. *Water Research, 36*(4), 859–870. https://doi.org/10.1016/s0043-1354(01)00283-4

Jansson, M., Persson, L., De Roos, A. M., Jones, R. I., & Tranvik, L. J. (2007). Terrestrial carbon and intraspecific size-variation shape lake ecosystems. *Trends in Ecology & Evolution, 22*, 316. https://doi.org/10.1016/j.tree.2007.02.015

Kaiser, K., & Benner, R. (2012). Characterization of lignin by gas chromatography and mass spectrometry using a simplified CuO oxidation method. *Analytical Chemistry, 84*, 459–464. https://doi.org/10.1021/ac202004r

Kalra, S., Madguni, O., & Rana, D. (2023). Sustainable waste management of man-made dams and reservoirs—A review. In N. A. Siddiqui, A. S. Baxtiyarovich, A. Nandan, & P. Mondal (Eds.), *Advances in Waste Management. AIR 2021. Lecture notes in civil engineering, vol 301*. Springer. https://doi.org/10.1007/978-981-19-7506-6_11

Kassem, A., Abbas, L., Coutinho, O., Opara, S., Najaf, H., Kasperek, D., Li, X., & Tiquia-Arashiro, S. M. (2023). Applications of Fourier transform-infrared spectroscopy in microbial cell biology and environmental microbiology: Advances, challenges, and future perspectives. *Frontiers in Microbiology, 14*, 1304081. https://doi.org/10.3389/fmicb.2023.1304081

Kellerman, A. M., Dittmar, T., Kothawala, D. N., & Tranvik, L. J. (2014). Chemodiversity of dissolved organic matter in lakes driven by climate and hydrology. *Nature Communications, 5*, 3804. https://doi.org/10.1038/ncomms4804

Kiersztyn, B., Chróst, R., Kaliński, T., et al. (2019). Structural and functional microbial diversity along a eutrophication gradient of interconnected lakes undergoing anthropopressure. *Scientific Reports, 9*, 11144. https://doi.org/10.1038/s41598-019-47577-8

Kuang, B., Xiao, R., Hu, Y., Wang, Y., Zhang, L., Wei, Z., Bai, J., Zhang, K., Acuña, J. J., Jorquera, M. A., & Pan, W. (2023 Jan). Metagenomics reveals biogeochemical processes carried out by sediment microbial communities in a shallow eutrophic freshwater lake. *Frontiers in Microbiology, 11*(13), 1112669. https://doi.org/10.3389/fmicb.2022.1112669

Kurek, M. R., Harir, M., Shukle, J. T., Schroth, A. W., Schmitt-Kopplin, P., & Druschel, G. K. (2020 Sep). Chemical fractionation of organic matter and organic phosphorus extractions from freshwater lake sediment. *Analytica Chimica Acta, 15*(1130), 29–38. https://doi.org/10.1016/j.aca.2020.07.013

Landa, M., Burns, A. S., Durham, B. P., Esson, K., Nowinski, B., Sharma, S., Vorobev, A., Nielsen, T., Kiene, R. P., & Moran, M. A. (2019). Sulfur metabolites that facilitate oceanic phytoplankton–bacteria carbon flux. *ISME Journal, 13*, 2536–2550. https://doi.org/10.1038/s41396-019-0455-3

Larson, J. H., Frost, P. C., Xenopoulos, M. A., Williams, C. J., Morales-Williams, A. M., Vallazza, J. M., Nelson, J. C., & Richardson, W. B. (2014). Relationships between land cover and dissolved organic matter change along the river to Lake transition. *Ecosystems, 17*, 1413–1425. https://doi.org/10.1007/s10021-014-9804-2

Lee, M. H., Lee, Y. K., Derrien, M., Choi, K., Shin, K. H., Jang, K. S., & Hur, J. (2019). Evaluating the contributions of different organic matter sources to urban river water during a storm event via optical indices and molecular composition. *Water Research, 165*, 115006. https://doi.org/10.1016/j.watres.2019.115006

Li, X. M., Sun, G. X., Chen, S. C., Fang, Z., Yuan, H. Y., Shi, Q., & Zhu, Y. G. (2018). Molecular Chemodiversity of dissolved organic matter in Paddy soils. *Environmental Science & Technology, 52*, 963–971. https://doi.org/10.1021/acs.est.7b00377

Liu, X., Tiquia, S. M., Holguin, G., Wu, L., Nold, S. C., Devol, A. H., Luo, K., Palumbo, A. V., Tiedje, J. M., & Zhou, J. (2003 Jun). Molecular diversity of denitrifying genes in continental margin sediments within the oxygen-deficient zone off the Pacific coast of Mexico.

Applied and Environmental Microbiology, 69(6), 3549–3560. https://doi.org/10.1128/AEM.69.6.3549-3560.2003

Liu, S., He, Z., Tang, Z., Liu, L., Hou, J., Li, T., Zhang, Y., Shi, Q., Giesy, J. P., & Wu, F. (2020). Linking the molecular composition of autochthonous dissolved organic matter to source identification for freshwater lake ecosystems by combination of optical spectroscopy and FT-ICR-MS analysis. *Science of the Total Environment, 703*, 134764. https://doi.org/10.1016/j.scitotenv.2019.134764

Ma, Y., Li, J., Wu, J., et al. (2018). Bacterial and fungal community composition and functional activity associated with Lake wetland water level gradients. *Scientific Reports, 8*, 760. https://doi.org/10.1038/s41598-018-19153-z

Malhi, Y., Franklin, J., Seddon, N., Solan, M., Turner, M. G., Field, C. B., & Knowlton, N. (2020 Mar 16). Climate change and ecosystems: Threats, opportunities and solutions. *Philosophical Transactions of the Royal Society of London. Series B, Biological Sciences, 375*(1794), 20190104. https://doi.org/10.1098/rstb.2019.0104

McKindles, K. M., & Tiquia-Arashiro, S. M. (2012). Functional gene arrays for analysis of microbial communities on ocean platform. In S. Tiquia-Arashiro (Ed.), *Molecular biological technologies for ocean sensing* (Springer Protocols Handbooks). Humana Press. https://doi.org/10.1007/978-1-61779-915-0_9

Nassereddine, Z., Opara, S., Coutinho, O., Qyteti, F., Book, R., Heinicke, M. P., Napieralski, J., & Tiquia-Arashiro, S. M. (2024). Critical perspectives on advancing antibiotic resistant gene (ARG) detection technologies in aquatic ecosystems. *Science of the Total Environment, 957*(20), 177775. https://doi.org/10.1016/j.scitotenv.2024.177775

Nebbioso, A., & Piccolo, A. (2013). Molecular characterization of dissolved organic matter (DOM): A critical review. *Analytical and Bioanalytical Chemistry, 405*, 109–124. https://doi.org/10.1007/s00216-012-6363-2

Nielsen, A., Trolle, D., Søndergaard, M., Lauridsen, T. L., Bjerring, R., Olesen, J. E., & Jeppesen, E. (2012). Watershed land use effects on lake water quality in Denmark. *Ecological Applications, 22*, 1187–1200. https://doi.org/10.1890/11-1831.1

Oest, A., Alsaffar, A., Fenner, M., Azzopardi, D., & Tiquia-Arashiro, S. M. (2018). Patterns of change in metabolic capabilities of sediment microbial communities in river and lake ecosystems. *International Journal of Microbiology., 2018*, 1. https://doi.org/10.1155/2018/6234931

Pagnucco, G., Overfield, D., Chamlee, Y., Fortuna, A., Sulaiman, F., Farinas, J., Schittenhelm, R., Shuler, C., Kassem, A., Opara, S., Najaf, H., Abbas, L., Coutinho, O., Catalfano, B., Li, X., & Tiquia-Arashiro, S. M. (2023). Metal tolerance and biosorption capacities of bacterial strains isolated from an urban watershed. *Frontiers in Microbiology, 14*, 1278886. https://doi.org/10.3389/fmicb.2023.1278886

Pang, Y., Wang, K., Sun, Y., Zhou, Y., Yang, S., Li, Y., He, C., Shi, Q., & He, D. (2021). Linking the unique molecular complexity of dissolved organic matter to flood period in the Yangtze River mainstream. *Science of the Total Environment, 764*, 142803. https://doi.org/10.1016/j.scitotenv.2020.142803

Parr, T. B., Cronan, C. S., Ohno, T., Findlay, S. E. G., Smith, S. M. C., & Simon, K. S. (2015). Urbanization changes the composition and bioavailability of dissolved organic matter in headwater streams. *Limnology and Oceanography, 60*, 885–900. https://doi.org/10.1002/lno.10060

Patel, D., Gismondi, R., Alsaffar, A., & Tiquia-Arashiro, S. M. (2019). Applicability of API ZYM to capture seasonal and spatial variabilities in lake and river sediments. *Environmental Technology, 40*(24), 3227–3239. https://doi.org/10.1080/09593330.2018.1468492

Qin, B., Xu, P., Wu, Q., Luo, L., & Zhang, Y. (2007). Environmental issues of Lake Taihu, China. *Hydrobiologia., 581*, 3–14. https://doi.org/10.1007/s10750-006-0521-5

Raghul, M., & Porchelvan, P. (2024). A critical review of remote sensing methods for inland water quality monitoring: Progress, limitations, and future perspectives. *Water, Air, and Soil Pollution, 235*, 159. https://doi.org/10.1007/s11270-024-06957-1

Ratnadass, A., Fernandes, P., Avelino, J., et al. (2012). Plant species diversity for sustainable management of crop pests and diseases in agroecosystems: A review. *Agronomy for Sustainable Development, 32*, 273–303. https://doi.org/10.1007/s13593-011-0022-4

Raymond, P. A., Hartmann, J., Lauerwald, R., Sobek, S., McDonald, C., Hoover, M., Butman, D., Striegl, R., Mayorga, E., Humborg, C., Kortelainen, P., Dürr, H., Meybeck, M., Ciais, P., & Guth, P. (2014). Erratum: Global carbon dioxide emissions from inland waters. *Nature, 503*, 355–359. https://doi.org/10.1038/nature13142

Simon, C., Roth, V. N., Dittmar, T., & Gleixner, G. (2018). Molecular signals of heterogeneous terrestrial environments identified in dissolved organic matter: A comparative analysis of Orbitrap and ion cyclotron resonance mass spectrometers. *Frontiers in Earth Science (Lausanne), 6*. https://doi.org/10.3389/feart.2018.00138

Sobek, S., Tranvik, L. J., Prairie, Y. T., Kortelainen, P., & Cole, J. J. (2007). Patterns and regulation of dissolved organic carbon: An analysis of 7,500 widely distributed lakes. *Limnology and Oceanography, 52*, 1208–1219. https://doi.org/10.4319/lo.2007.52.3.1208

Soomro, S., Guo, J., Shi, X., et al. (2023). Climate change critique on dams and anthropogenic impact to Mediterranean Mountains for freshwater ecosystem – A review. *Polish Journal of Environmental Studies., 32*(4), 2981–2992. https://doi.org/10.15244/pjoes/161512

Tiquia, S. M. (2008). Diversity of sulfate-reducing genes (dsrAB) in sediments from Puget Sound. *Environmental Technology, 29*(10), 1095–1108. https://doi.org/10.1080/09593330802190608

Tiquia, S. M. (2010a). Metabolic diversity of the heterotrophic microorganisms and potential link to pollution of the Rouge River. *Environmental Pollution, 158*, 1435–1443. https://doi.org/10.1016/j.envpol.2009.12.035

Tiquia, S. M. (2010b). Salt-adapted bacteria isolated from the Rouge River and potential for degradation of contaminants and biotechnological applications. *Environmental Technology, 31*, 967–978. https://doi.org/10.1080/09593331003706226

Tiquia, S. M. (2011). Extracellular hydrolytic enzyme activities of the heterotrophic microbial communities of the Rouge River: An approach to evaluate ecosystem response to urbanization. *Microbial Ecology, 62*, 679–689. https://doi.org/10.1007/s00248-011-9871-2

Tiquia, S. M., Masson, S. A., & Devol, A. (2006a Dec). Vertical distribution of nitrite reductase genes (nir S) in continental margin sediments of the Gulf of Mexico. *FEMS Microbiology Ecology, 58*(3), 464–475. https://doi.org/10.1111/j.1574-6941.2006.00173.x

Tiquia, S. M., Gurczynski, S., Zholi, A., & Devol, A. (2006b Dec). Diversity of biogeochemical cycling genes from Puget Sound sediments using DNA microarrays. *Environmental Technology, 27*(12), 1377–1389. https://doi.org/10.1080/09593332708618756

Tiquia-Arashiro, S. M. (2012). *Molecular biological technologies for Ocean Sensing* (Springer protocols handbooks) (p. 307). Humana. https://doi.org/10.1007/978-1-61779-915-0

Tiquia-Arashiro, S. M., & Pant, D. (2020). *Microbial electrochemical technologies* (p. 518). CRC Press. https://doi.org/10.1201/9780429487118

Verpoorter, C., Kutser, T., Seekell, D. A., & Tranvik, L. J. (2014). A global inventory of lakes based on high-resolution satellite imagery. *Geophysical Research Letters, 41*, 6396–6402. https://doi.org/10.1002/2014GL060641

Wang, Y., Guo, M., Li, X., Liu, G., Hua, Y., Zhao, J., Huguet, A., & Li, S. (2022). Shifts in microbial communities in shallow lakes depending on trophic states: Feasibility as an evaluation index for eutrophication. *Ecological Indicators, 136*, 108691. https://doi.org/10.1016/j.ecolind.2022.108691

Wei, W., Gao, Y., Huang, J., & Gao, J. (2020). Exploring the effect of basin land degradation on lake and reservoir water quality in China. *Journal of Cleaner Production, 268*, 122249. https://doi.org/10.1016/j.jclepro.2020.122249

Wen, Z., Han, J., Shang, Y., Tao, H., Fang, C., Lyu, L., Li, S., Hou, J., Liu, G., & Song, K. (2024, March 15). Spatial variations of DOM in a diverse range of lakes across various frozen ground zones in China: Insights into molecular composition. *Water Research, 252*, 121204. https://doi.org/10.1016/j.watres.2024.121204

Wu, M., Li, P., Li, G., Petropoulos, E., Feng, Y., & Li, Z. (2021). The chemodiversity of paddy soil dissolved organic matter is shaped and homogenized by bacterial communities that are orchestrated by geographic distance and fertilizations. *Soil Biology and Biochemistry, 161*, 108374. https://doi.org/10.1016/j.soilbio.2021.108374

Xu, L., Hu, Q., Liao, L., Duan, Z., Liu, S., Chen, L., Zhu, Q., & Zhong, A. (2022). Hydrological isolation affected the chemo-diversity of dissolved organic matter in a large river-connected Lake (Poyang lake, China). *Science of the Total Environment, 851*, 158047. https://doi.org/10.1016/j.scitotenv.2022.158047

Xu, X., Wu, C., Xie, D., & Ma, J. (2023). Sources, migration, transformation, and environmental effects of organic carbon in Eutrophic Lakes: A critical review. *International Journal of Environmental Research and Public Health., 20*(1), 860. https://doi.org/10.3390/ijerph20010860

Ye, Q., Wang, Y. H., Zhang, Z. T., Huang, W. L., Li, L. P., Li, J., Liu, J., Zheng, Y., Mo, J. M., Zhang, W., & Wang, J. J. (2020). Dissolved organic matter characteristics in soils of tropical legume and non-legume tree plantations. *Soil Biology and Biochemistry, 148*, 107880. https://doi.org/10.1016/j.soilbio.2020.107880

Zhao, Z., Gonsior, M., Schmitt-Kopplin, P., Zhan, Y., Zhang, R., Jiao, N., & Chen, F. (2019). Microbial transformation of virus-induced dissolved organic matter from picocyanobacteria: Coupling of bacterial diversity and DOM chemodiversity. *ISME Journal, 13*, 2551–2565. https://doi.org/10.1038/s41396-019-0449-1

Zhao, B., Song, P., Yang, W., Mai, Y., Li, H., Liu, Q., Zeng, Y., Gao, Y., Du, W., & Wang, C. (2022). Bacterioplankton community indicators for seasonal variation in a fragmented subtropical river. *Environmental Monitoring and Assessment, 194*, 458. https://doi.org/10.1007/s10661-022-10101-z

Zhu, S., Peng, R., Wu, X., Ge, X., Yang, J., Lin, X., Peng, S., Feng, L., & Wang, N. (2023). The spatial distribution characteristics of dissolved organic matter (DOM) in sediments and water quality of Wuli Lake, Taihu Lake Basin, China. *Sustainability., 15*(13), 10097. https://doi.org/10.3390/su151310097

Chapter 6
Interactions Between the Microbiome, Virome, and Cyanobacterial Blooms in Tropical Urban Catchments

Shu Harn Te, Dong Zhang, Zhi Yang Sim, Jerome Wai Kit Kok, and Karina Yew-Hoong Gin

1 Introduction

The protection of freshwater resources is becoming increasingly urgent as global populations increase. There are two key sources of vulnerability from climate change and urbanization (Wiederkehr et al., 2020). However, these patterns are likely to differ across latitudes. Specifically, whereas temperate regions experience strong seasonality throughout the year, tropical waters tend to maintain a persistently warm temperature. This variability alters the ecosystem and therefore the environmental responses across temperate and tropical water bodies (Johnson et al., 2024). In comparison, tropical waters tend to be more poorly understood (Vieillard et al., 2020) although they are expected to account for >50% of the world's population in 2050 (Marcotullio et al., 2021). In many places especially within Southeast Asia, livelihoods depend heavily on freshwater bodies. For instance, the Mekong River starts from China and runs through Myanmar, Lao PDR, Cambodia, and Vietnam (Liu et al., 2022). The Johor River is also an important driver of productivity within the Johor Straits, which is the coastal channel between Singapore to the South and Malaysia to the North (Pak et al., 2021). In both cases, urbanization and construction activities (e.g., dams) within the freshwater system have severely altered freshwater hydrodynamics within the respective environments. This chapter

S. H. Te · Z. Y. Sim · K. Y.-H. Gin (✉)
Department of Civil and Environmental Engineering, National University of Singapore, Singapore, Singapore
e-mail: ceeginyh@nus.edu.sg

D. Zhang
Singapore Centre for Environmental Life Science Engineering, Nanyang Technological University, Singapore, Singapore

J. W. K. Kok
NUS College, National University of Singapore, Singapore, Singapore

S. M. Tiquia-Arashiro et al. (eds.), *Urban Watershed Microbiology, Volume 1*,
https://doi.org/10.1007/978-3-031-89071-0_6

therefore provides a review of the current knowledge in tropical freshwater systems, providing an important tool for the management of such environments within the region

One of the key challenges facing freshwater security is the occurrence of cyanobacterial blooms, which are prevalent in both freshwater and marine ecosystems, especially in tropical urban catchments. Typically, cyanobacteria play a fundamental role in aquatic ecosystems as primary producers, contributing to the cycling of carbon and nutrients (Nie et al., 2013; Winder & Sommer, 2012). Blooms occur when these single-celled phytoplankton are allowed to dominate the environments under conducive growth conditions. These blooms are driven by a combination of environmental factors such as nutrient availability, light intensity, and temperature, which promote the excessive growth of cyanobacteria

Historically, research on cyanobacterial harmful algal blooms (CHABs) has focused on managing the abiotic factors such as nutrient levels, pH, and water temperature as primary drivers. However, manipulating the environment has tended to favor certain bloom species over others instead of decreasing the number of bloom incidents, which is a particular problem for tropical freshwaters that tend to be biodiverse. Specifically, toxic HAB-causing cyanobacteria like *Raphidiopsis* sp., which is a species found in freshwaters all over the world, have been known to thrive at low-nutrient conditions (Shi et al., 2022). Broad strategies that, for instance, reduce eutrophication would therefore need to be carefully assessed and implemented. Instead, there has been recent interest in the more targeted and density-dependent treatment outcomes such as the interaction between cyanobacteria and other microbial communities, including bacteria and viruses (Needham & Fuhrman, 2016). The discovery of viruses, particularly cyanophages (i.e., viruses that infect cyanobacteria), has expanded our understanding of bloom regulation, with viral lysis accounting for a significant portion of cyanobacteria mortality in freshwater systems (Brussaard, 2004). Cyanophages were first reported in the 1960s, but interest has been steadily increasing over the years (Grasso et al., 2022). They present a potent solution to HAB phenomena because they exhibit varying specificity to the target species or to more general assemblages of taxa. Cyanophages play dual roles by controlling cyanobacterial populations through lytic and lysogenic cycles, which can significantly alter microbial community structure and influence nutrient cycling. Cyanophage infection occurs in four initial steps—attachment, penetration, biosynthesis, and maturation (Grasso et al., 2022). These four steps integrate the genetic material in the cyanophage into the host cell, altering protein expression patterns. In the case of lytic cycles, maturation will lead directly to cell lysis, releasing the newly expressed cyanophage into the environment. In the lysogenic process, genetic material is incorporated into the host genome without lysis of the host cell. The phage-introduced elements, known as prophages, continue to be expressed and passed down the lineage into daughter cells. Under stress conditions, these lysogenic prophages can switch into a lytic phase and cause lysis of the cell, leading to a release of cyanophage into the environment. Zhang et al. (2022a) observed that this switch between the lytic and the lysogenic can be affected by nutrient conditions but also suggested that the direction of the switch might be affected by the

extant microbiome. Other studies have observed similar drivers, identifying also that pH/pCO_2, salinity, as well as light can be important (Ni & Zeng, 2016; Stough et al., 2017; Cheng et al., 2019). Apart from controlling HAB-causative species populations, cyanophages play a wider ecological role when through cell lysis that releases intracellular nutrients while itself being an organic matter that sinks within the water column; they also drive carbon dynamics by slowing the carbon-fixing processes in infected cells (Fuchsman et al., 2019; Antosiak et al., 2022; Aranda et al., 2023)

Recent advances in molecular techniques, such as high-throughput sequencing, have allowed for rapid sequencing and data collection for detailed characterization of the microbiome and virome associated with cyanobacterial blooms. While cyanobacteria serve as primary producers, heterotrophic bacteria play vital roles in degrading organic matter and cycling nutrients, which further supports bloom dynamics (Zambrano et al., 2009; Tiquia, 2010; Flood et al., 2015; Yang et al., 2021; Ye et al., 2020; Nassereddine et al., 2024). Additionally, cyanophages contribute to the regulation of bloom formation and collapse, as well as the potential enhancement of toxin production through interactions with their cyanobacterial hosts. In general, the molecular tools have highlighted the intricate relationships between cyanobacteria, heterotrophic bacteria, and viruses in aquatic environments as well as the broader emergent patterns that can arise from such interactions. For instance, Chen and Zeng (2020) discovered synchronicity in gene expression for a population of cyanophages that they roughly categorized into early-, middle-, and late-stage infection phases. The ability to map such synchronicity supports bloom management efforts by highlighting the stage of cyanophage infection in a toxic HAB event. Another useful outcome of such high-throughput technologies has been in mapping the spread of cyanophage-induced novel traits through horizontal gene transfer (HGT) processes (Zhu et al., 2023a, b). In one study, Gin et al. (2021) documented a novel toxic strain of the freshwater *Synechococcus* sp. Although the study did not investigate HGT, the potential for novel traits to emerge means that we have needed faster technologies that can acquire more data across a wider spatial range. Understanding the rate and pattern of infection across a water body can better inform about future risks especially if novel harmful traits become more widespread. Importantly, there is an effort to improve the cost of such technologies (e.g., Cuber et al., 2023) so that they become more accessible to countries with limited resources, especially within tropical climates.

In general, tropical urban freshwater systems are important to address. Urban watersheds are typically very complex because there are many nutrient sources that drive eutrophication regimes (McDowell et al., 2019; Oest et al., 2018). This challenge is exacerbated by climate change trajectories that create unprecedented environmental conditions. It is generally unclear how the phytoplankton community and other organisms from higher trophic levels can respond. One expected trend in tropical waters is that phytoplankton communities will be dominated by small-sized pico-cyanobacteria (e.g., Gin et al., 2021). This uncertainty differs from the challenges facing temperate climates, which are expected to become more tropical-like (Zarzyczny et al., 2023). In other words, knowledge about tropical waters needs to

be studied separately from temperate systems, with appropriate monitoring programs to observe how tropical waters are responding to climate change and urbanization trends. In this chapter, we therefore have two broad objectives. The first is to explore the complex interplay between the microbiome, the virome, and cyanobacteria populations specifically within tropical urban freshwaters. Second, we seek to specifically investigate the role of cyanophages in these interactions

2 Cyanobacterial Bloom Microbiome

Heterotrophic bacteria and phytoplankton are intricately connected, forming a significant portion of the overall planktonic biomass in aquatic ecosystems (Patten et al., 2018; Yang et al., 2019). Phytoplankton community profoundly influence the abundance and community structure of heterotrophic bacteria, as the interactions are highly dynamic and dependent on the algal species present (Needham & Fuhrman, 2016).

In shallow freshwater environments, there is often a rich diversity of microorganisms. However, land use, human activities,climate or environmental changes can dramatically impact the composition and distribution of the microbiomes (Chopyk et al., 2020; Boix et al., 2012; Fuhrman et al., 2015). Nutrient pulses, a characteristic of cyanobacterial blooms, elicit varying responses from bacteria and viruses (Brum et al., 2016; Malits et al., 2014). Although much research has focused on the abiotic factors that trigger phytoplankton blooms, such as nutrients, pH, temperature, and light, earlier studies often overlooked the interactions between cyanobacteria and heterotrophic bacteria due to limitations in traditional identification methods. The advent of high-throughput sequencing has revolutionized the study of microbiomes associated with cyanobacterial blooms. While both cyanobacteria and heterotrophic bacteria are influenced by the same environmental factors, the resilience of bacteria to environmental changes and their effects on phytoplankton have been less explored. Recently, there has been growing interest in understanding the role of bacterial communities that coexist with cyanobacterial blooms, particularly in terms of their contributions to nutrient cycling, population succession, and toxin degradation (Te et al., 2023).

2.1 Bacterial Responses to Phytoplankton Blooms

The interaction between heterotrophic bacteria and phytoplankton becomes particularly relevant during bloom events. Studies have shown that during phytoplankton blooms, specific bacterial species consistently dominate, despite variations in phytoplankton composition and environmental conditions. Rhodobacteraceae and Gammaproteobacteria are frequently identified in 16S rRNA gene-based surveys (Buchan et al., 2005; Teeling et al., 2012) due to their ability to quickly respond to

nutrient pulses—a trait of phytoplankton blooms. *Roseobacters* and *Flavobacteria*, in particular, have become model organisms for studying bacteria–phytoplankton interactions (Buchan et al., 2014).

The structure of bacterial communities changes in response to shifts in phytoplankton abundance and species composition during bloom events (Pinhassi et al., 2004; Riemann et al., 2000; Te et al., 2017). Factors such as nutrient availability, temperature, and the type of phytoplankton species present can significantly influence the timing of these shifts. Different phytoplankton species release distinct types of organic matter, and bacteria vary in their ability to consume and remineralize these substrates (Alonso-Sáez & Gasol, 2007; Moran et al., 2022). For example, studies on two separate bloom events in the South Pacific Ocean found similar bacterial communities despite different cyanobacteria-to-diatom ratios (LeCleir et al., 2014). Although generalizations about bacterial community responses can be difficult due to variations in bloom location, phytoplankton succession and environmental factors, Bacteroidetes and Proteobacteria, nevertheless, frequently dominate bloom events (Pinhassi et al., 2004; Fandino et al., 2001; Beate et al., 2007).

Recent advances in functional community analysis, such as metagenomics, metatranscriptomics, and metaproteomics, provide a deeper understanding of bacterial activity during blooms (Campbell & Kirchman, 2013; Georges et al., 2014). Studies of *Roseobacter* and *Flavobacteria* strains have revealed high niche specialization, with closely related phylotypes exhibiting distinct responses during bloom progression (Teeling et al., 2012). High-throughput sequencing methods have uncovered the vast diversity of bacterial communities in freshwater environments, with Proteobacteria, Actinobacteria, Bacteroidetes, and Cyanobacteria being the most abundant phyla (Andersson et al., 2010; Galand et al., 2009; Sogin et al., 2006; Liu et al., 2018; Tang et al., 2020).

2.2 *Bacterial Conversion of Organic Matter*

Phytoplankton blooms generally create favorable conditions for bacterial growth, as they provide a habitat rich in organic material. This is evident in the positive correlation between bacterial and phytoplankton cell concentrations (Buchan et al., 2014). Measurements of bacterial generation and production rates often increase during bloom events (Arrieta & Herndl, 2002; Bird & Karl, 1999; Riemann et al., 2000). Interestingly, a decline in bacterioplankton abundance is typically observed at the start of blooms, possibly due to competition for nutrients or bacterial predation by protists (Castberg et al., 2001; Pernthaler, 2005). As blooms peak, high levels of organic matter released by phytoplankton support the recovery and subsequent growth of bacterial populations, allowing them to exploit organic material from decaying phytoplankton even after blooms collapse (Bratbak et al., 1998).

Both laboratory and field studies have demonstrated that similar groups of bacteria are often found associated with cyanobacterial aggregates, forming a microenvironment known as the phycosphere. In this interactive space, cyanobacteria release

compounds essential for heterotrophic bacteria, such as oxygen, nitrogen compounds, and dissolved organic carbon through photosynthesis and cell lysis. In return, heterotrophic bacteria provide essential services, such as removing reactive oxygen species and supplying readily available nutrients and CO_2 through mineralization. This reciprocal relationship within the phycosphere underscores the complex interactions between cyanobacteria and heterotrophic bacteria and highlights the need for a holistic approach to managing harmful algal blooms.

Phytoplankton species and abundance affect the composition and concentration of detrital pools throughout bloom events, altering bacterial activity. During early bloom stages, phytoplankton releases low molecular weight compounds that act as chemo-attractants for bacteria. Bacteria then synthesize compounds that support phytoplankton growth. Toward the bloom's peak, nutrient-limiting conditions often lead to the release of small molecules, further enhancing bacterial activity (Bratbak & Thingstad, 1985). In the later stages, higher molecular weight compounds, such as lipids, polysaccharides, and proteins are released due to phytoplankton cell lysis. Bacteria with the ability to solubilize these particles dominate, quickly colonizing senescent phytoplankton and exhibiting extracellular enzyme activity (Smith et al., 1992).

Organic matter from phytoplankton is assimilated and remineralized by bloom-associated heterotrophic bacteria to provide energy, involving complex processes (Arnosti, 2011;, Kujawinski, 2011). The initial step is the transformation of bioavailable particulate organic matter (POM) into dissolved organic matter (DOM), enabling it to pass through bacterial cell membranes—a process requiring cell surface and extracellular enzymes (Azam & Malfatti, 2007; Patel et al., 2019; Pagnucco et al., 2023). Bacteria with a full complement of hydrolytic enzymes rapidly colonize senescent phytoplankton cells and aggregates, showing significant extracellular enzyme activity (Smith et al., 1992; Patel et al., 2019). Once POM is converted into DOM, it is quickly absorbed into bacterial biomass, with some carbon respired as CO_2. However, about 30% of the organic matter released by phytoplankton is resistant to microbial degradation (Jiao et al., 2010). The microbial transformation of organic matter produces metabolic by-products distinct from the original material, with heterotrophic bacteria typically prioritizing the nitrogen components of both DOM and POM. This leads to an increase in the carbon-to-nitrogen ratio of organic matter resistant to degradation, which can persist for centuries to millennia in marine environments (Hopkinson & Vallino, 2005). Consequently, the metabolically altered organic matter from phytoplankton contributes to long-term carbon storage in marine waters, playing a crucial role in global carbon cycling over extended timescales (Ogawa et al., 2001).

3 Cyanophages

Cyanophages are viruses, which infect cyanobacteria. Given their parasitic nature, cyanophages primarily depend on host machinery and cellular functions for replication. Cyanophage infection can be divided into two modes, the lytic and lysogenic infection cycle. In the lytic infection cycle, cyanophage adsorption onto susceptible host receptors allows the injection of viral DNA into the host. The viral DNA in the host cell can hijack host cellular activity and synthesize phage DNA as well as proteins. Eventually, viral structural proteins and DNA are assembled to form progeny phage. The host cells are lysed by lytic enzymes, thereby releasing numerous cyanophages into the environment for the next round of infection (Dion et al., 2020). The lysogenic infection cycle represents a different lifestyle. Instead of altering host cellular processes which could incur antiviral responses, cyanophage DNA becomes integrated into the host chromosome upon injection during a lysogenic infection cycle. The incorporated cyanophage DNA will replicate with the host and be passed on to daughter cells (Brussow et al., 2004). The lysogenic cyanophage genome in the host chromosome can remain dormant until triggering factors result in dormant phages initiating the lytic cycle. Activation of dormant lysogenic phages has been linked to stressors including mitomycin C (Dillon & Parry, 2008), elevated levels of UV radiation (Jiang & Paul, 1998), high temperature (Ashy & Agustí, 2020), and heavy metal contamination (Lee et al., 2006). However, the mechanisms determining the lysogenic life cycles of cyanophages are unclear and different theories have been proposed. Some researchers believe in the rise in lytic activity with dense host populations, whereas more recent studies observed the opposite phenomenon (Knowles et al., 2016). Furthermore, the presence of signature genes, such as the integrase gene, was commonly seen as a sign of lysogenic potential, whereas a recent study found that T-7-like marine cyanophages infecting *Synechococcus* did not form a stable lysogeny despite the presence and active expression of the integrase gene (Shitrit et al., 2021). Although there is a lack of studies on cyanophage lysogeny in tropical catchments, elevated levels of UV radiation and temperatures within these environments suggest that there should be an active lytic cyanophage population.

The first cyanophage was isolated from a freshwater pond in 1964 (Schneider et al., 1964). Since then, cyanophages have been identified in wide range of environments from tropical oceans to polar freshwater lakes (Schneider et al., 1964; Safferman et al., 1983). Freshwater cyanophages are generally genetically distinct from their marine counterparts, possessing unique genes and infection strategies (Morimoto et al., 2018). Nevertheless, a few freshwater cyanophages show a closer evolutionary relationship with marine cyanophages, such as S-EIV1 infecting polar freshwater *Synechococcus* and S-CRM01 infecting estuarine *Synechococcus* (Dreher et al., 2011; Chenard et al., 2015). Decades of isolation efforts have yielded over 100 cyanophage isolates, primarily focusing on marine cyanophages infecting *Prochlorococcus* and *Synechococcus* (Morimoto et al., 2018; Padan et al., 1967). In the past 5 years, the isolation of cyanophage from the freshwater environment has

gained increased attention. Thirty studies on cyanophage isolation work were published from 2019 to 2023. Of these, 19 studies described the isolation of freshwater cyanophages, with only 1 study in India, 1 study in Thailand, and 3 studies in Singapore describing cyanophages isolated from tropical freshwaters (Zhang et al., 2020, 2021, 2022b; Noronha et al., 2022; Naknaen et al., 2021). Characterizing freshwater cyanophage isolates has revealed specific activities, such as cyanophage Ma-LMM01 degrading the host phycobilisome during infection to limit excess light energy absorption and thus, providing an advantage in surface water (Yoshida et al., 2008b). Similar activity was also found in cyanophage PaV-LD (Gao et al., 2012). Besides enhancing the understanding of cyanophages, isolation work creates a valuable pool of cyanophages for potential applications in treating cyanobacterial harmful algal blooms (CHABs). Although a limited portion of cyanophages come from the tropical urban catchment, the findings and isolations on cyanophage from different climates can still be relevant and applicable in the tropical urban context. For instance, cyanophage PA-SR01 first isolated in an urban freshwater lake in Singapore was proposed to be present in Lake Michigan based on metagenomic reads mapping. Later in 2022, researchers isolated cyanophage Pam2 sharing over 95% nucleotide identity to PA-SR01 in Lake Chaohu, China (Du et al., 2022).

3.1 Implications of Cyanophages in the Environment

Cyanophages play pivotal roles in the ecological dynamics of cyanobacteria and the biogeochemical cycling of nutrients (Zimmerman et al., 2020; Zheng et al., 2021). Common investigations of cyanophage groups based on PCR amplification of the g20 gene, DNA pol gene, and g91 gene can be used to monitor cyanomyoviruses, cyanopodoviruses and Ma-LMM01-type phages, respectively (Dorigo et al., 2004; Huang et al., 2010; Kimura et al., 2012; Jaskulska et al., 2021). The PCR-based environmental monitoring of cyanophages has identified the widespread presence of cyanophages throughout the world, highlighting their global prevalence and significant impacts on aquatic ecosystems (Jameson et al., 2011; Matteson et al., 2011; Loh, 2017; Xia et al., 2013). Quantification of real-time PCR (qPCR) of cyanophages further enables the examination of cyanophage associations with host cyanobacteria abundance.

In a tropical freshwater lake in Singapore, the dynamics of several free and host cell fraction cyanophage and cyanobacteria were quantified, demonstrating active infection from each of the phage targets despite differing abundances (Zhang et al., 2022a). Correlation of free cyanophage abundance and cyanobacteria populations indicated lytic infection of the dominant cyanobacteria by cyanophage S-SRP02 and S-SRM01 in the entire lake, supporting the hypothesis that they may exhibit greater host range than in in vitro experiments. Additionally, cyanophage infection has also been proposed to mediate cyanotoxin production through altering the host cyanobacteria population structure (Yoshida et al., 2008a). Although cyanophage lytic infection can kill harmful cyanobacteria rapidly, the mass release of

intracellular cyanotoxin could also raise concerns over water quality (Stough et al., 2017; McKindles et al., 2020). Viral metagenomic analysis in Lake Erie also pointed to the possibility of cyanophage infection to facilitate certain cyanotoxin production (Zhang et al., 2022a).

In the lytic infection cycle, cyanophage can reprogram the host cell in order to facilitate cyanophage infection and alter key metabolic pathways (Waldbauer et al., 2019). This can be achieved through the use of host-derived auxiliary metabolic genes (AMGs) (Gao et al., 2016). Common cyanobacterial AMGs include genes involved in light-dependent reactions of photosynthesis, the pentose phosphate pathway, nutrient acquisition, and nucleotide biosynthesis (Thompson et al., 2011). The presence of such AMGs holds significant implications for the ecosystem. For instance, transcriptome analysis of the marine *Prochlorococcus* demonstrated that a substantial portion of *psb*A expression, responsible for encoding the photosystem D1 protein crucial in oxygenic photosynthesis, originated from viral *psb*A genes rather than cyanobacterial sources (Sieradzki et al., 2019). Phages harboring PSI gene cassettes are prevalent in the tropical Pacific, Indian, and Atlantic Oceans. This prevalence implies that cyanophages carrying both PSI and PSII genes are inclined to sustain and potentially influence photosynthesis while infecting *Prochlorococcus* hosts in the tropical oceans (Fridman et al., 2017), emphasizing the substantial impact of viral contributions to primary production. Similar psbA genes were identified in cyanophages isolated from a tropical lake in Singapore (Zhang et al., 2021). Thus, a proportion of cyanobacterial photosynthetic activity in tropical environments might be controlled by or come from cyanophage-encoded photosynthetic genes.

Another key ecological process mediated by cyanophage infection is the release of intracellular nutrients into the surrounding water, setting free host biomass as DOM and particulate organic matter. At the end of a lytic infection cycle, host cyanobacteria are lysed, releasing viral progeny and cellular debris into the surrounding environment, a process known as "viral shunt" (Fuhrman, 1999). This provides nutrients for surrounding microbes and redirects nutrients to lower trophic levels (Steven et al., 1999). Additionally, cyanophage infection can alter the structure and population dynamics of the cyanobacteria and bacteria communities, thus affecting the whole microbiome in the system through killing of the dominant cyanobacteria population (Pound et al., 2021). This not only drives shifts in the microbial community but also promotes co-evolution of cyanophage and cyanobacteria. In Osaka Bay, Japan, metagenomic and transcriptomic characterization revealed that a majority of the viral population identified in the viral metagenome is active and rapidly disseminates upon lysis, influenced by ocean currents (Yoshida et al., 2018). Interestingly, cyanophage infection followed the circadian rhythm. Transcriptomic analysis of oceanic cyanobacteria suggested that cyanophage activity is synchronized with the light-dark cycle, exhibiting peak expression during nighttime (Chen & Zeng, 2020). Such results point to the synchronized peak release of intracellular nutrients at night, thereby changing the fate of other co-occurring microbes.

Cyanophages play a crucial role in shaping the ecological and biogeochemical dynamics of tropical urban catchments. By regulating cyanobacteria populations,

cyanophages prevent blooms that can lead to decreased dissolved oxygen, increased pH, and toxin production, thereby maintaining ecosystem resilience against environmental disturbances. The ecological impacts of cyanophages are multifaceted. They alter the composition of microbial communities by selectively targeting specific cyanobacteria strains, influencing the balance between beneficial and harmful bacteria. Additionally, cyanophages facilitate nutrient release from infected cyanobacteria, influencing nitrogen, phosphorus, and carbon availability. This nutrient cycling has significant implications for ecosystem health, as it can impact nutrient availability and potentially affect eutrophication. In tropical urban catchments, urban pollution and climate change create ideal conditions for cyanobacteria growth. The frequent rainfall and strong solar radiation also point to a high viral decay rate. Yet highly abundant and dynamic cyanophage infection has been observed in tropical urban environments, suggesting rapid turnover of cyanobacteria regulated by cyanophage infection (Zhang, Te, et al., 2022a).

3.2 Environmental Factors Affecting Cyanophage Infection Dynamics

Cyanophages in the environment face various factors that can modify their degradation, adsorption, and infection mechanisms. pH levels and water temperature can influence cyanophage stability and infection dynamics. Extreme pH conditions may lead to phage inactivation, with bacteriophage TSE1–3 exhibiting optimal stability at a pH of 7 (Khawaja et al., 2016). Similar findings were observed for cyanophage MaMV-DH01 (Meng et al., 2023). However, for cyanophage S-PM2, researchers have reported that stability remains largely unaffected by pH changes, although they did observe a reduction in burst size and latent period at low pH values (Traving et al., 2014). Water temperature, on the other hand, can have a significant impact on cyanophage infection dynamics. For example, the cyanophage CrV-01 showed an almost 90% reduction in the latent period when the temperature was increased from 22 °C to 30 °C, without compromising burst size (Steenhauer et al., 2016). Higher temperatures facilitate increased metabolic rates in host cells, providing ample nutrients and energy for phage replication. This may also explain why the polar freshwater cyanophage S-EIV1 exhibits an exceptionally long latent period of 5–7 days (Chenard et al., 2015). In a study by Niu et al., the combined effect of elevated temperature and CO_2 on cyanophage PP infection dynamics revealed synergistic effects, promoting cyanophage adsorption and reducing the latent period (Niu et al., 2012).

UV radiation is identified as a primary contributor to cyanophage degradation (Paul & Kellogg, 2000). Exposure to full sunlight conditions has shown UV to cause over 90% inactivation of coliphages, which are double-stranded DNA viruses similar to cyanophages in terms of protein protection of nucleic acids (Lester et al., 2002). Both coliphage and cyanophage are double-stranded DNA viruses with

similar protein protection of nucleic acids. Consequently, sunlight is likely to induce comparable decay levels in cyanophages. The mortality of cyanophages induced by UV radiation may also provide an explanation for the commonly observed synchronized cyanophage expression and release in the oceans (Yoshida et al., 2018; Chen & Zeng, 2020; Liu et al., 2019). Recently, researchers have developed predictive models simulating virus inactivation in natural settings, allowing for better incorporation of environmental viruses in modeling efforts (Rockey et al., 2021). In addition to cyanophage degradation, UV radiation can impact cyanophage behavior concerning adsorption, latent period, and burst size (Zhang et al., 2009; Clokie & Mann, 2006). Studies have indicated that appropriate sunlight levels are crucial for cyanophage activity. For instance, the adsorption rate of the marine cyanophage S-PM2 onto the host cyanobacterium, *Synechococcus* sp. WH7803, significantly decreased under dark conditions compared to sunlit conditions. Moreover, the burst size of certain cyanophages was found to be positively correlated with the duration of sunlight exposure (Jia et al., 2010). In addition, the burst size of some cyanophages was found to be positively correlated with the length of sunlight exposure (Liu et al., 2019). Furthermore, low levels of radiation were observed to extend the latent period of CrV-01 infecting *Raphidiopsis raciborskii*, potentially due to the limitation of host metabolism and energy production in low-light conditions (Steenhauer et al., 2016).

In tropical aquatic ecosystems with ample sunlight and warm water throughout the year, increased cyanophage degradation and facilitated infection cycles in terms of adsorption as well as progeny phage production, point to rapid replacement of cyanophages. Therefore, the temporal profile of cyanophage population in tropical urban catchments is likely to be highly dynamic over time, signifying rapid turnover and co-evolution of the cyanophage and cyanobacteria communities.

Notably, emerging contaminants such as microplastics were revealed to act as new carriers for viral adsorption and transport (Li et al., 2022). On the other hand, cyanophages could also be carriers of emerging contaminants. In the freshwater viral metagenome of the Han River, functional phage-encoded antimicrobial resistance genes (ARG) were detected and authenticated, highlighting the potential for phages to serve as vectors for antimicrobial resistance through horizontal gene transfer (Moon et al., 2020). This is of particular concern in tropical urban catchments with abundant cyanophage presence and proximity to human activities. More investigation into the ARG-carrying potential of cyanophages and their possible impact on the antimicrobial properties of aquatic microbiomes is necessary to evaluate the potential outcome of such processes on human health risk.

4 A Case Study in Singapore—CHAB Virome/Microbiome

Singapore has typically been seen as a leader in urban watershed management efforts because it has needed to manage an increasingly dense urban population without the benefit of large upstream watersheds that, for some places, can include

sources like snowmelt (e.g., the Mekong River). As a small island, Singapore's main source of water has been rainfall. A complex network of canals and drains captures most of the rainfall within 17 reservoirs (e.g., Gu et al., 2018), all of which have been constructed since the 1970s and did not previously exist. Importantly, reservoirs similarly also exist in surrounding countries (e.g., Subehi et al., 2018). The experience and knowledge gained by Singapore in managing its reservoirs can therefore provide useful information for such water resources in other tropical urban cities. One key vulnerability facing urban watersheds is the impact of land-use patterns. One study examining nutrient inputs associated with land-use patterns demonstrated the highest contribution in parkland areas (Song et al., 2019), coinciding with many studies demonstrating the importance of agricultural practices on watershed conditions. However, Pearce and Chertow (2017) also demonstrated that Singapore's increasing shift away from traditional soil-based agriculture has replaced these agricultural drivers of phosphorus eutrophication with more industrial- and household-based sources. Nutrient flows are therefore especially complex and important to document especially within rapidly urbanizing cities like Singapore where land-use patterns can vary widely.

In tropical cities like Singapore where there is weak variation in seasonality, biological activities within aquatic environments become dominated by community interactions and are less affected by environmental forcing although these conditions still play an important role in supporting the growth and survival of indigenous organisms (Xu et al., 2018). Specifically, it is expected that interactions between phytoplankton, bacteria, and viruses would have a more pronounced effect on bloom conditions in the tropics, affectinghigher trophic levels and biogeochemical export fluxes. Understanding the succession patterns of these organisms therefore becomes crucial (Dhib et al., 2016; Patten et al., 2018; Yang et al., 2019) in managing the health of water resources. The ability to detect and monitor these different organisms has been challenging but is being improved by technological advances. Specifically, model predictions and empirical observations that small-sized pico-phytoplankton would become more dominant have raised unique challenges because they are difficult to detect using typical methods like microscopy (e.g., Te et al., 2023). Instead, methods like qPCR, flow cytometry (FCM), and 16S rRNA gene amplicon sequencing can provide useful insights into the dynamics of these species.

4.1 Bloom Dynamics Influence Bacteria and Virus Abundances

Currently, knowledge about bacteria and virus dynamics in Singapore urban freshwater systems indicates often eutrophic conditions that support almost persistent year-round blooms. In a year-long study, Zhang et al. (2022a) observed average chlorophyll-*a* concentrations of 121.6 μg/mL, total nitrogen (N) of 2.57 mg/L, and

total phosphorus (P) of 0.099 mg/L. The N:P ratio of 28.3 suggested P-limited conditions, influencing cyanobacteria proliferation patterns. qPCR assays were used to quantify cyanobacteria abundance, *Microcystis*, *Raphidiopsis*, microcystin synthetase gene, cylindrospermopsin amidinotransferase, and 2-MIB (an off-flavor compound) cyclase gene. The bacterial and cyanobacterial communities were analyzed with amplicon sequencing based on the 16S rRNA V4 region. For cyanophage qPCR analysis, primer pairs were mostly adopted from previous studies (Yoshida et al., 2008a; Zhang et al., 2021, 2022b), except for primers C8F and C8R targeting cyanophage PA-SR01 which were newly designed. Phage abundance was quantified in two fractions—the host cell fraction cyanophages detected from cyanobacterial biomass as well as the free phage fraction which was detected from filtered samples without host cells. Environmental viruses ranged around 20–200 nm. For FCM, picocyanobacteria were enumerated according to three *Synechococcus* clusters based on their photosynthetic pigments—phycocyanin ("Syn_PC"), high phycoerythrin ("Syn_PE1"), and low phycoerythrin ("Syn_PE2"). Cyanophages were also analyzed using FCM with staining by SYBR gold (e.g., Kaletta et al., 2020). Other methodologies for measuring cyanotoxins, nutrients, and environmental parameters followed Te et al. (2023).

In the study, Te et al. (2023) reported *Microcystis* and *Synechococcus* as the most prevalent taxa in the tropical water body, with another 13 minor co-existing cyanobacteria genera. The cyanobacterial community showed distinct spatial variability, with higher coccoid cyanobacteria upstream and a mixed community downstream, including filamentous cyanobacteria like *Nodosilinea* and *Planktothricoides*. The coexistence of two bloom types was observed in the reservoir, where upstream blooms were dominated by *Microcystis* during blooms and by *Synechococcus* during non-bloom phases. Further downstream, a multi-genera bloom tended to be more prominent. This discrepancy was attributed to varying nutrient conditions between upstream and downstream sites. Upstream sites exhibited higher nutrient levels for nitrogen and phosphorus, where a high N/P ratio favored the dominance of *Microcystis*. In contrast, the downstream areas had a more diverse community composition likely due to a more heterogeneous environment caused by differences in water flow, depth, and nutrient availability. Water quality parameters such as temperature, salinity, dissolved oxygen, and total dissolved solids also likely influenced the spatial distribution of cyanobacteria. Notably, *Synechococcus* thrived in both upstream and downstream conditions, displaying wide tolerance to ionic strength and salinity.

In Singapore, bacteria and virus concentrations ranged 4.1×10^6–1.3×10^7 bacteria-cells/mL and 2.1×10^7–5.2×10^7 virus-cells/mL (Sim et al., 2025). These values were consistent across other water bodies. In a study of a different reservoirs in Singapore, virus concentrations were reported to range 1.5×10^7–1.2×10^8 virus-cells/mL (Yang et al., 2019). In Sri Lanka, two tropical reservoirs reported concentrations of around 3.4×10^6 bacteria-cells/mL and 4.9×10^7 virus-cells/mL (Peduzzi & Schlemer, 2004). The values were consistent even comparing with two temperate reservoirs in France—Grangent had concentrations ranging 1×10^6 bacteria-cells/mL and 8×10^6 virus-cells/mL in Autumn to 3×10^6 bacteria-cells/mL and 7×10^6

virus-cells/mL in Spring; whereas Villerest concentrations ranged 2×10^7 bacteria-cells/mL and 1×10^8 virus-cells/mL in Autumn to 3×10^7 bacteria-cells/mL and 2×10^8 virus-cells/mL in Spring (Pradeep Ram et al., 2019). The comparable bacteria concentrations across temperate and tropical waters slightly contrasted with Amado et al. (2013), who demonstrated that bacteria growth efficiency should be reduced in tropical waters compared to temperate waters. It is possible that seasonal forcing in temperate waters may significantly suppress populations especially during Winter conditions whereas bacteria and viruses can accumulate year-round within tropical waters. This outcome was reported by Sim et al. (2025), where bacteria generally maintained a moderately high abundance throughout the year, showing few deviations. Abundances increased with chlorophyll-*a* concentration during bloom events and remained elevated even after the bloom peaks. Consistent trait profiles determined using FCM coupled with nucleic acid staining methods suggested that populations in Singapore waters generally retained their composition throughout the year, which may reflect the retention of functional groups within the ecosystem (Mitra et al., 2014; Polz et al., 2006). The results however contrast with a study of a temperate lake, where nucleic acid content varied according to nutrient levels in the water (Pradeep Ram et al., 2020).

Overall, bacterial abundances observed in the Singapore waters were stable apart from a moderate positive correlation with the causative phytoplankton during the *Microcystis* bloom ($r = 0.56$, $P < 0.01$) and a downstream *Synechococcus* sp. "SYN_PE2" bloom ($r = 0.46$, $P < 0.01$). This trend contrasted with other studies where bacteria and viruses persistently increased in abundance along with phytoplankton (Buchan et al., 2014; Zimmerman et al., 2020). Viral dynamics also displayed different temporal patterns and significant correlations with potential hosts (cyanobacteria). This was particularly evident in some phytoplankton communities, which showed significant correlations with viruses. For instance, blooms in the upstream environments had positive correlations with *Synechococcus* ($r = 0.37$, $P < 0.01$) and the "Syn_PE1" group ($r = 0.33$, $P < 0.01$). However, weak correlations were also observed, suggesting that taxa-specific viral infections do not lead to the destruction of entire populations, as was previously reported in one study on *Prochlorococcus* (Kashtan et al., 2014). A higher frequency of daily sampling intervals could also track the correlations more closely since viral concentrations should generally follow host abundances (Needham et al., 2017). In a study carried out in Brazil, the correlation between virus-like particles and bacterioplankton was observed even at a 3-h sampling frequency, but not at 15-min intervals (de Araujo & Godinho, 2009).

There is clear potential for viruses to be an important control for phytoplankton communities (e.g., Bhatt et al., 2023). However, the mechanism for this control is not well understood. Although cell lysis reduces host cell density, the release of intracellular dissolved carbon can drive bacterial growth especially for non-target species (Pradeep Ram & Sime-Ngando, 2023). Furthermore, infection rates are known to be driven by environmental conditions such as light conditions, nutrient availability, and temperature (Ni & Zeng, 2016; Grasso et al., 2022). Specifically, a *Microcystis* cyanophage Ma-LMM01 was found to affect cellular pigments in the target host cell (Yoshida-Takashima et al., 2012), emphasizing the importance of

interaction between phage infection and light conditions. Across the studies cited in this chapter, viral loads have persistently occurred at a magnitude higher than bacteria cell densities. Consequently, the persistent high abundance of potential host cyanobacteria cells suggests that the function of cyanophages is not simply to infect and eradicate target host cells. Instead, intervening factors such as a density-dependent infection responses or environmental preferences may play a bigger role in supporting a healthy and biodiverse ecosystem than is presently understood. Understanding some of these natural feedback responses is needed to properly assess the impact of cyanophages as a treatment solution, complementing efforts by researchers like Zhu et al. (Zhu et al., 2023a, b) to manipulate the specificity of the cyanophage–host association for better alignment with environmental management outcomes.

4.2 Importance of Nutrients in Shaping the Phytoplankton Community

In the study by Te et al. (2023), blooms of *Synechococcus* and *Microcystis* were observed to coexist and interact with distinct microbial communities. During peak bloom periods, *Microcystis* formed dense colonies that facilitate access to light and nutrients while providing protection from environmental stresses. *Microcystis* sp. colonies were typically associated with heterotrophic bacteria such as Microscillaceae, Chthoniobacteraceae, and *Roseomonas*, which thrive in the nutrient-rich conditions provided by the dense cyanobacterial biomass. When the *Microcystis* sp. bloom subsided, the pico-cyanobacteria *Synechococcus* became more prominent and subsequently maintained high concentrations. *Synechococcus* blooms are linked to different bacterial taxa such as Phycisphaeraceae and Methylacidiphilaceae, representing a shift in the heterotrophic bacteria community associated with the prevalent bloom causative species. In the case of *Synechococcus*, Zhang et al. (2021) showed that long-term co-culturing can result in mutualistic interactions with the heterotrophic community, therefore demonstrating potential adaptations even for ecological associations. Such interactions between cyanobacteria and their associated microbiomes therefore exert influence on nutrient cycling and bloom dynamics patterns.

Specifically, heterotrophic bacteria associated with cyanobacterial blooms play critical roles in nutrient cycling by degrading organic matter and releasing inorganic nutrients. Heterotrophic bacteria can rapidly utilize the labile component—estimated to make up around 25–55%—of the organic carbon excreted by cyanobacteria (Hoikkala et al., 2016, Patriarca et al., 2021) or can be released when cyanophage-driven cell lysis occurs (Ataeian et al., 2022). A similar dynamic was observed for heterotrophic bacteria associated with *Microcystis* sp., where phosphonate-degrading enzymes within the microbiome helped to retain phosphorus nutrients that the cyanobacteria can utilize to grow (Zhao et al., 2023). An

association between *Microcystis* sp. and its associated heterotrophic bacteria communities is moderated by variations in the chemical composition of the polymeric exudates, supporting cell density equilibriums that balance nutrient release and uptake rates in both populations (Li et al., 2024). The cyanobacteria-heterotrophic bacteria association therefore not only supports cyanobacterial growth but also influences the overall nutrient dynamics in the aquatic ecosystem. In the eutrophic waters of Singapore, a succession pattern linked to nutrient availability was observed with three phases. When nutrients, particularly nitrate (NO_3) and ammonium (NH_4), were low, the system was dominated by pico-sized rapidly growing genera, i.e., *Synechococcus*, with *Microcystis* growing in parallel. As nutrients increased, long-chain filamentous phytoplankton like *Nodosilinea* emerged. When nutrient concentration further decreased, nitrogen-fixing cyanobacteria such as *Raphidiopsis* started flourishing in the system, which was similar to observations by other research studies (Burford et al., 2018; Yema et al., 2016).

Negative correlations were also observed between *Synechococcus* and nitrogen (N) sources both upstream and downstream. A random forest analysis showed that Syn_PC was an important regulator of NO_3 with PE_cyanobacteria, *Pseudanabaena*, *Microcystis*, and *Nodosilinea*; whereas Syn_PC, Syn_PE1, and *Microcystis* regulated NH_4. However, phosphate (PO_4) could not achieve a positive model score, indicating nutrient supply could not be easily predicted based on phytoplankton communities. PO_4 is typically considered a limiting nutrient in freshwater (Bestion et al., 2018; Dignum et al., 2005). Picocyanobacteria, sensitive to PO4 changes, are observed in many lakes, including oligotrophic lakes where they have a high affinity for PO_4 and rapid growth response due to their small size (Callieri et al., 2012; Schallenberg et al., 2001). This sensitivity partially explains the low PO_4 detected and the high *Synechococcus* abundances observed by Te et al. (2023). In contrast, the study demonstrated that the *Synechococcus*-virus association was stronger in driving the biogeochemical N cycle in this system, affected by limited PO_4 to some extent. Viruses appeared to play a more active role, with multiple linear regression predicting PO_4 accounting for 88.0% of the variance in surface waters.

For cyanophage dynamics, principal component analysis identified pH and nitrogen as key drivers in Singapore waters, with microbial compositions shifting during the Southwest Monsoon season around June–September coinciding with rainfall events that introduced nutrients into the watershed (Zhang et al., 2022a). This highlighted the connection between environmental changes and microbial population dynamics, emphasizing the influence of abiotic factors on cyanophages. Cyanophages such as S-SRM01 and S-SRP02 (with *Synechococcus* sp. being the host target) were particularly responsive, with high pH levels positively linked to free cyanophage abundance. This suggests that alkaline conditions enhance cyanophage activity and replication, influencing cyanobacterial bloom collapses by promoting viral infections in cyanobacteria. Host cell fraction cyanophage abundance showed similar trends to free cyanophages, particularly in relation to pH and nitrogen. Elevated nitrogen levels were associated with increased cyanophage infections, suggesting that nutrient availability influences cyanophage-host interactions. More broadly, *Synechococcus* sp. and *Microscystis* sp. phages are known to utilize more

than half of host resources to express phage proteins (Grasso et al., 2022). Similarly, a study on *Phormidium* sp. phage indicated that P-limitation greatly decreased its production rate (Cheng et al., 2019). Nutrients may also be important in switching between lytic and lysogenic infection patterns (Zhang et al., 2022a), demonstrating the role of environmental conditions in driving cyanophage dynamics.

4.2.1 Roles of Heterotrophic Bacteria in Toxin Regulation

In addition to altering growth conditions, the microbiome can also affect toxicity associated with causative cyanobacteria species. Recent studies suggest that microbial interactions, rather than the cyanobacteria alone, could influence the production of harmful metabolites. Such dynamics have been attributed to the occurrence of novel toxicity, for example in the case of tropical *Synechococcus* spp. strains which—for the first time—exhibited cylindrospermopsin (CYN) production (Gin et al., 2021). Subsequent molecular analysis indicated that although the *Synechococcus* sp., the genome lacked the complete molecular machinery for toxin production; many of these components were analogs found in the bins belonging to the microbiome community. This finding suggests that the microbiome could play a crucial role in providing the molecular components necessary for toxin production, acting as a support system for the host cyanobacterium. These results emphasize the importance of a "whole-system" approach to understanding toxin production, which involves not only the primary species but also its interactions with the surrounding microbial community.

Apart from directly supplementing the production machinery, the heterotrophic bacteria community is also known to metabolize cyanotoxins (Berg et al., 2009; Banerji & Benesh, 2022), thereby lowering bloom toxicity. Other species have been known to mitigate chemicals in the environment. For instance, the microbiome community associated with *Microcystis* sp. can break down hydrogen peroxide (H_2O_2) in the environment (Smith et al., 2022), thereby protecting the cyanobacteria from bloom treatment chemicals. Other communities have provided protection for *Microcystis* sp. colonies against osmotic shock especially within salinity changes that happen along a marine-freshwater gradient in estuaries (Reignier et al., 2024). Such benefits may not be significant for freshwater reservoirs where saltwater is excluded but may be important for riverine environments exposed to coastal waters. Much of the studies in this space have tended to use *Microcystis* sp. colonies because they form colonies and are known to have organic exudates that support a thriving microbiome community. Results from this model organism have however demonstrated that cyanobacteria can have important associations with its microbiome. Such processes can be important especially in biodiverse tropical environments where populations can spatially overlap. The interaction between different microbiomes is poorly understood but could hold significant ecological implications (e.g., Pound et al., 2021) for the management of freshwater ecosystems. Furthermore, many environmental models predict increasing dominance by pico-cyanobacteria populations, which may encounter a higher incidence of horizontal gene transfer

mechanisms (Park et al., 2024), highlighting novel dynamics that would need to be carefully studied.

5 Technological Advances in Study of Environmental Microbes

5.1 Emerging Molecular Techniques on Cyanophage Research

The metagenomic era has unraveled the intricacies of viral communities, showcasing their roles as environmental influencers. With metagenomic investigation of the cyanobacteria and cyanophage community in tropical urban catchments, the journey into the depths of cyanophage–host interactions holds promise for unveiling the complexity of microbial ecology and the potential applications in addressing ecological challenges. Hybrid assembly, complementing high-accuracy short-read sequencing with long-read technology, could substantially improve the completeness of viral genomes recovered and bring the study of environmental cyanophages to another level (Zaragoza-Solas et al., 2022). Additionally, advances in PCR-based technology point to new possibilities in the study of environmental cyanophages. The iPolony method developed and adapted by Baran and colleagues (Baran et al., 2018) combines flow-cytometric sorting with the PCR-based polony technique, thereby producing sensitive and high-throughput quantification of cyanophage infection by various viral populations. Leveraging on solid-phase single-molecule PCR polony method, this innovative tool can analyze thousands of individual cells, taxonomically identified, for intracellular viral DNA, allowing for sensitive, high-throughput detection and quantification of diverse viral lineages. The application of iPolony successfully tracked viral infection throughout the lytic cycle, distinguishing between varying infection levels. In North Pacific subtropical surface waters, it revealed that a minor fraction of *Prochlorococcus* cells (0.35–1.6%) were infected, primarily by T4-like cyanophages, with infection rates fluctuating in synchrony with the diel cycle.

Another emerging technique is EpicPCR (Emulsion, Paired Isolation, and Concatenation PCR). EpicPCR is a single-cell genomic technique that utilizes fusion-PCR to link viral marker sequences of interest with 16S rRNA gene fragments within emulsion droplets, enabling taxonomic identification of bacteria carrying specific functions through high-throughput amplicon sequencing. This is capable of reflecting virus–host interactions and tracking virus–host dynamics with higher precision and specificity than purely relying on implied virus–host relationship through co-abundance profiling (Sakowski et al., 2021). Furthermore, it could capture in situ virus–host interactions for viral clades with no closely related database representatives. Although there is a lack of application of EpicPCR directly on cyanophage, tracking virus–host interaction dynamics on actinophages revealed contrasting dynamics between multi-host and single-host viruses. Multi-host viruses

engaged in extended interactions, while single-host viruses demonstrated efficient interactions at minimal abundance levels.

5.2 Comparison of FCM and Amplicon Sequencing Data

Both FCM and 16S rRNA gene amplicon sequencing were employed to study the phytoplankton compositional dynamics in the tropical urban aquatic ecosystem. These methods are popular in microbiome research (Dashkova et al., 2017; Te et al., 2017; Yang et al., 2019). However, 16S amplicon sequencing typically reports relative abundances rather than absolute quantitation (Props et al., 2017). Without absolute quantitation, the underlying physiology and ecology of certain microbial taxa may be obscured by their relative abundance. For example, an increase in relative abundance does not always equate to an increase in absolute abundance. Conversely, there is no straightforward method to identify phytoplankton communities based on FCM data.

The high abundance of *Synechococcus* in this study indicates their importance in tropical water bodies, consistent with their prevalence in many freshwater ecosystems (Ruber et al., 2016; Ruber et al., 2018). *Synechococcus* is polyphyletic and can be divided into three taxonomic groups based on pigmentation observed in our study—Syn_PC, Syn_PE1, and Syn_PE2. Different ecotypes of Synechococcus generally co-occur in aquatic ecosystems with known competitiveness among them in both fresh and marine waters (Callieri, 2008; Haverkamp et al., 2009). Syn_PC dominated the phytoplankton community in our study (up to 8.2×10^5 cells mL − 1), consistent with observations that phycocyanin-contained cyanobacteria tend to dominate in eutrophic waters (Li et al., 2016; Vörös et al., 2009), while phycoerythrin-contained species thrive in clear, oligotrophic waters due to more efficient green light harvesting, which is prevalent in clear waters (Haverkamp et al., 2009; Stomp et al., 2007; Vörös et al., 1998). As observed in many studies, *Synechococcus* pigment compositions correspond to their adaptability to varying light intensities and depths (Callieri, 2017; Descy et al., 2005; Kolodny et al., 2022; Zhang et al., 2022a). It is interesting to note that the relative abundance of *Synechococcus* determined with 16S rRNA sequencing showed no correlation with *Synechococcus* absolute abundances quantified by FCM. This reflects the challenge relating 16S amplicon sequencing to FCM analysis in microbiome research, requiring various validation methods and often using machine learning tools. This area needs further exploration for more consistent results (Kinet et al., 2016; Özel Duygan et al., 2020; Props et al., 2017).

Both FCM and amplicon sequencing provide valuable, complementary insights into phytoplankton community structures in aquatic ecosystems. As demonstrated here and supported by other research findings, FCM's rapid, high-throughput quantification based on cell size and pigment fluorescence enables detailed tracking of population dynamics over time and space, effectively highlighting broad functional groups such as *Prochlorococcus* and *Synechococcus* (Ning et al., 2021; Sudek et al.,

2015). However, as other studies have also noted, FCM's limited taxonomic resolution restricts its ability to distinguish closely related species, and it cannot detect non-pigmented groups due to its reliance on fluorescence (Props et al., 2017). In contrast, 16S rRNA gene amplicon sequencing offers high taxonomic resolution, identifying specific clades and low-abundance taxa across prokaryotic and some eukaryotic groups. Still, its reliance on relative abundance data can mask true population densities, and PCR biases may distort community composition (Te et al., 2017; Özel Duygan et al., 2020). By combining the high-throughput cell counts of FCM with the detailed taxonomic profiling of amplicon sequencing, researchers gain a robust and comprehensive perspective on microbial dynamics. This combined approach aligns with findings in other studies and holds great promise for advancing our understanding of complex community interactions and ecological processes within diverse aquatic ecosystems.

5.3 *Could Cyanophages Be a Potential Method for Treating Harmful Algal Blooms?*

The common methods of treating CHABs, such as mechanical removal or chemical algaecides, have limitations in terms of cost-effectiveness and potential environmental repercussions (Xiong et al., 2014; Barrington & Ghadouani, 2008). Utilizing cyanophages as biological control agents appears promising for CHAB treatment due to their predatory nature on cyanobacteria and rapid replication. The application of specific cyanophages could effectively control cyanobacteria populations in eutrophic water bodies. Bacteriophages have been effectively used as early as 1929 to treat *Staphylococcus* infections (Larkum, 1929), and have been explored as antimicrobial agents in the food and medical industries. For instance, it has been successful in treating multidrug-resistant pathogenic *Escherichia coli* in mice with phages like T-4 and KEP10 (Nishikawa et al., 2008). Applying cyanophages to address CHABs in aquatic environments would require larger quantities due to the volume of water, which poses a challenge. Possible improvements could include encapsulating cyanophages in alginate beads for slow and constant release (Huang & Lin, 2017), as well as applying cyanophages in the early stages of a bloom to prevent widespread and unmanageable situations. Addressing the diversity of cyanobacteria populations necessitates isolating a wide range of cyanophages, requiring collaborative efforts among research groups worldwide. Developing more effective cyanophage isolation methods is essential, considering that the current method is time-consuming and there is also the potential emergence of resistant mutant cyanobacteria post-cyanophage infection. Genetic engineering technologies provide potential solutions to alter cyanophage genomes and overcome host defense systems (Shitrit et al., 2021). Although phage treatment is considered a low-side-effect approach, studies suggest that mass cyanophage lytic infection might lead to an increase in dissolved cyanotoxins, endangering drinking water supplies (Grasso

et al., 2022; Steffen et al., 2017). Therefore, in treating CHABs with high intracellular cyanotoxin levels, combining cyanophage treatment with cyanotoxin degraders may be necessary to mitigate released cyanotoxin concentrations.

A significant challenge in using cyanophages to treat CHABs is the development of cyanophage resistance after phage inoculation, but is a more targeted and therefore promising alternative to other chemical and antibiotic treatments. Cyanophages and cyanobacteria engage in constant co-evolutionary pathways in nature, where cyanophages select for resistant cyanobacteria hosts, and the resistant cyanobacteria population, in turn, selects for resistance-breaking cyanophage mutants (Schwartz & Lindell, 2017). The mechanisms of cyanophage resistance are not fully understood, but various antiviral defense systems play roles. One direct measure is preventing phage adsorption by blocking phage receptors, producing extracellular matrices, and generating competitive inhibitors (Riede & Eschbach, 1986; Stummeyer et al., 2006; Zaleski, 2005). Superinfection exclusion systems are also employed by some bacteria to block phage DNA entry, hindering infection (Labrie et al., 2010). If phage DNA enters the host cell, bacteria can cleave it through the restriction-modification system or CRISPR-Cas system (Pingoud et al., 2005; Cai et al., 2013). In cases where these systems fail, bacteria can prevent further phage spreading in a suicidal manner known as the abortive infection system, which halts intracellular activities and leads to the death of the infected cell, protecting the wider uninfected population (Molineux, 1991). The study of cyanophage resistance mechanisms is less comprehensive than that of other bacteriophages, but the presence of hypervariable genomic islands in marine *Prochlorococcus* genomes has been associated with resistance to cyanophage infection (Coleman et al., 2006). The CRISPR-Cas system and restriction modification system are commonly found in cyanobacteria genomes (Cai et al., 2013; Papoulis et al., 2021). Resistance to cyanophage infection appears to be primarily intracellular against broad-host range cyanophages and extracellular against specialist cyanophages (Zborowsky & Lindell, 2019).

However, the co-evolutionary nature of bacteria and phages means that phages have similarly also evolved countermeasures. For instance, phages infecting *E. coli* developed polysaccharide-degrading enzymes to remove extracellular polymeric substances, acting as a barrier to phage adsorption (Stummeyer et al., 2006). The highly variable nature of phage genomes allows them to evade the CRISPR-Cas system through genetic sequence changes (Doron et al., 2018; Dion et al., 2021). Additionally, cyanophages may evade host defense systems by not altering host gene expression, thereby avoiding detection by the host (Morimoto et al., 2018). Conversely, it has been suggested that a cyanophage's ability to infect multiple hosts depends on the effectiveness of the host defense system, as observed through unchanged cyanophage gene expression patterns during infection of different hosts (Doron et al., 2016).

Due to cyanophage host specificity and the development of cyanophage resistance in host populations, the currently available cyanophage isolates are unlikely to be in sufficient quantities to account for the complex permutations required for effectively targeting major CHAB-forming cyanobacteria groups in the

environment (Schwartz & Lindell, 2017; Doron et al., 2016; Wang & Chen, 2008). While significant strides have been made in cyanophage studies over the past two decades, with over 100 isolates sequenced and environmental cyanophage genetic diversity characterized across various habitats, limitations persist. There is a lack of international collaboration on bringing together various cyanophage isolates from diverse backgrounds. The diversity captured by isolation falls short of representing the entire natural population. Resistance to cyanophage infection, coupled with a limited understanding of the resistance mechanism, remains a major obstacle to cyanophage treatment of CHABs. The diversity of environmental waters further complicates the issue. Therefore, the ongoing need to obtain new cyanophage isolates and comprehend cyanophage infection and phage–host interactions in the environment remains crucial for advancing the cyanophage treatment of CHABs.

Furthermore, no real-world field trials have been conducted to date. To bridge this knowledge gap, researchers must first test cyanophage applications in laboratory microcosm settings that simulate environmental conditions. This controlled environment allows for monitoring cyanophage and host population dynamics using sensitive molecular techniques, such as quantitative polymerase chain reaction (qPCR), to assess the efficacy of cyanophage-mediated killing. Additionally, analytical chemistry methods like liquid chromatography-mass spectrometry (LC-MS) are necessary to detect potential cyanotoxin release. By conducting these microcosm experiments, researchers can gain valuable insights into the dynamics of cyanophage–host interactions, toxin production, and ecosystem responses, ultimately informing the development of safe and effective cyanophage-based management strategies for harmful algal blooms at larger environmental scales.

6 Conclusions and Future Perspectives

Monitoring of urban waterways is becoming increasingly important especially with increasing urban populations. Higher land-use intensity implies more complex pollution pathways which in turn elevates the threat to water resources within these watersheds. These trends are coupled with climate change scenarios of warming temperatures and more intense rainfall patterns. Accordingly, environmental models especially within the tropics predict an increasing shift toward favoring small-sized phytoplankton such as pico-cyanobacteria. This trend raises two urgent challenges—1) there is a paucity of information about the ecology of these communities; and 2) existing sampling methods may not have sufficient resolution to detect these organisms. This current review of the literature provides sufficient evidence that the ecology of pico-cyanobacteria should depend heavily upon microbiome associations. However, microbiomes present a unique dynamic that is also not well understood. At a minimum, microbiomes can alter cyanobacteria tolerance to environmental conditions and regulate nutrient conditions for the primary cyanobacteria species (e.g., Wang et al., 2021). Knowledge gained from controlled laboratory studies using axenic cultures therefore needs to account for these ecological

conditions. An increasing amount of research especially using non-axenic cultures (e.g., Ataeian et al., 2022, Zhao et al., 2023, and others in this chapter) has provided valuable information about the ecological function and robustness conferred by microbiome communities.

Concurrently, improving technologies—for instance, high throughput sequencing as well as staining methods coupled with FCM or epifluorescence microscopy—have provided opportunities for more rapid monitoring of larger numbers of samples, allowing observation of synchronicity or large-scale responses in microbiome communities triggered by environmental changes. Additionally, cyanophages present very different ecological interactions because they ultimately reduce cell populations. There is some early research on developing cyanophages into targeted bio-treatment systems for bloom management (i.e., Zhu et al., 2023a, b), but bloom occurrences may not be mono-specific (e.g., Te et al., 2023) and may also involve heterotrophic communities. Looking forward, the potential for cross-interaction between microbiome communities presents a very important challenge (e.g., Pound et al., 2021) because novel traits and responses can potentially emerge from the transference of heterotrophic community members. Furthermore, such interactions are expected to be especially pronounced in tropical waters where there is higher biodiversity and where environmental conditions remain conducive for microbial growth throughout the year. This potential has been suggested by Gin et al. (2021), but more research is required to establish such dynamics.

In general, increasing pressures upon our urban waterways have required shifts in our monitoring systems and research approaches. The importance placed upon microbiome and virus dynamics means that shifts in community conditions—potentially leading toward bloom conditions—can occur suddenly, driven by a more complex pattern of urban pollution within the watershed. One benefit has been the improvement of technologies and the proximity to urban infrastructure which supports laboratory equipment. However, such solutions may be challenging in many places especially across Southeast Asia that may not be able to afford it. Closer collaboration and sharing of resources across the countries can help to alleviate some of the cost. Ultimately, it is hoped that this chapter provides sufficient documentation and sharing of information to help facilitate such cooperation across different researchers.

References

Alonso-Sáez, L., & Gasol, J. M. (2007). Seasonal variations in the contributions of different bacterial groups to the uptake of low-molecular-weight compounds in northwestern Mediterranean coastal waters. *Applied and Environmental Microbiology, 73*, 3528–3535.

Amado, A. M., Meirelles-Pereira, F., Vidal, L. O., Sarmento, H., Suhett, A. L., Farjalla, V. F., Cotner, J. B., & Roland, F. (2013). Tropical freshwater ecosystems have lower bacterial growth efficiency than temperate ones. *Frontiers in Microbiology, 4*, 167.

Andersson, A. F., Riemann, L., & Bertilsson, S. (2010). Pyrosequencing reveals contrasting seasonal dynamics of taxa within Baltic Sea bacterioplankton communities. *The ISME Journal, 4*, 171–181.

Antosiak, A., Sulcius, S., Malec, P., Tokodi, N., Lobodzinska, A., & Dziga, D. (2022). Cyanophage infections reduce photosynthetic activity and expression of CO_2 fixation genes in the freshwater bloom-forming cyanobacterium *Aphanizomenon flos-aquae*. *Harmful Algae, 116*, 102215.

Aranda, Y. N., Bhatt, P., Ates, N., Engel, B. A., & Simsek, H. (2023). Cyanophage-cyanobacterial interactions for sustainable aquatic environment. *Environmental Research, 229*, 115728.

Arnosti, C. (2011). Microbial extracellular enzymes and the marine carbon cycle. *Annual Review of Marine Science, 3*, 401–425.

Arrieta, J. M., & Herndl, G. J. (2002). Changes in bacterial β-glucosidase diversity during a coastal phytoplankton bloom. *Limnol Ocean, 47*, 594–599.

Ashy, R. A., & Agustí, S. (2020). Low host abundance and high temperature determine switching from lytic to lysogenic cycles in planktonic microbial communities in a tropical sea (Red Sea). *Viruses, 12*, 761.

Ataeian, M., Liu, Y., Kouris, A., Hawley, A. K., & Strous, M. (2022). Ecological interactions of cyanobacteria and heterotrophs enhances the robustness of cyanobacterial consortium for carbon sequestration. *Frontiers in Microbiology, 13*, 780346.

Azam, F., Malfatti, F., 2007. Microbial structuring of marine ecosystems. Nat. Rev. Microbiol. 5, 782–791.

Banerji, A., & Benesh, K. (2022). Incorporating microbial species interaction in management of freshwater toxic cyanobacteria: A systems science challenge. *Ecologies, 3*, 570–587.

Baran, N., Goldin, S., Maidanik, I., & Lindell, D. (2018). Quantification of diverse virus populations in the environment using the polony method. *Nature Microbiology, 3*, 62–72.

Barrington, D. J., & Ghadouani, A. (2008). Application of hydrogen peroxide for the removal of toxic cyanobacteria and other phytoplankton from wastewater. *Environmental Science & Technology, 42*, 8916–8921.

Beate, R., Susanne, S., Torben, M., Claus-Dieter, D., Meinhard, S., & Thorsten, B. (2007). Effects of phytoplankton bloom in a coastal ecosystem on the composition of bacterial communities. *Aquatic Microbial Ecology, 48*, 47–60.

Berg, K. A., Lyra, C., Sivonen, K., Paulin, L., Suomalainen, S., Tuomi, P., & Rapala, J. (2009). High diversity of cultivable heterotrophic bacteria in association with cyanobacterial water blooms. *The ISME Journal, 3*, 314–325.

Bestion, E., Schaum, C.-E., & Yvon-Durocher, G. (2018). Nutrient limitation constrains thermal tolerance in freshwater phytoplankton. *Limnology and Oceanography Letters, 3*, 436–443.

Bhatt, P., Engel, B. A., Reuhs, M., & Simsek, H. (2023). Cyanophage technology in removal of cyanobacteria mediated harmful algal blooms: A novel and eco-friendly method. *Chemosphere, 315*, 137769.

Bird, D. F., & Karl, D. M. (1999). Uncoupling of bacteria and phytoplankton during the austral spring bloom in Gerlache Strait, Antarctic Peninsula. *Aquatic Microbial Ecology, 19*, 13–27.

Boix, D., Biggs, J., Céréghino, R., Hull, A. P., Kalettka, T., & Oertli, B. (2012). Pond research and management in Europe: "Small is Beautiful". *Hydrobiologia, 689*, 1–9.

Bratbak, G., & Thingstad, T. F. (1985). Phytoplankton-bacteria interactions: an apparant paradox? Analysis of a model system with both competition and commensalism. *Marine Ecology – Progress Series, 25*, 23–30.

Bratbak, G., Jacobsen, A., & Heldal, M. (1998). Viral lysis of Phaeocystis pouchetii and bacterial secondary production. *Aquatic Microbial Ecology, 16*, 11–16.

Brum, J. R., Hurwitz, B. L., Schofield, O., Ducklow, H. W., & Sullivan, M. B. (2016). Seasonal time bombs: dominant temperate viruses affect Southern Ocean microbial dynamics. *The ISME Journal, 10*, 437–449.

Brussaard, C. P. (2004). Viral control of phytoplankton populations–A review. *The Journal of Eukaryotic Microbiology, 51*, 125–138.

Brussow, H., Canchaya, C., & Hardt, W. D. (2004). Phages and the evolution of bacterial pathogens: From genomic rearrangements to lysogenic conversion. *Microbiology and Molecular Biology Reviews, 68*, 560–602. table of contents.

Buchan, A., González, J. M., & Moran, M. A. (2005). Overview of the marine roseobacter lineage. *Applied and Environmental Microbiology, 71*, 5665–5677.

Buchan, A., LeCleir, G. R., Gulvik, C. A., & González, J. M. (2014). Master recyclers: Features and functions of bacteria associated with phytoplankton blooms. *Nature Reviews Microbiology, 12*, 686–698.

Burford, M. A., Willis, A., Chuang, A., Man, X., & Orr, P. T. (2018). Recent insights into physiological responses to nutrients by the cylindrospermopsin producing cyanobacterium, Cylindrospermopsis raciborskii. *Journal of Oceanology and Limnology, 36*, 1032–1039.

Cai, F., Axen, S. D., & Kerfeld, C. A. (2013). Evidence for the widespread distribution of CRISPR-Cas system in the Phylum Cyanobacteria. *RNA Biology, 10*, 687–693.

Callieri, C. (2008). Picophytoplankton in freshwater ecosystems: The importance of small-sized phototrophs. *Freshwater Reviews, 1*(1–28), 28.

Callieri, C. (2017). Synechococcus plasticity under environmental changes. *FEMS Microbiology Letters, 364*.

Callieri, C., Cronberg, G., & Stockner, J. G. (2012). Freshwater picocyanobacteria: Single cells, microcolonies and colonial forms. In B. A. Whitton (Ed.), *Ecology of Cyanobacteria II: Their diversity in space and time*. Springer Netherlands.

Campbell, B. J., & Kirchman, D. L. (2013). Bacterial diversity, community structure and potential growth rates along an estuarine salinity gradient. *The ISME Journal, 7*, 210–220.

Castberg, T., Larsen, A., Sandaa, R. A., Brussaard, C. P. D., Egge, J. K., Heldal, M., Thyrhaug, R., van Hannen, E. J., & Bratbak, G. (2001). Microbial population dynamics and diversity during a bloom of the marine coccolithophorid Emiliania huxleyi (Haptophyta). *Marine Ecology Progress Series, 221*, 39–46.

Chen, Y., & Zeng, Q. (2020). Correction to: Temporal transcriptional patterns of cyanophage genes suggest synchronized infection of cyanobacteria in the oceans. *Microbiome, 8*, 109.

Chenard, C., Chan, A. M., Vincent, W. F., & Suttle, C. A. (2015). Polar freshwater cyanophage S-EIV1 represents a new widespread evolutionary lineage of phages. *The ISME Journal, 9*, 2046–2058.

Cheng, K., Frenken, T., Brussaard, C. P. D., & van de Waal, D. B. (2019). Cyanophage propagation in the freshwater cyanobacterium *Phormidium* is constrained by phosphorus limitation and enhanced by elevated pCO_2. *Frontiers in Microbiology, 10*, 617.

Chopyk, J., Nasko, D. J., Allard, S., Bui, A., Pop, M., Mongodin, E. F., & Sapkota, A. R. (2020). Seasonal dynamics in taxonomy and function within bacterial and viral metagenomic assemblages recovered from a freshwater agricultural pond. *Environmental Microbiomes, 15*(1), 18.

Clokie, M. R. J., & Mann, N. H. (2006). Marine cyanophages and light. *Environmental Microbiology, 8*, 2074–2082.

Coleman, M. L., Sullivan, M. B., Martiny, A. C., Steglich, C., Barry, K., DeLong, E. F., & Chisholm, S. W. (2006). Genomic islands and the ecology and evolution of *Prochlorococcus*. *Science, 311*, 1768–1770.

Cuber, P., Chooneea, D., Geeves, C., et al. (2023). Comparing the accuracy and efficiency of third generation sequencing technologies, Oxford Nanopore Technologies, and Pacific Biosciences, for DNA barcode sequencing applications. *Ecological Genetics and Genomics, 28*, 100181.

Dashkova, V., Malashenkov, D., Poulton, N., Vorobjev, I., & Barteneva, N. S. (2017). Imaging flow cytometry for phytoplankton analysis. *Methods, 112*, 188–200.

de Araujo, M. F. F., & Godinho, M. J. L. (2009). Short-term variations of virus-like particles in a tropical lake: relationship with microbial communities (bacteria, ciliates and flagellates). *Microbiological Research, 164*, 411–419.

Descy, J.-P., Hardy, M.-A., Sténuite, S., Pirlot, S., Leporcq, B., Kimirei, I., Sekadende, B., Mwaitega, S. R., & Sinyenza, D. (2005). Phytoplankton pigments and community composition in Lake Tanganyika. *Freshwater Biology, 50*, 668–684.

Dhib, A., Denis, M., Barani, A., Turki, S., & Aleya, L. (2016). Ultra- and microplankton assemblages as indicators of trophic status in a Mediterranean lagoon. *Ecological Indicators, 64*, 59–71.

Dignum, M., Matthijs, H. C. P., Pel, R., Laanbroek, H. J., & Mur, L. R. (2005). Nutrient limitation of freshwater Cyanobacteria. In J. Huisman, H. C. P. Matthijs, & P. M. Visser (Eds.), *Harmful Cyanobacteria*. Springer Netherlands.

Dillon, A., & Parry, J. D. (2008). Characterization of temperate cyanophages active against freshwater phycocyanin-rich Synechococcus species. *Freshwater Biology, 53*, 1253–1261.

Dion, M., Oechslin, F., & Moineau, S. (2020). Phage diversity, genomics and phylogeny. *Nature Reviews Microbiology, 18*, 125+.

Dion, M. B., Plante, P.-L., Zufferey, E., Shah, S. A., Corbeil, J., & Moineau, S. (2021). Streamlining CRISPR spacer-based bacterial host predictions to decipher the viral dark matter. *Nucleic Acids Research, 49*, 3127–3138.

Dorigo, U., Jacquet, S., & Humbert, J. F. (2004). Cyanophage diversity, inferred from g20 gene analyses, in the largest natural lake in France, Lake Bourget. *Applied and Environmental Microbiology, 70*, 1017–1022.

Doron, S., Fedida, A., Hernandez-Prieto, M. A., Sabehi, G., Karunker, I., Stazic, D., Feingersch, R., Steglich, C., Futschik, M., Lindell, D., & Sorek, R. (2016). Transcriptome dynamics of a broad host-range cyanophage and its hosts. *The ISME Journal, 10*, 1437–1455.

Doron, S., Melamed, S., Ofir, G., Leavitt, A., Lopatina, A., Keren, M., Amitai, G., & Sorek, R. (2018). Systematic discovery of antiphage defense systems in the microbial pangenome. *Science, 359*, eaar4120.

Dreher, T. W., Brown, N., Bozarth, C. S., Schwartz, A. D., Riscoe, E., Thrash, C., Bennett, S. E., Tzeng, S. C., & Maier, C. S. (2011). A freshwater cyanophage whose genome indicates close relationships to photosynthetic marine cyanomyophages. *Environmental Microbiology, 13*, 1858–1874.

Du, K., Yang, F., Zhang, J.-T., Yu, R.-C., Deng, Z., Li, W.-F., Chen, Y., Li, Q., & Zhou, C.-Z. (2022). Comparative genomic analysis of five freshwater cyanophages and reference-guided metagenomic data mining. *Microbiome, 10*, 128.

Fandino, L. B., Riemann, L., Steward, G., Long, R., & Azam, F. (2001). Variations in bacterial community structure during a dinoflagellate bloom analyzed by DGGE and 16S rDNA sequencing. *Aquatic Microbial Ecology, 23*, 119–130.

Flood, M., Frabutt, D., Floyd, D., Powers, E. U., Devol, A., & Tiquia-Arashiro, S. M. (2015). Ammonia-oxidizing bacteria and archaea in sediments of the Gulf of Mexico. *Environmental Technology, 36*, 124–135. https://doi.org/10.1080/09593330.2014.942385

Fridman, S., Flores-Uribe, J., Larom, S., et al. (2017). A myovirus encoding both photosystem I and II proteins enhances cyclic electron flow in infected *Prochlorococcus* cells. *Nature Microbiology, 2*, 1350–1357.

Fuchsman, C. A., Palevsky, H. I., Widner, B., et al. (2019). Cyanobacteria and cyanophage contributions to carbon and nitrogen cycling in an oligotrophic oxygen-deficient zone. *The ISME Journal, 13*, 2714–2726.

Fuhrman, J. A. (1999). Marine viruses and their biogeochemical and ecological effects. *Nature, 399*, 541–548.

Fuhrman, J. A., Cram, J. A., & Needham, D. M. (2015). Marine microbial community dynamics and their ecological interpretation. *Nature Reviews Microbiology, 13*, 133–146.

Galand, P. E., Casamayor, E. O., Kirchman, D. L., & Lovejoy, C. (2009). Ecology of the rare microbial biosphere of the Arctic Ocean. *Proceedings of the National Academy of Sciences, 106*, 22427–22432.

Gao, E. B., Gui, J. F., & Zhang, Q. Y. (2012). A novel cyanophage with a cyanobacterial nonbleaching protein A gene in the genome. *Journal of Virology, 86*, 236–245.

Gao, E. B., Huang, Y., & Ning, D. (2016). Metabolic Genes within Cyanophage Genomes: Implications for diversity and evolution. *Genes, 7*, 80.

Georges, A. A., El-Swais, H., Craig, S. E., Li, W. K., & Walsh, D. A. (2014). Metaproteomic analysis of a winter to spring succession in coastal northwest Atlantic Ocean microbial plankton. *The ISME Journal, 8*, 1301–1313.

Gin, K. Y., Sim, Z. Y., Goh, K. C., Kok, J. W. K., Te, S. H., Tran, N. H., Li, W., & He, Y. (2021). Novel cyanotoxin-producing Synechococcus in tropical lakes. *Water Research, 192*, 116828.

Grasso, C. R., Pokrzywinski, K. L., Waechter, C., Rycroft, T., Zhang, Y., Aligata, A., Kramer, M., & Lamsal, A. (2022). A review of cyanophage-host relationships: Highlighting cyanophages as a potential cyanobacteria control strategy. *Toxins, 14*, 385.

Gu, X., Tay, Q. X. M., Te, S. H., Saeidi, N., Goh, S. G., Kushmaro, A., Thompson, J. R., & Gin, K. Y.-H. (2018). Geospatial distribution of viromes in tropical freshwater ecosystems. *Water Research, 137*, 220–232.

Haverkamp, T. H. A., Schouten, D., Doeleman, M., Wollenzien, U., Huisman, J., & Stal, L. J. (2009). Colorful microdiversity of Synechococcus strains (picocyanobacteria) isolated from the Baltic Sea. *The ISME Journal, 3*, 397–408.

Hoikkala, L., Tammert, H., Lignell, R., Eronen-Rasimus, E., Spilling, K., and Kisand, V. (2016). Autochthonous Dissolved Organic Matter Drives Bacterial Community Composition during a Bloom of Filamentous Cyanobacteria. Front. Mar. Sci. 3:111. doi: https://doi.org/10.3389/fmars.2016.00111

Hopkinson, C. S., & Vallino, J. J. (2005). Efficient export of carbon to the deep ocean through dissolved organic matter. *Nature, 433*, 142–145.

Huang, S.-L., & Lin, Y.-S. (2017). The size stability of alginate beads by different ionic crosslinkers. *Advances in Materials Science and Engineering, 2017*, 9304592.

Huang, S., Wilhelm, S. W., Jiao, N., & Chen, F. (2010). Ubiquitous cyanobacterial podoviruses in the global oceans unveiled through viral DNA polymerase gene sequences. *The ISME Journal, 4*, 1243–1251.

Jameson, E., Mann, N. H., Joint, I., Sambles, C., & Mühling, M. (2011). The diversity of cyanomyovirus populations along a North–South Atlantic Ocean transect. *The ISME Journal, 5*, 1713–1721.

Jaskulska, A., Nájera, A. F., Czarny, P., Serwecińska, L., & Mankiewicz-Boczek, J. (2021). Daily dynamic of transcripts abundance of Ma-LMM01-like cyanophages in two lowland European reservoirs. *Ecohydrology & Hydrobiology, 21*, 543–548.

Jia, Y., Shan, J., Millard, A., Clokie, M. R., & Mann, N. H. (2010). Light-dependent adsorption of photosynthetic cyanophages to Synechococcus sp. WH7803. *FEMS Microbiology Letters, 310*, 120–126.

Jiang, S. C., & Paul, J. H. (1998). Significance of lysogeny in the marine environment: Studies with isolates and a model of lysogenic phage production. *Microbial Ecology, 35*, 235–243.

Jiao, N., Herndl, G. J., Hansell, D. A., et al. (2010). Microbial production of recalcitrant dissolved organic matter: long-term carbon storage in the global ocean. *Nature Reviews. Microbiology, 8*, 593–599.

Johnson, M. F., Albertson, L. K., Algar, A. C., et al. (2024). Rising water temperature in rivers: ecological impacts and future resilience. *WIREs Water, 11*(4), e1724.

Kaletta, J., Pickl, C., Grievler, C., Klingl, A., Kurmayer, R., & Deng, L. (2020). A rigorous assessment and comparison of enumeration methods for environmental viruses. *Scientific Reports, 10*, 18625.

Kashtan, N., Roggensack, S. E., Rodrigue, S., et al. (2014). Single-cell genomics reveals hundreds of coexisting subpopulations in wild Prochlorococcus. *Science, 344*, 416–420.

Khawaja, K. A., Abbas, Z., Rehman, S., & u. (2016). Isolation and characterization of lytic phages TSE1-3 against Enterobacter cloacae. *Open Life Sciences, 11*, 287–292.

Kimura, S., Yoshida, T., Hosoda, N., Honda, T., Kuno, S., Kamiji, R., Hashimoto, R., & Sako, Y. (2012). Diurnal infection patterns and impact of Microcystis cyanophages in a Japanese pond. *Applied and Environmental Microbiology, 78*, 5805–5811.

Kinet, R., Dzaomuho, P., Baert, J., Taminiau, B., Daube, G., Nezer, C., Brostaux, Y., Nguyen, F., Dumont, G., Thonart, P., & Delvigne, F. (2016). Flow cytometry community fingerprinting

and amplicon sequencing for the assessment of landfill leachate cellulolytic bioaugmentation. *Bioresource Technology, 214*, 450–459.

Knowles, B., Silveira, C. B., Bailey, B. A., et al. (2016). Lytic to temperate switching of viral communities. *Nature, 531*, 466–470.

Kolodny, Y., Avrahami, Y., Zer, H., Frada, M. J., Paltiel, Y., & Keren, N. (2022). Phycobilisome light-harvesting efficiency in natural populations of the marine cyanobacteria Synechococcus increases with depth. *Communications Biology, 5*, 727.

Kujawinski, E. B. (2011). The impact of microbial metabolism on marine dissolved organic matter. *Annual Review of Marine Science, 3*, 567–599.

Labrie, S. J., Samson, J. E., & Moineau, S. (2010). Bacteriophage resistance mechanisms. *Nature Reviews. Microbiology, 8*, 317–327.

Larkum, N. W. (1929). Bacteriophage treatement of staphylococcus infections. *The Journal of Infectious Diseases, 45*, 34–41.

LeCleir, G. R., DeBruyn, J. M., Maas, E. W., Boyd, P. W., & Wilhelm, S. W. (2014). Temporal changes in particle-associated microbial communities after interception by nonlethal sediment traps. *FEMS Microbiology Ecology, 87*, 153–163.

Lee, L. H., Lui, D., Platner, P. J., Hsu, S. F., Chu, T. C., Gaynor, J. J., Vega, Q. C., & Lustigman, B. K. (2006). Induction of temperate cyanophage AS-1 by heavy metal—copper. *BMC Microbiology, 6*, 17.

Lester, S., Carollyn, H., Philippa, L., & Davies-Colley, R. J. (2002). Sunlight inactivation of fecal indicator bacteria and bacteriophages from waste stabilization pond effluent in fresh and saline waters. *Applied and Environmental Microbiology, 68*, 1122.

Li, S., Shi, X., Lepère, C., Liu, M., Wang, X., & Kong, F. (2016). Unexpected predominance of photosynthetic picoeukaryotes in shallow eutrophic lakes. *Journal of Plankton Research, 38*, 830–842.

Li, R., Zhu, L., Cui, L., & Zhu, Y. G. (2022). Viral diversity and potential environmental risk in microplastic at watershed scale: Evidence from metagenomic analysis of plastisphere. *Environment International, 161*, 107146.

Liu, T., Zhang, A. N., Wang, J., Liu, S., Jiang, X., Dang, C., Ma, T., Liu, S., Chen, Q., Xie, S., Zhang, T., & Ni, J. (2018). Integrated biogeography of planktonic and sedimentary bacterial communities in the Yangtze River. *Microbiome, 6*, 16.

Liu, R., Liu, Y., Chen, Y., Zhan, Y., & Zeng, Q. (2019). Cyanobacterial viruses exhibit diurnal rhythms during infection. *Proceedings of the National Academy of Sciences of the United States of America, 116*, 14077–14082.

Liu, J., Chen, D., Mao, G., Irannezhad, M., & Pokhrel, Y. (2022). Past and future changes in climate and water resources in the Lancang-Mekong River Basin: current understanding and future research directions. *Engineering, 13*, 144–152.

Loh, L. (2017). *Dynamics of microcystis-infecting cyanophages in tropical freshwaters through qpcr and next generation sequencing*. Degree of doctor of philosophy. National University of Singapore.

Malits, A., Christaki, U., Obernosterer, I., & Weinbauer, M. G. (2014). Enhanced viral production and virus-mediated mortality of bacterioplankton in a natural iron-fertilized bloom event above the Kerguelen Plateau. *Biogeosciences, 11*, 6841–6853.

Marcotullio, P. J., Kebler, C., Gonzalez, R. Q., & Schmeltz, M. (2021). Urban growth and heat in tropical climates. *Frontiers in Ecology and Evolution, 9*.

Matteson, A. R., Loar, S. N., Bourbonniere, R. A., & Wilhelm, S. W. (2011). Molecular enumeration of an ecologically important cyanophage in a laurentian great lake. *Applied and Environmental Microbiology, 77*, 6772–6779.

McDowell, W. H., McDowell, W. G., Potter, J. D., & Ramírez, A. (2019 Mar). Nutrient export and elemental stoichiometry in an urban tropical river. *Ecological Applications, 29*(2), e01839. https://doi.org/10.1002/eap.1839

McKindles, K. M., Manes, M. A., DeMarco, J. R , McClure, A., McKay, R. M., Davis, T. W., Bullerjahn, G. S., & Schaffner, D. W. (2020). Dissolved microcystin release coincident with

lysis of a bloom dominated by *Microcystis* spp. in western lake erie attributed to a novel cyanophage. *Applied and Environmental Microbiology, 86*, e01397–e01320.

Meng, L.-H., Ke, F., Zhang, Q.-Y., & Zhao, Z. (2023). Biological and genomic characteristics of MaMV-DH01, a novel freshwater *Myoviridae* Cyanophage strain. *Microbiology Spectrum, 11*, e02888–e02822.

Mitra, A., Flynn, K. J., Burkholder, J. M., Berge, T., Calbet, A., Raven, J. A., Granéli, E., Glibert, P. M., Hansen, P. J., Stoecker, D. K., Thingstad, F., Tillmann, U., Våge, S., Wilken, S., & Zubkov, M. V. (2014). The role of mixotrophic protists in the biological carbon pump. *Biogeosciences, 11*, 995–1005.

Molineux, I. J. (1991). Host-parasite interactions: recent developments in the genetics of abortive phage infections. *The New Biologist, 3*, 230–236.

Moon, K., Jeon, J. H., Kang, I., Park, K. S., Lee, K., Cha, C.-J., Lee, S. H., & Cho, J.-C. (2020). Freshwater viral metagenome reveals novel and functional phage-borne antibiotic resistance genes. *Microbiome, 8*, 75–75.

Moran, M. A., Kujawinski, E. B., Schroer, W. F., et al. (2022). Microbial metabolites in the marine carbon cycle. *Nature Microbiology, 7*, 508–523.

Morimoto, D., Kimura, S., Sako, Y., & Yoshida, T. (2018). Transcriptome analysis of a bloom-forming cyanobacterium *Microcystis aeruginosa* during Ma-LMM01 phage infection. *Frontiers in Microbiology, 9*.

Naknaen, A., Suttinun, O., Surachat, K., Khan, E., & Pomwised, R. (2021). A novel jumbo phage PhiMa05 inhibits harmful Microcystis sp. *Frontiers in Microbiology, 12*, 660351.

Nassereddine, Z., Opara, S., Coutinho, O., Qyteti, F., Book, R., Heinicke, M. P., Napieralski, J., & Tiquia-Arashiro, S. M. (2024). Critical perspectives on advancing antibiotic resistant gene (ARG) detection technologies in aquatic ecosystems. *Science of the Total Environment, 957*(20), 177775. https://doi.org/10.1016/j.scitotenv.2024.177775

Needham, D. M., & Fuhrman, J. A. (2016). Pronounced daily succession of phytoplankton, archaea and bacteria following a spring bloom. *Nature Microbiology, 1*, 16005.

Needham, D. M., Sachdeva, R., & Fuhrman, J. A. (2017). Ecological dynamics and co-occurrence among marine phytoplankton, bacteria and myoviruses shows microdiversity matters. *The ISME Journal, 11*, 1614–1629.

Ni, T., & Zeng, Q. (2016). Diel infection of cyanobacteria by cyanophages. *Frontiers in Marine Science, 2*, 123.

Nie, X.-P., Liu, B.-Y., Yu, H.-J., Liu, W.-Q., & Yang, Y.-F. (2013). Toxic effects of erythromycin, ciprofloxacin and sulfamethoxazole exposure to the antioxidant system in *Pseudokirchneriella subcapitata. Environmental Pollution, 172*, 23–32.

Ning, M., Li, H., Xu, Z., Chen, L., & He, Y. (2021). Picophytoplankton identification by flow cytometry and high-throughput sequencing in a clean reservoir. *Ecotoxicology and Environmental Safety, 216*, 112216.

Nishikawa, H., Yasuda, M., Uchiyama, J., et al. (2008). T-even-related bacteriophages as candidates for treatment of Escherichia coli urinary tract infections. *Archives of Virology, 153*, 507–515.

Niu, X., Cheng, K., Rong, Q., Min, X. U., & Zhao, Y. (2012). The combined effects of elevated CO_2 and elevated temperature on proliferation of cyanophage PP. *Acta Ecologica Sinica.*

Noronha, J. M., Gauns, M. U., & Ghadi, S. C. (2022). Isolation of a novel cyanophage infecting potentially bloom-forming Limnothrix sp. *Limnologica, 93*, 125961.

Oest, A., Alsaffar, A., Fenner, M., Azzopardi, D., & Tiquia-Arashiro, S. M. (2018 May). Patterns of Change in Metabolic Capabilities of Sediment Microbial Communities in River and Lake Ecosystems. *International Journal of Microbiology, 27*(2018), 6234931. https://doi.org/10.1155/2018/6234931

Ogawa, H., Amagai, Y., Koike, I., Kaiser, K., & Benner, R. (2001). Production of refractory dissolved organic matter by bacteria. *Science, 292*, 917–920.

Özel Duygan, B. D., Hadadi, N., Babu, A. F., Seyfried, M., & van der Meer, J. R. (2020). Rapid detection of microbiota cell type diversity using machine-learned classification of flow cytometry data. *Communications Biology, 3*, 379.

Padan, E., Shilo, M., & Kislev, N. (1967). Isolation of "cyanophages" from freshwater ponds and their interaction with Plectonema boryanum. *Virology, 32*, 234–246.

Pagnucco, G., Overfield, D., Chamlee, Y., Fortuna, A., Sulaiman, F., Farinas, J., Schittenhelm, R., Shuler, C., Kassem, A., Opara, S., Najaf, H., Abbas, L., Coutinho, O., Catalfano, B., Li, X., & Tiquia-Arashiro, S. M. (2023). Metal tolerance and biosorption capacities of bacterial strains isolated from an urban watershed. *Frontiers in Microbiology, 14*, 1278886. https://doi.org/10.3389/fmicb.2023.1278886

Pak, H. Y., Chuah, C. J., Yong, E. L., & Snyder, S. A. (2021). Effects of land use configuration, seasonality and point source on water quality in a tropical watershed: a case study of the Johor River Basin. *Science of the Total Environment, 780*, 146661.

Papoulis, S. E., Wilhelm, S. W., Talmy, D., Zinser, E. R., & Moran, M. A. (2021). Nutrient loading and viral memory drive accumulation of restriction modification systems in bloom-forming Cyanobacteria. *MBio, 12*, e00873–e00821.

Park, H., Bulzu, P.-A., Shabarova, T., Kavagutti, V. S., Ghai, R., Kasalicky, V., & Jezberova, J. (2024). Uncovering the genomic basis of symbiotic interactions and niche adaptations in freshwater picocyanobacteria. *Microbiome, 12*.

Patel, D., Gismondi, R., Alsaffar, A., & Tiquia-Arashiro, S. M. (2019). Applicability of API ZYM to capture seasonal and spatial variabilities in lake and river sediments. *Environmental Technology, 40*(24), 3227–3239. https://doi.org/10.1080/09593330.2018.1468492

Patten, N. L., van Ruth, P. D., & Rodriguez, A. R. (2018). Spatial variability in picophytoplankton, bacteria and viruses in waters of the Great Australian Bight (southern Australia). *Deep Sea Research Part II: Topical Studies in Oceanography, 157–158*, 46–57.

Patriarca, C., Sedano-Núñez, V.T., Garcia, S.L., Bergquist, J., Bertilsson, S., Sjöberg, P.J.R., Tranvik, L.J. and Hawkes, J.A. (2021). Character and environmental lability of cyanobacteria-derived dissolved organic matter. *Limnol Oceanogr, 66*: 496-509. https://doi.org/10.1002/lno.11619

Paul, J. H., & Kellogg, C. A. (2000). Ecology of bacteriophages in nature. *Viral Ecology*.

Pearce, B. J., & Chertow, M. (2017). Scenarios for achieving absolute reductions in phosphorus consumption in Singapore. *Journal of Cleaner Production, 140*, 1587–1601.

Peduzzi, P., & Schlemer, F. (2004). Bacteria and viruses in the water column of tropical freshwater reservoirs. *Environmental Microbiology, 6*, 707–715.

Pernthaler, J. (2005). Predation on prokaryotes in the water column and its ecological implications. *Nature Reviews. Microbiology, 3*, 537–546.

Pingoud, A., Fuxreiter, M., Pingoud, V., & Wende, W. (2005). Type II restriction endonucleases: Structure and mechanism. *Cellular and Molecular Life Sciences, 62*, 685.

Pinhassi, J., Sala, M. M., Havskum, H., Peters, F., Guadayol, O., Malits, A., & Marrasé, C. (2004). Changes in bacterioplankton composition under different phytoplankton regimens. *Applied and Environmental Microbiology, 70*, 6753–6766.

Polz, M. F., Hunt, D. E., Preheim, S. P., & Weinreich, D. M. (2006). Patterns and mechanisms of genetic and phenotypic differentiation in marine microbes. *Philosophical Transactions of the Royal Society of London. Series B, Biological Sciences, 361*, 2009–2021.

Pound, H. L., Martin, R. M., Sheik, C. S., et al. (2021). Environmental Studies of Cyanobacterial Harmful Algal Blooms Should Include Interactions with the Dynamic Microbiome. *Environmental Science & Technology, 55*, 12776–12779.

Pradeep Ram, A. S., & Sime-Ngando, T. (2023). Differential effects of viruses on the growth efficiency of freshwater bacterioplankton in eutrophic relative to non-eutrophic lakes. *Microorganisms, 11*, 384.

Pradeep Ram, A. S., Keshri, J., & Sime-Ngando, T. (2019). Distribution patterns of bacterial communities and their potential link to variable viral lysis in temperate freshwater reservoirs. *Aquatic Sciences, 81*, 72.

Pradeep Ram, A. S., Keshri, J., & Sime-Ngando, T. (2020). Differential impact of top-down and bottom-up forces in structuring freshwater bacterial communities. *FEMS Microbiology Ecology, 96*.

Props, R., Kerckhof, F.-M., Rubbens, P., et al. (2017). Absolute quantification of microbial taxon abundances. *The ISME Journal, 11*, 584–587.

Reignier, O., Bormans, M., Herve, F., et al. (2024). Spatio-temporal connectivity of a toxic cyanobacterial community and its associated microbiome along a freshwater-marine continuum. *Harmful Algae, 134*, 102627.

Riede, I., & Eschbach, M.-L. (1986). Evidence that TraT interacts with OmpA of *Escherichia coli*. *FEBS Letters, 205*, 241–245.

Riemann, L., Steward, G. F., & Azam, F. (2000). Dynamics of bacterial community composition and activity during a mesocosm diatom bloom. *Applied and Environmental Microbiology, 66*, 578–587.

Rockey, N. C., Henderson, J. B., Chin, K., Raskin, L., & Wigginton, K. R. (2021). Predictive modeling of virus inactivation by UV. *Environmental Science and Technology, 55*.

Ruber, J., Bauer, F. R., Millard, A. D., Raeder, U., Geist, J., & Zwirglmaier, K. (2016). Synechococcus diversity along a trophic gradient in the Osterseen Lake District, Bavaria. *Microbiology (Reading), 162*, 2053–2063.

Ruber, J., Geist, J., Hartmann, M., Millard, A., Raeder, U., Zubkov, M., & Zwirglmaier, K. (2018). Spatio-temporal distribution pattern of the picocyanobacterium *Synechococcus* in lakes of different trophic states: a comparison of flow cytometry and sequencing approaches. *Hydrobiologia, 811*, 77–92.

Safferman, R. S., Cannon, R. E., Desjardins, P. R., Gromov, B. V., Haselkorn, R., Sherman, L. A., & Shilo, M. (1983). Classification and nomenclature of viruses of Cyanobacteria. *Intervirology, 19*, 61–66.

Sakowski, E. G., Arora-Williams, K., Tian, F., Zayed, A. A., Zablocki, O., Sullivan, M. B., & Preheim, S. P. (2021). Interaction dynamics and virus–host range for estuarine actinophages captured by epicPCR. *Nature Microbiology, 6*, 630–642.

Schallenberg, M., Burns, W., & C. (2001). Tests of autotrophic picoplankton as early indicators of nutrient enrichment in an ultra-oligotrophic lake. *Freshwater Biology, 46*, 27–37.

Schneider, I. R., Diener, T. O., & Safferman, R. S. (1964). Blue-green algal virus LPP-1: purification and partial characterization. *Science, 144*, 1127–1130.

Schwartz, D. A., & Lindell, D. (2017). Genetic hurdles limit the arms race between Prochlorococcus and the T7-like podoviruses infecting them. *The ISME Journal, 11*, 1836–1851.

Shi, J., He, S., Zhao, L., Ji, L., Yang, S., & Wu, Z. (2022). Physiological and molecular responses of invasive cyanobacterium *Raphidiopsis raciborskii* to ambient phosphorus deficiency. *Journal of Oceanology and Limnology, 40*, 1792–1803.

Shitrit, D., Hackl, T., Laurenceau, R., et al. (2021). Genetic engineering of marine cyanophages reveals integration but not lysogeny in T7-like cyanophages. *The ISME Journal, 16*, 488–499.

Sieradzki, E. T., Ignacio-Espinoza, J. C., Needham, D. M., Fichot, E. B., & Fuhrman, J. A. (2019). Dynamic marine viral infections and major contribution to photosynthetic processes shown by spatiotemporal picoplankton metatranscriptomes. *Nature Communications, 10*, 1169.

Sim, Z. Y., Goh, K. C., Sukarji, N. H. B., Mao, F., He, Y., & Gin, K. Y.-H. (2025). Influence of phytoplankton, bacteria and viruses on nutrient supply in tropical waters. *Journal of Environmental Sciences, 151*, 174–186.

Smith, D. C., Simon, M., Alldredge, A. L., & Azam, F. (1992). Intense hydrolytic enzyme activity on marine aggregates and implications for rapid particle dissolution. *Nature, 359*, 139–142.

Smith, D. J., Berry, M. A., Cory, R. M., Johengen, T. H., Kling, G. W., Davis, T. W., & Dick, G. J. (2022). Heterotrophic bacteria dominate catalase expression during *Microcystis* blooms. *Applied and Environmental Microbiology, 88*, e02544–e02521.

Sogin, M. L., Morrison, H. G., Huber, J. A., Welch, D. M., Huse, S. M., Neal, P. R., Arrieta, J. M., & Herndl, G. J. (2006). Microbial diversity in the deep sea and the underexplored "rare

biosphere". *Proceedings of the National Academy of Sciences of the United States of America, 103*, 12115–12120.

Song, H., Qin, T., Wang, J., & Wong, T. H. F. (2019). Characteristics of stormwater quality in Singapore catchments in 9 different types of land use. *Water, 11*, 1089.

Steven, et al. (1999). Viruses and Nutrient Cycles in the Sea Viruses play critical roles in the structure and function of aquatic food webs. Bioscience.

Steenhauer, L. M., Wierenga, J., Carreira, C., Limpens, R., Koster, A. J., Pollard, P. C., & Brussaard, C. P. D. (2016). Isolation of cyanophage CrV infecting Cylindrospermopsis raciborskii and the influence of temperature and irradiance on CrV proliferation. *Aquatic Microbial Ecology, 78*, 11–23.

Steffen, M. M., Davis, T. W., McKay, R. M. L., et al. (2017). Ecophysiological examination of the lake erie microcystis bloom in 2014: Linkages between biology and the water supply shutdown of Toledo, OH. *Environmental Science & Technology, 51*, 6745–6755.

Stomp, M., Huisman, J., Vörös, L., Pick, F. R., Laamanen, M., Haverkamp, T., & Stal, L. J. (2007). Colourful coexistence of red and green picocyanobacteria in lakes and seas. *Ecology Letters, 10*, 290–298.

Stough, J. M. A., Tang, X., Krausfeldt, L. E., Steffen, M. M., Gao, G., Boyer, G. L., & Wilhelm, S. W. (2017). Molecular prediction of lytic vs lysogenic states for *Microcystis* phage: Metatranscriptomic evidence of lysogeny during large bloom events. *PLoS One, 12*, e0184146.

Stummeyer, K., Schwarzer, D., Claus, H., Vogel, U., & Mühlenhoff, M. (2006). Evolution of bacteriophages infecting encapsulated bacteria: Lessons from *Escherichia coli* K1-specific phages. *Molecular Microbiology, 60*, 1123–1135.

Subehi, L., Ismail, S. N., Ridwansyah, I., Hamid, M. A., & Mansor, M. (2018). Analysis of the influence of reservoirs utilisation to water quality profiles in Indonesia (Saguling–Jatiluhur) and Malaysia (Temengor–Chenderoh) with special references to cascade reservoirs. *IOP Conference Series: Earth and Environmental Science, 118*, 012025.

Sudek, S., Everroad, R. C., Gehman, A. L., Smith, J. M., Poirier, C. L., Chavez, F. P., & Worden, A. Z. (2015). Cyanobacterial distributions along a physico-chemical gradient in the Northeastern Pacific Ocean. *Environmental Microbiology, 17*, 3692–3707.

Tang, X., Xie, G., Shao, K., Hu, Y., Cai, J., Bai, C., Gong, Y., & Gao, G. (2020). Contrast diversity patterns and processes of microbial community assembly in a river-lake continuum across a catchment scale in northwestern China. *Environmental Microbiomes, 15*, 10.

Te, S. H., Tan, B. F., Thompson, J. R., & Gin, K. Y.-H. (2017). Relationship of microbiota and cyanobacterial secondary metabolites in planktothricoides-dominated bloom. *Environmental Science & Technology, 51*, 4199–4209.

Te, S. H., Kok, J. W. K., Luo, R., et al. (2023). Coexistence of *Synechococcus* and *Microcystis* blooms in a tropical urban reservoir and their links with microbiomes. *Environmental Science & Technology, 57*, 1613–1624.

Teeling, H., Fuchs, B. M., Becher, D., et al. (2012). Substrate-controlled succession of marine bacterioplankton populations induced by a phytoplankton bloom. *Science, 336*, 608–611.

Thompson, L. R., Zeng, Q., Kelly, L., Huang, K. H., Singer, A. U., Stubbe, J., & Chisholm, S. W. (2011). Phage auxiliary metabolic genes and the redirection of cyanobacterial host carbon metabolism. *Proceedings of the National Academy of Sciences of the United States of America, 108*, E757–E764.

Tiquia, S. M. (2010). Salt-adapted bacteria isolated from the Rouge River and potential for degradation of contaminants and biotechnological applications. *Environmental Technology, 31*, 967–978.

Traving, S. J., Clokie, M. R. J., & Middelboe, M. (2014). Increased acidification has a profound effect on the interactions between the cyanobacterium *Synechococcus* sp. WH7803 and its viruses. *FEMS Microbiology Ecology, 87*, 133–141.

Vieillard, A. M., Newell, S. E., & Thrush, S. F. (2020). Recovering from bias: a call for futher study of underrepresented tropical and low-nutrient estuaries. *Journal of Geophysical Research: Biogeosciences, 125*(7), e2020JG005766.

Vörös, L., Callieri, C., Balogh, K. V., & Bertoni, R. (1998). *Freshwater picocyanobacteria along a trophic gradient and light quality range* (pp. 117–125). Springer.

Vörös, L., Mózes, A., & Somogyi, B. (2009). A five-year study of autotrophic winter picoplankton in Lake Balaton, Hungary. *Aquatic Ecology, 43*, 727–734.

Waldbauer, J. R., Coleman, M. L., Rizzo, A. I., Campbell, K. L., Lotus, J., & Zhang, L. (2019). Nitrogen sourcing during viral infection of marine cyanobacteria. *Proceedings of the National Academy of Sciences of the United States of America, 116*, 15590–15595.

Wang, K., & Chen, F. (2008). Prevalence of highly host-specific cyanophages in the estuarine environment. *Environmental Microbiology, 10*, 300–312.

Wang, K., Mou, X., Cao, H., Struewing, I., Allen, J., & Lu, J. (2021). Co-occurring microorganisms regulate the succession of cyanobacterial harmful algal blooms. *Environmental Pollution, 288*, 117682.

Wiederkehr, F., Wilkinson, C. L., Zeng, Y., Yeo, D. C. J., Ewers, R. M., & O'Gorman, E. J. (2020). Urbanisation affects ecosystem functioning more than structure in tropical streams. *Biological Conservation, 249*, 108634.

Winder, M., & Sommer, U. (2012). Phytoplankton response to a changing climate. *Hydrobiologia, 698*, 5–16.

Xia, H., Wang, M., Ge, X., Wu, Y., Yang, X., Zhang, Y., Li, T., & Shi, Z. (2013). Study of the dynamics of *Microcystis aeruginosa* and its cyanophage in East Lake using quantitative PCR. *Virologica Sinica, 28*, 309–311.

Xiong, H., Li, Y., & Zhang, Q. (2014). Mechanical removal and dewatering of cyanobacteria blooms in Chaohu Lake. *Chinese Journal of Environmental Engineering, 8*, 599–604.

Xu, Z., Te, S. H., He, Y., & Gin, K. Y.-H. (2018). The characteristics and dynamics of cyanobacteria-heterotrophic bacteria between two estuarine reservoirs—tropical versus sub-tropical regions. *Frontiers in Microbiology, 9*.

Yang, Y., Gu, X., Te, S. H., Goh, S. G., Mani, K., He, Y., & Gin, K. Y. (2019). Occurrence and distribution of viruses and picoplankton in tropical freshwater bodies determined by flow cytometry. *Water Research, 149*, 342–350.

Yang, Y., Chen, J., Chen, X., Jiang, Q., Liu, Y., & Xie, S. (2021 Sep 1). Cyanobacterial bloom induces structural and functional succession of microbial communities in eutrophic lake sediments. *Environmental Pollution, 284*, 117157. https://doi.org/10.1016/j.envpol.2021.117157

Ye, T., Zhao, Z., Bai, L., Song, N., & Jiang, H. (2020 Dec). Characteristics and bacterial community dynamics during extracellular polymeric substance (EPS) degradation of cyanobacterial blooms. *Science of the Total Environment, 15*(748), 142309. https://doi.org/10.1016/j.scitotenv.2020.142309

Yema, L., Litchman, E., & de Tezanos Pinto, P. (2016). The role of heterocytes in the physiology and ecology of bloom-forming harmful cyanobacteria. *Harmful Algae, 60*, 131–138.

Yoshida, M., Yoshida, T., Kashima, A., Takashima, Y., Hosoda, N., Nagasaki, K., & Hiroishi, S. (2008a). Ecological dynamics of the toxic bloom-forming cyanobacterium Microcystis aeruginosa and its cyanophages in freshwater. *Applied and Environmental Microbiology, 74*, 3269–3273.

Yoshida, T., Nagasaki, K., Takashima, Y., Shirai, Y., Tomaru, Y., Takao, Y., Sakamoto, S., Hiroishi, S., & Ogata, H. (2008b). Ma-LMM01 infecting toxic Microcystis aeruginosa illuminates diverse cyanophage genome strategies. *Journal of Bacteriology, 190*, 1762–1772.

Yoshida, T., Nishimura, Y., Watai, H., Haruki, N., Morimoto, D., Kaneko, H., Honda, T., Yamamoto, K., Hingamp, P., Sako, Y., Goto, S., & Ogata, H. (2018). Locality and diel cycling of viral production revealed by a 24 h time course cross-omics analysis in a coastal region of Japan. *The ISME Journal, 12*, 1287–1295.

Yoshida-Takashima, Y., Yoshida, M., Ogata, H., Nagasaki, K., Hiroishi, S., & Yoshida, T. (2012). Cyanophage infection in the bloom-forming cyanobacteria *Microcystis aeruginosa* in surface freshwater. *Microbes and Environments, 27*, 350–355.

Zaleski, P. (2005). The role of Dam methylation in phase variation of Haemophilus influenzae genes involved in defence against phage infection. *Microbiology, 151*, 3361–3369.

Zambrano, L., Contreras, V., Mazari-Hiriart, M., et al. (2009). Spatial heterogeneity of water quality in a highly degraded tropical freshwater ecosystem. *Environmental Management, 43*, 249–263. https://doi.org/10.1007/s00267-008-9216-1

Zaragoza-Solas, A., Haro-Moreno, J. M., Rodriguez-Valera, F., & López-Pérez, M. (2022). Long-read metagenomics improves the recovery of viral diversity from complex natural marine samples. *mSystems, 7*, e00192–e00122.

Zarzyczny, Karolina & Rius, Marc & Williams, Suzanne & Fenberg, Phillip. (2023). The ecological and evolutionary consequences of tropicalisation. Trends in Ecology & Evolution, 39. https://doi.org/10.1016/j.tree.2023.10.006

Zborowsky, S., & Lindell, D. (2019). Resistance in marine cyanobacteria differs against specialist and generalist cyanophages. *Proceedings of the National Academy of Sciences of the United States of America, 116*, 16899–16908.

Zhang, Y. H., Jiang, H., Cheng, K., & Zhao, Y. J. (2009). *The influence of different kinds of sunlight conditions on the adsorption rate, latent period and burst size of Cyanophage PP infecting its wild host*. Huazhong Normal University Journal of Postgraduates.

Zhang, D., You, F., He, Y., Te, S. H., & Gin, K. Y.-H. (2020). Isolation and characterization of the first freshwater cyanophage infecting *Pseudanabaena*. *Journal of Virology, 94*. https://doi.org/10.1128/jvi.00682-20

Zhang, D., He, Y., & Gin, K. Y.-H. (2021). Novel freshwater cyanophages provide new insights into evolutionary relationships between freshwater and marine cyanophages. *Microbiology Spectrum, 9*, e00593–e00521.

Zhang, D., Te, S. H., He, Y., & Gin, K. Y.-H. (2022a). Cyanophage dynamics in a tropical urban freshwater lake. *Ecological Indicators, 142*, 109257.

Zhang, D., He, Y., & Gin, K. Y.-H. (2022b). Genomic characterization of a novel freshwater cyanophage reveals a new lineage of cyanopodovirus. *Frontiers in Microbiology, 12*.

Zhao, L., Lin, L. Z., Zeng, Y., Teng, W. K., Chen M. Y, Brand, J. J, Zheng, L. L., Gan, N. Q., Gong Y. H., Li, X. Y., Lv, J., Chen. T., Han, B. P., Song, L. R., Shu, W.S. (2023). The facilitating role of phycospheric heterotrophic bacteria in cyanobacterial phosphonate availability and Microcystis bloom maintenance. Microbiome, 26;11(1):142. doi: https://doi.org/10.1186/s40168-023-01582-2. PMID: 37365664; PMCID: PMC10291797.

Zheng, Q., Lin, W., Wang, Y., Xu, D., Liu, Y., & Jiao, N. (2021). Top-down controls on nutrient cycling and population dynamics in a model estuarine photoautotroph–heterotroph co-culture system. *Molecular Ecology, 30*, 592–607.

Zhu, X., Li, Z., Tong, Y., Chen, L., Sun, T., & Zhang W. (2023a). From natural to artificial cyanophages: current progress and application prospects. *Environmental Research, 223*, 115428.

Zhu, J., Yang, F., Du, K., et al. (2023b). Phylogenomics of five *Pseudanabaena* cyanophages and evolutionary traces of horizontal gene transfer. *Environmental Microbiomes, 18*, 3.

Zimmerman, A. E., Howard-Varona, C., Needham, D. M., John, S. G., Worden, A. Z., Sullivan, M. B., Waldbauer, J. R., & Coleman, M. L. (2020). Metabolic and biogeochemical consequences of viral infection in aquatic ecosystems. *Nature Reviews. Microbiology, 18*, 21–34.

Part II
Urban Development and the Occurrence and Persistence of Antimicrobial Resistance and Microbial Source Tracking Technologies

Chapter 7
Quanti-Tray-Based Amplicon Sequencing (QT-AMP) Application for Fecal Indicator Bacteria Identification After Hurricane Ian

Hidetoshi Urakawa and Taylor L. Hancock

1 Introduction

Enterococcus is a gram-positive non-spore-forming facultative anaerobic bacteria. Because of their ubiquity in human feces and persistence in the environment, *Enterococcus* has been adopted as a useful indicator of human fecal pollution in water (Byappanahalli et al., 2012; Devane et al., 2020; Gilmore et al., 2014). Traditionally, detection of enterococci has been made by the membrane filtration method (e.g., US EPA method 1600 membrane filtration technique) with a combination of selective medium and incubation temperature. More recently, enterococci have also been quantified using an enzymatic reaction and the most probable number (MPN) format and quantitative polymerase chain reaction (qPCR) (Boehm & Sassoubre, 2014). Now, a semi-automated MPN method for enumeration of enterococci is standardized and commercially available as Enterolert (IDEXX, Westbrook, ME, USA), a defined substrate methodology that measures a fluorescent endpoint based on enterococci metabolizing 4-methylumbelliferone-β-D-glucoside. The method has been widely adopted worldwide (Abbott et al., 1998; Eckner, 1998; Ramoutar, 2020) and greatly improved the standardization of the counting procedure, which may cause less contamination and counting errors, and greatly reduces labor and time (Budnick et al., 1996).

H. Urakawa (✉)
Department of Marine and Ecological Sciences,
Florida Gulf Coast University, Fort Myers, FL, USA
e-mail: hurakawa@fgcu.edu

T. L. Hancock
Department of Ecology and Environmental Studies, The Water School,
Florida Gulf Coast University, Fort Myers, FL, USA

School of Geosciences, University of South Florida, Tampa, FL, USA

S. M. Tiquia-Arashiro et al. (eds.), *Urban Watershed Microbiology, Volume 1*,
https://doi.org/10.1007/978-3-031-89071-0_7

Although these counting methods are standardized and accepted broadly, the use of enterococci as an indicator of fecal contamination could be problematic (Budnick et al., 1996). Identification of origins of enterococci in water bodies is confounded by multiple sources as well as by the many factors that influence the ultimate fate of microorganisms once they are released into the environment. Enterococci are present in human feces (Devane et al., 2020; Elmir et al., 2009), as well as in wild and domestic animal feces (Harwood et al., 2000; Layton et al., 2010), birds (Devriese et al., 1992), river beds, soils (Byappanahalli & Fujioka, 2004), beaches (Abdelzaher et al., 2010; Ran et al., 2013; Yamahara et al., 2007), and aquatic and terrestrial plants (Byappanahalli et al., 2003; Imamura et al., 2011). Due to the uncertainty about how long enterococci are persistent, many researchers have focused on studying their dynamics and persistence in both freshwater and saltwater environments, which can be changed by multiple environmental factors controlled by seasons, weather, hydrology, water quality, and microbial communities (Ran et al., 2013; Yamahara et al., 2007).

Identification of human-specific enterococci species or genotypes could aid in the discrimination of human fecal contamination from other environmental sources of the organisms. However, regular enterococci monitoring only addresses their numbers (Lipp et al., 2001). In the past, isolation, purification, and characterization of *Enterococcus* strains have been used to identify this bacterium at the species level (Ferguson et al., 2013; Ran et al., 2013). As an alternative approach, a cloning method targeting the 16S rRNA gene has been used to characterize enterococci without cultivation efforts (Sercu et al., 2011). However, these two approaches are difficult to be incorporated into regular water monitoring only possible in the microbiology laboratory, and are not applicable in many water chemistry labs.

To solve this issue and quantify and identify enterococci in a streamlined fashion, we recently developed a new method named QT-AMP (Quanti-Tray-Based Amplicon Sequencing), in which first the Enterolert Quanti-Tray is used for MPN counting and then pools of positive wells are used as a source for high-throughput 16S rRNA gene amplicon sequencing (Fig. 7.1) (Urakawa et al., 2022). This method can improve the limitations of the existing MPN counting method using Enterolert, particularly in the context of the characterization of *Enterococcus* populations and identifying key *Enterococcus* species, which could be used in various water management projects and human health risk assessment in the future.

Hurricane Ian made landfall on September 28, 2022, as a Category 4 storm in southwest Florida with a significant impact, bringing strong winds, heavy rainfall, and widespread flooding to the region. The storm surge led to coastal inundation, causing damage to homes, infrastructure, and natural habitats (Bucci et al., 2023). Many areas experienced power outages and transportation disruptions due to fallen trees and debris. Emergency response teams worked diligently to assist affected communities and mitigate further damage. The aftermath of Hurricane Ian underscored the importance of preparedness and resilience in the face of extreme weather events. Water contamination by pathogenic bacteria was one of the major concerns of citizens, particularly when the recovery efforts were made, and many workers were exposed to potentially contaminated standing stormwater. Hurricane-induced

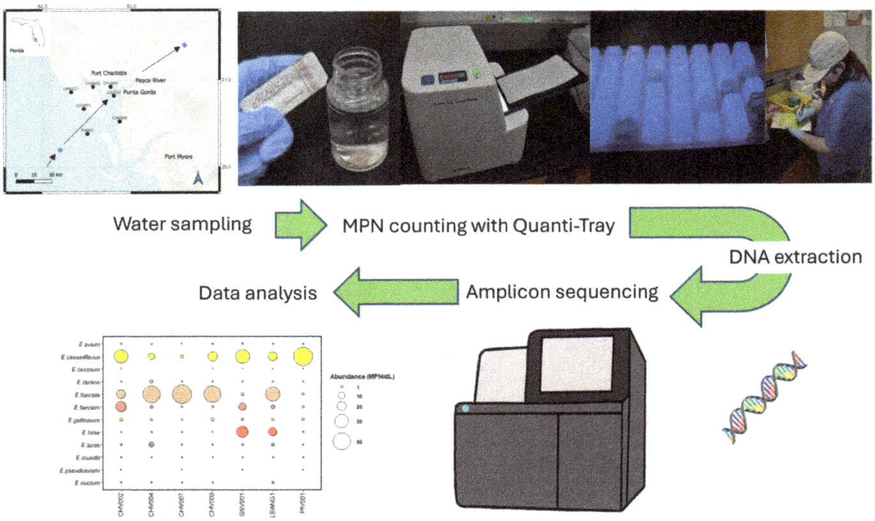

Fig. 7.1 Procedure of QT-AMP (Quanti-Tray Based Amplicon Sequencing). A flowchart illustrating the process of water sampling and analysis. The top row shows images of a map with sampling locations, mixing Enterolert in a water sample, a Quanti-Tray Sealer PLUS for MPN counting, blue fluorescing positive wells, and a lab technician working with samples. After counting the number of cells, a culture medium is used for DNA extraction. The bottom row shows Illumina amplicon sequencing and data analysis. The relative abundance of each *Enterococcus* species is determined and may be visualized as a heatmap, stuck bars, or a bubble plot.

stormwater runoffs carry various terrestrial pollutants, negatively affecting the microbiological quality of water and introducing fecal and other pathogens into coastal environments (Bae & Hou, 2013; Saingam et al., 2021). In these conditions, MPN counts of *Enterococcus* were used to assess the health risk of microbial contaminations.

The purpose of this study was to determine if QT-AMP can specifically detect storm-affected *Enterococcus* genotypes in hurricane-induced waters, providing better insights into contamination sources post-hurricane.

2 Materials and Methods

2.1 Sampling Sites

Surface water samples were collected at 7 sampling sites in October 2022 from the greater Charlotte Harbor estuary in Charlotte and Lee counties, Florida (Fig. 7.2). The water samples were collected 25 and 33 days after Hurricane Ian, which landed on southwest Florida on September 28, 2022, as a strong Category 4.

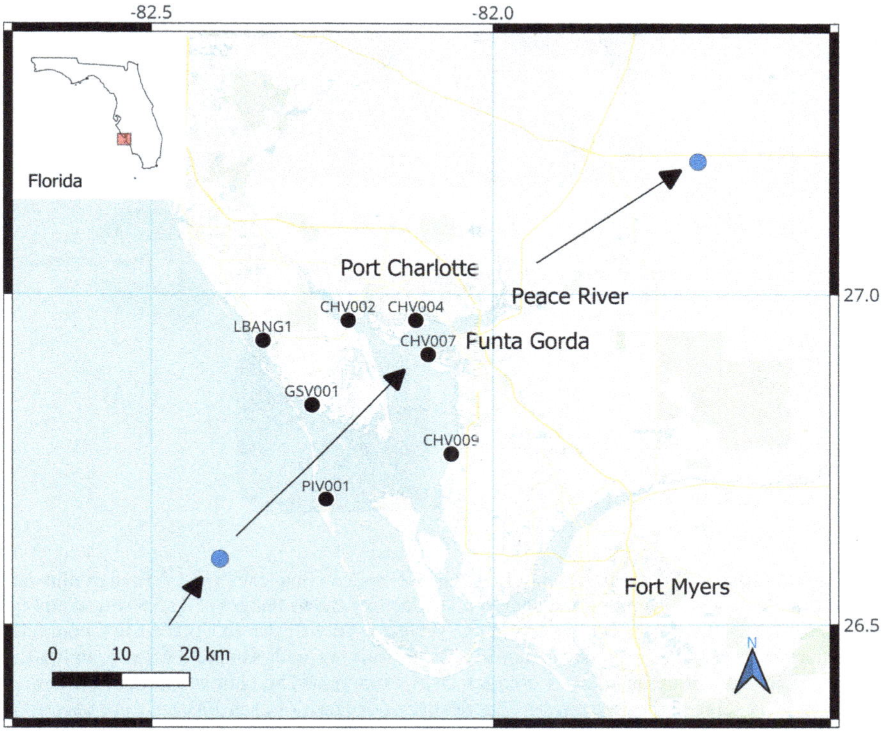

Fig. 7.2 Map of water sampling in southwest Florida after Hurricane Ian. Two blue dots connected by black arrows indicate the path of Hurricane Ian. An inset shows Florida with the mapped area shaded. A scale bar and compass rose are present for reference.

2.2 *Water Sampling and Water Quality Analyses*

Surface water samples (0.5 m depth) were collected in high-density polyethylene (HDPE) bottles in volumes of 250 mL for acid-preserved samples and two 950 mL bottles for other measurements. The collected water samples were kept below 8 °C (iced cooler) and transported to the Sanders Laboratories (Fort Myers, FL, USA) for microbiological analyses and the Florida Department of Environmental Protection Laboratory (Tallahassee, FL, USA) for chemical analyses. Water quality (temperature, salinity, pH, dissolved oxygen [DO], specific conductivity, and turbidity) readings were conducted on-site by a YSI multiparameter probe. Nutrients were analyzed using EPA approved methods: total Kjeldahl nitrogen (TKN) and total phosphate (TP) were measured after heated acid digestion. Total nitrogen (TN) was calculated as a total of TKN and nitrite + nitrate. Chlorophyll *a* measurement was conducted at the FDEP laboratory following Standard Method 10200H (mod.). Biological oxygen demand (BOD) was measured as a 5-day test using a standard method SM5210 B at Eurofins Tampa.

Table 7.1 qPCR gene markers and their targets and methods

Gene marker	Target bacteria	Method	Reference
HF183	Human-specific *Bacteroides*	SOP-PCR-1.0	Bernhard and Field (2000)
Gull2	Coastal bird-*specific Catellicoccus marimammalium*	SOP-PCR-1.2	Lu et al. (2008)
GFD	Bird-specific *Helicobacter*	SOP-PCR-1.3	Green et al. (2012)
BacR	Ruminant-specific Bacteroidetes	SOP-PCR-1.4	Reischer et al. (2006)
DG3	Canine-specific *Bacteroides*	SOP-PCR-1.5	Green et al. (2014)

Method: Detailed PCR methods are available on the DEP Laboratory Standard Operating Procedures website

2.3 Quantitative PCR Analysis

Five quantitative PCR (qPCR) gene markers were used in the present study. The target organism of each gene marker and corresponding methods are summarized in Table 7.1.

2.4 Chemical Tracer Analysis

All chemical tracers were analyzed by EPA method 8321B using HPLC-MS/MS at the Florida Department of Environmental Protection Laboratory, Tallahassee, FL.

2.5 MPN Counting Using the Enterolert Quanti-Tray Method

Water samples for *Enterococcus* were aseptically collected and brought back to the laboratory on ice (below 8 °C). All samples were analyzed within eight hours of collection following US EPA standards and the Enterolert manufacturer's instructions. Enterolert ampules (IDEXX Laboratories, Westbrook, ME, USA) were aseptically added into 100 mL of water samples (10 times diluted with sterilized distilled water) and thoroughly mixed. The sample-reagent mixture was poured into a Quanti-Tray/2000 and sealed with the Quanti-Tray Sealer PLUS (IDEXX) (Fig. 7.1). The trays were incubated at 41 ± 0.5 °C for 24 h. Under UV light (365 nm) in a dark room, blue fluorescing positive wells were counted, and cell density was calculated from the IDEXX MPN table.

2.6 DNA Extraction from Quanti-Trays

For DNA extraction, the rear surface of the Quanti-Tray was sterilized with 70% alcohol, after which media from up to five prominently fluorescing (positive) wells were aspirated using sterile syringes. The volume aspirated was meticulously recorded for each operation, with approximately 10 mL of culture medium from positive wells subsequently filtered through disposable 0.22 μm polysulfone filters (Millipore Sigma). DNA extraction was performed utilizing a modified phenol-chloroform method as described by Urakawa et al. (2010). Subsequently, the purity and concentration of the extracted DNA were assessed using a NanoDrop spectrophotometer (Thermo Fisher Scientific, Foster City, CA, USA) and a Qubit 2.0 Fluorometer with the Qubit dsDNA HS Assay Kit (Thermo Fisher Scientific), respectively. The extracted DNA was then stored at −20 °C until further analysis.

2.7 High-Throughput DNA Sequencing

A PCR primer set consisting of 27f (AGRGTTTGATCMTGGCTCAG) and 519r (GTNTTACNGCGGCKGCTG) was employed, with the 27f primer incorporating a DNA index for sample identification. Amplification of samples was carried out using the HotStarTaq Plus Master Mix Kit (Qiagen, Valencia, CA, USA) under the following cycling conditions: initial denaturation at 94 °C for 3 min, followed by 30 cycles of denaturation at 94 °C for 30 s, annealing at 53 °C for 40 s, and extension at 72 °C for 1 min, with a final extension step at 72 °C for 10 min. PCR amplification success was confirmed through agarose gel electrophoresis (2%). The quantified amplicons were pooled equitably and paired-end 2 × 300 cycle sequencing was conducted utilizing the Illumina MiSeq platform with a MiSeq Reagent Kit v3 at MR DNA (Shallowater, TX, USA). Sequence data underwent processing using the MR DNA taxonomic analysis pipeline. Initially, sequence data underwent trimming to eliminate index sequences shorter than <200 bp, and ambiguous base calls were eliminated. Sequences were quality-filtered using a maximum expected error threshold of 1.0 and then dereplicated. Unique sequences were denoised to identify and remove sequencing and/or PCR point errors, followed by chimera removal (Edgar et al., 2011), resulting in denoised amplicon sequence variants (ASVs). Final ASVs were taxonomically classified using BLASTn against a curated database sourced from the National Center for Biotechnology Information (NCBI). All data associated with 16S rRNA gene amplicon sequencing reads are available in NCBI under BioProject PRJNA1227612.

2.8 Estimation of the Abundance of Each Enterococcus *Species*

Estimation of the abundance of each *Enterococcus* species was performed based on the multiplication of the MPN counts of Quanti-Tray and the relative abundance of *Enterococcus* species.

2.9 Statistical Analysis

A Pearson correlation and descriptive statistical analyses were made using XLSTAT (Addinsoft, NY, USA). Box plots were generated and non-parametric analysis (Kruskal-Wallis test and Dunn's multiple comparisons test) was performed using GraphPad Prism 9.3.1 (GraphPad Software, CA, USA). A bubble plot was made using R and R-studio (ver. 4.3.1. R Core Team, 2023; Posit Team, 2023).

3 Results

3.1 Environmental Factors

The environmental parameters of each sampling site are listed in Table 7.2. The mean and standard deviation (SD) of temperature was 26.5 ± 1.3 °C and ranged between 23.8 and 27.8 °C. A steep gradient of salinity was found among estuary sites. The mean and SD of salinity was 17.3 ± 9.1 ppt and ranged between 3.65 and 27.7 ppt. The observed DO concentration was lower than the normal condition. The mean and SD of DO was 3.3 ± 1.4 mg/L and ranged between 2.0 and 5.8 mg/L. The mean and SD of BOD was 2.6 ± 0.8 mg/L and ranged between 2.0 and 3.8 mg/L.

3.2 Enterococcus *Compositions*

The QT-AMP was conducted for seven samples and a bubble plot was used to show the spatiotemporal distribution of *Enterococcus* species among samples (Fig. 7.3). The method successfully detected 12 *Enterococcus* species: four rare species (*E. cecorum*, *E. lactis*, *E. pseudoavium*, and *E. rivorum*) and eight cosmopolitan species (*E. faecalis*, *E. faecium*, *E. casseliflavus*, *E. hirae*, *E. mundtii*, *E. gallinarum*, *E. avium*, and *E. durans*), which have been commonly documented in various *Enterococcus* isolation studies (Table 7.3). Furthermore, the abundance of each *Enterococcus* species was estimated based on the MPN count and the relative abundance of species composition determined by QT-AMP (Fig. 7.4).

Table 7.2 Physiochemical parameters, nutrients, and Enterolert MPN counts of sampling sites

Site ID	Date	Location	Temp (°C)	Salinity (ppt)	pH	DO%	DO (mg/L)	Sp. Cond (mS/cm)	Turbidity (NTU)	Total N (mg-N/L)	Total *P* (mg-P/L)	NOx (mg-N/L)	Chl-a (μg/L)	BOD (mg/L)	Entero (MPN/dL)
CHV002	10/17/2022	Myakka River	27.8	3.65	7.18	56.6	4.36	6.70	2.3	1.415	0.56	0.015	9.4	2.0	10
CHV004	10/17/2022	Charlotte Harbor	26.7	8.31	7.35	36.2	2.48	14.38	2.2	1.215	0.56	0.015	3.5	2.0	10
CHV007	10/17/2022	Ponce de Leon Inlet	26.3	13.20	7.49	41.5	3.11	21.58	3.2	1.212	0.52	0.012	4.5	2.0	30
CHV009	10/17/2022	Burnt Store Entrance	27.0	21.05	7.35	31.0	2.20	33.67	1.2	1.003	0.25	0.013	13	2.0	20
GSV001	10/17/2022	Coral Creek	26.8	25.87	7.42	29.0	2.01	40.79	3.6	0.704	0.094	0.004	9.6	3.8	134
LBANG1	10/17/2022	Ainger Creek	27.2	20.99	7.60	46.4	3.28	33.68	2.6	1.004	0.18	0.004	28	3.3	109
PIV001	10/25/2022	Pine Island Sound	23.8	27.69	7.97	80.6	5.80	42.92	3.1	0.667	0.11	0.007	12	3.0	20

Environmental parameters: Sp. Cond, Specific conductivity, NOx, nitrate + nitrite, Entero, Enterolert MPN counts

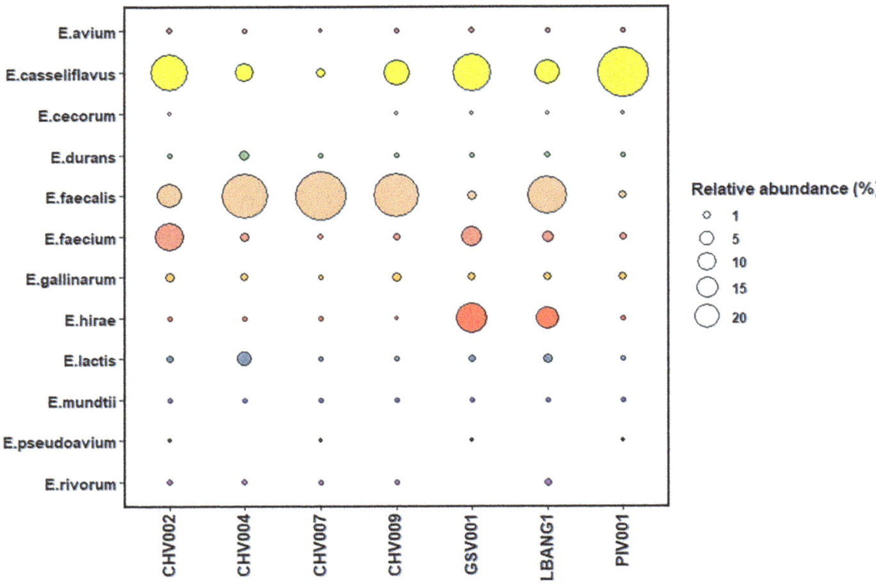

Fig. 7.3 Species composition of enterococci found in QT-AMP analysis of water samples collected after Hurricane Ian. Bubble chart showing the relative abundance of different *Enterococcus* species across six samples: CHV002, CHV004, CHV007, CHV009, GSV001, LBANG1, and PIV001. Each species is represented by a row of bubbles, with bubble size indicating abundance percentage. The largest bubbles, representing higher abundance, are seen for *E. casseliflavus* in PIV001 and *E. faecalis* in CHV007. A legend on the right indicates bubble sizes corresponding to 1, 5, 10, 15, and 20% abundance.

3.3 Enterococcus *Abundance*

The mean and SD of MPN counting using Enterolert was 47.6 ± 51.5 MPN/dL and ranged between 10.0 and 134 MPN/dL. Only one tested water sample (GSV001, Coral Creek) exceeded USEPA's statistical threshold value (STV) of 130 CFU/100 mL. Two tested water samples (GSV001, Coral Creek, and LBANG1, Ainger Creek) exceeded USEPA's geometric mean criterion of 35 CFU/100 mL. Estimation of the abundance of each *Enterococcus* species was performed based on the multiplication of the MPN counts of Quanti-Tray and the relative abundance of *Enterococcus* species. The analysis revealed the dominance of two species, *E. casseliflavus* and *E. faecalis*, and their numbers were nearly identical. The mean and SD of the MPN count of *E. casseliflavus* was 17.2 ± 24.7 cells/dL and ranged between 0.4 and 69.9 cells/dL. The mean and SD of the MPN count of *E. faecalis* was 17.2 ± 22.8 cells/dL and ranged between 0.2 and 63.3 cells/dL. These two species were followed by *E. hirae*, attributed to multiple animal and plant sources (Gilmore et al., 2014). The mean and SD of the MPN count of *E. hirae* was 8.8 ± 16.7 cells/dL and ranged between 0.005 and 43.5 cells/dL. The fourth species was *E. faecium*, typically a common counterpart to *E. faecalis* in sewage and other fecal sources (Gilmore et al., 2014). The mean and SD of the MPN count of *E. faecium* was

Table 7.3 List of major *Enterococcus* species found in each study

Source	Sample type	Country	Number of isolates	Number of species detected	*E. casseliflavus*	*E. faecalis*	*E. faecium*	*E. hirae*	*E. mundtii*	*E. gallinarum*
Ran et al. (2013)	Water, sand	USA	2441	8	+	+	+	+	+	+
Ferguson et al. (2013)	Multiple	USA	1279	9	+	+	+	+	+	+
Badgley et al. (2010)	Sed, SAV	USA	50	5	+	+	+	+	+	
Nayak et al. (2011)	Water, sed, SAV	USA	61	6	+		+	+	+	
Müller et al. (2001)	Plant (grass)	Germany	54	5	+	+	+		+	
Graves et al. (2009)	Livestock manure	USA	640	9	+	+	+	+	+	+
Švec and Sedláček (1999)	Surface fresh water	Czech Republic	630	8	+	+	+	+	+	+
Monticelli et al. (2019)	Sewage, coastal water	Italy	152	6	+	+	+			+
Urakawa et al. (2022)	River water	USA	QT-AMP	11	+	+	+	+	+	+
This study	Coastal water	USA	QT-AMP	12	+	+	+	+	+	+

Sample type: Sed, Sediment; SAV, submerged aquatic vegetation. Multiple: wastewater, beach water, urban runoff

E. avium	*E. durans*	*E. lactis*	*E. rivorum*	*E. malodoratus*	*E. saccharolyticus*	*E. gilvus*	*E. cecorum*	*E. silesiacus/ moraviensis/ caccae*	*E. sulfureus*	*E. flavescens*	*E. pseudoavium*
+	+										
+	+				+						
						+		+			
									+		
	+									+	+
+	+										
+	+										
+	+	+	+	+							
+	+	+	+				+				+

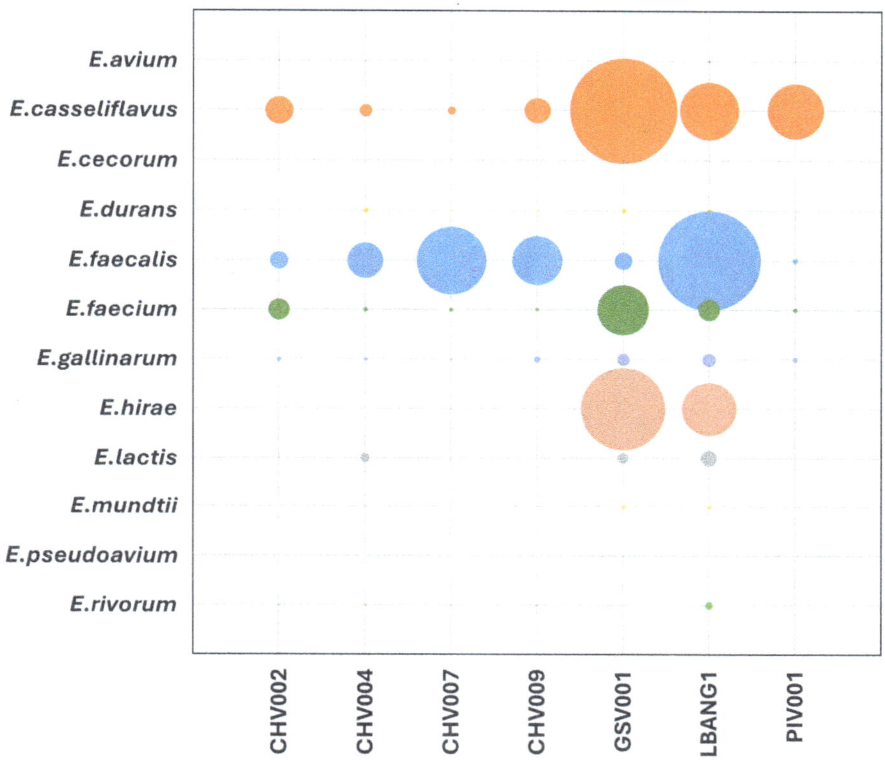

Fig. 7.4 Estimated cell counts of *Enterococcus* species using QT-AMP. A bubble chart displays the presence of different *Enterococcus* species across various samples. Each species, such as *E. avium*, *E. casseliflavus*, and others, is represented by bubbles of varying sizes and colors, indicating the abundance in each sample. The unit is cells per dL.

3.2 ± 5.9 cells/dL and ranged between 0.08 and 16.3 cells/dL. The means of all other eight species were less than 1 cell/dL and considered minor species.

3.4 qPCR Assays

The results of qPCR assays are presented in Fig. 7.5. No significant difference was found in gene copy numbers among all tested samples. All gene markers were found in all tested samples. The gene distribution patterns were similar among three markers (GULL2, BacR, and DG3). The mean and standard SD of HF183 was 2533 ± 981 copies/dL and ranged between 830 and 3700 copies/dL (Table 7.4). The mean and standard SD of GULL2 was 2357 ± 1045 copies/dL and ranged between 900 and 3600 copies/dL. The mean and standard SD of GFD was 1959 ± 3186 copies/dL and ranged between 480 and 9130 copies/dL. An extremely high gene count (9130 copies/dL) was found for GFD at PIV001 (Pelican Bay, Pine Island Sound). The mean

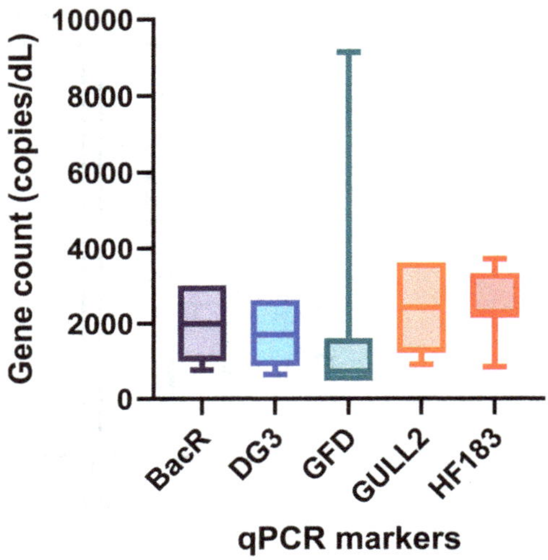

Fig. 7.5 Box plot of gene copy numbers of qPCR gene markers in water samples collected after Hurricane Ian. Box plot illustrating gene count in copies per deciliter for five qPCR markers: BacR, DG3, GFD, GULL2, and HF183. Each marker is represented by a colored box, showing the distribution of gene counts. BacR and DG3 have lower median values, while GFD shows a wider range with a higher maximum value. GULL2 and HF183 have similar distributions with moderate gene counts. The y-axis ranges from 0 to 10,000 gene copies per deciliter. None of the pairs were significantly different ($p > 0.05$)

Table 7.4 Gene copy abundance of qPCR makers

Site ID	Location	Enterolert (MPN/dL)	Salinity (ppt)	HF183 (copy/dL)	GULL2 (copy/dL)	GFD (copy/dL)	BacR (copy/dL)	DG3 (copy/dL)
CHV002	Myakka River	10	3.7	3300	3600	720	3000	2600
CHV004	Charlotte Harbor	10	8.3	2200	2400	580	2000	1700
CHV007	Ponce de Leon Inlet	30	13.2	2200	2400	480	2000	1700
CHV009	Burnt Store Entrance	20	21.1	3700	1200	1600	1000	860
GSV001	Coral Creek	134	25.9	2200	2400	480	2000	1700
LBANG1	Ainger Creek	109	21.0	3300	3600	720	3000	2600
PIV001	Pine Island Sound	20	27.7	830	900	9130	760	640

and standard SD of BacR was 1966 ± 869 copies/dL and ranged between 760 and 3000 copies/dL. The mean and standard SD of DG3 was 1686 ± 758 copies/dL and ranged between 640 and 2600 copies/dL.

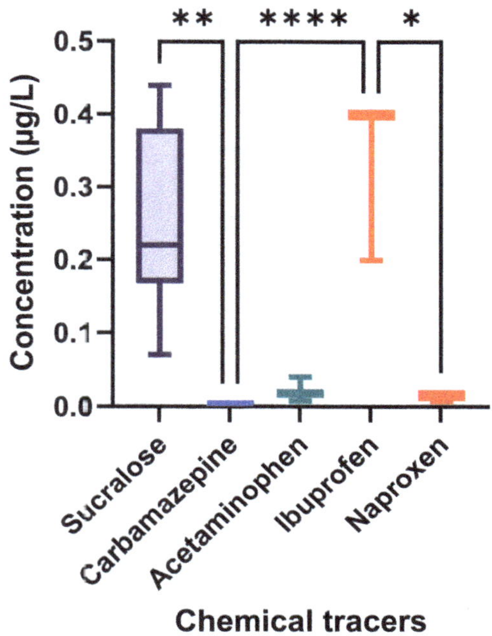

Fig. 7.6 Box plot of the concentration of five chemical tracers in water samples collected after Hurricane Ian. Statistical significance is indicated with asterisks above the comparisons. *P*-values: * ($p = 0.0310$), **($p = 0.002$), ****($p < 0.0001$)

3.5 Chemical Tracer Measurements

Box plot of the concentration of chemical tracers in water samples collected after Hurricane Ian is shown in Fig. 7.6. The significant differences among chemical tracers were found between sucralose and carbamazepine ($p = 0.002$), Carbamazepine and ibuprofen ($p < 0.0001$), and ibuprofen and naproxen ($p = 0.0310$). The mean and standard SD of sucralose was 255.7 ± 129.8 ng/L and ranged between 70.0 and 440.0 ng/L (Table 7.5). The mean and standard SD of carbamazepine was 1.5 ± 0.3 ng/L and ranged between 0.8 and 1.6 ng/L. The mean and standard SD of acetaminophen was 19.4 ± 10.2 ng/L and ranged between 8.0 and 40.0 ng/L. The mean and standard SD of ibuprofen was 371.4 ± 75.6 ng/L and ranged between 200.0 and 400.0 ng/L. The mean and standard SD of naproxen was 14.9 ± 3.0 ng/L and ranged between 8.0 and 16.0 ng/L. All tracers showed the lowest concentrations at PIV001 (Pine Island Sound), which also had the highest salinity.

3.6 Pearson Correlation Matrix

The Pearson correlation matrix among environmental factors, qPCR assays, chemical tracers, and *Enterococcus* species detected by QT-AMP is shown in Table 7.6. A high correlation was found between Enterolert MPN counts and BOD ($r = 0.88$).

Table 7.5 Concentrations of chemical tracer markers

Site ID	Location	Enterolert (MPN/dL)	Salinity (ppt)	Sucralose (ng/L)	Carbamazepine (ng/L)	Acetaminophen (ng/L)	Ibuprofen (ng/L)	Naproxen (ng/L)
CHV002	Myakka River	10	3.7	192	1.6	16	400	16
CHV004	Charlotte Harbor	10	8.3	380	1.6	16	400	16
CHV007	Ponce de Leon Inlet	30	13.2	220	1.6	16	400	16
CHV009	Burnt Store Entrance	20	21.1	168	1.6	40	400	16
GSV001	Coral Creek	134	25.9	320	1.6	24	400	16
LBANG1	Ainger Creek	109	21.0	440	1.6	16	400	16
PIV001	Pine Island Sound	20	27.7	70	0.8	8	200	8

Table 7.6 Correlation matrix (Pearson) of *Enterococcus* species, counts, qPCR gene markers, **chemical tracer, BOD, and salinity**

Variables	ENTERO	Sucralose	BacR	DG3	GFD	GULL2	HF183	BOD	*E. avium*	*E. casseliflavus*	*E. cecorum*	*E. durans*	*E. faecalis*	*E. faecium*
ENTERO	**1.00**	0.57	0.32	0.32	−0.27	0.32	0.07	0.88	0.11	0.02	0.63	−0.29	−0.32	0.02
Sucralose	0.57	**1.00**	0.65	0.64	−0.67	0.65	0.33	0.28	−0.19	−0.58	0.11	0.47	0.29	−0.07
BacR	0.32	0.65	**1.00**	1.00	−0.66	1.00	0.48	0.05	0.30	−0.38	−0.10	0.06	0.05	0.57
DG3	0.32	0.64	1.00	**1.00**	−0.66	1.00	0.49	0.05	0.31	−0.37	−0.09	0.05	0.05	0.58
GFD	−0.27	−0.67	−0.66	−0.66	**1.00**	−0.67	−0.70	0.20	−0.01	0.81	−0.16	−0.23	−0.49	−0.27
GULL2	0.32	0.65	1.00	1.00	−0.67	**1.00**	0.49	0.05	0.30	−0.38	−0.10	0.06	0.06	0.57
HF183	0.07	0.33	0.48	0.49	−0.70	0.49	**1.00**	−0.29	0.05	−0.57	0.43	−0.12	0.37	0.30
BOD	0.88	0.28	0.05	0.05	0.20	0.05	−0.29	**1.00**	0.22	0.46	0.54	−0.32	−0.64	−0.02
E. avium	0.11	−0.19	0.30	0.31	−0.01	0.30	0.05	0.22	**1.00**	0.55	0.23	−0.22	−0.78	0.92
E. casseli flavus	0.02	−0.58	−0.38	−0.37	0.81	−0.38	−0.57	0.46	0.55	**1.00**	0.12	−0.36	−0.90	0.25
E. cecorum	0.63	0.11	−0.10	−0.09	−0.16	−0.10	0.43	0.54	0.23	0.12	**1.00**	−0.44	−0.32	0.09
E. durans	−0.29	0.47	0.06	0.05	−0.23	0.06	−0.12	−0.32	−0.22	−0.36	−0.44	**1.00**	0.36	−0.21
E. faecalis	−0.32	0.29	0.05	0.05	−0.49	0.06	0.37	−0.64	−0.78	−0.90	−0.32	0.36	**1.00**	−0.51
E. faecium	0.02	−0.07	0.57	0.58	−0.27	0.57	0.30	−0.02	0.92	0.25	0.09	−0.21	−0.51	**1.00**
E. gallina rum	−0.18	−0.13	0.07	0.09	0.10	0.07	0.52	−0.09	0.44	0.29	0.40	−0.09	−0.31	0.42
E. hirae	0.97	0.52	0.28	0.27	−0.28	0.28	0.03	0.87	0.28	0.10	0.66	−0.21	−0.43	0.14
E. lactis	−0.13	0.63	0.23	0.22	−0.29	0.23	−0.03	−0.19	−0.21	−0.38	−0.38	0.97	0.33	−0.17
E. mundtii	0.65	0.79	0.37	0.37	−0.36	0.37	0.34	0.54	0.07	−0.14	0.53	0.31	−0.13	0.01
E. pseudo avium	0.49	−0.22	−0.18	−0.19	0.28	−0.18	−0.51	0.69	0.60	0.67	0.35	−0.34	−0.81	0.31
E. rivorum	0.02	0.71	0.65	0.65	−0.47	0.65	0.55	−0.20	−0.29	−0.58	−0.18	0.43	0.48	−0.03
Salinity	0.54	−0.16	−0.56	−0.56	0.53	−0.56	−0.36	0.74	−0.18	0.47	0.57	−0.44	−0.39	−0.48
Temp	0.19	0.56	0.72	0.73	−0.91	0.73	0.88	−0.22	0.23	−0.63	0.28	0.10	0.31	0.50
DO	−0.38	−0.59	−0.14	−0.13	0.79	−0.14	−0.50	0.01	0.23	0.71	−0.45	−0.29	−0.48	0.16
Sp. Cond	0.55	−0.14	−0.56	−0.56	0.51	−0.56	−0.35	0.74	−0.20	0.45	0.58	−0.43	−0.38	−0.49
pH	0.11	−0.30	−0.50	−0.50	0.84	−0.50	−0.72	0.46	−0.33	0.56	−0.13	−0.24	−0.30	−0.56
Total *N*	−0.51	0.17	0.58	0.58	−0.59	0.58	0.52	−0.78	0.04	−0.60	−0.50	0.30	0.54	0.40
NO*x*	−0.86	−0.25	−0.01	−0.01	−0.24	−0.01	0.26	−0.96	0.00	−0.37	−0.50	0.43	0.48	0.19
Total P	−0.63	0.07	0.39	0.38	−0.46	0.39	0.22	−0.83	−0.02	−0.54	−0.69	0.48	0.54	0.27
Chl-a	0.52	0.33	0.29	0.30	0.06	0.29	0.37	0.50	−0.12	0.08	0.46	−0.40	−0.14	−0.09
Turbidity	0.52	0.07	0.12	0.10	0.17	0.11	−0.66	0.63	0.18	0.34	−0.14	−0.22	−0.44	0.06

E. gallinarum	*E. hirae*	*E. lactis*	*E. mundtii*	*E. pseudoavium*	*E. rivorum*	*Salinity*	Temp	DO	Sp. Cond	pH	Total N	NO*x*	Total *P*	Chl-a	Turbidity
−0.18	0.97	−0.13	0.65	0.49	0.02	0.54	0.19	−0.38	0.55	0.11	−0.51	−0.86	−0.63	0.52	0.52
−0.13	0.52	0.63	0.79	−0.22	0.71	−0.16	0.56	−0.59	−0.14	−0.30	0.17	−0.25	0.07	0.33	0.07
0.07	0.28	0.23	0.37	−0.18	0.65	−0.56	0.72	−0.14	−0.56	−0.50	0.58	−0.01	0.39	0.29	0.12
0.09	0.27	0.22	0.37	−0.19	0.65	−0.56	0.73	−0.13	−0.56	−0.50	0.58	−0.01	0.38	0.30	0.10
0.10	−0.28	−0.29	−0.36	0.28	−0.47	0.53	−0.91	0.79	0.51	0.84	−0.59	−0.24	−0.46	0.06	0.17
0.07	0.28	0.23	0.37	−0.18	0.65	−0.56	0.73	−0.14	−0.56	−0.50	0.58	−0.01	0.39	0.29	0.11
0.52	0.03	−0.03	0.34	−0.51	0.55	−0.36	0.88	−0.50	−0.35	−0.72	0.52	0.26	0.22	0.37	−0.66
−0.09	0.87	−0.19	0.54	0.69	−0.20	0.74	−0.22	0.01	0.74	0.46	−0.78	−0.96	−0.83	0.50	0.63
0.44	0.28	−0.21	0.07	0.60	−0.29	−0.18	0.23	0.23	−0.20	−0.33	0.04	0.00	−0.02	−0.12	0.18
0.29	0.10	−0.38	−0.14	0.67	−0.58	0.47	−0.63	0.71	0.45	0.56	−0.60	−0.37	−0.54	0.08	0.34
0.40	0.66	−0.38	0.53	0.35	−0.18	0.57	0.28	−0.45	0.58	−0.13	−0.50	−0.50	−0.69	0.46	−0.14
−0.09	−0.21	0.97	0.31	−0.34	0.43	−0.44	0.10	−0.29	−0.43	−0.24	0.30	0.43	0.48	−0.40	−0.22
−0.31	−0.43	0.33	−0.13	−0.81	0.48	−0.39	0.31	−0.48	−0.38	−0.30	0.54	0.48	0.54	−0.14	−0.44
0.42	0.14	−0.17	0.01	0.31	−0.03	−0.48	0.50	0.16	−0.49	−0.56	0.40	0.19	0.27	−0.09	0.06
1.00	−0.14	−0.03	0.33	−0.19	0.27	−0.09	0.29	0.16	−0.09	−0.25	0.11	0.17	−0.12	0.41	−0.68
−0.14	**1.00**	−0.08	0.66	0.63	−0.11	0.49	0.21	−0.41	0.50	0.00	−0.52	−0.78	−0.60	0.35	0.53
−0.03	−0.08	**1.00**	0.48	−0.37	0.59	−0.42	0.19	−0.30	−0.41	−0.24	0.30	0.29	0.41	−0.20	−0.19
0.33	0.66	0.48	**1.00**	0.02	0.52	0.20	0.41	−0.47	0.21	−0.16	−0.22	−0.45	−0.37	0.54	−0.10
−0.19	0.63	−0.37	0.02	**1.00**	−0.75	0.49	−0.32	0.12	0.48	0.24	−0.64	−0.54	−0.53	−0.17	0.72
0.27	−0.11	0.59	0.52	−0.75	**1.00**	−0.46	0.55	−0.19	−0.44	−0.29	0.54	0.15	0.35	0.47	−0.43
−0.09	0.49	−0.42	0.20	0.49	−0.46	**1.00**	−0.57	0.06	1.00	0.71	−0.96	−0.81	−0.96	0.42	0.32
0.29	0.21	0.19	0.41	−0.32	0.55	−0.57	**1.00**	−0.62	−0.56	−0.90	0.64	0.28	0.42	0.12	−0.38
0.16	−0.41	−0.30	−0.47	0.12	−0.19	0.06	−0.62	**1.00**	0.04	0.60	−0.11	−0.05	−0.08	0.11	0.22
−0.09	0.50	−0.41	0.21	0.48	−0.44	1.00	−0.56	0.04	**1.00**	0.70	−0.96	−0.81	−0.96	0.42	0.31
−0.25	0.00	−0.24	−0.16	0.24	−0.29	0.71	−0.90	0.60	0.70	**1.00**	−0.71	−0.59	−0.61	0.30	0.43
0.11	−0.52	0.30	−0.22	−0.64	0.54	−0.96	0.64	−0.11	−0.96	−0.71	**1.00**	0.78	0.93	−0.27	−0.42
0.17	−0.78	0.29	−0.45	−0.54	0.15	−0.81	0.28	−0.05	−0.81	−0.59	0.78	**1.00**	0.86	−0.62	−0.60
−0.12	−0.60	0.41	−0.37	−0.53	0.35	−0.96	0.42	−0.08	−0.96	−0.61	0.93	0.86	**1.00**	−0.57	−0.27
0.41	0.35	−0.20	0.54	−0.17	0.47	0.42	0.12	0.11	0.42	0.30	−0.27	−0.62	−0.57	**1.00**	−0.11
−0.68	0.53	−0.19	−0.10	0.72	−0.43	0.32	−0.38	0.22	0.31	0.43	−0.42	−0.60	−0.27	−0.11	**1.00**

4 Discussion

Although a dozen *Enterococcus* spp. were found, only two species, *E. faecalis* and *E. casseliflavus*, mainly dominated in the QT-AMP samples. Four samples (CHV004, CHV007, CHV009, and LBANG1) were led by *E. faecalis*, and three samples (CHV002, GSV001, and PIV001) were led by *E. casseliflavus* (Fig. 7.3). The prevalence of these two species was previously documented by our study conducted in three rivers in southwest Florida (Urakawa et al., 2022). In the present study, the most abundant species was *E. faecalis* and this species showed a large variety among samples. The mean and SD of the relative abundance of *E. faecalis* was 47.9 ± 40.1% and ranged between 0.9 and 97.8%. The lowest abundance was reported at PIV001 (Pine Island Sound) which was the highest salinity site (27.7 ppt). The highest abundance was reported at CHV007 (Ponce de Leon Inlet). This site exhibited moderate salinity at 13.2 ppt and no conspicuous environmental anomalies indicative of wastewater influence were observed (Table 7.2). The secondary abundant species was *E. casseliflavus* and this species also showed a great range among samples. The mean and SD of the relative abundance of *E. casseliflavus* was 35.8 ± 33.1% and ranged between 1.4% and 97.3%. The highest abundance was reported at PIV001 (Pine Island Sound) which was the highest salinity site (27.7 ppt). The lowest abundance was reported at CHV007 (Ponce de Leon Inlet). Thus, the abundance of these two species was inversive in relation to salinity. The correlation matrix data showed that the relative abundance of these two species was negatively correlated ($r = -0.90$, $n = 7$) (Table 7.6). These differences could be attributed to the difference in salt stress tolerance and UV resistance between two *Enterococcus* species (Byappanahalli et al., 2012).

The relationship between fecal indicator bacteria (FIB) density and environmental conditions has been a focal point of previous research, enhancing our ability to interpret the association between FIB densities and human health risks (Byappanahalli et al., 2012). Previous studies in southwest Florida have examined the relationship between enterococci density and water body types, concluding that larger water volumes afford greater dilution and buffer against high *Enterococcus* densities (Badgley et al., 2010). Consistent with these findings, our study revealed elevated *Enterococcus* densities in two small creeks, Coral Creek and Ainger Creek, echoing the patterns observed in Charlotte Harbor, which is linked to the Gulf of Mexico.

This research was conducted after Hurricane Ian. The study area received devastating damage from winds and storm surges. The strong input of freshwater inflow from the watershed was reported as well as up to 26 inches (66 cm) of rain in Placida, FL. Consequently, river and estuary *Enterococcus* communities were also anticipated to be influenced by this natural disturbance. Although we do not have pre-hurricane QT-AMP data for this study area, we were able to compare the data from three tidal rivers in southwest Florida: the Caloosahatchee River, Estero River, and Imperial River (Urakawa et al., 2022). The species composition (i.e., mean relative abundance or read frequency) was compared between pre-hurricane (dry and wet seasons) and post-hurricane (Table 7.7). The wet season pre and post-hurricane

Table 7.7 Species composition of *Enterococcus* in pre- and post-hurricane water samples collected from southwest Florida

	This study	Urakawa et al. (2022)	
Species	Post (%)	Pre-wet (%)	Pre-dry (%)
E. faecalis	47.9	42.0	36.6
E. casseliflavus	35.8	34.9	30.1
E. hirae	7.1	1.7	0.8
E. faecium	6.5	6.5	9.4
E. lactis	1.1	0.7	1.0
E. gallinarum	0.9	6.6	1.6
E. durans	0.3	4.8	1.8
E. rivorum	0.14	0.1	1.1
E. mundtii	0.09	1.5	16.7
E. avium	0.08	1.3	0.5
E. pseudoavium	0.01	0	0
E. cecorum	0.005	0	0
E. malodoratus	0	0.04	0

data were very similar in the composition of three species, *E. faecalis*, *E. casseliflavus*, and *E. faecium*.

The major dissimilarity was caused by the difference between *E. hirae* and *E. gallinarum*. *E. hirae* was less abundant in pre-hurricane samples but abundant in the post-hurricane samples. *E. gallinarum* showed the opposite relationship. In post-hurricane Charlotte Harbor samples, a notably strong positive correlation ($r = 0.97$) was detected between MPN counts and the abundance of *E. hirae*, attributed to multiple animal and plant sources (Byappanahalli et al., 2012). Thus, this *Enterococcus* species may be a good indicator demonstrating an association between the hurricane event and FIB contamination in the post-hurricane samples.

In recent years, there has been a growing utilization of source-specific molecular markers, such as HF183 and BacR targeting human fecal wastes, and Gull2 targeting fecal inputs from shoreline birds, alongside traditional fecal indicators like *E. coli* or *Enterococcus*, to enhance the characterization of water quality in monitoring programs (Ahmed et al., 2008; Koskey et al., 2014; Ryu et al., 2012; Sinigalliano et al., 2013). In our study, we employed five gene markers in qPCR assays, revealing a notable correlation ($r = 1.0$) among Gull2, BacR, and DG3, despite their expected diverse origins: Gull2 from shoreline birds, BacR from human fecal wastes, and DG3 from canine-specific *Bacteroides*. At the location with the highest salinity (PIV001), these markers demonstrated their lowest abundance, implying a terrestrial origin for these bacteria. This observation also indicates a decrease in gene copy numbers attributed to the dilution of freshwater by seawater. In southwest Florida, *Enterococcus* MPN counts consistently surpass the criteria of the US EPA, indicating persistent high levels. Typical point sources that could account for such high levels in the freshwater area are not evident in the data, suggesting that enterococci originate from various unidentified sources beyond wastewater (Byappanahalli

et al., 2012). Despite HF183 being a widely employed human-specific gene marker, it did not exhibit a strong correlation with *Enterococcus* species. There was a modest positive association between HF183 and five species, *E. secorum*, *E. faecalis*, *E. faecium*, *E. gallinarum*, and *E. rivorum*, while negative correlations were observed with *E. casseliflavus* and *E. pseudoavium*. No correlation was found between HF183 and species like *E. avium*, *E. durans*, *E. hirae*, and *E. lactis*. The most significant correlations between gene markers and *Enterococcus* species were observed with BacR and *E. faecium* ($r = 0.57$), DG3 and *E. rivorum* ($r = 0.65$), GFD and *E. casseliflavus* ($r = 0.81$), and GULL2 and *E. rivorum* ($r = 0.65$).

Urakawa et al. (2022) delineated the relative prevalence of *Enterococcus* spp. in QT-AMP analysis via a heat map. On the contrary, in this study, akin data were represented using a bubble plot (Fig. 7.3). The depiction of the relative abundance of species composition serves as a pragmatic tool for cross-referencing with datasets from fellow researchers. Yet, our endeavor in this study was to amalgamate MPN counts with the aforementioned relative abundance data to approximate the theoretical cellular quantities of individual species (Fig. 7.4). Water management practitioners may favor such an approach for its potential applicability in water resource oversight. The highest MPN counts were found in two sites, GSV001 (134 cells/dL) and LBANG1 (109 cells/dL). While the MPN counts exhibit similarity, the QT-AMP analysis revealed notable disparities in the composition of *Enterococcus* species: GSV001 exhibited dominance by *E. casseliflavus*, contrasting with *E. faecalis* prevalence in LBANG1 (Fig. 7.4). This difference suggests distinct microbial origins at the two sampling sites following the impact of Hurricane Ian on the area (Fig. 7.2). *E. faecalis* is commonly identified as a member of the normal intestinal flora of multiple animal species (Gilmore et al., 2014; Staley et al., 2014). Although *E. casseliflavus* could be found in the human intestinal tract, however, it is considered a rare human pathogen, naturalized populations persisting in freshwater environments may greatly contribute to the *Enterococcus* population in estuary in the absence of fecal pollution events (Ferguson et al., 2013; Staley et al., 2014). Thus, water managers should pay more attention to locations where a high abundance of *E. faecalis* is often reported. In this case, the microbial risk factor of LBANG1 was greater than GSV001 even though both showed similar MPN counts.

In recent years, there has been a growing trend toward utilizing source-specific molecular markers, such as HF183 and BacR targeting human fecal wastes, and Gull2 targeting fecal inputs from shoreline birds, alongside traditional fecal indicators like *E. coli* or enterococci, to improve water quality characterization in monitoring programs (Ahmed et al., 2008; Koskey et al., 2014; Ryu et al., 2012; Sinigalliano et al., 2013). However, while high-throughput 16S rRNA gene amplicon sequencing offers advanced insights, its widespread use is hindered by the need for specialized expertise and laboratory infrastructure, posing challenges for regular water quality monitoring (Brewton et al., 2022; Nassereddine et al., 2024). Addressing this challenge requires collaborative efforts between government/state agencies, water managers, and academic institutions (Brewton et al., 2022). The adoption of QT-AMP could offer a practical solution to augment existing water quality monitoring efforts. In the proposed workflow, as shown in Fig. 7.1, water

samples are collected by the local environmental laboratories as part of routine monitoring, with subsequent MPN counting and freezing of Quanti-Tray at −20 °C. These trays are then shipped to a university or commercial molecular research laboratory for sequencing analysis. The resulting DNA sequence data, containing species names of *Enterococcus* and corresponding sequence read numbers, can be conveniently provided in a spreadsheet. At a cost of US $50–100 per Illumina sequence read, with potential cost reductions for bulk analysis and future technological advancements (Van Dijk et al. 2018), local water quality managers could feasibly send Quanti-Tray samples directly to contracted commercial molecular research laboratories for high-throughput sequencing. This streamlined approach would enable them to receive enterococci abundance data in an easily interpretable format, facilitating informed decision-making in water quality management.

This study was conducted to assess the microbiological quality of estuary water post-Hurricane Ian using QT-AMP targeting enterococci. QT-AMP method was coevaluated with quantitative PCR analysis and chemical tracer analysis and this process extremely limited the sample size. Thus, more extensive research is required to validate our findings.

Future research directions may include the establishment of shared species composition libraries for various environmental samples and the introduction of artificial intelligence to determine the potential source of contamination. The accuracy of *Enterococcus* species identification should improve if we can use nearly complete 16S rRNA gene sequences or other evolutionarily conserved molecular markers (e.g., *gyrB*, *dnaK*). We could retrieve more genomic information (e.g., antibiotic resistance genes) if we combine metagenomic approaches.

5 Conclusion

In summary, we documented the usefulness of a new approach to characterize enterococci using an amplicon sequencing platform from Quanti Trays after being used for counting the MPN of enterococci. While the current Enterolert MPN method offers quantitative data alone, the integration of amplicon sequencing allows for an examination of the relative abundance of *Enterococcus* species composition. QT-AMP may have the ability to trace back contamination sources more accurately or understand the potential risks associated with each species. Through QT-AMP, a diverse array of *Enterococcus* species can be detected in post-hurricane water samples. Seven surface water samples collected after the hurricane landing from the Charlotte Harbor estuary were tested, revealing 12 *Enterococcus* species post-hurricane, including four rare and eight cosmopolitan species. *E. hirae* showed increased abundance post-hurricane, notably correlating with MPN counts, attributed to multiple animal and plant sources. QT-AMP analysis identified *E. hirae* as correlated with elevated MPN counts, indicating its importance as an indicator of post-hurricane samples. The two sites with the highest MPN counts, out of the seven sites examined, exhibited similar numerical values. However, the QT-AMP analysis

revealed distinct compositions of *Enterococcus* species in these sites, with one site being dominated by *E. faecalis* and the other by *E. casseliflavus*. Our recommendation is for water managers to prioritize locations with frequent reports of high *E. faecalis* abundance. We propose targeted testing for specific *Enterococcus* species at high-risk sites to further investigate and mitigate contamination. Despite similar MPN counts, LBANG1 exhibited a greater microbial risk factor than GSV001 in this instance. The relative abundance data of enterococci obtained via QT-AMP enabled the differentiation of estuary water samples in our study. This highlights the potential of QT-AMP to enhance the quantification and identification of *Enterococcus*, thus making it applicable to a range of water management projects and future assessments of human health risks.

Acknowledgments Thank you to the FDEP-Charlotte Harbor Estuaries Volunteer Water Quality Monitoring Network (CHEVWQMN) volunteers for the collection of samples and the FDEP Laboratory and Sanders Laboratory staff for the analyses. We also thank Melynda Brown at FDEP and Brandon Moody at Charlotte County for helpful discussion and critical reading of the early draft, and Michael Kratz and Haruka Urakawa at FGCU for QT-AMP sample processing. This research was supported by the FGCU Wings for the Future Award and Florida Sea Grant.

Author Contributions Hidetoshi Urakawa contributed to the conception and design of this study. Material preparation, data collection, and analysis were performed by Hidetoshi Urakawa and Taylor Hancock. The first draft of the manuscript was written by Hidetoshi Urakawa, and Taylor Hancock commented on previous versions of the manuscript. All authors read and approved the final manuscript.

Data Availability All data associated with 16S rRNA gene amplicon sequencing reads are available in NCBI under BioProject PRJNA1227612.

Conflict of Interest The authors have no competing interests to declare that are relevant to the content of this article.

References

Abbott, S., Caughley, B., & Scott, G. (1998). Evaluation of Enterolert® for the enumeration of enterococci in the marine environment. *New Zealand Journal of Marine and Freshwater Research, 32*(4), 505–513.

Abdelzaher, A. M., Wright, M. E., Ortega, C., Solo-Gabriele, H. M., Miller, G., Elmir, S., Newman, X., Shih, P., Bonilla, J. A., Bonilla, T. D., & Palmer, C. J. (2010). Presence of pathogens and indicator microbes at a non-point source subtropical recreational marine beach. *Applied and Environmental Microbiology, 76*(3), 724–732.

Ahmed, W., Stewart, J., Powell, D., & Gardner, T. (2008). Evaluation of Bacteroides markers for the detection of human faecal pollution. *Letters in Applied Microbiology, 46*(2), 237–242.

Badgley, B. D., Nayak, B. S., & Harwood, V. J. (2010). The importance of sediment and submerged aquatic vegetation as potential habitats for persistent strains of enterococci in a subtropical watershed. *Water Research, 44*(20), 5857–5866.

Bae, H. S., & Hou, A. (2013). 23S rRNA gene-based enterococci community signatures in Lake Pontchartrain, Louisiana, USA, following urban runoff inputs after Hurricane Katrina. *Microbial Ecology, 65*, 289–301. https://doi.org/10.1007/s00248-012-0166-z

Bernhard, A. E., & Field, K. G. (2000). A PCR assay to discriminate human and ruminant feces on the basis of host differences in Bacteroides-Prevotella genes encoding 16S rRNA. *Applied and Environmental Microbiology, 66*(10), 4571–4574.

Brewton, R. A., Kreiger, L. B., Tyre, K. N., Baladi, D., Wilking, L. E., Herren, L. W., & Lapointe, B. E. (2022). Septic system–groundwater–surface water couplings in waterfront communities contribute to harmful algal blooms in Southwest Florida. *Science of The Total Environment, 837*, 155319.

Boehm, A. B., & Sassoubre, L. M. (2014). Enterococci as indicators of environmental fecal contamination. In M. S. Gilmore, D. B. Clewell, Y. Ike, & N. Shankar (Eds.), *Enterococci: From commensals to leading causes of drug resistant infection*. Eye and Ear Infirmary.

Bucci, L., Alaka, L., Hagen, A., Delgado, S., & Beven, J. (2023). Hurricane Ian. In *Report AL092022*. NOAA National Hurricane Center Tropical Cyclone Report. https://www.nhc.noaa.gov/data/tcr/AL092022_Ian.pdf

Budnick, G. E., Howard, R. T., & Mayo, D. R. (1996). Evaluation of Enterolert for enumeration of enterococci in recreational waters. *Applied and Environmental Microbiology, 62*(10), 3881–3884.

Byappanahalli, M., & Fujioka, R. (2004). Indigenous soil bacteria and low moisture may limit but allow faecal bacteria to multiply and become a minor population in tropical soils. *Water Science and Technology, 50*(1), 27–32.

Byappanahalli, M. N., Shively, D. A., Nevers, M. B., Sadowsky, M. J., & Whitman, R. L. (2003). Growth and survival of *Escherichia coli* and enterococci populations in the macro-alga Cladophora (Chlorophyta). *FEMS Microbiology Ecology, 46*(2), 203–211.

Byappanahalli, M. N., Nevers, M. B., Korajkic, A., Staley, Z. R., & Harwood, V. J. (2012). Enterococci in the environment. *Microbiology and Molecular Biology Reviews, 76*(4), 685–706.

Devane, M., Moriarty, E., Weaver, L., Cookson, A., & Gilpin, B. (2020). Fecal indicator bacteria from environmental sources; strategies for identification to improve water quality monitoring. *Water Research*, 116204.

Devriese, L., Cruz Colque, J., Haesebrouck, F., Desmidt, M., Uyttebroek, E., & Ducatelle, R. (1992). *Enterococcus hirae* in septicaemia of psittacine birds. *Veterinary Record, 130*(25), 558–559.

Eckner, K. F. (1998). Comparison of membrane filtration and multiple-tube fermentation by the Colilert and Enterolert methods for detection of waterborne coliform bacteria, *Escherichia coli*, and enterococci used in drinking and bathing water quality monitoring in southern Sweden. *Applied and Environmental Microbiology, 64*(8), 3079–3083.

Edgar, R. C., Haas, B. J., Clemente, J. C., Quince, C., & Knight, R. (2011). UCHIME improves sensitivity and speed of chimera detection. *Bioinformatics, 27*(16), 2194–2200.

Elmir, S. M., Shibata, T., Solo-Gabriele, H. M., Sinigalliano, C. D., Gidley, M. L., Miller, G., Plano, L. R., Kish, J., Withum, K., & Fleming, L. E. (2009). Quantitative evaluation of enterococci and Bacteroidales released by adults and toddlers in marine water. *Water Research, 43*(18), 4610–4616.

Ferguson, D. M., Griffith, J. F., McGee, C. D., Weisberg, S. B., & Hagedorn, C. (2013). Comparison of *Enterococcus* species diversity in marine water and wastewater using Enterolert and EPA Method 1600. *Journal of Environmental and Public Health 2013, 2013*, 848049., 6 pages. https://doi.org/10.1155/2013/848049

Gilmore, M. S., Clewell, D. B., Ike, Y., & Shankar, N. (2014). *Enterococci: From commensals to leading causes of drug resistant infection*. Eye and Ear Infirmary.

Graves, A., Weaver, R. W., & Entry, J. (2009). Characterization of enterococci populations in livestock manure using BIOLOG. *Microbiological Research, 164*(3), 260–266.

Green, H. C., Dick, L. K., Gilpin, B., Samadpour, M., & Field, K. G. (2012). Genetic markers for rapid PCR-based identification of gull, Canada goose, duck, and chicken fecal contamination in water. *Applied and Environmental Microbiology, 78*(2), 503–510.

Green, H. C., White, K. M., Kelty, C. A., & Shanks, O. C. (2014). Development of rapid canine fecal source identification PCR-based assays. *Environmental Science & Technology, 48*(19), 11453–11461.

Harwood, V. J., Whitlock, J., & Withington, V. (2000). Classification of antibiotic resistance patterns of indicator bacteria by discriminant analysis: Use in predicting the source of fecal contamination in subtropical waters. *Applied and Environmental Microbiology, 66*(9), 3698–3704.

Imamura, G. J., Thompson, R. S., Boehm, A. B., & Jay, J. A. (2011). Wrack promotes the persistence of fecal indicator bacteria in marine sands and seawater. *FEMS Microbiology Ecology, 77*(1), 40–49.

Koskey, A. M., Fisher, J. C., Traudt, M. F., Newton, R. J., & McLellan, S. L. (2014). Analysis of the gull fecal microbial community reveals the dominance of *Catellicoccus marimammalium* in relation to culturable enterococci. *Applied and Environmental Microbiology, 80*(2), 757–765.

Layton, B., Walters, S., Lam, L., & Boehm, A. (2010). *Enterococcus* species distribution among human and animal hosts using multiplex PCR. *Journal of Applied Microbiology, 109*(2), 539–547.

Lipp, E. K., Farrah, S. A., & Rose, J. B. (2001). Assessment and impact of microbial fecal pollution and human enteric pathogens in a coastal community. *Marine Pollution Bulletin, 42*(4), 286–293.

Lu, J., Santo Domingo, J. W., Lamendella, R., Edge, T., & Hill, S. (2008). Phylogenetic diversity and molecular detection of bacteria in gull feces. *Applied and Environmental Microbiology, 74*(13), 3969–3976.

Monticelli, L. S., Decembrini, F., Bergamasco, A., & Caruso, G. (2019). Water quality assessment of transitional and coastal marine Sicilian waters (Italy): Ecological and epidemiological significance of multiple antimicrobial resistant *Enterococcus* spp. *Estuarine, Coastal and Shelf Science, 217*, 173–184.

Müller, T., Ulrich, A., Ott, E. M., & Müller, M. (2001). Identification of plant-associated enterococci. *Journal of Applied Microbiology, 91*(2), 268–278.

Nassereddine, Z., Opara, S., Coutinho, O., Qyteti, F., Book, R., Heinicke, M. P., Napieralski, J., & Tiquia-Arashiro, S. M. (2024). Critical perspectives on advancing antibiotic resistant gene (ARG) detection technologies in aquatic ecosystems. *Science of the Total Environment, 957*, 177775. https://doi.org/10.1016/j.scitotenv.2024.177775

Nayak, B. S., Badgley, B., & Harwood, V. J. (2011). Comparison of genotypic and phylogenetic relationships of environmental *Enterococcus* isolates by BOX-PCR typing and 16S rRNA gene sequencing. *Applied and Environmental Microbiology, 77*(14), 5050–5055.

Posit Team. (2023). *RStudio: Integrated development environment for R. Posit Software*. PBC. http://www.posit.co/

R Core Team. (2023). *R: A language and environment for statistical computing*. R Foundation for Statistical Computing. https://www.R-project.org/.

Ramoutar, S. (2020). The use of Colilert-18, Colilert and Enterolert for the detection of faecal coliform, *Escherichia coli* and *Enterococci* in tropical marine waters, Trinidad and Tobago. *Regional Studies in Marine Science, 40*, 101490.

Ran, Q., Badgley, B. D., Dillon, N., Dunny, G. M., & Sadowsky, M. J. (2013). Occurrence, genetic diversity, and persistence of enterococci in a Lake Superior watershed. *Applied and Environmental Microbiology, 79*(9), 3067–3075.

Reischer, G. H., Kasper, D. C., Steinborn, R., Mach, R. L., & Farnleitner, A. H. (2006). Quantitative PCR method for sensitive detection of ruminant fecal pollution in freshwater and evaluation of this method in alpine karstic regions. *Applied and Environmental Microbiology, 72*(8), 5610–5614.

Ryu, H., Griffith, J. F., Khan, I. U., Hill, S., Edge, T. A., Toledo-Hernandez, C., Gonzalez-Nieves, J., & Santo Domingo, J. (2012). Comparison of gull feces-specific assays targeting the 16S rRNA genes of *Catellicoccus marimammalium* and *Streptococcus* spp. *Applied and Environmental Microbiology, 78*(6), 1909–1916.

Saingam, P., Di, D. Y., & Yan, T. (2021). Diversity and health risk potentials of the *Enterococcus* population in tropical coastal water impacted by Hurricane Lane. *Journal of Water and Health, 19*(6), 990–1001. https://doi.org/10.2166/wh.2021.209

Schang, C., Henry, R., Kolotelo, P. A., Prosser, T., Crosbie, N., Grant, T., Cottam, D., O'Brien, P., Coutts, S., & Deletic, A. (2016). Evaluation of techniques for measuring microbial hazards in bathing waters: A comparative study. *PLoS One, 11*(5), e0155848.

Sercu, B., Van De Werfhorst, L. C., Murray, J. L., & Holden, P. A. (2011). Cultivation-independent analysis of bacteria in IDEXX Quanti-Tray/2000 fecal indicator assays. *Applied and Environmental Microbiology, 77*(2), 627–633.

Staley, C., Dunny, G. M., & Sadowsky, M. J. (2014). Environmental and animal-associated enterococci. *Advances in Applied Microbiology, 87*, 147–186.

Sinigalliano, C. D., Ervin, J. S., Van De Werfhorst, L. C., Badgley, B. D., Ballesté, E., Bartkowiak, J., Boehm, A. B., Byappanahalli, M., Goodwin, K. D., & Gourmelon, M. (2013). Multi-laboratory evaluations of the performance of *Catellicoccus marimammalium* PCR assays developed to target gull fecal sources. *Water Research, 47*(18), 6883–6896.

Švec, P., & Sedláček, I. (1999). Occurrence of *Enterococcus* spp. in waters. *Folia Microbiologica, 44*, 3–10.

Urakawa, H., Martens-Habbena, W., & Stahl, D. A. (2010). High abundance of ammonia-oxidizing Archaea in coastal waters, determined using a modified DNA extraction method. *Applied and Environmental Microbiology, 76*(7), 2129–2135.

Urakawa, H., Kratz, M. A., Hancock, T. L., & Armstrong, R. A. (2022). QT-AMP: Sequencing PCR amplicons from Quanti-Tray wells to analyze enterococci communities. *Science of the Total Environment, 839*, 156188.

Van Dijk, E. L., Jaszczyszyn, Y., Naquin, D., & Thermes, C. (2018). The third revolution in sequencing technology. *Trends in Genetics, 34*(9), 666–681.

Yamahara, K. M., Layton, B. A., Santoro, A. E., & Boehm, A. B. (2007). Beach sands along the California coast are diffuse sources of fecal bacteria to coastal waters. *Environmental Science & Technology, 41*(13), 4515–4521.

Chapter 8
A Review of Metal and Antibiotic Resistance of Microbial Strains and Their Bioremediation Ability

Subhomita Ghosh Roy and Joseph C. Mester

1 Introduction

Freshwater is a scarce and valuable resource (Walker et al., 2019) but can get easily contaminated by surface water pollution (Singh et al., 2022). Nutrients, plastics, heavy metals, pesticides, fertilizers, antibiotics, and industrial waste discharges into waterbodies like rivers, streams, wetlands, ponds, and oceans generally cause surface water pollution (Singh et al., 2022; Nassereddine et al., 2024). Stormwater runoff also majorly contributes to surface water pollution (Singh et al., 2022). Over time as public awareness about the deterioration of freshwater quality has increased, several water pollution control acts have been passed globally. For example, the US Clean Water Act (CWA) 1972 (USEPA, 2023b), the Water Framework Directive of the EU (EU, 2023), the Water Pollution Prevention and Control Law of the People's Republic of China—Amended in 2017 (Ministry of Ecology and Environment The People's Republic of China, 2017), and The Water (Prevention & Control of Pollution) Act 1974, India (CPCB, 2019). The common goal of all these laws is to control pollution discharge into waterbodies and maintain water quality standards.

The two main ways surface water gets polluted are through urban runoff and wastewater (treated, untreated, or sewage overflow), USEPA, 2023a). According to the United Nations, 42% of household wastewater is improperly treated, damaging ecosystems and human health (UN, 2023). Wastewater from industry, domestic sources, or pharmaceuticals that comes to a wastewater treatment plant (WWTP) has heavy metals (e.g., Cu, Pb, Cd, Zn, Ag, Ni, Fe, and Cr) and antibiotics (e.g., Beta-Lactams, Tetracyclines, Quinolones, Sulfonamides, and Macrolides). If traces of these metals and antibiotics make their way through the treatment process

S. Ghosh Roy (✉) · J. C. Mester
Department of Biological Sciences, Northern Kentucky University,
Highland Heights, KY, USA
e-mail: subhomitagr@gmail.com; mesterj1@nku.edu

S. M. Tiquia-Arashiro et al. (eds.), *Urban Watershed Microbiology, Volume 1*,
https://doi.org/10.1007/978-3-031-89071-0_8

they can end up in surface water. Metals and antibiotics can affect the biota or even enter our households, as surface waters are also a primary source of consumable water after treatment (USEPA, 2017). For example, in 2014 a water crisis began in Flint, Michigan, when the city switched its water source to the Flint River from the Detroit Water and Sewage Department (DWSD) (Masten et al., 2016). Soon after the switch, elevated Pb (lead) levels were detected in the drinking water, even after the water went through water treatment. The decision was made to save money, but it resulted in a public health emergency due to inadequate treatment. The incident raised concerns about water safety if a heavy metal pollutant like Pb leaked into drinking water facilities (Masten et al., 2016). Similarly, beginning in 2016, high Pb contamination was observed in Newark, New Jersey schools. It was found that this was a result of corrosion of old lead pipes and insufficient water treatment. By January 2021, Newark was on its way to removing a significant number of lead service lines, implementing a new treatment process designed to reduce the lead corrosion from pipes, and also provide water filters to many affected households (NRDC, 2021). The Centers for Disease Control and Prevention has explained that antibiotics can be traced to hospitals, farms, or sewage discharges, and sometimes wastewater treatment can fall short of removing antibiotic-resistant microbial pathogens or their genes (CDC, 2021).

Therefore, the need to treat incoming wastewater with efficient treatment facilities is evident. In this scenario, emerging technologies have given rise to the advancement of "Bioremediation" as a potent alternative to mitigate the toxic effects of contaminant accumulation (Bhatnagar & Kumari, 2013; Tiquia-Arashiro, 2018). Bioremediation is a process of biodegradation either to eliminate or mitigate the impacts of environmental contaminants (Cho et al., 2012; Seagren, 2023). Bioremediation can degrade or eliminate toxic contaminants such as heavy metals or antibiotics present in wastewater by reducing, detoxifying, transforming, or mineralizing them into less toxic forms (Liu et al., 2021; Seagren, 2023).

Specifically, while addressing heavy metal (e.g., Cu, Pb, Cd, Zn, Ag, Ni, Fe, and Cr) contamination bioremediation with bacteria, fungi, and algae has been found as a cost-effective, efficient, and environmentally safe process to use (Bowman et al., 2018; Ameen et al., 2020; Pagnucco et al., 2023). Studies have reported the bioremediation potential of bacterial species such as *Pseudomonas aeruginosa* with a high bioremediation potential of 58.80% for 50 mg Cd L^{-1} and 33.67% for 300 mg Pb L^{-1}. Additionally, *Klebsiella edwardsii* exhibited a remediation potential of 79.87% and 92.41% for Cd and Pb respectively (Oziegbe et al., 2021).

Similarly, antibiotics have also been observed to be remediated from wastewater. Antibiotics such as Triclosan, enrofloxacin, metronidazole, tetracycline, levofloxacin, and sulfamethoxazole have shown 100%, 99.8%, 99%, 98.8%, 98.4%, 96.69%, and 91.9% removal respectively within a setting of constructed wetland wastewater treatment process with the help of residing microbes (Ohore et al., 2022). Bioaugmentation has been used as a very effective bioremediation technology for specific antibiotics (Hong et al., 2020). Bioaugmentation is adding specific microorganisms to efficiently biodegrade pollutants from the environment (Handore et al., 2023). For example, the novel bacterium *Burkholderia cepacia* (BC) was

observed to degrade tetracycline antibiotics with the help of bioaugmentation (inoculation dose 15% v/w, temperature 28–43 °C, pH 4.5–6.5) (Hong et al., 2020).

This indicates the potential of bioremediation in treating contaminants like antibiotics and heavy metals, commonly found in wastewater. Bioremediation is a highly efficient, and environment-friendly process that does not cause any secondary pollutants and only relies on the action of the microbes for pollutant removal from wastewater (Liu et al., 2021; Bwapwa, 2022). Bioremediation can be done in the original location of contamination (in situ) or off-site (ex situ). Several molecular mechanisms in microbes can result in bioremediation.

The goal of this review is threefold. First, this review will discuss the processes and molecular mechanisms of bioremediation, followed by the role of microorganisms in metal and antibiotic bioremediation with specific examples. Finally, the significance of bioremediation in metal and antibiotic reduction will be summarized along with possible future research directions.

2 The Process of Bioremediation

Bioremediation is a "biodegradation" process where toxins in a polluted environment are biologically degraded or eliminated. Biodegradation happens through biological activities, which can be either aerobic (when oxygen is present) or anaerobic (when oxygen is absent) (Sayqal & Ahmed, 2021). Bioremediation can also be classified as in situ or ex situ. The in situ bioremediation methods treat contaminants at their original location, on the other hand in ex situ bioremediation technologies involve excavating the soil or the contaminated materials and then treating them off-site (Sayqal & Ahmed, 2021; Seagren, 2023).

2.1 Ex Situ Bioremediation Types

Biopile Excavated polluted soil is piled above ground and then nutrients are added. Sometimes aeration is part of the process to increase microbial reactions and thus bioremediation (Azubuike et al., 2016). Biopiled soil has been successfully applied to cold regions and polluted areas where the chances of volatilization of pollutants are high (Dias et al., 2015; Gomez & Sartaj, 2013; Whelan et al., 2015).

Windrows In this technique, the polluted soil is regularly turned around over, and water is added. This increases microbial activity and aeration and enhances the bioremediation process (Azubuike et al., 2016). When used for bioremediation, Windrowed soil is often seen to remove more hydrocarbons (Coulon et al., 2010).

Bioreactors This is a vessel where a series of biochemical reactions take place with the addition of enzymes, microorganisms, animal cells, and plant cells. These vessels have been observed to successfully bioremediate a wide range of pollutants

(e.g., crude oil-polluted sediment, waste lubricating oil, laundry wastewater, pharmaceutical sludge, effluents, and hydrocarbon-rich contaminated soil) (Bhattacharya et al., 2015; Chikere et al., 2016; Delforno et al., 2015; Mustafa et al., 2015; Saravanan et al., 2015; Shannon et al., 2015). Bioreactors are often reported to have a high rate of bioremediation (Tiquia-Arashiro, 2019; Sayqal & Ahmed, 2021).

Land Farming This is the simplest and low-cost bioremediation technique. In this technique, contaminated soil is excavated over a pre-existing prepared site along with tilling. The degradation of pollutants is achieved with the help of microorganisms. This is a very effective technique for hydrocarbon-contaminated soil remediation (Sayqal & Ahmed, 2021; Sivakumar et al., 2014).

2.2 *In Situ Bioremediation Types*

Bioventing and Biosparging Bioventing involves limited airflow and additional nutrients being supplied to stimulate microbial activity (Sayqal & Ahmed, 2021). Biosparging is the injection of low-pressure air below the water table to increase oxygen in the groundwater. This increased oxygen increases bioremediation activity at the pollution site. Both are low-cost technologies for the bioremediation process (Vidali, 2001).

Biostimulation In this process, the polluted soil is added with additional nutrients (phosphorus, nitrogen, oxygen, or carbon) to increase or stimulate microbial activity, to initiate bioremediation (Medina-Bellver et al., 2005).

Phytoremediation This is an emerging technology where plants are used to remove pollutants from water and soil. This technology works best with low-level contaminants. However, this technology has some concerns such as the stress tolerance of plants, bioaccumulation, toxicity, and the type of plant to be used (Farraji et al., 2016).

Bioaugmentation With this technology, specific types of efficient microorganisms are added to the site of pollution. This technology can be helpful and highly cost-effective in wastewater treatment rich with complicated pollutants (Handore et al., 2023). But this technology has some disadvantages as well. For example, it has been observed that as new microbes are introduced, they tend to outcompete the native microbial population living at the site of contamination (Bouchez et al., 2000; Stroo et al., 2013).

3 Role of Microorganisms in Metal Bioremediation

All forms of microbial organisms such as bacteria, fungi, and algae are capable of bioremediation, a process of pollution treatment. Microorganisms are present universally and can remove environmental pollutants using a variety of mechanisms (Ayilara & Babalola, 2023). These mechanisms are either by mobilization or immobilization of the pollutants (Ndeddy Aka & Babalola, 2016; Verma & Kuila, 2019). Microbes carry diverse sets of genes for metabolic processes that enable them for immobilization or mobilization (Fig. 8.1). However, the complete degradation depends on the interaction of microbes, the substrates, and the environmental conditions (Thakkar & Vincent, 2021; Pagnucco et al., 2023).

In mobilization, contaminants are converted into a form that allows them to move around the environment (Ayilara & Babalola, 2023) (Fig. 8.1). Contaminants such as metals can be mobilized through processes like protonation, chelation, and chemical transformation, which can volatilize the metal. This process can break down insoluble metal compounds like phosphates, oxides, or sulfates (Gadd, 2004). Specifically, the biochemical processes of mobilization are enzymatic oxidation, bioleaching, biostimulation, bioaugmentation, and enzymatic reduction (Ayangbenro & Babalola, 2017; Tak et al., 2012) (Fig. 8.1).

As the terms indicate in immobilization the contaminants are converted into a form that is not available to the environment (Fig. 8.1), for example, protein-heavy metal complexes, without disturbing their functions (Jeyakumar et al., 2023). Hence, heavy metals are remediated using immobilization (Ayilara & Babalola, 2023) (Fig. 8.1). Microbes use immobilization for both in situ and ex situ methods of bioremediation (Pratush et al., 2018). Immobilization decreases the number of free metal species (Gadd, 2004). Specifically, the biochemical processes of immobilization are bioaccumulation, biosorption, and precipitation (Ayangbenro & Babalola, 2017; Tak et al., 2012) (Fig. 8.1). However, some reactions can

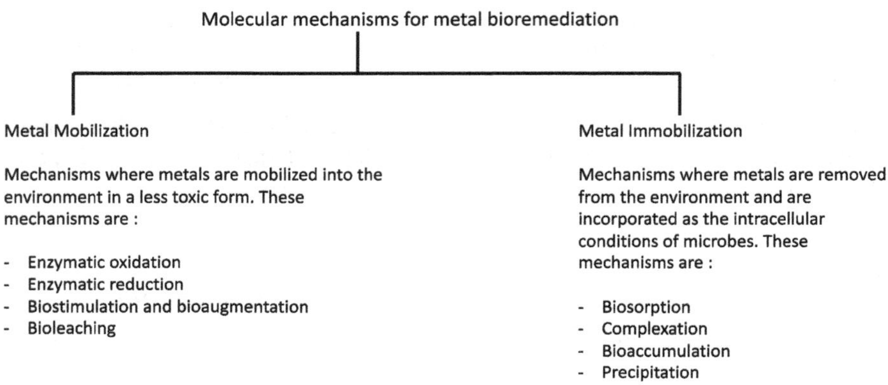

Fig. 8.1 The two main molecular mechanisms for metal bioremediation with specific processes for each (Ayilara & Babalola, 2023)

immobilize or mobilize metals, depending on the metal in question (Gadd, 2004). Microbes also have a way of protecting themselves from contaminants by making hydrophobic or solvent efflux pumps that protect their outer cell membrane (Verma & Kuila, 2019).

3.1 The Molecular Mechanisms of Microbial Remediation for Metal-Mobilization

Enzymatic Oxidation In this mobilization process pollutant compounds are oxidized from a higher to a lower oxidation state causing the heavy metals to lose an electron and become less toxic. This method relies on an enzyme (oxidoreductase) produced by microbes. The oxidative enzymes generate radicals that fragment into various fractions, forming compounds with high molecular weights. It is particularly effective in remediation scenarios involving dyes, phenols, and other pollutants that are resistant to bacterial degradation (Unuofin et al., 2021).

Enzymatic Reduction This mobilization process is the reverse of enzymatic oxidation and transforming pollutants into a state of reduced oxidation, making them insoluble. Obligate and facultative anaerobic microbes can carry out this process. For example, chrome reductase facilitates the conversion of hexavalent chromium to trivalent chromium, while azoreductase is reduced to azo compounds by breaking the azo bonds by this process (Saxena et al., 2020; Pagnucco et al., 2023).

Biostimulation and Bioaugmentation These two mobilization techniques are used by microbes in in situ bioremediation mechanisms. In biostimulation, additional nutrients (phosphorus, nitrogen, oxygen, or carbon) are added to the polluted sites to stimulate more microbial activity (Medina-Bellver et al., 2005). Whereas, in Bioaugmentation efficient microorganisms are added to the site of the pollution (Handore et al., 2023). Microbes have been observed to mobilize and remediate pesticides in less toxic forms using bioaugmentation (Goswami et al., 2018; Ojuederie et al., 2021). Similarly, microbes also have been observed to mobilize and remediate heavy metals using biostimulation (Ayilara & Babalola, 2023).

Bioleaching In this process, acidophilic microbes solubilize solid-state heavy metals from sediment, particularly iron or sulfur pollutants (Bhandari et al., 2023; Sun et al., 2019).

3.2 The Molecular Mechanisms of Microbial Remediation for Metal-Immobilization

Biosorption: Heavy metal pollutants are removed in this process, by replacing ions and protons resulting in chelates, complexes, and physical interaction with electrostatic forces (Mahmoud, 2021). Biosorption involves the transfer of ions

between metal cations (positive charge) and negatively charged active groups on the outer cell surface of microorganisms (Michalak et al., 2013; Mathew & Krishnamurthy, 2018; Baltazar et al., 2019; Sedlakova-Kadukova et al., 2019).

Complexation is an important mechanism within biosorption. It uses ligands (ligands are ions or neutral molecules that bond to a central metal atom or ion) to create complexes with inorganic metals, which are environmental pollutants, particularly in solid wastes (Ayangbenro & Babalola, 2017). Extracellular polymeric substances on microbial surfaces, complex with polluting heavy metals (Xiao et al., 2019). For example, in a study by Xiao et al. (2019), biochar was used to remove copper (II) oxide and hexavalent chromium from wastewater through a mechanism involving complexation.

Bioaccumulation In this process, the absorption rate of a compound exceeds its elimination rate. Potentially toxic compounds are accumulated in the intracellular environment of microbes (Sharma et al., 2022a). Heavy metals are transported across microbial membranes through various mechanisms such as carrier-mediated transport, protein channels, and ion pumps (Mir-Tutusaus et al., 2018).

Precipitation Involves the transformation of heavy metals or pollutants into precipitates or crystals, reducing their toxicity. This can happen during biogeochemical cycles, leading to the deposition of metals like iron and manganese (Sharma et al., 2022a, b). For example, in alkaline pH (level above 7) conditions sulfate-reducing bacteria can convert organo-phosphate to ortho-phosphate (Pratush et al., 2018).

While both metal mobilization and immobilization can be used for bioremediation, and both can be equally efficient and cost-effective. The applicability or advantage is largely dependent on the chemical state of the contaminating metal. For example, if the metal is in an insoluble state, then the metal mobilization technique can be used to transform the metal, as insoluble metal species are not biodegradable (Jeyakumar et al., 2023). Whereas the immobilization technique can be used for metal removal as microorganisms create protein-heavy metal complexes and trap the contaminant metal without affecting their own internal function (Jeyakumar et al., 2023).

4 Environmental Factors Impacting Microbial Bioremediation of Metals and Examples of Microorganisms

4.1 Environmental Factors

Environmental factors such as acid conditions, carbon content, moisture content, temperature, chemical structure and concentration of pollutants and soil contents can affect bioremediation potential. Excessive acid, low environmental moisture, cold temperatures, and highly saline soil can adversely affect the survivability of the

Table 8.1 List of environmental conditions and the effects on metal bioremediation of microbes with references

Environmental conditions	Effect	Reference
1. Excessive acid and carbon content		
A. Excessive acid	Survivability of microbes is low	Kharangate-Lad and D'Souza (2021)
B. Carbon	Main energy source. Affects survivability and competition among microbial species	Ayilara and Babalola (2023) and Patel et al. (2022)
2. Moisture content		
A. Low moisture	Survivability of microbes is low	Ayilara and Babalola (2023)
B. High moisture	Increases anaerobic activity suppresses growth of aerobic microbes	Ayilara and Babalola (2023)
3. Temperature		
A. High temperatures	Metal solubility and availability increase bioremediation in the form of degradation increases.	Mahmoud (2021)
B. Cold temperatures	Survivability of microbes is low, metabolic activity decreases, metal reduction decreases	Bala et al. (2022) and Ren et al. (2018)
4. Chemical structure and concentration		
Simple chemical structure and low concentration of metal	Better bioremediation potential; easy for microbes to degrade	Kebede et al. (2021)
5. Soil conditions		
Top layer soil (0–10 cm)	Higher oxygen and organic matter so more microbes present. Bioremediation in the form of degradation increases	Ndeddy Aka and Babalola (2016)
Well-drained soil with fine texture	Habitat of many microbes so better possibilities of bioremediation	Huang et al. (2019)
High salinity	Decreases metabolic activity thus decreasing bioremediation potential	Imron et al. (2020) and Kebede et al. (2021)

microbial population and decrease the metal bioremediation rate (Table 8.1). But conditions like higher temperature, higher moisture, top layer of soil, well-drained soil, simple chemical structure, and low concentration of metal increase the possibility of metal bioremediation rate (Table 8.1). The presence of Carbon (the main energy source) in the environment can also affect the bioremediation rate as this increases competition among microbial populations (Table 8.1). A detailed description of the effect of environmental conditions on microbial metal bioremediation with all references are listed below:

Table 8.2 Microorganisms effective for metal bioremediation

Organism	Metal	Bioremediation potential	Reference
Bacillus megaterium *Rhizopus stolonifera*	Heavy metal	Effective bioremediation	Njoku et al. (2020)
Klebsiella *Enterobacter*	As, Pb, Cd	Effective bioremediation	González Henao and Ghneim-Herrera (2021)
Alcaligenes faecalis *Bacillus pumilus* *Bacillus* sp. *Pseudomonas aeruginosa* *Brevibacterium iodinium*	Hg, Cd, Pb	Detox efficiency. Can remove 70% Cadmium and 98% Lead within 72 and 96 h.	De et al. (2008)
Micrococcus luteus	Pb	Can remove or bioremediate large amount of Lead from synthetic material	Puyen et al. (2012)
Bacillus megaterium *Bacillus subtilis*	Pb	High lead reduction	Abioye et al. (2018)
Aspergillus niger *Penicillium* sp.	Cr	High chromium reduction	Abioye et al. (2018)
Bacillus subtilis *Bacillus megaterium*	Cd	High cadmium reduction after 20 days	Abioye et al. (2018)
Bacterial Consortia	Cr, Zn, Cd, Pb, Cu, Co	Removes 75–85% Chromium, Zinc, Cadmium, Lead, Copper, and Cobalt in less than two hours	Abbas et al. (2014)

4.2 Examples of Microorganisms Involved in Metal Bioremediation

Several researchers have investigated the metal bioremediation potential of microorganisms. The examples of microorganisms along with the metals are described in Table 8.2.

5 Role of Microorganisms in Antibiotic Bioremediation

Following the initial discovery of penicillin in the early 1900s, antibiotics have been produced on a large scale and extensively utilized in medical, veterinary, and agricultural practices. Antibiotics are natural, synthetic, or partly synthetic molecules with potent antibacterial properties (Catteau et al., 2018). In addition to widespread use in humans and animals to treat infectious diseases, antibiotics have been used

globally as growth-promoting agents for agricultural animal production (Dibner & Richards, 2005; Gong et al., 2024; Kemper, 2008). After application, antibiotics make their way into wastewaters, WWTPs, biosolids, soil, groundwater, surface water, and even drinking water (Barancheshme & Munir, 2018; Fayaz et al., 2024; Zhang et al., 2018). In addition to disrupting aquatic and soil ecosystems by depleting microbial communities, antibiotic exposure can be toxic to plants and animals at low doses (ng/L to mg/L concentrations) (Bhatt & Chatterjee, 2022; Cheong et al., 2020; Liu et al., 2009; Yang et al., 2020; Yang et al., 2021).

Bacterial populations have rapidly adapted to prevalent antibiotics by developing and sharing antibiotic resistance genes (ARGs) (Allen et al., 2010; Economou & Gousia, 2015; Gong et al., 2024). Bacteria may acquire resistance via mutation in existing genes or by the transmission of ARGs via horizontal gene transfer (HGT) (Darby et al., 2023; Wellington et al., 2013). Resistance genes may be transferred between related or divergent taxonomic groups of bacteria and between pathogenic and non-pathogenic organisms (Wellington et al., 2013; Zhang et al., 2022). The widespread distribution of ARGs in bacterial populations is concerning due to the increased risk of untreatable infections by pathogenic organisms (Zhang et al., 2022). Many medically relevant bacteria are now resistant to multiple antibiotics and are a major public health concern (Kumar et al., 2019; Wellington et al., 2013).

From an environmental perspective, bacteria with ARGs may perform a critical function by neutralizing and/or degrading antibiotics commonly found in ground and wastewater. Focusing bioremediation on the effective removal of antibiotics would lessen the selective pressure for HGT and potentially decrease the spread of ARGs (Nicoloff & Andersson, 2016; Sorg et al., 2016). Several organisms have been identified that metabolize antibiotics as their primary carbon and energy sources (Barnhill et al., 2010; Bhatt et al., 2022; Dantas et al., 2008; Gao et al., 2024; Wang, Han, et al., 2024b). Hence, it is critically important to determine the full potential of microbial bioremediation by investigating and optimizing the molecular mechanisms of antibiotic removal.

6 The Molecular Mechanisms of Microbial Remediation of Antibiotics

There are several mechanisms by which microbes protect themselves from antibiotics. These include the following: (i) barriers and reduced access to targets, (ii) alteration of the drug target, (iii) efflux pumps, (iv) sequestration/biosorption, and (v) degradation, alteration, and inactivation of the drug (Darby et al., 2023). Microbial barriers, alterations of cellular drug targets, and efflux pumps protect the individual microbial cell from antibiotics but by themselves may have limited bioremediation potential. The ability to bioabsorb and/or chemically neutralize the drug will be considered in greater detail.

Enzymatic Pathways and Targets Biochemical inactivation of antibiotics involves chemical modification of the drug by cellular hydrolases, transferases, and redox enzymes (reviewed in (Egorov et al., 2018)). Bacterial hydrolases include

beta-lactamases (carbapenemases) and macrolide esterases (Egorov et al., 2018). Thousands of beta-lactamases have been described and are catalogued in the beta-lactamase database (BLDB) (Naas et al., 2017). Transferases and redox enzymes include aminoglycoside-modifying enzymes, monooxygenases, and lyases (Egorov et al., 2018). Specific biodegradation and biotransformation pathways have been described to varying levels of detail for most antibiotic classes, including the sulfonamides, aminoglycosides, amphenicols, tetracyclines, beta-lactams, macrolides, and quinolones (Reis et al., 2020a, b).

Biofilms Microbes may produce extracellular polymers (biofilms) that bind metals and antibiotics, adsorb them to their surface (biosorption), transport them internally and bioaccumulate (store) them, or modify and inactivate the chemicals by chelation, precipitation, or other detoxification pathways (Mahto et al., 2022; Wang et al., 2024b; Zhao et al., 2022). Several types of biofilm-based bioreactors are currently being used for antibiotic removal from wastewater (Mishra et al., 2023). Environmental conditions and the embedded organisms can be modified to optimize bioremediation (Ali et al., 2023). Biofilm formation enhances the ability of microalgae to form consortia with bacteria to sequester and optimally biodegrade antibiotics (Eheneden et al., 2023; Fayaz et al., 2024).

Microalgae, Microalgae Consortiums, and Fungi High-rate algal ponds (HRAPs) have been used for wastewater treatment since the 1950s (Oswald et al., 1957). Microalgae and mixed algae-bacteria consortiums have been shown to effectively remove antibiotics in wastewater (Hena et al., 2021; Matamoros et al., 2015). Algae initiate stress response mechanisms when exposed to antibiotics and remove or degrade antibiotics by bioadsorption, biodegradation, bioaccumulation, photodegradation, hydrolysis, and volatilization (Abdelfattah et al., 2022; Li et al., 2022). The close association of algae and bacteria in biofilms allows for synergistic metabolic interactions and enhanced clearance of antibiotics (Eheneden et al., 2023). Algae-fungi consortia have also been shown to be effective in removing antibiotics from wastewater (Li et al., 2022). Fungi are effective bioremediators due to their ability to secrete a wide range of enzymes, including laccases, peroxidases, and oxidases that degrade antibiotics by chemical reduction, oxidation, hydroxylation, dehalogenation, dehydrogenation, deamination, and other mechanisms (Tiquia-Arashiro & Grube, 2019; Russell & Yost, 2021; Amobonye et al., 2023).

7 Environmental Factors Impacting Microbial Bioremediation of Antibiotics and Examples of Microorganisms

7.1 Environmental Factors

Environmental factors such as pH, temperature, oxygen, salinity, the presence of ammonia, and the presence of additional carbon sources affect the rate of microbial inactivation of antibiotics. Optimal pH and temperature vary depending on the

organism and antibiotic (Bhatt et al., 2022; Wang & Wang, 2018). Temperature and pH levels affect enzyme activity, microbial physiology, and biofilm formation. The solubility and ionic state of antibiotic molecules are also affected by pH levels (Mishra et al., 2023; Wang & Wang, 2018). The addition of carbon sources such as glucose, acetate, or succinate can augment antibiotic bioremediation in some cases, possibly by a co-metabolic effect, or may have no effect or an inhibitory effect if the organism prioritizes the added carbon source over the antibiotic (Li et al., 2022; Mishra et al., 2023; Wang & Wang, 2018). The addition of exogenous factors such as vitamin B6, vitamin C, and oxidized glutathione have been shown to enhance bacterial bioremediation of antibiotics (Zhang et al., 2016). For algae, pH, temperature, CO_2 levels, light intensity, presence of nutrients, and photocatalysts all affect antibiotic bioremediation (Amaro et al., 2023; Eheneden et al., 2023; Li et al., 2022). In natural systems, unique environmental conditions will select, and determine the performance of, individual organisms. In engineered systems, environmental conditions can be fully optimized for the selected organism(s).

7.2 Examples of Microorganisms Involved in Antibiotic Bioremediation

Bacteria Bacterial populations have rapidly adapted to antibiotics in the environment by modifying their metabolism to block or remove antibiotics. As a result, many different bacterial species have bioremediation abilities (Reis et al., 2020a; Reis et al., 2020b). Bacterial species from 19 different genera were observed to biodegrade sulfonamide antibiotics with removal efficiencies ranging from 3 to 100% (Reis et al., 2020a; Wang & Wang, 2018). Seven different bacterial genera, *Klebsiella*, *Streptomyces*, *Eschericia*, *Salmonella*, *Haemophilus, Clostridium*, and *Flavobacterium*, have been shown to degrade or transform amphenicol antibiotics (Reis et al., 2020a). Additional examples of bioremediation microorganisms and their antibiotic targets are shown in Table 8.3. These bioremediating bacteria were isolated from patients, sediments, sludge, soil, WWTPs, animal feces, and seawater.

Current approaches for identifying the molecular mechanisms of bioremediation bacteria utilize whole genome sequencing, transcriptome profiling, and biochemical analysis of the degraded products. Three such approaches found that the isolated organisms (*Serratia marcescens* MSM2304, *Alcaligenes aquatillis* FA, and *Achromobacter mucicolens* JD417, shown in Table 8.3) had multiple genes and metabolic pathways for degrading not only the target antibiotic but many additional ARGs and enzymes for degrading multiple antibiotic types (Du et al., 2023; Wang et al., 2024b; Zeng et al., 2024). For example, *Alcaligenes aquatillis* FA was shown to have approximately 100 genes capable of xenobiotic degradation (representing 4.9% of all metabolic genes) and 17 ARGs for resistance to 11 different antibiotics (Du et al., 2023). *Serratia marcescens* MSM2304 was isolated with tetracycline as its sole carbon and energy source and made a biofilm to enhance clearance of the

Table 8.3 Bacteria demonstrated to be effective for antibiotic bioremediation

Antibiotic target.	Organism(s)	Reference
Cephalosporin	*Bacillus cereus* P41	Erickson et al. (2014)
	Azospirillum sp., Bacillus sp., *Roseomonas* sp.	Rafii et al. (2009)
Chloramphenicol	*Sphingobium* WTD-1	Gao et al. (2024)
	Klebsiella pneumoniae, Escherichia fergusonii	Xin et al. (2012)
Doxycycline	*Escherichia* sp.	Wen et al. (2018)
Fluoroquinolones	*Microbacterium* sp.	Kim et al. (2011)
	Consortiums of *Acinetobacter baumannii* + *Klebsiella pneumoniae* and *Elizabethkingia miricola* + *Klebsiella pneumoniae*	Shaker et al. (2022)
	Labrys portucalensis F11	Amorim et al. (2014)
Penicillin & Neomycin	*Achromobacter* sp., *Bradyrhizobium* sp., *Burkholderia* sp., *Caulobacter* sp., *Chitinophaga arvensicoloa, Klebsiella sp., Pseudomonas* sp., *Sphingobacterium* sp., *Strenotrophomonas* sp.	Zhang and Dick (2014)
Sulfonamides	*Achromobacter mucicolens* JD417	Zeng et al. (2024)
	Microbacterium sp.	Topp et al. (2013)
	Escherichia sp. HS21, *Acinetobacter* sp. HS51	Zhang et al. (2012)
	Gordonia sp. SCD14, *Labrys* sp. SC11, *Onchrobactrum* sp. SA1	Mulla et al. (2018)
	Alcaligenes aquatillis FA	Du et al. (2023)
Tetracycline	*Alcaligenes* sp. T17	Chen et al. (2022)
	Achromobacter aegrifaciens SG-4	Zhao et al. (2018)
	Serratia marcescens MSM2304	Wang, Han, et al. (2024b)
	Serratia marcescens WW1	Bhatt and Chatterjee (2022)
	Stenotrophomonas maltophilia	Leng et al. (2016)

antibiotic. Whole genome sequencing of *S. marcescens* MSM2304 revealed the presence of dehydrogenases, monooxygenases, and methyltransferases (enzymes involved in antibiotic biotransformation/degradation) and 418 ARGs (Wang et al., 2024a, b). *Achromobacter mucicolens* JD417 had 106 genes related to xenobiotic degradation and metabolism, including oxidoreductases, transferases, hydrolases, lyases, isomerases, and synthases, and 393 ARGs targeting 9 major antibiotic classes (Zeng et al., 2024). Genetic knock-out experiments coupled with downstream analysis of any changes in biodegradation will precisely identify which genes are directedly associated with degradation of a specific antibiotic target.

Microalgae, Consortiums, and Fungi Algae activate stress response mechanisms to remove or degrade antibiotics. Algal species from genera *Chlamydomonas*, *Spyrogira*, *Scenedesmus*, *Chorella*, *Haematococcus*, *Microcystis*, *Nannochloris* and others have been shown to effectively remove a variety of antibiotics with efficiencies of 20–100% (Eheneden et al., 2023; Li et al., 2022; Russell & Yost, 2021; Xiong et al., 2021; Zhou et al., 2023). Consortia of algae and bacteria in biofilms allow for synergistic metabolic interactions and enhanced clearance of antibiotics (Eheneden et al., 2023; Xiong et al., 2021). Algae-fungi consortia have also been shown to be effective in removing antibiotics from wastewater (Li et al., 2022).

Fungi are natural decomposers and produce a varied and unique spectrum of enzymes capable of degrading complex chemical structures (Amobonye et al., 2023). Fungal genera shown to degrade or transform tetracyclines include *Trichosporon*, *Trichoderma*, *Fusarium*, *Xylaria*, *Penicillium*, *Paecilomyces*, and *Pleurotus* (Reis et al., 2020a). Transcriptome and metabolomic analysis of the ability of *Apiotrichum loubieri* strain MFZ-16 identified 9 ARGs and 137 genes involved in tetracycline degradation (Wang et al., 2024a). In a bioreactor system, *Trametes versicolor* was able to remove quinolones, beta-lactams, tetracyclines, metronidazoles, and trimethoprim type antibiotics with efficiencies ranging from 8 to 96% (Lucas et al., 2016). Simultaneous removal of ciprofloxacin, norfloxacin, and sulfamethoxazole was documented for fungal species *Phanerochaete chrysosporium* and *Pycnoporus sanguineus* (Gao et al., 2018). *Irpex lacteus* and *Leptosphaerulina* species were also shown to be highly efficient in removing a variety of antibiotics from test cultures (Russell & Yost, 2021). *Bjerkandera adusta*, *Porosterum spadiceum*, *Trametes versicolor*, *Cladosporium cladosporioides* 11, and *Pleuroteus ostreatus* demonstrated efficient biodegradation of fluoroquinolones that are commonly used in human and veterinary medicine via fungal-specific mechanisms such as heme-peroxidase and laccase activity (Akrout et al., 2024; Ma et al., 2024).

8 Conclusion

The persistence of heavy metals and antibiotics in the environment continues to pose significant threats to Earth's ecosystems and jeopardize human health. Bioremediation is a process by which environmental contaminants such as heavy metals and antibiotics are degraded or eliminated by microorganisms. This review discussed techniques that are currently being used for bioremediation, various microorganisms capable of metals and antibiotics bioremediation, and the molecular mechanisms and processes utilized by microbes for bioremediation. Chemical reduction, oxidation, and biosorption are some of the microbial mechanisms observed for both metal and antibiotic bioremediation. Environmental factors including pH, temperature, oxygen, salinity, and the presence of additional carbon sources affect the rate of microbial inactivation of heavy metals and antibiotics.

Bioremediation is a cost-effective, flexible, and environment-friendly method for reducing contaminants from wastewater, and a crucial upcoming technology from

the perspective of environmental management and public health. Various species of bacteria, microalgae, and fungi capable of effective bioremediation of heavy metals and antibiotics, and the optimum conditions for their activity, have been defined. At the same time, there are still gaps in our understanding of the molecular mechanisms of individual microorganisms that are involved in bioremediation, and how these organisms would interact when introduced into biological communities in various environments. Much of what is known has been documented with isolated organisms in laboratory/pilot scale applications and has yet to be demonstrated in natural systems where there are many potential environmental variables including the presence of other pollutants and other active organisms.

9 Future Directions

Bioremediating organisms can be added to natural systems to enhance the removal of antibiotics and heavy metals; however, the potential level of disruption caused by the addition of a known bioremediating organism to a natural environmental system is largely unknown. The range of degradation by-products and their potential toxicities needs to be fully explored. Systems biology approaches such as metagenomic DNA sequencing and single-cell RNA sequencing (transcriptome profiling) technologies will be necessary to monitor the impact of the bioremediating organism(s) on the survival and function of resident organisms. Bioreactors and other designed/engineered systems will allow for more precise control of operational conditions and input organisms but at a potentially higher cost than with natural systems. Mechanisms for monitoring and controlling the spread of ARGs from environmental organisms to pathogenic organisms are needed. This need supports the use of bioreactors/designed systems where the bioremediating organisms can be more easily contained and controlled.

Another challenge for the application of microbial bioremediation is that target substrates are complex chemical mixtures of various pollutants whose composition will vary from region to region and from day to day within one location. Still, bioremediating organisms that have been well-characterized to date have extensive and varied mechanisms for antibiotic and heavy metal degradation. Future investigations should look more broadly at the ability of individual organisms to bioremediate multiple antibiotics and heavy metals. This review found that even though antibiotics and heavy metals are very different kinds of contaminants, there are common mechanisms such as shared enzymatic pathways and biosorption that microorganisms use to eliminate both contaminant types. Future research should be directed toward understanding the application of these molecular mechanisms in natural and engineered systems so that efficient technologies can be designed to prevent contaminants like metal and antibiotics from entering surface water from wastewater.

Acknowledgments We thank our colleague Sarah Griffith, Laboratory and Pretreatment Manager, Northern Kentucky Sanitation District No. 1, and Hailey Ausbrooks, a student at the Biological Science Department, at Northern Kentucky University for their support and feedback. We would also like to thank Dr. Timothy Ehlinger, Emeritus faculty, University of Wisconsin-Milwaukee for this opportunity.

Author Contributions Dr. Subhomita Ghosh Roy performed the conceptualization, research, manuscript draft preparation, writing, reviewing, and manuscript edits for the introduction, conclusion, future directions, and the metal sections. Dr. Joseph Mester performed the conceptualization, research, manuscript draft preparation, writing, reviewing, and manuscript edits of the conclusion, future directions, and antibiotic sections.

Declarations Conflict of Interest
The authors declare no conflict of interest.

Funding No funding was received to assist with the preparation of this manuscript.

Data Availability Not Applicable

References

Abbas, S. H., Ismail, I. M., Mostafa, T. M., & Sulaymon, A. H. (2014). Biosorption of heavy metals: A review. *Journal of Chemical Science and Technology, 3*(4), 74–19.

Abdelfattah, A., Ali, S. S., Ramadan, H., El-Aswar, E., Eltawab, R., Shih-Hsin, H., Elsamahy, T., Li, S., El-Sheekh, M. M., Schagerl, M., Kornaros, M., & Sun, J. (2022). Microalgae-based wastewater treatment: Mechanisms, challenges, recent advances, and future prospects. *Environmental Science and Ecotechnology, 13*(100205), 100205. https://doi.org/10.1016/j.ese.2022.100205

Abioye, O. P., Oyewole, O. A., Oyeleke, S. B., Adeyemi, M. O., & Orukotan, A. A. (2018). Biosorption of lead, chromium and cadmium in tannery effluent using indigenous microorganisms. *Brazilian Journal of Biological Sciences, 5*(9), 25–32.

Akrout, I., Staita, K., Zouari-Mechichi, H., Ghariani, B., Khmaissa, M., Navarro, D., Doan, A., Albert, Q., Faulds, C., Sciara, G., Record, E., & Mechichi, T. (2024). Valorizing fungal diversity for the degradation of fluoroquinolones. *Heliyon, 10*(10), e30611.

Ali, A., Zahra, A., Kamthan, M., Husain, F., Albalawi, T., Zubair, M., Alatawy, R., Abid, M., & Noorani, M. (2023). Microbial biofilms: Applications, clinical consequences, and alternative therapies. *Microorganisms, 11*(8), 1934. https://doi.org/10.3390/microorganisms11081934

Allen, H., Donato, J., Wang, H. H., Cloud-Hansen, K. A., & Handelsman, J. D. J. (2010). Call of the wild: Antibiotic resistance genes in natural environments. *Nature Reviews. Microbiology, 8*(8), 251–259. https://doi.org/10.1038/nrmicro2312

Amaro, H., Salgado, E., Nunes, O., Pires, J., & Esteves, A. (2023). Microalgae systems – Environmental agents for wastewater treatment and further potential biomass valorisation. *Journal of Environmental Management, 337*, 117678. https://doi.org/10.1016/j.jenvman.2023.117678

Ameen, F. A., Hamdan, A. M., & El-Naggar, M. Y. (2020). Assessment of the heavy metal bioremediation efficiency of the novel marine lactic acid bacterium, Lactobacillus plantarum MF042018. *Scientific Reports, 10*(1), 1–11. https://doi.org/10.1038/s41598-019-57210-3

Amobonye, A., Aruwa, C., Aransiola, S., Omame, J., Alabi, T., & Lalung, J. (2023). The potential of fungi in the bioremediation of pharmaceutically active compounds: A comprehensive review. *Frontiers in Microbiology, 14*, 1207792. https://doi.org/10.3389/fmicb.2023.1207792

Amorim, C., Moreira, I., Maia, A., Tiritan, M., & Castro, P. (2014). Biodegradation of ofloxacin, norfloxacin, and ciprofloxacin as single and mixed substrates by Labrys portucalensis

F11. *Applied Microbiology and Biotechnology, 98*(7), 3181–3190. https://doi.org/10.1007/s00253-013-5333-8

Ayangbenro, A. S., & Babalola, O. O. (2017). A new strategy for heavy metal polluted environments: A review of microbial biosorbents. *International Journal of Environmental Research and Public Health, 14*(1), 94. https://doi.org/10.3390/ijerph14010094

Ayilara, M. S., & Babalola, O. O. (2023). Bioremediation of environmental wastes: The role of microorganisms. *Frontiers in Agronomy, 5*(May), 1–15. https://doi.org/10.3389/fagro.2023.1183691

Azubuike, C. C., Chikere, C. B., & Okpokwasili, G. C. (2016). Bioremediation techniques–classification based on site of application: Principles, advantages, limitations and prospects. *World Journal of Microbiology and Biotechnology, 32*(11), 1–18. https://doi.org/10.1007/s11274-016-2137-x

Bala, S., Garg, D., Thirumalesh, B. V., Sharma, M., Sridhar, K., & Inbaraj, B. S. (2022). Recent strategies for bioremediation of emerging pollutants: A review for a green and sustainable environment. *Toxics, 10*(8), 484. https://doi.org/10.3390/toxics10080484

Baltazar, M., Gracioso, L. H., Avanzi, I. R., Karolski, B., Tenório, J. A. S., & do, & Nascimento, C. A. O. (2019). Copper biosorption by rhodococcus erythropolis isolated from the sossego mine–PA–Brazil. *Journal of Materials Research and Technology, 8*(1), 475–483. https://doi.org/10.1016/j.jmrt.2018.04.006

Barancheshme, F., & Munir, M. (2018). Strategies to combat antibiotic resistance in the wastewater treatment plants. *Frontiers in Microbiology, 8*, 2603.

Barnhill, A., Weeks, K., Xiong, N., Day, T., & Carlsona, S. (2010). Identification of multiresistant salmonella isolates capable of subsisting on antibiotics. *Applied and Environmental Microbiology, 76*(8), 2678–2680. https://doi.org/10.1128/AEM.02516-09

Bhandari, G., Gupta, S., Chaudhary, P., Chaudhary, S., & Gangola, S. (2023). Bioleaching: A sustainable resource recovery strategy for urban mining of e-waste. In P. Debbarma, S. Kumar, D. C. Suyal, & R. Soni (Eds.), *Microbial technology for sustainable e-waste management*. Cham.

Bhatnagar, S., & Kumari, R. (2013). Bioremediation: A sustainable tool for environmental management—A review. *Annual Research & Review in Biology, 3*(4), 974–993. www.sciencedomain.org

Bhatt, S., & Chatterjee, S. (2022). Fluoroquinolone antibiotics: Occurrence, mode of action, resistance, environmental detection, and remediation – A comprehensive review. *Environmental Pollution, 315*, 120440. https://doi.org/10.1016/j.envpol.2022.120440

Bhatt, P., Jeon, C., & Kim, W. (2022). Tetracycline bioremediation using the novel Serratia marcescens strain WW1 isolated from a wastewater treatment plant. *Chemosphere, 298*, 134344. https://doi.org/10.1016/j.chemosphere.2022.134344

Bhattacharya, M., Guchhait, S., Biswas, D., & Datta, S. (2015). Waste lubricating oil removal in a batch reactor by mixed bacterial consortium: A kinetic study. *Bioprocess and Biosystems Engineering, 38*, 2095–2106. https://doi.org/10.1007/s00449-015-1449-9

Bouchez, T., Patureau, D., & Dabert, P. (2000). Ecological study of a bioaugmentation failure. *Environmental Microbiology, 2*(2), 179–190.

Bowman, N., Patel, P., Xu, W., Alsaffar, A., & Tiquia-Arashiro, S. M. (2018). Enrichment and isolation of Pb-resistant bacteria from Saint Clair River sediments and their potential for Pb removal in aqueous solutions. *Applied Microbiology and Biotechnology, 102*, 2391–2398. https://doi.org/10.1007/s00253-018-8772-4

Bwapwa, J. K. (2022). Chapter 23 – Factors affecting the bioremediation of industrial and domestic wastewaters. In S. Das & H. R. Dash (Eds.), *Microbial biodegradation and bioremediation* (2nd ed., pp. 461–472). Elsevier. https://doi.org/10.1016/B978-0-323-85455-9.00030-8

Catteau, L., Zhu, L., Van Bambeke, F., & Quetin-Leclercq, J. (2018). Natural and hemi-synThetic pentacyclic triterpenes as antimicrobials and resistance modifying agents against Staphylococcus aureus: A review. *Phytochemistry Reviews, 17*, 1129–1163.

CDC. (2021). Understanding antibiotic resistance in water: A one health approach. .

Chen, X., Shen, W., Chen, J., Zhu, Y., Chen, C., & Xie, S. (2022). Tetracycline biotransformation by a novel bacterial strain Alcaligenes sp. T17. *Science of the Total Environment, 832*, 155130. https://doi.org/10.1016/j.scitotenv.2022.155130

Cheong, M. S., Yoon, Y.-E., Kim, J. W., Young, K. H., Kim, S. C., & Yong, B. L. (2020). Chlortetracycline inhibits seed germination and seedling growth in Brassica campestris by disrupting H_2O_2 signaling. *Applied Biological Chemistry, 63*, 1. https://doi.org/10.1186/s13765-019-0484-7

Chikere, C., Okoye, A., & Okpokwasili, G. (2016). Microbial community profiling of active oleophilic bacteria involved in bioreactor based crude-oil polluted sediment treatment. *Journal of Applied & Environmental Microbiology, 4*, 1–20. https://doi.org/10.12691/jaem-4-1-1

Cho, K., Zholi, A., Frabutt, D., Flood, M., Floyd, D., & Tiquia, S. M. (2012). Linking bacterial diversity and geochemistry of uranium-contaminated groundwater. *Environmental Technology, 33*, 1629–1640. https://doi.org/10.1080/09593330.2011.641036

Coulon, F., Al Awadi, M., Cowie, W., Mardlin, D., Pollard, S., Cunningham, C., Risdon, G., Arthur, P., Semple, K., & Paton, G. (2010). When is a soil remediated? Comparison of biopiled and windrowed soils contaminated with bunker-fuel in a full-scale trial. *Environmental Pollution, 158*, 3032–3040. https://doi.org/10.1016/j.envpol.2010.06.001

CPCB. (2019). *Water pollution*. https://cpcb.nic.in

Dantas, G., Sommer, M., Oluwasegun, R., & Church, G. (2008). Bacteria subsisting on antibiotics. *Science, 320*(5872), 100–103. https://doi.org/10.1126/science.1155157

Darby, E., Trampari, E., Siasat, P., Gaya, M., Alav, I., Webber, M. A., & Blair, J. M. A. (2023). Molecular mechanisms of antibiotic resistance revisited. *Nature Reviews. Microbiology, 21*(5), 280–295. https://doi.org/10.1038/s41579-022-00820-y

De, J., Ramaiah, N., & Vardanyan, L. (2008). Detoxification of toxic heavy metals by marine bacteria highly resistant to mercury. *Marine Biotechnology, 10*, 471–477. https://doi.org/10.1007/s10126-008-9083-z

Delforno, T., Moura, A., Okada, D., Sakamoto, I., & Varesche, M. (2015). Microbial diversity and the implications of sulfide levels in an anaerobic reactor used to remove an anionic surfactant from laundry wastewater. *Bioresource Technology, 192*, 37–45. https://doi.org/10.1016/j.biortech.2015.05.050

Dias, R., Ruberto, L., Calabro, A., Balbo, A., Del Panno, M., & MacCormack, W. (2015). Hydrocarbon removal and bacterial community structure in on-site biostimulated biopile systems designed for bioremediation of diesel-contaminated Antarctic soil. *Polar Biology, 38*, 677–687. https://doi.org/10.1007/s00300-014-1630-7

Dibner, J. J., & Richards, J. D. (2005). Antibiotic growth promoters in agriculture: History and mode of action. *Poultry Science, 84*, 634–643.

Du, Y., Cheng, Q., Qian, M., Liu, Y., Wang, F., Ma, J., Zhang, X., & Lin, H. (2023). Biodegradation of sulfametoxydiazine by Alcaligenes aquatillis FA: Performance, degradation pathways, and mechanisms. *Journal of Hazardous Materials, 452*, 131186. https://doi.org/10.1016/j.jhazmat.2023.131186

Economou, V., & Gousia, P. (2015). Agriculture and food animals as a source of antimicrobial-resistant bacteria. *Infection and Drug Resistance, 8*, 49–61. https://doi.org/10.2147/IDR.S55778

Egorov, A., Ulyashova, M., & Rubtsova, M. (2018). Bacterial enzymes and antibiotic resistance. *Acta Naturae., 10*(4), 33–48.

Eheneden, I., Wang, R., & Zhao, J. (2023). Antibiotic removal by microalgae-bacteria consortium: Metabolic pathways and microbial responses. *Science of the Total Environment, 15*(164489), 891. https://doi.org/10.1016/j.scitotenv.2023.164489

Erickson, B., Elkins, C., Mullis, L., Heinze, T., Wagner, R., & Cerniglia, C. (2014). A metallo-β-lactamase is responsible for the degradation of ceftiofur by the bovine intestinal bacterium Bacillus cereus P41. *Veterinary Microbiology, 172*(3–4), 499–504. https://doi.org/10.1016/j.vetmic.2014.05.032

EU. (2023). *Water framework directive*. https://environment.ec.europa.eu/topics/water/water-framework-directive_en

Farraji, H., Zaman, N. Q., Tajuddin, R., & Faraji, H. (2016). Advantages and disadvantages of phytoremediation: A concise review. *International journal of Environmental Science and Technology, 2*, 69–75.

Fayaz, T., Renuka, N., & Ratha, S. (2024). Antibiotic occurrence, environmental risks, and their removal from aquatic environments using microalgae: Advances and future perspectives. *Chemosphere, 140822*, 349. https://doi.org/10.1016/j.chemosphere.2023.140822

Gadd, G. M. (2004). Microbial influence on metal mobility and application for bioremediation. *Geoderma, 122*(2–4 SPEC. IIS), 109–119. https://doi.org/10.1016/j.geoderma.2004.01.002

Gao, N., Liu, C., Xu, Q., Cheng, J., & Yuan, Y. (2018). Simultaneous removal of ciprofloxacin, norfloxacin, sulfamethoxazole by co-producing oxidative enzymes system of Phanerochaete chrysosporium and Pycnoporus sanguineus. *Chemosphere, 195*, 146–155. https://doi.org/10.1016/j.chemosphere.2017.12.062

Gao, Y., Chen, Y., Zhu, F., Pan, D., Huang, J., & Wu, X. (2024). Revealing the biological significance of multiple metabolic pathways of chloramphenicol by Sphingobium sp. WTD-1. *Journal of Hazardous Materials, 469*, 134069. https://doi.org/10.1016/j.jhazmat

Gomez, F., & Sartaj, M. (2013). Field scale ex situ bioremediation of petroleum contaminated soil under cold climate conditions. *International Biodeterioration & Biodegradation, 85*, 375–382. https://doi.org/10.1016/j.ibiod.2013.08.003

Gong, W., Guo, L., Huang, C., Xie, B., Jiang, M., Zhao, Y., Zhang, H., Wu, Y., & Liang, H. (2024). A systematic review of antibiotics and antibiotic resistance genes (ARGs) in mariculture wastewater: Antibiotics removal by microalgal-bacterial symbiotic system (MBSS), ARGs characterization on the metagenomic. *Science of the Total Environment, 930*, 172601. https://doi.org/10.1016/j.scitotenv.2024.172601

González Henao, S., & Ghneim-Herrera, T. (2021). Heavy metals in soils and the remediation potential of bacteria associated with the plant microbiome. *Frontiers in Environmental Science, 9*(April), 1–17. https://doi.org/10.3389/fenvs.2021.604216

Goswami, M., Chakraborty, P., Mukherjee, K., Mitra, G., Bhattacharyya, P., & Dey, S. (2018). Bioaugmentation and biostimulation: A potential strategy for environmental remediation. *Journal of Microbiology & Experimentation, 6*(5), 223–231.

Handore, A. V., Jagtap, V. S., Ghangale, S. S., Khandelwal, S. R., Bholay, A. D., Karmakar, R., & Handore, D. V. (2023). Chapter 1 – Bioaugmentation: An emerging ecofriendly technology for wastewater treatment. In *Integrative strategies of bioremediation of environmental contaminants, volume two* (pp. 1–16).

Hena, S., Gutierrez, L., & Croué, J. (2021). Removal of pharmaceutical and personal care products (PPCPs) from wastewater using microalgae: A review. *Journal of Hazardous Materials, 403*, 124041. https://doi.org/10.1016/j.jhazmat.2020.124041

Hong, X., Zhao, Y., Zhuang, R., Liu, J., Guo, G., Chen, J., & Yao, Y. (2020). Bioremediation of tetracycline antibiotics-contaminated soil by bioaugmentation. *RSC Advances, 10*(55), 33086–33102. https://doi.org/10.1039/d0ra04705h

Huang, Y., Pan, H., Wang, Q., Ge, Y., Liu, W., & Christie, P. (2019). Enrichment of the soil microbial community in the bioremediation of a petroleum-contaminated soil amended with rice straw or sawdust. *Chemosphere, 224*, 265–271. https://doi.org/10.1016/j.chemosphere.2019.02.148

Imron, M. F., Kurniawan, S. B., Ismail, N. I., & Abdullah, S. R. S. (2020). Future challenges in diesel biodegradation by bacteria isolates: A review. *Journal of Cleaner Production, 119716*, 251. https://doi.org/10.1016/j.jclepro.2019.119716

Jeyakumar, P., Debnath, C., Vijayaraghavan, R., & Muthuraj, M. (2023). Trends in bioremediation of heavy metal contaminations. *Environmental Engineering Research, 28*(4), 220630–220631. https://doi.org/10.4491/eer.2021.631

Kebede, G., Tafese, T., Abda, E. M., Kamaraj, M., & Assefa, F. (2021). Factors influencing the bacterial bioremediation of hydrocarbon contaminants in the soil: Mechanisms and impacts. *Journal of Chemistry*, 1–17. https://doi.org/10.1155/2021/9823362

Kemper, N. (2008). Veterinary antibiotics in the aquatic and terrestrial environment. *Ecological Indicators, 8*, 1–13.

Kharangate-Lad, A., & D'Souza, N. C. (2021). Current approaches in bioremediation of toxic contaminants by application of microbial cells; biosurfactants and bioemulsifiers of microbial origin. In V. Kumar, R. Prasad, & M. Kumar (Eds.), *Rhizobiont in Bioremediation of Hazardous Waste* (pp. 217–263). Springer. https://doi.org/10.1007/978-981-16-0602-1_11

Kim, D., Heinze, T., Kim, B., Schnackenberg, L., Woodling, K., & Sutherland, J. (2011). Modification of norfloxacin by a microbacterium sp. strain isolated from a wastewater treatment plant. *Applied and Environmental Microbiology, 77*(17), 6100–6108.

Kumar, M., Jaiswal, S., Sodhi, K. K., Shree, P., Singh, D. K., Agrawal, P. K., & Shukla, P. (2019). Antibiotics bioremediation: Perspectives on its ecotoxicity and resistance. *Environment International, 124*(October 2018), 448–461. https://doi.org/10.1016/j.envint.2018.12.065

Leng, Y., Bao, J., Chang, G., Zheng, H., Li, X., Du, J., Snow, D., & Li, X. (2016). Biotransformation of tetracycline by a novel bacterial strain Stenotrophomonas maltophilia DT1. *Journal of Hazardous Materials, 318*, 125–133. https://doi.org/10.1016/j.jhazmat.2016.06.053

Li, S., Show, P. L., Ngo, H. H., & SH., H. (2022). Algae-mediated antibiotic wastewater treatment: A critical review. *Environmental Science and Ecotechnology, 9*, 100145.

Liu, F., Ying, G. G., Tao, R., Zhao, J.-L., Yang, J. F., & Zhao, L. F. (2009). Effects of six selected antibiotics on plant growth and soil microbial and enzymatic activities. *Environmental Pollution, 157*, 1636–1642.

Liu, C., Li, T., Liming, Z., Weiqian, T., & Lanqing, M. (2021). A review of the distribution of antibiotics in water in different regions of China and current antibiotic degradation pathways. *Frontiers in Environmental Science, 9, 692298*. https://doi.org/10.3389/fenvs.2021.692298

Lucas, D., Badia-Fabregat, M., Vicent, T., Caminal, G., Rodríguez-Mozaz, S., Balcázar, J., & Barceló, D. (2016). Fungal treatment for the removal of antibiotics and antibiotic resistance genes in veterinary hospital wastewater. *Chemosphere, 152*, 301–308. https://doi.org/10.1016/j.chemosphere.2016.02.113

Ma, N., Zhang, H., Yuan, L., Li, Y., Yang, W., & Huang, Y. (2024). Characterization and removal mechanism of fluoroquinolone-bioremediation by fungus Cladosporium cladosporioides 11 isolated from aquacultural sediments. *Environmental Science and Pollution Research International, 31*(20), 29525–29535. https://doi.org/10.1007/s11356-024-33142-y

Mahmoud, G. A.-E. (2021). Microbial scavenging of heavy metals using bioremediation strategies. In K. Vivek, P. Ram, & K. Manoj (Eds.), *Rhizobiont in bioremediation of hazardous waste* (pp. 265–289). Springer.

Mahto, K., Kumari, S., & Das, S. (2022). Unraveling the complex regulatory networks in biofilm formation in bacteria and relevance of biofilms in environmental remediation. *Critical Reviews in Biochemistry and Molecular Biology, 57*(3), 305–332.

Masten, S. J., Davies, S. H., & McElmurry, S. P. (2016). Flint water crisis: What happened and why? *Journal – American Water Works Association, 108*(12), 22–34. https://doi.org/10.5942/jawwa.2016.108.0195

Matamoros, V., Gutiérrez, R., Ferrer, I., García, J., & Bayona, J. M. (2015). Capability of microalgae-based wastewater treatment systems to remove emerging organic contaminants: A pilot-scale study. *Journal of Hazardous Materials, 288*, 34–42. https://doi.org/10.1016/j.jhazmat.2015.02.002

Mathew, B. B., & Krishnamurthy, N. B. (2018). Screening and identification of bacteria isolated from industrial area groundwater to study lead sorption: Kinetics and statistical optimization of biosorption parameters. *Groundwater for Sustainable Development, 7*, 313–327. https://doi.org/10.1016/j.gsd.2018.07.007

Medina-Bellver, J. I., Marín, P., & Delgado, A. (2005). Evidence for in situ crude oil biodegradation after the prestige oil spill. *Environmental Microbiology, 7*(6), 773–779.

Michalak, I., Chojnacka, K., & Witek-Krowiak, A. (2013 Jul). State of the art for the biosorption process--a review. *Applied Biochemistry and Biotechnology, 170*(6), 1389–1416. https://doi.org/10.1007/s12010-013-0269-0

Ministry of Ecology and Environment The People's Republic of China. (2017). *Environmental laws: Water pollution prevention and control law of the people's Republic of China (Amended in 2017).* https://english.mee.gov.cn/Resources/laws/environmental_laws/202012/t20201211_812662.shtml

Mir-Tutusaus, J. A., Baccar, R., Caminal, G., & Sarrà, M. (2018). Can white-rot fungi be a real wastewater treatment alternative for organic micropollutants removal? A review. *Water Research, 138*, 137–151. https://doi.org/10.1016/j.watres.2018.02.056

Mishra, S., Cheng, L., & Lian, Y. (2023). Response of biofilm-based systems for antibiotics removal from wastewater: Resource efficiency and process resiliency. *Chemosphere, 139878*, 340. https://doi.org/10.1016/j.chemosphere.2023.139878

Mulla, S., Hu, A., Sun, Q., Li, J., Suanon, F., Ashfaq, M., & Yu, C. (2018). Biodegradation of sulfamethoxazole in bacteria from three different origins. *Journal of Environmental Management, 206*, 93–102. https://doi.org/10.1016/j.jenvman.2017.10.029

Mustafa, Y., Abdul-Hameed, H., & Razak, Z. (2015). Biodegradation of 2,4-dichlorophenoxyacetic acid contaminated soil in a roller slurry bioreactor. *Clean-Soil Air Water, 43*, 1115–1266. https://doi.org/10.1002/clen.201400623

Naas, T., Oueslati, S., Bonnin, R., Dabos, M., Zavala, A., Dortet, L., Retailleau, P., & Iorga, B. (2017). Beta-lactamase database (BLDB) – Structure and function. *Journal of Enzyme Inhibition and Medicinal Chemistry, 32*(1), 917–919. https://doi.org/10.1080/14756366.2017.1344235

Nassereddine, Z., Opara, S., Coutinho, O., Qyteti, F., Book, R., Heinicke, M. P., Napieralski, J., & Tiquia-Arashiro, S. M. (2024). Critical perspectives on advancing antibiotic resistant gene (ARG) detection technologies in aquatic ecosystems. *Science of the Total Environment, 957*(20), 177775. https://doi.org/10.1016/j.scitotenv.2024.177775

Ndeddy Aka, R. J., & Babalola, O. O. (2016). Effect of bacterial inoculation of strains of pseudomonas aeruginosa, alcaligenes feacalis and bacillus subtilis on germination, growth and heavy metal (cd, cr, and Ni) uptake of brassica juncea. *International Journal of Phytoremediation, 18*(2), 200–209.

Nicoloff, H., & Andersson, D. (2016). Indirect resistance to several classes of antibiotics in cocultures with resistant bacteria expressing antibiotic-modifying or -degrading enzymes. *The Journal of Antimicrobial Chemotherapy, 71*(1), 100–110. https://doi.org/10.1093/jac/dkv312

Njoku, K. L., Akinyede, O. R., & Obidi, O. F. (2020). Microbial remediation of heavy metals contaminated media by bacillus megaterium and Rhizopus stolonifer. *Scientific African, 10*, e00545. https://doi.org/10.1016/j.sciaf.2020.e00545

NRDC. (2021). *Fighting for safe drinking water in newark.* https://www.nrdc.org/resources/fighting-safe-drinking-water-newark

Ohore, O. E., Qin, Z., Sanganyado, E., Wang, Y., Jiao, X., Liu, W., & Wang, Z. (2022). Ecological impact of antibiotics on bioremediation performance of constructed wetlands: Microbial and plant dynamics, and potential antibiotic resistance genes hotspots. *Journal of Hazardous Materials, 424*(August 2021), 127495. https://doi.org/10.1016/j.jhazmat.2021.127495

Ojuederie, O. B., Chukwuneme, F., Olanrewaju, O., Ayilara, M., Adegboyega, T. T., & Babalola, O. O. (2021). Contribution of microbial inoculants in sustainable maintenance of human health, including test methods and evaluation of safety of microbial pesticide microorganisms. In *Biopesticides: Botanicals microorganisms improving agriculture and human health* (pp. 207–240).

Oswald, W. J., Gotaas, H. B., Golueke, C. G., Kellen, W. R., Gloyna, E. F., & Hermann, E. R. (1957). Algae in waste treatment [with discussion]. *Sewage and Industrial Wastes, 29*(4), 437–457.

Oziegbe, O., Oluduro, A. O., Oziegbe, E. J., Ahuekwe, E. F., & Olorunsola, S. J. (2021). Assessment of heavy metal bioremediation potential of bacterial isolates from landfill soils. *Saudi Journal of Biological Sciences, 28*(7), 3948–3956. https://doi.org/10.1016/j.sjbs.2021.03.072

Pagnucco, G., Overfield, D., Chamlee, Y., Fortuna, A., Sulaiman, F., Farinas, J., Schittenhelm, R., Shuler, C., Kassem, A., Opara, S., Najaf, H., Abbas, L., Coutinho, O., Catalfano, B., Li, X., & Tiquia-Arashiro, S. M. (2023). Metal tolerance and biosorption capacities of bacterial

strains isolated from an urban watershed. *Frontiers in Microbiology, 14*, 1278886. https://doi.org/10.3389/fmicb.2023.1278886

Patel, A. B., Jain, K. R., Manvar, T., Desai, C., & Madamwar, D. (2022). Enriched bacterial community efficiently degrade polycyclic aromatic hydrocarbons in soil ecosystem: Insights from a mesocosms study. *Biochemical Engineering Journal, 108516*, 185. https://doi.org/10.1016/j.bej.2022.108516

Pratush, A., Kumar, A., & Hu, Z. (2018). Adverse effect of heavy metals (As, Pb, Hg, and cr) on health and their bioremediation strategies: A review. *International Microbiology, 21*(3), 97–106. https://doi.org/10.1007/s10123-018-0012-3

Puyen, Z. M., Villagrasa, E., Maldonado, J., Elia Diestra, I., Esteve, S., & Solé, A. (2012). Biosorption of lead and copper by heavy-metal tolerant Micrococcus luteus DE2008. *Bioresource Technology, 126*, 233–237. https://doi.org/10.1016/j.biortech.2012.09.036

Rafii, F., Williams, A., Park, M., Sims, L., Heinze, T., Cerniglia, C., & Sutherland, J. (2009). Isolation of bacterial strains from bovine fecal microflora capable of degradation of ceftiofur. *Veterinary Microbiology, 139*(1–2), 89–96. https://doi.org/10.1016/j.vetmic.2009.04.023

Reis, A., Kolvenbach, B., Nunes, O., & Corvini, P. (2020a). Biodegradation of antibiotics: The new resistance determinants – Part I. *New Biotechnology, 54*, 34–51.

Reis, A., Kolvenbach, B., Nunes, O., & Corvini, P. (2020b). Biodegradation of antibiotics: The new resistance determinants – Part II. *New Biotechnology, 54*, 13–27.

Ren, X., Zeng, G., Tang, L., Wang, J., Wan, J., & Wang, J. (2018). The potential impact on the biodegradation of organic pollutants from composting technology for soil remediation. *Waste Management, 72*, 138–149. https://doi.org/10.1016/j.wasman.2017.11.032

Russell, J., & Yost, C. (2021). Alternative, environmentally conscious approaches for removing antibiotics from wastewater treatment systems. *Chemosphere, 263*, 128177. https://doi.org/10.1016/j.chemosphere.2020.128177

Saravanan, V., Rajasimman, M., & Rajamohan, N. (2015). Performance of packed bed biofilter during transient operating conditions on removal of xylene vapour. *International journal of Environmental Science and Technology, 12*, 1625–1634. https://doi.org/10.1007/s13762-014-0521-3

Saxena, G., Kishor, R., & Bharagava, R. N. (2020). *Application of microbial enzymes in degradation and detoxification of organic and inorganic pollutants, in bioremediation of industrial waste for environmental safety*. Springer.

Sayqal, A., & Ahmed, O. B. (2021). Advances in heavy metal bioremediation: An overview. *Applied Bionics and Biomechanics, 2021*, 1. https://doi.org/10.1155/2021/1609149

Seagren, E. A. (2023). Bioremediation. In *Encyclopedia of toxicology* (Vol. 3, 4th ed., pp. 863–868) https://www.sciencedirect.com/science/article/pii/B9780128243152003870

Sedlakova-Kadukova, J., Kopcakova, A., Gresakova, L., Godany, A., & Pristas, P. (2019). Bioaccumulation and biosorption of zinc by a novel streptomyces K11 strain isolated from highly alkaline aluminium brown mud disposal site. *Ecotoxicology and Environmental Safety, 167*, 204–211. https://doi.org/10.1016/j.ecoenv.2018.09.123

Shaker, R., Nagy, Y., Adly, M., Khattab, R., & Ragab, Y. (2022). Acinetobacter baumannii, Klebsiella pneumoniae and Elizabethkingia miricola isolated from wastewater have biodegradable activity against fluoroquinolone. *World Journal of Microbiology and Biotechnology, 38*(11), 187. https://doi.org/10.1007/s11274-022-03367-5

Shannon, J., Hauser, L., Liu, X., Parkin, G., Mattes, T., & Just, G. (2015). Partial nitritation ANAMMOX in submerged attached growth bioreactors with smart aeration at 20. *Environmental Science: Process & Impacts, 17*, 81–89. https://doi.org/10.1039/c4em00481g

Sharma, P., Dutta, D., Udayan, A., Nadda, A. K., Lam, S. S., & Kumar, S. (2022a). Role of microbes in bioaccumulation of heavy metals in municipal solid waste: Impacts on plant and human being. *Environmental Pollution, 119248*, 305. https://doi.org/10.1016/j.envpol.2022.119248

Sharma, P., Singh, S. P., Parakh, S. K., & Tong, Y. W. (2022b). Health hazards of hexavalent chromium (Cr (VI)) and its microbial reduction. *Bioengineered, 13*(3), 4923–4938. https://doi.org/10.1080/21655979.2022.2037273

Singh, R., Andaluri, G., & Pandey, V. C. (2022). Chapter 1 – Cities water pollution — Challenges and controls. In *Algae and aquatic Macrophytes in cities bioremediation, biomass, biofuels and bioproducts* (pp. 3–22). https://doi.org/10.1016/B978-0-12-824270-4.00015-8

Sivakumar, D., Kandaswamy, A., Gomathi, V., Rajeshwaran, R., & Murugan, N. (2014). Bioremediation studies on reduction of heavy metals toxicity. *Pollution Research, 33*(3), 553–558.

Sorg, R., Lin, L., van Doorn, G., Sorg, M., Olson, J., Nizet, V., & Veening, J. (2016). Collective resistance in microbial communities by intracellular antibiotic deactivation. *PLoS Biology, 14*(12), e2000631. https://doi.org/10.1371/journal.pbio.2000631

Stroo, H. F., Leeson, A., & Ward, C. H. (2013). *Bioaugmentation for groundwater remediation*. Spinger.

Sun, L., Guo, D., Liu, K., Meng, H., Zheng, Y., & Yuan, F. (2019). Levels, sources, and spatial distribution of heavy metals in soils from a typical coal industrial city of Tangshan, China. *Catena, 175*, 101–109. https://doi.org/10.1016/j.catena.2018.12.014

Tak, H. I., Ahmad, F., & Babalola, O. O. (2012). Advances in the application of plant growth-promoting rhizobacteria in phytoremediation of heavy metals. *Reviews of Environmental Contamination and Toxicology, 223*(1), 33–52.

Thakkar, H., & Vincent, V. (2021). Application of Microorganisms in Bioremediation. https://doi.org/10.1002/9781119717317

Tiquia-Arashiro, S. M. (2018). Lead absorption mechanisms in bacteria as strategies for lead bioremediation. *Applied Microbiology and Biotechnology, 102*(13), 5437–5444. https://doi.org/10.1007/s00253-018-8969-6

Tiquia-Arashiro, S. M. (2019). Thermophilic fungi in composts: Their role in composting and industrial processes. In Fungi in extreme environments: Ecological role and biotechnological significance (pp. 587–606). Springer. https://doi.org/10.1007/978-3-030-19030-9_29

Tiquia-Arashiro, S. M., & Grube, M. (2019). Fungi in extreme environments: Ecological role and biotechnological significance (p. 626). Springer Nature. https://doi.org/10.1007/978-3-030-19030-9.

Topp, E., Chapman, R., Devers-Lamrani, M., Hartmann, A., Marti, R., Martin-Laurent, F., Sabourin, L., Scott, A., & Sumarah, M. (2013). Accelerated biodegradation of veterinary antibiotics in agricultural soil following long-term exposure, and isolation of a sulfamethazine-degrading microbacterium sp. *Journal of Environmental Quality, 42*(1), 173–178. https://doi.org/10.2134/jeq2012.0162

UN. (2023). *Blueprint for acceleration: Sustainable development goal 6 synthesis report on water and sanitation 2023*. https://www.unwater.org/publications/sdg-6-synthesis-report-2023

Unuofin, J. O., Falade, A. O., & Aladekoyi, O. J. (2021). Applications of microbial laccases in bioremediation of environmental pollutants: Potential issues, challenges, and prospects. *Bioremediation for Environmental Sustainability, 1*, 519–540. https://doi.org/10.1016/B978-0-12-820524-2.00021-3

USEPA. (2017). *Determining the prevalence of contaminants in treated and untreated drinking water*. https://www.epa.gov/water-research/determining-prevalence-contaminants-treated-and-untreated-drinking-water

USEPA. (2023a). *Sanitary Sewer Overflows (SSOs)*. https://www.epa.gov/npdes/sanitary-sewer-overflows-ssos

USEPA. (2023b). *Summary of the Clean Water Act*. https://www.epa.gov/laws-regulations/summary-clean-water-act

Verma, S., & Kuila, A. (2019). Bioremediation of heavy metals by microbial process. *Environmental Technology & Innovation, 14*, 100369. https://doi.org/10.1016/j.eti.2019.100369

Vidali, M. (2001). Bioremediation. An overview. *Pure and Applied Chemistry, 73*(7), 1163–1172.

Walker, D. B., Baumgartner, D. J., Gerba, C. P., & Fitzsimmons, K. (2019). Chapter 16 Surface Water Environment. In *Environmental and pollution science* (3rd ed., pp. 261–292). https://doi.org/10.1016/B978-0-12-814719-1.00016-1

Wang, J., & Wang, S. (2018). Microbial degradation of sulfamethoxazole in the environment. *Applied Microbiology and Biotechnology, 102*(8), 3573–3582. https://doi.org/10.1007/s00253-018-8845-4

Wang, J., Zhang, M., Cheng, J., Han, Y., Ma, H., Jiang, X., & Liu, Y. (2024a). Molecular mechanism underlying the degradation of tetracycline by Apiotrichum loubieri MFZ-16. *Environmental Technology & Innovation, 33*, 103489. https://doi.org/10.1016/j.eti.2023.103489

Wang, S., Han, J., Ge, Z., Su, X., Chen, Y., & Meng, J. (2024b). Biotransformation characteristics of tetracycline by strain Serratia marcescens MSM2304 and its mechanism evaluation based on products analysis and genomics. *Journal of Environmental Management, 356*, 120684. https://doi.org/10.1016/j.jenvman.2024.120684

Wellington, E., Boxall, A., Cross, P., Feil, E., Gaze, W., Hawkey, P., Johnson-Rollings, A., Jones, D., Lee, N., Otten, W., Thomas, C., & Williams, A. (2013). The role of the natural environment in the emergence of antibiotic resistance in gram-negative bacteria. *The Lancet Infectious Diseases, 13*(2), 155–165. https://doi.org/10.1016/S1473-3099(12)70317-1

Wen, X., Wang, Y., Zou, Y., Ma, B., & Wu, Y. (2018). No evidential correlation between veterinary antibiotic degradation ability and resistance genes in microorganisms during the biodegradation of doxycycline. *Ecotoxicology and Environmental Safety, 147*, 759–766. https://doi.org/10.1016/j.ecoenv.2017.09.025

Whelan, M., Coulon, F., Hince, G., Rayner, J., McWatters, R., Spedding, T., & Snape, I. (2015). Fate and transport of petroleum hydrocarbons in engineered biopiles in polar regions. *Chemosphere, 131*, 232–240. https://doi.org/10.1016/j.chemosphere.2014.10.088

Xiao, F., Cheng, J., Cao, W., Yang, C., Chen, J., & Luo, Z. (2019). Removal of heavy metals from aqueous solution using chitosan-combined magnetic biochars. *Journal of Colloid and Interface Science, 540*, 579–584. https://doi.org/10.1016/j.jcis.2019.01.068

Xin, Z., Fengwei, T., Gang, W., Xiaoming, L., Qiuxiang, Z., Hao, Z., & Wei, C. (2012). Isolation, identification and characterization of human intestinal bacteria with the ability to utilize chloramphenicol as the sole source of carbon and energy. *FEMS Microbiology Ecology, 82*(3), 703–712. https://doi.org/10.1111/j.1574-6941.2012.01440.x

Xiong, Q., Hu, L., Liu, Y., Zhao, J., He, L., & Ying, G. (2021). Microalgae-based technology for antibiotics removal: From mechanisms to application of innovational hybrid systems. *Environment International, 155*, 106594. https://doi.org/10.1016/j.envint.2021.106594

Yang, C., Song, G., & Lim, W. (2020). A review of the toxicity in fish exposed to antibiotics. *Comparative Biochemistry and Physiology Part C: Pharmacology, Toxicology, 108840*, 237.

Yang, Q., Gao, Y., Ke, J., Show, P., Ge, Y., Liu, Y., Guo, R., & Chen, J. (2021). Antibiotics: An overview on the environmental occurrence, toxicity, degradation, and removal methods. *Bioengineered., 12*(1), 7376–7416.

Zeng, L., Du, H., Lin, X., Liao, R., Man, Y., Fang, H., Yang, Y., & Tao, R. (2024). Isolation, identification and whole-genome analysis of an Achromobacter strain with a novel sulfamethazine resistance gene and sulfamethazine degradation gene cluster. *Bioresource Technology, 399*, 130598. https://doi.org/10.1016/j.biortech.2024.130598

Zhang, Q., & Dick, W. A. (2014). Growth of soil bacteria, on penicillin and neomycin, not previously exposed to these antibiotics. *Science of the Total Environment, 493*, 445–453. https://doi.org/10.1016/j.scitotenv.2014.05.114

Zhang, W., Wen, Y., Niu, Z., Yin, K., Xu, D., & Chen, L. (2012). Isolation and characterization of sulfonamide-degrading bacteria Escherichia sp. HS21 and Acinetobacter sp. HS51. *World Journal of Microbiology and Biotechnology, 28*(2), 447–452. https://doi.org/10.1007/s11274-011-0834-z

Zhang, Y., Zhou, J., Xu, Q., Cheng, J., Luo, Y., & Yuan, Y. (2016). Exogenous cofactors for the improvement of bioremoval and biotransformation of sulfamethoxazole by Alcaligenes faecalis. *Science of the Total Environment, 565*, 547–556. https://doi.org/10.1016/j.scitotenv.2016.05.063

Zhang, H., Jia, Y., Khanal, S. K., Lu, H., Fang, H., & Zhao, Q. (2018). Understanding the role of extracellular polymeric substances (EPS) on ciprofloxacin (CIP) adsorption in aerobic sludge,

anaerobic sludge and sulfate-reducing bacteria (SRB) sludge systems. *Environmental Science & Technology, 52*(11), 6476–6486.

Zhang, Z., Zhang, Q., Wang, T., Xu, N., Lu, T., Hong, W., Penuelas, J., Gillings, M., Wang, M., Gao, W., & Qian, H. (2022). Assessment of global health risk of antibiotic resistance genes. *Nature Communications, 13*(1), 1553. https://doi.org/10.1038/s41467-022-29283-8

Zhao, X., Li, Y., Yang, L., Wang, X., Chen, Z., & Shen, J. (2018). Screen and study of tetracycline-degrading bacteria from activated sludge and granular sludge. *Clean—Soil, Air Water, 46*, 1700411. https://doi.org/10.1002/clen.201700411

Zhao, W., You, J., Yin, S., Yang, H., He, S., Feng, L., Li, J., Zhao, Q., & Wei, L. (2022). Extracellular polymeric substances-antibiotics interaction in activated sludge: A review. *Environmental Science and Ecotechnology, 13*, 100212. https://doi.org/10.1016/j.ese.2022.100212

Zhou, T., Zhang, Z., Liu, H., Dong, S., Nghiem, L., Gao, L., Chaves, A., Zamyadi, A., Li, X., & Wang, Q. (2023). A review on microalgae-mediated biotechnology for removing pharmaceutical contaminants in aqueous environments: Occurrence, fate, and removal mechanism. *Journal of Hazardous Materials, 443*(*Part A*), 130213. https://doi.org/10.1016/j.jhazmat.2022.130213

Chapter 9
Aquatic Environmental Influence on Antimicrobial Resistance in Various Water Sources: A Hydrological Perspective

Yazeed van Wyk and Eunice Ubomba-Jaswa

1 Introduction

The issue of antimicrobial resistance has gained increasing attention worldwide due to its potential impact on public health and the environment. In this context, understanding the role of hydrology in combating the spread and development of antimicrobial resistance in aquatic environments is crucial. The main purpose is to examine from a hydrological perspective how various water sources influence antimicrobial resistance. By examining the interplay of factors such as antibiotic use, poor sanitation, and misuse of antibiotics in the context of aquatic environments, we can gain insights into the complexities of antimicrobial resistance proliferation. Thus far research has shown that natural water sources, such as rivers, lakes, groundwater, and coastal areas, can serve as reservoirs for antimicrobial-resistant bacteria. The discharge of untreated sewage, agricultural runoff, and industrial effluents into these water bodies carries a considerable number of antimicrobial agents and resistant bacteria, further exacerbating the problem.

The United Nations projects that by 2025, approximately 1.8 billion people worldwide will confront severe water stress and scarcity (Koncagül et al., 2018). According to the latest World Economic Forum Global Risk Report (World Economic Forum, 2024), antimicrobial-resistant bacteria (ARB) in aquatic environments warrant categorization under "Biological, chemical, or nuclear hazards," recognizing its substantial biological hazard potential. This classification is fitting as

Y. van Wyk
Water Research Commission, Pretoria, South Africa

Department of Geology, University of Pretoria, Pretoria, South Africa

E. Ubomba-Jaswa (✉)
Water Research Commission, Pretoria, South Africa
e-mail: euniceuj@wrc.org.za

S. M. Tiquia-Arashiro et al. (eds.), *Urban Watershed Microbiology, Volume 1*,
https://doi.org/10.1007/978-3-031-89071-0_9

ARB not only affects ecosystems but also poses a threat to human health through the proliferation of antibiotic-resistant bacteria and genes in water sources. Consequently, urgent measures are required to explore and establish alternative freshwater resources through economic and effective water reclamation, recycling, and preservation techniques. Moreover, safeguarding existing water reservoirs from further contamination is imperative for national governments worldwide. Antimicrobial resistance is widely acknowledged as a global public health crisis necessitating mitigative interventions that are no longer restricted to just surface water and wastewater, but to groundwater with the entire hydrological cycle being at risk (Lapworth et al., 2023; Nassereddine et al., 2024). The high incidence of antimicrobial-resistant bacteria (ARB) in groundwater, as evidenced by Andrade et al. (2020), suggests that this vital resource serves as a substantial reservoir for ARB on a global scale. The study, synthesizing findings from 70 peer-reviewed studies, revealed alarming statistics regarding ARB prevalence in groundwater. A substantial percentage of groundwater isolates displayed resistance to at least one antimicrobial (80.2% ± 29.0) and to at least three antimicrobials (57.2% ± 36.8), indicating widespread resistance across various regions and time periods. Moreover, the presence of ARB was identified in a substantial proportion of individual boreholes and springs (76.9% ± 33.7) where bacterial presence was detected. The calculated mean MAR index, reflecting the level of ARB presence in global groundwater, further highlights the severity of the issue. Notably, MAR index values were notably high in regions such as Kenya, Nepal, China, Morocco, Saudi Arabia, and South Africa, with Africa and Asia exhibiting the highest continental MAR indices.

Pharmaceutical residues on the other hand, especially antibiotics and antiretroviral drugs, can play a substantial role in the prevalence of antimicrobial resistance in aquatic environments. The persistence of pharmaceutical residues in health-care waste containing pharmaceutical residues can lead to groundwater contamination and the spread of pathogens, further contributing to the presence of antibiotic-resistant bacteria in water sources (Ngqwala & Muchesa, 2020; Nassereddine et al., 2024). A study conducted in an informal settlement in Kampala City, Uganda, sheds light on the startling link between on-site sanitation practices and antibiotic contamination in shallow groundwater aquifers (Twinomucunguzi et al., 2023). Over 65% of the sampled locations showed an elevated risk for antibiotic resistance during the dry season (1400 ng/L) as opposed to the wet season (730 ng/L). The lower concentrations during the wet season can be attributed to the dilution effect caused by increased recharge and flow in the groundwater system due to higher precipitation levels during this period. The seven different antibiotics found in groundwater sources are likely attributed to anthropogenic activities. In addition, a substantial positive correlation between sulfathiazole and *Escherichia coli* was identified ($p < 0.001$), suggesting a direct influence of sanitation practices on antibiotic resistance exposure. These findings emphasize the pressing need to address sanitation practices and antibiotic contamination in groundwater to protect public health and mitigate the spread of antibiotic resistance in vulnerable communities. From a surface water perspective, the impact of antibiotic pollution on water sources in mountain areas, particularly from tourism-related sewage and agricultural practices,

poses a substantial threat to microbial populations and public health. Studies have shown the presence of antibiotic residues in groundwater and river water, with antibiotics like sulfamethazine detected in surface water (Tiquia et al., 2007, 2008; Tamtam et al., 2008; Ngigi et al., 2020). The persistence of cephalosporin antibiotics in lake water, with half-lives ranging from 3.1 to 18.7 days, emphasizes the environmental challenges posed by antibiotic contamination. Advanced wastewater treatment methods utilizing metal-organic frameworks and membrane bioreactors are crucial for reducing antibiotic pollution and ensuring the safe reuse of water resources in sensitive environments like aquatic ecosystems (Wu et al., 2005; Gentile et al., 2006; Tiquia-Arashiro & Pant, 2020; Khalidi-Idrissi et al., 2023; Nassereddine et al., 2024). Monitoring antibiotic contamination in water sources and implementing effective removal processes are essential for protecting aquatic ecosystems and mitigating the risks of antibiotic resistance development (Kulik et al., 2023; Nassereddine et al., 2024).

In response to the growing awareness of threatened water resources and the associated risks to ecosystems and human health, there is a universal demand for restoring water resources to safe and acceptable standards for various purposes, including drinking. The relationship between hydrological changes driven by climate change and the spread of antimicrobial resistance (AMR) is multifaceted. Increasing temperatures associated with climate change can enhance bacterial growth rates and infection rates, leading to more antibiotic-resistant infections (Burnham, 2021; Nassereddine et al., 2024). Extreme weather events like flooding can result in the spread of waterborne infections and contamination by pollutants like nitrogen fertilizers and composts, which increase antibiotic resistance (Cabral, 2010; Tiquia, 2010a, b; Palaniappan et al., 2012; Nguyen et al., 2013; Pomaranski & Tiquia-Arashiro, 2016; Bashir et al., 2020; Magnano San Lio et al., 2023; Nassereddine et al., 2024). Water scarcity during droughts can reduce sanitation and increase population densities sharing water sources, priming the conditions for outbreaks of waterborne infections and the acquisition of antibiotic-resistant enteric pathogens. Additionally, pollution-induced expression of antibiotic-resistance genes and bacterial mutagenesis due to increased runoff and higher pollution levels in water systems contribute to the dissemination of antibiotic resistance. Overall, the combination of hydrological changes, temperature increases, and pollution in aquatic environments driven by climate change creates a scenario where the proliferation and spread of antimicrobial resistance pose a substantial threat to public health (Patz & Olson, 2006; Patz & Hahn, 2012; Bowman et al., 2018; Oest et al., 2018; Patel et al., 2019; Nassereddine et al., 2024).

This chapter will explore the intricate relationship between hydrology and the proliferation of antimicrobial resistance (AMR) in aquatic environments. This will include the influence of water sources such as rivers, lakes, groundwater, and coastal areas on AMR dissemination. Additionally, we will examine the impact of factors like antibiotic use, poor sanitation, and pharmaceutical residues on the prevalence of antibiotic-resistant bacteria in water sources. Furthermore, this chapter will discuss the need for mitigating interventions to safeguard water reservoirs from further contamination and to establish alternative freshwater resources through effective

water reclamation and preservation techniques. By addressing these aspects, this chapter aims to provide insights into the complexities of AMR in aquatic environments and highlight the pressing need for proactive measures to protect public health and preserve water resources.

2 Methods

We conducted a comprehensive mini-review using a range of online resources, including Google Scholar, PubMed, Springer, Web of Science, Science Direct, and John Wiley. Our aim was to gather the most recent and relevant information available on antimicrobial resistance in water sources, ensuring the preparation of an up-to-date review article.

2.1 Literature Search

Literature review searches were carried out on 30 April 2024 using multiple keywords as a basis to identify relevant research articles with Boolean combinations ("AND", "OR", "WITH", "SAME") to further constrain the search results. Additionally, we scanned the reference lists of selected studies and relevant review articles that might have been missed during the initial search. We collected and organized the search results to ensure comprehensive coverage across various databases, by employing the following search terms:

Wastewater treatment (Keyword) OR wastewater OR effluent OR liquid waste OR animal waste.

AND

Hospitals (Keyword) OR health care setting OR health-care facility.

AND

Water resources (Keyword) OR groundwater OR surface water.

AND

Antibiotics (Keyword) OR) antimicrobial resistance OR antibiotic resistance genes.

The screening process was conducted by two authors (YvW and EUJ), who independently reviewed and evaluated the literature based on the criteria mentioned above. We used Endnote™ 20 software for record identification and removing duplicates. The initial screening of titles and abstracts was carried out using Microsoft Excel, resulting in the selection of 121 studies for full-text review (Fig. 9.1). Following the full-text screening, we excluded 40 articles according to the above-mentioned criteria and considered the remaining 81 studies for analysis.

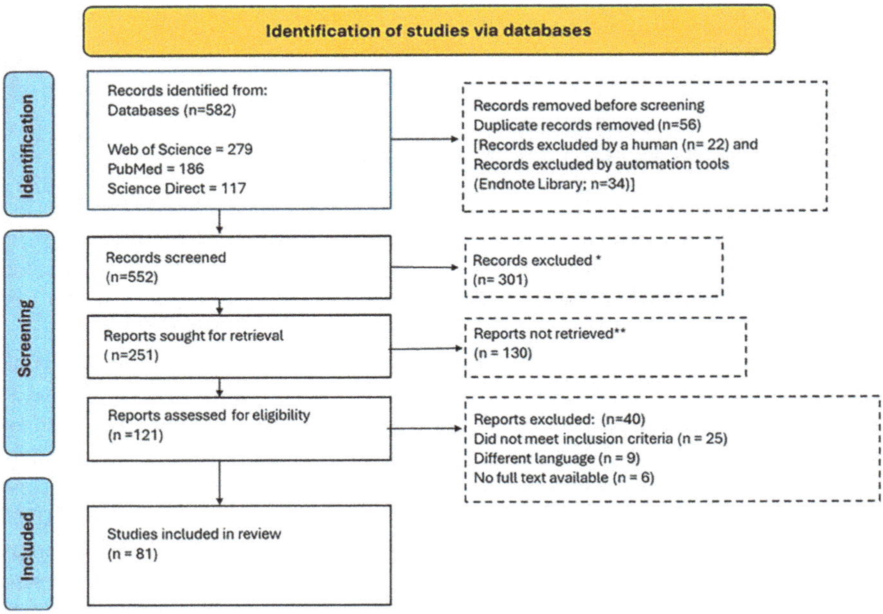

Fig. 9.1 An outline of the process that was used for the literature search and screening for eligibility of included studies

We performed a quality evaluation of the methods, sampling approach and number of samples collected. However, studies were not excluded based on the quality assessment.

2.2 *Inclusion and Exclusion Criteria*

The criteria for inclusion of studies in this research were as follows: studies published between 2005 and 2024 that described the presence and impact of antibiotics in water bodies, wastewater treatment plants, and hospital wastewaters; studies that discussed the role of aquaculture facilities, untreated wastewater discharge, and agricultural practices in spreading antibiotic-resistant bacteria; availability of original peer-reviewed journal articles; retrieval of articles from indexed journals; and publication in the English language.

Exclusion criteria involved omitting articles that did not specifically address the environmental dimensions of antibiotic resistance or those that lacked empirical data on the prevalence and dissemination of antibiotic-resistant bacteria and genes. The search aimed to encompass a multidisciplinary perspective, incorporating insights from veterinarians, medical professionals, environmental scientists, and policymakers, and emphasizing the need for global cooperation and

interdisciplinary strategies to mitigate the spread of antimicrobial resistance in aquatic environments.

3 Sources of Antibiotics in Aquatic Environments

The pervasive presence of antibiotics in aquatic environments represents a complex challenge that threatens both aquatic balance and public health. The importance of antibiotics cannot be taken for granted; however, they now permeate water bodies across the globe through diverse pathways, fostering the proliferation of antimicrobial resistance (AMR). It is essential to comprehend the origins and intricacies of antibiotic contamination in aquatic ecosystems to develop robust mitigation strategies and uphold worldwide health standards. Anthropogenic contaminants, such as antibiotics, antimicrobial compounds, heavy metals, pesticides, biocides, and nutrients, are considered selective pressures for antibiotic resistance in aquatic environments (Varela & Manaia, 2013; Kassem et al., 2023; Pagnucco et al., 2023; Nassereddine et al., 2024). Wastewater treatment plants (WWTPs) and agricultural waste have been identified as substantial sources and reservoirs for antibiotic resistance genes (ARGs) due to the excessive selective pressures they contain (Table 9.1). Additionally, urban stormwater runoff has been recognized as a source of selective pressures and ARGs into aquatic environments, although it has not been as extensively researched as WWTPs. These sources contribute to the diversification of the resistome, increasing the risk of ARB dissemination to human pathogens and exacerbating the global issue of antibiotic resistance (Benjamin et al., 2024). In this context, a resistome is the collection of all antibiotic-resistance genes in a particular environment, including those in both pathogenic and non-pathogenic bacteria (Paul & Das, 2022). It encompasses not only the genes that provide resistance to antibiotics currently in use but also those that could potentially develop into resistance mechanisms in the future. This concept is crucial in understanding how resistance can spread and evolve, affecting the efficacy of antibiotics.

Effluents from wastewater treatment plants, chemical manufacturing facilities, and agricultural and aquacultural activities emerge as primary vectors for the introduction of antibiotics such as Ciprofloxacin, which can bypass water treatment processes and enter the environment, directly impacting water bodies (Morin-Crini et al., 2022). This raises concerns about the potential for these antibiotics to contribute to the development of antibiotic-resistant pathogens, posing risks to both ecosystems and human health. Monitoring and regulating the presence of antibiotics in aquatic environments are crucial steps to mitigate these risks and protect water quality. On a global scale, wastewater treatment plants contribute substantially to antibiotic pollution. Pharmaceutical manufacturing facilities also play a substantial role, in releasing antibiotic residues into open waters. Additionally, antibiotics from husbandry and hospitals, where a high percentage of inpatients are prescribed antibiotics, further exacerbate aquatic contamination. Household consumption adds to the issue, with antibiotic residues found in urban drainage, city canals, and rivers.

Table 9.1 Various sources of antibiotics in aquatic environments

Source	Description
Wastewater treatment plants (WWTPs)	Effluents containing antibiotics from human and veterinary use, as well as pharmaceutical manufacturing discharges
Agricultural runoff	Runoff from agricultural lands carrying antibiotics used in animal husbandry or aquaculture
Aquaculture	Antibiotics used in aquaculture to prevent and treat diseases in farmed fish and shellfish
Hospital and healthcare facility discharges	Effluents from healthcare facilities containing antibiotics from patient treatment and medical waste disposal
Pharmaceutical manufacturing	Discharges from pharmaceutical manufacturing processes and improper waste disposal
Urban runoff	Runoff from urban areas carrying antibiotics from household use, pharmaceuticals, and personal care products
Land application of biosolids	Biosolids applied as fertilizer to agricultural land containing residual antibiotics
Septic systems	Discharges from septic systems containing antibiotics from household use
Animal waste	Runoff and direct deposition of antibiotics from animal agriculture and manure management practices
Natural sources	Minimal contribution from naturally occurring antibiotics produced by microorganisms in aquatic environments

Focusing on a specific case study, Vietnam illustrates the severity of this issue. The study revealed that antibiotics like enrofloxacin have been detected at concentrations as high as 680 ng L^{-1} in aquaculture wastewater in Vietnam (Vu Ngan et al., 2018). This regional data highlights the impact of antibiotic pollution in specific contexts, demonstrating the importance of localized monitoring and regulation efforts to address and mitigate the contamination of aquatic environments. Efforts to address this widespread contamination, both globally and regionally, are essential to maintaining environmental and public health.

The presence of antibiotics in these water bodies poses a risk to human and animal health, as well as the ecosystem, by promoting the development of antibiotic-resistant bacteria. The interdisciplinary collaboration encouraged by the "One Health" concept is vital in tackling this complex issue, and further research is needed to fully understand the extent of antibiotic contamination in Vietnam's aquatic environments. The interconnectedness of human, animal, and environmental health by way of advocating for a collaborative approach to tackle complex health issues like antibiotic resistance cannot be overstated. Interdisciplinary cooperation is crucial in this context, as it brings together expertise from various fields to develop comprehensive strategies for combating antibiotic resistance. For instance, veterinarians, medical professionals, environmental scientists, and policymakers can work together to identify and mitigate the sources of antibiotic contamination in aquatic environments. A practical example of such collaboration is the implementation of integrated surveillance systems that monitor antibiotic use and resistance patterns across human healthcare, agriculture, and aquaculture sectors. By sharing

data and insights, these systems can identify emerging resistance trends and hotspots, enabling targeted interventions. Global agencies, including the WHO's Global Antimicrobial Resistance Surveillance System (GLASS), the CDC, the Food and Agriculture Organization (FAO), and the World Organisation for Animal Health (OIE), are diligently working to control antibiotic resistance. Additional initiatives addressing the global threat of antibiotic resistance include the Global Health Security Agenda (GHSA) and its Antimicrobial Resistance Action Package (Belay et al., 2017; Sirijatuphat et al., 2018). The European Centre for Disease Prevention and Control (ECDC) also plays a crucial role in collecting, analyzing, and reporting data related to antimicrobial resistance (AMR) through a network of national surveillance systems across EU/EEA countries as defined in the EARS-Net protocol, to enable action against AMR (WHO, 2022). Furthermore, initiatives like the European Union's Joint Programming Initiative on Antimicrobial Resistance (JPIAMR) illustrate the potential of interdisciplinary cooperation. This initiative brings together researchers from diverse fields to conduct studies on the environmental dimensions of antibiotic resistance, promote best practices in antibiotic stewardship, and develop new technologies for wastewater treatment (Berg, 2020). In agricultural settings, collaboration between agronomists and environmental engineers can lead to the development of sustainable farming practices that reduce antibiotic runoff, such as optimizing the use of manure and implementing buffer zones around water bodies. Similarly, partnerships between public health officials and pharmaceutical companies can lead to the creation of antibiotic disposal programs to prevent improper disposal of unused medications (Samreen et al., 2021). By leveraging the strengths and expertise of different disciplines, the "One Health" approach can create robust, multifaceted strategies to combat antibiotic resistance, protecting both environmental and public health.

Furthermore, the global perspective reveals a complex history of human activities exacerbating the issue. Agricultural runoff, driven by practices such as the application of manure and irrigation with contaminated water, introduces antibiotics and antibiotic resistance genes (ARGs) into surface waters, significantly impacting aquatic ecosystems. A study highlighted that agricultural practices contribute to the spread of antibiotic resistance, with antibiotic residues and ARGs being detected in rivers and streams. Additionally, improper disposal of unused medications and inadequate wastewater treatment facilities can result in the release of antibiotics and ARGs into rivers and streams, further intensifying the issue of antimicrobial resistance in aquatic environments (Sala-Comorera et al., 2021). Similarly, inadequate wastewater treatment facilities and improper disposal of unused medications serve as additional conduits for the release of antibiotics and ARGs into water bodies, perpetuating the cycle of antibiotic resistance propagation. Research has shown that substantial quantities of antibiotics consumed by humans and animals remain unmetabolized, entering aquatic environments through various pathways, including wastewater treatment plants and direct deposition (Abia et al., 2015; Chukwu et al., 2023). Moreover, the presence of pathogens like Arcobacter further compounds the issue, originating from sources such as animal feces and water bodies, posing risks to human and animal health. The study highlights the potential risks posed by

antibiotic pollution to public health, highlighting the importance of identifying and treating bacteria like Arcobacter to reduce exposure and adverse effects (Ghaju Shrestha et al., 2022).

Antibiotics enter aquatic environments primarily through a multitude of sources such as animal husbandry, plant production, and aquaculture, contributing to diffuse concentrations of antimicrobial chemicals and their transformation products (TPs) within water bodies (Löffler et al., 2023). Notably, the volume of antibiotics used in food-producing animals surpasses that of human consumption in certain European countries, accentuating the substantial contribution of animal agriculture to antibiotic pollution (Löffler et al., 2023). A pan-European study found sulfonamide antibiotics in 24% of groundwater samples, with concentrations reaching a maximum of 38 ng/L and an average of 2 ng/L (Singer et al., 2016). In Spain, higher concentrations of sulfonamides, up to 3.4 μg/L, were recorded in groundwater (García-Galán et al., 2010). Additionally, triclosan, a biocide, has been detected in UK aquifers at levels exceeding 2 μg/L over several years (Stuart et al., 2012; Manamsa et al., 2016). In the United States, veterinary pharmaceuticals from livestock wastewater impoundments have infiltrated groundwater, emphasizing the substantial role of livestock wastewater lagoons in groundwater pollution (Bartelt-Hunt et al., 2011). Antibiotics present in manure- or sludge-amended agricultural soils can enter groundwater due to rainfall, irrigation, and other human activities. The magnitude of this impact is further exemplified by the UK's reported sale of 2 tons of antibiotics specifically for fish production in 2014 alone, underlining the prominent role aquaculture plays in antibiotic dissemination (Singer et al., 2016). Moreover, Tiquia (2002), Tiquia (2011), Cho et al. (2012), and Apreja et al. (2022) highlight various other sources of antibiotic contamination in groundwater, encompassing hospital and industrial discharge, livestock farming, aquaculture runoff, leaky sewerage, and wastewater effluent. Collectively, these findings highlight the diverse and persistent nature of antibiotic pollution in aquatic environments, emphasizing the urgent need for comprehensive mitigation strategies to safeguard water quality and mitigate the risks associated with antibiotic resistance.

Antibiotics enter aquatic environments through various transport processes, primarily leaching, runoff, and direct discharge. Leaching occurs when antibiotics present in soils, particularly those treated with manure or biosolids, dissolve in water and percolate through the soil, eventually reaching groundwater. This process is influenced by the properties of the soil, rainfall, and irrigation practices (Du & Liu, 2012). Runoff can be considered as another major pathway, where antibiotics applied on agricultural fields or urban landscapes are washed away by heavy rains or irrigation water, flowing over the land surface and into rivers, lakes, and streams (Oest et al., 2018). This is especially prevalent in areas with heavy rainfall or inadequate vegetation cover (Zheng et al., 2020), and is expected to become more prevalent with increasing climate change impacts and urbanization. Direct discharge involves the release of antibiotics from point sources, such as wastewater treatment plants (WWTPs), pharmaceutical manufacturing facilities, hospitals, and aquaculture operations, directly into water bodies. WWTPs, for example, may not fully remove antibiotics during the treatment process as noted earlier, allowing residues

to be discharged into receiving waters. These transport processes collectively contribute to the widespread presence of antibiotics in aquatic environments, posing risks to both ecosystems and human health by promoting the development of antibiotic-resistant bacteria (Kim et al., 2011; Carvalho & Santos, 2016).

4 Pathways of Antibiotics, ARB, and ARGs in Water Environments

Understanding the intricate pathways through which antibiotics, antibiotic-resistant bacteria (ARB), and antibiotic genes (ARGs) traverse aquatic environments is essential in comprehending the dynamics of antimicrobial resistance (AMR) proliferation. These pathways, which represent a complex interplay of several factors, play a crucial role in shaping the landscape of AMR dissemination within aquatic ecosystems. By unraveling these multifaceted pathways, we gain insights into the mechanisms driving the spread of AMR, from its origins in anthropogenic activities such as landfill leachate, wastewater treatment plants, agriculture activities, aquaculture, urban runoff, and pharmaceutical manufacturing to its impact on water quality and public health. These compounds and microorganisms interact within water bodies, creating a complex network of transmission routes that contribute to the spread of AMR (Fig. 9.2).

Wastewater treatment plants (WWTPs) serve as major conduits for the release of antibiotics into aquatic environments, as highlighted by (Sorn et al., 2022). These facilities receive a diverse array of contaminants from sources such as households, slaughterhouses, and hospitals, with antibiotics constituting a substantial portion. Despite undergoing treatment processes, a considerable proportion of these antibiotics persist throughout the treatment steps and ultimately end up in the final effluent, contributing to the degradation of water quality. Conventional WWTPs are not equipped to completely remove all types of chemical and biological contaminants, thus allowing the release of these pollutants into various water bodies. Moreover, alongside antibiotics, emerging pollutants like polyaromatic hydrocarbons (PAHs), polychlorinated biphenyls (PCBs), and per- and poly-fluoroalkyl substances (PFAS) also accumulate in WWTPs, creating conducive environments for the proliferation of antibiotic-resistant bacteria (ARB) and antibiotic resistance genes (ARGs). The frequent detection of antibiotics, ARB, and ARGs throughout treatment stages again displays the inefficiency of WWTPs in effectively removing these contaminants. This persistence poses severe risks to both aquatic ecosystems and human health, as it can lead to the development and spread of antimicrobial resistance, compromising the effectiveness of antibiotics, and threatening public health.

Locally, in South Africa, the operation and maintenance of wastewater treatment plant (WWTPs) infrastructure faces numerous challenges, with many facilities not complying with national discharge standards. Studies indicate that a big portion of WWTPs, particularly those along the Vaal River, do not satisfactorily treat

Fig. 9.2 Sources and pathways of antimicrobial resistance in water environments

wastewater, highlighting systemic issues in wastewater management (Gani et al., 2021). Despite initiatives like the Green Drop program aiming to improve treatment operation and management, recent reports show a concerning decline in wastewater management standards. A study assessing the prevalence, distribution, and antibiotic resistance patterns of *Escherichia coli* and *Enterococcus* spp. in raw and treated wastewater from a major (WWTP) in KwaZulu-Natal, South Africa, as well as in the receiving river water upstream and downstream from the WWTP discharge point, revealed interesting findings. Quantitative polymerase chain reaction (PCR)-confirmed isolates were assessed for antibiotic susceptibility against multiple antibiotics. Results showed that most *E. coli* were susceptible to meropenem (94.8%) and piperacillin-tazobactam (92.9%), while most *Enterococcus* were susceptible to ampicillin (97.8%) and vancomycin (96.7%). However, a high proportion of

isolates exhibited multidrug resistance (64.8% for *E. coli* and 80.8% for *Enterococcus*), with many showing multiple antibiotic resistance indices >0.2 (42.4% for *E. coli* and 65.1% for *Enterococcus*). The log removal values (LRVs), a measure of the effectiveness of the treatment process in removing contaminants, for *E. coli* ranged from 2.97 to 3.99, and for *Enterococci*, the range was observed from 1.83 to 3.98 (Mbanga et al., 2021). LRVs indicate the percentage of a contaminant removed during treatment, with higher values reflecting greater removal efficiency. These findings suggest inadequate removal of multidrug-resistant bacteria during wastewater treatment, highlighting the public health risks associated with bacterial contamination of environmental waters from WWTPs. The occurrence of multidrug-resistant bacteria in treated wastewater effluent signifies that these bacteria have the potential to infiltrate natural water bodies like groundwater, rivers, lakes, and streams. Once released into these environments, they can disseminate resistance genes to other bacteria, heightening the risk of infections that are challenging to manage with conventional antibiotics. This poses a substantial public health threat, particularly in communities with compromised immune systems, especially those residing in rural areas. Therefore, improving the efficiency of WWTPs and addressing the systemic issues in wastewater management are crucial steps in mitigating the spread of AMR.

In addition to wastewater treatment plants, agricultural activities are another major source of antibiotic pollution. Agricultural practices significantly contribute to environmental contamination, with the application of manure and biosolids spreading antibiotics, ARB, and ARGs into soil, groundwater, and surface water through leaching and runoff, exacerbating AMR proliferation (Singer et al., 2016; Sverdrup et al., 2020; Mansi et al., 2022). The application of biosolids in agricultural fields can lead to the release of antibiotics and antibiotic residues into the soil, eventually reaching water bodies through surface runoff or leaching, thereby contributing to the spread of antimicrobial resistance. These sources highlight the complex pathways through which antibiotics enter aquatic ecosystems, emphasizing the need for comprehensive strategies to mitigate the impact of antimicrobial resistance in these environments (Calero-Cáceres et al., 2022). While the case study of South African municipal WWTPs provides a detailed assessment of local challenges, it also mirrors broader global trends in wastewater management and AMR proliferation. Similar inefficiencies in wastewater treatment and resultant environmental contamination have been reported worldwide, highlighting the universal need for improved management practices and robust regulatory frameworks to address this pressing public health issue (Hiller et al., 2019; Mukherjee et al., 2021).

Similarly, aquaculture operations are another major source of antibiotic contamination in aquatic environments. These facilities use antibiotics extensively to prevent disease and promote growth in fish and other aquatic organisms. However, the runoff from these facilities introduces large quantities of antibiotics, ARB, and ARGs into nearby water bodies, thereby further amplifying the spread of AMR (Singer et al., 2016; Uluseker et al., 2021). Studies have shown that aquaculture activities contribute to the presence of antibiotics and resistant bacteria in surrounding aquatic environments, impacting water quality and posing risks to public health

(Preena et al., 2020; Miranda et al., 2024). The antibiotics used in aquaculture not only persist in the environment but also promote the selection of resistant bacteria, which can transfer resistance genes to other bacteria, exacerbating the spread of AMR (Suyamud et al., 2023). The presence of antibiotics and resistant bacteria in aquaculture effluents has been documented worldwide, indicating a global issue that requires international cooperation and stringent regulations to mitigate. For instance, research in China has shown that aquaculture ponds are hotspots for the dissemination of ARGs, with considerable amounts of antibiotics and resistant bacteria detected in both pond water and sediments (Lu et al., 2022). Similarly, studies in European and Latin American countries have reported high levels of antibiotics and ARGs in water bodies affected by aquaculture activities (Marti et al., 2013; Tapia-Arreola et al., 2022).

While the focus has been on specific sources such as WWTPs, agricultural practices, and aquaculture, it is crucial to recognize that direct discharge of untreated wastewater remains an effective pathway for introducing antibiotics and resistance elements into surface waters, particularly in regions with inadequate wastewater treatment infrastructure. Seasonal hydrological dynamics also influence ARG behavior in water bodies, stressing the interconnectedness between environmental factors and AMR dissemination (Uluseker et al., 2021; Adyari et al., 2023). Moreover, the infiltration of antibiotics from land-applied manure and sewage sludge, coupled with effluent discharge from WWTPs, contaminates groundwater, further exacerbating AMR spread (Sverdrup et al., 2020). The persistence of these pollutants, including emerging contaminants like PAHs, PCBs, and PFAS, compounds the challenge by creating environments that foster the proliferation of ARB and ARGs. Addressing AMR in water environments necessitates a comprehensive approach encompassing stringent agricultural practices, and aquaculture management, with collaborative efforts imperative to mitigate AMR's environmental dimensions and safeguard public health (EMA/CVMP/ERA/632109/2014, 2021). It is therefore important to look at innovative ways of enhancing wastewater treatment efficiency to mitigate the risks posed by emerging pollutants. This can include implementing a combination of technological advancements, operational improvements, and regulatory measures. Advanced treatment technologies such as advanced oxidation processes (AOPs), membrane filtration, and activated carbon adsorption can effectively remove emerging pollutants, including antibiotics and other contaminants. Additionally, upgrading existing wastewater treatment plants (WWTPs) with additional treatment units or retrofitting them with advanced treatment technologies can enhance removal efficiency for emerging pollutants. Optimizing biological treatment processes, implementing source control measures, and strengthening regulatory frameworks are also essential components of a comprehensive approach to wastewater treatment (Bradley et al., 2002; Tiquia & Mormile, 2010; Tiquia-Arashiro & Mormile, 2013; Chettri et al., 2024). Furthermore, robust monitoring and surveillance programs and public awareness campaigns play critical roles in ensuring compliance with effluent quality standards and promoting responsible behavior in pharmaceutical use and disposal.

5 Impact of Antibiotics on Aquatic Ecosystems and Human Health

The widespread use and misuse of antibiotics in aquatic environments have created a complex web of ecological and health impacts, where the spread of antibiotic resistance in water bodies threatens both the health of ecosystems and human populations. The ecological impacts of antibiotics in aquatic environments are multifaceted and far-reaching. Antibiotics can alter the composition of microbial communities, leading to changes in nutrient cycling and ecosystem function. They can also select for ARBs, which can then spread to other environments and potentially cause infections in humans and animals. Furthermore, the persistence of antibiotics in the environment can lead to chronic exposure of organisms, which can have long-term ecotoxicological effects. The health impacts of antibiotics in aquatic environments are equally concerning. Antibiotic-resistant bacteria can cause infections that are difficult or impossible to treat, leading to increased morbidity and mortality rates. The overuse and misuse of antibiotics can also contribute to the development of antimicrobial resistance, which is a major public health concern worldwide. Additionally, the presence of antibiotics in drinking water and food can pose health risks to humans, particularly for vulnerable populations such as the elderly and young children. To address these complex issues, it is essential to adopt a multidisciplinary approach that considers both ecological and health impacts. This includes improving wastewater treatment practices, reducing antibiotic use in agriculture and aquaculture, and developing new technologies for removing antibiotics from the environment. Additionally, there is a need for increased public awareness and education about the risks associated with antibiotic use and the importance of responsible antibiotic stewardship.

5.1 Impact on Aquatic Life

Antibiotics, hailed as life-saving medications, have become pervasive pollutants in aquatic environments, posing a complex threat to both ecosystems and human health. The widespread use of antibiotics in various sectors, coupled with inadequate disposal and treatment practices, has led to their infiltration into water bodies worldwide. From their origins in anthropogenic activities such as agriculture, aquaculture, and healthcare, to their consequential effects on water quality and microbial ecology, the influence of antibiotics extends far beyond their intended therapeutic purposes. Using a multidisciplinary lens, the mechanisms driving antibiotic contamination in aquatic environments, the proliferation of antibiotic-resistant bacteria, and the transmission of antibiotic resistance genes can be assessed. Antibiotics enter aquatic environments through various pathways, including runoff from agricultural fields where manure containing antibiotics is used as fertilizer, leaching from

aquaculture, and improper disposal of unused medications. This contamination significantly affects microbial communities, which play crucial roles in biogeochemical cycling and organic contaminant degradation (Chhabra et al., 2006; Tiquia et al., 2006a, b; Tiquia, 2008; McKindles & Tiquia-Arashiro, 2012; Flood et al., 2015). Antibiotics can exert bactericidal and bacteriostatic effects, leading to the disappearance of sensitive microbial populations and disrupting ecological functions. Additionally, the presence of sub-lethal concentrations of antibiotics accelerates the proliferation of antibiotic-resistant bacteria and the transmission of antibiotic resistance genes (ARGs), compromising the efficacy of antibiotics in treating infections. Aquatic organisms, from microorganisms to higher trophic levels, are adversely impacted, with phytoplankton and cyanobacteria being particularly sensitive, leading to inhibited growth and affected primary production. Zooplankton also experience sub-lethal effects, impacting growth and reproductive success, which can cascade through the food web, ultimately jeopardizing ecosystem stability (Tiquia et al., 2008; Pan et al., 2017; Islam & Midya, 2023).

5.2 *Consequence for Human Health*

Research conducted by (Salas et al., 2022) on the antimicrobial pharmacodynamics of fluoroquinolones and cephalosporins against Salmonella strains with varying levels of acquired resistance provides valuable insights into the impact of antibiotics on both aquatic ecosystems and human health. By analyzing the pharmacodynamic models and parameter values for these antibiotics, the study sheds light on how bacterial strains with acquired resistance influence the effectiveness of treatment regimens. Specifically, the study highlights the importance of understanding resistance mechanisms in designing effective treatment strategies, emphasizing the need to consider various pharmacodynamic parameters beyond just the minimum inhibitory concentration (MIC). These findings are crucial in addressing the global challenge of antimicrobial resistance, which not only affects human health but also has implications for the ecological balance of aquatic ecosystems where these antibiotics may be present. Furthermore, antibiotics have a huge impact on aquatic ecosystems and human health due to their extensive use and incomplete elimination, leading to their detection in various environmental waters, including surface water and groundwater. The persistence of antibiotics in the environment and chronic exposure of organisms to these chemical stressors has ecotoxicological effects. The sub-therapeutic doses of antibiotics present risks to humans and other animals, contributing to the development of antibiotic-resistant bacteria and genes, which is a major health concern worldwide (Ngqwala & Muchesa, 2020). Overuse and misuse of antibiotics may cause a risk to human health by promoting antibiotic-resistant bacteria and antibiotic resistance genes in aquatic environments. As mentioned previously, poor management of wastewater treatment plants may release antibiotics, antibiotic-resistant bacteria, and antibiotic resistance genes into the environment,

presenting a potential environmental health risk. Antimicrobial resistance presents a formidable challenge in clinical therapy, compromising antibiotic effectiveness and resulting in therapeutic failure, elevated health costs, and increased morbidity and mortality rates.

5.3 *Progress in Modeling and Surveillance*

While microbial water quality modeling has attained a solid foundation, the inclusion of antibiotic-resistant bacteria (ARB) and antibiotic resistance genes (ARG) is still in its early stages (Kleczkowski et al., 2019). The fate and transport of antibiotics, ARB and ARGs in the environment and their impact on human health are still poorly understood. Given the intricacies involved in the dissemination of ARB from source to human exposure, a modular modeling framework emerges as the most adept approach to confront this challenge (Jampani et al., 2022). Developing modeling scenarios and management strategies based on the framework could also contribute to achieving Sustainable Development Goals 3 (good health and wellbeing) and 6 (clean water and sanitation). Gothwal and Thatikonda (2018) attempted to develop a mathematical model to understand the fate and transport of fluoroquinolone-resistant bacteria in the Musi River (India); however, the model could not predict the observed occurrences accurately. A few recent case studies have analyzed the fate and transport of ARB in aquatic environments by developing research-specific regional mathematical models (Gothwal & Thatikonda, 2018, 2020; Ikuma & Rehmann, 2020; Barrios et al., 2021), but many details on the evolution of resistance were missing in those models. Other limitations include a lack of data and a limited understanding of fate and transport dynamics. Improving wastewater and freshwater quality monitoring and surveillance in watersheds for ARB and ARGs, especially in low and middle-income countries (LMICs), could help evaluate fate and transport approach and mitigate public health risks moving forward. Larsson et al. (2018) present key findings explaining the multifaceted dynamics of antibiotic resistance. The team showed the importance of assessing and implementing social interventions to comprehend the environmental role in resistance development, particularly regarding anthropogenic inputs of selective agents. Furthermore, they advocate for a global perspective in identifying high-risk environments for resistance evolution and emergence. Additionally, the study emphasizes the necessity of quantitative modeling to grasp the evolutionary processes driving resistance emergence across diverse environments, despite the challenges in acquiring relevant input data. Understanding the molecular mechanisms underpinning resistance mobilization and transfer, alongside the involved organisms and environmental transmission arenas, is deemed crucial for estimating risks associated with the emergence of novel resistant pathogens.

5.4 Antibiotic Pollution and Ecosystem Health

Costanzo et al. (2005) investigated the presence and effects of antibiotics, namely ciprofloxacin, norfloxacin, and cephalexin, in sewage effluent and downstream environmental waters. Antibiotics in both effluent and receiving waters were detected, with concentrations ranging from 90.0 to 2000.0 ng/L for cephalexin, 41.5 to 132.2 ng/L for ciprofloxacin, and 15.4 to 210.0 ng/L for norfloxacin. Bacteria cultured from these waters exhibited resistance to multiple antibiotics, with all strains displaying resistance to erythromycin, and varying resistance to other antibiotics such as ampicillin, tetracycline, and trimethoprim. The study also examined the impact of short-term antibiotic exposure on bacterial denitrification rates, revealing variable responses depending on the antibiotic type and concentration. These findings, based on actual data, paint a clear picture of the significance of antibiotic pollution in aquatic ecosystems and its potential threat to ecosystem function and human health. This contamination extends to surface waters, groundwater, and effluents, impacting non-target species and selecting resistant bacteria, particularly affecting cyanobacteria (Kovalakova et al., 2020). Sensitive microbial groups, such as cyanobacteria and ammonium oxidizing bacteria, play crucial roles in aquatic ecosystems, including nutrient cycling and ecosystem function. Antibiotics can alter the composition of microbial communities, leading to changes in these processes and potentially disrupting ecosystem balance (Paumelle et al., 2021). The mechanisms of antibiotic impact on microbial communities involve several factors. Antibiotics can inhibit the growth of sensitive microorganisms, leading to changes in community composition and potentially altering ecosystem processes. Additionally, antibiotics can be selected for antibiotic-resistant bacteria, which can then spread to other environments and potentially cause infections in humans and animals (Wang et al., 2022). The impact of antibiotics on microbial communities links directly to overall ecosystem health. Changes in microbial communities can have cascading effects on ecosystem processes, such as nutrient cycling and decomposition. These changes can also affect the health of other organisms in the ecosystem, including plants and animals. To better understand the impact of antibiotics on microbial communities and overall ecosystem health, it is essential to consider the ecological context in which these interactions occur. This includes factors such as the type and concentration of antibiotics, the composition of microbial communities, and the presence of other pollutants in the environment.

Furthermore, Grenni et al. (2018) showed that antibiotics can affect natural microbial communities, which play a crucial role in maintaining soil and water quality by being involved in biogeochemical cycling and organic contaminant degradation. Antibiotics can directly impact microbial communities by causing bactericide and bacteriostatic actions, leading to the disappearance of certain microbial populations and their ecological functioning. Indirect impacts include the development of antibiotic-resistant bacteria and bacterial strains capable of degrading

antibiotics through metabolic or co-metabolic processes. The implication of these direct and indirect effects is influenced by several factors, such as antibiotic concentration, exposure time, receiving ecosystem, and the co-occurrence of other antibiotics and/or contaminants. Gunathilaka et al. (2023) have shown that when antibiotics are inadequately metabolized and discharged into the environment, they fuel the dissemination of antibiotic resistance among bacterial pathogens, posing grave environmental and public health problems. This pollution profoundly impacts planktonic communities, crucial primary producers in aquatic ecosystems, inducing sublethal effects that ripple through higher trophic levels. Notably, antibiotics exhibit toxicity toward phytoplankton, notably inhibiting growth, while also eliciting sublethal responses in zooplankton, ultimately jeopardizing individual growth and community stability (Pan et al., 2017; Islam & Midya, 2023). In this sense, antibiotic resistance poses a formidable challenge, impeding disease treatment and exacerbating public health risks. Crucially, as planktonic organisms underpin essential ecosystem services like oxygen production and nutrient cycling, antibiotic pollution threatens to disrupt these vital functions, necessitating a multifaceted approach encompassing reduced consumption, enhanced wastewater treatment, and rigorous environmental monitoring for effective policy and management interventions.

The persistence of antibiotics in aquatic environments is a major concern due to their potential for chronic, long-term impacts on ecosystems. Antibiotics can be considered persistent substances because their entry rate into the environment is often higher than their elimination rate (Polianciuc et al., 2020). This leads to the accumulation of antibiotics in water bodies over time, exposing organisms to continuous, low-level doses (Kulik et al., 2023). Chronic exposure of aquatic organisms to antibiotic residues can have severe ecological consequences, as antibiotics can bioaccumulate in the tissues of aquatic life, leading to sublethal effects that ripple through food webs. Välitalo et al. (2017) emphasize the critical role of aquatic microorganisms in ecosystems and their susceptibility to antibiotic exposure. Cyanobacteria and ammonium oxidizing bacteria are among the most sensitive microorganisms to antibiotics, while proteobacteria exhibited lower sensitivity, highlighting the need for chronic toxicity tests over acute tests. The study also points out the challenges associated with antibiotic resistance, complicating the application of chronic tests. The review further discusses the importance of considering the ecotoxicological effects of antibiotics on aquatic microorganisms, given their substantial role in various ecosystems. The data presented, including effect concentrations for distinct types of microorganisms, provide valuable insights into the potential impacts of antibiotics on aquatic ecosystems. The accumulation of antibiotics and the resulting disruption of sensitive microbial communities and ecosystem functions can have cascading impacts on the overall health and resilience of aquatic environments.

6 Role of Hospital Wastewaters in AMR Dissemination

Hospital wastewaters are a substantial source of antimicrobial resistance (AMR) in aquatic environments, warranting specific attention in the context of AMR mitigation. Unlike municipal wastewaters, which also contribute to the spread of AMR, hospital wastewaters have a unique combination of antibiotic-resistant bacteria (ARB) and antimicrobial resistance genes (ARGs) that pose a heightened risk to public health (Hiller et al., 2019). For instance, studies have shown that hospital wastewaters contain higher concentrations of antibiotics and ARGs compared to municipal wastewaters (Li et al., 2015; Lien et al., 2016). HWWs often contain 5% and 15% more pollution load compared to wastewater from other sources (Kumari et al., 2020). Healthcare facilities, which play a crucial role in patient treatment and medical waste disposal, generate effluents that carry a potent mix of antibiotics and other pharmaceutical residues into water bodies. In addition, the water demand in hospitals is estimated to range from 200 to 1200 liters per bed per day, with an equal amount of wastewater discharge accompanying this consumption (Khan et al., 2021). This highlights the significant impact of healthcare facilities on the environmental pollution load, emphasizing the need for effective treatment and disposal strategies to mitigate this issue.

Despite the vital role of wastewater treatment plants (WWTPs) in mitigating pollution, studies indicate that a considerable proportion of antibiotics persist throughout treatment processes, ultimately entering aquatic ecosystems. Consequently, hospital wastewaters serve as vectors for the transmission of antibiotic-resistant bacteria (ARB) and antibiotic resistance genes (ARGs), amplifying the global challenge of AMR proliferation (Perry et al., 2021). Within hospital settings, the widespread use of antibiotics among inpatients further exacerbates the contamination of wastewaters with antimicrobial agents. This practice not only contributes to the direct release of antibiotics but also fosters the development and dissemination of antibiotic-resistant pathogens. As a result, hospital wastewaters emerge as reservoirs of ARB and ARGs, posing substantial risks to both environmental health and public health.

A study conducted by the CDC revealed the considerable role of hospital wastewater in the dissemination of AMR. Through insights provided by CDC's National Wastewater Surveillance System (NWSS), it showed that hospitals contribute substantially to the presence of antibiotic-resistant pathogens in water systems. These pathogens find ideal repositories in biofilms within hospital plumbing systems, where they thrive and share their resistance genes (Control & Prevention, 2022). Efforts to address this issue include innovative approaches such as applying Phages to hospital plumbing surfaces to combat biofilm structures and prevent the spread of resistant bacteria.

In related studies, investigations reveal that hospital wastewaters harbor a plethora of antibiotic-resistant bacteria (ARB) and antimicrobial resistance genes (ARGs), including clinically substantial strains such as extended-spectrum β-lactamase (ESBL) and carbapenemase-producing *Enterobacteriaceae*

(Hassoun-Kheir et al., 2020). Building on these findings, research from Romania (Surleac et al., 2020) demonstrates the transmission of multidrug-resistant *Klebsiella* pneumoniae clones from hospitals to wastewater, underlining the risk of inadequate sewage disposal. Similarly, studies from Brazil highlight the genomic and virulence potential of *Pseudomonas aeruginosa* strains carrying clinically relevant resistance mechanisms, suggesting dissemination from hospital settings to urban rivers. These findings point to the need for robust wastewater treatment technologies as mentioned and risk assessment protocols to mitigate the spread of AMR from hospitals to the environment (Calero-Cáceres et al., 2022).

In a comprehensive metagenomic analysis of wastewater samples from seven hospitals and one community sewage sample, Petrovich et al. (2020) revealed an average read pair count of 38.4 million, with approximately 62% allocated to bacteria. The study found a higher abundance of ARGs in hospital wastewater compared to community sewage, with 0.25% of reads mapping to ARGs in hospital samples versus 0.1% in the community sample. Hospital samples also exhibited a higher predominance of gut-associated bacteria, such as *Faecalibacterium, Bacteroides, Bifidobacterium*, and *Escherichia,* compared to the community sample. This analysis detected 1154 unique bacterial genera across all samples, with the top 19 genera, including *Pseudomonas* and *Acinetobacter*, accounting for over 70% of bacterial abundance.

Further studies (Perry et al., 2021) support these findings, showing that ARG abundance and composition varied across different hospital collection points, with 502 different resistance genes identified, predominantly belonging to the aminoglycoside and macrolide antimicrobial classes. Key ARGs of interest for infection control, including *blaOXA, blaIMP*, and genes of the vanA cluster, were also identified. Despite no substantial relationship found between total antimicrobial usage or length of stay and ARG abundance in sewage, increased phenotypically matched antimicrobial usage positively affected resistance gene abundance, suggesting a small role of direct selection. Positive associations were observed between vancomycin ARGs and vancomycin usage, and a trend toward significance was noted for carbapenem ARG abundance and carbapenem antimicrobial usage. However, no association was found between amoxicillin usage and amoxicillin ARGs. Additionally, ARG abundance did not mirror resistance patterns in clinical isolates, but *Enterococcaceae* and *Staphylococcaceae* showed substantial positive associations with ARG abundance for the same antimicrobial class. Analysis of antibiotic residues revealed a 12-fold increase in concentration in hospital effluent compared to the community, reflecting high antimicrobial usage within the hospital.

Complementary studies in the Middle East (Daoud et al., 2017; Alaali & Thani, 2020) have demonstrated that hospital wastewaters host antibiotic-resistant bacteria and antibiotic residues, which can spread through wastewater treatment plants to various water bodies, contributing to the dissemination of different bacterial species. Specifically, the study found that 66.7% of E*nterobacteriaceae* isolates from hospital wastewaters showed resistance to gentamicin and amikacin, while no resistance was observed in *Klebsiella* isolates towards these antibiotics from the second hospital. Moreover, the presence of resistance genes like CTX-M and SHV was

notable, with 94.4% (17 out of 18 isolates) of *E. coli* and 100% (5 out of 5 isolates) of *Klebsiella* species carrying the CTX-M gene and 55.6% (10 out of 18 isolates) of *E. coli* have SHV gene which was absent in *Klebsiella* species. Additionally, AmpC producers reached 25% and 28.9% in the first and second hospital respectively, and New Delhi metallo-beta-lactamase-1 (NDM-1) gene was detected in two strains of *Enterobacter cloacae* but only one of them expressed the phenotypic resistance.

A systematic review and pooled analysis conducted by Andrade et al. (2020) explored the occurrence of antimicrobial-resistant bacteria (ARB) in global groundwater, a critical drinking water source for 2.2 billion people. Their findings revealed that 80.2% ± 29.0 of groundwater isolates were resistant to at least one antimicrobial, with 57.2% ± 36.8 displaying resistance to three or more antimicrobials. ARB were detected in 76.9% ± 33.7 of individual groundwater sources, highlighting the widespread nature of the issue. The analysis emphasized multidrug resistance and the prevalence of critical ARB as listed by the World Health Organization, with certain bacterial genera like *Pseudomonas, Klebsiella*, and *Enterobacter* spp. consistently exhibiting high resistance rates. Additionally, specific antimicrobial classes such as first generation *Cephalosporins* and *Glycopeptides* were associated with elevated MAR indices. Across 70 relevant studies spanning diverse settings, high occurrence rates of ARB were observed, particularly in urban areas and regions with human waste sources, including those near hospitals. Furthermore, the study indicated that one-off sampling regimes yielded higher resistance percentages compared to repeat sampling, and MAR indices were notably higher in arid regions, during dry seasons, and in urban areas. These findings emphasize the need for refined methodologies, comprehensive data collection, and more representative studies to effectively address the global concern of antimicrobial resistance in groundwater.

The dissemination of antimicrobial resistance (AMR) through hospital wastewater (HWW) varies significantly across different regions, influenced by factors such as regulations, treatment technologies, and hospital practices. In regions like South India and the Democratic Republic of Congo, tropical conditions exacerbate the prevalence of antimicrobial-resistant bacteria in hospital effluents. Studies have shown that wastewater from different sources, including hospitals, significantly impacts the prevalence of antimicrobial-resistant *Escherichia coli* in sewage treatment plants (Mubedi et al., 2013). The removal rates of antibiotics in these regions are also notably variable, with some treatment plants achieving higher removal efficiencies than others (Prabhasankar et al., 2016). In contrast, regions with well-regulated water and wastewater treatment infrastructure, such as parts of Europe and North America, tend to have more effective control over the dissemination of AMR. These regions often employ advanced treatment technologies that can degrade pharmaceuticals and other contaminants more efficiently. However, even in these areas, certain substances like antitumor agents and organohalogen compounds can pass through wastewater treatment plants (WWTPs) without degradation, leading to environmental contamination (Khan et al., 2021). In low-resource settings, the situation is more dire. These regions often lack the infrastructure for effective wastewater treatment, leading to higher concentrations of pharmaceuticals and

AMR bacteria in the environment. The presence of multidrug-resistant bacteria in the environment poses a substantial risk to public health, as these bacteria can transfer their resistance genes to human pathogens (Khan et al., 2021). Additionally, the lack of stringent regulations and monitoring exacerbates the problem, making it difficult to assess and mitigate the impact of HWW on human and environmental health (Kumari et al., 2020).

Hospital wastewater plays a considerable role in the dissemination of antimicrobial resistance (AMR) due to the high concentrations of antibiotic-resistant bacteria (ARB) and antimicrobial resistance genes (ARGs) they contain. This issue is exacerbated by the widespread use of antibiotics in hospital settings, leading to the direct release of these agents into wastewater systems. Consequently, hospital wastewaters have been found to harbor clinically substantial strains, such as extended-spectrum β-lactamase (ESBL) and carbapenemase-producing *Enterobacteriaceae*, posing substantial risks to both environmental and public health. Key takeaways include the fact that hospital wastewaters contain higher concentrations of antibiotics and ARGs compared to municipal wastewaters, making them potent vectors for the spread of AMR. Additionally, a considerable proportion of antibiotics persist through wastewater treatment processes, ultimately entering aquatic ecosystems and contributing to the spread of resistant bacteria. The impact of hospital wastewater on AMR dissemination varies significantly across regions, influenced by factors such as local regulations, treatment technologies, and hospital practices. Policymakers, healthcare facilities, and water treatment professionals must address these issues. Policymakers need to enforce robust infection control practices, stringent regulations, and enhanced monitoring of hospital wastewaters to mitigate the spread of ARB and ARGs into the environment. Policies should focus on improving wastewater treatment technologies and implementing region-specific strategies to address the varying factors influencing AMR dissemination. Healthcare facilities should adopt practices to reduce antibiotic use and improve the management of pharmaceutical residues, considering innovative approaches such as applying phages to hospital plumbing surfaces to combat biofilm structures and prevent the spread of resistant bacteria. Water treatment professionals need to develop and implement advanced treatment technologies that can more efficiently degrade pharmaceuticals and other contaminants, including substances that pass through current treatment processes without degradation, such as antitumor agents and organohalogen compounds.

These investigations collectively underline the pressing need for robust infection control practices, reduced antibiotic use, and enhanced monitoring of hospital wastewaters to mitigate the spread of antibiotic-resistant bacteria (ARB) and antimicrobial resistance genes (ARGs) into the environment. Addressing the impact of hospital wastewater on AMR dissemination is crucial to prevent the spread of these resistant strains and to protect environmental and public health. Overall, the varying factors influencing AMR dissemination highlight the need for region-specific strategies. In well-regulated regions, the focus should be on improving treatment technologies and monitoring systems, while in low-resource settings, there is a need for infrastructure development and stringent regulatory frameworks to control the spread of AMR.

7 Conclusion

The exploration of antimicrobial resistance (AMR) in aquatic environments has revealed a complex web of factors contributing to the widespread presence of antibiotic-resistant bacteria (ARB) and resistance genes (ARGs) in water sources globally. This study highlights the pressing need for a unified and integrated strategy to tackle the multifaceted challenges posed by AMR in aquatic ecosystems. From the origins of antibiotics in water bodies to the intricate pathways through which resistance disseminates, a thorough understanding of AMR dynamics is imperative for protecting public health, ensuring environmental sustainability, and fortifying global health security. The research has shown high occurrence rates of ARB, particularly in urban areas and regions with human waste sources, including those near hospitals. There are significant differences in the efficiency of wastewater treatment plants (WWTPs) across different regions, leading to variable removal rates of antibiotics. Hospital effluents are seen as significant reservoirs of ARB and ARGs, especially in regions with inadequate treatment infrastructure. Additionally, higher resistance percentages were observed in arid regions, during dry seasons, and in urban areas. It therefore becomes, a crucial investment in interdisciplinary research to deepen our understanding of the intricate interactions fueling AMR, including the influence of climate change, hydrological dynamics, and seasonal variations on the dissemination of resistance genes. To effectively combat antimicrobial resistance (AMR) in aquatic environments, fostering interdisciplinary collaborations that bring together diverse expertise is crucial. One such collaboration should involve microbiologists and environmental scientists to investigate the mechanisms of antibiotic resistance gene (ARG) transfer in aquatic ecosystems. This collaboration should ideally focus on understanding how environmental factors such as temperature, pH, and salinity influence the survival and transfer of ARGs in water bodies. Microbiologists can study the predominant microbial communities in different aquatic environments and their role in the spread of AMR. Environmental scientists on the other hand can examine how natural and anthropogenic activities impact the prevalence and distribution of ARGs in various water sources. By combining their expertise, we can gain an understanding of the complex interactions between environmental conditions, microbial communities, and the dissemination of ARGs in aquatic environments. Integrating climate change into AMR strategies and models is crucial because it helps in understanding and mitigating the compounded effects of environmental changes on AMR dissemination. By considering climate change, we can develop more robust and adaptive strategies to safeguard water resources and public health. This integrated approach ensures that mitigation efforts are comprehensive and resilient to future climatic variations, ultimately contributing to sustainable water management and reduced AMR risks.

8 Recommendations

- Enhanced Monitoring and Surveillance Programs.

Implement continuous and comprehensive monitoring of antibiotic contamination levels in various water sources by establishing nationwide surveillance networks. These networks should monitor antibiotic levels in surface water, groundwater, and wastewater, utilizing advanced analytical techniques such as liquid chromatography-mass spectrometry (LC-MS) for precise detection and quantification. Additionally, develop a centralized database to collect and share data on antibiotic contamination and resistance patterns across different regions, ensuring a coordinated and informed response to the issue.

- Advanced Wastewater Treatment Methods.

Upgrade existing wastewater treatment plants (WWTPs) with membrane bioreactors (MBRs) to enhance the removal efficiency of antibiotics and other micropollutants. Implement advanced oxidation processes (AOPs) such as ozonation and photocatalysis to break down resistant compounds. Also, by exploring the use of metal-organic frameworks (MOFs) due to their high adsorption capacity and selectivity towards antibiotics.

- Targeted Mitigation Strategies.

Develop and enforce specific strategies tailored to diverse pollution sources, including aquaculture facilities, untreated wastewater discharge, and agricultural practices. Implement best management practices (BMPs) in aquaculture to minimize antibiotic use and discharge. Enforce stricter regulations on the discharge of untreated wastewater from industrial and agricultural sources. Promote the use of alternative, non-antibiotic treatments in agriculture to reduce reliance on antibiotics.

- Improved Hospital Wastewater Management.

As shown hospital wastewaters are major reservoirs of ARB and ARGs, necessitating robust infection control practices. Mandate the installation of on-site treatment units in hospitals to pre-treat wastewater before discharge into municipal systems. Utilize technologies such as activated carbon adsorption and advanced oxidation processes to effectively remove antibiotics from hospital effluents. Finally, conduct regular audits and compliance checks to ensure hospitals adhere to wastewater treatment standards.

- Collaborative Efforts Among Stakeholders.

Establishing multi-stakeholder task forces at local, national, and international levels fosters coordination and facilitates knowledge exchange and capacity-building workshops for professionals in these sectors. Furthermore, public awareness campaigns are crucial for educating communities about the risks of AMR and promoting responsible antibiotic use, thereby enhancing community engagement and support for mitigation efforts. On a global scale, international agencies such as

the WHO's Global Antimicrobial Resistance Surveillance System (GLASS), the CDC, the Food and Agriculture Organization (FAO), and the World Organisation for Animal Health (OIE) play pivotal roles in harmonizing efforts to combat AMR across borders. Initiatives like the European Union's Joint Programming Initiative on Antimicrobial Resistance (JPIAMR) exemplify interdisciplinary cooperation, bringing together researchers from various fields to study the environmental dimensions of antibiotic resistance and advocate for best practices in antibiotic stewardship. By leveraging these international collaborations, both high-income countries (HICs) and low- and middle-income countries (LMICs) can benefit from shared data, advanced technologies, and best practices.

- Reduction of Antibiotic Use in Agriculture.

Implement sustainable farming practices that reduce antibiotic runoff, such as optimizing the use of manure and implementing protective buffer zones around water bodies.

Author Contributions EUJ and YvW conceived the idea for this chapter. Both authors contributed to the literature search and data analysis. YvW and EUJ drafted the initial manuscript, and EUJ provided critical revisions. Both authors approved the final version of the manuscript for submission.

Conflict of Interest On behalf of all authors, the corresponding author states that there is no conflict of interest.

References

Abia, A. L. K., Ubomba-Jaswa, E., & Momba, M. N. B. (2015). Impact of seasonal variation on Escherichia coli concentrations in the riverbed sediments in the Apies River, South Africa. *Science of the Total Environment, 537*, 462–469.

Adyari, B., Hou, L., Zhang, L., Chen, N., Ju, F., Zhu, L., & Hu, A. (2023). Seasonal hydrological dynamics govern lifestyle preference of aquatic antibiotic resistome. *Environmental Science and Ecotechnology, 13*, 100223.

Alaali, Z., & Thani, A. S. B. (2020). Patterns of antimicrobial resistance observed in the Middle East: Environmental and health care retrospectives. *Science of the Total Environment, 740*, 140089.

Andrade, L., Kelly, M., Hynds, P., Weatherill, J., Majury, A., & O'Dwyer, J. (2020). Groundwater resources as a global reservoir for antimicrobial-resistant bacteria. *Water Research, 170*, 115360.

Apreja, M., Sharma, A., Balda, S., Kataria, K., Capalash, N., & Sharma, P. (2022). Antibiotic residues in environment: Antimicrobial resistance development, ecological risks, and bioremediation. *Environmental Science and Pollution Research International, 29*(3), 3355–3371. https://doi.org/10.1007/s11356-021-17374-w

Barrios, R. E., Bartelt-Hunt, S. L., Li, Y., & Li, X. (2021). Modeling the vertical transport of antibiotic resistance genes in agricultural soils following manure application. *Environmental Pollution, 285*, 117480.

Bartelt-Hunt, S., Snow, D. D., Damon-Powell, T., & Miesbach, D. (2011). Occurrence of steroid hormones and antibiotics in shallow groundwater impacted by livestock waste control facilities. *Journal of Contaminant Hydrology, 123*(3–4), 94–103.

Bashir, I., Lone, F. A., Bhat, R. A., Mir, S. A., Dar, Z. A., & Dar, S. A. (2020). Concerns and threats of contamination on aquatic ecosystems. In K. Hakeem, R. Bhat, & H. Qadri (Eds.), *Bioremediation and biotechnology*. Springer. https://doi.org/10.1007/978-3-030-35691-0_1

Belay, E. D., Kile, J. C., Hall, A. J., Barton-Behravesh, C., Parsons, M. B., Salyer, S., & Walke, H. (2017). Zoonotic disease programs for enhancing global health security. *Emerging Infectious Diseases, 23*(Suppl 1), S65.

Benjamin, B., Kassidy, O. M., Greg, D., Charitha, G., & Walter, M. (2024). Review of emerging contaminants in green stormwater infrastructure: Antibiotic resistance genes, microplastics, tire wear particles, PFAS, and temperature. *Science of the Total Environment, 906*, 167195. https://doi.org/10.1016/j.scitotenv.2023.167195

Berg, A. (2020). The roadmap for action on AMR. *European Journal of Public Health, 30*(Supplement_5), ckaa165. 1152.

Bowman, N., Patel, P., Xu, W., Alsaffar, A., & Tiquia-Arashiro, S. M. (2018). Enrichment and isolation of Pb-resistant bacteria from Saint Clair River sediments and their potential for Pb removal in aqueous solutions. *Applied Microbiology and Biotechnology, 102*, 2391–2398. https://doi.org/10.1007/s00253-018-8772-4

Bradley, B. R., Daigger, G. T., Rubin, R., et al. (2002). Evaluation of onsite wastewater treatment technologies using sustainable development criteria. *Clean Technologies and Environmental Policy, 4*, 87–99. https://doi.org/10.1007/s10098-001-0130-y

Burnham, J. P. (2021). Climate change and antibiotic resistance: A deadly combination. *Therapeutic Advances in Infectious Disease, 8*. https://doi.org/10.1177/2049936121991374

Cabral JPS. Water Microbiology. (2010). Bacterial pathogens and water. *International Journal of Environmental Research and Public Health, 7*(10), 3657–3703. https://doi.org/10.3390/ijerph7103657

Calero-Cáceres, W., Marti, E., Olivares-Pacheco, J., & Rodriguez-Rubio, L. (2022). Antimicrobial resistance in aquatic environments. *Frontiers in Microbiology, 13*, 866268.

Carvalho, I. T., & Santos, L. (2016). Antibiotics in the aquatic environments: A review of the European scenario. *Environment International, 94*, 736–757.

Chettri, D., Verma, A. K., & Verma, A. K. (2024). Bioaugmentation: An approach to biological treatment of pollutants. *Biodegradation, 35*, 117–135. https://doi.org/10.1007/s10532-023-10050-5

Chhabra, A., et al. (2006). Multiple impacts of land-use/cover change. In E. F. Lambin & H. Geist (Eds.), *Land-use and land-cover change. Global change – The IGBP series*. Springer. https://doi.org/10.1007/3-540-32202-7_4

Cho, K., Zholi, A., Frabutt, D., Flood, M., Floyd, D., & Tiquia, S. M. (2012). Linking bacterial diversity and geochemistry of uranium-contaminated groundwater. *Environmental Technology, 33*, 1629–1640. https://doi.org/10.1080/09593330.2011.641036

Chukwu, K. B., Abafe, O. A., Amoako, D. G., Essack, S. Y., & Abia, A. L. K. (2023). Environmental concentrations of antibiotics, biocides, and heavy metals fail to induce phenotypic antimicrobial resistance in Escherichia coli. *Science of the Total Environment, 899*, 165721. https://doi.org/10.1016/j.scitotenv.2023.165721

Costanzo, S. D., Murby, J., & Bates, J. (2005). Ecosystem response to antibiotics entering the aquatic environment. *Marine Pollution Bulletin, 51*(1–4), 218–223.

Daoud, Z., Farah, J., Sokhn, E. S., El Kfoury, K., Dahdouh, E., Masri, K., & Matar, G. M. (2017). Multidrug-resistant Enterobacteriaceae in Lebanese hospital wastewater: Implication in the one health concept. *Microbial Drug Resistance, 24*(2), 166–174. https://doi.org/10.1089/mdr.2017.0090

Du, L., & Liu, W. (2012). Occurrence, fate, and ecotoxicity of antibiotics in agro-ecosystems. A review. *Agronomy for Sustainable Development, 32*, 309–327.

EMA/CVMP/ERA/632109/2014. (2021). *Reflection paper on antimicrobial resistance in the environment: Considerations for current and future risk assessment of veterinary medicinal products*. European Medicines Agency.

Flood, M., Frabutt, D., Floyd, D., Powers, E. U., Devol, A., & Tiquia-Arashiro, S. M. (2015). Ammonia-oxidizing bacteria and archaea in sediments of the Gulf of Mexico. *Environmental Technology, 36*, 124–135. https://doi.org/10.1080/09593330.2014.942385

Gani, K. M., Hlongwa, N., Abunama, T., Kumari, S., & Bux, F. (2021). Emerging contaminants in South African water environment-a critical review of their occurrence, sources and ecotoxicological risks. *Chemosphere, 269*, 128737.

García-Galán, M. J., Garrido, T., Fraile, J., Ginebreda, A., Díaz-Cruz, M. S., & Barceló, D. (2010). Simultaneous occurrence of nitrates and sulfonamide antibiotics in two ground water bodies of Catalonia (Spain). *Journal of Hydrology, 383*(1–2), 93–101.

Gentile, M., Yan, T., Tiquia, S. M., Fields, M. W., Nyman, J., Zhou, J., & Criddle, C. S. (2006). Stability and resilience in a denitrifying fluidized bed reactor. *Microbial Ecology, 52*(2), 311–321. https://doi.org/10.1007/s00248-006-9024-1

Ghaju Shrestha, R., Tanaka, Y., & Haramoto, E. (2022). A review on the prevalence of Arcobacter in aquatic environments. *Water, 14*(8), 1266.

Gothwal, R., & Thatikonda, S. (2018). Mathematical model for the transport of fluoroquinolone and its resistant bacteria in aquatic environment. *Environmental Science and Pollution Research, 25*, 20439–20452.

Gothwal, R., & Thatikonda, S. (2020). Modeling transport of antibiotic resistant bacteria in aquatic environment using stochastic differential equations. *Scientific Reports, 10*(1), 15081.

Grenni, P., Ancona, V., & Caracciolo, A. B. (2018). Ecological effects of antibiotics on natural ecosystems: A review. *Microchemical Journal, 136*, 25–39.

Gunathilaka, M. L., Bao, S., Liu, X., Li, Y., & Pan, Y. (2023). Antibiotic pollution of planktonic ecosystems: A review focused on community analysis and the causal chain linking individual- and community-level responses. *Environmental Science & Technology, 57*(3), 1199–1213.

Hassoun-Kheir, N., Stabholz, Y., Kreft, J.-U., De La Cruz, R., Romalde, J. L., Nesme, J., & Paul, M. (2020). Comparison of antibiotic-resistant bacteria and antibiotic resistance genes abundance in hospital and community wastewater: A systematic review. *Science of the Total Environment, 743*, 140804.

Hiller, C. X., Hübner, U., Fajnorova, S., Schwartz, T., & Drewes, J. E. (2019). Antibiotic microbial resistance (AMR) removal efficiencies by conventional and advanced wastewater treatment processes: A review. *Science of the Total Environment, 685*, 596–608. https://doi.org/10.1016/j.scitotenv.2019.05.315

Ikuma, K., & Rehmann, C. R. (2020). Importance of extracellular DNA in the fate and transport of antibiotic resistance genes downstream of a wastewater treatment plant. *Environmental Engineering Science, 37*(2), 164–168.

Islam, S. S., & Midya, S. (2023). Growth regulatory pattern of zooplankton in herbicide and antibiotic contaminated aquatic ecosystem: An overview. *Watershed Ecology and the Environment, 5*, 153.

Jampani, M., Gothwal, R., Mateo-Sagasta, J., & Langan, S. (2022). Water quality modelling framework for evaluating antibiotic resistance in aquatic environments. *Journal of Hazardous Materials Letters, 3*, 100056.

Kassem, A., Abbas, L., Coutinho, O., Opara, S., Najaf, H., Kasperek, D., Li, X., & Tiquia-Arashiro, S. M. (2023). Applications of Fourier transform-infrared spectroscopy in microbial cell biology and environmental microbiology: Advances, challenges, and future perspectives. *Frontiers in Microbiology, 14*, 1304081. https://doi.org/10.3389/fmicb.2023.1304081

Khalidi-Idrissi, A., Madinzi, A., Anouzla, A., Pala, A., Mouhir, L., Kadmi, Y., & Souabi, S. (2023 Mar 16). Recent advances in the biological treatment of wastewater rich in emerging pollutants produced by pharmaceutical industrial discharges. *International Journal of Environmental Science and Technology, 20*, 11719–11740. https://doi.org/10.1007/s13762-023-04867-z

Khan, M. T., Shah, I. A., Ihsanullah, I., Naushad, M., Ali, S., Shah, S. H. A., & Mohammad, A. W. (2021). Hospital wastewater as a source of environmental contamination: An overview of management practices, environmental risks, and treatment processes. *Journal of Water Process Engineering, 41*, 101990.

Kim, K.-R., Owens, G., Kwon, S.-I., So, K.-H., Lee, D.-B., & Ok, Y. S. (2011). Occurrence and environmental fate of veterinary antibiotics in the terrestrial environment. *Water, Air, & Soil Pollution, 214*, 163–174.

Kleczkowski, A., Hoyle, A., & McMenemy, P. (2019). One model to rule them all? Modelling approaches across one health for human, animal and plant epidemics. *Philosophical Transactions of the Royal Society B, 374*(1775), 20180255.

Koncagül, E., Tran, M., Connor, R., & Uhlenbrook, S. (2018). The United Nations world water development report 2018: Nature-based solutions for water; facts and figures.

Kovalakova, P., Cizmas, L., McDonald, T. J., Marsalek, B., Feng, M., & Sharma, V. K. (2020). Occurrence and toxicity of antibiotics in the aquatic environment: A review. *Chemosphere, 251*, 126351.

Kulik, K., Lenart-Boroń, A., & Wyrzykowska, K. (2023). Impact of antibiotic pollution on the bacterial population within surface water with special focus on mountain Rivers. *Water, 15*(5), 975.

Kumari, A., Maurya, N. S., & Tiwari, B. (2020). Hospital wastewater treatment scenario around the globe. In *Current developments in biotechnology and bioengineering* (pp. 549–570). Elsevier.

Lapworth, D., Boving, T., Brauns, B., Dottridge, J., Hynds, P., Kebede, S., & Vargas, C. R. (2023). Groundwater quality: Global challenges, emerging threats and novel approaches. *Hydrogeology Journal, 31*(1), 15–18. https://doi.org/10.1007/s10040-022-02542-0

Larsson, D. J., Andremont, A., Bengtsson-Palme, J., Brandt, K. K., de Roda Husman, A. M., Fagerstedt, P., & Kuroda, M. (2018). Critical knowledge gaps and research needs related to the environmental dimensions of antibiotic resistance. *Environment International, 117*, 132–138.

Li, J., Cheng, W., Xu, L., Strong, P., & Chen, H. (2015). Antibiotic-resistant genes and antibiotic-resistant bacteria in the effluent of urban residential areas, hospitals, and a municipal wastewater treatment plant system. *Environmental Science and Pollution Research, 22*, 4587–4596.

Lien, L. T. Q., Hoa, N. Q., Chuc, N. T. K., Thoa, N. T. M., Phuc, H. D., Diwan, V., & Lundborg, C. S. (2016). Antibiotics in wastewater of a rural and an urban hospital before and after wastewater treatment, and the relationship with antibiotic use—A one year study from Vietnam. *International Journal of Environmental Research and Public Health, 13*(6), 588.

Löffler, P., Escher, B. I., Baduel, C., Virta, M. P., & Lai, F. Y. (2023). Antimicrobial transformation products in the aquatic environment: Global occurrence, Ecotoxicological risks, and potential of antibiotic resistance. *Environmental Science & Technology, 57*(26), 9474–9494. https://doi.org/10.1021/acs.est.2c09854

Lu, J., Zhang, Y., Wu, J., & Wang, J. (2022). Intervention of antimicrobial peptide usage on antimicrobial resistance in aquaculture. *Journal of Hazardous Materials, 427*, 128154.

Magnano San Lio, R., Favara, G., Maugeri, A., Barchitta, M., & Agodi, A. (2023). How antimicrobial resistance is linked to climate change: An overview of two intertwined global challenges. *International Journal of Environmental Research and Public Health., 20*(3), 1681. https://doi.org/10.3390/ijerph20031681

Manamsa, K., Crane, E., Stuart, M., Talbot, J., Lapworth, D., & Hart, A. (2016). A national-scale assessment of micro-organic contaminants in groundwater of England and Wales. *Science of the Total Environment, 568*, 712–726.

Mansi, A., Sharma, A., Balda, S., Kataria, K., Capalash, N., & Sharma, P. (2022). Antibiotic residues in environment: Antimicrobial resistance development, ecological risks, and bioremediation. *Environmental Science and Pollution Research, 29*(3), 3355–3371. https://doi.org/10.1007/s11356-021-17374-w

Marti, R., Scott, A., Tien, Y.-C., Murray, R., Sabourin, L., Zhang, Y., & Topp, E. (2013). Impact of manure fertilization on the abundance of antibiotic-resistant bacteria and frequency of detection of antibiotic resistance genes in soil and on vegetables at harvest. *Applied and Environmental Microbiology, 79*(18), 5701–5709. https://doi.org/10.1128/AEM.01682-13

Mbanga, J., Abia, A. L. K., Amoako, D. G., & Essack, S. Y. (2021). Longitudinal surveillance of antibiotic resistance in Escherichia coli and enterococcus spp. from a wastewater treatment plant and its associated waters in KwaZulu-Natal, South Africa. *Microbial Drug Resistance, 27*(7), 904–918.

McKindles, K. M., & Tiquia-Arashiro, S. M. (2012). Functional gene arrays for analysis of microbial communities on ocean platform. In S. Tiquia-Arashiro (Ed.), *Molecular biological Technologies for Ocean Sensing* (Springer Protocols Handbooks). Humana Press. https://doi.org/10.1007/978-1-61779-915-0_9

Miranda, C. D., Concha, C., Hurtado, L., Urtubia, R., Rojas, R., & Romero, J. (2024). Occurrence of antimicrobial-resistant bacteria in intestinal contents of wild marine fish in Chile. *Antibiotics, 13*(4), 332.

Morin-Crini, N., Lichtfouse, E., Liu, G., Balaram, V., Ribeiro, A. R. L., Lu, Z., & Picos-Corrales, L. A. (2022). Worldwide cases of water pollution by emerging contaminants: A review. *Environmental Chemistry Letters, 20*(4), 2311–2338.

Mubedi, J. I., Devarajan, N., Le Faucheur, S., Mputu, J. K., Atibu, E. K., Sivalingam, P., & Poté, J. (2013). Effects of untreated hospital effluents on the accumulation of toxic metals in sediments of receiving system under tropical conditions: Case of South India and Democratic Republic of Congo. *Chemosphere, 93*(6), 1070–1076.

Mukherjee, M., Laird, E., Gentry, T. J., Brooks, J. P., & Karthikeyan, R. (2021). Increased antimicrobial and multidrug resistance downstream of wastewater treatment plants in an urban watershed. *Frontiers in Microbiology, 12*, 657353.

Nassereddine, Z., Opara, S., Coutinho, O., Qyteti, F., Book, R., & Heinicke, M. P. (2024). Napieralski J and Tiquia-Arashiro SM (2024) critical perspectives on advancing antibiotic resistant gene (ARG) detection technologies in aquatic ecosystems. *Science of the Total Environment, 957*(20), 177775. https://doi.org/10.1016/j.scitotenv.2024.177775

Ngigi, A. N., Magu, M. M., & Muendo, B. M. (2020). Occurrence of antibiotics residues in hospital wastewater, wastewater treatment plant, and in surface water in Nairobi County, Kenya. *Environmental Monitoring and Assessment, 192*, 18. https://doi.org/10.1007/s10661-019-7952-8

Ngqwala, N. P., & Muchesa, P. (2020). Occurrence of pharmaceuticals in aquatic environments: A review and potential impacts in South Africa. *South African Journal of Science, 116*(7–8), 1–7.

Nguyen, S., Ala, F., Cardwell, C., Cai, D., McKindles, K. M., Lotvola, A., Hodges, S., Deng, Y., & Tiquia-Arashiro, S. M. (2013). Isolation and screening of carboxydotrophs isolated from composts and their potential for butanol synthesis. *Environmental Technology, 34*, 1995–2007. https://doi.org/10.1080/09593330.2013.795987

Oest, A., Alsaffar, A., Fenner, M., Azzopardi, D., & Tiquia-Arashiro, S. M. (2018, May 27). Patterns of change in metabolic capabilities of sediment microbial communities in river and Lake ecosystems. *International Journal of Microbiology, 2018*, 6234931. https://doi.org/10.1155/2018/6234931

Pagnucco, G., Overfield, D., Chamlee, Y., Fortuna, A., Sulaiman, F., Farinas, J., Schittenhelm, R., Shuler, C., Kassem, A., Opara, S., Najaf, H., Abbas, L., Coutinho, O., Catalfano, B., Li, X., & Tiquia-Arashiro, S. M. (2023). Metal tolerance and biosorption capacities of bacterial strains isolated from an urban watershed. *Frontiers in Microbiology, 14*, 1278886. https://doi.org/10.3389/fmicb.2023.1278886

Palaniappan, M., Gleick, P. H., Allen, L., Cohen, M. J., Christian-Smith, J., & Smith, C. (2012). Water quality. In P. H. Gleick (Ed.), *The world's water. The world's water*. Island Press. https://doi.org/10.5822/978-1-59726-228-6_3

Pan, Y., Yan, S.-W., Li, R.-Z., Hu, Y.-W., & Chang, X.-X. (2017). Lethal/sublethal responses of Daphnia magna to acute norfloxacin contamination and changes in phytoplankton-zooplankton interactions induced by this antibiotic. *Scientific Reports, 7*(1), 40385.

Patel, D., Gismondi, R., Alsaffar, A., & Tiquia-Arashiro, S. M. (2019). Applicability of API ZYM to capture seasonal and spatial variabilities in lake and river sediments. *Environmental Technology, 40*(24), 3227–3239. https://doi.org/10.1080/09593330.2018.1468492

Patz, J. A., & Hahn, M. B. (2012). Climate change and human health: A one health approach. In J. Mackenzie, M. Jeggo, P. Daszak, & J. Richt (Eds.), *One health: The human-animal-environment interfaces in emerging infectious diseases. Current topics in microbiology and immunology* (Vol. 366). Springer. https://doi.org/10.1007/82_2012_274

Patz, J. A., & Olson, S. H. (2006). Climate change and health: Global to local influences on disease risk. *Annals of Tropical Medicine & Parasitology, 100*(5–6), 535–549. https://doi.org/10.1179/136485906X97426

Paul, D., & Das, B. (2022). Gut microbiome in the emergence of antibiotic-resistant bacterial pathogens. *Progress in Molecular Biology and Translational Science, 192*(1), 1–31.

Paumelle, M., Donnadieu, F., Joly, M., Besse-Hoggan, P., & Artigas, J. (2021). Effects of sulfonamide antibiotics on aquatic microbial community composition and functions. *Environment International, 146*, 106198.

Perry, M. R., Lepper, H. C., McNally, L., Wee, B. A., Munk, P., Warr, A., & de Roda Husman, A. M. (2021). Secrets of the hospital underbelly: Patterns of abundance of antimicrobial resistance genes in hospital wastewater vary by specific antimicrobial and bacterial family. *Frontiers in Microbiology, 12*, 703560.

Petrovich, M. L., Zilberman, A., Kaplan, A., Eliraz, G. R., Wang, Y., Langenfeld, K., & Avisar, D. (2020). Microbial and viral communities and their antibiotic resistance genes throughout a hospital wastewater treatment system. *Frontiers in Microbiology, 11*, 153.

Polianciuc, S. I., Gurzău, A. E., Kiss, B., Ştefan, M. G., & Loghin, F. (2020). Antibiotics in the environment: Causes and consequences. *Medicine and Pharmacy Reports, 93*(3), 231.

Pomaranski, E., & Tiquia-Arashiro, S. M. (2016). Butanol tolerance of carboxydotrophic bacteria isolated from manure composts. *Environmental Technology, 37*(15), 1970–1982. https://doi.org/10.1080/09593330.2015.1137360

Prabhasankar, V. P., Joshua, D. I., Balakrishna, K., Siddiqui, I. F., Taniyasu, S., Yamashita, N., & Guruge, K. S. (2016). Removal rates of antibiotics in four sewage treatment plants in South India. *Environmental Science and Pollution Research, 23*, 8679–8685.

Preena, P. G., Swaminathan, T. R., Kumar, V. J. R., & Singh, I. S. B. (2020). Antimicrobial resistance in aquaculture: A crisis for concern. *Biologia, 75*, 1497–1517.

Sala-Comorera, L., Nolan, T. M., Reynolds, L. J., Venkatesh, A., Cheung, L., Martin, N. A., & Meijer, W. G. (2021). Bacterial and bacteriophage antibiotic resistance in marine bathing waters in relation to rivers and urban streams. *Frontiers in Microbiology, 12*, 718234.

Salas, J. R., Gaire, T., Quichocho, V., Nicholson, E., & Volkova, V. V. (2022). Modelling the antimicrobial pharmacodynamics for bacterial strains with versus without acquired resistance to fluoroquinolones or cephalosporins. *Journal of Global Antimicrobial Resistance, 28*, 59–66.

Samreen, Ahmad, I., Malak, H. A., & Abulreesh, H. H. (2021). Environmental antimicrobial resistance and its drivers: A potential threat to public health. *Journal of Global Antimicrobial Resistance, 27*, 101–111. https://doi.org/10.1016/j.jgar.2021.08.001

Singer, A. C., Shaw, H., Rhodes, V., & Hart, A. (2016). Review of antimicrobial resistance in the environment and its relevance to environmental regulators. *Frontiers in Microbiology, 7*, 1728. https://doi.org/10.3389/fmicb.2016.01728

Sirijatuphat, R., Sripanidkulchai, K., Boonyasiri, A., Rattanaumpawan, P., Supapueng, O., Kiratisin, P., & Thamlikitkul, V. (2018). Implementation of global antimicrobial resistance surveillance system (GLASS) in patients with bacteremia. *PLoS One, 13*(1), e0190132.

Sorn, S., Sulfikar, Lin, M., Shuto, M., Noguchi, M., Honda, R., & Watanabe, T. (2022). Potential impact factors on the enhancement of antibiotic resistance in a lake environment. *Journal of Water and Health, 20*(6), 1017–1026.

Stuart, M., Lapworth, D., Crane, E., & Hart, A. (2012). Review of risk from potential emerging contaminants in UK groundwater. *Science of the Total Environment, 416*, 1–21.

Surleac, M., Czobor Barbu, I., Paraschiv, S., Popa, L. I., Gheorghe, I., Marutescu, L., & Nita, M. (2020). Whole genome sequencing snapshot of multi-drug resistant Klebsiella pneumoniae strains from hospitals and receiving wastewater treatment plants in southern Romania. *PLoS One, 15*(1), e0228079.

Suyamud, B., Chen, Y., Dong, Z., Zhao, C., & Hu, J. (2023). Antimicrobial resistance in aquaculture: Occurrence and strategies in Southeast Asia. *Science of the Total Environment, 907*, 167942.

Sverdrup, H. U., Frolova, L. L., & Sverdrup, A. E. (2020). Using a system dynamics model for investigating potential levels of antibiotics pollution in the Volga River. *Water, Air, & Soil Pollution, 231*, 1–32.

Tamtam, F., Mercier, F., Le Bot, B., Eurin, J., Tuc Dinh, Q., Clément, M., & Chevreuil, M. (2008, April 1). Occurrence and fate of antibiotics in the Seine River in various hydrological conditions. *Science of the Total Environment, 393*(1), 84–95. https://doi.org/10.1016/j.scitotenv.2007.12.009

Tapia-Arreola, A. K., Ruiz-Garcia, D. A., Rodulfo, H., Sharma, A., & De Donato, M. (2022). High frequency of antibiotic resistance genes (ARGs) in the Lerma River basin, Mexico. *International Journal of Environmental Research and Public Health, 19*(21), 13988.

The Centers for Disease Control and Prevention. (2022). Understanding antibiotic resistance in water: A one health approach. Available at: https://www.cdc.gov/onehealth/in-action/combating-ar-in-people-and-animals.html (Accessed 10/08/2023).

Tiquia, S. M. (2002). Microbial transformation of nitrogen during composting. In H. Insam, N. Riddech, & S. Klammer (Eds.), *Microbiology of composting*. Springer. https://doi.org/10.1007/978-3-662-08724-4_20

Tiquia, S. M. (2008). Diversity of sulfate-reducing genes (dsrAB) in sediments from Puget Sound. *Environmental Technology, 29*(10), 1095–1108. https://doi.org/10.1080/09593330802190608

Tiquia, S. M. (2010a). Salt-adapted bacteria isolated from the Rouge River and potential for degradation of contaminants and biotechnological applications. *Environmental Technology, 31*, 967–978. https://doi.org/10.1080/09593331003706226

Tiquia, S. M. (2010b). Metabolic diversity of the heterotrophic microorganisms and potential link to pollution of the Rouge River. *Environmental Pollution, 158*, 1435–1443. https://doi.org/10.1016/j.envpol.2009.12.035

Tiquia, S. M. (2011). Extracellular hydrolytic enzyme activities of the heterotrophic microbial communities of the Rouge River: An approach to evaluate ecosystem response to urbanization. *Microbial Ecology, 62*, 679–689. https://doi.org/10.1007/s00248-011-9871-2

Tiquia, S. M., & Mormile, M. (2010). Extremophiles: A source of innovation for industrial and environmental applications. *Environmental Technology, 31*(10), 1061–1181. https://doi.org/10.1080/09593330.2010.489333

Tiquia, S. M., Gurczynski, S., Zholi, A., & Devol, A. (2006a, December). Diversity of biogeochemical cycling genes from Puget Sound sediments using DNA microarrays. *Environmental Technology, 27*(12), 1377–1389. https://doi.org/10.1080/09593332708618756

Tiquia, S. M., Masson, S. A., & Devol, A. (2006b, December). Vertical distribution of nitrite reductase genes (nirS) in continental margin sediments of the Gulf of Mexico. *FEMS Microbiology Ecology, 58*(3), 464–475. https://doi.org/10.1111/j.1574-6941.2006.00173.x

Tiquia, S. M., Davis, D., Hadid, H., Kasparian, S., Ismail, M., & Murray, K. S. (2007). Halophilic and halotolerant bacteria from river waters and shallow groundwater along the Rouge River of southeastern Michigan. *Environmental Technology, 28*, 297–307. https://doi.org/10.1080/09593332808618789

Tiquia, S. M., Schleibak, M., Schlaff, J., Floyd, J. C., Benipal, B., Zakhem, E., & Murray, K. S. (2008). Microbial community profiling and characterization of some heterotrophic bacterial isolates from river waters and shallow groundwater wells along the Rouge River, Southeast Michigan. *Environmental Technology, 29*, 651–663. https://doi.org/10.1080/09593330801986998

Tiquia-Arashiro, S. M., & Mormile, M. (2013). Sustainable technologies: Bioenergy and biofuel from biowaste and biomass. *Environmental Technology, 34*(13-14), 1637–1638. https://doi.org/10.1080/09593330.2013.834162

Tiquia-Arashiro, S. M., & Pant, D. (2020). *Microbial electrochemical technologies* (p. 518). CRC Press. https://doi.org/10.1201/9780429487118

Twinomucunguzi, F. R. B., Nyenje, P. M., Semiyaga, S., Kebirungi, P., Kulabako, R. N., & Kansiime, F. (2023). Antibiotics in shallow groundwater underlying urban informal settlements in developing countries: Influence of on-site sanitation practices and risk assessment. *Urban Water Journal, 20*(10), 1731–1743. https://doi.org/10.1080/1573062X.2022.2059688

Uluseker, C., Kaster, K. M., Thorsen, K., Basiry, D., Shobana, S., Jain, M., & Pala-Ozkok, I. (2021). A review on occurrence and spread of antibiotic resistance in wastewaters and in wastewater treatment plants: Mechanisms and perspectives. *Frontiers in Microbiology, 12*, 717809.

Välitalo, P., Kruglova, A., Mikola, A., & Vahala, R. (2017). Toxicological impacts of antibiotics on aquatic micro-organisms: A mini-review. *International Journal of Hygiene and Environmental Health, 220*(3), 558–569.

Varela, A. R., & Manaia, C. M. (2013). Human health implications of clinically relevant bacteria in wastewater habitats. *Environmental Science and Pollution Research, 20*, 3550–3569. https://doi.org/10.1007/s11356-013-1594-0

Vu Ngan, B., Nhung, D., Nguyen Thi Kieu, A., Le Xuan, K., & Phong, K. T. (2018). Antibiotics in the aquatic environment of Vietnam: Sources, concentrations, risk and control strategy. *Chemosphere, 197*, 438–450. https://doi.org/10.1016/j.chemosphere.2018.01.061

Wang, X., Lin, Y., Zheng, Y., & Meng, F. (2022). Antibiotics in mariculture systems: A review of occurrence, environmental behavior, and ecological effects. *Environmental Pollution, 293*, 118541.

WHO. (2022). *Antimicrobial resistance surveillance in Europe 2022–2020 data*. World Health Organization. Regional Office for Europe.

World Economic Forum. (2024). *The global risks report. In* (19th ed.). World Economic Forum.

Wu, W., Gu, B., Fields, M. W., Gentile, M., Ku, Y.-K., Yan, H., Tiquia, S. M., Yan, T., Nyman, J., Zhou, J., Jardine, P. M., & Criddle, C. S. (2005). Uranium (VI) reduction by denitrifying biomass. *Bioremediation Journal, 9*, 1–13. https://doi.org/10.1080/10889860590929628

Zheng, W., Guo, M., & Czapar, G. (2020). Environmental fate and transport of veterinary antibiotics derived from animal manure. *Animal Manure: Production, Characteristics, Environmental Concerns, and Management, 67*, 409–430.

Chapter 10
Diverse Aminoglycoside Gene Cassettes (*aad*A1, *aad*B, and *aac*A4) in Class 1 Integrons in *Escherichia coli* Strains Isolated from Three Urban Watersheds

Bryan Catalfano, Yanesa Chamlee, Victoria Khamis, Diana Kasperek, Erica Parmenter, Reis Schittenhelm, Zainab Nassereddine, and Sonia M. Tiquia-Arashiro

1 Introduction

Antibiotic-resistant microorganisms are increasingly being found in various environmental compartments, including wastewater, sewage, sediments, wastewater treatment plants, surface waters, groundwater, and even drinking water sources (Wu et al., 2005; Henriques et al., 2006; Gentile et al., 2006; Tiquia et al., 2008; Zhang et al., 2009a; Flood et al., 2015; Bowman et al., 2018; Zainab et al., 2020; Li et al., 2021a). This resistance phenomenon is closely linked to the presence of antibiotic resistance genes (ARGs), which are either acquired through mutations within bacterial populations or through horizontal gene transfer (HGT) mechanisms (Peak et al., 2007; Roberts, 2005; Robicsek et al., 2005, 2006a, b, c; Zainab et al., 2020; Li et al., 2021a; Nassereddine et al., 2024). The selective pressure exerted by antibiotic usage contributes significantly to the proliferation of resistant strains within bacterial communities. The emergence and spread of antibiotic resistance pose significant challenges to human medicine, as it diminishes the efficacy of antibiotics and exacerbates the prevalence of antibiotic-resistant bacteria (Kemper, 2008). Thus, it is imperative to implement effective strategies to curb the dissemination of ARGs and antibiotic-resistant bacteria. Despite the critical need for such measures, our understanding of the genetic makeup and distribution patterns of antibiotic-resistant bacteria in aquatic environments remains limited (Yang et al., 2018; Amarasiri et al., 2019; Nassereddine et al., 2024). As antibiotic resistance continues to increase, comprehensive knowledge about the mechanisms facilitating the transfer of resistance genes among bacterial strains is crucial. This understanding will aid in the

B. Catalfano · Y. Chamlee · V. Khamis · D. Kasperek · E. Parmenter · R. Schittenhelm
Z. Nassereddine · S. M. Tiquia-Arashiro (✉)
Department of Natural Sciences, University of Michigan-Dearborn, Dearborn, MI, USA
e-mail: smtiquia@umich.edu

S. M. Tiquia-Arashiro et al. (eds.), *Urban Watershed Microbiology, Volume 1*,
https://doi.org/10.1007/978-3-031-89071-0_10

development of targeted interventions to mitigate the spread of antibiotic resistance in aquatic ecosystems.

One established method for studying antibiotic resistance involves investigating gene cassettes within class 1 integrons (Zhang et al., 2009b; Deng et al., 2015; Domingues et al., 2015; Oliveira-Pinto et al., 2016). Integrons are a new set of mobile elements that specifically integrate and express resistance genes. They adhere to transposons or plasmids and transfer within or between strains, enabling the spread of resistance genes. The integron contains an integrase (*int*I1) gene, a specific recombination site (*att*I1), and a promoter P (Stokes & Hall, 1989). The role of integrons in antimicrobial resistance (AMR) transmission is consistently being recognized, especially in gram-negative bacteria (Chen et al., 2019). They can be classified into five types (class 1–5) depending on their integrase genes. Class 1 integrons are strongly associated with bacterial resistance (Wang et al., 2021). Integrons usually carry antibiotic resistance genes that contribute to resistance against aminoglycosides, macrolides, β-lactams, chloramphenicol, and sulfonamides (Ye et al., 2020).

Aminoglycosides are used to treat infections caused by Gram-negative bacteria such as *Escherichia coli* (*E. coli*). Their mechanism of action involves binding specifically to the 16S rRNA of the 30S ribosomal subunit, thereby disrupting protein synthesis (Zárate et al., 2018). Among aminoglycosides, gentamicin, tobramycin, and amikacin exhibit notable efficacy against staphylococci and are frequently used alongside other antibiotic classes such as β-lactams (Schmitz et al., 1999). The primary mechanism of resistance to aminoglycosides involves the inactivation of the drug by aminoglycoside-modifying enzymes (AMEs), which diminish the affinity of aminoglycosides for their natural target, the 16S rRNA. One crucial group of AMEs is the acetyltransferase (AAC) enzymes (Kapoor et al., 2017). AACs catalyze the acetylation of an amino group and are often found as fusion proteins adjacent to other AMEs like APH and ANT (Ramirez & Tolmasky, 2017; Zárate et al., 2018). AACs are categorized into various groups based on where the acetyl group is inserted in the acceptor aminoglycoside drug. These include AAC (1), AAC (3), AAC (2), or AAC (6). Among them, AAC (6) enzymes are the most prevalent, with over 40 known enzymes (Ramirez & Tolmasky, 2010). One noteworthy plasmid-mediated acetyltransferase, AAC(6)-Ib, also known as aacA4, confers resistance to several aminoglycosides, including amikacin (Ramirez & Tolmasky, 2017). AAC (6) modifies aminoglycosides like gentamicin, tobramycin, and amikacin by acetylating the amino group at the 6′-position, rendering the drug inactive by releasing the 16S-rRNA receptor (labeled green) from its ligand (Robicsek et al., 2006a, b, c; Serio et al., 2018; El-Baz et al., 2021).

Given the concern over antibiotic resistance in bacteria and the propensity of local watersheds to serve as breeding grounds for organisms like *E. coli* (Wight et al., 2024), this study aimed to identify various genes associated with *E. coli* resistance to specific antibiotics. The target genes under investigation are *aad*A, *aad*B, and *aac*A4, each conferring varying levels of resistance to specific antibiotics. Both *aa*dA and *aa*dB confer resistance to tobramycin (Mostafa et al., 2015; Gallagher et al., 2017; Sabbagh et al., 2021), with *aad*A also affecting streptomycin and

spectinomycin, while *aad*B additionally correlates with kanamycin and gentamicin . On the other hand, *aac*A4 deactivates antibiotics such as kanamycin (Vourli et al., 2003; Yan et al., 2007; Nap et al., 1992; Li, 2016; Ahmadian et al., 2020), gentamicin (Nap et al., 1992; Yan et al., 2007; Partridge et al., 2018; Ahmadian et al., 2020), and to some extent, neomycin (Nap et al., 1992; Azizi et al., 2016; Naderi et al., 2023). All these antibiotics belong to the class of aminoglycosides, which inhibit bacterial protein synthesis, thereby thwarting infection. *E. coli* strains, particularly virulent ones, pose significant challenges in treatment.

In this study, we investigated the presence of *E. coli* within three distinct watersheds: Lake St. Clair, the St. Clair River, and the Rouge River. These aquatic environments are known to periodically experience outbreaks of *E. coli* (Tiquia et al., 2007; Tiquia, 2010a; Oest et al., 2018; Patel et al., 2019). Across these three locations, a total of 66 strains of the bacterium were collected and analyzed. Polymerase chain reaction (PCR) amplification tests were conducted to detect the presence of the target class 1 integron gene cassettes, providing insights into the prevalence of these genes within the strains. The significance of such gene surveillance extends beyond merely assessing the infectious potential of *E. coli* strains (Tiquia-Arashiro, 2012; McKindles & Tiquia-Arashiro, 2012). it is also crucial due to the local impact of the watersheds, frequented by numerous visitors annually. For instance, industrial activities near the Rouge and St. Clair Rivers often lead to pollution, while Lake St. Clair, though more stagnant, remains accessible to the public (Tiquia et al., 2007; Tiquia, 2010a, b, 2011; Oest et al., 2018; Patel et al., 2019; Pagnucco et al., 2023). Understanding the genetic makeup of bacteria in these environments aids in implementing effective management strategies to mitigate the spread of antibiotic resistance and safeguard public health.

2 Materials and Methods

2.1 Sample Collection and Water Quality Testing

Water samples were collected from the Rouge River, Lake Saint Clair, and Saint Clair River using 1-L sterile borosilicate glass bottles. These samples were stored at 6 °C until processing. Surface water samples were obtained from three locations spanning a 50 m^2 area across the water flow at each site, with an approximate depth of 0.5 m. Upon collection, all water samples underwent characterization within 1 day. The following water quality parameters were assessed: temperature, pH, turbidity, electrical conductivity, total organic carbon (TOC), total suspended solids (TSS), and *E. coli* presence. The pH levels were measured using an electrode meter (Hanna Instruments, HI2002-01, Ann Arbor, Michigan, USA). Turbidity was assessed using a turbidimeter (Hanna Instruments, H198703-01 Ann Arbor, Michigan, USA). The electrical conductivity, TSS, and water temperature were measured using a Hann conductivity/TSS meter (Hanna Instruments, HI98703-01, Ann Arbor, Michigan, USA). Dissolved oxygen was measured using a dissolved

oxygen meter (FiveGo F2 portable meter, Mettler Toledo, Columbus, Ohio, USA). TOC levels were measured utilizing a Hach low-range TOC reagent set and a portable spectrophotometer (Hach DR/2400, Loveland, Colorado, USA).

2.2 *Recovery of* E. coli *Strains in Water Samples*

Enumeration of *E. coli* was conducted through membrane filtration using a modified membrane-thermotolerant *E. coli* agar method, following EPA method 1603 (USEPA, 2005). Within 3 h of collection, samples were processed. Each sample was passed through 0.45 μm polyethylene sulfonate membrane filters (Millipore Sigma, St. Louis, Missouri, USA) in appropriate volumes, and these filters were then placed on the medium. Subsequently, plates were incubated in a 35 °C environment for 2 h for resuscitation. Following this, the plates were transferred to water-tight plastic bags and submerged in a 44.5 °C water bath. After precisely 20 h, the plates were removed from the water bath. Countable filters were then transferred to pads saturated with urea substrate for an in situ urease test (Dufour & Cabelli, 1975). After 15 min of exposure to the urea substrate, all yellow colonies were counted and recorded.

Following the in situ urease test, yellow colonies underwent verification as *E. coli* through subculturing. Each colony was subcultured onto nutrient agar (BD Biosciences, San Jose, California, USA) and Simmons citrate plates, which were then incubated at 35 °C. Twenty-four hours post-inoculation, nutrient agar cultures were tested for ferroxidase (Paik, 1970; Tiquia-Arashiro & Pant, 2020) and urease (Barry et al., 1969; Patel et al., 2019; Tiquia-Arashiro, 2019a, b, 2020) activities. Additionally, growth on citrate plates was observed. Yellow colonies negative for oxidase, urease, and citrate were confirmed as *E. coli*. Non-yellow colonies observed prior to the in situ urease test were subcultured to lauryl sulfate-lactose broth (BD Biosciences, San Jose, California, USA). Gas-positive tubes underwent the same testing process as the target colonies. Colonies that were positive for lactose and gas but negative for oxidase, urease, and citrate were designated as *E. coli*. The percentages of false-positive and false-negative colonies were calculated relative to the total number of target colonies examined and the total number of verified *E. coli* colonies, respectively. Sixty-six *E. coli* colonies were purified on mTech agar (BD Biosciences, San Jose, California, USA). These strains include 33 from the Rouge River, 18 from Lake Saint Clair, and 15 from Saint Clair River.

2.3 *Antibiotic Susceptibility Pattern*

All *E. coli* isolates were tested for antimicrobial resistance using the method of Bauer et al. (1966) on standard antibiotic discs (Carolina Biological Supply Company, NC, USA). Twelve antimicrobial agents were selected for the drug susceptibility test in this experiment. Susceptibility for ampicillin (AMP, 20 μg per

disk), ciprofloxacin (CIP, 5 μg per disk), ceftriaxone (CEF, 30 μg per disk), chloramphenicol (CHL, 30 μg per disk), gentamicin (GEN, 10 μg per disk), imipenem (IPM, 10 μg per disk), nitrofurantoin (NIT, 10 μg per disk), rifampin (RIF, 5 μg per disk), streptomycin (STR, 20 μg per disk), sulfamethoxazole (SUL, 25 μg per disk), trimethoprim (SXT, 25 μg per disk), tetracycline (TET, 25 μg per disk), and tobramycin (TOB, 10 μg per disk) in accordance with the Clinical and Laboratory Standards Institute (CLSI) (Clinical and Laboratory Standards Institute, 2021). Overnight broth cultures of the test strains were used and the turbidity of the inoculum was adjusted in phosphate-buffered saline (pH 7.4) to a 0.5 McFarland opacity standard (Tiquia et al., 1998; Tiquia-Arashiro, 2014a, b, c, d). These cultures (0.1 mL) were then aseptically spread onto Mueller–Hinton agar plates. Standard antibiotic-impregnated discs (6-mm diameter, Carolina Biological Supply Company, NC, USA) were placed on freshly prepared lawns of each isolate on Mueller–Hinton plates. After 24 h of incubation at 30 °C, the diameter of the inhibition zones around the discs was measured after incubation and the bacterial strains were classified as resistant, intermediate, and susceptible. Multiple antibiotic resistance (MAR) index was determined for each isolate by using the formula MAR = *a*/*b*, where a represents the number of antibiotics to which the test isolate depicted resistance and b represents the total number of antibiotics to which the test isolate has been evaluated for susceptibility (Krumperman, 1983).

2.4 Incidence of Class 1 Integron

The isolates were screened for the integrase genes of class 1 and 2 integrons using PCR. The DNA templates needed for polymerase chain reaction (PCR) were prepared as described previously (Bass et al., 1999; Nguyen et al., 2013; Plecha et al., 2013; Pomaranski & Tiquia-Arashiro, 2016). Integrons were detected using PCR with specific primers, *int*I1 (Forward primer: 5′-CCTCCCGCACGATGATC-3′, Reverse primer: 5′-TCCACG-CATCGTCAGGC-3′), designed to target conserved regions of the integron-encoded integrase genes intI1 (Zhao et al., 2001) and *int*I2 specific primers targeting the 5′-conserved (5′-ATGTCTAACAGTCATTTTTAAATTCTA-3′) and 3′-conserved segments (5′-AAATCTTTAACCCGCAAACGC-3′) (Fling et al., 1985) (Table 10.1). Strains containing the *int*I1 were subsequently subjected to PCR amplification of the variable regions of class 1 integrons.

2.5 Identifying the Variable Regions in Class 1 Integrons

Amplification of the gene cassettes in class 1 integrons by PCR using primers in the 5′ conserved segment (5′-CS) and the 3′-CS (Lévesque & Roy, 1993; Lévesque et al., 1995) has been used in hundreds of studies to identify integron-associated resistance genes. To assess the potential insertion of antimicrobial-resistant gene

Table 10.1 PCR primers used for analysis of integrons

Primer	DNA sequence	Nucleotide position	Accession number	Target gene	Reference
Int1F	5′-CCTCCCGCACGATGATC-3′	786–766	U49101	*Intl*1	Zhao et al. (2001)
Int1R	5′-TCCACG-CATCGTCAGGC-3′	303–324			
Int2A	5′-ATGTCTAACAGTCATTTTTAAATTCTA-3′	1471–1495	AJ002782	*Intl*2	Fling et al. (1985)
Int2B	5′-AAATCTTTAACCCGCAAACGC-3′	1917–1887			
5′-CS	5′-GGCATCCAAGCACAAGC-3′	1190–1206	M73819	Class 1 integron variable region	Lévesque et al. (1995)
3′CS	5′-AAGCAGACTTGACTGAT-3′	1341–1327			

Table 10.2 PCR primers used for aminoglycoside gene cassettes

Resistant gene	Gene product	Forward sequence	Reverse sequence	Reference
aacA4	Aminoglycoside N(6)- acetyltransferase	*aacA4-F* [5′ TGT GAC GGA ATC GTT GC 3′]	*aacA4-R* [5′ GTT TGA GAG GCA AGG TA 3′]	Plante et al. (2003)
aac(3) IV	Aminoglycoside 3-n-acetyltransferase	*aac(3)IV-F* [5′ TGC TGG TCC ACA GCT CCT TC]	*aac(3)IV-R* [5′ CGG ATG CAG GAA GAT CAA]	Boerlin et al. (2005)
aadB 2′	Aminoglycoside nucleotidyltransferase	*aadB-F* [5′ CCG CAG CTA GAA TTT TG 3′]	*aadB-R* [5′ AGG TTG AGG TCT TGC GT 3′]	Vanhoof et al. (1992)
aadA1	3″-adenylyltransferase	*aadA1-F* [5′ GAG AAC ATA GCG TTG CCT TGG 3′]	*aadA1-R* [5′ TCG GCG CGA TTT TGC CGG TTA C 3′]	Sunde et al. (2015)

cassettes, specific primers targeting the 5′-conserved (5′-GGCATCCAAGCACAAGC-3′) and 3′-conserved segments (5′-AAGCAGACTTGACTGAT-3′) (Lévesque et al., 1995) were employed to amplify the entire integron cassette insertion region (Table 10.1). PCR reactions were carried out at 94 °C for 1 min, followed by 30 cycles at 94 °C for 30 s, 58 °C for 20 s, and 72 °C for 30 s. As a control, plasmid pUB2401, which carries transposon Tn21 containing the Intl2 class 1 integron, was utilized (Hall & Collis, 1995). Following PCR, the products were purified using the Qiagen PCR purification kit (Qiagen, Germantown, Maryland, USA). Purified products were then sequenced by primer walking, starting with the 3CS and 5CS primers (Table 10.1) (Lévesque et al., 1995) and then the aminoglycoside primers (Table 10.2).

3 Results

3.1 Water Quality in the Rouge River, Lake Saint Clair, and Saint Clair River

The water temperatures across all three watersheds were similar, ranging from 22 to 26.7 °C (Table 10.3). The Rouge River watershed had the lowest pH at 7.00, followed by the Saint Clair River at 7.04, and Lake Saint Clair at 7.64. The Rouge River watershed also had a much higher turbidity of 17.87 NTU, compared to the other two watersheds, which both had turbidity levels around 3 NTU. The Rouge River watershed had a significantly higher electrical conductivity at 493.33 μS cm^{-1}, compared to 192.22 μS cm^{-1} in the Saint Clair River and 207.73 μS cm^{-1} in Lake Saint Clair. The Rouge River watershed had a total suspended solids concentration of 244 mg L^{-1}, compared to 90 mg L^{-1} in the Saint Clair River and 92 mg L^{-1} in Lake

Table 10.3 Water quality data

Sample source	Water temperature (°C)	pH	Turbidity (NTU)	Electrical conductivity (μS cm^{-1})	Total suspended solids (mg L^{-1})	Dissolved oxygen (mg L^{-1})	TOC (mg L^{-1} of C)	*E. coli* (CFU per 100 ml)
Rouge River	22.1	7.00	17.87	493.33	244	8.58	7.68	38.77
Lake Saint Clair	24.6	7.64	3.04	192.22	90	10.89	9.96	18.40
Saint Clair River	26.7	7.04	3.61	207.73	92	7.58	6.25	22.31

Saint Clair. The dissolved oxygen readings were 8.58 mg L^{-1} in the Rouge River, 7.58 mg L^{-1} in the Saint Clair River, and 10.89 mg L^{-1} in Lake Saint Clair. The water samples showed the lowest TOC content in the Saint Clair River at 6.25 mg L^{-1}. In comparison, the TOC content was 7.68 mg L^{-1} in the Rouge River and 9.96 mg L^{-1} in Lake Saint Clair. The *E. coli* concentration, measured in CFU/100 mL, was found to be highest in the Rouge River at 38.77 CFU/100 mL. This was followed by the Saint Clair River with 22.31 CFU/100 mL, and the lowest concentration was in Lake Saint Clair at 18.40 CFU/100 mL (Table 10.3).

3.2 *Antibiotic Resistance of* E. coli *Strains*

Figure 10.1a provides a comprehensive analysis of the susceptibility of strains sourced from the Rouge River to a spectrum of antibiotics. Among the tested antibiotics, the strains demonstrated the greatest resistance to rifampin. The data indicates that the resistance is substantial, with an insignificant minority (less than or equal to 5%) showing an intermediate level of response to this antibiotic. In contrast, the strains showed a high level of susceptibility to tetracycline, as over 90% of them were affected by this antibiotic, suggesting tetracycline is highly effective against these strains. In more detail regarding the antibiotic imipenem, the results are promising yet show some level of resistance. Approximately 3% of the strains were categorized as resistant to imipenem, with about the same percentage exhibiting an intermediate response, while the vast majority (around 90%) were deemed susceptible. When comparing the efficacy of different antibiotics, the order of susceptibility from most to least effective is TET > IPM> SXT> CHL > AMP > CRO > GEN > TOB > NIT > SUL > CIP > RIF (Fig. 10.1a).

Figure 10.1b reports the antibiotic susceptibility profile of bacterial strains from Lake Saint Clair. The results showed complete susceptibility to imipenem, with all strains tested showing effective inhibition with no recorded instances of intermediate resistance to this antibiotic. In addition, the effectiveness of chloramphenicol

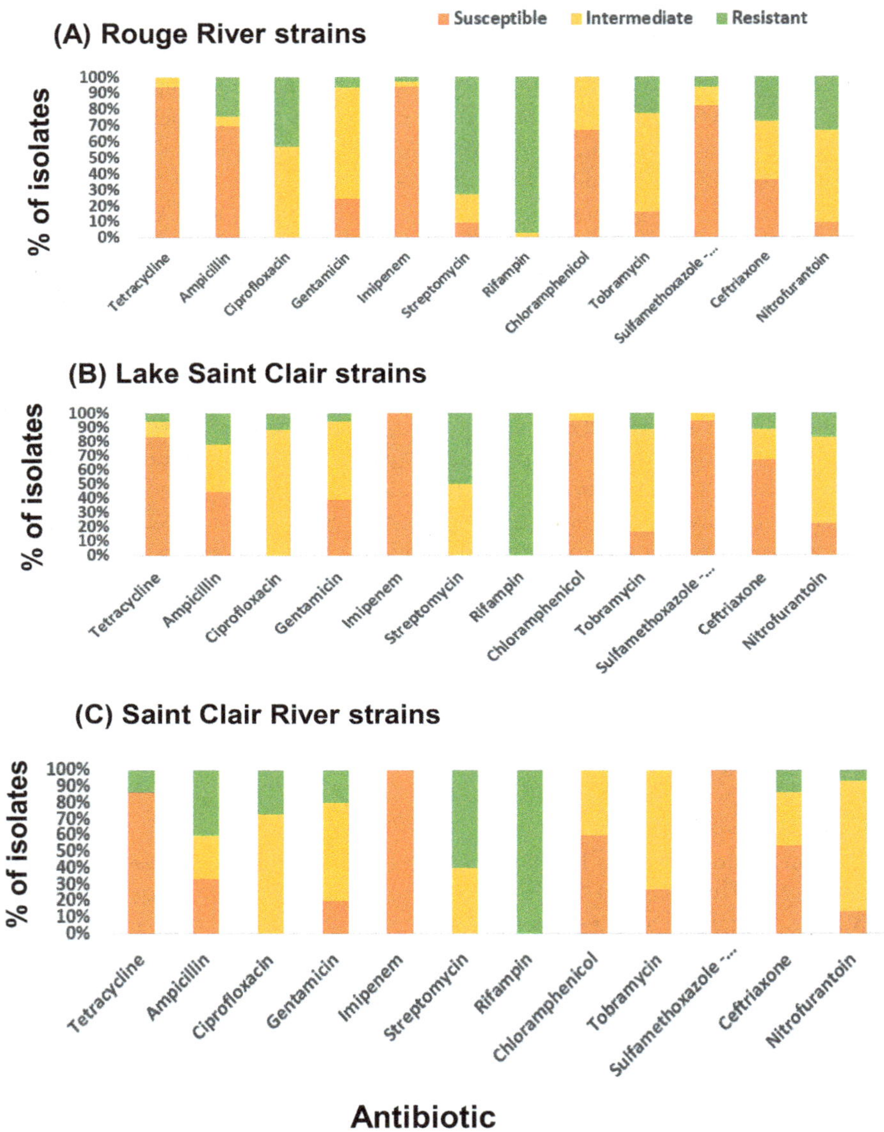

Fig. 10.1 Percentage of distribution of *E. coli* strains categorized as susceptible, intermediately resistant, and strains from (**a**) Rouge River, (**b**) Lake Saint Clair, and (**c**) Saint Clair River

was notable, with over 90% of strains being susceptible. Sulfamethoxazole-trimethoprim also demonstrated high efficacy, although with a marginally higher proportion of strains exhibiting an intermediate response when compared to chloramphenicol. Interestingly, the strains from Lake Saint Clair displayed universal resistance to rifampin, with an observed resistance rate of 100% across all tested isolates. This indicates a significant concern for the effectiveness of rifampin against these strains. Regarding streptomycin, the bacterial isolates were evenly split in their response: 50% were resistant, and the remaining 50% were classified with intermediate susceptibility, suggesting considerable variability in response to this antibiotic. The hierarchy of antibiotic susceptibility for the Lake Saint Clair strains is listed as follows, moving from the most effective to the least effective: IMI > IPM> CHL = SXT> TET > CRO > AMP > GEN > NIT > TOB > CIP > SUL > RIF (Fig. 10.1b).

Figure 10.1c illustrates the antibiotic susceptibility profile of bacterial strains from Saint Clair River. The results depicted total susceptibility to both imipenem and sulfamethoxazole-Trimethoprim with all isolates eliciting effective inhibition with no occurrence of intermediate resistance to these antibiotics. Next, the effectiveness of tetracycline was exceptional, with around 85% of strains being susceptible. Similarly, Chloramphenicol demonstrated some efficiency, although with a much higher percentage of isolates (40%) tested exhibiting an intermediate response. This suggests high variability in response to this antibiotic from these isolates. Notably, the strains from Saint Clair River displayed complete resistance to rifampin, showing total resistance of all tested isolates. This raises concern for the efficacy of rifampin when used to treat these strains. The hierarchy of antibiotic susceptibility for the Saint Clair River strains is listed as follows, listed from the most effective to the least effective: IPM = SXT > TET > CHL > CRO > AMP > TOB > GEN > NIT > CIP > SUL > RIF (Fig. 10.1c).

The multiple antibiotic resistance (MAR) index values of various bacterial strains provide insights into the potential risk of contamination based on their exposure to antibiotics. A critical threshold has been set at an MAR index of 0.2; values above this threshold suggest that the bacterial samples originate from environments where antibiotics are frequently used, thereby designating them as high-risk sources of contamination. Figure 10.2a focused on the isolates from the Rouge River; the MAR indices are depicted for each strain, indicating their respective risks for antibiotic resistance. Out of the total isolates evaluated, a significant 66.7% (22 strains) showed MAR index values greater than 0.2, which classifies them as originating from high-risk sources of contamination. Notably, the strain 6RR2 exhibited the highest MAR index in this group, with a recorded value of approximately 0.65, far exceeding the threshold and underscoring its significant exposure to antibiotic environments. Conversely, the remaining 33.3% (11 strains) of the Rouge River isolates had MAR index values below 0.2, suggesting these strains are likely not from areas where antibiotic use is prevalent and consequently represent a lower risk of contamination in this context. Among the lower-risk isolates, the strains 6RR5 and 7RR16 were identified as having particularly low MAR indexes, approximately

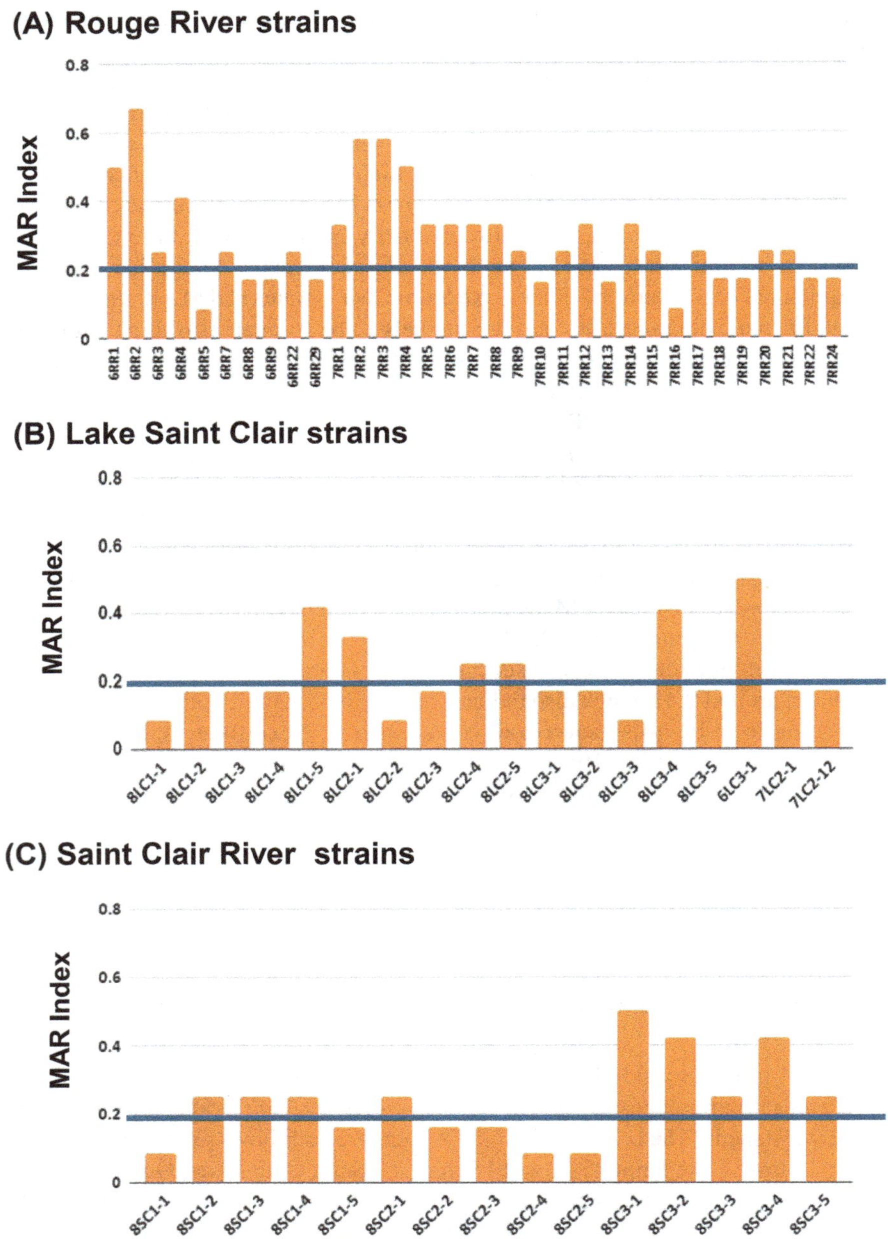

Fig. 10.2 Multiple Antibiotic Resistance (MAR) index values for bacterial strains isolated from (**A**) Rouge River, (**B**) Lake Saint Clair, and (**C**) Saint Clair River. The horizontal line marks the threshold of 0.2; values at or below this suggest the bacteria likely originated from low-risk environments with minimal antibiotic exposure, such as natural or uncontaminated waters. In contrast, values above 0.2 indicate potential contamination from high-risk sources where antibiotics are frequently used, including hospitals, livestock operations, or urban wastewater, posing possible public health concerns

around 0.08, which underscore their minimal exposure to antibiotic environments (Fig. 10.2a).

Figure 10.2b depicts the MAR indices of isolates from Lake Saint Clair indicating their respective risks for antibiotic resistance. Of all isolates evaluated, an exceptional 33.3% (6 strains) had MAR index values greater than the threshold of 0.2 categorizing them as originating from high-risk sources of contamination. Notably, the strain 6LC3-1 recorded the highest MAR index in this tested group, with a value of approximately 0.5, exceptionally higher than the threshold highlighting its significant exposure to antibiotic environments. On the other hand, the remaining 66.7% (12 strains) of the Lake Saint Clair isolates had MAR index values below 0.2, indicating these isolates are unlikely to be from areas of high antibiotic use and consequently represent a lower risk of contamination in this context. Among the lower-risk isolates, the strains 8LC1-1, 8LC2-2, and 8LC3-3 were identified as having particularly low MAR indices, approximately around 0.08, which underscore their minimal exposure to antibiotic environments (Fig. 10.2b).

Figure 10.2c lists the MAR indices of *E.coli* isolates from Saint Clair River illustrating their respective risks for antibiotic resistance. Of all isolates studied, a majority of 60% (9 strains) had MAR index values above the threshold of 0.2 which identifies them as originating from high-risk sources of contamination. The isolate 8SC3-1 recorded the highest MAR index with a value around 0.5, which is significantly higher than the threshold. This highlights the strain's notable exposure to antibiotic rich environments. The second highest index was of strains 8SC3-2 and 8SC3-4 at around 0.41 which also indicates the high probability of these two strains originating from high-risk sources of contamination. Next, the remaining 40% (6 strains) of the Saint Clair River strains had MAR index values smaller than 0.2, meaning these isolates are likely NOT from areas of high antibiotic use and likely represent a lower risk of contamination in this context. Among these lower-risk isolates are 8SC1-1, 8SC2-4, and 8SC2-5 having especially low MAR index readings of 0.1 (Fig. 10.2c).

3.3 Incidence of Class 1 Integron

Table 10.4 provides a side-by-side comparison of the presence of class 1 and class 2 integrons, referred to as Int1 and Int2, in 66 isolated *E. coli* strains from three different watersheds. The breakdown of these strains is as follows: 33 from the Rouge River, 18 from Lake Saint Clair, and 15 from the Saint Clair River. Out of the total, 43 strains were found to contain Int1. Within this subset, the Rouge River contributed the largest number with 23 strains, followed by Lake Saint Clair with 11 strains, and the Saint Clair River with 9 strains. On the other hand, Int2 was found much less frequently, appearing in only 7 out of the 66 *E. coli* strains. Of these, six strains containing Int2 were from the Rouge River, while the Saint Clair River had only a single strain with Int2. None of the strains from Lake Saint Clair showed the presence of Int2. Meanwhile, both integrons coexisted in 16 of the 66 *E. coli* strains

Table 10.4 Presence of Int1 and Int2 in *E. coli* strains

Sites	Number of strains	Int1	Int2	Int + In2
Rouge River	33	23 (70%)	6 (18%)	4 (12%)
Lake Saint Clair	18	11 (61%)	0 (0%)	7 (39%)
Saint Clair River	15	9 (60%)	1 (7)	5 (33%)
Total	66	43 (65%)	7 (11%)	16 (24%)

strains, documenting a significant incidence of concurrent resistance mechanisms, with noted variations among the sources: 4 strains in the Rouge River, 7 strains in Lake Saint Clair, and 5 strains in the Saint Clair River strains (Table 10.4). This distribution underscores a notable environmental footprint of antibiotic resistance genes across these watersheds.

Of the 66 strains screened, 43 were found to contain Int1. These 43 strains were subsequently subjected to PCR amplification of the variable region of class 1 integrons. Table 10.5 displays the distribution of class 1 integron cassette arrays found within 43 Int1-positive *E coli* strains from three different watersheds: Rouge River (23 strains), Lake Saint Clair (11 strains), and Saint Clair River (9 strains). Amplification, sequencing, and analysis of these Int-1 containing strains revealed the presence of 11 gene cassettes. Notably, there was a more diverse set of gene arrays found in Rouge River when compared to the other two watersheds, with combinations like *aad*A1, *aac*(3) IV, *aac*A4, and several multi-gene arrays including *aad*A1-*aac*(3) IV and *aad*A1-*aac*A4. Gene cassette *aad*A1-*aac*(3) IV was most abundant in Rouge River as it was present in six Int1-containing strains and not found in any strains from Lake Saint Clair and Saint Clair River. The Lake Saint Clair strains most frequently contained the *aad*B-*aac*A1-*aac*(3) IV cassette array with four occurrences. In the Saint Clair River strains, both *aad*B-*aad*A4 and *aad*B-*aac*A1-*aac*(3) IV arrays were most common, each found in three strains. The arrays *aad*A1, "*aac*(3) IV, *aac*A4, *aad*B, *aad*A1-*aac*A4, and *aac*A4-aa(3) IV showed varied presence across the watersheds, while some arrays such as *aad*A1-*aac*(3) IV," *aad*A1-*aac*A4, and *aad*B-*aac*(3) IV were exclusively found in the Rouge River strains. Overall, the gene cassette *aad*B-*aad*A4 frequently occurred in strains from all three watershed surfaces: 3 from Rouge River, 4 from Lake Saint Clair, and 4 from Saint Clair River (Table 10.5), which illustrate a diverse and distinct spread of resistance gene arrangements among the sampled *E. coli* strains in these aquatic environments.

Table 10.6 details the antimicrobial resistance phenotypes of 43 *E. coli* strains carrying various aminoglycoside gene cassettes. Strains with the aadA1 cassette, specifically strains 6RR4, 7RR12, and 7RR15, exhibit resistance to a range of antibiotics including antibiotics ampicillin, rifampin, ceftriaxone, nitrofurantoin, ciprofloxacin, and streptomycin. Strains with the "*aac*(3) IV" cassette, specifically strains 7RR16 and 8LC1-5, exhibit resistance to a range of antibiotics including rifampin and combinations including ciprofloxacin, streptomycin, and nitrofurantoin. Strains possessing the *aac*A4 gene cassette, including 6RR1, 7RR2, 7RR3, 7RR7, 7RR17, and 8LC1-5 have a wider resistance profile, including agents like ampicillin,

Table 10.5 Class 1 integron cassette arrays occurring in the 43 Int1-containing genes in *Escherichia coli* originating from the Rouge River, Lake Saint Clair and Saint Clair River Surface waters

Cassette arrays	Rouge River strains with Int1 ($n = 23$)	Lake Saint Clair strains with Int1 ($n = 11$)	Saint Clair River strains with Int1 ($n = 9$)
aadA1	3	0	0
aac(3) IV	1	1	0
aacA4	4	1	0
aadB	1	0	2
aadA1-aac(3) IV	6	0	0
aadA1-aacA4	3	1	1
aacA4-aa(3) IV	1	0	0
aadB-aadA4	4	3	3
aadB-aac(3) IV	1	0	0
aadB-aacA4-aac(3) IV	0	2	1
aadB-aacA1-aac(3) IV	0	4	3
Total	23	11	9

ciprofloxacin, streptomycin, rifampin, trimethoprim/sulfamethoxazole, tobramycin, and nitrofurantoin. The aadB gene cassette was associated with resistance to ampicillin, streptomycin, rifampin, and nitrofurantoin, with particular varieties also resistant to chloramphenicol. Strains 7RR1, 7RR2, 7RR5, 7RR8, 7RR9, and 7RR10, which have the "*aad*A1-*aac*(3) IV" cassette, are resistant to several antibiotics: strain 7RR1 is resistant to streptomycin, rifampin, tobramycin, and nitrofurantoin; strain 7RR2 is resistant to ampicillin, ciprofloxacin, streptomycin, rifampin, tobramycin, and nitrofurantoin; strain 7RR5 is resistant to ciprofloxacin, streptomycin, rifampin, and tobramycin; strains 7RR8 and 7RR9 are resistant to streptomycin, rifampin, tobramycin, and ceftriaxone; and strain 7RR10 is resistant to streptomycin and rifampin. Strains with the "*aad*A1-*aac*A4" cassette, specifically strains 7RR13, 7RR21, 7LC2-12, and 8SC2-1, exhibit resistance to a range of antibiotics including streptomycin and ceftriaxone for strain 7RR13; ciprofloxacin, streptomycin, and rifampin for strain 7RR21; antibiotics streptomycin, rifampin, tobramycin, and ceftriaxone for strain 7LC2-12; and tetracycline, streptomycin, and rifampin for strain 8SC2-1. The "*aac*A4-aa(3) IV" cassette, present in strain 7RR18 confers resistance to streptomycin and rifampin. Strains with the "*aad*B-*aad*A4" cassette, specifically strains 6RR2, 7RR4, 7RR16, 7RR19, 8SC2-2, 8SC2-4, 8SC3-2, 6LC3-1, 8LC2-4, and 8LC2-5 show resistance to a wide range of antibiotics, including ampicillin, ciprofloxacin, rifampin, tobramycin, trimethoprim/sulfamethoxazole, and nitrofurantoin. The "*aad*B-*aac*A4-*aac*(3) IV" cassette specifically strains 8SC2-1, 7LC2-12, and 8LC1-1, confer resistance to tetracycline, streptomycin, rifampin, tobramycin, and ceftriaxone. Finally, the "*aad*B-*aac*A1-*aac*(3) IV" cassette, strains 8SC3-3, 8SC3-4, 8SC3-5, 8LC2-3, 8LC3-3, 8LC3-1, and 8LC3-2 are

Table 10.6 Antimicrobial phenotypes among strains carrying aminoglycoside gene cassettes

Cassette arrays	Strains	Antimicrobial resistance phenotype
aadA1	6RR4	AMP, RIF, CEF, NIT
	7RR12	CIP, STR, RIF, CEF
	7RR15	CIP, STR, RIF
aac(3) IV	7RR16	RIF
	8LC1-5	CIP, STR, RIF, CEF, NIT
aacA4	6RR1	AMP, CIP, STR, RIF, SXT, NIT
	7RR2	AMP, CIP, STR, RIF, TOB, CEF, NIT
	7RR3	CIP, GEN, STR, RIF, TOB, CEF, NIT
	7RR7	STR, RIF, TOB, CEF
	7RR17	CIP, STR, RIF
	8LC1-5	STR, RIF
aadB	6RR3	AMP, STR, RIF, NIT
	8SC1-2	AMP, STR, RIF,
	8SC2-3	STE, RIF
aadA1-aac(3) IV	7RR1	STR, RIF, TOB, NIT
	7RR2	AMP, CIP, STR, RIF, TOB, NIT
	7RR5	CIP, STR, RIF, TOB
	7RR8	STR, RIF, TOB, CEF
	7RR9	STR, RIF, TOB, CEF
	7RR10	STR, RIF
aadA1-aacA4	7RR13	STR, CEF
	7RR21	CIP, STR, RIF
	7LC2-12	STR, RIF, TOB, CEF
	8SC2-1	TET, STR, RIF
aacA4-aa(3) IV	7RR18	STR, RIF
aadB-aadA4	6RR2	AMP, CIP, RIF, TOB, SXT, CEF, NIT
	7RR4	AMP, CIP, GEN, STR, RIF, TOB
	7RR16	RIF
	7RR19	STR, RIF
	8SC2-2	TET, STR, RIF
	8SC2-4	RIF
	8SC3-2	AMP, CIP, GEN,STR, RIF
	6LC3-1	AMP, RIF, NIT
	8LC2-4	AMP, RIF, NIT
	8LC2-5	
aadB-aacA4-aac(3) IV	8SC2-1	TET, STR, RIF
	7LC2-12	STR, RIF, TOB, CEF
	8LC1-1	RIF
aadB-aacA1-aac(3) IV	8SC3-3	AMP, RIF, NIT
	8SC3-4	AMP, CIP, RIF, CEF
	8SC3-5	CIP, GEN, RIF
	8LC2-3	STR, RIF
	8LC3-3	RIF
	8LC3-1	AMP, RIF
	8LC3-2	AMP, RIF

AMP Ampicillin, *CIP* ciprofloxacin, *CRO* ceftriaxone, *CHL* chloramphenicol, *GEN* gentamicin, *IPM* imipenem, *NIT* nitrofurantoin, *RIF* rifampin, *STR* streptomycin, *SXT* trimethoprim/sulfamethoxazole, *TET* tetracycline, and *TOB* tobramycin

linked to resistance to ampicillin, ciprofloxacin, rifampin, ceftriaxone, gentamicin, and nitrofurantoin. Notably, resistance to rifampin and streptomycin is widespread among these strains, whereas no strains are resistant to chloramphenicol and imipenem, and only two strains (6RR1 and 6RR2) are resistant to trimethoprim/sulfamethoxazole (Table 10.6).

4 Discussion

4.1 Water Quality Parameters Are Indicators of Proliferation of Antibiotic Resistance in Three Watersheds

The water quality parameters of three watersheds (Table 10.3) reveal that despite similar temperatures, the Rouge River exhibits higher turbidity, electrical conductivity, total suspended solids, and organic carbon content, alongside a lower pH level compared to the Saint Clair River and Lake Saint Clair. These conditions may create a more conducive environment for the survival and propagation of antibiotic-resistant bacteria, as evidenced by higher *E. coli* concentrations in the Rouge River. Dissolved oxygen levels were highest in Lake Saint Clair, suggesting better water quality, but does not preclude the presence of antibiotic resistance. Overall, these various environmental factors combined with fecal contamination across all watersheds could be contributing to the spread of antibiotic resistance genes within these aquatic ecosystems. High conductivity in the Rouge River, indicative of greater ion content, may suggest larger inputs of pollutants, which can include heavy metals that have been shown to co-select for antibiotic resistance (Tiquia-Arashiro & Rodrigues, 2016a, b; Cytryn et al., 2017; Tiquia-Arashiro, 2018; Bytesnikova et al., 2019). The relatively high TOC indicates more organic matter, which can enrich microbial communities that carry resistance genes (Tiquia et al., 2006a; Tiquia, 2008, 2010a; Cheng et al., 2021; Li et al., 2021a). This might correlate with the higher *E. coli* concentrations observed, raising concerns about the potential presence of resistant bacterial strains. Interestingly, Lake Saint Clair presented with the highest levels of dissolved oxygen, which, despite being generally indicative of better water quality, could also support a robust microbial community (Tiquia et al., 2006b, 2008; Tiquia-Arashiro, 2012; Cho et al., 2012; Qin et al., 2018; Oest et al., 2018; Patel et al. 2019; Shahraki et al., 2021) where resistant bacteria might persist. The lower bacterial concentrations in Lake Saint Clair and the Saint Clair River, along with their generally lower levels of pollutants compared to the Rouge River, suggest less fecal contamination and possibly fewer antibiotic-resistant bacteria present. However, the presence of even a moderate level of *E. coli* across all watersheds indicates fecal contamination and an inherent risk of antibiotic-resistant bacteria being present. Integrated approaches considering both the physicochemical parameters and direct identification of resistance genes are necessary to provide a more comprehensive understanding of the environmental dissemination of antibiotic resistance in these aquatic ecosystems.

4.2 *Antibiotic Resistance Profiles of the* E. coli *Strains in the Three Watersheds*

In this study aimed to first determine whether the 66 *E. coli* strains from three watersheds exhibited antibiotic resistance. To achieve this, we first screened these isolates for resistance to 12 antibiotics across 10 different classes, including aminoglycosides, fluoroquinolones, tetracyclines, penicillins, carbapenems, antimycobacterials, chloramphenicol, sulfonamides, cephalosporins, and nitrofurans. Our results confirm high resistance levels of *E. coli* strains to various antibiotics in the three watersheds (Figs. 10.1 and 10.2). In the Rouge River, the strains showed the greatest resistance to rifampin, with nearly all the strains resistant to it and less than or equal to 5% exhibiting an intermediate response. For Lake Saint Clair, there was universal resistance to rifampin and significant resistance to streptomycin, with 50% of strains being resistant and the other 50% showing intermediate susceptibility. In the Saint Clair River, the strains demonstrated complete resistance to rifampin. This indicates that rifampin is consistently ineffective across all three watersheds, while streptomycin resistance is a particular concern in Lake Saint Clair. Other antibiotics such as chloramphenicol, sulfamethoxazole-trimethoprim, and imipenem generally exhibit higher effectiveness across different watersheds, with lower rates of resistance observed. The resistance patterns of *E. coli* to antibiotics like rifampin in the Rouge River, Lake Saint Clair, and the Saint Clair River are not isolated phenomena and have global parallels. Published studies from other watersheds indicate similar trends in antibiotic resistance. A study by Pruden et al. (2013) in the Cache la Poudre River watershed in Colorado found that antibiotic resistance genes were pervasive, particularly for sulfonamides and tetracyclines. Research by Servais and Passerat (2009) in the Seine River in France cited high levels of antibiotic-resistant *E. coli*, with particularly high resistance rates to ampicillin and trimethoprim, which are commonly used in human and veterinary medicine. In the Haihe River in China, a study showed a significant prevalence of antibiotic-resistant bacteria, often displaying multi-drug resistance, including resistance to antibiotics such as ampicillin, tetracycline, and chloramphenicol (Zhang et al., 2009a). These findings align with previous studies indicating high levels of antibiotic resistance in *E. coli* from aquatic environments exposed to human activities. Studies have noted that human activities contribute significantly to antibiotic pollution, which in turn fosters the development and spread of antibiotic-resistant bacteria in the environment (Martinez, 2009). The high resistance to rifampin observed across all sites is consistent with findings from studies in aquatic cultures, where antibiotics are frequently detected in water sources (Liao et al., 2021). The significant rifampin resistance observed in the three watersheds presents a challenge for treating bacterial infections, notably given rifampin's role in treating various bacteria including the causative agent of tuberculosis (Beloor Suresh et al., 2024). The diminished effectiveness of rifampin could impact treatment regimens. On the other hand, imipenem and tetracycline maintain high effectiveness against *E. coli*, with susceptibility rates over 90% in the studied areas, consistent with imipenem's broad-spectrum efficacy against

β-lactamase-producing bacteria (Scholar, 2007) and tetracycline's role as a protein synthesis inhibitor (Ziółkowski, 2023; Shutter & Akhondi, 2024). Resistance to rifampin commonly involves rpoB gene mutations (Li et al., 2021b), while imipenem resistance may involve the blaVM1 gene and blaOXA913, although their rarity suggests limited dissemination (Mirbagheri et al., 2015; Moon et al., 2021).

The MAR index values reveal varying levels of antibiotic contamination across the watersheds (Fig. 10.2). The Rouge River exhibits the highest proportion of high-risk strains (67%), likely due to industrial discharges, a common source of antibiotics. Lake Saint Clair shows lower MAR indices (33%), suggesting less contamination, possibly due to improved management practices or fewer pollution sources. Similarly, the Saint Clair River's MAR indices (60%) indicate exposure to antibiotic pollution sources. The variability in resistance, including intermediate susceptibility, suggests ongoing resistance evolution driven by antibiotic concentrations seeping into watersheds, promoting the selection of resistant mutants (Nassereddine et al., 2024). Intermediate responses indicate the changing nature of resistance development, influenced by environmental factors. The trend of universal rifampin resistance, particularly in Lake Saint Clair and Saint Clair River strains, is concerning, limiting treatment options and underscoring the need for strict antibiotic policies and enhanced water contamination monitoring.

4.3 Gene Cassettes in the Class I Integrons

Integrons, particularly class 1 integrons (Int1), play a crucial role in the spread of antibiotic resistance among bacterial populations. Integrons are genetic elements that capture and express antibiotic resistance genes, allowing bacteria to spread resistance traits across different species and environments (Gillings et al., 2008; Bhat et al., 2023; Nassereddine et al., 2024). Class 1 integrons are widely recognized for their association with antibiotic resistance in clinical settings. They are a major concern due to their ability to capture multiple resistance genes, leading to complex resistance species (Gillings, 2014, 2017; Zheng et al., 2020). In this study, we first determined the resistance profiles of each strain and then screened them for the presence of class 1 integrons (Int1). We also assessed whether these integrons contained multiple resistant genes within their gene cassettes. Our results show that the majority of the *E. coli strains* (43 out of 66) carry Int1: 23 from Rouge River, 11 from Lake Saint Clair, and 9 from Saint Clair River (Table 10.4). This widespread occurrence aligns with other studies that have documented high rates of Int1 in diverse environments. For example, Gillings et al. (2015) reported a high prevalence of Int1 in whole watersheds, attributing this to human activities such as wastewater discharge and agriculture. The high frequency of Int1 in this study highlights the nature of antibiotic resistance genes in environments influenced by human activity. In contrast to the high presence of *Int*1, class 2 integrons (*Int*2) were scarce, detected in only 18% of Rouge River strains and 7% of Saint Clair River strains, with no

presence in Lake Saint Clair strains. This discrepancy may reflect different evolutionary pressures favoring *Int*1 over *Int*2. The varied prevalence of *Int*1 and *Int*2 among different water sources can be attributed to ecological dynamics, including water quality, human activity, and geographical differences (Luo et al., 2010; Tiquia-Arashiro, 2014a, b, c, d; Yang et al., 2021; Racewicz et al., 2022; Shin et al., 2023). Rivers like the Rouge River, which may be subjected to high levels of pollution and human waste, provide a suitable environment for the proliferation of *Int*1 integrons due to the constant influx of antibiotics and other pollutants. In contrast, the lower prevalence of integrons in Lake Saint Clair may reflect differences in pollution levels, water treatment practices, and local human activities. Potential environmental factors contributing to the high prevalence of Int1 include contamination with pharmaceutical substances, heavy metals, pesticides, and other pollutants that exert selective pressure on bacterial populations (Lu et al., 2015; Tiquia & Mormile, 2010; Tiquia-Arashiro & Rodrigues, 2016b, c; Jing et al., 2023; Pagnucco et al., 2023; Liu et al., 2024). These contaminants can promote the maintenance and spread of resistance genes, including those carried by Int1 integrons (Gillings et al., 2015). The widespread use of antibiotics in both human medicine and agriculture from surface run-off, leaching, livestock wastes/compost (Tiquia et al., 1997; Richard & Tiquia, 1999; Tam & Tiquia, 1999; Tiquia & Tam, 1999; Garrison et al., 2001; Tiquia & Michel, 2002) likely contributes to the selection of *E. coli* strains carrying *Int*1 integrons, enhancing their survival and propagation in contaminated environments. The observed co-occurrence of *Int*1 and *Int*2 in 24% of the total strains is significant, as it indicates the presence of multiple resistance mechanisms within single bacterial populations. This co-occurrence suggests that these strains have been exposed to diverse selective pressures, leading to the acquisition of various resistance elements. The identification of class 1 integron in 43 out of 66 tested strains highlights its significant role in mediating antibiotic resistance. This high prevalence indicates that *Int*1 is a major factor for the spread of resistance genes among bacterial populations in the Rouge River, for example. Integrons, particularly *Int*1, are known to capture and spread gene cassettes that encode antibiotic resistance, facilitating rapid adaptation to selective pressures such as antibiotic use in clinical and agricultural settings. This makes them a concern for public health and environmental safety due to their potential to spread widely in the environment, ultimately posing a risk to human health by compromising the effectiveness of antibiotics (Gillings et al., 2008, 2015; Tiquia-Arashiro & Mormile, 2013; Gillings, 2014).

4.4 Aminoglycosides Gene Cassettes

Further screening of the 43 *E. coli* strains containing Int-1 revealed the presence of 11 different aminoglycoside gene cassettes (Table 10.5). The diversity of the gene aminoglycoside cassettes observed among the Rouge River strains suggests that this environment is subject to various ecological and selective pressures that drive

genetic variation. Factors such as the presence of different pollutants, varying levels of antibiotics, and interactions with various environmental factors could contribute to this diversity. Studies have shown that environments subjected to human activities, such as urban rivers, often exhibit high levels of microbial diversity and an increase in the spread of resistance genes due to constant exposure to pollutants and antibiotics (Gaze et al., 2011; Gillings et al., 2015; Oest et al., 2018). When considering the strains from Lake Saint Clair and the Saint Clair River, it is crucial to note whether these environments exhibit a similar diversity of aminoglycoside gene cassettes. According to the data collected, there were 9 out of 11 gene cassettes present in the Rouge River, 6 in Lake Saint Clair, and 5 in the Saint Clair River. This discrepancy suggests that the Rouge River may be subject to more intense or varied selective pressures than the other two environments. The higher diversity of gene cassettes in the Rouge River could be attributed to its significantly higher levels of pollutants, as indicated by higher turbidity (17.87 NTU), electrical conductivity (493.33 μS cm^{-1}), and total suspended solids (244 mg L^{-1}) (Table 10.3). These factors, combined with the higher *E. coli* concentration (38.77 CFU/100 mL), suggest a more polluted and complex environment that could promote greater genetic variation and the spread of resistance genes. In contrast, Lake Saint Clair, and the Saint Clair River, with lower levels of turbidity (both around 3 NTU), electrical conductivity (207.73 μS cm^{-1} in Lake Saint Clair and 192.22 μS cm^{-1} in Saint Clair River), and total suspended solids (92 mg L^{-1} in Lake Saint Clair and 90 mg L^{-1} in Saint Clair River), might experience less intense selective pressures, resulting in fewer gene cassettes being present (Table 10.3). The water quality data show that Lake Saint Clair and the Saint Clair River have similar, and relatively lower, levels of pollutants compared to the Rouge River, which might account for the reduced diversity in gene cassettes observed in these environments. The lower *E. coli* concentrations in Lake Saint Clair (18.40 CFU/100 mL) and the Saint Clair River (22.31 CFU/100 mL) also support the notion of less favorable conditions for the proliferation and spread of antibiotic resistance genes (Table 10.3).

The unique presence of specific gene arrays, such as *aad*A1-aac (3) IV in the Rouge River, raises questions about the mechanisms driving the localization of these cassettes. One hypothesis could be that these gene cassettes are introduced through sources, such as discharge from nearby healthcare facilities, agricultural runoff, or specific industrial discharges that release antibiotics (Gillings et al., 2015). Another possibility is that these gene cassettes have evolved due to selective pressures in the Rouge River environment, such as specific antibiotic residues or heavy metals that select for these resistance genes (Gillings, 2013). Strains with multiple resistance genes have significant implications for treatment options and the spread of resistance. Such strains are more difficult to treat, as they can withstand a broad range of antibiotics, leading to limited therapeutic options. This can result in longer illness durations, higher medical costs, and increased mortality rates (Davies & Davies, 2010; Bartoletti et al., 2018). Furthermore, bacteria with multiple resistance genes can serve as reservoirs for these genes, facilitating their transfer to other bacteria through HGT mechanisms (Gaze et al., 2011; Gillings, 2014). This

facilitates the spread of resistance within and between different environments, making it more difficult to control and prevent antibiotic resistance.

The widespread occurrence of the *aad*B-*aad*A4 gene cassette across all three watersheds may suggest a highly efficient mechanism of spreading for this resistance element. The presence of this gene cassette in the three watersheds may indicate that it may be successful at spreading through water systems, possibly due to mobile genetic elements such as plasmids or transposons that facilitate its transfer between bacteria (Partridge et al., 2009). This finding implies that there are significant cross-environment transfers of resistance genes, which can be traced back to common human sources such as wastewater treatment or agricultural runoff (Tiquia, 2002; Gillings et al., 2015; Nassereddine et al., 2024). The findings from this study have important ecological and health implications. The widespread presence of antibiotic resistance genes in environmental bacteria highlights the impact of human activities on antibiotic resistance. In terms of human health, the accumulation of antibiotic resistance genes in pathogens leads to increasing difficulties in treating bacterial infections and higher mortality rates. Environmentally, ecosystems serve as reservoirs for antibiotic resistance genes, playing a crucial role in the evolution and spread of resistance. High antibiotic pollution in the environment, particularly from manufacturing emissions creates strong selection pressures for resistance, facilitating the movement of antibiotic resistance genes and resistant bacteria across various ecological niches (Larsson & Flach, 2022).

4.5 Antimicrobial Phenotypes Among Strains Carrying Aminoglycoside Gene Cassettes

The comprehensive analysis presented in Table 10.6 shows the antimicrobial resistance (AMR) profiles observed across 43 *E. coli* strains carrying diverse aminoglycoside gene cassettes. Each aminoglycoside gene cassette demonstrated specific resistance patterns. For instance, strains with the *aac*A4 gene cassette exhibited resistance to multiple aminoglycosides, such as gentamicin, tobramycin, netilmicin, and amikacin (El-Baz et al., 2021; Pagano et al., 2016). This broader resistance profile suggests that the *aac*A4 cassette may be linked to several resistance mechanisms, facilitated by horizontal gene transfer (HGT). Studies have detected the presence of *aac*A4 gene in different species of *E. coli* and non-fermentative bacteria indicating the occurrence of HGT (Xia et al., 2016). Other cassettes, like aadA1, primarily conferred resistance to streptomycin and spectinomycin (Pagano et al., 2016), indicating a more targeted resistance mechanism. The resistance extends to crucial antibiotics such as ceftriaxone and nitrofurantoin, indicating that the possession of these gene cassettes enables the bacteria to withstand a broad spectrum of therapeutic agents.

Certain cassettes when combined, such as "*aad*A1-*aac*(3) IV," "*aad*A1-*aac*A4," and "*aad*B-*aac*A4-*aac*(3) IV," exhibit an even wider range of drug resistance,

illustrating the complexity of AMR mechanisms within these strains. For instance, the "*aad*A1-*aac*(3) IV" cassette provides resistance to essential antibiotics like streptomycin, rifampin, tobramycin, and nitrofurantoin, with some strains also resistant to ampicillin, ciprofloxacin, and ceftriaxone. Strains carrying the "*aad*A1-*aac*A4" cassette exhibit varying resistance patterns, encompassing antibiotics such as streptomycin, ceftriaxone, rifampin, and ciprofloxacin, among others. Similarly, the "*aad*B-*aac*A4-*aac*(3) IV" cassette provides a broad range of resistance, including tetracycline, streptomycin, rifampin, tobramycin, and ceftriaxone.

Notably, resistance to rifampin and streptomycin is common across many of these gene configurations, which poses a significant challenge to treatment regimens. It's equally important to note the absence of resistance to chloramphenicol and imipenem in these strains, and a limited resistance to trimethoprim/sulfamethoxazole, recognizing potential vulnerabilities in the bacteria's defense that may be exploited therapeutically. Such detailed AMR profiling is pivotal for the development of targeted treatment strategies to combat antibiotic resistance in *E. coli*, a major public health threat. This study also noted widespread resistance to rifampin and streptomycin across multiple gene cassettes. This pattern is consistent with existing literature (Liao et al., 2021) and may be due to the excessive use of these antibiotics leading to significant environmental exposure, possibly from agricultural runoff or other human sources. Additionally, no resistance to chloramphenicol and imipenem was observed in the strains studied. This absence may indicate that these antibiotics are either not extensively used in the three watersheds or that the resistance mechanisms against them are not easily acquired or maintained by *E. coli* in these settings.

The observed resistance phenotypes present significant clinical challenges in treating infections caused by these *E. coli* strains. The presence of multiple resistance genes, especially those with resistance to commonly used antibiotics, limits the available treatment options, potentially leading to higher healthcare costs, longer hospital stays, and increased mortality rates (Livermore, 2012; Davies & Davies, 2010; Bartoletti et al., 2018). The resistance to aminoglycosides, combined with resistance to other antibiotic classes, can complicate therapeutic regimens and necessitate the use of more toxic alternatives. Environmental factors likely contribute to the observed distribution of antibiotic resistance. Sources such as agricultural runoff, wastewater treatment plants, and hospital discharges can introduce antibiotics and resistant bacteria into aquatic environments, creating reservoirs of resistance genes. These environments facilitate the horizontal gene transfer among bacterial populations, promoting the spread of resistance (Gillings et al., 2015; Larsson & Flach, 2022; Tiquia-Arashiro et al., 2025; Nassereddine et al., 2024).

5 Conclusions

The analysis conducted in this study highlights several crucial findings regarding antibiotic resistance in *E. coli* strains within three distinct watersheds. First, the examination of water quality parameters indicates varying environmental conditions across the Rouge River, Saint Clair River, and Lake Saint Clair, with the Rouge River exhibiting higher levels of pollutants and organic matter. These conditions, coupled with fecal contamination, potentially contribute to the proliferation and dissemination of antibiotic-resistant bacteria within these aquatic ecosystems. Furthermore, the antibiotic resistance profiles of the *E. coli* strains reveal significant resistance to various antibiotics, with rifampin resistance being particularly widespread across all three watersheds. This resistance trend aligns with global observations and underscores the challenges in treating bacterial infections, especially those caused by *E. coli* strains from aquatic environments influenced by human activities. Additionally, the presence of class 1 integrons (Int1) in the majority of tested strains suggests their role in mediating antibiotic resistance dissemination, highlighting the impact of human activities on resistance gene prevalence in the environment. Moreover, the diversity of aminoglycoside gene cassettes observed among the *E. coli* strains underscores the complexity of antibiotic resistance mechanisms and the potential for horizontal gene transfer (HGT) to facilitate resistance gene dissemination. The combination of different gene cassettes further expands the resistance profile of these strains, posing challenges for effective antibiotic treatment.

Despite the widespread resistance to certain antibiotics, vulnerabilities in the bacteria's defense, such as the absence of resistance to chloramphenicol and imipenem, suggest potential therapeutic opportunities. However, the observed resistance phenotypes present significant clinical challenges, emphasizing the urgent need for targeted treatment strategies and enhanced surveillance to combat antibiotic resistance in aquatic environments. These results emphasize the complex relationship between environmental conditions, bacterial genetic traits, and the dynamics of antibiotic resistance and highlight the importance of multifaceted approaches to address this pressing public health concern.

Acknowledgments The authors like to express their sincere thanks and deep gratitude to the University of Michigan-Dearborn Office of Research and Sponsored Programs for financially supporting this work. Furthermore, Bryan Catalfano and Yanesa Chamlee received support for this participation in this research through the Summer Undergraduate Research Experience (SURE) Program at the University of Michigan-Dearborn.

Author Contributions All authors contributed to the study conception and design. Material preparation, data collection and analysis were performed by Bryan Catalfano, Yanesa Chamlee, Victoria Khamis, Diana Kasperek, Erica Parmenter, Reis Schittenhelm, and Zainab Nassereddine. The first draft of the manuscript was written by Zainab Nassereddine, and all authors commented on previous versions of the manuscript. All authors read and approved the final manuscript.

Conflict of Interest The authors declare they have no financial interests. Sonia Tiquia-Arashiro is the editor of the book project and was not involved in the editorial review, or the decision to publish, this article. All authors declare that there are no competing interests.

Data Availability The authors confirm that the data supporting the findings of this study are available within the article [and/or] its supplementary materials.

Compliance with Ethical Standards The authors affirm that they have no conflicts of interest to disclose. Furthermore, this article does not involve any studies conducted by the authors with human or animal participants.

References

Ahmadian, L., Haghshenas, M. R., Mirzaei, B., Norouzi Bazgir, Z., & Goli, H. R. (2020). Distribution and molecular characterization of resistance gene cassettes containing class 1 integrons in multi-drug resistant (MDR) clinical isolates of Pseudomonas aeruginosa. *Infection and Drug Resistance, 13*, 2773–2781. https://doi.org/10.2147/IDR.S263759

Amarasiri, M., Sano, D., & Suzuki, S. (2019). Understanding human health risks caused by antibiotic resistant bacteria (ARB) and antibiotic resistance genes (ARG) in water environments: Current knowledge and questions to be answered. *Critical Reviews in Environmental Science and Technology, 50*(19), 2016–2059. https://doi.org/10.1080/10643389.2019.1692611

Azizi, O., Shakibaie, M. R., Badmasti, F., Modarresi, F., Ramazanzadeh, R., Mansouri, S., & Shahcheraghi, F. (2016). Class 1 integrons in non-clonal multidrug-resistant *Acinetobacter baumannii* from Iran, description of the new blaIMP-55 allele in In1243. *Journal of Medical Microbiology, 65*(9), 928–936. https://doi.org/10.1099/jmm.0.000315

Barry, A. L., Bernsohn, K. L., & Thrupp, L. D. (1969). Four-hour urease test for distinguishing between *Klebsiella* and *Enterobacter*. *Applied Microbiology, 18*(2), 156–158. https://doi.org/10.1128/am.18.2.156-158.1969

Bartoletti, M., Giannella, M., Tedeschi, S., & Viale, P. (2018). Multidrug-resistant bacterial infections in solid organ transplant candidates and recipients. *Infectious Disease Clinics of North America, 32*(3), 551–580. https://doi.org/10.1016/j.idc.2018.04.004

Bass, L., Liebert, C. A., Lee, M. D., Summers, A. O., White, D. G., Thayer, S. G., & Maurer, J. J. (1999). Incidence and characterization of integrons, genetic elements mediating multiple-drug resistance, in avian *Escherichia coli*. *Antimicrobial Agents and Chemotherapy, 43*(12), 2925–2929. https://doi.org/10.1128/AAC.43.12.2925

Bauer, A. W., Kirby, W. M., Sherris, J. C., & Turck, M. (1966). Antibiotic susceptibility testing by a standardized single disk method. *American Journal of Clinical Pathology, 45*(4), 493–496.

Beloor Suresh, A., Rosani, A., Patel, P., et al. (2024). Rifampin. [Updated 2023 Nov 12]. In *StatPearls [Internet]*. StatPearls Publishing. Available from: https://www.ncbi.nlm.nih.gov/books/NBK557488/

Bhat, B. A., Mir, R. A., Qadri, H., Dhiman, R., Almilaibary, A., Alkhanani, M., & Mir, M. A. (2023). Integrons in the development of antimicrobial resistance: Critical review and perspectives. *Frontiers in Microbiology, 14*, 1231938. https://doi.org/10.3389/fmicb.2023.1231938

Boerlin, P., Travis, R., Gyles, C. L., Reid-Smith, R., Janecko, N., Lim, H., Nicholson, V., McEwen, S. A., Friendship, R., & Archambault, M. (2005). Antimicrobial resistance and virulence genes of *Escherichia coli* isolates from swine in Ontario. *Applied and Environmental Microbiology, 71*(11), 6753–6761. https://doi.org/10.1128/AEM.71.11.6753-6761.2005

Bowman, N., Patel, P., Xu, W., Alsaffar, A., & Tiquia-Arashiro, S. M. (2018). Enrichment and isolation of Pb-resistant bacteria from Saint Clair River sediments and their potential for Pb

removal in aqueous solutions. *Applied Microbiology and Biotechnology, 102*, 2391–2398. https://doi.org/10.1007/s00253-018-8772-4

Bytesnikova, Z., Richtera, L., Smerkova, K., et al. (2019). Graphene oxide as a tool for antibiotic-resistant gene removal: A review. *Environmental Science and Pollution Research, 26*, 20148–20163. https://doi.org/10.1007/s11356-019-05283-y

Chen, M., Wu, Y., Yu, S., Liu, S., Wang, Y., Huang, D., Xu, X., & Lin, F. (2019). Drug resistance and integron genes in *Escherichia coli* isolated from urinary tract infection. *Journal of Nanoscience and Nanotechnology, 19*(9), 5989–5993. https://doi.org/10.1166/jnn.2019.16501

Cheng, X., Lu, Y., Song, Y., Zhang, R., ShangGuan, X., Xu, H., Liu, C., & Liu, H. (2021). Analysis of antibiotic resistance genes, environmental factors, and microbial community from aquaculture farms in five provinces, China. *Frontiers in Microbiology, 12*, 679805. https://doi.org/10.3389/fmicb.2021.679805

Cho, K., Zholi, A., Frabutt, D., Flood, M., Floyd, D., & Tiquia, S. M. (2012). Linking bacterial diversity and geochemistry of uranium-contaminated groundwater. *Environmental Technology, 33*, 1629–1640. https://doi.org/10.1080/09593330.2011.641036

Clinical and Laboratory Standards Institute. (2021). *Performance Standards for Antimicrobial Susceptibility Testing*. CLSI Documents M100 (31st ed.). CLSI.

Cytryn, E., Markiewicz, Z., & Popowska, M. (2017). Antibiotics and antibiotics resistance genes dissemination in soils. In M. Hashmi, V. Strezov, & A. Varma (Eds.), *Antibiotics and antibiotics resistance genes in soils. Soil biology* (Vol. 51). Springer. https://doi.org/10.1007/978-3-319-66260-2_9

Davies, J., & Davies, D. (2010). Origins and evolution of antibiotic resistance. *Microbiology and Molecular Biology Reviews, 74*(3), 417–433. https://doi.org/10.1128/MMBR.00016-10

Deng, Y., Bao, X., Ji, L., Chen, L., Liu, J., Miao, J., Chen, D., Bian, H., Li, Y., & Yu, G. (2015). Resistance integrons: Class 1, 2 and 3 integrons. *Annals of Clinical Microbiology and Antimicrobials, 14*, 45. https://doi.org/10.1186/s12941-015-0100-6

Domingues, S., da Silva, G. J., & Nielsen, K. M. (2015). Global dissemination patterns of common gene cassette arrays in class 1 integrons. *Microbiology (Reading), 161*(7), 1313–1337. https://doi.org/10.1099/mic.0.000099

Dufour, A. P., & Cabelli, V. J. (1975). Membrane filter procedure for enumerating the component genera of the coliform group in seawater. *Applied Microbiology, 29*(6), 826–833. https://doi.org/10.1128/am.29.6.826-833.1975

El-Baz, A. M., Yahya, G., Mansour, B., El-Sokkary, M. M., Alshaman, R., Alattar, A., & El-Ganiny, A. M. (2021). The link between occurrence of class I integron and acquired aminoglycoside resistance in clinical MRSA isolates. *Antibiotics, 10*(5), 488. https://doi.org/10.3390/antibiotics10050488

Fling, M. E., Kopf, J., & Richards, C. (1985). Nucleotide sequence of the transposon Tn7 gene encoding an aminoglycoside-modifying enzyme, 3″(9)-O-nucleotidyltransferase. *Nucleic Acids Research, 13*(19), 7095–7106. https://doi.org/10.1093/nar/13.19.7095

Flood, M., Frabutt, D., Floyd, D., Powers, E. U., Devol, A., & Tiquia-Arashiro, S. M. (2015). Ammonia-oxidizing bacteria and archaea in sediments of the Gulf of Mexico. *Environmental Technology, 36*, 124–135. https://doi.org/10.1080/09593330.2014.942385

Gallagher, L. A., Lee, S. A., & Manoil, C. (2017). Importance of core genome functions for an extreme antibiotic resistance trait. *MBio, 8*(6), 10–1128. https://doi.org/10.1128/mbio.01655-17

Garrison, M., Richard, T. L., Tiquia, S. M., & Honeyman, M. S. (2001). *Nutrient losses from unlined bedded swine hoop structures and an associated windrow composting site*. ASAE Paper No. 012238. American Society of Agricultural Engineers (ASAE) International Meeting, Sacramento, July 29–August 1, 2001.

Gaze, W. H., Zhang, L., Abdouslam, N. A., Hawkey, P. M., Calvo-Bado, L., Royle, J., Brown, H., Davis, S., Kay, P., Boxall, A. B. A., & Wellington, E. M. H. (2011). Impacts of anthropogenic activity on the ecology of class 1 integrons and integron-associated genes in the environment. *The ISME Journal, 5*(8), 1253–1261. https://doi.org/10.1038/ismej.2011.15

Gentile, M., Yan, T., Tiquia, S. M., Fields, M. W., Nyman, J., Zhou, J., & Criddle, C. S. (2006). Stability and resilience in a denitrifying fluidized bed reactor. *Microbial Ecology, 52*(2), 311–321. https://doi.org/10.1007/s00248-006-9024-1

Gillings, M. R. (2013). Evolutionary consequences of antibiotic use for the resistome, mobilome and microbial pangenome. *Frontiers in Microbiology, 4*, 4. https://doi.org/10.3389/fmicb.2013.00004

Gillings, M. R. (2014). Integrons: Past, present, and future. *Microbiology and Molecular Biology Reviews, 78*(2), 257–277. https://doi.org/10.1128/MMBR.00056-13

Gillings, M. R. (2017). Class 1 integrons as invasive species. *Current Opinion in Microbiology, 38*, 10–15. https://doi.org/10.1016/j.mib.2017.03.002

Gillings, M., Boucher, Y., Labbate, M., Holmes, A., Krishnan, S., Holley, M., & Stokes, H. W. (2008). The evolution of class 1 integrons and the rise of antibiotic resistance. *Journal of Bacteriology, 190*(14), 5095–5100. https://doi.org/10.1128/JB.00152-08

Gillings, M. R., Gaze, W. H., Pruden, A., Smalla, K., Tiedje, J. M., & Zhu, Y.-G. (2015). Using the class 1 integron-integrase gene as a proxy for anthropogenic pollution. *The ISME Journal, 9*(6), 1269–1279. https://doi.org/10.1038/ismej.2014.226

Hall, R. M., & Collis, C. M. (1995). Mobile gene cassettes and integrons: Capture and spread of genes by site-specific recombination. *Molecular Microbiology, 15*(4), 593–600. https://doi.org/10.1111/j.1365-2958.1995.tb02368.x

Henriques, I. S., Fonseca, F., Alves, A., Saavedra, M. J., & Correia, A. (2006). Occurrence and diversity of integrons and beta-lactamase genes among ampicillin-resistant isolates from estuarine waters. *Research in Microbiology, 157*(10), 938–947. https://doi.org/10.1016/j.resmic.2006.09.003

Jing, K., Li, Y., Yao, C., Jiang, C., & Li, J. (2023). Towards the fate of antibiotics and the development of related resistance genes in stream biofilms. *Science of the Total Environment, 898*, 165554. https://doi.org/10.1016/j.scitotenv.2023.165554

Kapoor, G., Saigal, S., & Elongavan, A. (2017). Action and resistance mechanisms of antibiotics: A guide for clinicians. *Journal of Anaesthesiology Clinical Pharmacology, 33*(3), 300–305. https://doi.org/10.4103/joacp.JOACP_349_15

Kemper, N. (2008). Veterinary antibiotics in the aquatic and terrestrial environment. *Ecological Indicators, 8*, 1–13. https://doi.org/10.1016/j.ecolind.2007.06.002

Krumperman, P. H. (1983). Multiple antibiotic resistance indexing of *Escherichia coli* to identify high-risk sources of fecal contamination of foods. *Applied and Environmental Microbiology, 46*(1), 165–170. https://doi.org/10.1128/aem.46.1.165-170.1983

Larsson, D. G. J., & Flach, C.-F. (2022). Antibiotic resistance in the environment. *Nature Reviews Microbiology, 20*(5), 257–269. https://doi.org/10.1038/s41579-021-00649-x

Lévesque, C., & Roy, P. H. (1993). PCR analysis of integrons. In D. H. Persing, T. F. Smith, F. C. Tenover, & T. J. White (Eds.), *Diagnostic molecular microbiology: Principles and applications* (pp. 590–594). Mayo Foundation.

Lévesque, C., Piché, L., Larose, C., & Roy, P. H. (1995). PCR mapping of integrons reveals several novel combinations of resistance genes. *Antimicrobial Agents and Chemotherapy, 39*(1), 185–191. https://doi.org/10.1128/AAC.39.1.185

Li, X. Z. (2016). Antimicrobial resistance in bacteria: An overview of mechanisms and role of drug efflux pumps. In X. Z. Li, C. Elkins, & H. Zgurskaya (Eds.), *Efflux-mediated antimicrobial resistance in bacteria*. Adis. https://doi.org/10.1007/978-3-319-39658-3_6

Li, S., Liu, J., Yao, Q., Yu, Z., Li, Y., Jin, J., Liu, X., & Wang, G. (2021a). Potential role of organic matter in the transmission of antibiotic resistance genes in black soils. *Ecotoxicology and Environmental Safety, 227*, 112946. https://doi.org/10.1016/j.ecoenv.2021.112946

Li, M., Lu, J., Lu, Y., Xiao, T., Liu, H., Lin, S., Xu, D., Li, G., Zhao, X., Liu, Z., Zhao, L., & Wan, K. (2021b). rpoB mutations and effects on rifampin resistance in mycobacterium tuberculosis. *Infection and Drug Resistance, 14*, 4119–4128. https://doi.org/10.2147/IDR.S333433

Liao, C.-Y., Balasubramanian, B., Peng, J.-J., Tao, S.-R., Liu, W.-C., & Ma, Y. (2021). Antimicrobial resistance of *Escherichia coli* from aquaculture farms and their environment in Zhanjiang, China. *Frontiers in Veterinary Science, 8*, 806653. https://doi.org/10.3389/fvets.2021.806653

Liu, K., Gan, C., Peng, Y., Gan, Y., He, J., Du, Y., Tong, L., Shi, J., & Wang, Y. (2024). Occurrence and source identification of antibiotics and antibiotic resistance genes in groundwater surrounding urban hospitals. *Journal of Hazardous Materials, 465*, 133368. https://doi.org/10.1016/j.jhazmat.2023.133368

Livermore, D. M. (2012). Fourteen years in resistance. *International Journal of Antimicrobial Agents, 39*(4), 283–294. https://doi.org/10.1016/j.ijantimicag.2011.12.012

Lu, Z., Na, G., Gao, H., Wang, L., Bao, C., & Yao, Z. (2015). Fate of sulfonamide resistance genes in estuary environment and effect of anthropogenic activities. *Science of the Total Environment, 527–528*, 429–438. https://doi.org/10.1016/j.scitotenv.2015.04.101

Luo, Y., Mao, D., Rysz, M., Zhou, Q., Zhang, H., Xu, L., & Alvarez, P. J. J. (2010). Trends in antibiotic resistance genes occurrence in the Haihe River, China. *Environmental Science & Technology, 44*(19), 7220–7225. https://doi.org/10.1021/es100233w

Martinez, J. L. (2009). Environmental pollution by antibiotics and by antibiotic resistance determinants. *Environmental Pollution, 157*(11), 2893–2902. https://doi.org/10.1016/j.envpol.2009.05.051

McKindles, K. M., & Tiquia-Arashiro, S. M. (2012). Functional gene arrays for analysis of microbial communities on ocean platform. In S. Tiquia-Arashiro (Ed.), *Molecular biological technologies for ocean sensing* (Springer protocols handbooks). Humana Press. https://doi.org/10.1007/978-1-61779-915-0_9

Mirbagheri, S. Z., Meshkat, Z., Naderinasab, M., Rostami, S., Nabavinia, M. S., & Rahmati, M. (2015). Study on imipenem resistance and prevalence of blaVIM1 and blaVIM2 metallo-beta lactamases among clinical isolates of Pseudomonas aeruginosa from Mashhad, Northeast of Iran. *Iranian Journal of Microbiology, 7*(2), 72–78.

Moon, D.-C., Mechesso, A. F., Kang, H.-Y., Kim, S.-J., Choi, J.-H., Song, H.-J., Yoon, S.-S., & Lim, S.-K. (2021). Imipenem resistance mediated by blaOXA-913 gene in Pseudomonas aeruginosa. *Antibiotics, 10*(10), 1188. https://doi.org/10.3390/antibiotics10101188

Mostafa, M., Siadat, S. D., Shahcheraghi, F., Vaziri, F., Japoni-Nejad, A., Vand Yousefi, J., & Rahbar, M. (2015). Variability in gene cassette patterns of class 1 and 2 integrons associated with multi drug resistance patterns in Staphylococcus aureus clinical isolates in Tehran-Iran. *BMC Microbiology, 15*, 1–9. https://doi.org/10.1186/s12866-015-0488-3

Naderi, G., Talebi, M., Gheybizadeh, R., Seifi, A., Ghourchian, S., Rahbar, M., et al. (2023). Mobile genetic elements carrying aminoglycoside resistance genes in *Acinetobacter baumannii* isolates belonging to global clone 2. *Frontiers in Microbiology, 14*, 1172861. https://doi.org/10.3389/fmicb.2023.1172861

Nap, J. P., Bijvoet, J., & Stiekema, W. J. (1992). Biosafety of kanamycin-resistant transgenic plants. *Transgenic Research, 1*, 239–249. https://doi.org/10.1007/BF02525165

Nassereddine, Z., Opara, S., Coutinho, O., Qyteti, F., Book, R., Heinicke, M. P., Napieralski, J., & Tiquia-Arashiro, S. M. (2024). Critical perspectives on advancing antibiotic resistant gene (ARG) detection technologies in aquatic ecosystems. *Science of the Total Environment, 957*, 177775. https://doi.org/10.1016/j.scitotenv.2024.177775

Nguyen, S., Ala, F., Cardwell, C., Cai, D., McKindles, K. M., Lotvola, A., Hodges, S., Deng, Y., & Tiquia-Arashiro, S. M. (2013). Isolation and screening of carboxydotrophs isolated from composts and their potential for butanol synthesis. *Environmental Technology, 34*, 1995–2007. https://doi.org/10.1080/09593330.2013.795987

Oest, A., Alsaffar, A., Fenner, M., Azzopardi, D., & Tiquia-Arashiro, S. M. (2018). Patterns of change in metabolic capabilities of sediment microbial communities in river and lake ecosystems. *International Journal of Microbiology, 2018*, 6234931. https://doi.org/10.1155/2018/6234931

Oliveira-Pinto, C., Costa, P. S., Reis, M. P., et al. (2016). Diversity of gene cassettes and the abundance of the class 1 integron-integrase gene in sediment polluted by metals. *Extremophiles, 20*, 283–289. https://doi.org/10.1007/s00792-016-0820-3

Pagano, M., Martins, A. F., & Barth, A. L. (2016). Mobile genetic elements related to carbapenem resistance in Acinetobacter baumannii. *Brazilian Journal of Microbiology, 47*(4), 785–792. https://doi.org/10.1016/j.bjm.2016.06.005

Pagnucco, G., Overfield, D., Chamlee, Y., Fortuna, A., Sulaiman, F., Farinas, J., Schittenhelm, R., Shuler, C., Kassem, A., Opara, S., Najaf, H., Abbas, L., Coutinho, O., Catalfano, B., Li, X., & Tiquia-Arashiro, S. M. (2023). Metal tolerance and biosorption capacities of bacterial strains isolated from an urban watershed. *Frontiers in Microbiology, 14*, 1278886. https://doi.org/10.3389/fmicb.2023.1278886

Paik, G. (1970). Reagents, strains, and test procedures. In J. E. Blair, E. H. Lennette, & J. P. Truant (Eds.), *Manual of clinical microbiology* (pp. 675–692). American Society for Microbiology.

Partridge, S. R., Tsafnat, G., Coiera, E., & Iredell, J. R. (2009). Gene cassettes and cassette arrays in mobile resistance integrons. *FEMS Microbiology Reviews, 33*(4), 757–784. https://doi.org/10.1111/j.1574-6976.2009.00175.x

Partridge, S. R., Kwong, S. M., Firth, N., & Jensen, S. O. (2018). Mobile genetic elements associated with antimicrobial resistance. *Clinical Microbiology Reviews, 31*, 10–1128. https://doi.org/10.1128/cmr.00088-17

Patel, D., Gismondi, R., Alsaffar, A., & Tiquia-Arashiro, S. M. (2019). Applicability of API ZYM to capture seasonal and spatial variabilities in lake and river sediments. *Environmental Technology, 40*(24), 3227–3239. https://doi.org/10.1080/09593330.2018.1468492

Peak, N., Knapp, C. W., Yang, R. K., Hanfelt, M. M., Smith, M. S., Aga, D. S., & Graham, D. W. (2007). Abundance of six tetracycline resistance genes in wastewater lagoons at cattle feedlots with different antibiotic use strategies. *Environmental Microbiology, 9*(1), 143–151. https://doi.org/10.1111/j.1462-2920.2006.01123.x

Plante, I., Centrón, D., & Roy, P. H. (2003). An integron cassette encoding erythromycin esterase, ere(A), from Providencia stuartii. *The Journal of Antimicrobial Chemotherapy, 51*(4), 787–790. https://doi.org/10.1093/jac/dkg169

Plecha, S., Hall, D., & Tiquia-Arashiro, S. M. (2013). Screening for novel bacteria from the bioenergy feedstock switchgrass (Panicum virgatum L.). *Environmental Technology, 34*, 1895–1904. https://doi.org/10.1080/09593330.2013.818701

Pomaranski, E., & Tiquia-Arashiro, S. M. (2016). Butanol tolerance of carboxydotrophic bacteria isolated from manure composts. *Environmental Technology, 37*(15), 1970–1982. https://doi.org/10.1080/09593330.2015.1137360

Pruden, A., Larsson, D. G., Amézquita, A., Collignon, P., Brandt, K. K., Graham, D. W., Lazorchak, J. M., Suzuki, S., Silley, P., Snape, J. R., Topp, E., Zhang, T., & Zhu, Y. G. (2013). Management options for reducing the release of antibiotics and antibiotic resistance genes to the environment. *Environmental Health Perspectives, 121*(8), 878–885. https://doi.org/10.1289/ehp.1206446

Qin, W., Amin, S. A., Lundeen, R. A., Heal, K. R., Martens-Habbena, W., Turkarslan, S., Urakawa, H., Costa, K. C., Hendrickson, E. L., Wang, T., Beck, D. A., Tiquia-Arashiro, S. M., Taub, F., Holmes, A. D., Vajrala, N., Berube, P. M., Lowe, T. M., Moffett, J. W., Devol, A. H., Baliga, N. S., Arp, D. J., Sayavedra-Soto, L. A., Hackett, M., Armbrust, E. V., Ingalls, A. E., & Stahl, D. A. (2018). Stress response of a marine ammonia-oxidizing archaeon informs physiological status of environmental populations. *The ISME Journal, 12*(2), 508–519. https://doi.org/10.1038/ismej.2017.186

Racewicz, P., Majewski, M., Biesiada, H., et al. (2022). Prevalence and characterisation of antimicrobial resistance genes and class 1 and 2 integrons in multiresistant *Escherichia coli* isolated from poultry production. *Scientific Reports, 12*, 6062. https://doi.org/10.1038/s41598-022-09996-y

Ramirez, M. S., & Tolmasky, M. E. (2010). Aminoglycoside modifying enzymes. *Drug Resistance Updates, 13*(6), 151–171. https://doi.org/10.1016/j.drup.2010.08.003

Ramirez, M. S., & Tolmasky, M. E. (2017). Amikacin: Uses, resistance, and prospects for inhibition. *Molecules, 22*(12), 2267. https://doi.org/10.3390/molecules22122267

Richard, T. L., & Tiquia, S. M. (1999). In-barn composting system plus a warm bed for pigs. *Biocycle, 40*(10), 44.

Roberts, M. C. (2005). Update on acquired tetracycline resistance genes. *FEMS Microbiology Letters, 245*(2), 195–203. https://doi.org/10.1016/j.femsle.2005.02.034

Robicsek, A., Sahm, D. F., Strahilevitz, J., Jacoby, G. A., & Hooper, D. C. (2005). Broader distribution of plasmid-mediated quinolone resistance in the United States. *Antimicrobial Agents and Chemotherapy, 49*(7), 3001–3003. https://doi.org/10.1128/AAC.49.7.3001-3003.2005

Robicsek, A., Strahilevitz, J., Jacoby, G. A., Macielag, M., Abbanat, D., Park, C. H., Bush, K., & Hooper, D. C. (2006a). Fluoroquinolone-modifying enzyme: A new adaptation of a common aminoglycoside acetyltransferase. *Nature Medicine, 12*(1), 83–88. https://doi.org/10.1038/nm1347

Robicsek, A., Strahilevitz, J., Sahm, D. F., Jacoby, G. A., & Hooper, D. C. (2006b). qnr prevalence in ceftazidime-resistant Enterobacteriaceae isolates from the United States. *Antimicrobial Agents and Chemotherapy, 50*(8), 2872–2874. https://doi.org/10.1128/AAC.01647-05

Robicsek, A., Jacoby, G. A., & Hooper, D. C. (2006c). The worldwide emergence of plasmid-mediated quinolone resistance. *The Lancet Infectious Diseases, 6*(10), 629–640. https://doi.org/10.1016/S1473-3099(06)70599-0

Sabbagh, P., Rajabnia, M., Maali, A., & Ferdosi-Shahandashti, E. (2021). Integron and its role in antimicrobial resistance: A literature review on some bacterial pathogens. *Iranian Journal of Basic Medical Sciences, 24*(2), 136. https://doi.org/10.22038/ijbms.2020.48905.11208

Schmitz, F. J., Fluit, A. C., Gondolf, M., Beyrau, R., Lindenlauf, E., Verhoef, J., Heinz, H. P., & Jones, M. E. (1999). The prevalence of aminoglycoside resistance and corresponding resistance genes in clinical isolates of staphylococci from 19 European hospitals. *The Journal of Antimicrobial Chemotherapy, 43*(2), 253–259. https://doi.org/10.1093/jac/43.2.253

Scholar, E. (2007). Imipenem. In S. J. Enna & D. B. Bylund (Eds.), *xPharm: The comprehensive pharmacology reference* (pp. 1–5). Elsevier. https://doi.org/10.1016/B978-008055232-3.61923-3

Serio, A. W., Keepers, T., Andrews, L., & Krause, K. M. (2018). Aminoglycoside revival: Review of a historically important class of antimicrobials undergoing rejuvenation. *EcoSal Plus, 8*(1), 10–1128. https://doi.org/10.1128/ecosalplus.esp-0002-2018

Servais, P., & Passerat, J. (2009). Antimicrobial resistance of fecal bacteria in waters of the Seine river watershed (France). *Science of the Total Environment, 408*(2), 365–372. https://doi.org/10.1016/j.scitotenv.2009.09.042

Shahraki, A. H., Chaganti, S. R., & Heath, D. (2021). Spatio-temporal dynamics of bacterial communities in the shoreline of Laurentian great Lake Erie and Lake St. Clair's large freshwater ecosystems. *BMC Microbiology, 21*, 253. https://doi.org/10.1186/s12866-021-02306-y

Shin, H., Kim, Y., Han, S., & Hur, H. G. (2023). Resistome study in aquatic environments. *Journal of Microbiology and Biotechnology, 33*(3), 277–287. https://doi.org/10.4014/jmb.2210.10044

Shutter, M. C., & Akhondi, H. (2024). Tetracycline. [Updated 2023 Jun 5]. In *StatPearls [Internet]*. StatPearls Publishing. Available from: https://www.ncbi.nlm.nih.gov/books/NBK549905/

Stokes, H. W., & Hall, R. M. (1989). A novel family of potentially mobile DNA elements encoding site-specific gene-integration functions: Integrons. *Molecular Microbiology, 3*(12), 1669–1683. https://doi.org/10.1111/j.1365-2958.1989.tb00153.x

Sunde, M., Simonsen, G. S., Slettemeås, J. S., Böckerman, I., & Norström, M. (2015). Integron, plasmid and host strain characteristics of *Escherichia coli* from humans and food included into the Norwegian antimicrobial resistance monitoring programs. *PLoS One, 10*(6), e0128797. https://doi.org/10.1371/journal.pone.0128797

Tam, N. F. Y., & Tiquia, M. (1999). Nitrogen transformation during co-composting of spent pig manure, sawdust litter and sludge under forced-aerated system. *Environmental Technology, 20*(3), 259–267. https://doi.org/10.1080/09593332008616816

Tiquia, S. M. (2002). Microbial transformation of nitrogen during composting. In H. Insam, N. Riddech, & S. Klammer (Eds.), *Microbiology of composting*. Springer. https://doi.org/10.1007/978-3-662-08724-4_20

Tiquia, S. M. (2008). Diversity of sulfate-reducing genes (dsrAB) in sediments from Puget Sound. *Environmental Technology, 29*(10), 1095–1108. https://doi.org/10.1080/09593330802190608

Tiquia, S. M. (2010a). Salt-adapted bacteria isolated from the Rouge River and potential for degradation of contaminants and biotechnological applications. *Environmental Technology, 31*, 967–978. https://doi.org/10.1080/09593331003706226

Tiquia, S. M. (2010b). Metabolic diversity of the heterotrophic microorganisms and potential link to pollution of the Rouge River. *Environmental Pollution, 158*, 1435–1443. https://doi.org/10.1016/j.envpol.2009.12.035

Tiquia, S. M. (2011). Extracellular hydrolytic enzyme activities of the heterotrophic microbial communities of the Rouge River: An approach to evaluate ecosystem response to urbanization. *Microbial Ecology, 62*, 679–689. https://doi.org/10.1007/s00248-011-9871-2

Tiquia, S. M., & Michel, F. C. (2002). Bacterial diversity in livestock manure composts as characterized by terminal restriction fragment length polymorphisms (T-RFLP) of PCR-amplified 16s rRNA gene sequences. In H. Insam, N. Riddech, & S. Klammer (Eds.), *Microbiology of composting*. Springer. https://doi.org/10.1007/978-3-662-08724-4_6

Tiquia, S. M., & Mormile, M. (2010). *Extremophiles: A source of innovation for industrial and environmental applications* (Vol. 31, pp. 1061–1181). Taylor and Francis. https://doi.org/10.1080/09593330.2010.489333

Tiquia, S. M., & Tam, N. F. Y. (1999). Chemical parameters for maturity determination of pig manure disposed from the pig-on-litter (POL) system in Hong Kong. In *Manure management* (99th ed., pp. 449–463). Extension Division-University of Saskatchewan.

Tiquia, S. M., Tam, N. F., & Hodgkiss, I. J. (1997). Composting of spent pig litter at different seasonal temperatures in subtropical climate. *Environmental Pollution, 98*(1), 97–104. https://doi.org/10.1016/s0269-7491(97)00116-4

Tiquia, S. M., Tam, N. F. Y., & Hodgkiss, I. J. (1998). *Salmonella* elimination during composting of spent pig litter. *Bioresource Technology, 63*(2), 193–196. https://doi.org/10.1016/S0960-8524(97)00113-2

Tiquia, S. M., Gurczynski, S., Zholi, A., & Devol, A. (2006a). Diversity of biogeochemical cycling genes from Puget Sound sediments using DNA microarrays. *Environmental Technology, 27*(12), 1377–1389. https://doi.org/10.1080/09593332708618756

Tiquia, S. M., Masson, S. A., & Devol, A. (2006b). Vertical distribution of nitrite reductase genes (nirS) in continental margin sediments of the Gulf of Mexico. *FEMS Microbiology Ecology, 58*(3), 464–475. https://doi.org/10.1111/j.1574-6941.2006.00173.x

Tiquia, S. M., Davis, D., Hadid, H., Kasparian, S., Ismail, M., & Murray, K. S. (2007). Halophilic and halotolerant bacteria from river waters and shallow groundwater along the Rouge River of Southeastern Michigan. *Environmental Technology, 28*, 297–307. https://doi.org/10.1080/09593332808618789

Tiquia, S. M., Schleibak, M., Schlaff, J., Floyd, J. C., Benipal, B., Zakhem, E., & Murray, K. S. (2008). Microbial community profiling and characterization of some heterotrophic bacterial isolates from river waters and shallow groundwater wells along the Rouge River, Southeast Michigan. *Environmental Technology, 29*, 651–663. https://doi.org/10.1080/09593330801986998

Tiquia-Arashiro, S. M. (2012). *Molecular biological technologies for ocean sensing* (Springer protocols handbooks) (p. 307). Humana Press. https://doi.org/10.1007/978-1-61779-915-0

Tiquia-Arashiro, S. M. (2014a). Biotechnological applications of thermophilic carboxydotrophs. In *Thermophilic carboxydotrophs and their applications in biotechnology* (SpringerBriefs in microbiology). Springer. https://doi.org/10.1007/978-3-319-11873-4_4

Tiquia-Arashiro, S. M. (2014b). Microbial CO metabolism. In *Thermophilic carboxydotrophs and their applications in biotechnology* (Chapter 2. SpringerBriefs in microbiology) (pp. 5–9). Springer. https://doi.org/10.1007/978-3-319-11873-4_2

Tiquia-Arashiro, S. M. (2014c). CO-oxidizing microorganisms. In *Thermophilic carboxydotrophs and their applications in biotechnology* (Chapter 3. Springer SpringerBriefs in microbiology) (pp. 11–28). Springer. https://doi.org/10.1007/978-3-319-11873-4_3

Tiquia-Arashiro, S. M. (2014d). *Thermophilic carboxydotrophs and their applications in biotechnology* (SpringerBriefs in microbiology) (p. 131). Springer. https://doi.org/10.1007/978-3-319-11873-4

Tiquia-Arashiro, S. M. (2018). Lead absorption mechanisms in bacteria as strategies for lead bioremediation. *Applied Microbiology and Biotechnology, 102*(13), 5437–5444. https://doi.org/10.1007/s00253-018-8969-6

Tiquia-Arashiro, S. M. (2019a). Synthesis of metallic nanoparticles by halotolerant fungi. In *Fungi in extreme environments: Ecological role and biotechnological significance* (pp. 383–394). Springer International Publishing. https://doi.org/10.1007/978-3-030-19030-9_19

Tiquia-Arashiro, S. M. (2019b). Thermophilic fungi in composts: Their role in composting and industrial processes. In *Fungi in extreme environments: Ecological role and biotechnological significance* (pp. 587–606). Springer International Publishing. https://doi.org/10.1007/978-3-030-19030-9_29

Tiquia-Arashiro, S. M. (2020). Electricity production from carbon monoxide and syngas in a thermophilic microbial fuel cell. In S. M. Tiquia-Arashiro & D. Pant (Eds.), *Microbial electrochemical technologies* (pp. 288–299). CRC Press. https://www.taylorfrancis.com/chapters/edit/10.1201/9780429487118-19/electricity-production-carbon-monoxide-syngas-thermophilic-microbial-fuel-cell-sonia-tiquia-arashiro

Tiquia-Arashiro, S. M., & Mormile, M. (2013). Sustainable technologies: Bioenergy and biofuel from biowaste and biomass. *Environmental Technology, 34*(13–14), 1637–1638. https://doi.org/10.1080/09593330.2013.834162

Tiquia-Arashiro, S. M., & Pant, D. (2020). *Microbial electrochemical technologies* (p. 518). CRC Press. https://doi.org/10.1201/9780429487118

Tiquia-Arashiro, S. M., & Rodrigues, D. (2016a). *Extremophiles: Applications in nanotechnology* (Springerbriefs in microbiology: Extremophilic microorganisms) (p. 193). Springer International Publishing. https://doi.org/10.1007/978-3-319-45215-9

Tiquia-Arashiro, S. M., & Rodrigues, D. (2016b). Nanoparticles synthesized by microorganisms. In *Extremophiles: Applications in nanotechnology* (pp. 1–51). Springer International Publishing. https://doi.org/10.1007/978-3-319-45215-9_1

Tiquia-Arashiro, S. M., & Rodrigues, D. (2016c). Applications of nanoparticles. In *Extremophiles: Applications in nanotechnology* (pp. 163–193). Springer International Publishing.

Tiquia-Arashiro, S. M., Medema, G., Ubomba-Jaswa, E., Urakawa, H., Ella, V., & Libardi, N. (2025). *Watershed microbiology: Biodiversity composition, ecological function, and technological advances*. Springer Nature Switzerland AG. (In press).

USEPA (U.S. Environmental Protection Agency). (2005). *Method 1603: Escherichia coli (E. coli) in water by membrane filtration using modified membrane-thermotolerant Escherichia coli agar (modified mTEC), EPA-821-R-04-025*. Office of Water, U.S. Environmental Protection Agency.

Vanhoof, R., Content, J., Van Bossuyt, E., Dewit, L., & Hannecart-Pokorni, E. (1992). Identification of the aadB gene coding for the aminoglycoside-2″-O-nucleotidyltraiisferase, ANT(2″), by means of the polymerase chain reaction. *Journal of Antimicrobial Chemotherapy, 29*(4), 365–374. https://doi.org/10.1093/jac/29.4.365

Vourli, S., Tzouvelekis, L. S., Tzelepi, E., Lebessi, E., Legakis, N. J., & Miriagou, V. (2003). Characterization of In111, a class 1 integron that carries the extended-spectrum beta-lactamase gene blaIBC-1. *FEMS Microbiology Letters, 225*(1), 149–153. https://doi.org/10.1016/S0378-1097(03)00510-X

Wang, X., Kong, N., Cao, M., Zhang, L., Sun, M., Xiao, L., Li, G., & Wei, Q. (2021). Comparison of class 2 integron integrase activities. *Current Microbiology, 78*(3), 967–978. https://doi.org/10.1007/s00284-021-02352-9

Wight, J., Byrne, A. S., Tahlan, K., & Lang, A. S. (2024). Anthropogenic contamination sources drive differences in antimicrobial-resistant *Escherichia coli* in three urban lakes. *Applied and Environmental Microbiology, 90*(3), e0180923. https://doi.org/10.1128/aem.01809-23

Wu, W., Gu, B., Fields, M. W., Gentile, M., Ku, Y.-K., Yan, H., Tiquia, S. M., Yan, T., Nyman, J., Zhou, J., Jardine, P. M., & Criddle, C. S. (2005). Uranium (VI) reduction by denitrifying biomass. *Bioremediation Journal, 9*, 1–13. https://doi.org/10.1080/10889860590929628

Xia, W., Xu, T., Qin, T., Li, P., Liu, Y., Kang, H., Gu, B., & Ma, P. (2016). Characterization of integrons and novel cassette arrays in bacteria from clinical isloates in China, 2000–2014. *Journal of Biomedical Research, 30*(4), 292–303. https://doi.org/10.7555/JBR.30.20150153

Yan, H., Shi, L., Yamasaki, S., Li, X., Cao, Y., Li, L., et al. (2007). A plasmidic class 1 integron from five Pseudomonas aeruginosa clinical strains harbored aacA4 and nonsense-mutated cmlA1 gene cassettes. *Journal of Health Science, 53*(6), 750–755. https://doi.org/10.1248/jhs.53.750

Yang, Y., Song, W., Lin, H., Wang, W., Du, L., & Xing, W. (2018). Antibiotics and antibiotic resistance genes in global lakes: A review and meta-analysis. *Environment International, 116*, 60–73. https://doi.org/10.1016/j.envint.2018.04.011

Yang, S., Liang, M., Qin, Z., Qian, Y., Li, M., & Cao, Y. (2021). A novel assessment considering spatial and temporal variations of water quality to identify pollution sources in urban rivers. *Scientific Reports, 11*(1), 8714. https://doi.org/10.1038/s41598-021-87671-4

Ye, C., Hou, F., Xu, D., Huang, Q., Chen, X., Zeng, Z., Peng, Y., & Fang, R. (2020). Prevalence and characterisation of class 1 and 2 integrons in multi-drug resistant Staphylococcus aureus Isolates from pig farms in Chongqing, China. *Journal of Veterinary Research, 64*(3), 381–386. https://doi.org/10.2478/jvetres-2020-0061

Zainab, S. M., Junaid, M., Xu, N., & Malik, R. N. (2020). Antibiotics and antibiotic resistant genes (ARGs) in groundwater: A global review on dissemination, sources, interactions, environmental and human health risks. *Water Research, 187*, 116455. https://doi.org/10.1016/j.watres.2020.116455

Zárate, S. G., De la Cruz Claure, M. L., Benito-Arenas, R., Revuelta, J., Santana, A. G., & Bastida, A. (2018). Overcoming aminoglycoside enzymatic resistance: Design of novel antibiotics and inhibitors. *Molecules, 23*(2), 284. https://doi.org/10.3390/molecules23020284

Zhang, X., Wu, B., Zhang, Y., Zhang, T., Yang, L., Fang, H. H., Ford, T., & Cheng, S. (2009a). Class 1 integronase gene and tetracycline resistance genes tetA and tetC in different water environments of Jiangsu Province, China. *Ecotoxicology, 18*(6), 652–660. https://doi.org/10.1007/s10646-009-0332-3

Zhang, X. X., Zhang, T., Zhang, M., et al. (2009b). Characterization and quantification of class 1 integrons and associated gene cassettes in sewage treatment plants. *Applied Microbiology and Biotechnology, 82*, 1169–1177. https://doi.org/10.1007/s00253-009-1886-y

Zhao, S., White, D. G., Ge, B., Ayers, S., Friedman, S., English, L., Wagner, D., Gaines, S., & Meng, J. (2001). Identification and characterization of integron-mediated antibiotic resistance among Shiga toxin-producing *Escherichia coli* isolates. *Applied and Environmental Microbiology, 67*(4), 1558–1564. https://doi.org/10.1128/AEM.67.4.1558-1564.2001

Zheng, W., Huyan, J., Tian, Z., Zhang, Y., & Wen, X. (2020). Clinical class 1 integron-integrase gene–A promising indicator to monitor the abundance and elimination of antibiotic resistance genes in an urban wastewater treatment plant. *Environment International, 135*, 105372. https://doi.org/10.1016/j.envint.2019.105372

Ziółkowski, H. (2023). Bioavailability of tetracyclines is substantially increased by administration of cyclosporine A, a non-specific efflux-pump blocker. *Drug Metabolism and Pharmacokinetics, 50*, 100493. https://doi.org/10.1016/j.dmpk.2023 100493

Chapter 11
Anthropogenic Environments and Antimicrobial-Resistant Dynamics: A General Overview of a Complex Network

Tania Henriquez and Massimiliano Marvasi

1 Introduction

Pollution is defined as the introduction of harmful substances or contaminants into the environment, leading to adverse effects on ecosystems, wildlife, and human health (Noël et al., 2022). *Anthropogenic environment* refers to the environment significantly impacted by human activities, leading to changes in various ecological systems. These activities include industrial processes, urbanization, agriculture, and waste mismanagement (Jasinavičiūtė & Veteikis, 2022). The origins of pollution can be traced back to ancient civilizations, where early urbanization and industrial activities began generating waste. For instance, cities like Babylon in ancient Mesopotamia and Mohenjo-Daro (circa 3000 B.C.) developed rudimentary sewage systems (Jansen, 1989), with wastewater from latrines being directed into nearby rivers.

The industrial revolution in the eighteenth century marked a pivotal shift, significantly accelerating pollution levels (Panchuk, 2021; Wuaten, 2022). Mass production powered by fossil fuels led to substantial air pollution from factories, while the growing need for agricultural productivity in the twentieth century introduced pesticides and fertilizers (Labaran et al., 2021), contributing to water pollution through runoff and leaching. Rapid urbanization and industrialization during this period also led to the dumping of solid, industrial, and domestic waste into water bodies (Flood et al., 2015; Bowman et al., 2018; Lin et al., 2022), severely impacting aquatic ecosystems and air quality. Developing countries, experiencing similar industrial booms, also saw declines in environmental quality.

T. Henriquez
Department of Medical Biotechnologies, University of Siena, Siena, Italy

M. Marvasi (✉)
Department of Biology, University of Florence, Sesto Fiorentino, Florence, Italy
e-mail: Massimiliano.marvasi@unifi.it

S. M. Tiquia-Arashiro et al. (eds.), *Urban Watershed Microbiology, Volume 1*,
https://doi.org/10.1007/978-3-031-89071-0_11

The increase in greenhouse gas emissions, particularly CO_2, began during the industrial revolution (Labaran et al., 2021; Hadipoor et al., 2021) which is often cited as the starting point of modern climate change (Ahmed et al., 2022). Since then, human activities have been the primary drivers of rising greenhouse gas levels, contributing to environmental degradation and public health concerns. In recent decades, studies have consistently linked pollution to adverse health effects, stretching back to the nineteenth century (Hanlon, 2015). Anthropogenic activities continue to shape the environment today, with microplastic contamination in oceans (Barrows et al., 2018), alterations in water quality (Oest et al., 2018; Patel et al., 2019; Rahman et al., 2020), and changes in the carbon and nitrogen flux in rivers (Tiquia, 2010a, b; Neu et al., 2023), and the accumulation of potentially toxic elements in groundwater (Cho et al., 2012; Sacchi et al., 2021; Marengo et al., 2022; Tiquia et al., 2008; Pagnucco et al., 2023). Atmospheric pollution, particularly from aerosols like particulate matter ($PM_{2.5}$) (Jin et al., 2023; Zhou et al., 2023), has global consequences, affecting ecosystems and human health (Ma et al., 2021). These environmental changes, driven by human activities, have far-reaching implications not only for conservation and environmental management (Kennish, 2021) but also for the spread of antimicrobial resistances (AMRs). Antimicrobial agents, which include antibiotics, antivirals, antifungals, and antiparasitics, are crucial in fighting infections. However, the overuse and misuse of these agents have led to the rise of multidrug-resistant organisms, posing a severe threat to public health (Ahmed et al., 2024).

Antimicrobial resistance primarily manifests through two key elements: antimicrobial-resistant bacteria (ARB) and antimicrobial-resistant genes (ARGs). ARB are bacteria that can withstand the effects of antibiotics, making infections more difficult to treat (Amarasiri et al., 2020), while ARG are the genetic components that confer this resistance and can be transferred between bacteria, facilitating the spread of resistance (Smith et al., 2015; Nassereddine et al., 2024). The proliferation of ARB and ARGs is driven by factors such as antibiotic use in healthcare, agriculture, and animal husbandry, as well as pollution. Their presence has been documented across diverse environments, from hospitals and farms to aquatic ecosystems and human microbiomes (Hu et al., 2013, 2014; Thamlikitkul et al., 2019; Heard Matthew et al., 2021).

Wastewater treatment plants and untreated veterinary healthcare waste have been identified as significant sources of ARB and ARGs pollution (Ahmed, 2013; Lupan et al., 2017). The rising concern over these elements, now considered contaminants of emerging concern (e-CEC), highlights the need to incorporate evolutionary perspectives into risk assessments (Vassallo et al., 2021). This is especially pertinent for practices like urban wastewater reuse in agriculture (Rizzo et al., 2020), where controlling the release of ARB and ARGs is essential for ensuring safe and sustainable food production.

The evolutionary dynamics of ARB and ARG are critical to understanding their persistence and spread in the environment (this important aspect is discussed at the end of this chapter). Addressing these challenges requires a multidisciplinary approach, with a particular focus on how pollution influences the evolution and

dissemination of AMR. By gaining a deeper understanding of these complex interactions, we can develop more effective strategies to mitigate the environmental and public health risks associated with antimicrobial resistance.

2 Antibiotic Resistance Development and Its Implications for Human Health

Antibiotic resistance occurs when bacteria evolve and become resistant to the drugs that are designed to kill them or inhibit their growth (Willey et al., 2011). The resistance can be simplified by comprising two primary elements: the antibiotic or antimicrobial medication, which by inhibiting the growth of the susceptible organisms, poses a selective pressure, favoring the selection of resistant ones. The second component is the genetic resistance factor within microorganisms which can arise through several mechanisms, including the acquisition of resistance genes from other bacteria, mutations in the DNA that prevent antibiotics from interacting with the target, and the formation of biofilms that protect bacteria from antibiotics (Willey et al., 2011). *Selective pressure* refers to environmental conditions that favor organisms with traits that enhance their survival, such as antibiotic resistance (Uddin et al., 2021). Pollutants can create selective pressure by killing or inhibiting susceptible bacteria, leaving resistant strains to thrive. This process drives the evolution and spread of resistance in bacterial populations (Uddin et al., 2021).

The identified resistance genes and their hosts proliferate and disseminate under ongoing antimicrobial pressure, exacerbating and spreading the issue to various hosts and geographical areas.

The collection of all antimicrobial resistance genes that is found in both pathogenic and non-pathogenic microbes is known as the *resistome* (Wright, 2010). This comprises resistance genes that are actively involved in resistance as well as those that have the capacity to induce resistance to antimicrobial agents. The genes present in bacteria, viruses, fungi, and parasites are included in the resistome. These genes comprise both expressed and mobilizable genes that may be transferred to other organisms.

The range of antimicrobial resistance genes present in the environment, including soil, water, air, and natural ecosystems, is referred to as the *environmental resistome* (Davies & Davies, 2010). These genes might multiply in hosts residing in anthropogenic sources such as industrial pollution, wastewater, and agricultural runoff, or they could come from natural microbial populations (Bobate et al., 2023). The environmental resistome plays a critical role in the evolution and spread of resistance, as it acts as a reservoir from which resistance genes can transfer to pathogenic organisms, contributing to the global AMR crisis (Bobate et al., 2023).

Globally, millions of kilograms of antimicrobials are utilized annually (Levy & Marshall, 2004) as prophylaxis and treatment of human, animal, and agricultural

food and feed, intensifying the resistance dilemma by eliminating susceptible strains and favoring the survival of resistant ones (Levy & Marshall, 2004).

Bacteria possess remarkable genetic adaptability enabling them to replicate themselves during various environmental challenges, including the presence of antibiotics that threaten their survival. Bacteria inhabiting the same ecological niche as antimicrobial-producing organisms have developed ancient defense mechanisms to withstand the impact of these molecules, thereby enabling their thriving in their presence (Dantas et al., 2008). The acquisition of foreign DNA material through horizontal gene transfer stands as a pivotal driver of bacterial evolution and is frequently accountable for the emergence of antimicrobial resistance (Munita & Arias, 2016). Most antimicrobial agents used in clinical settings originate from natural sources, primarily soil. Bacteria coexisting with these molecules harbor intrinsic genetic resistance determinants, constituting a part of the environmental resistome and helping in the acquisition of antibiotic resistance genes by clinically significant bacteria. Moreover, this genetic exchange could be implicated in disseminating resistance to commonly used antibiotics (Dantas et al., 2008; Nassereddine et al., 2024).

Bacteria assimilate external genetic material via four principal strategies: transformation (incorporation of naked DNA), transduction (phage-mediated transfer), conjugation (bacterial "mating"), and membrane vesicles. While transformation is a straightforward form of horizontal gene transfer, only a limited number of bacterial species can naturally integrate exogenous DNA to develop resistance (Johnsborg & Håvarstein, 2009). Resistance emergence in hospital environments often involves conjugation, an efficient gene transfer mechanism occurring through cell-to-cell contact, particularly prevalent in the gastrointestinal tract of humans undergoing antibiotic treatment (Schwarzenlander & Averhoff, 2006). Generally, conjugation employs mobile genetic elements as carriers to disseminate valuable genetic information, although direct chromosome transfer is also documented (Wang et al., 2005). Key mobile genetics elements include plasmids and transposons, pivotal in facilitating antimicrobial resistance development and spread among clinically significant organisms (Reid et al., 2015; Feßler et al., 2018; Jiang et al., 2020; Nassereddine et al., 2024).

Bacteria possess remarkable capability to respond to anthropogenic antibiotic pressure, such as the responding to overused antibiotics in prescribing practices in healthcare settings, the overuse of antibiotics in livestock and aquaculture, and the ease of access to antibiotics without a prescription. This has led to the widespread occurrence of antibiotic-resistant strains, not only in hospitals but also in communities worldwide. These resistant strains are causing an increase in the prevalence of infections that are difficult, and sometimes impossible, to treat effectively. Infections such as tuberculosis, gonorrhea, and foodborne illnesses, to cite some examples, have become more challenging to manage due to the decreasing effectiveness of antibiotics (Barkan et al., 2009; Colavecchio et al., 2017; Tao et al., 2022). Without prompt and effective action, we are rapidly approaching a *post-antibiotic era* where common infections and minor injuries can become life-threatening once again. The consequences of antibiotic resistance are multifaceted and far-reaching. In terms of

human health, the impact of antibiotic resistance is profound. It leads to longer and more severe illnesses, increased mortality rates, and higher healthcare costs. Additionally, antibiotic resistance undermines the effectiveness of medical procedures that rely on the ability to prevent or treat infections, such as surgeries, chemotherapy, and organ transplantation.

In general, human behaviors such as improper disposal of antibiotics and inadequate sanitation infrastructure significantly contribute to the exacerbation of antibiotic resistance (Anwar et al., 2020). Improper disposal of antibiotics introduces sub-therapeutic levels of these drugs into the environment, creating selective pressure that promotes the development and spread of AMR genes among bacterial populations (Anwar et al., 2020). Inadequate sanitation infrastructure leads to the contamination of water and soil with human and animal waste containing ARB and ARG, facilitating their dissemination in the environment. Additionally, the excessive use of antibiotics in agriculture and animal husbandry introduces resistant bacteria into soil and water systems (Heuer et al., 2011), further increasing the risk of resistance gene transfer. A *lack of public awareness* and *weak regulatory controls* exacerbate these issues by allowing for the continued misuse and improper management of antibiotics, contributing to the global rise of antimicrobial resistance.

3 General Impact of Pollution on Antibiotic Resistance

While the use of antibiotics is the main factor driving antibiotic resistance, environmental pollution can also significantly contribute to the development and spread of ARB and ARG on a global scale.

Polluted *water bodies*, for example, that could be charged with residues from pharmaceutical manufacturing, agriculture, healthcare facilities, and/or community sewage, serve as reservoirs for ARB (Singh et al., 2022; Tiquia-Arashiro & Pant, 2020). Wastewater treatment plants have been identified as hotspots for the exchange of mobile genetic elements, allowing ARGs to mix between pathogens, opportunistic pathogens, and environmental bacteria (Chen et al., 2017). Current treatments in wastewater treatment plants are ineffective in removing bacteria that could contain ARB and ARGs (Matesun et al., 2024), causing the release of large quantities of ARB and ARGs into the aquatic environment via the effluent of the facilities (Singh et al., 2022; Su et al., 2023). Studies have demonstrated that, although up to 99.9% of antibiotic-resistant *Escherichia coli* and related genes are removed in wastewater treatment plants, they are still detectable in the final effluent (Matesun et al., 2024). This suggests that the mechanical and biological treatment processes applied to *E. coli* have varying degrees of effectiveness (Raboni et al., 2016; Le et al., 2023; Matesun et al., 2024). In other scenarios, the pollutants in the environment, including polycyclic aromatic hydrocarbons (PAHs), may compete with antibiotics, such as tetracycline (Maruzani et al., 2018). Recent studies have shown that the concentration of bioavailable PAHs has a stronger influence on the relative abundance of tetracycline resistance genes when compared with the total concentration of

tetracycline (Sun et al., 2015; Maruzani et al., 2018). In this context, it is important to understand the whole process and to develop management strategies to control the dissemination of ARGs.

Soil is considered a fertile reservoir of ARGs due to its intricate microbial community and the diverse array of antibiotic-producing microorganisms (Su et al., 2014). In this context, the contamination of soil with agricultural runoff containing antibiotics and pesticides or animal wastes can significantly increase its potential as a reservoir of ARGs (Manyi-Loh et al., 2018; Nguyen et al., 2013; Tiquia et al., 1998). This process can lead to the proliferation of antibiotic-resistant bacteria in the soil microbiome, potentially impacting plant health, food safety, and even spreading resistance through the food chain (Manyi-Loh et al., 2018).

Airborne pollutants, including particulate matter and chemicals from industrial processes and vehicle emissions, can be deposited onto soil and water bodies. These pollutants may contribute to the selection pressure favoring antibiotic-resistant bacteria and genes. For example, it was recently described that the levels of particulate matter $PM_{2.5}$ and antibiotic resistances were significantly correlated in airborne samples (Zhou et al., 2023). According to Zhou and colleagues, globally, a 10% increase in annual $PM_{2.5}$ could lead to about 1% increase in aggregate antibiotic resistance and thousands of premature deaths linked to antibiotic resistance (Zhou et al., 2023).

Pollution knows no boundaries and it can travel across continents through air and water currents. Consequently, antibiotic-resistant bacteria and resistance genes can disseminate globally, posing further challenges.

4 How Do Pollutants Contribute to the Horizontal Transfer of Genes?

Pollutants can promote horizontal gene transfer (HGT) by creating stressful environments that encourage bacteria to exchange genetic material, including antibiotic resistance genes (Sobecky & Coombs, 2009; Chen et al., 2023). Certain pollutants, such as heavy metals, antibiotics, and other chemical contaminants, can induce oxidative stress, DNA damage, and increased mutation rates in bacteria, all of which can enhance HGT. These pollutants also promote biofilm formation, which facilitates close contact between bacteria, making gene transfer more likely (Michaelis & Grohmann, 2023). Additionally, pollutants can disrupt bacterial communities and ecosystems, further accelerating the spread of resistance genes through various HGT mechanisms like transformation, transduction, and conjugation. HGT is one of the main mechanisms for the dissemination of ARG (Vinayamohan et al., 2022).

The HGT between *Escherichia coli* and susceptible strains (not resistant to antibiotics) has been observed in agricultural runoff and animal feces, emphasizing the role of anthropogenic activities in disseminating antibiotic resistances (Nagachinta & Chen, 2008). The presence of pollutants in water, such as disinfectants,

nanomaterials, and organic contaminants, has been shown to promote the horizontal transfer of ARGs among bacterial communities (Qiu et al., 2012; Karkman et al., 2018; Jin et al., 2020). For instance, chlorine disinfection has been found to enrich ARGs in bacteria and promote their transmission in water, posing a potential risk to public health (Jin et al., 2020). This is because chemicals like chlorine, quaternary ammonium compounds, or other disinfectants can promote resistance by selecting bacteria that also have mechanisms to expel antibiotics (co-selection) (Wales & Davies, 2015; Vinayamohan et al., 2022). In a similar way, pesticides and herbicides used in agriculture, such as glyphosate, can alter gene abundance in microbial communities and promote the survival of resistant bacteria even in non-antibiotic environments (Liao et al., 2021). Additionally, nanomaterials in water have been demonstrated to significantly promote the horizontal conjugative transfer of multidrug-resistant genes mediated by plasmids (Qiu et al., 2012). Moreover, changes in water quality and point-source pollution in urban rivers have been identified as factors that may lead to the spread of ARGs, highlighting the need for greater attention to control the transfer of ARGs between environmental bacteria and pathogens (Zhou et al., 2017). The dissemination of ARGs in water environments is primarily facilitated by horizontal gene transfer via mobile genetic elements, including plasmids, transposons, and integrons (Chen et al., 2017; Fernanda et al., 2022; Nassereddine et al., 2024). These genetic elements play a crucial role in the transfer of ARGs between bacteria, contributing to the spread of antibiotic resistance in water systems (Nassereddine et al., 2024).

Furthermore, the selective pressure caused by antibiotics associated with heavy metals in wastewater has been identified as an inducer for the existence of ARGs in potential pathogens, further emphasizing the role of environmental factors in promoting the transfer of ARGs (Marathe et al., 2016; Tangwa et al., 2019; Zhao et al., 2021; Bai et al., 2022; Nassereddine et al., 2024). Studies have shown that poorly treated sewage can facilitate the transfer of resistant genes to human pathogens, contributing to the dissemination of antibiotic-resistant bacteria (Kunhikannan et al., 2021). Additionally, low-intensity ultrasound and the presence of catabolic genes for pollutant degradation, have been implicated in promoting mutation accumulation or horizontal transfer of antibiotic-resistant genes between bacteria, further contributing to the spread of antibiotic resistance (Wang et al., 2014; Song et al., 2018).

Additionally, the presence of antibiotic-resistant genes in bacteriophages and their potential role in transduction further highlights the significance of horizontal gene transfer in the dissemination of ARGs (Han et al., 2016; Karkman et al., 2019). Furthermore, reclaimed water irrigation has been shown to impact the soil resistome, indicating the potential for the horizontal gene transfer to influence antibiotic resistance in agricultural environments (Thakali et al., 2021; Nassereddine et al., 2024). The abundance of ARGs in aquatic environments has been linked to fecal pollution, and the correlation between fecal pollution and ARG abundance suggests the role of horizontal gene transfer in the dissemination of antibiotic resistance. Overall, horizontal gene transfer in water environments serves as a mechanism for the spread of antibiotic resistance, *emphasizing the need for effective strategies to mitigate the*

transfer of ARGs and control the environmental dissemination of antibiotic resistance. In conclusion, the transfer of ARGs between bacteria in water environments is influenced by a multitude of factors, including the presence of pollutants, selective pressures, and the role of mobile genetic elements. Understanding these dynamics is crucial for developing effective strategies to control pollution, mitigating the spread of ARG and preserving the utility of antibiotics.

5 Current Policies Addressing Pollution and ARGs/ARB

It is imperative to establish policies and strategies aimed at tackling pollution and antimicrobial resistance across various regions to safeguard both human health and the environment. These initiatives encompass a range of measures, such as regulating industrial emissions, advocating for renewable energy adoption, enhancing waste management protocols, endorsing sustainable farming techniques, and instituting antimicrobial stewardship programs to curb unnecessary antibiotic usage in healthcare settings. Effective implementation and enforcement of these policies require collaborative efforts among governments, healthcare providers, and regulatory agencies.

According to research conducted by the UN Environment (Gaze & Depledge, 2017), the escalating threat of antimicrobial resistance, attributed to the release of drugs and specific chemicals into the environment, stands out as one of the most alarming health concerns of our time. Recognized as a priority at the World Health Assembly since 1998 (WHO, 1998), AMR has gained increasing attention over the years, leading to a series of resolutions on the topic. The culmination of these efforts was witnessed at the Sixty-eighth World Health Assembly in May 2015, where a global action plan to combat antimicrobial resistance as the most pressing trend was endorsed (WHO, 2015).

Although 194 countries committed to generating multisectoral national action plans by the 2017 World Health Assembly to overcome antimicrobial resistance (WHO, 2015; Patel et al., 2023), only a fraction of them had such a plan by 2022 (Patel et al., 2023). In this context, Patel and colleagues described that the efforts to control antimicrobial resistance vary considerably between countries and regions (Patel et al., 2023). This could be partially due to differences in resources, infrastructure, and governance. For example, in their study, Norway had the highest governance score, while the Federated States of Micronesia had the lowest governance score (Patel et al., 2023). Of the 20 nations with the highest overall governance scores, 17 of them belonged to the high-income group of countries and 11 were from the European zone (Patel et al., 2023). These results highlight the need for international cooperation to assist less-resourced countries by providing financial aid, technology transfer, capacity building, and global policy frameworks. Collaborative efforts are essential for addressing these global challenges effectively.

More recently, the Thirteenth General Programme of Work (2019–2023) by WHO underscored the imperative to confront this emerging threat, specifically

highlighting the need to address waste and wastewater management from pharmaceutical production under the section titled "Tackling antimicrobial resistance" (WHO, 2019). Notably, it was only in the recent past that the necessity to tackle waste and wastewater management from the production of critically important antimicrobials was explicitly acknowledged. This was evidenced by the decision made during the World Health Organization's Executive Board meeting on November 30, 2018, wherein it was decided to provide technical input for good manufacturing practices (GMP) on waste and wastewater management (WHO, 2016, 2018).

These efforts seek to diminish pollution levels and are linked to the proliferation of antimicrobial resistance, fostering a sustainable future for forthcoming generations. Currently strategies and policies to address these challenges include the following: (1) Enforcing stricter regulations on industrial emissions to mitigate air and water pollution. (2) Encouraging the adoption of renewable energy sources to mitigate carbon emissions and combat climate change. (3) Enhancing waste management systems to minimize pollution from improper disposal and promote recycling while ensuring proper treatment of hazardous materials. (4) Promoting sustainable agricultural practices to diminish pollution arising from pesticide and fertilizer usage and to bolster biodiversity. (5) Implementing antimicrobial stewardship programs within healthcare settings to foster judicious antibiotic usage and prevent the emergence of antimicrobial resistance. In this context, alternative treatments to antibiotics, such as phage therapy, probiotics, and vaccination strategies, are gaining attention in both human and veterinary medicine as a way to combat AMR (Gupta & Sharma, 2022). Phage therapy uses bacteriophages—viruses that specifically target bacteria—to kill resistant strains, offering a precise approach that minimizes harm to beneficial microbes. Probiotics, which promote healthy microbiota, can enhance immunity and outcompete harmful bacteria, reducing the need for antibiotics. Vaccination strategies also provide a preventative approach by strengthening the immune system, lowering infection rates, and curbing the overuse of antibiotics. These alternatives represent promising avenues to reduce reliance on traditional antibiotics and mitigate the growing threat of AMR (Gupta & Sharma, 2022).

Emerging technologies, particularly artificial intelligence (AI), hold significant potential in monitoring and addressing pollution-related antimicrobial resistance (AMR) (Ali et al., 2023). AI can process vast amounts of environmental data to identify patterns of pollution that contribute to AMR, such as the presence of antibiotic residues in water sources or soil (Iftikhar et al., 2023). Machine learning algorithms can enhance predictive models to assess future AMR risks and guide targeted interventions (Ali et al., 2023). Additionally, AI-driven bio-surveillance systems could monitor microbial mutations in real time (Alsulimani et al., 2024), allowing for faster detection of resistance hotspots and informing policies to mitigate pollution's role in the spread of AMR. Therefore, AI's ability to analyze environmental data and predict AMR risks highlights its vital role in combating pollution-driven resistance, offering promising avenues for more proactive and informed solutions in the fight against AMR.

6 Integrating Evolutionary Perspectives into Risk Assessment: A Necessary Step?

As seen in this chapter, the anthropogenic environments, the selective pressure exerted by them, and the antimicrobial resistance dynamics are strictly connected in an evolving path. A comprehensive risk evaluation of ARG/ARB should include both the "evolution component" in addition to the "dose-response factor." Chemo-centric environmental risk assessment involves the identification of hazard(s) in a dose-response manner in animal or in human models. However, it should be noted that the dose-response of a chemical cannot be used in the case of ARG/ARB evolution. Both dose-response and transmission route assessments should be merged to provide an immediate understanding of the risk associated with the presence of ARGs/ARB in polluted environments. The evolutionary component, on the other hand, offers insights into the long-term trajectory, potential evolutionary pathways, and variations due to environmental changes. We have recently provided a number of experimental procedures to implement evolution in risk assessment (Vassallo et al., 2021). The experiments and the models proposed certainly will require adaptation to accurately reflect real environmental conditions. It is unlikely that all the evolutionary tools proposed by Vassallo et al. (2021) can be fully integrated; however, incorporating a few of the experimental evidence will enhance the identification of the evolutionary trajectory of ARGs/ARB in real environments under anthropogenic selective pressure. As our understanding improves, new algorithms can be developed and integrated to better evaluate the contribution of all factors to overall risk.

How can these concepts be practically integrated into risk assessment? A conceptual framework for risk assessments of antibiotic resistance and anthropogenic environments, by incorporating the new evolutionary component, is proposed in Fig. 11.1, adapted from Ben and colleagues (2019).

As microorganisms adapt to specific polluted environments, their evolution influences the long-term selection and persistence of ARB and ARG. It would be important to classify ARG and ARB as evolving anthropogenic pollutants. When these pollutants are released into the environment, they have the potential to give rise to new, potentially hazardous, and selection-responsive populations. Therefore, it is essential to recognize these evolving traits and incorporate evolutionary principles into modern risk assessments. This should include implementing evolutionary fitness tests to evaluate the resilience and adaptability of ARB.

7 Conclusion

The study of links between anthropogenic environments and antimicrobial resistance dynamics reveals a complex interplay between human activities and the emergence and spread of antimicrobial resistances: the clinical related selective pressure

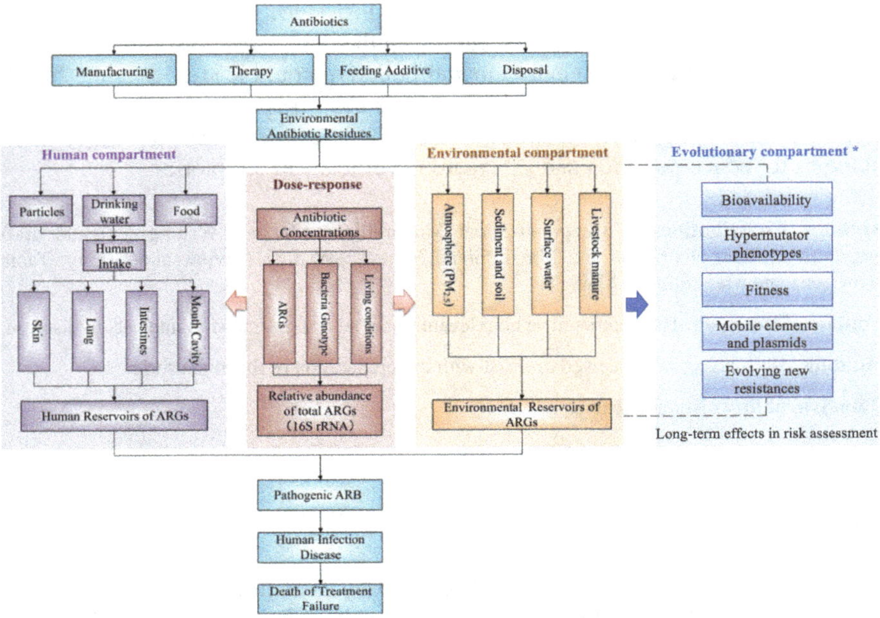

Fig. 11.1 Conceptual framework for human health risk assessment of antibiotic resistance, featuring the new "Evolutionary Compartment" (depicted on the right in blue). This component forms connections with the "Environmental Component" and influences the "Human Compartment," thereby shaping the overall system's fate. The "Evolutionary Component" facilitates assessment of microbial evolution in specific environments, which helps evaluate long-term impacts on ARB and ARGs selection and persistence. (Adapted from Ben et al., 2019; Vassallo et al., 2021). ARGs and ARB, much like other evolving entities such as viruses, bacteria, or genes, can evade environmental controls. When this occurs, even a single remaining molecule at a control point can become the seed of a new, potentially hazardous and evolutionarily responsive population. Consequently, it may be time to recognize this characteristic and integrate evolutionary principles into the current framework for pollutant risk assessment. Terms used in the blue section (deeply explained by Vassallo et al. (2021)): *Bioavailability*: Fraction of a chemical that reaches the target or induces a selective pressure. *Hypermutator phenotypes*: Microorganisms with clearly distinct mutation frequencies. *Distribution of fitness effects (DFE) and fitness*: Measuring the diverse subpopulation of ARG/ARB to co-exists and fluctuate. *Mobile elements and plasmids*: Measuring the capacity of strains to generate resistance phenotypes. *Evolving resistance to newest antimicrobial*: Measuring the capacity of strains to generate resistance phenotypes to new antibiotics in the system

is not the only concern. Mitigation of persistence and spread of ARGs and ARB goes through various pathways such as agricultural practices, wastewater disposal, and antibiotic usage in healthcare. Moreover, anthropogenic environments serve as reservoirs for resistant bacteria and resistance genes, facilitating their dissemination into natural ecosystems and potentially impacting human and animal health. Understanding these links is crucial for implementing effective strategies to mitigate the spread of antimicrobial resistance, including prudent antibiotic use, improved sanitation practices, and environmental management strategies.

The link between pollution and antibiotic resistance is a pressing issue that demands immediate attention. If we fail to act, the unchecked spread of antibiotic-resistant bacteria fueled by environmental pollution could lead to a public health crisis where common infections become untreatable. It is crucial that we invest in cleaner practices, tighter regulations, better risk assessment, and innovative technologies to reduce pollution and safeguard the efficacy of antibiotics.

Authors' Contributions Conceptualization: Massimiliano Marvasi; Writing—original draft preparation: Tania Henriquez and Massimiliano Marvasi; Writing—review and editing: Tania Henriquez and Massimiliano Marvasi.

Conflict of Interest The authors have no relevant financial or non-financial interests to disclose.

Funding No funding was received to assist with the preparation of this manuscript.

Data Availability Not applicable.

References

Ahmed, S. (2013). The Central Cattle Breeding and Dairy Farm, Bangladesh waste contributes in emergence and spread of aminoglycoside-resistant bacteria. *Advances in Bioscience and Biotechnology, 04*, 278–282.

Ahmed, F., Ali, I., Kousar, S., & Ahmed, S. (2022). The environmental impact of industrialization and foreign direct investment: Empirical evidence from Asia-Pacific region. *Environmental Science and Pollution Research International, 29*(20), 29778–29792.

Ahmed, S. K., Hussein, S., Qurbani, K., Ibrahim, R. H., Fareeq, A., Mahmood, K. A., & Mohamed, M. G. (2024). Antimicrobial resistance: Impacts, challenges, and future prospects. *Journal of Medicine, Surgery, and Public Health, 2*, 100081.

Ali, T., Ahmed, S., & Aslam, M. (2023). Artificial intelligence for antimicrobial resistance prediction: Challenges and opportunities towards practical implementation. *Antibiotics (Basel), 12*(3).

Alsulimani, A., Akhter, N., Jameela, F., Ashgar, R. I., Jawed, A., Hassani, M. A., & Dar, S. A. (2024). The impact of artificial intelligence on microbial diagnosis. *Microorganisms, 12*(6).

Amarasiri, M., Sano, D., & Suzuki, S. (2020). Understanding human health risks caused by antibiotic resistant bacteria (ARB) and antibiotic resistance genes (ARG) in water environments: Current knowledge and questions to be answered. *Critical Reviews in Environmental Science and Technology, 50*(19), 2016–2059.

Anwar, M., Iqbal, Q., & Saleem, F. (2020). Improper disposal of unused antibiotics: An often overlooked driver of antimicrobial resistance. *Expert Review of Anti-Infective Therapy, 18*(8), 697–699.

Bai, Y., Ruan, X., Li, R., Zhang, Y., & Wang, Z. (2022). Metagenomics-based antibiotic resistance genes diversity and prevalence risk revealed by pathogenic bacterial host in Taihu Lake, China. *Environmental Geochemistry and Health, 44*(8), 2531–2543.

Barkan, D., Liu, Z., Sacchettini, J. C., & Glickman, M. S. (2009). Mycolic acid cyclopropanation is essential for viability, drug resistance, and cell wall integrity of Mycobacterium tuberculosis. *Chemistry & Biology, 16*(5), 499–509.

Barrows, A. P. W., Cathey, S. E., & Petersen, C. W. (2018). Marine environment microfiber contamination: Global patterns and the diversity of microparticle origins. *Environmental Pollution, 237*, 275–284.

Ben, Y., Fu, C., Hu, M., Liu, L., Wong, M. H., & Zheng, C. (2019). Human health risk assessment of antibiotic resistance associated with antibiotic residues in the environment: A review. *Environmental Research, 169*, 483–493.

Bobate, S., Mahalle, S., Dafale, N. A., & Bajaj, A. (2023). Emergence of environmental antibiotic resistance: Mechanism, monitoring and management. *Environmental Advances, 13*, 100409.

Bowman, N., Patel, P., Xu, W., Alsaffar, A., & Tiquia-Arashiro, S. M. (2018). Enrichment and isolation of Pb-resistant bacteria from Saint Clair River sediments and their potential for Pb removal in aqueous solutions. *Applied Microbiology and Biotechnology, 102*, 2391–2398. https://doi.org/10.1007/s00253-018-8772-4

Chen, B., He, R., Yuan, K., Chen, E., Lin, L., Chen, X., Sha, S., Zhong, J., Lin, L., Yang, L., Yang, Y., Wang, X., Zou, S., & Luan, T. (2017). Polycyclic aromatic hydrocarbons (PAHs) enriching antibiotic resistance genes (ARGs) in the soils. *Environmental Pollution, 220*, 1005–1013.

Chen, X., Du, Z., Song, X., Wang, L., Wei, Z., Jia, L., & Zhao, R. (2023). Evaluating the occurrence frequency of horizontal gene transfer induced by different degrees of heavy metal stress. *Journal of Cleaner Production, 382*, 135371.

Cho, K., Zholi, A., Frabutt, D., Flood, M., Floyd, D., & Tiquia, S. M. (2012). Linking bacterial diversity and geochemistry of uranium-contaminated groundwater. *Environmental Technology, 33*, 1629–1640. https://doi.org/10.1080/09593330.2011.641036

Colavecchio, A., Cadieux, B., Lo, A., & Goodridge, L. D. (2017). Bacteriophages contribute to the spread of antibiotic resistance genes among foodborne pathogens of the Enterobacteriaceae family – A review. *Frontiers in Microbiology, 8*, 1108.

Dantas, G., Sommer, M. O., Oluwasegun, R. D., & Church, G. M. (2008). Bacteria subsisting on antibiotics. *Science, 320*(5872), 100–103.

Davies, J., & Davies, D. (2010). Origins and evolution of antibiotic resistance. *Microbiology and Molecular Biology Reviews, 74*(3), 417–433.

Fernanda, P. A., Liu, S., Yuan, T., Ramalingam, B., Lu, J., & Sekar, R. (2022). Diversity and abundance of antibiotic resistance genes and their relationship with nutrients and land use of the inflow rivers of Taihu Lake. *Frontiers in Microbiology, 13*, 1009297.

Feßler, A., Kadlec, K., Wang, Y., Zhang, W.-J., Wu, C., Shen, J., & Schwarz, S. (2018). Small antimicrobial resistance plasmids in livestock-associated methicillin-resistant *Staphylococcus aureus* CC398. *Frontiers in Microbiology, 9*, 2063.

Flood, M., Frabutt, D., Floyd, D., Powers, A., Ezegwe, U., Devol, A., & Tiquia-Arashiro, S. M. (2015). Ammonia-oxidizing bacteria and archaea in sediments of the Gulf of Mexico. *Environmental Technology, 36*, 124–135. https://doi.org/10.1080/09593330.2014.942385

Gaze, W., & Depledge, M. (2017). *Chapter 1. Antimicrobial resistance: Investigating the environmental dimension – frontiers 2017: Emerging issues of environmental concern* (pp. 12–22). United Nations Environment Programme 2017. https://wedocs.unep.org/handle/20.500.11822/22263?show=full

Gupta, R., & Sharma, S. (2022). Role of alternatives to antibiotics in mitigating the antimicrobial resistance crisis. *The Indian Journal of Medical Research, 156*(3), 464–477.

Hadipoor, M., Keivanimehr, F., Baghban, A., Ganjali, M. R., & Habibzadeh, S. (2021). Chapter 24 – Carbon dioxide as a main source of air pollution: Prospective and current trends to control. In A. Núñez-Delgado (Ed.), *Sorbents materials for controlling environmental pollution* (pp. 623–688). Elsevier.

Han, X.-M., Hu, H.-W., Shi, X.-Z., Wang, J.-T., Han, L.-L., Chen, D., & He, J.-Z. (2016). Impacts of reclaimed water irrigation on soil antibiotic resistome in urban parks of Victoria, Australia. *Environmental Pollution, 211*, 48–57.

Hanlon, W. W. (2015). *Pollution and mortality in the 19th century*. National Bureau of Economic Research. Working Paper 21647. https://doi.org/10.3386/w21647

Heard Matthew, J., Barton Christopher, E., Frost Victoria, J., & Hongo, R. (2021). Sourcing antibiotic-resistant *Escherichia coli* in aquatic ecosystems: A combined laboratory and field module. *Journal of Microbiology & Biology Education, 22*(3), e00207–e00221.

Heuer, H., Schmitt, H., & Smalla, K. (2011). Antibiotic resistance gene spread due to manure application on agricultural fields. *Current Opinion in Microbiology, 14*(3), 236–243.

Hu, Y., Yang, X., Qin, J., Lu, N., Cheng, G., Wu, N., Pan, Y., Li, J., Zhu, L., Wang, X., Meng, Z., Zhao, F., Liu, D., Ma, J., Qin, N., Xiang, C., Xiao, Y., Li, L., Yang, H., Wang, J., Yang, R., Gao, G. F., Wang, J., & Zhu, B. (2013). Metagenome-wide analysis of antibiotic resistance genes in a large cohort of human gut microbiota. *Nature Communications, 4*(1), 2151.

Hu, Y., Yang, X., Lu, N., & Zhu, B. (2014). The abundance of antibiotic resistance genes in human guts has correlation to the consumption of antibiotics in animal. *Gut Microbes, 5*(2), 245–249.

Iftikhar, S., Karim, A. M., Karim, M. A., Aslam, M. Rubab, F., Malik, S. K., Kwon, J. E., Hussain, I., Azhar, E. I., Kang, S. C., & Yasir, M. (2023). Prediction and interpretation of antibiotic-resistance genes occurrence at recreational beaches using machine learning models. *Journal of Environmental Management, 328*, 116969.

Jansen, M. (1989). Water supply and sewage disposal at Mohenjo-Daro. *World Archaeology, 21*(2), 177–192.

Jasinavičiūtė, A., & Veteikis, D. (2022). Assessing landscape instability through land-cover change based on the hemeroby index (Lithuanian example). *Land, 11*(7), 1056.

Jiang, Y., Wang, Y., Hua, X., Qu, Y., Peleg, A. Y., & Yu, Y. (2020). Pooled plasmid sequencing reveals the relationship between mobile genetic elements and antimicrobial resistance genes in clinically isolated *Klebsiella pneumoniae. Genomics, Proteomics & Bioinformatics, 18*(5), 539–548.

Jin, M., Liu, L., Wang, D. N., Yang, D., Liu, W. I., Yin, J., Yang, Z. W., Wang, H. R., Qiu, Z. G., Shen, Z. Q., Shi, D. Y., Li, H. B., Guo, J. H., & Li, J. W. (2020). Chlorine disinfection promotes the exchange of antibiotic resistance genes across bacterial genera by natural transformation. *The ISME Journal, 14*(7), 1847–1856.

Jin, S., Ma, Y., Huang, Z., Huang, J., Gong, W., Liu, B., Wang, W., Fan, R., & Li, H. (2023). A comprehensive reappraisal of long-term aerosol characteristics, trends, and variability in Asia. *Atmospheric Chemistry and Physics, 23*(14), 8187–8210.

Johnsborg, O., & Håvarstein, L. S. (2009). Regulation of natural genetic transformation and acquisition of transforming DNA in *Streptococcus pneumoniae. FEMS Microbiology Reviews, 33*(3), 627–642.

Karkman, A., Do, T. T., Walsh, F., & Virta, M. P. J. (2018). Antibiotic-resistance genes in waste water. *Trends in Microbiology, 26*(3), 220–228.

Karkman, A., Pärnänen, K., & Larsson, D. G. J. (2019). Fecal pollution can explain antibiotic resistance gene abundances in anthropogenically impacted environments. *Nature Communications, 10*(1), 80.

Kennish, M. (2021). Drivers of change in estuarine and coastal marine environments: An overview. *Open Journal of Ecology, 11*, 224–239.

Kunhikannan, S., Thomas, C. J., Franks, A. E., Mahadevaiah, S., Kumar, S., & Petrovski, S. (2021). Environmental hotspots for antibiotic resistance genes. *MicrobiologyOpen, 10*(3), e1197.

Labaran, Y. H., Mathur, V. S., & Farouq, M. M. (2021). The carbon footprint of construction industry: A review of direct and indirect emission. *Journal of Sustainable Construction Materials and Technologies, 6*, 101–115.

Le, T. H., Tong, D. K., Pham, T. T. V., Hoang, T. L., Tran, T. K., & Ng, C. (2023). Occurrence and removal of antibiotic resistant Escherichia coli and antibiotic resistance genes in an urban wastewater treatment plant. *International Journal of Environmental Science and Technology, 20*(12), 13739–13752.

Levy, S. B., & Marshall, B. (2004). Antibacterial resistance worldwide: Causes, challenges and responses. *Nature Medicine, 10*(12 Suppl), S122–S129.

Liao, H., Li, X., Yang, Q., Bai, Y., Cui, P., Wen, C., Liu, C., Chen, Z., Tang, J., Che, J., Yu, Z., Geisen, S., Zhou, S., Friman, V. P., & Zhu, Y. G. (2021). Herbicide selection promotes antibiotic resistance in soil microbiomes. *Molecular Biology and Evolution, 38*(6), 2337–2350.

Lin, L., Yang, H., & Xu, X. (2022). Effects of water pollution on human health and disease heterogeneity: A review. *Frontiers in Environmental Science, 10*.

Lupan, I., Carpa, R., Oltean, A., Kelemen, B. S., & Popescu, O. (2017). Release of antibiotic resistant bacteria by a waste treatment plant from Romania. *Microbes and Environments, 32*(3), 219–225.

Ma, R., Li, L., & Zhang, B. (2021). Impact assessment of anthropogenic activities on the ecological systems in the Xiongan New Area in the North China Plain. *Integrated Environmental Assessment and Management, 17*(4), 866–876.

Manyi-Loh, C., Mamphweli, S., Meyer, E., & Okoh, A. (2018). Antibiotic use in agriculture and its consequential resistance in environmental sources: Potential public health implications. *Molecules, 23*(4).

Marathe, N. P., Gaikwad, S. S., Vaishampayan, A. A., Rasane, M. H., Shouche, Y. S., & Gade, W. N. (2016). Mossambicus tilapia (Oreochromis mossambicus) collected from water bodies impacted by urban waste carries extended-spectrum beta-lactamases and integron-bearing gut bacteria. *Journal of Biosciences, 41*(3), 341–346.

Marengo, M., Fullgrabe, L., Fontaine, Q., Boissery, P., Cancemi, M., Lejeune, P., & Gobert, S. (2022). Ecological and human health risk assessment of potentially toxic element contamination in waters of a former asbestos mine (Canari, Mediterranean Sea): Implications for management. *Environmental Monitoring and Assessment, 195*(1), 150.

Maruzani, R., Canali, A., Serafim, V., Munoz, L. P., Shah, A. J., Perito, B., & Marvasi, M. (2018). Effect of anthropogenic pollution on the fitness of tetracycline sensitive *Shigella flexneri* in Thames river water. *Journal of Environmental Chemical Engineering, 6*(1), 19–27.

Matesun, J., Petrik, L., Musvoto, E., Ayinde, W., & Ikumi, D. (2024). Limitations of wastewater treatment plants in removing trace anthropogenic biomarkers and future directions: A review. *Ecotoxicology and Environmental Safety, 281*, 116610.

Michaelis, C., & Grohmann, E. (2023, February 4). Horizontal gene transfer of antibiotic resistance genes in biofilms. *Antibiotics (Basel, Switzerland), 12*(2), 328. https://doi.org/10.3390/antibiotics12020328

Munita, J. M., & Arias, C. A. (2016). Mechanisms of antibiotic resistance. *Microbiology Spectrum, 4*(2).

Nagachinta, S., & Chen, J. (2008). Transfer of class 1 integron-mediated antibiotic resistance genes from Shiga toxin-producing Escherichia coli to a susceptible *E. coli* K-12 strain in storm water and bovine feces. *Applied and Environmental Microbiology, 74*(16), 5063–5067.

Nassereddine, Z., Opara, S., Coutinho, O., Qyteti, F., Book, R., Heinicke, M. P., Napieralski, J., & Tiquia-Arashiro, S. M. (2024, December 20). Critical perspectives on advancing antibiotic resistant gene (ARG) detection technologies in aquatic ecosystems. *Science of the Total Environment, 957*, 177775. https://doi.org/10.1016/j.scitotenv.2024.177775

Neu, V., Araújo, M. G. da S., Guedes, V. M., Ward, N. D., Ribeiro, M. M., Richey, J. E., & Krusche, A. V. (2023). Composition and flux of dissolved and particulate carbon and nitrogen in the lower tocantins river. *Journal of Geophysical Research: Biogeosciences, 128*(6), e2022JG006846.

Nguyen, S., Ala, F., Cardwell, C., Cai, D., McKindles, K. M., Lotvola, A., Hodges, S., Deng, Y., & Tiquia-Arashiro, S. M. (2013). Isolation and screening of carboxydotrophs isolated from composts and their potential for butanol synthesis. *Environmental Technology, 34*, 1995–2007. https://doi.org/10.1080/09593330.2013.795987

Noël, C., Van Landschoot, L., Vanroelen, C., & Gadeyne, S. (2022). The public's perceptions of air pollution. What's in a name? *Environmental Health Insights, 16*, 11786302221123563.

Oest, A., Alsaffar, A., Fenner, M., Azzopardi, D., & Tiquia-Arashiro, S. M. (2018, May 27). Patterns of change in metabolic capabilities of sediment microbial communities in river and Lake ecosystems. *International Journal of Microbiology, 2018*, 6234931. https://doi.org/10.1155/2018/6234931

Pagnucco, G., Overfield, D., Chamlee, Y., Fortuna, A., Sulaiman, F., Farinas, J., Schittenhelm, R., Shuler, C., Kassem, A., Opara, S., Najaf, H., Abbas, L., Coutinho, O., Catalfano, B., Li, X., & Tiquia-Arashiro, S. M. (2023). Metal tolerance and biosorption capacities of bacterial strains isolated from an urban watershed. *Frontiers in Microbiology, 14*, 1278886. https://doi.org/10.3389/fmicb.2023.1278886

Panchuk, K. (2021). *Physical geology*. BCcampus.
Patel, D., Gismondi, R., Alsaffar, A., & Tiquia-Arashiro, S. M. (2019). Applicability of API ZYM to capture seasonal and spatial variabilities in lake and river sediments. *Environmental Technology, 40*(24), 3227–3239. https://doi.org/10.1080/09593330.2018.1468492
Patel, J., Harant, A., Fernandes, G., Mwamelo, A. J., Hein, W., Dekker, D., & Sridhar, D. (2023). Measuring the global response to antimicrobial resistance, 2020–21: A systematic governance analysis of 114 countries. *The Lancet Infectious Diseases, 23*(6), 706–718.
Qiu, Z., Yu, Y., Chen, Z., Jin, M., Yang, D., Zhao, Z., Wang, J., Shen, Z., Wang, X., Qian, D., Huang, A., Zhang, B., & Li, J.-W. (2012). Nanoalumina promotes the horizontal transfer of multiresistance genes mediated by plasmids across genera. *Proceedings of the National Academy of Sciences, 109*(13), 4944–4949.
Raboni, M., Gavasci, R., & Torretta, V. (2016). Assessment of the fate of Escherichia coli in different stages of wastewater treatment plants. *Water, Air, & Soil Pollution, 227*(12), 455.
Rahman, M. M., Howladar, M. F., Hossain, M. A., Shahidul Huqe Muzemder, A. T. M., & Al Numanbakth, M. A. (2020). Impact assessment of anthropogenic activities on water environment of Tillai River and its surroundings, Barapukuria Thermal Power Plant, Dinajpur, Bangladesh. *Groundwater for Sustainable Development, 10*, 100310.
Reid, C. J., Chowdhury, P. R., & Djordjevic, S. P. (2015). Tn6026 and Tn6029 are found in complex resistance regions mobilised by diverse plasmids and chromosomal islands in multiple antibiotic resistant *Enterobacteriaceae*. *Plasmid, 80*, 127–137.
Rizzo, L., Gernjak, W., Krzeminski, P., Malato, S., McArdell, C. S., Perez, J. A. S., Schaar, H., & Fatta-Kassinos, D. (2020). Best available technologies and treatment trains to address current challenges in urban wastewater reuse for irrigation of crops in EU countries. *Science of the Total Environment, 710*, 136312.
Sacchi, E., Bergamini, M., Lazzari, E., Musacchio, A., Mor, J.-R., & Pugliaro, E. (2021). Natural background levels of potentially toxic elements in groundwater from a former Asbestos mine in Serpentinite (Balangero, North Italy). *Water, 13* https://doi.org/10.3390/w13050735
Schwarzenlander, C., & Averhoff, B. (2006). Characterization of DNA transport in the thermophilic bacterium Thermus thermophilus HB27. *The FEBS Journal, 273*(18), 4210–4218.
Singh, A. K., Kaur, R., Verma, S., & Singh, S. (2022). Antimicrobials and antibiotic resistance genes in water bodies: Pollution, risk, and control. *Frontiers in Environmental Science, 10*.
Smith, R. A., M'ikanatha, N. M., & Read, A. F. (2015). Antibiotic resistance: A primer and call to action. *Health Communication, 30*(3), 309–314.
Sobecky, P. A., & Coombs, J. M. (2009). Horizontal gene transfer in metal and radionuclide contaminated soils. *Methods in Molecular Biology (Clifton, N.J.), 532*, 455–472. https://doi.org/10.1007/978-1-60327-853-9_26
Song, L., Wang, X., Zhang, W., Ye, L., & Feng, X. (2018). Low-intensity ultrasound promotes the horizontal transfer of resistance genes mediated by plasmids in E. coli. *3 Biotech, 8*(5), 224.
Su, J. Q., Wei, B., Xu, C. Y., Qiao, M., & Zhu, Y. G. (2014). Functional metagenomic characterization of antibiotic resistance genes in agricultural soils from China. *Environment International, 65*, 9–15.
Su, Z., Chen, L., & Wen, D. (2023). Impact of wastewater treatment plant effluent discharge on the antibiotic resistome in downstream aquatic environments: A mini review. *Frontiers of Environmental Science & Engineering, 18*(3), 36.
Sun, M., Ye, M., Wu, J., Feng, Y., Shen, F., Tian, D., Liu, K., Hu, F., Li, H., Jiang, X., Yang, L., & Kengara, F. O. (2015). Impact of bioaccessible pyrene on the abundance of antibiotic resistance genes during Sphingobium sp.- and sophorolipid-enhanced bioremediation in soil. *Journal of Hazardous Materials, 300*, 121–128.
Tangwa, B., Keubou, H., Ngwang, N., & Albert, N. (2019). Antimicrobial resistance profile of bacteria isolated from boreholes and hand dug wells water in Ngaoundere municipality of Adamawa Region in Cameroon. *Advances in Microbiology, 09*, 629–645.

Tao, Q., Wu, Q., Zhang, Z., Liu, J., Tian, C., Huang, Z., Malakar, P. K., Pan, Y., & Zhao, Y. (2022). Meta-analysis for the global prevalence of foodborne pathogens exhibiting antibiotic resistance and biofilm formation. *Frontiers in Microbiology, 13*.

Thakali, O., Malla, B., Tandukar, S., Sthapit, N., Raya, S., Furukawa, T., Sei, K., Sherchand, J. B., & Haramoto, E. (2021). Release of antibiotic-resistance genes from hospitals and a wastewater treatment plant in the Kathmandu Valley, Nepal. *Water, 13*. https://doi.org/10.3390/w13192733

Thamlikitkul, V., Tiengrim, S., Thamthaweechok, N., Buranapakdee, P., & Chiemchaisri, W. (2019). Contamination by antibiotic-resistant bacteria in selected environments in Thailand. *International Journal of Environmental Research and Public Health, 16*. https://doi.org/10.3390/ijerph16193753

Tiquia, S. M. (2010a). Salt-adapted bacteria isolated from the Rouge River and potential for degradation of contaminants and biotechnological applications. *Environmental Technology, 31*, 967–978. https://doi.org/10.1080/09593331003706226

Tiquia, S. M. (2010b). Metabolic diversity of the heterotrophic microorganisms and potential link to pollution of the Rouge River. *Environmental Pollution, 158*, 1435–1443. https://doi.org/10.1016/j.envpol.2009.12.035

Tiquia, S. M., Tam, N. F. Y., & Hodgkiss, I. J. (1998). Salmonella elimination during composting of spent pig litter. *Bioresource Technology, 63*(2), 193–196. https://doi.org/10.1016/S0960-8524(97)00113-2

Tiquia, S. M., Schleibak, M., Schlaff, J., Floyd, J. C., Benipal, B., Zakhem, E., & Murray, K. S. (2008). Microbial community profiling and characterization of some heterotrophic bacterial isolates from river waters and shallow groundwater wells along the Rouge River, Southeast Michigan. *Environmental Technology, 29*, 651–663. https://doi.org/10.1080/09593330801986998

Tiquia-Arashiro, S. M., & Pant, D. (2020). *Microbial electrochemical technologies* (p. 518). CRC Press. https://doi.org/10.1201/9780429487118

Uddin, T. M., Chakraborty, A. J., Khusro, A., Zidan, B. M. R. M., Mitra, S., Emran, T. B., Dhama, K., Ripon, M. K. H., Gajdács, M., Sahibzada, M. U. K., Hossain, M. J., & Koirala, N. (2021). Antibiotic resistance in microbes: History, mechanisms, therapeutic strategies and future prospects. *Journal of Infection and Public Health, 14*(12), 1750–1766.

Vassallo, A., Kett, S., Purchase, D., & Marvasi, M. (2021). Antibiotic-resistant genes and bacteria as evolving contaminants of emerging concerns (e-CEC): Is it time to include evolution in risk assessment? *Antibiotics, 10*. https://doi.org/10.3390/antibiotics10091066

Vinayamohan, P. G., Pellissery, A. J., & Venkitanarayanan, K. (2022). Role of horizontal gene transfer in the dissemination of antimicrobial resistance in food animal production. *Current Opinion in Food Science, 47*, 100882.

Wales, A. D., & Davies, R. H. (2015). Co-selection of resistance to antibiotics, biocides and heavy metals, and its relevance to foodborne pathogens. *Antibiotics (Basel), 4*(4), 567–604.

Wang, J., Karnati, P. K., Takacs, C. M., Kowalski, J. C., & Derbyshire, K. M. (2005). Chromosomal DNA transfer in *Mycobacterium smegmatis* is mechanistically different from classical Hfr chromosomal DNA transfer. *Molecular Microbiology, 58*(1), 280–288.

Wang, Y., Jiang, Q., Zhou, C., Chen, B., Zhao, W., Song, J., Fang, R., Chen, J., & Xiao, M. (2014). In-situ remediation of contaminated farmland by horizontal transfer of degradative plasmids among rhizosphere bacteria. *Soil Use and Management, 30*(2), 303–309.

WHO. (1998, May 11–16). Emerging and other communicable diseases: Antimicrobial resistance. In *Fifty-first world health assembly, Geneva*. World Health Organization.

WHO. (2015). *Global action plan on antimicrobial resistance*. World Health Organization.

WHO. (2016). *Critically important antimicrobials for human medicine: 5th Revision*.

WHO. (2018, November 30). Follow-up to the high-level meetings of the United Nations General Assembly on health-related issues. Antimicrobial resistance. Report by the Director-General. In *Executive Board 144th session, Geneva*. World Health Organization.

WHO. (2019). *Thirteenth General Programme of Work 2019–2023. Promote health, keep the world safe, serve the vulnerable*. World Health Organization.

Willey, J. M., Sherwood, L., & Woolverton, C. J. (2011). *Prescott's microbiology*. McGraw-Hill. ISBN 0070172595.

Wright, G. D. (2010). Antibiotic resistance in the environment: A link to the clinic? *Current Opinion in Microbiology, 13*(5), 589–594.

Wuaten, F. M. (2022). The role of sustainable finance and technology at Bank BJB in supporting the "sustainable development goals". *Aptisi Transactions on Technopreneurship (ATT), 5*, 1SP. https://doi.org/10.34306/att.v5i1Sp.318

Zhao, X. L., Qi, Z., Huang, H., Tu, J., Song, X. J., Qi, K. Z., & Shao, Y. (2021). Diversity and distribution of potential pathogens and antibiotic resistance genes in anthropogenic disturbances aquatic environment and their relationship with microbial indicators. *Research Square*. PREPRINT (Version 1). https://doi.org/10.21203/rs.3.rs-601610/v1

Zhou, Z.-C., Zheng, J., Wei, Y.-Y., Chen, T., Dahlgren, R. A., Shang, X., & Chen, H. (2017). Antibiotic resistance genes in an urban river as impacted by bacterial community and physicochemical parameters. *Environmental Science and Pollution Research, 24*(30), 23753–23762.

Zhou, Z., Shuai, X., Lin, Z., Yu, X., Ba, X., Holmes, M. A., Xiao, Y., Gu, B., & Chen, H. (2023). Association between particulate matter (PM)(2·5) air pollution and clinical antibiotic resistance: A global analysis. *Lancet Planet Health, 7*(8), e649–e659.

Chapter 12
Fecal-Derived Microbiota and Antimicrobial-Resistant Gene Profiles Along an Urban Watershed Using Microbial DNA Arrays

Florent Qyteti, Vaughn Karaim, Reeghan Book, and Sonia M. Tiquia-Arashiro

1 Introduction

Urban watersheds, such as the Rouge River in the Metropolitan Detroit area, are critical focal points for understanding the impacts of human activities on natural water systems (Tiquia, 2011; Bowman et al., 2018; Oest et al., 2018; Patel et al., 2019). Extending over 127 miles, the Rouge River has been highly urbanized and industrialized, serving as a key waterway that bisects a densely populated region. The degradation of its water quality reflects broader challenges faced by aquatic ecosystems worldwide, including both freshwater and marine environments (Tiquia, 2008; Nassereddine et al., 2024; Liu et al., 2025; Kaphle et al., 2025). Freshwater systems, such as rivers, lakes, and wetlands, provide essential ecosystem services, including water purification, habitat for biodiversity, and nutrient cycling, yet they are increasingly threatened by urbanization, pollution, and climate change (Liu et al., 2003; Tiquia et al. 2006a, b). Similarly, marine ecosystems, particularly coastal regions that receive freshwater inputs, are affected by nutrient loading, chemical contaminants, and habitat degradation, exacerbating issues such as eutrophication, harmful algal blooms, and declines in fish populations (Lan et al., 2024). The Rouge River, like many urban watersheds, receives an influx of stormwater runoff, combined sewer overflows (CSOs), and effluent from wastewater treatment plants, leading to significant biological and chemical contamination (Tiquia et al., 2007, 2008; Tiquia, 2010a, b). High concentrations of heavy metals, polycyclic aromatic hydrocarbons (PAHs), and persistent organic pollutants, legacies of industrial pollution, (Wụ et al., 2005; Gentile et al., 2006; Tiquia-Arashiro & Mormile, 2013; Pagnucco et al., 2023) threaten not only the river's biodiversity but also the health

F. Qyteti · V. Karaim · R. Book · S. M. Tiquia-Arashiro (✉)
Department of Natural Sciences, University of Michigan-Dearborn, Dearborn, MI, USA
e-mail: smtiquia@umich.edu

S. M. Tiquia-Arashiro et al. (eds.), *Urban Watershed Microbiology, Volume 1*,
https://doi.org/10.1007/978-3-031-89071-0_12

of connected freshwater and marine ecosystems. Additionally, untreated or inadequately treated sewage, agricultural runoff, and urban waste contribute excessive nutrients, pathogens, and emerging contaminants such as pharmaceuticals and personal care products (Tiquia, 2011; Pomaranski & Tiquia-Arashiro, 2016). The complex interplay between urban watersheds, freshwater, and marine environments underscores the need for targeted research to address the environmental and public health implications of these pollutants and to develop sustainable remediation strategies.

The microbiological health of the Rouge River is an urgent concern, particularly regarding fecal contamination and the associated microbial pollution. Urban runoff and combined sewer overflows (CSOs) introduce fecal bacteria into the river, signaling water quality degradation and posing severe health risks. These risks are exacerbated by the propagation of antibiotic-resistant genes (ARGs) (Nassereddine et al., 2024). ARGs, found increasingly in aquatic environments, represent a significant global health threat (Tiquia-Arashiro, 2012; Tiquia-Arashiro et al., 2025). Their ability to be transferred easily among bacterial populations complicates efforts to manage antibiotic-resistant pathogens, which heighten public health costs and challenge medical treatments (Tiquia et al., 1998).

Extensive research has highlighted the challenges of managing microbial contamination in urban water bodies. For instance, a study by Al Salah et al. (2020) and Wu et al. (2019) underscores the role of urban waterways as reservoirs of resistance genes, exacerbated by the discharge of untreated sewage. Additionally, Czatzkowska et al. (2022) reveal that environments impacted by human activity, such as rivers in urban areas, show heightened levels of ARGs due to the influx of anthropogenic waste. The integration of advanced molecular techniques, like quantitative polymerase chain reaction (qPCR) and high-throughput sequencing, has been pivotal in elucidating the diversity and abundance of ARGs in these environments (Nassereddine et al., 2024; Ouyang et al., 2024). The current study seeks to expand on these findings by specifically evaluating fecal-derived microbial communities and ARG profiles along the Rouge River. Through methodologies like quantitative PCR (qPCR) and microbial DNA arrays, we will analyze water samples from six distinct sites, aiming to identify the dominant fecal-associated bacteria and assess the prevalence of ARGs in these communities. This approach aligns with previous research strategies that prioritize pinpointing hot spots of microbial and chemical pollution to mitigate potential threats to human and ecological health (Tiquia & Mormile, 2010; McKindles & Tiquia-Arashiro, 2012; Tiquia-Arashiro & Pant, 2020; Awewomom et al., 2024; Tiquia-Arashiro et al., 2025).

Furthermore, understanding these dynamics in the Rouge River watershed is critical for the development of effective management strategies. As suggested by Li et al. (2020), addressing microbial pollution requires integrating molecular analysis with environmental management to reduce ARG dissemination and improve water quality. By mapping the microbial and chemical pollution landscapes, our study contributes to identifying contamination hotspots and informing further mitigation strategies. This research is essential not only for preserving the health of the Rouge River ecosystem but also for protecting public health by limiting the spread of antibiotic resistance in urban environments.

2 Methods

2.1 *Site Description*

Our study focused on six sites along the lower Rouge River: Canton (42.28139° N, 83.46586° W), Venoy (42.28854° N, 83.35928° W), Inkster (42.29872° N, 83.30706° W), Ford Field (42.31189° N, 83.24300° W), Dearborn (42.31068° N, 83.23165° W), and Rouge Park (42.32550° N, 83.21400° W). These sites generally follow an eastward trend, except for Rouge Park, which is situated further north. These sites capture the range of landscapes within the Rouge River watershed, from urban settings to rural floodplains, providing a comprehensive basis for ecological study.

2.2 *Field Analysis and Water Sampling*

On the sampling day, ambient air temperature, relative humidity, water temperature, water depth, and air pressure were recorded on-site. Air temperature ranged from 18 °C at Venoy to 25 °C at Ford Field, while relative humidity was highest in Inkster (99%) and lowest in Dearborn (62%). Water temperature varied from 19 °C in Canton to 22 °C at Ford Field. Water depth showed significant variation, with the shallowest measurement at Venoy (131 mm) and the deepest at Rouge Park (2304 mm), likely reflecting differences in river flow and topography. Barometric pressure remained relatively stable, ranging from 29.5 inHg (1000 mb) in Inkster to 30.1 inHg (1013 mb) in Dearborn. Relative humidity data for Rouge Park was unavailable. These environmental factors may influence water quality and ecosystem dynamics along the Rouge River.

Surface water samples were collected manually by lowering a sterile glass bottle to a depth of 1 m. At each site, a 1-hour composite sample was created by combining four individual 400-mL grab samples taken every 20 minutes (Tiquia, 2011; Oest et al., 2018; Patel et al., 2019).

For microbial analysis, water samples were transported on ice and stored in a climate-controlled cold room. Since bacterial populations could change during transport and storage, all samples were analyzed within 8 hours of collection. For qPCR analysis, 300–500 mL of water was filtered in triplicate at each site using Sterivex filters (Millipore, Billerica, MA, USA). The filters were then placed in a dewar with liquid nitrogen and stored at −20 °C until further use.

Total coliform and *Escherichia coli* counts were determined using membrane filtration with Chromocult agar. Water was filtered through a 0.45-micrometer membrane, which was then placed on Chromocult coliform agar (Millipore, Billerica, MA, USA), a chromogenic medium that differentiates bacteria based on enzyme activity. Coliforms, exhibiting β-galactosidase activity, appeared as pink to red colonies, while *E. coli*, with β-glucuronidase activity, formed dark blue to violet

colonies. Plates were incubated at 35–37 °C for 18–24 hours before colony counts were performed. Additional confirmation tests were conducted as needed. Bacterial concentrations were reported as colony-forming units (CFU) per mL using the geometric mean (Caudill et al., 2007). For detailed procedures, refer to EPA Method 1604 or Method 2000/015.

Laboratory analyses included measuring pH (Hanna Edge H12002 meter), electrical conductivity and total dissolved solids (Hanna 9810 pH/EC/TDS meter), and turbidity (Hanna HI98703-01 portable turbidity meter), all conducted in triplicate. Key water quality parameters varied notably across sites. The lowest pH was recorded in Canton (6.63) and Ford Field (6.66), indicating slightly acidic conditions. Electrical conductivity, indicative of dissolved ion content, was highest in Dearborn (1617 $\mu S\ cm^{-1}$) and Ford Field (1263 $\mu S\ cm^{-1}$), suggesting potential pollutant influx. Total suspended solids were highest in Dearborn (750 mg/L), with elevated levels also observed at Ford Field (640 mg/L) and Canton (527 mg/L). Rouge Park had the highest turbidity (56 NTU), reflecting increased suspended particles. Dissolved oxygen was lowest in Dearborn (4.0 mg/L), suggesting potential organic pollution. Microbial contamination was most severe at Ford Field, which recorded the highest total coliform (1465 CFU/100 mL) and *E. coli* (1229 CFU/100 mL) levels, followed by Venoy (*E. coli*: 1100 CFU/100 mL) and Inkster (*E. coli*: 990 CFU/100 mL). These findings indicate that Dearborn, Ford Field, and Rouge Park exhibit multiple indicators of pollution, while Ford Field and Venoy pose the highest microbial contamination risks, raising concerns about water quality and public health in these areas.

2.3 DNA Extraction

DNA extraction was performed using the Qiagen DNeasy PowerWater Kit (Qiagen, Germantown, MD) according to the manufacturer's instructions to ensure consistency and reliability of the results. Prior to extraction, the Sterivex filters were removed from storage and thawed on ice. Each filter was carefully opened using sterile techniques to prevent contamination. The filter membrane was then placed into a PowerWater Bead Tube, which contains specialized beads to aid in mechanical disruption of cells. To begin the extraction process, Solution PW1 was added to the bead tube to facilitate cell lysis and release DNA. The tube was then subjected to vortexing to enhance the efficiency of the lysis step. Following vortexing, the sample was centrifuged to separate the lysate from the bead debris. The resulting supernatant, which contained the extracted DNA and cellular components, was carefully transferred to a clean microcentrifuge tube. The supernatant then underwent a series of purification steps involving multiple rounds of centrifugation and washing to remove contaminants. Specifically, inhibitors and unwanted cellular materials were removed using Solution IRS, followed by additional washing steps with Solutions PW3 and PW4 to further purify the DNA. After the final wash, the DNA was eluted in Solution EB, ensuring high-quality nucleic acid recovery. To

assess the quality and concentration of the extracted genomic DNA (gDNA), initial quantification was performed using a Nanodrop 1000 spectrophotometer (Thermo Fisher Scientific, Waltham, MA). Additionally, the Quantus fluorometer (Promega Corporation, Madison, WI) was used to obtain precise DNA concentration measurements, providing more accurate quantification for downstream applications such as qPCR. Only samples with sufficient DNA yield and quality were selected for further analysis. All extracted DNA samples were stored at −20 °C to preserve their integrity for future molecular analyses. Throughout the procedure, strict precautions were taken to minimize contamination, including the use of sterile equipment and adherence to best laboratory practices.

2.4 *qPCR Assays*

Quantitative polymerase chain reaction (qPCR) assays were conducted to detect and quantify both microbial species and antibiotic-resistant genes (ARGs) present in the environmental samples. Two specialized qPCR arrays were utilized: the Qiagen Fecal-Derived Microbes Array and the Qiagen Antibiotic Resistance Gene Microbes Array (Qiagen GmbH, Hilden, Germany). These arrays enabled the simultaneous identification of over 30 bacterial species commonly found in the gastrointestinal tracts of humans and animals, as well as the detection of 86 potential ARGs associated with antibiotic resistance in microbial communities. The Microbial DNA qPCR Kit facilitated the amplification and detection of specific ARGs, while the fecal-derived microbes array provided a comprehensive profile of fecal-associated microbial species. Together, these assays allowed for a thorough analysis of the presence and relative abundance of ARGs and fecal-derived bacteria in environmental samples collected from the Rouge River.

2.4.1 Sample Preparation

Prior to qPCR analysis, extracted DNA samples were thawed on ice to prevent degradation. To ensure representative detection across samples, DNA extracts were pooled into a single Eppendorf tube (Cho et al., 2012; Nguyen et al., 2013; Plecha et al., 2013; Flood et al., 2015; Pomaranski & Tiquia-Arashiro, 2016). The pooled DNA was gently mixed by pipetting up and down several times to achieve homogeneity, while vortexing was strictly avoided to prevent shearing of the DNA. For qPCR setup, a minimum of 500 ng of DNA was required per reaction. The microbial qPCR master mix was prepared by combining the kit's qPCR master mix reagent with the DNA template to obtain a total reaction volume of 1280 μL. The prepared reaction mixture was then aliquoted into a 96-well qPCR plate, ensuring even distribution. Each well contained 25 μL of the DNA:master mix solution, which was loaded into wells pre-coated with gene-specific primers and hydrolysis probes designed for target ARGs and microbial species. Once all wells were

prepared, the qPCR plate was sealed with an optically clear sealing film, preventing evaporation and ensuring accurate fluorescence readings. Care was taken to use a film that did not interfere with fluorescence detection within the 450–750 nm wavelength range. To eliminate any bubbles or inconsistencies in well volumes, the sealed plate was subjected to a brief centrifugation (quick spin-down) before loading into the qPCR instrument.

2.4.2 qPCR Thermal Cycling Conditions

qPCR reactions were performed using the Bio-Rad CFX Connect Thermal Cycler Real-time PCR Detection System (Bio-Rad, Hercules, CA), following the manufacturer's recommended protocol. The thermal cycling conditions were set as follows: (1) initial denaturation at 95 °C for 10 minutes to ensure complete DNA strand separation; (2) denaturation at 95 °C for 15 seconds to break apart double-stranded DNA; and (3) annealing and extension at 60 °C for 2 minutes to allow primers to bind to target sequences and enable DNA polymerase activity. Steps 2 and 3 were repeated for 39 additional cycles, ensuring sufficient amplification for detection. Upon completion of the thermal cycling process, fluorescence data were collected in real time and analyzed using Bio-Rad CFX Manager software.

2.4.3 Data Analysis and Interpretation

The qPCR data were analyzed based on cycle threshold (Ct) values, which indicate the number of cycles required for a detectable fluorescence signal to surpass the threshold. Ct values were used to assess the presence and relative abundance of target ARGs and microbial species in the samples. A lower Ct value corresponded to a higher concentration of the target gene or species, suggesting a greater abundance in the environmental sample. To determine the presence or absence of a specific microbial species or ARG, Ct values were compared against predefined threshold values specific to each assay (Nassereddine et al., 2024). If the Ct value was below the assay-specific threshold, the gene or microbial species was considered present (positive detection) in the sample. If the Ct value exceeded the threshold, it was deemed absent or at very low abundance (negative detection).

To validate the accuracy of the results, Ct values were also compared to positive controls and known standards included in the qPCR assay. The qPCR analysis provided critical insights into the microbial composition and antibiotic-resistant gene distribution within the Rouge River ecosystem. The detection of fecal-derived microbes helped assess potential fecal contamination, while the presence of ARGs indicated the possible dissemination of antibiotic-resistant bacteria in the aquatic environment. The findings from this study contribute to a better understanding of environmental health risks and microbial pollution dynamics, aiding in the evaluation of public health concerns related to waterborne contamination and antibiotic resistance (Nassereddine et al., 2024).

2.5 *Data and Statistical Analysis*

To analyze fecal-derived microbes, a heatmap was generated to visualize the distribution of bacterial strains across different sites. This heatmap was then used to create a frequency distribution table, which provided a clearer view of the specific bacterial strains or species detected at each site. Additionally, analysis of variance (ANOVA) tables were constructed to assess the variation in cycle threshold (Ct) values and the presence or absence of microbes at each site. A similar approach was applied to antibiotic-resistant genes (ARGs). After conducting qPCR, a heatmap was created to illustrate the distribution of ARGs across sites, followed by ANOVA tables to evaluate Ct values and the presence/absence of these genes. For both fecal-derived microbes and ARGs, cluster analysis maps were generated to identify patterns in microbial distribution, and paired comparison tests were conducted to assess significant differences between sites. For both fecal-derived microbes and ARGs, cluster analysis maps were generated to identify patterns in microbial distribution, and paired comparison tests were conducted to assess significant differences between sites.

3 Results

3.1 *qPCR Profiles*

3.1.1 Fecal-Derived-Microbes

Figure 12.1 (heatmap) displays the distribution of fecal-derived microbial gene profiles across six environmental sites. Ford Field and Rouge Park show the highest presence of microbial genes, suggesting significant contamination from urban runoff or wastewater discharge. In contrast, Dearborn and Inkster exhibit moderate levels of microbial presence, with some species frequently detected, while others are inconclusive or absent, indicating variable contamination sources. Venoy and Canton have the lowest microbial presence, with many genes marked as absent, suggesting better water quality compared to other sites. Comparing this with the antibiotic-resistant gene (ARG) heatmap in Fig. 12.2, locations with higher microbial diversity, particularly Ford Field and Rouge Park, also show increased ARG abundance, especially in beta-lactamase and efflux pump genes. This correlation suggests that greater microbial loads from fecal contamination may contribute to higher ARG presence. Overall, the data reveal variations in fecal contamination and microbial diversity, with Ford Field and Rouge Park showing the most contamination and Venoy and Canton the least. The presence of opportunistic pathogens and antibiotic-resistant bacteria in more contaminated sites highlights potential public health risks and underscores the environmental impact of urban wastewater discharge.

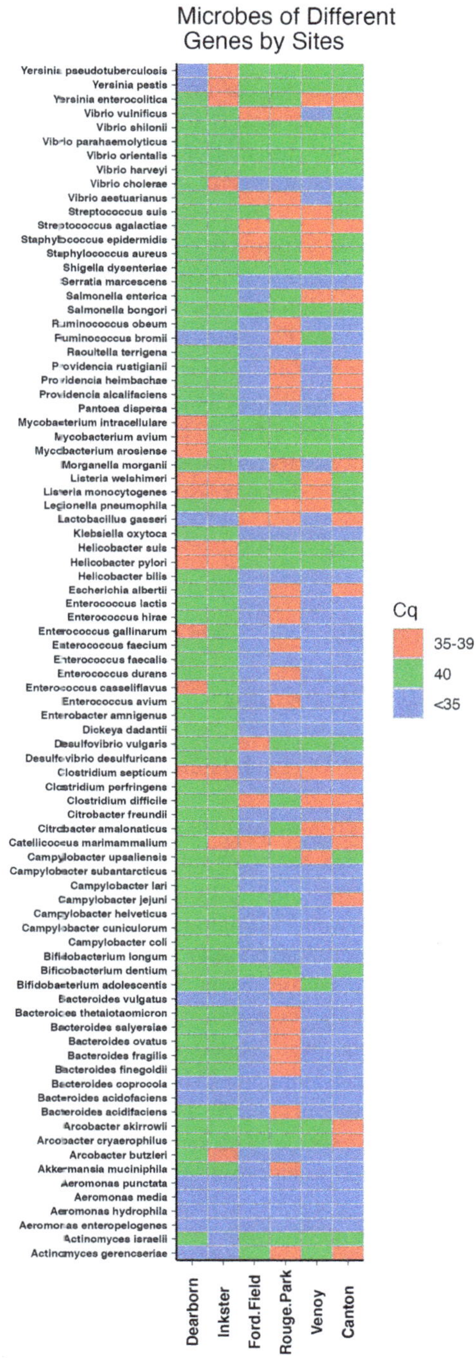

Fig. 12.1 Heatmap of fecal-derived microbial gene detection across sampling sites based on Cq Values. The color coding indicates the status of microbial gene detection: blue ($Cq < 35$) for presence, red (Cq 35–39) for inconclusive detection, and green ($Cq = 40$) for absence. The x-axis displays the sampling locations, and the y-axis lists the microbial species detected at each site

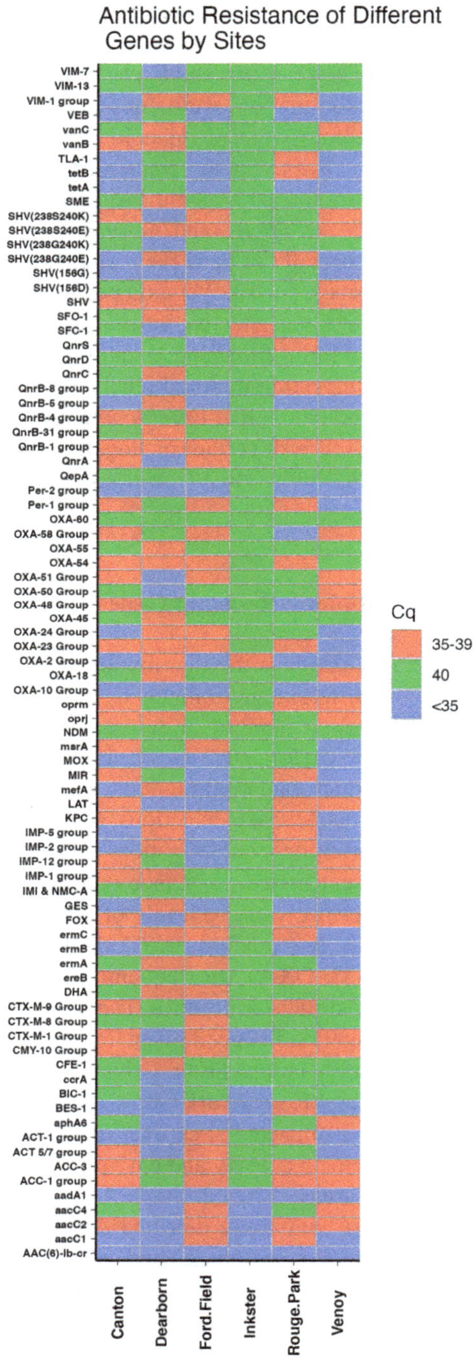

Fig. 12.2 Heatmap of antibiotic resistance genes (ARGs) detected across sampling sites based on Cq values. The color code indicates the detection status: blue (Cq < 35) for presence, red (Cq 35–39) for inconclusive, and green (Cq = 40) for absence. The x-axis displays the sampling locations, while the y-axis lists the ARGs detected at each site

The most commonly detected fecal-derived microbial genes across all sites include *Enterococcus faecalis*, *Enterococcus faecium*, *Bacteroides fragilis*, *Clostridium perfringens*, and *Escherichia albertii*. These bacteria, known indicators of fecal contamination, were consistently present or inconclusive across multiple sites, suggesting widespread contamination. Additionally, human gut-associated *Bacteroides ovatus* and *Bacteroides acidifaciens* were frequently found. Opportunistic and pathogenic bacteria like *Salmonella enterica*, *Campylobacter jejuni*, *Vibrio vulnificus*, and *Aeromonas hydrophila* were also present in varying amounts, especially in Ford Field and Rouge Park. Conversely, some fecal-derived microbial genes, such as Yersinia pestis, *Arcobacter skirrowii*, and *Mycobacterium arupense*, were not detected, indicating either their absence or concentrations below the detection limit. *Providencia* and *Citrobacter* species were also absent in several locations, suggesting minimal fecal contamination from these sources. While common fecal indicators like *Enterococcus* and *Bacteroide*s species were present at all sites, the absence of certain pathogens suggests variability in contamination levels and sources, with some locations experiencing more severe fecal pollution than others.

3.1.2 Antibiotic-Resistant Genes

Figure 12.2 displays the distribution of various antibiotic-resistant genes (ARGs) across six different environmental sites. Among these, Ford Field and Rouge Park have the highest number of detected ARGs, indicating strong selective pressures likely due to factors such as wastewater contamination, urban runoff, or human-associated microbial communities. Conversely, Venoy and Canton show a higher number of inconclusive detections, suggesting lower levels of ARGs and potentially reduced antibiotic contamination or weaker selection pressures. Dearborn and Inkster present mixed resistance patterns, with both positive and no ARG detections, suggesting varying levels of antibiotic resistance. Notably, genes like carbapenemases (e.g., NDM, KPC, IMP) and extended-spectrum beta-lactamases (e.g., CTX-M, OXA) are frequently detected at all sites, pointing to widespread resistance to critical antibiotics. The abundance and diversity of ARGs differ across locations, likely influenced by environmental factors, pollution sources, and microbial community composition. Urban areas with more wastewater influence, such as Ford Field and Rouge Park, show a higher prevalence and variety of ARGs compared to less impacted sites like Venoy and Canton. These findings highlight the impact of human activity and environmental contamination on antibiotic resistance in different areas.

The most commonly detected ARGs across all six sites involve resistance to beta-lactams, tetracyclines, and aminoglycosides, as illustrated in Fig. 12.2. Frequently observed genes include those related to bla (beta-lactamase resistance), tet (tetracycline resistance), and aad (aminoglycoside resistance), which were widely present (Cq < 35) or inconclusive (Cq 35–39) across multiple locations, indicating the prevalence of antibiotic-resistant bacteria. Specifically, genes like

*bla*CTX-M and *bla*TEM, responsible for beta-lactam resistance, were consistently detected, pointing to contamination from human and animal waste. Additionally, multidrug efflux pump genes, such as oqxA and qnrS (related to fluoroquinolone resistance), were often found, implying antibiotic pressure in these environments. On the other hand, some ARGs, such as certain carbapenem resistance genes (*bla*KPC, *bla*NDM) and colistin-resistant genes (*mcr*-1), were not detected at any site (Cq = 40). This suggests either their absence or concentrations below the detection limit, indicating that the presence of highly resistant pathogens like carbapenem-resistant Enterobacteriaceae (CRE) may be limited in these areas. Overall, while ARGs related to beta-lactams, tetracyclines, and aminoglycosides are common at all sites, more clinically significant resistance genes for carbapenem or colistin were not detected, reflecting variability in selective pressures and contamination sources across these locations.

3.2 *Frequency Distribution*

3.2.1 Fecal-Derived Microbes

The frequency distribution of fecal-derived bacteria varied significantly across sites, with some locations displaying completely different microbial compositions (Fig. 12.3). Each pie chart represents the relative abundance of various fecal-derived bacterial species detected at each site, providing a comparative analysis of microbial contamination in the Rouge River ecosystem. Several microbial genera, including *Aeromonas* spp. (8.2–25.3%), *Clostridium* spp. (8.1–19.5%), *Enterococcus* spp. (5.1–16.8%), *Bacteroides* spp. (6.7–20.3%), and *Vibrio* spp. (6.3–17.2%), appear across multiple sites, suggesting their widespread presence in the Rouge River. However, there is variation in dominant microbes at each location. Dearborn shows a noticeable proportion of *Bifidobacterium* spp. (9.1%) and *Clostridium* spp. (8.2%) (Fig. 12.3a). Venoy has a diverse microbial community, with *Aeromonas* spp. (11.5%), *Enterococcus* spp. (10.3%), and *Clostridium* spp. (9.8%) being the most abundant group (Fig. 12.3b). Inkster exhibits a relatively balanced distribution, with *Desulfovibrio* spp. (12.6%) and *Aeromonas* spp. (10.9%) being the most dominant (Fig. 12.3c). Ford Field has a strong presence of *Aeromonas* spp. (25.3%), *Bacteroides* spp. (20.3%), and *Clostridium* spp. (15.1%) (Fig. 12.3d). Rouge Park features notable contributions from *Morganella* spp. (7.8%), *Vibrio* spp. (16.7%), and *Bacteroides* spp. (11.2%) (Fig. 12.3e). Canton shows a high prevalence of *Aeromonas* spp. (18.7%) and *Bacteroides* spp. (13.9%), similar to other sites (Fig. 12.3f).

Site-specific differences are also evident, with Venoy displaying the most evenly distributed microbial diversity, indicating a highly contaminated or complex microbial environment. Rouge Park has a distinct microbial signature, with *Morganella* spp. appearing more frequently than at other locations. Inkster and Canton share similarities in microbial diversity, particularly with the presence of *Desulfovibrio*

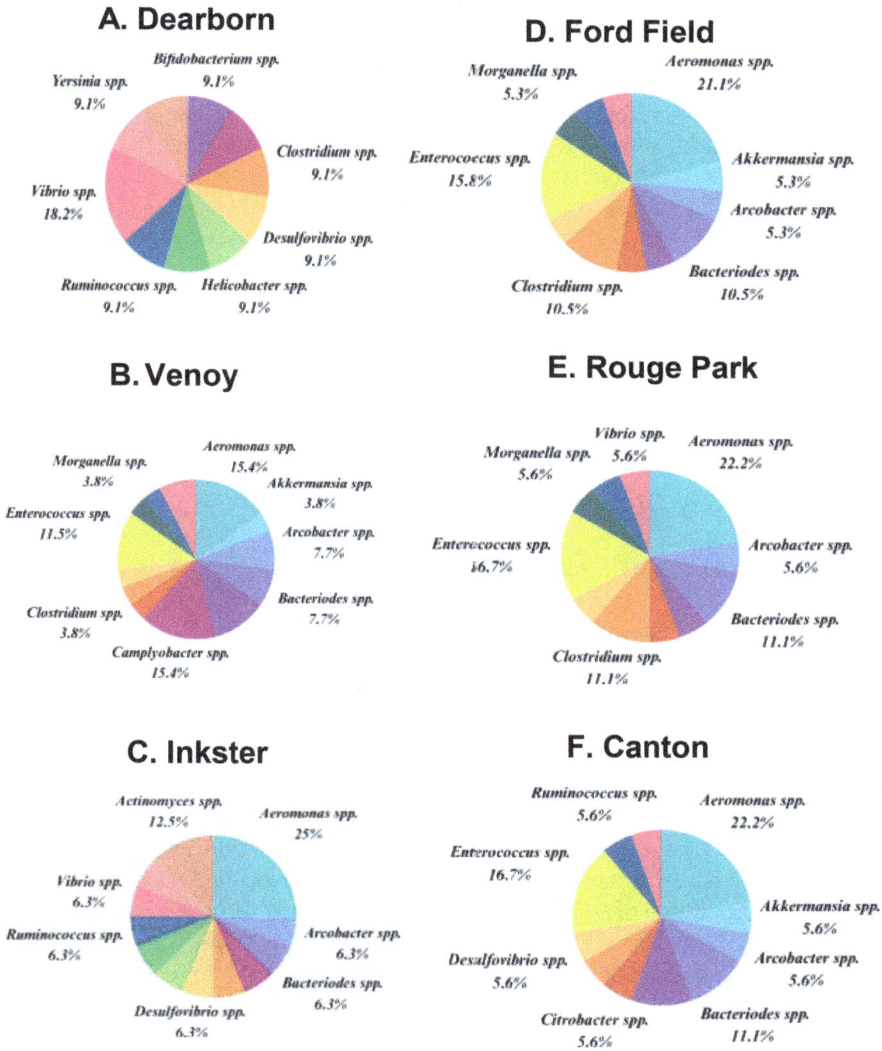

Fig. 12.3 Relative frequency distribution of fecal-derived microbes across six sampling sites: (**a**) Dearborn, (**b**) Venoy, (**c**) Inkster, (**d**) Ford Field, (**e**) Rouge Park, and (**f**) Canton

spp. Overall, this figure visually represents the microbial composition of the six sampling sites, highlighting both shared and site-specific distributions. The differences in microbial communities likely reflect variations in pollution sources, environmental conditions, and anthropogenic impacts at each location.

Several bacterial strains were not detected at any of the sampling sites including *Catellicoccus*, *Streptococcus*, and *Listeria*. The percentage of positive detections for the tested bacterial strains varied across sites: Inkster had 30%, Canton had 34%,

Ford Field had 38%, and Rouge Park had 24%, making it the closest to Dearborn in terms of overall detection rates. *Aeromonas* was present at each site, with four distinct strains detected at all locations except Dearborn, where no *Aeromonas* strains were found. In contrast, *Clostridium* was detected at every site, though in varying amounts (Fig. 12.3).

Based on the data in Table 12.1, two antibiotic-resistant gene (ARG) classes, fluoroquinolone resistance and aminoglycoside resistance were detected at all six sites. Fluoroquinolone resistance was found in varying quantities across locations, while aminoglycoside resistance was present at all sites except Rouge Park. Several ARG classes were not detected at any site, including erythromycin resistance, multidrug resistance efflux pump, vancomycin resistance, *Staphylococcus aureus* (MRSA) resistance, and Panton-Valentine leukocidin chain F precursor. These genes were absent in all qPCR analyses. The distribution of ARGs varied significantly across sites. Inkster had the highest detection rate, with a strong presence of multiple ARGs, particularly tetracycline efflux pump and aminoglycoside resistance, and the highest positivity rate overall (81.60%). Rouge Park had the fewest ARG detections (8) and identified only a limited number of gene classes. Venoy had the most diverse array of detected ARGs, with particularly high levels of fluoroquinolone resistance and Class A beta-lactamase, contributing to its relatively high total ARG count (26). Dearborn and Ford Field exhibited moderate levels of ARGs, with some overlapping gene classes but generally lower overall prevalence compared to Venoy and Inkster. The lower ARG prevalence at these sites is reflected in the fewer total gene classes detected and the lower positivity rates (43.67% for Dearborn and 47.13% for Ford Field) (Table 12.1). The two most dominant ARG groups that were found were Tetracycline efflux pump-resistant genes and Aminoglycoside-resistant genes, coming in at 58% and 53%, respectively. Results also highlight the variability in antibiotic-resistant gene distribution, with certain resistance classes being more prevalent across sites. The presence of Class A beta-lactamase (20.45%), Class B beta-lactamase (14.81%), Class C beta-lactamase (16.67%), and Class D beta-lactamase (15.38%) indicates notable resistance to beta-lactam antibiotics.

3.3 ANOVA

3.3.1 Fecal-Derived Microbes

The ANOVA results indicate highly significant differences in Ct values and presence/absence scores across sites, with *p*-values ≤ 0.0001, though the differences are not extreme overall (Table 12.2). A total of 56 samples were analyzed for both parameters. Overall, the ANOVA values for fecal-derived microbes do not show major variations among sites, as most locations share statistical groupings. Superscripts (*a, b, ab*) denote sites that are not significantly different from each other in Ct values. Venoy is the most distinct site, marked with a *b* designation,

Table 12.1 Distribution and relative frequency of antibiotic resistance genes (ARGs) across six sites

Antibiotic classification	Number of genes	Canton	Dearborn	Inkster	Venoy	Ford Field	Rouge Park	Relative frequency (%)
Fluoroquinolone resistance	11	2	3	1	3	3	1	19.70
Aminoglycoside-resistance	5	2	5	5	2	2	0	53.33
Class A beta-lactamase	22	5	6	3	5	7	1	20.45
Class B beta-lactamase	9	3	2	0	3	0	0	14.81
Class C beta-lactamase	11	2	3	0	4	2	0	16.67
Class D beta-lactamase	13	3	1	0	4	2	2	15.38
Erythromycin resistance	1	0	0	0	0	0	0	0.00
Macrolide Lincosamide Streptogramin_b	5	2	0	0	3	2	2	30.00
Multidrug resistance efflux pump	2	0	0	0	0	0	0	0.00
Tetracycline efflux pump	2	2	0	0	2	2	1	58.33
Vancomycin resistance	2	0	0	0	0	0	0	0.00
Staphylococcus aureus (MRSA)	1	0	0	0	0	0	0	0.00
Beta-lactam resistance	1	0	1	0	0	0	0	16.67
Panton-Valentine leukocidin chain F precursor	1	0	0	0	0	0	0	0.00
Immunoglobulin G binding protein A precursor	1	0	0	0	0	0	1	16.67
Total number of genes	87	21	21	9	26	20	8	
% of genes detected per site		56.32	43.67	81.60	42.52	47.13	63.22	

Table 12.2 ANOVA results for fecal-derived Ct values and presence/absence across sampling locations

Sites	Ct values (Mean ± Std. deviation)	Presence/absence (Mean ± Std. deviation)
Dearborn	37.25 ± 4.49[a]	0.28 ± 0.42[a]
Venoy	32.11 ± 8.65[b]	0.61 ± 0.43[a]
Inkster	36.87 ± 5.56[a]	0.28 ± 0.40[b]
Ford Field	33.99 ± 6.43[ab]	0.57 ± 0.47[ab]
Rouge Park	35.26 ± 5.87[ab]	0.46 ± 0.4369[b]
Canton	34.01 ± 6.12[ab]	0.55 ± 0.44[a]
P-value	≥0.0001	0.0001
Significance	***	***
n	56	56

Superscripts ([a, b, ab]): Sites sharing the same letter are not significantly different from each other
Significance Levels: ***($p \leq 0.0001$): Highly significant differences exist among sampling sites
Mean and standard deviation of 96 replicates are shown

indicating significantly higher fecal contamination. In contrast, Dearborn and Inkster (*a*) have the highest Ct values, suggesting lower contamination (i.e., fewer fecal-derived microorganisms). Ford Field, Rouge Park, and Canton show intermediate Ct values of 33.99 ± 6.43, 35.26 ± 5.87, and 34.01 ± 6.12, respectively. For presence/absence measurements, values range from 0.28 ± 0.40 in Inkster to 0.61 ± 0.43 in Venoy, with significant differences across locations ($p \leq 0.0001$). However, unlike Ct values, Inkster and Rouge Park are the most distinct in presence/absence, both marked with a *b* designation, whereas Venoy and the other sites show less variation. These results highlight Venoy as the most contaminated site based on Ct values, while Inkster and Rouge Park exhibit the greatest variation in the presence/absence of fecal-derived microbes.

3.3.2 Antibiotic-Resistant Genes

Table 12.3 presents the ANOVA results for antibiotic-resistant gene (ARG) Ct values and variability across different sampling locations along the Rouge River. The data indicate significant variability in antibiotic-resistant gene (ARG) contamination across the Rouge River sampling locations. Inkster shows significantly higher Ct values (38.52 ± 4.43), suggesting lower concentrations of ARGs compared to other sites. The remaining locations,(34.83 ± 6.24), Venoy (34.20 ± 6.37), Ford Field (34.85 ± 5.99), Rouge Park (35.91 ± 6.93), and Canton (35.06 ± 6.31), demonstrate similar ARG concentrations, indicating relatively consistent levels of contamination across these sites. Overall, the findings highlight notable spatial differences in ARG contamination along the river. For presence/absence data, Inkster stands out with a much lower value (0.12 ± 0.31), indicating significantly lower presence of ARGs compared to the other sites. The other sites, Dearborn (0.49 ± 0.40), Venoy (0.53 ± 0.42), Ford Field (0.50 ± 0.40), Rouge Park

Table 12.3 ANOVA results for ARG Ct values and variability across sampling locations

Sites	Ct values (Mean ± std. deviation)	Presence/absence (Mean ± Std. deviation)
Dearborn	34.83 ± 6.24^{b}	0.4943 ± 0.3998^{a}
Venoy	34.20 ± 6.37^{b}	0.5345 ± 0.4162^{a}
Inkster	38.52 ± 4.43^{a}	0.1207 ± 0.3141^{c}
Ford field	34.845 ± 5.99^{b}	$0.5000 \pm .4035^{a}$
Rouge Park	35.914 ± 6.93^{b}	0.3161 ± 0.3744^{b}
Canton	35.060 ± 6.31^{b}	$0.4540 \pm .4008^{ab}$
P-value	≥0.0001	≥0.0001
Significance	***	***
n	96	96

Superscripts ($^{a, b, ab}$): Sites sharing the same letter are not significantly different from each other
Significance Levels: ***($p \leq 0.0001$): Highly significant differences exist among sampling sites
Mean and standard deviation of 96 replicates are shown

(0.32 ± 0.37), and Canton (0.45 ± 0.40), show similar values, with Dearborn, Venoy, Ford Field, and Canton showing no significant differences from each other (marked with *a* and *ab* superscripts). Rouge Park has an intermediate presence/absence value, marked with *b*, indicating slightly lower ARG presence than other locations, except for Inkster.

3.4 *Paired* t-*Test*

3.4.1 Fecal-Derived Microbes

A paired *t*-test was conducted to compare the presence and absence of fecal-derived microbes and their corresponding Ct values across different sampling sites along the Rouge River. The left panel represents the mean presence and absence of fecal-derived microbes across different sites (Fig. 12.4a), while the right panel displays mean Ct values for fecal-derived microbes (Fig. 12.4b). Results showed significant differences in Ct values between certain sites, particularly between Inkster and Venoy ($p < 0.01$), Venoy and Dearborn ($p < 0.001$), and Rouge Park and Ford Field ($p < 0.05$), suggesting that microbial loads varied considerably between locations, with Venoy exhibiting the highest fecal-derived microbial concentrations and Dearborn displaying the lowest. For the presence and absence of fecal-derived microbes, statistical differences were observed across several sites, particularly between Venoy and Inkster ($p < 0.01$) and Dearborn and Venoy ($p < 0.01$), indicating distinct bacterial compositions. These findings align with the broader results of this study, reinforcing that Venoy had the highest fecal contamination, while Dearborn had the lowest.

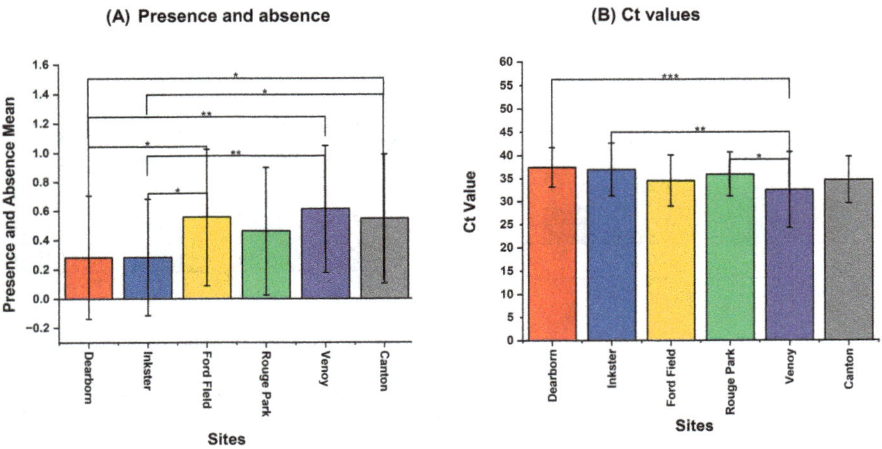

Fig. 12.4 Paired *t*-test results for fecal-derived microbes based on (**a**) presence and absence and (**b**) Ct values across six sampling sites. Asterisks represent statistical significance: $*p < 0.05$, $**p < 0.01$, $**p < 0.001$

3.4.2 Antibiotic-Resistant Genes

In terms of the ARG, the presence and absence analysis (Fig. 12.5a) ARG revealed significant differences between Canton-Inkster ($p < 0.001$), Dearborn-Inkster ($p < 0.001$), Ford Field-Inkster ($p < 0.001$), and Venoy-Inkster ($p < 0.001$), suggesting that the microbial ARG composition at Inkster significantly differs from these locations. Notably, Inkster exhibited the lowest mean presence and absence value, indicating a distinct ARG profile compared to other sites. Similarly, the paired *t*-test on Ct values (Fig. 12.5b) demonstrated statistically significant differences among Dearborn-Inkster ($p < 0.001$), Ford Field-Inkster ($p < 0.001$), and Venoy-Inkster ($p < 0.001$), further supporting the trend observed in the presence/absence analysis. Since lower Ct values correspond to higher ARG concentrations, these findings suggest that ARG distribution varies significantly across locations, with Inkster standing out as an outlier in terms of both presence/absence and Ct values, indicating a unique microbial and ARG composition at this site.

3.5 *Cluster Analysis*

3.5.1 Fecal-Derived Microbes

Cluster analysis was performed to evaluate the similarity between sites based on Ct values and the presence/absence of fecal-derived microbes. The results aligned with prior expectations given the historically documented pollution levels in the Rouge River. All locations tested positive for multiple strains of fecal-derived microbes,

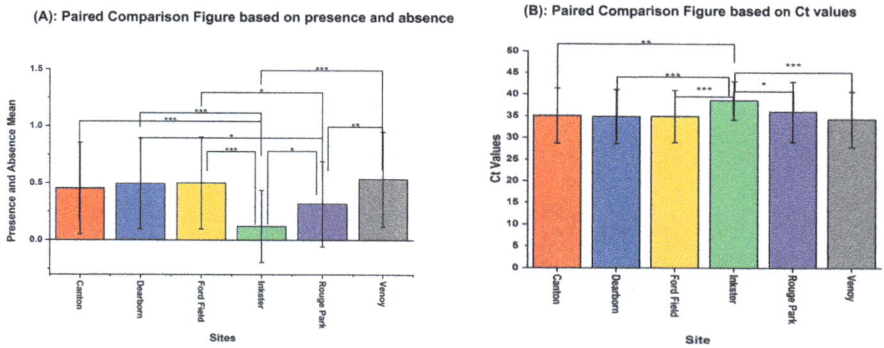

Fig. 12.5 Paired *t*-test results for ARGs based on (**a**) presence and absence and (**b**) Ct values across sampling sites. Asterisks represent statistical significance: $*p < 0.05$, $**p < 0.01$, $**p < 0.001$

and all sites exhibited some level of ARGs. The presence/absence-based cluster analysis (Fig. 12.6a) displayed a similar pattern but with a few notable differences. Ford Field and Canton again exhibited the highest similarity (90%), followed by Rouge Park (80%). Dearborn and Inkster also showed ~80% similarity, suggesting a shared microbial presence. However, Venoy displayed a much higher similarity (70%) in this analysis compared to its Ct value similarity (25%), indicating that while certain bacterial strains were present at this site, their relative abundance differed significantly from the other locations. The Ct value-based cluster analysis (Fig. 12.6b) revealed that Ford Field and Canton were the most similar (95%), despite their geographic distance. Rouge Park closely followed (90%), suggesting that these three sites share a relatively similar microbial profile. On the other hand, Venoy was the most distinct site (25% similarity), making it the least similar to the other locations. This result was somewhat unexpected given its central location between Canton and Inkster, but it indicates that Venoy's microbial community composition differs significantly from other sites. Dearborn and Inkster exhibited moderate similarity (65%) to each other and the rest of the sites. These findings demonstrate that while both Ct values and presence/absence provide insights into microbial distribution, they do not always yield identical clustering patterns. The variation in Venoy's similarity across the two analyses highlights the importance of considering both microbial abundance (Ct values) and species presence/absence when assessing water quality and contamination patterns in the Rouge River.

3.5.2 Antibiotic-Resistant Genes

Cluster analysis was also performed to evaluate site similarity based on Ct values and presence/absence of ARGs, revealing distinct microbial-resistant gene distributions across the Rouge River sampling locations (Fig. 12.7). The presence/absence-based cluster analysis (Fig. 12.7a) revealed a similar pattern, where Dearborn and Inkster again clustered separately, but with lower similarity (~43%), indicating that

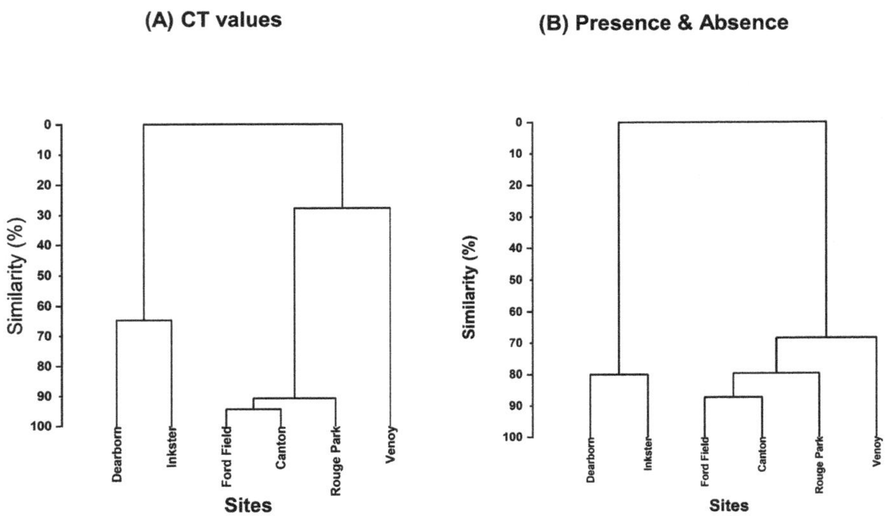

Fig. 12.6 Cluster analysis of fecal-Derived microbes based on (**a**) presence/absence and (**b**) Ct values across the six sampling sites

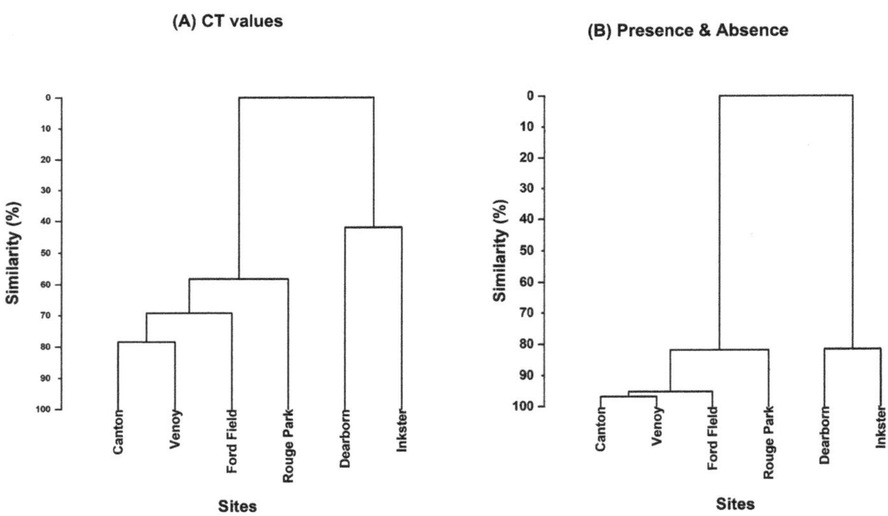

Fig. 12.7 Cluster Analysis of ARGs Based on (**a**) Ct values and (**b**) presence/absence across the six sampling sites

these sites harbor unique ARG profiles. Meanwhile, Venoy and Canton continued to show the highest similarity (~78%), reinforcing the strong similarity in their ARG composition. Ford Field and Rouge Park followed a clustering pattern similar to the Ct value analysis but exhibited slightly lower similarity percentages, highlighting differences in ARG presence rather than just abundance. The comparison between

Ct values and presence/absence analysis underscores that while certain sites may share the same ARGs, their expression levels or relative abundances can differ. The Ct value-based cluster analysis (Fig. 12.7b) revealed that Venoy and Canton exhibited the highest similarity (~97%), suggesting that these sites share highly comparable ARG abundance levels. Dearborn and Inkster formed a separate cluster with 80% similarity, indicating that while these two sites have distinct ARG compositions from the rest, they share a high degree of similarity with each other. Ford Field and Rouge Park displayed moderate similarity to the other sites, with Rouge Park appearing as the most distinct location. These findings suggest that ARG abundance levels vary significantly between sites, with Venoy and Canton being the most alike and Dearborn and Inkster forming their own unique grouping. These results suggest variability in antibiotic-resistant gene distribution across urban and suburban sites of the Rouge River watershed, with Dearborn and Inkster forming a distinct cluster, while Venoy and Canton displayed the highest similarity in both analyses.

4 Discussion

Recent studies on microbiological and antimicrobial-resistant gene (ARG) profiles have increasingly highlighted the need to understand their spatial variation, particularly in urban watershed environments such as the Rouge River. This study provides novel insights into these variations by employing advanced microbial DNA array technologies, a method that offers enhanced resolution and specificity compared to conventional techniques. Our findings demonstrate significant spatial differences in microbial communities and ARG distribution, influenced by a combination of environmental and anthropogenic factors. This aligns with recent literature emphasizing the critical role of urbanization in altering microbial landscapes (Tiquia-Arashiro & Grube, 2019; Nassereddine et al., 2024; Doyle et al., 2025).

4.1 Fecal-Derived Microbes

The fecal-derived microbial gene profiles across six environmental sites offer critical insights into the extent and variability of fecal contamination. Ford Field and Rouge Park emerge as the most contaminated sites, indicated by a high presence of microbial genes. This level of contamination is likely attributable to urban runoff or wastewater discharge, suggesting substantial environmental impact due to human activities in these areas (Bongiovanni et al., 2025). The high presence of these microbes at Venoy and Ford Field suggests intense urban runoff and possibly insufficient sewage treatment processes in these locations. Such conditions are fertile grounds for microbial proliferation, reflecting similar observations in other urban watersheds where human activities significantly impact water quality. In contrast, Dearborn and Inkster show moderate microbial presence, with a mix of frequently

detected, inconclusive, and absent species. This variability reflects differing contamination sources and fluctuating environmental conditions. Meanwhile, Venoy and Canton exhibit the lowest microbial presence, where many genes are absent, pointing towards better water quality relative to other sites. This pattern aligns with established research linking elevated microbial counts to urbanized areas with significant anthropogenic activities (Bongiovanni et al., 2025).

Commonly detected fecal-derived microbes such as *Enterococcus faecalis*, *Enterococcus faecium*, *Bacteroides fragilis*, *Clostridium perfringens*, and *Escherichia albertii* serve as indicators of widespread contamination across all sites. These findings are bolstered by the frequent detection of gut-associated *Bacteroides species*. Ford Field and Rouge Park also harbor opportunistic and pathogenic bacteria like *Salmonella enterica*, *Campylobacter jejuni*, *Vibrio vulnificus*, and *Aeromonas hydrophila*, which raise public health concerns due to their potential to cause disease. Conversely, some microbial genes associated with fecal contamination, such as *Yersinia pestis*, *Arcobacter skirrowii*, and *Mycobacterium arupense*, were not detected across the sites, indicating either their absence or concentrations below detectable levels. The absence of Providencia and Citrobacter species further suggests minimal contamination from these particular sources. The variability in pathogen presence reflects differing contamination severity and origins, highlighting that some sites face more intense fecal pollution than others. The consistent detection of fecal indicator species such as *Aeromonas* spp. across all sites supports existing literature that describes urban watersheds as prone to fecal contamination due to stormwater runoff and sewage overflows (Tiquia, 2002; Oest et al., 2018; Nassereddine et al., 2024; Xiao et al., 2025). This pervasive presence of fecal indicators highlights the ongoing challenges in managing water quality in urban environments, where even areas with fewer contamination sources are not immune to microbial pollution.

4.2 Antibiotic-Resistant Genes

The examination of antibiotic-resistant gene (ARG) distribution across six environmental sites reveals critical insights into the varying levels of antibiotic resistance influenced by environmental and anthropogenic factors. This comprehensive analysis underscores the differential impact of urbanization and pollution on ARG prevalence and diversity. Ford Field and Rouge Park exhibit the highest number of detected ARGs, suggesting intense selective pressures likely driven by elevated human activity, such as wastewater contamination and urban runoff. The consistent presence of high-impact ARGs in these areas indicates a significant environmental burden, pointing to both human-associated microbial communities and persistent pollution as major contributors to the resistance landscape. Conversely, Venoy and Canton show a greater number of inconclusive ARG detections, suggesting a lower level of antibiotic-resistant genes and weaker selection pressures. This could be attributed to reduced contamination sources or effective environmental

management in these areas. The lower ARG prevalence highlights potential differences in land use, infrastructure, and ecological management compared to more urbanized sites like Ford Field and Rouge Park. Dearborn and Inkster present mixed resistance profiles, indicating site-specific variations in ARG presence. This uneven distribution reflects the complex interplay of local factors influencing microbial populations, such as varied pollution inputs or differing levels of treatment in water discharge, affecting the persistence and dissemination of resistance genes.

The widespread detection of genes such as bla (beta-lactamase resistance), tet (tetracycline resistance), and aad (aminoglycoside resistance) across all sites indicates a baseline level of antibiotic-resistant bacteria pervasive in the environment. The consistent presence of genes like *bla*CTX-M and blaTEM suggests significant contributions from both human and animal waste, emphasizing the role of anthropogenic contamination in shaping resistance trends (Zhang et al., 2022; Nassereddine et al., 2024). The frequent detection of multidrug efflux pump genes, such as *oqx*A and *qnr*S, further underscores the antibiotic pressure within these environments, potentially driven by residues of antimicrobial agents. Interestingly, the absence of certain high-profile ARGs, such as carbapenem-resistant genes (e.g., *bla*KPC, *blaN*DM) and colistin-resistant genes (*mcr*-1), suggests that while there is a notable presence of common resistance genes, the distribution of clinically important resistance determinants is limited. This absence could reflect localized factors that prevent the proliferation or persistence of these highly resistant pathogens, such as specific environmental conditions or limited introduction of such genes through contamination (Sharma et al., 2024).

Comparing this with the antibiotic-resistant gene (ARG) profiles with the fecal-derived microbes profiles, locations with higher microbial diversity, particularly Ford Field and Rouge Park, also show increased ARG abundance, especially in beta-lactamase and efflux pump genes. This correlation suggests that greater microbial loads from fecal contamination may contribute to higher ARG presence. Overall, the data reveal variations in fecal contamination and microbial diversity, with Ford Field and Rouge Park showing the most contamination and Venoy and Canton the least. The presence of opportunistic pathogens and antibiotic-resistant bacteria in more contaminated sites highlights potential public health risks and underscores the environmental impact of urban wastewater discharge.

4.3 Physicochemical Parameters

Our study revealed that environmental parameters like water depth and electrical conductivity significantly correlate with the presence of both microbial communities and antibiotic-resistant genes (ARGs) across the Rouge River sites. For instance, Rouge Park, which demonstrated deeper water measurements and higher electrical conductivity, exhibited distinct pollution dynamics compared to shallower sites like Venoy. Rouge Park's higher conductivity suggests an influx of dissolved ions, potentially from urban runoff or natural sources, which could influence the

microbial and ARG profiles detected there. In contrast, Venoy, characterized by its relatively shallow water depth, presented with elevated microbial counts and ARG diversity. This may indicate a higher susceptibility to direct pollution through surface runoff and localized anthropogenic activities, resulting in a more concentrated contamination profile. The variation between these sites underscores the role of physical and chemical environmental factors in shaping microbial and ARG distribution patterns (Tiquia et al., 2007; Kassem et al., 2023). Statistical analyses, including ANOVA and paired *t*-tests, demonstrated significant differences in cycle threshold (Ct) values and microbial presence/absence scores across the sites. For example, Venoy's Ct values for fecal-derived microbes were notably lower than those observed at Dearborn, suggesting higher concentrations of these microbes and reinforcing the spatial heterogeneity of microbial pollution. Such differences are likely driven by variations in local land use practices, the density of urban infrastructure, and distinct hydrological patterns unique to each site, as suggested by O'Driscoll et al. (2010). This spatial heterogeneity emphasizes the complexity of pollution dynamics in urban watersheds, where site-specific environmental conditions and human activities interplay to influence microbial and ARG presence.

4.4 Cluster Analysis

The cluster analysis performed in this study yielded valuable insights into the similarities and differences between sites, effectively grouping locations with comparable microbial and ARG profiles. For example, our analysis revealed that Ford Field and Canton exhibited the highest similarity in both Ct values and presence/absence data, with 95% similarity based on Ct value clustering. This indicates that these sites might share common pollution sources or environmental conditions, such as nearby industrial activities or similar urban runoff patterns, influencing their microbial community and ARG distribution. Similarly, the cluster analysis highlighted that Venoy and Canton were notably alike in terms of ARG abundance, showing a 97% similarity in Ct values. Across all sites, certain ARGs, especially those for beta-lactam, tetracyclines, and sulfonamides, were detected with low Cq values, indicating widespread contamination shared across locations, possibly from common sources such as wastewater or agricultural runoff. Core fecal indicator microbes, including *Escherichia coli* and *Enterococcus faecalis*, were consistently detected at low Cq values, confirming their persistence in aquatic environments. Some ARGs and microbial genes showed intermediate Cq values (35–39), pointing to their presence at low abundance or inconsistent detection. This strong similarity suggests that these sites may be influenced by similar factors contributing to ARG presence, such as wastewater discharges or agricultural runoff, further underscoring the interconnected nature of pollution within the watershed (Tiquia, 2010a, b).

Conversely, sites such as Rouge Park and Inkster displayed lower similarity to other locations, indicating distinct microbial and ARG profiles. For instance, Rouge Park was identified as the least similar site in the Ct value-based cluster analysis

with only 25% similarity, suggesting that its unique environmental parameters, such as deeper water and higher conductivity, create different pollution dynamics compared to the more urbanized sites. Ford Field displayed the highest ARG abundance with Cq values <35, suggesting heightened antibiotic resistance contamination. In contrast, sites like Canton and Venoy had fewer ARGs, with many showing absence (Cq = 40), indicating less contamination and differing pollution sources or conditions affecting ARG persistence. Inkster had a more diverse and abundant fecal microbial profile, with numerous fecal-associated microbes detected at low Cq values. Meanwhile, Rouge Park and Canton showed lower fecal microbe prevalence, with several species absent or inconclusive, suggesting variable contamination levels and sources. Certain pathogens, like *Vibrio species* and *Listeria monocytogenes*, were detected at specific sites but absent elsewhere. Dearborn and Ford Field featured a broader range of microbial genes, hinting at higher contamination levels from wastewater or stormwater inputs. These clustering patterns emphasize the importance of integrated watershed management as a strategy to tackle the complex and multifaceted nature of river pollution. As suggested by Vriend et al. (2020), understanding these site-specific similarities and differences can guide targeted efforts in pollution control and remediation.

5 Conclusion

This study examines microbial communities and antibiotic-resistant gene (ARG) distribution along the Rouge River, revealing significant spatial variations driven by local environmental and human factors. Using advanced microbial DNA array technologies, we detected high levels of fecal microbes and ARGs at urbanized sites, such as Venoy and Ford Field, indicating significant impacts from urban runoff and inadequate sewage management. In contrast, Dearborn and Inkster showed lower contamination levels, suggesting more effective water management. The widespread detection of certain ARGs underscores the river's role as a conduit for antimicrobial resistance, posing potential public health risks. These findings highlight the necessity for targeted water management strategies and ongoing monitoring to protect water quality and public health in urban watersheds. Continued research is essential for adapting these strategies to changing environmental and urban conditions. Overall, our analysis reveals significant site-to-site differences in microbial contamination and ARG abundance. This highlights the urgent need for site-specific environmental management strategies. Addressing contamination, particularly in heavily impacted areas like Ford Field and Rouge Park, is crucial for mitigating public health risks and preserving water quality. While ARGs linked to beta-lactams, tetracyclines, and aminoglycosides are widespread, the distribution of more resistant genes for carbapenems and colistin is limited, emphasizing the complex interactions between environmental and human influences on antibiotic resistance. This complexity necessitates tailored strategies for both urbanized and less affected areas.

Acknowledgments The University of Michigan-Dearborn provided financial support for this project through the Office of Research and Sponsored Programs, specifically through the Research Initiation and Development Grant.

Authors' Contributions All authors contributed to the study conception and design. Material preparation, data collection and analysis were performed by Florent Qyteti, Vaughn Karaim, Reeghan Book, and Sonia M. Tiquia-Arashiro. The first draft of the manuscript was written by Florent Qyteti and Reeghan Book, and all authors commented on previous versions of the manuscript. All authors read and approved the final manuscript.

Conflict of Interest The authors declare they have no financial interests. Sonia Tiquia-Arashiro is the editor of the book project and was not involved in the editorial review, or the decision to publish, this article. All authors declare that there are no competing interests.

Data Availability Statement The authors confirm that the data supporting the findings of this study are available within the article [and/or] its supplementary materials.

Compliance with Ethical Standards The authors declare that there is no conflict of interest. This article does not contain any studies with human or animal participants performed by the authors.

References

Al Salah, D. M. M., Ngweme, G. N., Laffite, A., Otamonga, J. P., Mulaji, C., & Poté, J. (2020, September 1). Hospital wastewaters: A reservoir and source of clinically relevant bacteria and antibiotic resistant genes dissemination in urban river under tropical conditions. *Ecotoxicology and Environmental Safety, 200*, 110767. https://doi.org/10.1016/j.ecoenv.2020.110767

Awewomom, J., Dzeble, F., Takyi, Y. D., et al. (2024). Addressing global environmental pollution using environmental control techniques: A focus on environmental policy and preventive environmental management. *Discover Environment, 2*, 8. https://doi.org/10.1007/s44274-024-00033-5

Bongiovanni, D., Masson, S., Chialva, M., et al. (2025). Impact of urbanization on antimicrobial resistance in soil microbial communities. *Scientific Reports, 15*, 633. https://doi.org/10.1038/s41598-024-84945-5

Bowman, N., Patel, P., Xu, W., Alsaffar, A., & Tiquia-Arashiro, S. M. (2018). Enrichment and isolation of Pb-resistant bacteria from Saint Clair River sediments and their potential for Pb removal in aqueous solutions. *Applied Microbiology and Biotechnology, 102*, 2391–2398. https://doi.org/10.1007/s00253-018-8772-4

Caudill, S. P., Turner, W. E., & Patterson, D. G., Jr. (2007, September). Geometric mean estimation from pooled samples. *Chemosphere, 69*(3), 371–380. https://doi.org/10.1016/j.chemosphere.2007.05.061

Cho, K., Zholi, A., Frabutt, D., Flood, M., Floyd, D., & Tiquia, S. M. (2012). Linking bacterial diversity and geochemistry of uranium-contaminated groundwater. *Environmental Technology, 33*, 1629–1640. https://doi.org/10.1080/09593330.2011.641036

Czatzkowska, M., Wolak, I., Harnisz, M., & Korzeniewska, E. (2022). Impact of anthropogenic activities on the dissemination of ARGs in the environment—A review. *International Journal of Environmental Research and Public Health., 19*(19), 12853. https://doi.org/10.3390/ijerph191912853

Doyle, C., Wall, K., Fanning, S., & McMahon, B. J. (2025, January 6). Making sense of sentinels: Wildlife as the One Health bridge for environmental antimicrobial resistance surveillance. *Journal of Applied Microbiology, 136*(1), lxaf017. https://doi.org/10.1093/jambio/lxaf017

Flood, M., Frabutt, D., Floyd, D., Powers, A., Ezegwe, U., Devol, A., & Tiquia-Arashiro, S. M. (2015). Ammonia-oxidizing bacteria and archaea in sediments of the Gulf of Mexico. *Environmental Technology, 36*, 124–135. https://doi.org/10.1080/09593330.2014.942385

Gentile, M., Yan, T., Tiquia, S. M., et al. (2006). Stability in a denitrifying fluidized bed reactor. *Microbial Ecology, 52*, 311–321. https://doi.org/10.1007/s00248-006-9024-1

Kaphle, B., Wang, J., Ju, J., et al. (2025). Environmental risk assessment of the surface sediments based on trace elements analysis from the largest freshwater lake in the southern slope of the Himalaya, Nepal. *Environmental Monitoring and Assessment, 197*, 97.

Kassem, A., Abbas, L., Coutinho, O., Opara, S., Najaf, H., Kasperek, D., Li, X., & Tiquia-Arashiro, S. M. (2023). Applications of Fourier Transform-Infrared spectroscopy in microbial cell biology and environmental microbiology: Advances, challenges, and future perspectives. *Frontiers in Microbiology, 14*, 1304081. https://doi.org/10.3389/fmicb.2023.1304081

Lan, J., Liu, P., Hu, X., & Zhu, S. (2024). Harmful algal blooms in eutrophic marine environments: Causes, monitoring, and treatment. *Water, 16*(17), 2525. https://doi.org/10.3390/w16172525

Li, L. G., Huang, Q., Yin, X., & Zhang, T. (2020, October 15). Source tracking of antibiotic resistance genes in the environment – Challenges, progress, and prospects. *Water Research, 185*, 116127. https://doi.org/10.1016/j.watres.2020.116127

Liu, X., Tiquia, S. M., Holguin, G., Wu, L., Nold, S. C., Devol, A. H., Luo, K., Palumbo, A. V., Tiedje, J. M., & Zhou, J. (2003, June). Molecular diversity of denitrifying genes in continental margin sediments within the oxygen-deficient zone off the Pacific coast of Mexico. *Applied and Environmental Microbiology, 69*(6), 3549–3560. https://doi.org/10.1128/AEM.69.6.3549-3560.2003

Liu, Z., Gan, Y., Luo, J., Luo, X., Ding, C., & Cui, Y. (2025). Current status of emerging contaminant models and their applications concerning the aquatic environment: A review. *Water, 17*(1), 85. https://doi.org/10.3390/w17010085

McKindles, K. M., & Tiquia-Arashiro, S. M. (2012). Functional gene arrays for analysis of microbial communities on ocean platform. In S. Tiquia-Arashiro (Ed.), *Molecular biological technologies for ocean sensing. Springer protocols handbooks*. Humana Press. https://doi.org/10.1007/978-1-61779-915-0_9

Nassereddine, Z., Opara, S., Coutinho, O., Qyteti, F., Book, R., Heinicke, M. P., Napieralski, J., & Tiquia-Arashiro, S. M. (2024, 20 December). Critical perspectives on advancing antibiotic resistant gene (ARG) detection technologies in aquatic ecosystems. *Science of the Total Environment, 957*, 177775. https://doi.org/10.1016/j.scitotenv.2024.177775

Nguyen, S., Ala, F., Cardwell, C., Cai, D., McKindles, K. M., Lotvola, A., Hodges, S., Deng, Y., & Tiquia-Arashiro, S. M. (2013). Isolation and screening of carboxydotrophs isolated from composts and their potential for butanol synthesis. *Environmental Technology, 34*, 1995–2007. https://doi.org/10.1080/09593330.2013.795987

O'Driscoll, M., Clinton, S., Jefferson, A., Manda, A., & McMillan, S. (2010). Urbanization effects on watershed hydrology and in-stream processes in the southern United States. *Water, 2*(3), 605–648. https://doi.org/10.3390/w2030605

Oest, A., Alsaffar, A., Fenner, M., Azzopardi, D., & Tiquia-Arashiro, S. M. (2018, May 27). Patterns of change in metabolic capabilities of sediment microbial communities in river and lake ecosystems. *International Journal of Microbiology, 2018*, 6234931. https://doi.org/10.1155/2018/6234931

Ouyang, B., Yang, C., Lv, Z., Chen, B., Tong, L., & Shi, J. (2024, September). Recent advances in environmental antibiotic resistance genes detection and research focus: From genes to ecosystems. *Environment International, 191*, 108989. https://doi.org/10.1016/j.envint.2024.108989

Pagnucco, G., Overfield, D., Chamlee, Y., Fortuna, A., Sulaiman, F., Farinas, J., Schittenhelm, R., Shuler, C., Kassem, A., Opara, S., Najaf, H., Abbas, L., Coutinho, O., Catalfano, B., Li, X., & Tiquia-Arashiro, S. M. (2023). Metal tolerance and biosorption capacities of bacterial strains isolated from an urban watershed. *Frontiers in Microbiology, 14*, 1278886. https://doi.org/10.3389/fmicb.2023.1278886

Patel, D., Gismondi, R., Alsaffar, A., & Tiquia-Arashiro, S. M. (2019). Applicability of API ZYM to capture seasonal and spatial variabilities in lake and river sediments. *Environmental Technology, 40*(24), 3227–3239. https://doi.org/10.1080/09593330.2018.1468492

Plecha, S., Hall, D., & Tiquia-Arashiro, S. M. (2013). Screening for novel bacteria from the bioenergy feedstock switchgrass (*Panicum virgatum* L.). *Environmental Technology, 34*, 1895–1904. https://doi.org/10.1080/09593330.2013.818701

Pomaranski, E., & Tiquia-Arashiro, S. M. (2016). Butanol tolerance of carboxydotrophic bacteria isolated from manure composts. *Environmental Technology, 37*(15), 1970–1982. https://doi.org/10.1080/09593330.2015.1137360

Sharma, P., Pal, N., Kumawat, M., Mishra, S., Verma, V., Tiwari, R., Nagpal, R., Sarma, D., Singh, S., & Kumar, M. (2024). Antibiotic and non-antibiotic determinants of antimicrobial resistance: Insights from water ecosystems. *ACS ES&T Water, 4*(11), 4671–4689. https://doi.org/10.1021/acsestwater.4c00305

Tiquia, S. M. (2002). Microbial transformation of nitrogen during composting. In H. Insam, N. Riddech, & S. Klammer (Eds.), *Microbiology of composting*. Springer. https://doi.org/10.1007/978-3-662-08724-4_20

Tiquia, S. M. (2008, October). Diversity of sulfate-reducing genes (dsrAB) in sediments from Puget Sound, Environmental Technology, *29*(10), 1095–1108. https://doi.org/10.1080/09593330802190608

Tiquia, S. M. (2010a). Salt-adapted bacteria isolated from the Rouge River and potential for degradation of contaminants and biotechnological applications. *Environmental Technology, 31*, 967–978. https://doi.org/10.1080/09593331003706226

Tiquia, S. M. (2010b). Metabolic diversity of the heterotrophic microorganisms and potential link to pollution of the Rouge River. *Environmental Pollution, 158*, 1435–1443. https://doi.org/10.1016/j.envpol.2009.12.035

Tiquia, S. M. (2011). Extracellular hydrolytic enzyme activities of the heterotrophic microbial communities of the Rouge River: An approach to evaluate ecosystem response to urbanization. *Microbial Ecology, 62*, 679–689. https://doi.org/10.1007/s00248-011-9871-2

Tiquia, S. M., & Mormile, M. (2010). *Extremophiles: A source of innovation for industrial and environmental applications* (Vol. 31(10), pp. 1061–1181). Taylor and Francis. https://doi.org/10.1080/09593330.2010.489333

Tiquia, S. M., Tam, N. F. Y., & Hodgkiss, I. J. (1998). *Salmonella* elimination during composting of spent pig litter. *Bioresource Technology, 63*(2), 193–196. https://doi.org/10.1016/S0960-8524(97)00113-2

Tiquia, S. M., Gurczynski, S., Zholi, A., & Devol, A. (2006a, December). Diversity of biogeochemical cycling genes from Puget Sound sediments using DNA microarrays. *Environmental Technology, 27*(12), 1377–1389. https://doi.org/10.1080/09593332708618756

Tiquia, S. M., Masson, S. A., & Devol, A. (2006b, December). Vertical distribution of nitrite reductase genes (nirS) in continental margin sediments of the Gulf of Mexico. *FEMS Microbiology Ecology, 58*(3), 464–475. https://doi.org/10.1111/j.1574-6941.2006.00173.x

Tiquia, S. M., Davis, D., Hadid, H., Kasparian, S., Ismail, M., & Murray, K. S. (2007). Halophilic and halotolerant bacteria from river waters and shallow groundwater along the Rouge River of Southeastern Michigan. *Environmental Technology, 28*, 297–307. https://doi.org/10.1080/09593332808618789

Tiquia, S. M., Schleibak, M., Schlaff, J., Floyd, J. C., Benipal, B., Zakhem, E., & Murray, K. S. (2008). Microbial community profiling and characterization of some heterotrophic bacterial isolates from river waters and shallow groundwater wells along the Rouge River, Southeast Michigan. *Environmental Technology, 29*, 651–663. https://doi.org/10.1080/09593330801986998

Tiquia-Arashiro, S. M. (2012). *Molecular biological technologies for ocean sensing. Springer protocols handbooks* (p. 307). Humana. https://doi.org/10.1007/978-1-61779-915-0

Tiquia-Arashiro, S. M., & Grube, M. (2019). *Fungi in extreme environments: Ecological role and biotechnological significance* (p. 626). Springer Nature Switzerland AG. https://doi.org/10.1007/978-3-030-19030-9

Tiquia-Arashiro, S. M., & Mormile, M. (2013). Sustainable technologies: Bioenergy and biofuel from biowaste and biomass. *Environmental Technology, 34*(13–14), 1637–1638. https://doi.org/10.1080/09593330.2013.834162

Tiquia-Arashiro, S. M., & Pant, D. (2020). *Microbial electrochemical technologies* (p. 518). CRC Press. https://doi.org/10.1201/9780429487118

Tiquia-Arashiro, S. M., Medema, G., Ubomba-Jaswa, E., Urakawa, H., Ella, V., & Libardi, N. (2025). *Watershed microbiology: Biodiversity composition, ecological function, and technological advances*. Springer Nature Switzerland AG. (In press).

Vriend, P., Roebroek, C. T. J., & van Emmerik, T. (2020). Same but different: A framework to design and compare riverbank plastic monitoring strategies. *Frontiers in Water, 2*, 563791. https://doi.org/10.3389/frwa.2020.563791

Wu, W., Gu, B., Fields, M. W., Gentile, M., Ku, Y.-K., Yan, H., Tiquia, S. M., Yan, T., Nyman, J., Zhou, J., Jardine, P. M., & Criddle, C. S. (2005). Uranium (VI) reduction by denitrifying biomass. *Bioremediation Journal, 9*, 1–13. https://doi.org/10.1080/10889860590929628

Wu, D., Su, Y., Xi, H., Chen, X., & Xie, B. (2019, July 1). Urban and agriculturally influenced water contribute differently to the spread of antibiotic resistance genes in a mega-city river network. *Water Research, 158*, 11–21. https://doi.org/10.1016/j.watres.2019.03.010

Xiao, C., Ide, K., Matsunaga, H., Kogawa, M., Wagatsuma, R., & Takeyama, H. (2025, February). Metagenomic profiling of antibiotic resistance genes and their associations with the bacterial community along the Kanda River, an urban river in Japan. *Journal of Bioscience and Bioengineering, 139*(2), 147–155. https://doi.org/10.1016/j.jbiosc.2024.09.006

Zhang, Y., Li, K., Wu, Y., Liu, Y., Wu, R., Zhong, Y., Xiao, S., Mao, H., Li, G., Wang, Y., & Li, W. (2022, November 15). Distribution and correlation between antibiotic resistance genes and host-associated markers before and after swine fever in the longjiang watershed. *Environmental Pollution, 313*, 120101. https://doi.org/10.1016/j.envpol.2022.120101

Chapter 13
Integrated Approaches to Tracking Antibiotic Resistance in Aquatic Ecosystems

Reeghan Book, Florent Qyteti, Zainab Nassereddine, and Sonia M. Tiquia-Arashiro

1 Introduction

The emergence and dissemination of antibiotic resistance genes (ARGs) in aquatic ecosystems have become pressing environmental and public health concerns. Antibiotic resistance occurs when bacteria evolve mechanisms to resist the effects of antibiotics, rendering standard treatments ineffective and leading to persistent infections and increased mortality (Hasan et al., 2022). ARGs are the genetic components that encode this resistance, and their presence in the environment signifies an enhanced potential for the spread of antibiotic-resistant bacteria (ARBs). Aquatic habitats such as rivers, lakes, and oceans serve as significant reservoirs for ARGs due to their ability to harbor diverse microbial communities (Regina et al., 2021). These environments can act as conduits, facilitating the dissemination of ARGs across different microbial populations and potentially transferring these resistance traits to human pathogens (Salam et al., 2023; Nassereddine et al., 2024). The movement of ARGs in water bodies is influenced by various natural and anthropogenic factors. Natural processes, such as horizontal gene transfer among bacteria, allow ARGs to spread even in the absence of selective pressure from antibiotics. However, human activities have considerably exacerbated the issue. Anthropogenic activities, including agricultural runoff, wastewater discharge, and pharmaceutical contamination, have significantly contributed to the proliferation of ARGs in aquatic ecosystems (Akhter et al., 2024). Agricultural practices often involve the use of antibiotics to prevent disease in livestock, leading to the release of antibiotic residues and resistant bacteria into nearby water bodies through runoff (Oest et al., 2018; Iwu et al., 2020; Nassereddine et al., 2024; Patel et al., 2019; Tiquia, 2011; Tiquia et al., 2008). Similarly, wastewater discharge from hospitals and residential areas introduces a

R. Book · F. Qyteti · Z. Nassereddine · S. M. Tiquia-Arashiro (✉)
Department of Natural Sciences, University of Michigan-Dearborn, Dearborn, MI, USA
e-mail: smtiquia@umich.edu

S. M. Tiquia-Arashiro et al. (eds.), *Urban Watershed Microbiology, Volume 1*,
https://doi.org/10.1007/978-3-031-89071-0_13

multitude of antibiotics and ARG-containing bacteria into natural waters (Talat et al., 2023). Pharmaceutical contamination, often resulting from manufacturing and improper disposal of medications, further compounds the problem by continuously introducing high concentrations of antibiotics and resistant strains into aquatic environments (Felis et al., 2020). Given the vital role of aquatic ecosystems in sustaining life and the complex interplay of factors contributing to ARG spread, there is an urgent need for robust methodologies to detect and monitor these genes. Advanced detection methods are essential to understand the distribution and dynamics of ARGs, predict potential outbreaks of antibiotic-resistant infections, and devise effective intervention strategies to protect public health and environmental integrity.

Advancements in molecular biology, bioinformatics, and environmental science have propelled the development of sophisticated techniques for ARG detection and monitoring (Zhao et al., 2023; Nassereddine et al., 2024). Traditional methods like culture-based assays and polymerase chain reaction (PCR) remain foundational due to their reliability and specificity. However, these methods are increasingly being supplemented and, in some cases, replaced by high-throughput sequencing technologies that provide comprehensive insights into ARG diversity and distribution. High-throughput sequencing techniques, including metagenomics and whole-genome sequencing (WGS), allow researchers to analyze entire microbial communities and identify both known and novel ARGs without the need for pre-culturing. Despite these advancements, several challenges persist in the effective application of these technologies. Sample processing complexity is a significant hurdle, given the diverse and often harsh conditions of aquatic environments that can affect the integrity of DNA samples. Additionally, managing and interpreting the vast amounts of data generated by high-throughput sequencing requires robust bioinformatic tools and expertise, presenting another layer of complexity in ARG analysis.

This review aims to synthesize recent innovations in ARG detection and monitoring, critically evaluate their strengths and limitations, and explore future strategies for enhancing our ability to track and manage ARG dissemination in aquatic ecosystems. By addressing the challenges associated with current methodologies and highlighting potential future directions, this review seeks to contribute to the development of more effective approaches for combating the spread of antibiotic resistance, ultimately protecting both environmental and public health.

2 ARGs in Aquatic Ecosystems

Anthropogenic activities play a significant role in the dissemination of antibiotic resistance genes (ARGs) in water bodies. Various human-induced sources contribute to the introduction of ARGs into freshwater systems. Effluents from wastewater treatment plants (WWTPs), which often contain both antibiotic residues and resistant bacteria from hospitals, residential areas, and industrial sources, are a primary contributor (Omuferen et al., 2022; Nassereddine et al., 2024). Despite treatment processes, many WWTPs are unable to eliminate ARGs (Wang et al., 2021),

resulting in their discharge into natural water bodies (Wang et al., 2021; Tiquia-Arashiro, 2012; Tiquia et al., 2007). Agricultural runoff is another significant source of ARGs in aquatic ecosystems. The use of antibiotics in livestock farming to prevent disease and promote growth leads to the excretion of unmetabolized antibiotics and resistant bacteria into the environment (Flood et al., 2015; Tiquia et al., 1998; Tiquia-Arashiro & Pant, 2020; Zhu et al., 2013). Rain and irrigation can wash these contaminants from agricultural fields into nearby rivers and lakes, exacerbating the spread of ARGs (Anand et al., 2021; Richard & Tiquia, 1999; Bowman et al., 2018; Oest et al., 2018; Tiquia, 2010a, b; Nassereddine et al., 2024). Pharmaceutical manufacturing facilities also contribute to ARG contamination (Felis et al., 2020). During the production of antibiotics, waste products containing high concentrations of antibiotic compounds and resistant microorganisms can be released into waterways. This not only promotes the proliferation of ARGs but also facilitates the emergence of new resistance mechanisms due to continual exposure to sub-lethal doses of antibiotics (Mutuku et al., 2022).

The presence of ARGs in aquatic ecosystems poses several threats. First, it undermines biodiversity by altering microbial community structures and functions, which can have cascading effects on entire ecosystems (Quill et al., 2024; Tiquia et al. 2006a, b). Second, ARGs compromise ecosystem functions, such as nutrient cycling and water purification, by disrupting the natural microbial processes that underpin these services (McKindles & Tiquia-Arashiro, 2012; Quill et al., 2024). Third, from a public health perspective, the spread of ARGs in water bodies increases the risk of hard-to-treat infections. Humans can be exposed to resistant bacteria through recreational activities, consumption of contaminated water, and the food chain, leading to infections that are difficult or impossible to treat with available antibiotics. Environmental factors such as pollutants, temperature, and climate change further influence the dynamics of ARGs in aquatic ecosystems. Pollutants, including heavy metals and organic compounds, can co-select for resistance by creating environments that favor the survival of resistant bacteria (Bowman et al., 2018; Cho et al., 2012; Kassem et al., 2023; Pagnucco et al., 2023; Tiquia, 2002; Tiquia-Arashiro, 2018, 2019a, b; Wu et al., 2005). Temperature changes can affect bacterial growth rates and gene expression, potentially enhancing the transmission of ARGs. Climate change, through its impact on weather patterns and water cycles, can alter the distribution and concentration of ARGs, making it a critical factor in understanding and managing ARG spread. In summary, the presence and proliferation of ARGs in aquatic ecosystems represent a multifaceted challenge that threatens biodiversity, ecosystem functionality, and public health. Understanding the pathways and mechanisms of ARG dissemination, as well as the influence of environmental factors, is crucial for developing effective strategies to mitigate their impact. This necessitates ongoing research, robust monitoring systems, and interdisciplinary collaboration to address the environmental and public health risks posed by antibiotic resistance in aquatic environments (Tiquia-Arashiro et al., 2025).

3 Molecular Tools for ARGs and ARBs Studies

Traditional culture-based methods have been the cornerstone of identifying antimicrobial resistance. These methods isolate and grow bacteria under laboratory conditions, allowing scientists to study their resistance profiles. However, recent innovations in molecular methods have significantly improved detection speed, accuracy, and the breadth of what can be detected. For example, Lab-on-a-Chip devices miniaturize laboratory processes onto a single chip (Mark et al., 2010), enabling rapid and on-site testing of antibiotic resistance. Automated systems like BD Phoenix™ use advanced algorithms and digital imaging to quickly identify microbial growth and determine antibiotic susceptibility (Idelevich & Becker, 2019). Despite their advancements, these culture-based methods still have limitations, particularly their inability to detect non-cultivable bacteria, which make up a large proportion of microbial communities (Nguyen et al., 2013; Plecha et al., 2013; Pomaranski & Tiquia-Arashiro, 2016). This limitation necessitates the use of molecular techniques, such as PCR, metagenomics, and single-cell genomics, to provide a more comprehensive understanding of ARGs and antibiotic-resistant bacteria (ARBs). Strategies like new media formulations or co-cultures can enhance bacterial cultivation (Lagier et al., 2015; Tiquia-Arashiro, 2014a, b, c, d; Nassereddine et al., 2024).

3.1 *PCR and Quantitative PCR (qPCR) for Targeted ARG Detection*

PCR (Polymerase Chain Reaction) and qPCR (Quantitative PCR or Real-Time PCR) are foundational techniques for detecting specific ARGs. PCR amplifies DNA from small samples, making it possible to identify the presence of ARGs by targeting specific sequences (Galhano et al., 2021). It has been used extensively in environmental microbiology and microbial ecology (Tiquia, 2008, 2010c; Tiquia-Arashiro & Grube, 2019; Tiquia & Michel, 2002; Tiquia & Mormile, 2010; Tiquia-Arashiro & Mormile, 2013; Tiquia et al., 2004). qPCR enhances this capability by quantifying the amount of DNA, providing real-time measurements of ARG abundance (Galhano et al., 2021). Recent innovations in PCR techniques have significantly enhanced their precision and sensitivity. For instance:

- Digital PCR (dPCR) allows for absolute quantification of DNA, offering higher sensitivity and specificity by partitioning the sample into thousands of micro-reactions (Elder et al., 2021; Ferraro et al., 2024; Maestre-Carballa et al., 2024).
- Multiplex PCR enables the simultaneous amplification of multiple targets in a single reaction, increasing efficiency and throughput (Lata et al., 2016; Paulus et al., 2020; Strommenger et al., 2003).

- Loop-Mediated Isothermal Amplification (LAMP) offers a rapid and cost-effective alternative, often usable in field conditions without extensive laboratory equipment (Hassan et al., 2023; Meng et al., 2020; Mu et al., 2016).
- High-Throughput Quantitative PCR (HT-qPCR) allows the analysis of hundreds of samples simultaneously, making large-scale surveillance more feasible (Daw Elbait et al., 2024; Kasuga et al., 2022; Srathongneam et al., 2024; Stedtfeld et al., 2018).

However, the standardization of these methodologies remains a challenge. Variability in protocols, reagents, and instrumentation can lead to inconsistent results. Therefore, establishing uniform guidelines and standards is crucial for the reliable application of these techniques across different laboratories and studies.

3.2 Metagenomics for Resistome Analysis

Metagenomics involves sequencing the collective genetic material from environmental samples, providing comprehensive profiles of microbial communities and their ARGs. Unlike PCR, which targets specific genes, metagenomics offers a broader overview by analyzing all genetic material present, allowing for the discovery of unknown and novel ARGs (Ferrario et al., 2017; Galhano et al., 2021; Nassereddine et al., 2024). Advances in sequencing technologies, such as next-generation sequencing (NGS), have significantly expanded the capabilities of metagenomics. High-throughput sequencing generates large volumes of data, revealing the diversity and abundance of ARGs within complex microbial communities. Bioinformatics tools are then used to analyze these datasets, identifying and annotating resistance genes. However, managing these large datasets and ensuring accurate gene annotation are ongoing challenges. The complexity of the data requires sophisticated computational tools and expertise in bioinformatics. Additionally, ensuring that databases used for ARG identification are comprehensive and up to date is crucial for accurate analysis. Comparative metagenomics involves comparing metagenomic data across different samples, environments, or time points to track ARG distribution, evolution, and environmental responses (Fitzpatrick & Walsh, 2016; Forsberg et al., 2014; Pal et al., 2016). This approach has illuminated how human activities, such as agricultural practices and wastewater discharge, impact ARG dynamics (Louca et al., 2016). By integrating multi-omics approaches—such as metatranscriptomics (RNA), metaproteomics (proteins), and metabolomics (metabolites)—researchers can obtain a dynamic picture of microbial communities (Caporaso et al., 2012; Gloor et al., 2017; Quince et al., 2017). These combined data provide insights into the expression and regulation of ARGs, helping to understand their functional roles and environmental triggers. Despite the richness of information provided by comparative metagenomics, challenges in data analysis and standardization persist. Integrating different types of omics data

requires sophisticated analytical frameworks and consistent methodologies to ensure comparability across studies.

Functional genomics aims to understand the functions and interactions of genes within microbial communities. This field employs various techniques to study gene expression, regulation, and activity, offering in-depth insights into how ARGs operate and interact within their environments (Amos et al., 2014; Eduardo-Correia et al., 2020; Parsley et al., 2010).

- CRISPR-Cas systems provide powerful tools for gene editing and functional studies. These systems can be used to selectively knock out or modify specific ARGs, helping to elucidate their roles and mechanisms (Gholizadeh et al., 2020; Gupta et al., 2024; Lino et al., 2018; Pickar-Oliver & Gersbach, 2019; Tao et al., 2022; Zhang & Showalter, 2020).
- Single-cell sequencing allows for the examination of gene expression at the individual cell level, overcoming limitations associated with bulk sequencing approaches. This technique is particularly valuable for studying rare or unculti-vable bacteria (Blainey, 2013; Lan et al., 2017; Ma et al., 2023).
- Multi-omics integration combines data from genomics, transcriptomics, proteomics, and metabolomics to provide a holistic view of microbial communities and their resistance profiles (Liu et al., 2019).

However, functional genomics studies require extensive sampling and high-quality data preservation to ensure accurate results. The analysis of multi-omics data also demands sophisticated bioinformatics resources and expertise, underscoring the need for continued investment in computational tools and training.

3.3 Whole-Genome Sequencing (WGS)

Whole-genome sequencing (WGS) involves sequencing the entire genome of an organism, providing detailed genetic analysis. In the context of antibiotic resistance, WGS can identify ARGs, elucidate resistance mechanisms, and track genetic changes that confer resistance (Mardis, 2017). Recent advancements in third-generation sequencing technologies, such as Oxford Nanopore and PacBio, provide long reads that improve the accuracy of genome assembly (van der Helm et al., 2017; Wick et al., 2017). Hybrid approaches combining short-read and long-read sequencing further enhance the resolution and completeness of WGS (Ruan et al., 2020). However, WGS relies on comprehensive reference databases for accurate annotation of resistance genes. The analysis of WGS data is also resource-intensive, requiring significant computational infrastructure and bioinformatics expertise. These factors present challenges in the widespread application of WGS for ARG detection and characterization. In summary, molecular methods for detecting and characterizing ARGs and ARBs have evolved significantly, offering powerful tools for understanding and managing antibiotic resistance. Each method has its strengths and limitations, and ongoing advancements aim to address existing challenges,

improve accuracy, and facilitate broader application in both research and environmental monitoring contexts.

3.4 Bioinformatic Tools

Bioinformatic tools play a crucial role in processing and analyzing the vast datasets generated by high-throughput sequencing technologies. These tools are essential for extracting meaningful information from complex and voluminous data, which is particularly important for identifying and characterizing antibiotic resistance genes (ARGs). Key bioinformatic tools such as Antibiotic Resistance Genes Online Analysis Pipeline (ARGs-OAP), ResFinder, and the Comprehensive Antibiotic Resistance Database (CARD) have been developed to facilitate detailed ARG annotation and analysis.

1. *ARGs-OAP (Antibiotic Resistance Genes Online Analysis Pipeline)*: ARGs-OAP is designed specifically to handle metagenomic data, enabling the extraction and detailed annotation of ARG sequences from environmental samples. It utilizes a curated database of verified ARG sequences to provide accurate identification and functional annotation. This tool is particularly useful for studies focusing on the resistome—the collection of all ARGs in a given sample—allowing researchers to assess the diversity and abundance of resistance genes in microbial communities (Yang et al., 2016).
2. *ResFinder*: ResFinder is a web-based tool that identifies acquired antimicrobial resistance genes in whole-genome sequencing data from bacterial isolates. By comparing input sequences against a comprehensive database of known ARGs, ResFinder can accurately determine the presence and type of resistance genes. This tool is widely used in clinical microbiology to monitor resistance patterns and inform treatment strategies (Florensa et al., 2022).
3. *Comprehensive Antibiotic Resistance Database (CARD)*: CARD is an extensive, manually curated database that includes experimentally verified ARG sequences and detailed annotations on their resistance mechanisms, gene families, and associated antibiotics. The database's Resistance Gene Identifier (RGI) tool allows users to predict ARGs from nucleotide sequences. CARD is continually updated, ensuring that it remains a reliable resource for identifying both known and emerging resistance genes (Alcock et al., 2023).

Despite the significant advancements and capabilities of these bioinformatic tools, several challenges persist:

1. *Database Dependencies*: The accuracy of ARG identification and annotation relies heavily on the quality and comprehensiveness of reference databases. Many databases may not capture all existing ARGs, especially novel or rare variants, leading to potential misclassifications or missed detections. Regular updates

and expansion of these databases are essential to maintain their utility and accuracy.
2. *Computational Demands*: Analyzing high-throughput sequencing data requires significant computational resources. Large datasets necessitate powerful processors, extensive memory, and substantial storage capacity. Additionally, the complexity of the data demands advanced bioinformatics expertise to manage and interpret the results effectively.
3. *Standardization Needs*: Variability in data processing methods, software versions, and user-defined parameters can lead to inconsistencies in results. Standardizing bioinformatic pipelines across different research groups and studies is crucial for ensuring reproducibility and comparability of findings. Developing uniform protocols and guidelines can help mitigate these issues and enhance the reliability of ARG analyses.

To address these challenges and further improve the efficacy of bioinformatic tools in ARG research, several future directions are proposed:

1. *Improved Algorithms*: Developing more sophisticated algorithms can enhance the precision and efficiency of ARG detection and annotation. Machine learning and artificial intelligence approaches hold great potential for improving the accuracy of predictions and identifying novel resistance genes. These advanced algorithms can learn from large datasets and adapt to the evolving landscape of antibiotic resistance.
2. *Multi-Omics Integration*: Integrating data from various omics technologies, such as genomics, transcriptomics, proteomics, and metabolomics, can provide a more comprehensive understanding of ARGs. Combining these datasets can reveal insights into gene expression, regulation, and the functional impact of resistance genes within microbial communities. Multi-omics integration can also help identify environmental triggers and interactions that drive resistance dynamics.
3. *Standardized Pipelines*: Developing standardized bioinformatic pipelines can ensure consistency and reproducibility in ARG analyses. These pipelines should include uniform procedures for data processing, annotation, and interpretation, along with consensus guidelines for reporting results. Standardization can facilitate collaboration across research groups and enable more effective comparisons of findings.

In summary, bioinformatic tools are indispensable for processing and analyzing the complex data generated by high-throughput sequencing in ARG research. While challenges remain in database dependencies, computational demands, and standardization, ongoing advancements in algorithms, multi-omics integration, and standardized pipelines offer promising solutions. These efforts will enhance the accuracy, efficiency, and reliability of ARG detection and characterization, ultimately contributing to better management and mitigation of antibiotic resistance in diverse environments.

4 Environmental Monitoring Tools

Effectively monitoring antibiotic resistance genes (ARGs) in aquatic ecosystems requires advanced technologies that provide rapid, accurate, and comprehensive data. This section explores two primary methods: biosensors and Geographic Information Systems (GIS) combined with remote sensing technologies. Each method offers unique advantages and faces specific challenges.

4.1 *Biosensors*

Biosensors are analytical devices designed to detect specific biological molecules or microorganisms in real time (Tiquia-Arashiro, 2012, 2020; Tiquia-Arashiro & Pant, 2020). These devices are highly valued for their rapid response, sensitivity, and specificity, making them ideal for detecting ARGs in various environments.

4.1.1 Types of Biosensors

1. *Optical Biosensors*: Optical biosensors detect ARGs by measuring changes in light properties, such as fluorescence or surface plasmon resonance, upon binding with the target DNA or proteins. These sensors can provide real-time data with high sensitivity and specificity (Peltomaa et al., 2018).
2. *Electrochemical Biosensors*: These biosensors measure electrical signals generated by the interaction between the biosensor and ARGs. They offer advantages such as portability, cost-effectiveness, and quick results. Different types include amperometric (current-based), potentiometric (voltage-based), and impedimetric (impedance-based) biosensors (Wang et al., 2023).
3. *Piezoelectric Biosensors*: These sensors detect ARGs by measuring changes in the mass of the sensor's surface, resulting in shifts in crystal oscillation frequency. This method provides direct, real-time quantification of ARGs without the need for labeling (Tiquia-Arashiro, 2012; Zhao et al., 2013).

4.1.2 Recent Innovations

1. *Portable Biosensors*: Recent advancements have led to the development of portable biosensors that allow for on-site testing, reducing the need for extensive sample preparation. This innovation is particularly useful for real-time monitoring of ARGs in remote or field locations (Mao et al., 2024).
2. *Advanced Optical and Electrochemical Sensors*: Innovations such as nano-chip technology and hybridization chain reaction have improved the sensitivity and specificity of biosensors. For example, the Evanescent Wave Dual-color

fluorescence Fiber-embedded Optofluidic Nanochip (EDFON) has been demonstrated to detect ARGs with high precision (Chen et al., 2024).

4.1.3 Challenges

1. *Environmental Contaminants*: Biosensors often face interference from substances in aquatic samples, such as salts, organic matter, and other pollutants, which can affect the accuracy of detection. Developing sensors that are resistant to such interference is crucial (McKindles & Tiquia-Arashiro, 2012; Peltomaa et al., 2018).
2. *Need for Multiplexing*: Current biosensors generally detect single targets, limiting their efficiency. Efforts are needed to develop multiplexing capabilities, allowing simultaneous detection of multiple ARGs, which is essential for comprehensive monitoring (Zhao et al., 2013).
3. *Durability and Stability*: Ensuring that biosensors are durable and stable under various environmental conditions is necessary for their widespread application in field settings. Advances in materials science can contribute to improving sensor longevity and performance (Mao et al., 2024; Chen et al., 2024).

4.2 Geographic Information Systems (GIS)

Geographic Information Systems (GIS) are powerful tools for integrating and analyzing spatial data, which can be leveraged to map the distribution and dynamics of antibiotic resistance genes (ARGs) across various aquatic ecosystems. By combining ARG data with diverse geographic and environmental layers, GIS can provide invaluable insights into the sources, spread, and hotspots of ARG contamination (Chique et al., 2019; Zhu, 2016; Nassereddine et al., 2024).

4.2.1 Integrating and Analyzing Spatial Data

GIS allows for the integration of ARG data with multiple geographic layers, such as:

1. *Land Use*: Information on agricultural zones, urban areas, and industrial sites.
2. *Population Density*: Data on human settlements and population distributions.
3. *Water Quality Parameters*: Variables such as chemical concentrations, pH levels, turbidity, and temperature.

This integration helps in painting a comprehensive picture of how different factors interact to influence the presence and proliferation of ARGs in aquatic environments. For example, areas with high agricultural activity are often linked to elevated ARG levels due to runoff containing antibiotics from livestock farming.

4.2.2 Spatial Analysis

One of the key functionalities of GIS is spatial analysis, which enables the visualization and examination of ARG prevalence in relation to various environmental variables (Van Boeckel et al., 2015; Chique et al., 2019). This capability helps in:

1. *Identifying Patterns and Correlations*: GIS can highlight geographical regions where ARG presence is unusually high, known as hotspots. By overlaying ARG data with environmental layers, researchers can identify patterns and correlations between ARG occurrence and factors like land use, proximity to wastewater treatment plants, and industrial discharge points.
2. *Informing Targeted Interventions*: By pinpointing specific areas with high ARG concentrations, GIS analysis can guide targeted interventions. For example, implementing stricter waste management practices in identified hotspots can reduce ARG contamination. This data-driven approach ensures that resources and efforts are focused where they are most needed to effectively combat the spread of antibiotic resistance.

4.2.3 Predictive Modeling

GIS-based predictive modeling leverages historical data and current environmental conditions to forecast future trends in ARG distribution (Freire et al., 2024). Predictive models in GIS can:

1. *Forecast ARG Spread*: By analyzing historical ARG data and environmental variables, GIS models can predict how ARGs are likely to spread under different scenarios. These forecasts can consider factors such as seasonal changes, climate variations, and potential human activities like agricultural expansion or changes in wastewater management practices.
2. *Assist Policymakers*: These predictive insights are invaluable for policymakers and public health officials. By understanding potential future scenarios, they can develop proactive measures to mitigate the risk of antibiotic resistance. For instance, policy measures may include enacting regulations to limit the use of antibiotics in agriculture, improving wastewater treatment protocols, or increasing monitoring efforts in predicted high-risk areas.

In summary, GIS is a crucial tool for understanding and managing the complexities of ARG spread in aquatic environments. By integrating spatial data from various sources, conducting detailed spatial analyses, and developing predictive models, GIS provides a robust framework for tracking and mitigating the impact of antibiotic resistance. While the effectiveness of GIS applications is contingent on the availability of high-quality data and expertise, ongoing advancements in this field promise to enhance our capabilities in environmental monitoring and public health management.

4.3 Remote Sensing

Remote sensing is a powerful technique that utilizes satellite or drone-based technologies to gather extensive environmental data over large geographical areas. This method offers a comprehensive view of aquatic ecosystems, making it an invaluable tool for ARG monitoring and environmental assessment (Callejas, 2023; Khan et al., 2018; Nassereddine et al., 2024). Remote sensing eliminates the need for extensive ground-based sampling, providing a more efficient and wide-reaching approach to data collection.

4.3.1 Data Collection

Remote sensing technologies can continuously monitor a range of water quality parameters that are often associated with the presence and concentration of ARGs. Key parameters include:

1. *Turbidity*: Measures the clarity of water, with higher turbidity often indicating increased concentrations of organic and inorganic matter, which can correlate with ARG levels.
2. *Temperature*: Influences microbial activity and growth rates, which in turn affect ARG proliferation.
3. *Chemical Concentrations*: Detects pollutants, including antibiotics and other chemicals, which may promote the development and spread of ARGs.

These parameters can be monitored in real time, allowing for early detection of changes in water quality that could signify increased ARG presence (Hanlon et al., 2022; Lausch et al., 2024; Sudriani et al., 2023; Xie et al., 2024; Nassereddine et al., 2024). This continuous monitoring capability is crucial for timely interventions and management strategies.

4.3.2 Integration with GIS

Combining remote sensing data with Geographic Information Systems (GIS) significantly enhances the ability to analyze and interpret environmental variables affecting ARG distribution. This integration allows for:

1. *Enhanced Spatial Analysis*: By overlaying remote sensing data on GIS maps, researchers can visualize and analyze the spatial distribution of ARG-related environmental factors. This integrated approach can identify correlations between water quality parameters and ARG hotspots.
2. *Holistic Understanding*: The combination provides a more comprehensive understanding of the factors influencing ARG spread, such as land use patterns, hydrological processes, and climatic variations. This holistic view is essential for developing effective management and mitigation strategies.

Despite the numerous advantages of using remote sensing for ARG monitoring, several challenges need to be addressed to maximize its effectiveness:

1. *High-Quality Data Requirements*: The accuracy and reliability of remote sensing and GIS analysis heavily depends on high-resolution, high-quality data. Inaccurate or incomplete data can lead to erroneous conclusions and ineffective management strategies. Ensuring data accuracy and completeness is paramount.
2. *Technical Constraints*: The technical complexity of remote sensing technologies and GIS analysis requires specialized expertise. Skilled personnel are needed to operate sophisticated equipment, process large datasets, and interpret the results accurately. Additionally, the cost of acquiring and maintaining remote sensing platforms, such as satellites and drones, can be prohibitive for some organizations.
3. *Indirect Detection*: Remote sensing methods infer the presence of ARGs through associated environmental indicators (e.g., turbidity, temperature, chemical concentrations) rather than directly detecting ARGs themselves. This indirect approach may necessitate validation through ground-truthing—on-site sampling and analysis to confirm remote sensing findings—and complementary methods to ensure accurate assessments.

In summary, remote sensing technologies, particularly when integrated with GIS, offer robust methodologies for monitoring ARGs in aquatic ecosystems. While they present unique advantages such as large-scale coverage and continuous monitoring, the challenges of data quality, technical expertise, indirect detection, and high operational costs must be addressed. Continued advancements in remote sensing technologies and methodologies, along with interdisciplinary collaboration, are essential for enhancing our capabilities in monitoring and managing antibiotic resistance in the environment effectively.

5 Conclusions

The management of antibiotic resistance genes (ARGs) in aquatic ecosystems requires comprehensive and well-coordinated strategies, leveraging advances in molecular techniques, bioinformatics, and environmental monitoring. Recent advancements in high-throughput sequencing, digital PCR, metagenomics, whole-genome sequencing, and functional genomics have significantly improved the detection and characterization of ARGs, revealing detailed insights into their diversity, abundance, and dynamics within microbial communities. However, for these technologies to reach their full potential, issues such as standardization of protocols, efficient data processing, and adequate resource allocation must be addressed.

Bioinformatics tools like ARGs-OAP, ResFinder, and the Comprehensive Antibiotic Resistance Database (CARD) play a crucial role in analyzing and interpreting sequencing data, but challenges remain in ensuring database completeness, accessing sufficient computational resources, and standardizing data analysis

protocols. Environmental monitoring technologies, including biosensors, Geographic Information Systems (GIS), and remote sensing, are invaluable for real-time, spatially resolved monitoring of ARG prevalence, yet they require advancements in sensor design, data integration, and technical expertise.

Effective ARG management is inherently interdisciplinary, necessitating collaboration among microbiologists, environmental scientists, bioinformaticians, public health professionals, and policymakers. Establishing global and regional collaborative networks and developing standardized global data frameworks and repositories are essential for coordinated monitoring and management efforts. In conclusion, while significant progress has been made, ongoing efforts to address these challenges and foster interdisciplinary collaboration and data sharing are crucial for developing robust strategies to mitigate the impact of antibiotic resistance and protect both environmental and public health.

6 Future Perspectives

Looking ahead, future research in the field of antibiotic resistance gene (ARG) monitoring and management is poised to leverage several cutting-edge advancements to substantially enhance our understanding and capabilities. Innovations in CRISPR-Cas technologies hold the promise of highly precise gene editing and detection, presenting new avenues for identifying and potentially mitigating ARGs. Machine learning and artificial intelligence will revolutionize the analysis of complex datasets by automating ARG detection, predicting resistance patterns, and integrating multi-omics data to provide a comprehensive understanding of resistance mechanisms. Developments in "omics" techniques, including genomics, transcriptomics, proteomics, and metabolomics, will offer deeper insights into the functional roles of ARGs within microbial communities and the environmental triggers that influence their expression and transmission. Additionally, advancements in Geographic Information Systems (GIS) and remote sensing technologies will enhance environmental monitoring and public health strategies by providing more detailed spatial and temporal analyses of ARG distribution and dynamics. These technologies will enable the creation of predictive models to forecast ARG spread, thereby informing proactive intervention measures.

Addressing current challenges related to data complexity and standardization will be pivotal, as establishing uniform protocols and robust bioinformatic frameworks will ensure consistency, reproducibility, and comparability of results across different studies and regions. The integration of these advanced technologies and methodologies, along with interdisciplinary collaboration and the development of global data sharing frameworks, will pave the way for robust ARG surveillance and mitigation efforts, ultimately protecting both environmental and public health from the growing threat of antibiotic resistance.

Acknowledgments The authors express their sincere thanks and deep gratitude to the University of Michigan-Dearborn, Office of Research and Sponsored Programs for financially supporting this work.

Author Contributions All authors were involved in the conception and design of this review paper. Each author contributed to the writing process by drafting sections of the manuscript and providing critical feedback on earlier versions. Throughout the revision process, all authors participated in refining the content, ensuring accuracy, and enhancing the overall quality of the paper. All authors have thoroughly reviewed and approved the final version of the manuscript for submission.

Conflict of Interest The authors declare they have no financial interests. Sonia Tiquia-Arashiro is the editor of the book project and was not involved in the editorial review, or the decision to publish this article. All authors declare that there are no competing interests.

Data Availability The authors confirm that the data supporting the findings of this study are available within the article.

Compliance with Ethical Standards The authors affirm that they have no conflicts of interest to disclose. Furthermore, this article does not involve any studies conducted by the authors with human or animal participants.

References

Akhter, S., Bhat, M.A., Ahmed, S. et al. (2024). Antibiotic residue contamination in the aquatic environment, sources and associated potential health risks. *Environ Geochem Health 46*, 387. https://doi.org/10.1007/s10653-024-02146-5

Alcock, B. P., Huynh, W., Chalil, R., Smith, K. W., Raphenya, A. R., et al. (2023, January 6). CARD 2023: Expanded curation, support for machine learning, and resistome prediction at the Comprehensive Antibiotic Resistance Database. *Nucleic Acids Research, 51*(D1), D690–D699. https://doi.org/10.1093/nar/gkac920

Amos, G. C., Zhang, L., Hawkey, P. M., Gaze, W. H., & Wellington, E. M. (2014, July 16). Functional metagenomic analysis reveals rivers are a reservoir for diverse antibiotic resistance genes. *Veterinary Microbiology, 171*(3–4), 441–447. https://doi.org/10.1016/j.vetmic.2014.02.017

Anand, U., Reddy, B., Singh, V. K., Singh, A. K., Kesari, K. K., Tripathi, P., Kumar, P., Tripathi, V., & Simal-Gandara, J. (2021). Potential environmental and human health risks caused by antibiotic-resistant bacteria (ARB), antibiotic resistance genes (ARGs) and emerging contaminants (ECs) from municipal solid waste (MSW) landfill. *Antibiotics, 10*(4), 374. https://doi.org/10.3390/antibiotics10040374

Blainey, P. C. (2013). The future is now: Single-cell genomics of bacteria and archaea. *FEMS Microbiology Reviews, 37*(3), 407–427. https://doi.org/10.1111/1574-6976.12015

Bowman, N., Patel, P., Xu, W., Alsaffar, A., & Tiquia-Arashiro, S. M. (2018). Enrichment and isolation of Pb-resistant bacteria from Saint Clair River sediments and their potential for Pb removal in aqueous solutions. *Applied Microbiology and Biotechnology, 102*, 2391–2398. https://doi.org/10.1007/s00253-018-8772-4

Callejas, I. (2023). *Emerging techniques in coastal water quality in the US and Belize: Remote sensing and metagenomics* (Dissertation). University of California, Los Angeles. ProQuest Dissertations & Theses; 30525616.

Caporaso, J. G., Lauber, C. L., Walters, W. A., et al. (2012). Global patterns of 16S rRNA diversity at a depth of millions of sequences per sample. *Proceedings of the National Academy of Sciences, 108*(Supplement 1), 4516–4522. https://doi.org/10.1073/pnas.1000080107

Chen, D., Xu, W., Lu, Y., Zhuo, Y., Ji, T., & Long, F. (2024). Rapid and sensitive parallel on-site detection of antibiotics and resistance genes in aquatic environments using evanescent

wave dual-color fluorescence fiber-embedded OPTOFLUIDIC nanochip. *Biosensors and Bioelectronics, 257*, 116281. https://doi.org/10.1016/j.bios.2024.116281

Chique, C., Cullinan, J., Hooban, B., & Morris, D. (2019). Mapping and analysing potential sources and transmission routes of antimicrobial resistant organisms in the environment using geographic information systems—An exploratory study. *Antibiotics, 8*(1), 16. https://doi.org/10.3390/antibiotics8010016

Chique C, Cullinan J, Hooban B, Morris D. Mapping and Analysing Potential Sources and Transmission Routes of Antimicrobial Resistant Organisms in the Environment using Geographic Information Systems—An Exploratory Study. *Antibiotics. 2019; 8*(1):16. https://doi.org/10.3390/antibiotics8010016

Cho, K., Zholi, A., Frabutt, D., Flood, M., Floyd, D., & Tiquia, S. M. (2012). Linking bacterial diversity and geochemistry of uranium-contaminated groundwater. *Environmental Technology, 33*, 1629–1640. https://doi.org/10.1080/09593330.2011.641036

Daw Elbait, G., Daou, M., Abuoudah, M., et al. (2024, April 5). Comparison of qPCR and metagenomic sequencing methods for quantifying antibiotic resistance genes in wastewater. *PLoS One, 19*(4), e0298325. https://doi.org/10.1371/journal.pone.0298325

Eduardo-Correia, B., Morales-Filloy, H., & Abad, J. P. (2020, January 10). Bacteria from the multi-contaminated Tinto River Estuary (SW, Spain) show high multi-resistance to antibiotics and point to Paenibacillus spp. as antibiotic-resistance-dissemination players. *Frontiers in Microbiology, 10*, 3071. https://doi.org/10.3389/fmicb.2019.03071

Elder, F. C., Proctor, K., Barden, R., et al. (2021). Spatiotemporal profiling of antibiotics and resistance genes in a river catchment: Human population as the main driver of antibiotic and antibiotic resistance gene presence in the environment. *Water Research, 203*, 117533. https://doi.org/10.1016/j.watres.2021.117533

Felis, E., Kalka, J., Sochacki, A., Kowalska, K., Bajkacz, S., Harnisz, M., & Korzeniewska, E. (2020, January 5). Antimicrobial pharmaceuticals in the aquatic environment – Occurrence and environmental implications. *European Journal of Pharmacology, 866*, 172813. https://doi.org/10.1016/j.ejphar.2019.172813

Ferrario, C., Alessandri, G., Mancabelli, L., et al. (2017). Untangling the cecal microbiota of feral chickens by culturomic and metagenomic analyses. *Environmental Microbiology, 19*(11), 4771–4783. https://doi.org/10.1111/1462-2920.13943

Ferraro, G. B., Bonomo, C., Brandtner, D., et al. (2024). Characterisation of microbial communities and quantification of antibiotic resistance genes in Italian wastewater treatment plants using 16S rRNA sequencing and digital PCR. *Science of the Total Environment, 933*, 173217. https://doi.org/10.1016/j.scitotenv.2024.173217

Fitzpatrick, D., & Walsh, F. (2016). Antibiotic resistance genes across a wide variety of metagenomes. *FEMS Microbiology Ecology, 92*(2), fiw088. https://doi.org/10.1093/femsec/fiv168

Flood, M., Frabutt, D., Floyd, D., Powers, A., Ezegwe, U., Devol, A., & Tiquia-Arashiro, S. M. (2015). Ammonia-oxidizing bacteria and archaea in sediments of the Gulf of Mexico. *Environmental Technology, 36*, 124–135. https://doi.org/10.1080/09593330.2014.942385

Florensa, A. F., Kaas, R. S., Clausen, P. T. L. C., et al. (2022, January). ResFinder – An open online resource for identification of antimicrobial resistance genes in next-generation sequencing data and prediction of phenotypes from genotypes. *Microbial Genomics, 8*(1), 000748. https://doi.org/10.1099/mgen.0.000748

Forsberg, K., Patel, S., Gibson, M., et al. (2014). Bacterial phylogeny structures soil resistomes across habitats. *Nature, 509*, 612–616. https://doi.org/10.1038/nature13377

Freire, T. F. A., Hu, Z., Wood, K. B., & Gjini, E (2024, May 31). Modeling spatial evolution of multi-drug resistance under drug environmental gradients. *PLoS Computational Biology, 20*(5), e1012098. https://doi.org/10.1371/journal.pcbi.1012098

Galhano, B. S., Ferrari, R. G., Panzenhagen, P., et al. (2021). Antimicrobial resistance gene detection methods for bacteria in animal-based foods: A brief review of highlights and advantages. *Microorganisms, 9*(5), 923. https://doi.org/10.3390/microorganisms9050923

Gholizadeh, P., Köse, Ş., Dao, S., Ganbarov, K., Tanomand, A., Dal, T., Aghazadeh, M., Ghotaslou, R., Ahangarzadeh Rezaee, M., Yousefi, B., & Samadi Kafil, H. (2020, April 20). How CRISPR-Cas system could be used to combat antimicrobial resistance. *Infection and Drug Resistance, 13*, 1111–1121. https://doi.org/10.2147/IDR.S247271

Gloor, G. B., Macklaim, J. M., Pawlowsky-Glahn, V., & Egozcue, J. J. (2017). Microbiome datasets are compositional: And this is not optional. *Frontiers in Microbiology, 8*, 2224. https://doi.org/10.3389/fmicb.2017.02224

Gupta, S. S., Muneeb Hamza, K. H., Sones, C. L., et al. (2024). The CRISPR/Cas system as an antimicrobial resistance strategy in aquatic ecosystems. *Functional & Integrative Genomics, 24*, 110. https://doi.org/10.1007/s10142-024-01362-7

Hanlon, R., Jacquemin, S. J., Birbeck, J. A., Westrick, J. A., et al. (2022). Drone-based water sampling and characterization of three freshwater harmful algal blooms in the United States. *Frontiers in Remote Sensing, 3*, 949052. https://doi.org/10.3389/frsen.2022.949052

Hasan, C. M., Dutta, D., & Nguyen, A. N. T. (2022). Revisiting antibiotic resistance: Mechanistic foundations to evolutionary outlook. *Antibiotics, 11*(1), 40. https://doi.org/10.3390/antibiotics11010040

Hassan, M. M., van Vliet, A. H. M., Higgins, O., et al. (2023, May). Rapid culture-independent loop-mediated isothermal amplification detection of antimicrobial resistance markers from environmental water samples. *Microbial Biotechnology, 16*(5), 977–989. https://doi.org/10.1111/1751-7915.14227

Idelevich, E. A., & Becker, K. (2019, November). How to accelerate antimicrobial susceptibility testing. *Clinical Microbiology and Infection, 25*(11), 1347–1355. https://doi.org/10.1016/j.cmi.2019.04.025

Iwu, C. D., Korsten, L., & Okoh, A. I. (2020, September). The incidence of antibiotic resistance within and beyond the agricultural ecosystem: A concern for public health. *Microbiology, 9*(9), e1035. https://doi.org/10.1002/mbo3.1035

Kassem, A., Abbas, L., Coutinho, O., Opara, S., Najaf, H., Kasperek, D., Li, X., & Tiquia-Arashiro, S. M. (2023). Applications of Fourier Transform-Infrared spectroscopy in microbial cell biology and environmental microbiology: Advances, challenges, and future perspectives. *Frontiers in Microbiology, 14*, 1304081. https://doi.org/10.3389/fmicb.2023.1304081

Kasuga, I., Nagasawa, K., Suzuki, M., Kurisu, F., & Furumai, H. (2022). High-throughput screening of antimicrobial resistance genes and their association with class 1 integrons in urban rivers in Japan. *Frontiers in Environmental Science, 10*, 825372. https://doi.org/10.3389/fenvs.2022.825372

Khan, F. U., et al. (2018). Remote sensing technologies for enhancing forest inventories: A review. *Environmental Monitoring and Assessment, 190*(9), 1–23. https://doi.org/10.1080/07038992.2016.1207484

Lagier, J. C., Edouard, S., Pagnier, I., Mediannikov, O., Drancourt, M., & Raoult, D. (2015, January). Current and past strategies for bacterial culture in clinical microbiology. *Clinical Microbiology Reviews, 28*(1), 208–236. https://doi.org/10.1128/CMR.00110-14

Lan, F., Demaree, B., Ahmed, N., et al. (2017). Single-cell genome sequencing at ultra-high-throughput with microfluidic droplet barcoding. *Nature Biotechnology, 35*, 640–646. https://doi.org/10.1038/nbt.3880

Lata, P., Ram, S., & Shanker, R. (2016). Multiplex PCR based genotypic characterization of pathogenic vancomycin resistant Enterococcus faecalis recovered from an Indian river along a city landscape. *Springerplus, 5*, 1–11. https://doi.org/10.1186/s40064-016-2870-5

Lausch, A., Bannehr, L., Berger, S. A., et al. (2024). Monitoring water diversity and water quality with remote sensing and traits. *Remote Sensing, 16*(13), 2425. https://doi.org/10.3390/rs16132425

Lino, C. A., Harper, J. C., Carney, J. P., et al. (2018). Delivering CRISPR: A review of the challenges and approaches. *Drug Delivery, 25*(1), 1234–1257. https://doi.org/10.1080/10717544.2018.1474964

Liu, Z., Klümper, U., Liu, Y., Yang, Y., et al. (2019, August). Metagenomic and metatranscriptomic analyses reveal activity and hosts of antibiotic resistance genes in activated sludge. *Environment International, 129*, 208–220. https://doi.org/10.1016/j.envint.2019.05.036

Louca, S., Parfrey, L. W., & Doebeli, M. (2016, September 16). Decoupling function and taxonomy in the global ocean microbiome. *Science, 353*(6305), 1272–1277. https://doi.org/10.1126/science.aaf4507

Ma, P., Amemiya, H. M., He, L. L., Gandhi, S. J., Nicol, R., et al. (2023, February 16). Bacterial droplet-based single-cell RNA-seq reveals antibiotic-associated heterogeneous cellular states. *Cell, 186*(4), 877–891.e14. https://doi.org/10.1016/j.cell.2023.01.002

Maestre-Carballa, L., Navarro-López, V., & Martinez-Garcia, M. (2024). City-scale monitoring of antibiotic resistance genes by digital PCR and metagenomics. *Environmental Microbiomes, 19*, 16. https://doi.org/10.1186/s40793-024-00557-6

Mao, K., Zhang, H., Ran, F., Cao, H., Feng, R., Du, W., Li, X., & Yang, Z. (2024, January 15). Portable biosensor combining CRISPR/Cas12a and loop-mediated isothermal amplification for antibiotic resistance gene ermB in wastewater. *Journal of Hazardous Materials, 462*, 132793. https://doi.org/10.1016/j.jhazmat.2023.132793

Mardis, E. R. (2017, February). DNA sequencing technologies: 2006–2016. *Nature Protocols, 12*(2), 213–218. https://doi.org/10.1038/nprot.2016.182

Mark, D., Haeberle, S., Roth, G., et al. (2010). Microfluidic lab-on-a-chip platforms: Requirements, characteristics and applications. In S. Kakaç, B. Kosoy, D. Li, & A. Pramuanjaroenkij (Eds.), *Microfluidics based microsystems* (NATO science for peace and security series A: Chemistry and biology). Springer. https://doi.org/10.1007/978-90-481-9029-4_17

McKindles, K. M., & Tiquia-Arashiro, S. M. (2012). Functional gene arrays for analysis of microbial communities on ocean platform. In S. Tiquia-Arashiro (Ed.), *Molecular biological technologies for ocean sensing* (Springer protocols handbooks). Humana Press. https://doi.org/10.1007/978-1-61779-915-0_9

Meng, X., Zhang, G., Sun, B., Liu, S., et al. (2020) Rapid detection of mecA and femA genes by loop-mediated isothermal amplification in a microfluidic system for discrimination of different Staphylococcal species and prediction of methicillin resistance. *Frontiers in Microbiology, 11*, 1487. https://doi.org/10.3389/fmicb.2020.01487

Mu, X., Nakano, R., Nakano, A., Ubagai, T., et al. (2016). Loop-mediated isothermal amplification: Rapid and sensitive detection of the antibiotic resistance gene ISAba1-blaOXA-51-like in Acinetobacter baumannii. *Journal of Microbiological Methods, 121*, 36–40. https://doi.org/10.1016/j.mimet.2015.12.011

Mutuku, C., Gazdag, Z., & Melegh, S. (2022). Occurrence of antibiotics and bacterial resistance genes in wastewater: Resistance mechanisms and antimicrobial resistance control approaches. *World Journal of Microbiology and Biotechnology, 38*, 152. https://doi.org/10.1007/s11274-022-03334-0

Nassereddine, Z., Opara, S., Coutinho, O., Qyteti, F., Book, R., Heinicke, M. P., Napieralski, J., & Tiquia-Arashiro, S. M. (2024, December 20). Critical perspectives on advancing antibiotic resistant gene (ARG) detection technologies in aquatic ecosystems. *Science of the Total Environment, 957*, 177775. https://doi.org/10.1016/j.scitotenv.2024.177775

Nguyen, S., Ala, F., Cardwell, C., Cai, D., McKindles, K. M., Lotvola, A., Hodges, S., Deng, Y., & Tiquia-Arashiro, S. M. (2013). Isolation and screening of carboxydotrophs isolated from composts and their potential for butanol synthesis. *Environmental Technology, 34*, 1995–2007. https://doi.org/10.1080/09593330.2013.795987

Oest, A., Alsaffar, A., Fenner, M., Azzopardi, D., & Tiquia-Arashiro, S. M. (2018, May 27). Patterns of change in metabolic capabilities of sediment microbial communities in river and lake ecosystems. *International Journal of Microbiology, 2018*, 6234931. https://doi.org/10.1155/2018/6234931

Omuferen, L. O., Maseko, B., & Olowoyo, J. O. (2022). Occurrence of antibiotics in wastewater from hospital and convectional wastewater treatment plants and their impact on the effluent

receiving rivers: Current knowledge between 2010 and 2019. *Environmental Monitoring and Assessment, 194*, 306. https://doi.org/10.1007/s10661-022-09846-4

Pagnucco, G., Overfield, D., Chamlee, Y., Fortuna, A., Sulaiman, F., Farinas, J., Schittenhelm, R., Shuler, C., Kassem, A., Opara, S., Najaf, H., Abbas, L., Coutinho, O., Catalfano, B., Li, X., & Tiquia-Arashiro, S. M. (2023). Metal tolerance and biosorption capacities of bacterial strains isolated from an urban watershed. *Frontiers in Microbiology, 14*, 1278886. https://doi.org/10.3389/fmicb.2023.1278886

Pal, C., Bengtsson-Palme, J., Kristiansson, E., & Larsson, D. G. J. (2016). The structure and diversity of human, animal and environmental resistomes. *Microbiome, 4*(1), 54. https://doi.org/10.1186/s40168-016-0199-5

Parsley, L. C., Consuegra, E. J., Kakirde, K. S., Land, A. M., et al. (2010, June). Identification of diverse antimicrobial resistance determinants carried on bacterial, plasmid, or viral metagenomes from an activated sludge microbial assemblage. *Applied and Environmental Microbiology, 76*(11), 3753–3757. https://doi.org/10.1128/AEM.03080-09

Patel, D., Gismondi, R., Alsaffar, A., & Tiquia-Arashiro, S. M. (2019). Applicability of API ZYM to capture seasonal and spatial variabilities in lake and river sediments. *Environmental Technology, 40*(24), 3227–3239. https://doi.org/10.1080/09593330.2018.1468492

Paulus, G. K., Hornstra, L. M., & Medema, G. (2020). International tempo-spatial study of antibiotic resistance genes across the Rhine river using newly developed multiplex qPCR assays. *Science of the Total Environment, 706*, 135733. https://doi.org/10.1016/j.scitotenv.2019.135733

Peltomaa, R., Benito-Peña, E., & Moreno-Bondi, M. C. (2018). Bioinspired recognition elements for mycotoxin sensors. *Analytical Chemistry, 90*(1), 124–138. https://doi.org/10.1007/s00216-017-0701-3

Pickar-Oliver, A., & Gersbach, C. A. (2019). The next generation of CRISPR-Cas technologies and applications. *Nature Reviews. Molecular Cell Biology, 20*(8), 490–507. https://doi.org/10.1038/s41580-019-0131-5

Plecha, S., Hall, D., & Tiquia-Arashiro, S. M. (2013). Screening for novel bacteria from the bioenergy feedstock switchgrass (*Panicum virgatum* L.). *Environmental Technology, 34*, 1895–1904. https://doi.org/10.1080/09593330.2013.818701

Pomaranski, E., & Tiquia-Arashiro, S. M. (2016). Butanol tolerance of carboxydotrophic bacteria isolated from manure composts. *Environmental Technology, 37*(15), 1970–1982. https://doi.org/10.1080/09593330.2015.1137360

Quill, L., Ferreira, D., Joyce, B., Coleman, G., Harper, C., Martins, M., Hodkinson, T., Trimble, D., Gill, L., & O'Connell, D. W. (2024). An integrated mitigation approach to diffuse agricultural water pollution–a scoping review. *Frontiers in Environmental Science, 12*, 1340565. https://doi.org/10.3389/fenvs.2024.134056

Quince, C., Walker, A. W., Simpson, J. T., Loman, N. J., & Segata, N. (2017). Shotgun metagenomics, from sampling to analysis. *Nature Biotechnology, 35*(9), 833–844. https://doi.org/10.1038/nbt.3935

Regina, A. L. A., Medeiros, J. D., Teixeira, F. M., Côrrea, R. P., Santos, F. A. M., Brantes, C. P. R., Pereira, I. A., Stapelfeldt, D. M. A., Diniz, C. G., & da Silva, V. L. (2021, November 1). A watershed impacted by anthropogenic activities: Microbial community alterations and reservoir of antimicrobial resistance genes. *Science of the Total Environment, 793*, 148552. https://doi.org/10.1016/j.scitotenv.2021.148552

Richard, T. L., & Tiquia, S. M. (1999). In-barn composting system plus a warm bed for pigs. *BioCycle, 40*(10), 44.

Ruan, Z., Wu, J., Chen, H., Draz, M. S., Xu, J., & He, F. (2020, January 21). Hybrid genome assembly and annotation of a pandrug-resistant Klebsiella pneumoniae strain using nanopore and illumina sequencing. *Infection and Drug Resistance, 13*, 199–206. https://doi.org/10.2147/IDR.S240404

Salam, M. A., Al-Amin, M. Y., Salam, M. T., Pawar, J. S., Akhter, N., Rabaan, A. A., & Alqumber, M. A. A. (2023, July 5). Antimicrobial resistance: A growing serious threat for global public health. *Healthcare (Basel), 11*(13), 1946. https://doi.org/10.3390/healthcare11131946

Srathongneam, T., Sresung, M., Paisantham, P., et al. (2024). High throughput qPCR unveils shared antibiotic resistance genes in tropical wastewater and river water. *Science of the Total Environment, 908*, 167867. https://doi.org/10.1016/j.scitotenv.2023.167867

Stedtfeld, R. D., Guo, X., Stedtfeld, T. M., Sheng, H., et al. (2018, September 1). Primer set 2.0 for highly parallel qPCR array targeting antibiotic resistance genes and mobile genetic elements. *FEMS Microbiology Ecology, 94*(9), fiy130. https://doi.org/10.1093/femsec/fiy130

Strommenger, B., Kettlitz, C., Werner, G., & Witte, W. (2003). Multiplex PCR assay for simultaneous detection of nine clinically relevant antibiotic resistance genes in Staphylococcus aureus. *Journal of Clinical Microbiology, 41*(9), 4089–4094. https://doi.org/10.1128/jcm.41.9.4089-4094.2003

Sudriani, Y., Sebestyén, V., & Abonyi, J. (2023). Surface water monitoring systems—The importance of integrating information sources for sustainable watershed management. *IEEE Access, 11*, 36421–36451. https://doi.org/10.1109/ACCESS.2023.3263802

Talat, A., Blake, K. S., Dantas, G., & Khan, A. U. (2023, February 14). Metagenomic insight into microbiome and antibiotic resistance genes of high clinical concern in urban and rural hospital wastewater of northern India origin: A major reservoir of antimicrobial resistance. *Microbiology Spectrum, 11*(2), e0410222. https://doi.org/10.1128/spectrum.04102-22

Tao, S., Chen, H., Li, N., & Liang, W. (2022, August 2). The application of the CRISPR-Cas system in antibiotic resistance. *Infection and Drug Resistance, 15*, 4155–4168. https://doi.org/10.2147/IDR.S370869

Tiquia, S. M. (2002). Microbial transformation of nitrogen during composting. In H. Insam, N. Riddech, & S. Klammer (Eds.), *Microbiology of composting*. Springer. https://doi.org/10.1007/978-3-662-08724-4_20

Tiquia, S. M. (2008, October). Diversity of sulfate-reducing genes (dsrAB) in sediments from Puget Sound. *Environmental Technology, 29*(10), 1095–1108. https://doi.org/10.1080/09593330802190608

Tiquia, S. M. (2010a). Salt-adapted bacteria isolated from the Rouge River and potential for degradation of contaminants and biotechnological applications. *Environmental Technology, 31*, 967–978. https://doi.org/10.1080/09593331003706226

Tiquia, S. M. (2010b). Metabolic diversity of the heterotrophic microorganisms and potential link to pollution of the Rouge River. *Environmental Pollution, 158*, 1435–1443. https://doi.org/10.1016/j.envpol.2009.12.035

Tiquia, S. M. (2010c). Using terminal restriction fragment length polymorphism (T-RFLP) analysis to assess microbial community structure in compost systems. In S. Cummings (Ed.), *Bioremediation* (Methods in molecular biology) (Vol. 599). Humana Press. https://doi.org/10.1007/978-1-60761-439-5_6

Tiquia, S. M. (2011). Extracellular hydrolytic enzyme activities of the heterotrophic microbial communities of the Rouge River: An approach to evaluate ecosystem response to urbanization. *Microbial Ecology, 62*, 679–689. https://doi.org/10.1007/s00248-011-9871-2

Tiquia, S. M., & Michel, F. C. (2002). Bacterial diversity in livestock manure composts as characterized by terminal restriction fragment length polymorphisms (T-RFLP) of PCR-amplified 16s rRNA gene sequences. In H. Insam, N. Riddech, & S. Klammer (Eds.), *Microbiology of composting*. Springer. https://doi.org/10.1007/978-3-662-08724-4_6

Tiquia, S. M., & Mormile, M. (2010). *Extremophiles: A source of innovation for industrial and environmental applications* (Vol. 31(10), pp. 1061–1181). Taylor and Francis. https://doi.org/10.1080/09593330.2010.489333

Tiquia, S. M., Tam, N. F. Y., & Hodgkiss, I. J. (1998). *Salmonella* elimination during composting of spent pig litter. *Bioresource Technology, 63*(2), 193–196. https://doi.org/10.1016/S0960-8524(97)00113-2

Tiquia, S. M., Chong, S. C., Fields, M. W., & Zhou, J. (2004). Oligonucleotide-based functional gene arrays for analysis of microbial communities in the environment. In G. G. Kowalchuk, F. J. de Bruijn, I. M. Head, A. D. Akkermans, & J. D. van Elsas (Eds.), *Molecular microbial ecology manual* (pp. 1743–1763). The Netherlands.

Tiquia, S. M., Gurczynski, S., Zholi, A., & Devol, A. (2006a, December). Diversity of biogeochemical cycling genes from Puget Sound sediments using DNA microarrays. *Environmental Technology, 27*(12), 1377–1389. https://doi.org/10.1080/09593332708618756

Tiquia, S. M., Masson, S. A., & Devol, A. (2006b, December). Vertical distribution of nitrite reductase genes (nirS) in continental margin sediments of the Gulf of Mexico. *FEMS Microbiology Ecology, 58*(3), 464–475. https://doi.org/10.1111/j.1574-6941.2006.00173.x

Tiquia, S. M., Davis, D., Hadid, H., Kasparian, S., Ismail, M., & Murray, K. S. (2007). Halophilic and halotolerant bacteria from river waters and shallow groundwater along the Rouge River of Southeastern Michigan. *Environmental Technology, 28*, 297–307. https://doi.org/10.1080/09593332808618789

Tiquia, S. M., Schleibak, M., Schlaff, J., Floyd, J. C., Benipal, B., Zakhem, E., & Murray, K. S. (2008). Microbial community profiling and characterization of some heterotrophic bacterial isolates from river waters and shallow groundwater wells along the Rouge River, Southeast Michigan. *Environmental Technology, 29*, 651–663. https://doi.org/10.1080/09593330801986998

Tiquia-Arashiro, S. M. (2012). *Molecular biological technologies for ocean sensing* (Springer protocols handbooks) (p. 307). Humana. https://doi.org/10.1007/978-1-61779-915-0

Tiquia-Arashiro, S. M. (2014a). Biotechnological applications of thermophilic carboxydotrophs. In *Thermophilic carboxydotrophs and their applications in biotechnology* (SpringerBriefs in microbiology). Springer. https://doi.org/10.1007/978-3-319-11873-4_4

Tiquia-Arashiro, S. M. (2014b). Microbial CO metabolism. In *Thermophilic Carboxydotrophs and their applications in biotechnology* (SpringerBriefs in microbiology). Springer. https://doi.org/10.1007/978-3-319-11873-4_2

Tiquia-Arashiro, S. M. (2014c). Chapter 3: CO-oxidizing microorganisms. In *Thermophilic carboxydotrophs and their applications in biotechnology* (SpringerBriefs in microbiology) (pp. 11–28). Springer. https://doi.org/10.1007/978-3-319-11873-4_3

Tiquia-Arashiro, S. M. (2014d). *Thermophilic carboxydotrophs and their applications in biotechnology* (SpringerBriefs in microbiology) (p. 131). Springer. https://doi.org/10.1007/978-3-319-11873-4

Tiquia-Arashiro, S. M. (2018). Lead absorption mechanisms in bacteria as strategies for lead bioremediation. *Applied Microbiology and Biotechnology, 102*(13), 5437–5444. https://doi.org/10.1007/s00253-018-8969-6

Tiquia-Arashiro, S. M. (2019a). Synthesis of metallic nanoparticles by halotolerant fungi. In *Fungi in extreme environments: Ecological role and biotechnological significance* (pp. 383–394). Springer International Publishing. https://doi.org/10.1007/978-3-030-19030-9_19

Tiquia-Arashiro, S. M. (2019b). Thermophilic fungi in composts: Their role in composting and industrial processes. In *Fungi in extreme environments: Ecological role and biotechnological significance* (pp. 587–606). Springer International Publishing. https://doi.org/10.1007/978-3-030-19030-9_29

Tiquia-Arashiro, S. M. (2020). Electricity production from carbon monoxide and syngas in a thermophilic microbial fuel cell. In S. M. Tiquia-Arashiro & D. Pant (Eds.), *Microbial electrochemical technologies* (pp. 288–299). CRC Press.

Tiquia-Arashiro, S. M., & Grube, M. (2019). *Fungi in extreme environments: Ecological role and biotechnological significance* (p. 626). Springer Nature Switzerland AG. https://doi.org/10.1007/978-3-030-19030-9

Tiquia-Arashiro, S. M., & Mormile, M. (2013). Sustainable technologies: Bioenergy and biofuel from biowaste and biomass. *Environmental Technology, 34*(13–14), 1637–1638. https://doi.org/10.1080/09593330.2013.834162

Tiquia-Arashiro, S. M., & Pant, D. (2020). *Microbial electrochemical technologies* (p. 518). CRC Press. https://doi.org/10.1201/9780429487118

Tiquia-Arashiro, S. M., Medema, G., Ubomba-Jaswa, E., Urakawa, H., Ella, V., & Libardi, N. (2025). *Watershed microbiology: Biodiversity composition, ecological function, and technological advances*. Springer Nature Switzerland AG. (In press).

Van Boeckel, T. P., Brower, C., Gilbert, M., et al. (2015, May 5). Global trends in antimicrobial use in food animals. *Proceedings of the National Academy of Sciences of the United States of America, 112*(18), 5649–5654. https://doi.org/10.1073/pnas.1503141112

van der Helm, E., Imamovic, L., Hashim Ellabaan, M. M., et al. (2017, May 5). Rapid resistome mapping using nanopore sequencing. *Nucleic Acids Research, 45*(8), e61. https://doi.org/10.1093/nar/gkw1328

Wang, R., Ji, M., Zhai, H., Guo, Y., & Liu, Y. (2021, November 20). Occurrence of antibiotics and antibiotic resistance genes in WWTP effluent-receiving water bodies and reclaimed wastewater treatment plants. *Science of the Total Environment, 796*, 148919. https://doi.org/10.1016/j.scitotenv.2021.148919

Wang, B., Wang, H., Lu, X., Zheng, X., & Yang, Z. (2023). Recent advances in electrochemical biosensors for the detection of foodborne pathogens: Current perspective and challenges. *Foods, 12*(14), 2795. https://doi.org/10.3390/foods12142795

Wick, R. R., Judd, L. M., Gorrie, C. L., & Holt, K. E. (2017, June 8). Unicycler: Resolving bacterial genome assemblies from short and long sequencing reads. *PLoS Computational Biology, 13*(6), e1005595. https://doi.org/10.1371/journal.pcbi.1005595

Wu, W., Gu, B., Fields, M. W., Gentile, M., Ku, Y.-K., Yan, H., Tiquia, S. M., Yan, T., Nyman, J., Zhou, J., Jardine, P. M., & Criddle, C. S. (2005). Uranium (VI) reduction by denitrifying biomass. *Bioremediation Journal, 9*, 1–13. https://doi.org/10.1080/10889860590929628

Xie, Z., Zhao, Y., Jiang, R., Zhang, M., et al. (2024). Seasonal dynamics of fallow and cropping lands in the broadacre cropping region of Australia. *Remote Sensing of Environment, 305*, 114070. https://doi.org/10.1016/j.rse.2024.114070

Yang, Y., Jiang, X., Chai, B., et al. (2016, August 1). ARGs-OAP: Online analysis pipeline for antibiotic resistance genes detection from metagenomic data using an integrated structured ARG-database. *Bioinformatics, 32*(15), 2346–2351. https://doi.org/10.1093/bioinformatics/btw136

Zhang, Y., & Showalter, A. M. (2020, November 20). CRISPR/Cas9 genome editing technology: A valuable tool for understanding plant cell wall biosynthesis and function. *Frontiers in Plant Science, 11*, 589517. https://doi.org/10.3389/fpls.2020.589517

Zhao, J., Yan, Y., Zhu, L., Li, X., & Li, G. (2013). An amperometric biosensor for the detection of hydrogen peroxide released from human breast cancer cells. *Biosensors and Bioelectronics, 41*, 815–819. https://doi.org/10.1016/j.bios.2012.10.019

Zhao, Y., Huang, F., Wang, W., Gao, R., Fan, L., Wang, A., & Gao, S. H. (2023, November 20). Application of high-throughput sequencing technologies and analytical tools for pathogen detection in urban water systems: Progress and future perspectives. *Science of the Total Environment, 900*, 165867. https://doi.org/10.1016/j.scitotenv.2023.165867

Zhu, X. (2016). *GIS for environmental applications: A practical approach* (1st ed., p. 490). Routledge. https://doi.org/10.4324/9780203383124

Zhu, Y.-G., Johnson, T. A., Su, J.-Q., et al. (2013). Diverse and abundant antibiotic resistance genes in Chinese swine farms. *Proceedings of the National Academy of Sciences, 110*(9), 3435–3440. https://doi.org/10.1073/pnas.1222743110

Part III
Advances in Biological Monitoring and Modelling in Urban Watersheds and Community Science Monitoring

Chapter 14
Remote Sensing and Drivers of Harmful Cyanobacterial Algal Blooms in Freshwater Reservoirs

Carolina Acuña-Alonso, Elena Seoane-Martínez, Fernando António Leal Pacheco, Luís Filipe Sanches Fernandes, and Xana Álvarez

1 Introduction

Eutrophication is a natural process characterized by the growth of microalgae and cyanobacteria due to the concentration of nutrients such as nitrogen and phosphorus in water bodies. It becomes a critical problem when aggravated by human activity (Smith, 2009; Khan, 2014; Naeem et al., 2014; Oest et al., 2018; Igwaran et al., 2024). The pressure generated by human activity which contributes to the discharge of untreated wastewater and the excessive use of fertilizers in agriculture introduces significant amounts of nutrients into water bodies, resulting in rapid, uncontrolled growth of algae and aquatic plants (Lee & Jones-Lee, 2005; Pagnucco et al., 2023; Igwaran et al., 2024). In addition, factors such as climate change have resulted in

C. Acuña-Alonso
Universidade de Vigo, Hydro-Forestry Geomodeling Research Group, School of Forestry Engineering, Pontevedra, Spain

Centre for the Research and Technology of Agro-Environmental and Biological Sciences – CITAB, University of Trás-os-Montes and Alto Douro (UTAD), Vila Real, Portugal
e-mail: carolina.alonso@uvigo.es

E. Seoane-Martínez · X. Álvarez (✉)
Universidade de Vigo, Hydro-Forestry Geomodeling Research Group, School of Forestry Engineering, Pontevedra, Spain
e-mail: elena.seoane.martinez@uvigo.es; xaalvarez@uvigo.es

F. A. L. Pacheco
Chemistry Centre of Vila Real – CQVR, University of Trás-os-Montes and Alto Douro (UTAD), Vila Real, Portugal
e-mail: fpacheco@utad.pt

L. F. S. Fernandes
Centre for the Research and Technology of Agro-Environmental and Biological Sciences – CITAB, University of Trás-os-Montes and Alto Douro (UTAD), Vila Real, Portugal
e-mail: lfilipe@utad.pt

S. M. Tiquia-Arashiro et al. (eds.), *Urban Watershed Microbiology, Volume 1*,
https://doi.org/10.1007/978-3-031-89071-0_14

increased temperatures, thus generating a greater presence of nutrients due to increased runoff from agricultural activities (Rajamanickam et al., 2023). This inordinate response leads to the formation of "algal blooms"known as Cyanobacterial Harmful Algal Blooms (Cyano-HABs) which, as they die and decompose, deplete the dissolved oxygen in the water, reducing its quality and threatening aquatic life (Paerl & Otten, 2013; Flood et al., 2015; Patel et al., 2019; Khatoon et al., 2023). In this context the proliferation of cyanobacteria, particularly of the genus *Microcystis*, entails an additional risk due to their ability to produce cyanotoxins harmful to human health and the environment (Acuña-Alonso et al., 2020a, b; Acuña-Alonso et al., 2021).

The application of predictive and explanatory models is essential to anticipate and understand Cyano-HABs. Since 2013, this field of research has experienced a significant increase, driven by the adoption of new monitoring technologies and satellite remote sensing (Rousso et al., 2020). These models focus mainly on lakes that have been extensively studied, such as Lake Erie in Canada (Jetoo et al., 2015) and Lake Taihu in China (Guo, 2007), where severe episodes of Cyano-HABs have occurred in recent years, causing negative impacts on drinking water supply. These studies serve as examples of well-researched lakes that have been successfully modeled (Rousso et al., 2020). The research conducted by Rousso et al. (2020) summarizes and synthesizes a wide range of publications related to this topic, underscoring the importance and increasing attention being paid to this issue globally. On the other hand, several techniques have been explored, including the use of Bayesian Networks (Borsuk et al., 2004) and Artificial Neural Networks (ANN) (Huo et al., 2013; Srisuksomwong & Pekkoh, 2020), which have been applied in previous studies that have sought to predict the presence of cyanobacteria. The methodology used most frequently is PLS-PM (Partial Least Square Path Modeling). This methodology, based on statistical data analysis, is presented as the convergence of regression models, structural equation models, and multivariate analysis methods (González Huelva, 2018). It has been widely and predominantly applied in the field of information systems and its use has subsequently spread to other domains, including environmental sciences (Benitez et al., 2017; Acuña-Alonso et al., 2022), where it has been employed in several studies (McKindles & Tiquia-Arashiro, 2012; Fernandes et al., 2019; Oliveira et al., 2019; Tiquia-Arashiro & Pant, 2020; Gao et al., 2021; Zhao et al., 2022; Kassem et al., 2023; Zheng et al., 2024). Models created with the PLS-PM methodology not only permit accurate identification of the factors influencing the variable of interest but also open the door to the assessment of how climate change affects the proliferation of cyanobacteria in water bodies. This provides a long-term perspective that helps to anticipate future challenges and makes for more informed water management decisions. Such models are fundamental tools for planning and sustainable management of water resources in an ever-changing climate change context. It is important to note that few studies have been conducted to date using this model in relation to eutrophication and Cyano-HABs (Gao et al., 2021; Acuña-Alonso et al., 2022). However, Gao et al. (2021) analyze the influence of particulate organic matter on nutrient cycling and obtain adjustment coefficients ranging from 0.60 to 0.65. Moreover, Acuña-Alonso et al.

(2022) examine the impact of eutrophication, physicochemical parameters, and weather conditions on ecological integrity and obtains an adjustment coefficient of 0.515 for the variable "Eutrophication." While Gao's study focuses on the physicochemical composition of water, Acuña-Alonso's research is similar to the present investigation with the difference that it examines the effects of this phenomenon on ecological integrity but neither of these investigations incorporate the use of satellite imagery (Gao et al., 2021; Acuña-Alonso et al., 2022). However, in comparison to both, the current study aims to analyze the main factors that influence this phenomenon in a more comprehensive manner, in addition to introducing data from satellite imagery. Specifically, the new series of Sentinel satellites of the Copernicus European Union Earth Observation Program is used, especially Sentinel-2 (S2-A and B), which carries an MSI (Multispectral Instrument) sensor. These satellite images have been used to assess water quality (Toming et al., 2016; Ambrose-Igho et al., 2021; Nassereddine et al., 2024) and vegetation (Liang et al., 2022; Pu et al., 2022). This advanced system not only provides high-resolution satellite images but also offers open source information, increasing capacity for analysis in the estimation of phytoplankton (Tiquia-Arashiro, 2012; Ambrose-Igho et al., 2021; Pu et al., 2022) and cyanobacteria from the measurement of their main pigments, such as chlorophyll a and phycocyanin (Kutser et al., 2016; Pu et al., 2022).

Under this premise, the objective of this study has been to evaluate different predictive models of the Cyano-HABs phenomenon in the A Baxe reservoir, located in the Umia River Basin (Galicia, NW Spain) during the period 2017–2021. This study area was chosen because of the need to understand the factors influencing the numerous episodes of Cyano-HABs that have periodically affected the reservoir since 2006. However, the uniqueness of this approach lies not only in the prediction of eutrophication; it goes further by encompassing the estimation of the probability of Cyano-HABs proliferation. This study presents a multidimensional scope, as it is not limited only to the prediction of eutrophication events and cyanobacterial blooms but also focuses on the identification and detailed analysis of the critical variables that trigger and sustain these processes. For this purpose, cause–effect interactions between climatic conditions, water physicochemical parameters, and eutrophication indices (EVI, FAI, and NDVI) have been studied in relation to the eutrophication phenomenon and the proliferation of cyanobacteria. In particular, pressure scenarios have been simulated to understand the impact of future climatic conditions and different trophic states on the formation of cyanobacterial blooms in the reservoir. In addition, the present research introduces an innovative approach by integrating satellite images as a variable in the detection of cyanobacteria in the reservoir under study. The incorporation of these models allows for more informed decision making by providing a comprehensive view of the influential variables in eutrophication and cyanobacterial proliferation. This provides decision-makers with the ability to design more effective strategies and policies for water resource management, aquatic biodiversity conservation, and public health protection. This allows early warnings to be issued when conditions are conducive to the proliferation of toxic cyanobacteria. In this way, it is also possible to identify the areas with the highest probability of blooms, allowing a more focused management. These models also allow the evaluation of the control measures applied by simulating the impact of different strategies for informed decision making.

2 Materials and Methods

2.1 *Study Area*

The study area of this study is located in the Umia River Basin, precisely in the A Baxe reservoir. The Umia River is located in the region of Galicia, North-western Spain (Fig. 14.1) and has a total length of 70 km. An analysis of the main land uses that can be found in the Umia River Basin (upstream of the reservoir) shows that 35.13% corresponds to hardwood forests, 24.91% to cultivated areas, 15.48% to scrub and heathland, and 10.07% to coniferous forests. These land uses are strongly linked to the causes of blooms since agriculture and livestock are two of the industries that contribute most nutrients to lakes and reservoirs (Viso-Vázquez et al., 2021). The total surface area of the basin upstream of the reservoir is 440 km^2, while the average flow rate is 16.2 m^3/s. Construction of the Caldas de Reis dam, in the province of Pontevedra, began in 1988. The dam was built in the middle section of the Umia river and upstream of the town of Caldas de Reis. The construction of this reservoir was completed in 2000, thus creating the A Baxe reservoir with the intention of supplying drinking water to the 113,052 inhabitants (INE, 2023) of the O Salnés region and putting an end to the floods that inundated the thermal village almost every winter. Structurally, the reservoir has a volume of 6.16 hm^3 dammed for hydroelectric use in addition to supply use.

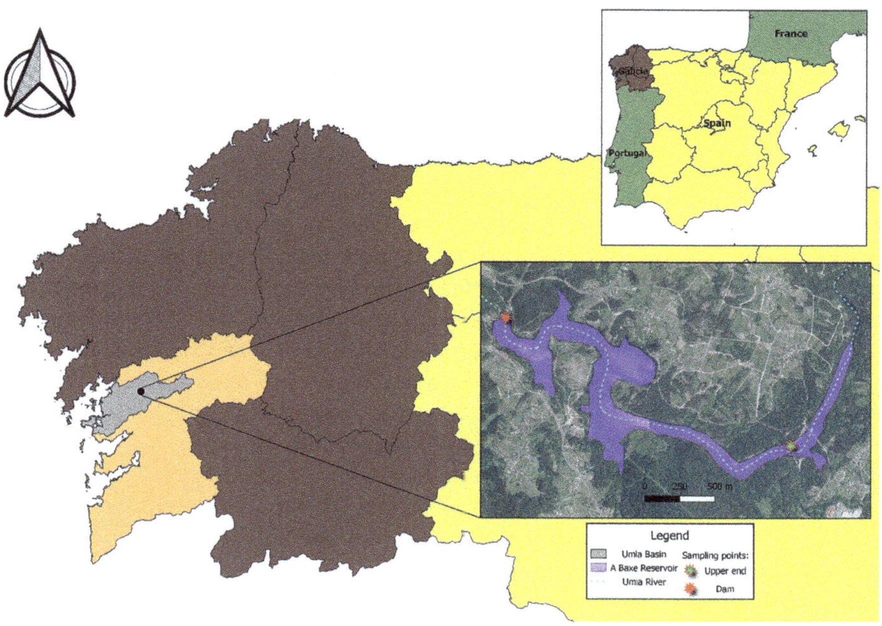

Fig. 14.1 Map of the study area showing the location of the A Baxe reservoir (EPSG: 25829)

The study area is classed as having a humid, oceanic climate, and within the varieties found in Galicia it falls under the Rías Baixas coastal subtype. This climate is characterized by high rainfall, with an annual total of over 1500 mm and precipitation distributed throughout the year but coming mainly in the autumn and winter months. Temperatures are mild in winter and warm in summer (Augas de Galicia, 2011). In 2022, the average temperature ranged from 9.2 °C in January to 21.8 °C in July. Average annual precipitation was found to be around 140.52 L/m^2. These variables are of great importance, since cyanobacterial blooms increase with lower precipitation and higher temperatures (Rosso & Giannuzzi, 2017).

Harmful algal blooms have become a persistent problem in the A Baxe reservoir. Many have been detected over the years, with the most striking episode being the bloom of Microcystis aeruginosa detected for the first time in 2006. Water supplies in the municipalities of Caldas de Reis and Vilagarcía de Arousa were compromised due to its toxicity, and the problem recurred every year up to 2012 (Lago Meijide et al., 2016). Various control methods have been evaluated to address this problem, such as the use of eucalyptus bark Xunta de Galicia – Augas de Galicia (2013), analysis of the influence of sediment (Acuña-Alonso et al., 2020a), and various ways of harvesting microalgae (Acuña-Alonso et al., 2021). In addition, the government has implemented a warning system for cyanobacteria blooms (GAIA, 2023), mainly the species *Microcystis* sp. For the period covered by the study reported here, this species accounts for 89.5% of the total, compared to *Pseudanabaena mucicola* with 8.9%, *Dolichospermum* sp. with 1.2%, and *Anabaena* sp., *Psedanabaena* sp., and *Aphanocaspa* sp. with less than 1% each.

2.2 Conceptual Model

In conducting this study, different interactions between data in the A Baxe reservoir were assessed. To construct the model, the interdependence between eutrophication indexes (NDVI, EVI, and FAI), weather conditions, reservoir flow, water physicochemical data, and eutrophication parameters was studied. These models have been studied for the period from 2017 to 2021, covering a total of 23 samples. To that end, samples were taken at two points in the reservoir: the dam (UTM 531475E 4717285 N) and the upper end of the reservoir (UTM 533645E 4716415 N) (Fig. 14.1).

Data collection began with the obtaining of the physicochemical parameters of the water. Following a request for data for research purposes, Augas de Galicia (the regional water authority) provided us with the data for the physicochemical parameters related to the A Baxe reservoir from 2017 to 2021. Data on reservoir flow during the period under study was obtained from the hydrological bulletins uploaded by Augas de Galicia (Augas de Galicia, 2023). To obtain data on meteorological conditions, we accessed the Meteogalicia online portal (Meteogalicia, 2023). In this context, the historical records of the climate station closest to the research area, that is, that in Caldas de Reis (UTM 531258E 4715846.2 N), were accessed. Moreover,

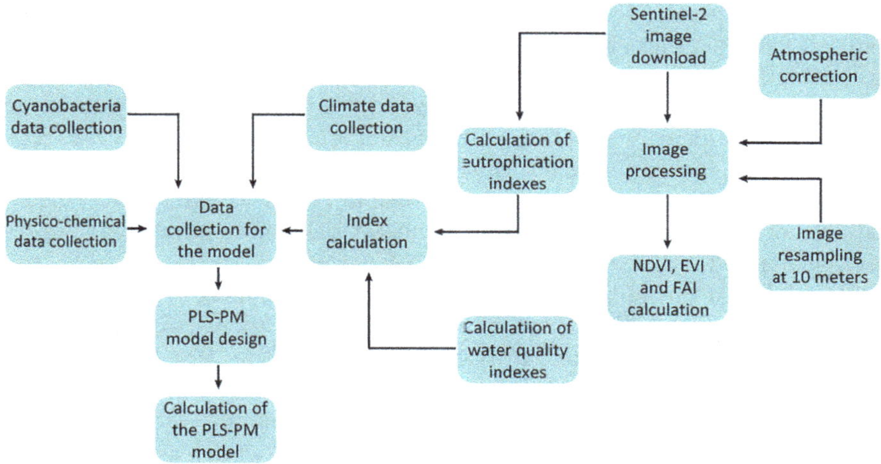

Fig. 14.2 Outline of the working methodology for the development of the model

data on the concentration of potentially toxic cyanobacteria was collected from the GAIA portal belonging to the Xunta de Galicia (GAIA, 2023). Finally, eutrophication indexes were calculated based on Sentinel-2 images (Fig. 14.2).

2.3 *Satellite Images*

The data used for the calculation of eutrophication indexes is from Sentinel-2 images taken between 2017 and 2021. The images downloaded are Level 1C and 2A products offered by Sentinel-2, and thus require atmospheric correction to eliminate interference caused by elements such as steam and suspended solids. This is done to counteract any reduction in clarity and brightness caused by such interference. In this study, a percentage of cloudiness <2% is used.

All these images were downloaded free of charge from the ESA website (https://scihub.copernicus.eu/). The images were taken on the same dates or on dates no more than 3 days apart, giving a total of 23 images for the period studied (Table 14.1).

2.4 *Sentinel-2 Image Processing*

Images were processed using Q-GIS software, with the Semi-automatic Classification Plug-in (SCP) (Augusto Mendonça Asciutti et al., 2019). For atmospheric corrections, the methodology used was Dark-Object Subtraction (DOS) (Chavez, 1988). DOS methods assume that non-zero signal values over supposedly

Table 14.1 Link between Augas de Galicia and Sentinel-2 images

Checkpoint	Sample data	Sentinel-2 images
Dam	20/06/2017	17/06/2017
Upper end	14/08/2017	13/08/2017
Dam	14/08/2017	13/08/2017
Upper end	09/10/2017	12/10/2017
Dam	09/10/2017	12/10/2017
Dam	26/02/2018	24/02/2018
Upper end	27/08/2018	26/08/2018
Dam	27/08/2018	26/08/2018
Upper end	22/10/2018	22/10/2018
Dam	22/10/2018	22/10/2018
Upper end	07/01/2019	05/01/2019
Dam	07/01/2019	05/01/2019
Dam	13/05/2019	13/05/2019
Upper end	19/08/2019	21/08/2019
Dam	19/08/2019	21/08/2019
Upper end	16/09/2019	15/09/2019
Dam	16/09/2019	15/09/2019
Dam	02/12/2019	04/12/2019
Dam	03/08/2020	05/08/2020
Dam	13/10/2020	16/10/2020
Dam	19/10/2020	16/10/2020
Dam	15/03/2021	15/03/2021
Dam	07/06/2021	08/06/2021

zero-valued dark shaded pixels are atmospheric scattering signals. This method is also fully automatic and requires no additional adjustments (Moravec et al., 2021). One of the main goals of this methodology is to obtain a cloud-free classification due to the influence of summer cloudiness and precipitation. In addition, the bands used in the classification were resampled to a pixel size of 10 m with the corresponding calibration and atmospheric correction. The 20 m bands were resampled to 10 m to harmonize all band data and ensure that the image classification matched this resolution.

2.5 Eutrophication Indexes

Eutrophication indexes were calculated from Sentinel-2 images using the Sentinel Application Platform (SNAP) v9.0.0 (SNAP-ESA Sentinel Application Platform, 2023) and the free software Q-GIS 3.16.16 (Q-GIS Development Team, 2023).

2.5.1 Enhanced Vegetation Index (EVI)

The Enhanced Vegetation Index (EVI) developed by Wardlow (Wardlow et al., 2007) was conceived with the purpose of improving vegetation assessment in regions with high biomass density. Its design focuses on improving detection sensitivity in areas of dense vegetation and increasing the accuracy of vegetation change monitoring. This is achieved by decoupling the signal obtained from the vegetation from the background canopy signal and by reducing interference caused by atmospheric factors. In essence, EVI seeks to provide a more accurate, more reliable view of vegetation conditions and changes by removing unwanted distortions in the data collected (Jiang et al., 2008). The range of values for the EVI index is −1 to 1, with healthy vegetation typically showing figures of between 0.20 and 0.80. The equation is applied via the following mathematical expression (Eq. (14.1)).

$$EVI = G \times \frac{(NIR - RED)}{(NIR + C1 \times RED - C2 \times Blue + L)} \tag{14.1}$$

where *NIR*, *RED,* and *Blue* are atmospherically corrected or partially corrected surface reflectances in the near-infrared, red, and blue-bands, respectively; *G* is a gain factor of 2,5; *C1* and *C2* are the coefficients correcting for atmospheric scattering effects ($C1 = 6$ and $C2 = 7.5$); and *L* functions as the ground adjustment factor.

For Sentinel-2, the index is calculated using the following expression (Eq. (14.2)):

$$EVI = 2.5 \times \frac{Band8 - Band4}{Band8 + 6 \times Band4 - 7.5 \times Band2 + 1} \tag{14.2}$$

2.5.2 Floating Algae Index (FAI)

The Floating Algae Index (FAI), first proposed by Hu (Hu, 2009), is a tool used to observe the spread of algae in water bodies. Fundamentally, the FAI is calculated by analyzing how the amount of reflected light changes at different wavelengths of near infrared, red, and shortwave infrared light. It is useful because factors such as atmosphere, observing conditions, or absorption of light by water have relatively little effect on it. FAI is often used to identify dense HABs in marine and inland waters. Therefore, spectral information from the red band, near infrared band, and shortwave infrared band can be used to correct atmospheric effects. According to Ma et al. (2021), when the FAI index is below −0.01 or above 0.02, there is no significant algal bloom. However, an index greater than −0.002 indicates harmful algal blooms. The algorithm is as follows (Eqs. (14.3) and (14.4)):

$$FAI = R_{NIR} - R'_{NIR} \tag{14.3}$$

$$R'_{NIR} = R_{RED} + \left(R_{SWIR} - R_{RED}\right) \times \frac{\lambda_{NIR} - \lambda_{RED}}{\lambda_{SWIR} - \lambda_{RED}} \quad (14.4)$$

where R_{RED}, R_{NIR}, and R_{SWIR} represent the reflectances of the red, near-infrared, and short-wave infrared bands; λ_{RED}, λ_{NIR}, and λ_{SWIR} represent the central wavelengths, which are 655, 865, and 1610 nm, respectively; and R'_{NIR} is the interpolation reflectance.

In the context of Sentinel-2, the FAI index is calculated via the following expression (Eq. (14.5)).

$$FAI = Band\,5 - \left(Band\,4 + \left(Band\,5 - Band\,6\right) \times \left(\frac{865 - 655}{1610 - 655} \right) \right) \quad (14.5)$$

2.5.3 Normalized Difference Vegetation Index (*NDVI*)

The Normalized Difference Vegetation Index (*NDVI*) by Kriegler (Kriegler et al., 1969) is a numerical indicator that uses information from the visible and near-infrared light bands within the electromagnetic spectrum. Its purpose is to analyze remotely sensed data and determine whether the observed area contains green and active vegetation. In other words, this numerical measure provides valuable information on the density and health of vegetation in an observed region. This index is defined by values ranging from −1 to 1, where negative values consist mainly of clouds, water, and snow, with negative values close to zero consisting mainly of rocks and bare soil. Very low values (0.1 or less) of the NDVI function correspond to areas without rocks, sand, or snow. Moderate values (0.2 to 0.3) represent shrubs and grasslands, while high values (0.6 to 0.8) indicate temperate and tropical forests. The equation is expressed as follows (Eq. (14.6)):

$$NDVI = \frac{NIR - RED}{NIR + RED} \quad (14.6)$$

For Sentinel-2, the index shows up as follows (Eq. (14.7)):

$$NDVI = \frac{Band\,8 - Band\,4}{Band\,8 + Band\,4} \quad (14.7)$$

2.6 *Water Quality Index (WQI)*

In the study reported here, a water quality index has been calculated in order to assess in depth the state of the water bodies in the reservoir of A Baxe. The Water Quality Index (*WQI*) calculated is a statistical tool that enables the quality of a body

Table 14.2 Rating rage for the WQI index

WQI value	Water Quality
91–100	Excellent
71–90	Good
51–70	Moderate
26–50	Poor
0–25	Very poor

Source: (Al-Tamimi & Braak, 2019)

of water to be estimated. It was developed in 1970 by the National Sanitation Foundation (*NSF*) of the United States, using the Rand Corporation's Delphi research technique (Parada & Solano, 2007). Nine parameters are used to determine the WQI (Limon & Webb, 1964): Fecal coliforms, pH, Biochemical Oxygen Demand in 5 days (BDO_5), nitrates, phosphates, temperature change, turbidity, and suspended solids. The index is calculated by applying the following formula (Eq. (14.8)).

$$WQI = \sum_{i=1}^{n} W_i \times I_i \tag{14.8}$$

Where W_i is the final weight of each parameter and I_i is the quality subindex of each parameter i. The functional curves determined by *NSF* are used to determine the subindexes of each parameter.

Once the quality index for each sampling date is determined, it must be compared with the water quality rating ranges (Table 14.2).

2.7 Partial Least Square-Paths

The Partial Least Squares (*PLS*) method seeks to model causal paths between blocks of variables called latent variables (*LV*) or constructs. It was originally developed by Wold (WOLD, 1980). In this method, paths are established between LV. The LV and paths form an internal model that is composed of endogenous variables and exogenous variables, and an external model represented by measured variables (*MV*). In the internal model, links between LV are quantified through path coefficients (β), while links between LV and MV in the external model are quantified through weights (*W*) (Edeh et al., 2023). All these interactions are summarized in equations. The measured score (*MS*) of an LV is the weighted sum of the associated MV scores (Eq. (14.8)). The predicted score (*PS*) of an endogenous LV is the weighted sum of all associated exogenous latent variables, where the weights are represented by path coefficients (Eqs. (14.9) and (14.10)) (Acuña-Alonso et al., 2022).

$$VL\,m = \sum_{i=1}^{n} \left(MVi \times Wi \right) \tag{14.9}$$

$$VL\,p,c = LV_{m,a} \times \beta_{cc} + LV_{m,b} \times \beta_{bc} \tag{14.10}$$

Table 14.3 Latent variables for the modeling

Latent variables (LV)	Measured variables (MV)	Units
Eutrophication	Potentially toxic cyanobacterias	Cell/mL
	Trophic state Index (Secchi disk)	
Flow	Occupation	%
Weather conditions	Precipitation	L/m^2
	Solar radiation	KJ/m^2*day
	Average temperature	°C
Eutrophication indexes	Enhanced vegetation index (EVI)	
	Floating algae index (FAI)	
	Normalized difference vegetation index (NDVI)	
Water quality	Alkalinity	mg/L
	Water temperature	°C
	Water quality index (WQI)	

This model focuses on analyzing the links between the LV "Flow", "Eutrophication indexes," "Weather conditions," and "Water quality" for the period between 2017 and 2021, providing a detailed understanding of the links and patterns that develop as a function of the variables involved. This analytical approach helps to identify significant relationships and capture essential dynamics within the context of the reservoir under study. Following the structure described above, two models were analyzed (models 1 and 2). Due to limitations in the number of samples, model 1 was tested with 9 MV and model 2 with 10, both with the same LV (Table 14.3).

Following the methodology proposed by Ketchen (Ketchen, 2013), the minimum sample size is calculated as ten times the maximum number of variables measured that make up a latent variable. In the context of this study, a total of 23 sampling points were analyzed, resulting in a maximum of two measured variables. With these constraints in mind, different models incorporating various combinations of measured variables were tested (Tables 14.4a and 14.4b). A set of variables was selected in which most of the cause and effect interactions aligned with the underlying theory.

In addition, Variance Inflation factors (*VIFs*) were evaluated (Martínez Ávila & Fierro Moreno, 2018). VIFs indicate potential failures in the training set fit, causing impacts on parameter estimation (weights and path coefficients). This statistical analysis measures how much larger the variance of an estimated coefficient is in the presence of collinearity, which makes it difficult to interpret PLS-PM results based on the training set. In this case, PLS-PM should be run with fewer variables until a non-collinearity group is identified (Oliveira et al., 2019) The SMART-PLS software (https://www.smartpls.com/) was used to develop and analyze the relationships between the variables using the PLS-PM methodology.

After running the models, we chose those with the highest R^2 values for the endogenous variables, because the higher the coefficient, the more robust the model is. Finally, the bootstrapping method integrated into the program was used to assess the statistical significance of the effects and weights obtained via the PLS-PM models run. This approach is an essential tool for assessing the statistical significance of the various results derived from PLS-PM analysis.

Table 14.4a Database used for the construction of the models

Checkpoint	Data	Cyanobacteria	TSI (DS)	Occupation	Temperature	Precipitation	Solar radiation
Tail	8/14/2017	7250	59.04	66.99	20.60	0	2477.5
Tail	10/9/2017	265,750	45.59896779	64.52	19.575	0	1744
Tail	8/27/2018	70,900	45.59896779	87.67	19.75	0.125	2285.75
Tail	10/22/2018	316,750	51.18	63.35	16.675	0	1206.75
Tail	1/7/2019	0	84.4030955	45.76	7.075	0	818.25
Tail	8/19/2019	83700.75	51.18	66.34	18.2	2.325	1901
Tail	9/16/2019	39,450	51.18	64.91	21.6	6.4	1895
Dam	6/20/2017	500	68.9560402	94.5	25.925	0	2701.75
Dam	8/14/2017	1000	59.0460231	66.99	20.6	0	2477.5
Dam	10/9/2017	73027.5	51.18	64.52	19.575	0	1744
Dam	2/26/2018	0	64.6270553	64	8.725	0	1509.75
Dam	8/27/2018	3750	55.5089849	87.67	19.75	0.125	2285.75
Dam	10/22/2018	23,750	59.0460231	63.35	16.675	0	1206.75
Dam	1/7/2019	8500	82.4030955	45.76	7.075	0	818.25
Dam	5/13/2019	0	68.9560402	93.11	17.7	1.3	2398.75
Dam	8/19/2019	65363.52	59.0460231	66.34	18.2	2.325	1901
Dam	9/16/2019	35,415	59.0460231	64.91	21.6	6.4	1895
Dam	12/2/2019	0	68.9560402	73.66	10.45	9.675	560.75
Dam	8/3/2020	348,550	66.47607279	66.73	19.45	0	2220.75
Dam	10/13/2020	39532.5	51.18	57.33	13.075	0.125	1144
Dam	10/19/2020	48,975	51.18	46.93	13.075	0.125	817.75
Dam	3/15/2021	0	82.4030955	42.6	10.25	1.05	1359.25
Dam	6/7/2021	7927.5	67.33000037	87.93	16.425	0	3020.25

Table 14.4b Database used for the construction of the models

Checkpoint	Data	EVI	NDVI	FAI	Alkalinity	Water temperaure	WQI
Tail	8/14/2017	−0.00648	0.00568	−0.14773	14.3	23.60	89.9
Tail	10/9/2017	0.0314	0.01211	0.20192	16.4	21.4	89.85
Tail	8/27/2018	−0.02136	−0.09965	−0.20707	14.5	24.2	87.2
Tail	10/22/2018	0.04794	0.01838	0.26596	8.97	20.9	92.9
Tail	1/7/2019	0.02665	0.01838	0.43548	5.93	11.8	90.7
Tail	8/19/2019	0.00286	0.00223	0.02011	8.95	24.7	84.8
Tail	9/16/2019	0.04279	0.01978	0.28719	8.95	24.7	84.9
Dam	6/20/2017	−0.00904	−0.00222	−0.064	8.17	21.7	92.1
Dam	8/14/2017	−0.00654	0.00175	−0.18571	14.6	23.5	91.45
Dam	10/9/2017	−0.00076	−0.00013	−0.00906	17	21.1	84.2
Dam	2/26/2018	0.00453	0.00155	0.09714	4.84	10.4	92.4
Dam	8/27/2018	−0.00078	−0.00021	−0.00984	15.8	24	82.05
Dam	10/22/2018	−0.00731	−0.00399	−0.06481	9.21	21	85.8
Dam	1/7/2019	0.00781	0.00363	0.33333	5.6	12	89.65
Dam	5/13/2019	0.00528	0.00238	0.12048	6.95	19.3	94.6
Dam	8/19/2019	−0.01163	−0.00557	−0.10945	8.29	24.6	85.4
Dam	9/16/2019	0.01396	0.00673	0.11368	8.29	24.6	85.4
Dam	12/2/2019	0.0134	0.00485	0.29944	6.95	19.3	92.35
Dam	8/3/2020	−0.04053	−0.01715	−0.19621	9.4	23.1	92.25
Dam	10/13/2020	−0.00051	−0.00026	−0.00741	5.58	10.8	91.6
Dam	10/19/2020	−0.00051	−0.00026	−0.00741	5.58	10.8	91.6
Dam	3/15/2021	0.00576	0.00358	0.22772	5.21	12.5	91.05
Dam	6/7/2021	−0.00404	−0.00127	−0.07109	7.52	20.8	91.5

Table 14.5 Trophic levels and values of the four key variables that define them

Trophic level	TSI	Chla (mg/m³)	SD (m)	TP (mg P/m³)	TN (mg N/m³)
Ultra-microtrophic	0.0 to 1.0	0.13–0.33	31–24	0.84–1.8	16–34
Microtrophic	1.0 to 2.0	0.33–0.82	24–15	1.8–4.1	34–73
Oligotrophic	2.0 to 3.0	0.82–2.0	15–7.8	4.1–9.0	73–157
Mesotrophic	3.0 to 4.0	2.0–5.0	7.8–3.6	9.0–20	157–337
Eutrophic	4.0 to 5.0	5.0–12	3.6–1.6	20–43	337–725
Supertrophic	5.0 to 6.0	12–31.0	1.6–0.7	43–96	725–1558
Hypertrophic	6.0 to 7.0	>31	<0.7	>96	>1558

Source: Burns et al. (2005)

2.8 Pressure Change Scenario

To help ensure suitable management, a simulation was calculated mainly to provide information for future decision making. This simulation analyzes scenarios of pressure change in the A Baxe reservoir taking into account the application of corrective measures. This methodology is applied to model 1, which analyzes the influence of weather factors (precipitation and temperature). This scenario is reformulated for the years 2030 and 2050, based on temperature and precipitation data provided by AdapteCCa (2023).

The simulation is based on the characteristic trophic classification of lakes, which is defined and grouped into seven categories, each of which is correlated with specific levels of nitrogen and phosphorus in the water column. However, for the scenario proposed here only one category was used: oligotrophic. From this physicochemical data, it is possible to calculate different trophic indexes (Fisher et al., 2021). In the course of this research, the trophic state index (TSI) was used in particular. In this sense, a scenario linked to pressure changes in the reservoir was considered. This scenario is based on the implementation of corrective measures based on the TSI with the aim of improving the state of the water towards an oligotrophic state. The measures that could be considered include reduction of discharges, more treatment stations, and improvement of riparian vegetation.

Changes in the levels of nitrogen, phosphorus, and the Secchi disk in the water must comply with recognized standards. In the process, TSI data for nitrogen, phosphorus, and the Secchi disk (Table 14.5) were used, along with data provided by AdapteCCa (RCP 8.5) for temperature and precipitation (20.50 °C and 4.33 mm/day, respectively, for 2030 and 21.88 °C and 4.45 mm/day for 2050).

3 Results

3.1 PLS-PM in the A Baxe Reservoir for 2017–2021

3.1.1 Model 1: Analysis of Influence of Weather Variables

This system in which the period 2017–2021 is studied led to the development of two different models. Within those models, the latent variable "Eutrophication indexes," which includes the EVI, FAI, and NDVI indexes, stands out. The consideration of

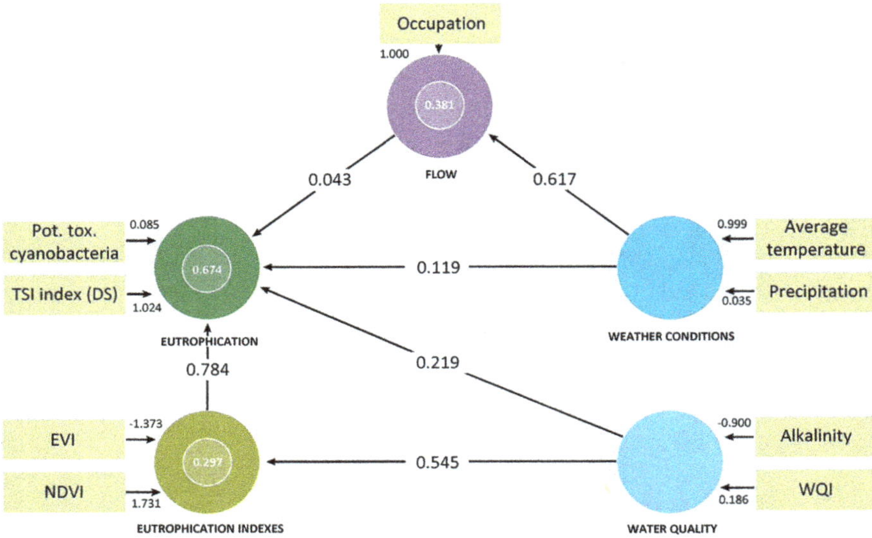

Fig. 14.3 PLS-PM results for model 1 with the influence of precipitation and temperature in the A Baxe reservoir

this latent variable seeks to integrate and represent more completely the eutrophication in the reservoir during this period. The two models resulting from this approach are presented as efforts to capture and further analyze the links between these indexes and their influence on eutrophication. First, model 1 (Fig. 14.3) shows visually the results obtained from the analysis via the influence of precipitation and temperature.

As shown in Fig. 14.3, the resulting R^2 values are 0.674 for "Eutrophication," 0.381 for "Flow," and 0.297 for "Eutrophication Indexes." In PLS-PM, statistical significance depends on all incoming arrows (from the internal and external models) and also on the number of samples, which in this case is 23 (Ketchen, 2013). For the variable "Eutrophication," the number of incoming arrows is 6. For "Eutrophication Indexes," it is 3 and for "Flow" it is 2. According to Ketchen (Ketchen, 2013), for a statistical significance of 5% the R^2 value in all these cases should exceed 0.75. Therefore, the model can be seen as adequately explaining "Eutrophication," but not the "Eutrophication Indexes" or "Flow" variables. Note also that the figures for these variables are low because they are not our target variable, that is, the variables that interfere with them were not studied. On the other hand, it is emphasized that this model predicts eutrophication at 67.4% via the independent variables of which it is composed.

To illustrate, an analysis of the variable "Water Quality" reveals that the weight assigned to alkalinity is negative (−0.900) and that assigned to the WQI index is (0.186). For "Weather Conditions" the weights for precipitation (0.035) and mean temperature (0.999) are both positive. For "Eutrophication" the weight of the TSI index for the Secchi Disk is highlighted (1.024), while in "Eutrophication Indexes" the weight of the EVI index is observed to be negative (−1.373), while the NDVI index is positive (1.731).

In regard to the path coefficients (β), in this modeling, the variables "Weather Conditions," "Flow," "Eutrophication Indexes," and "Water Quality" have positive influences on the variable "Eutrophication." "Eutrophication Indexes" in particular stands out as having a significant positive effect on "Eutrophication"($\beta = 0.784$). A positive, significant impact of the variable "Water Quality" on the "Eutrophication indices" ($\beta = 0.545$) is observed, and the same goes for the coefficient of "Flow" on "Eutrophication"($\beta = 0.043$), and the influence of "Weather Conditions" on "Flow" ($\beta = 0.617$).

The following equations show the measured score (*MS*) (Eqs. (14.11), (14.12), (14.13), (14.14), and (14.15)) and the predicted score (*PS*) (Eqs. (14.16), (14.17), and (14.18)) of each latent variable relative to this first modeling.

$$\begin{aligned} MS\,\text{Eutrophication} = {} & \text{Pot.tox.cyanobacteria} \times 0.085 \\ & + TSI\,\text{Index}\left(DS\right) \times 1.024 \end{aligned} \tag{14.11}$$

$$MS\,\text{Flow} = \text{Occupation} \times 1.000 \tag{14.12}$$

$$\begin{aligned} MS\,\text{Weather Conditions} = {} & \text{Precipitation} \times 0.035 \\ & + \text{Average temperature} \times 0.999 \end{aligned} \tag{14.13}$$

$$MS\,\text{Water Quality} = \text{Alkalinity} \times \left(-0.900\right) + WQI \times 0.186 \tag{14.14}$$

$$MS\,\text{Eutrophication Indexes} = EVI \times \left(-1.373\right) + NDVI \times 1.731 \tag{14.15}$$

$$\begin{aligned} PS\,\text{Eutrophication} = {} & MS\,\text{Flow} \times 0.043 + MS\,\text{Weather Conditions} \\ & \times 0.119 + MS\,\text{Water Quality} \times 0.545 \\ & + MS\,\text{Eutrophication Indexes} \times 0.784 \end{aligned} \tag{14.16}$$

$$PS\,\text{Flow} = MS\,\text{Weather Conditions} \times 0.617 \tag{14.17}$$

$$PS\,\text{Eutrophication Indexes} = MS\,\text{Water Quality} \times 0.545 \tag{14.18}$$

All the VIFs of the MVs are lower than 10 (Table 14.6). These results corroborate the absence of problematic collinearity and thus confirm the suitability of the PLS-PM approach.

Finally, to test the statistical significance of the effects and weights of the PLS-PM model, the bootstrapping algorithm included in the program was used. The results obtained are summarized in Table 14.7.

Table 14.7 reveals that most of the p-values obtained are higher than 0.05. This result suggests that most of the links analyzed are not statistically significant. However, it is interesting to note that the link between "Weather Conditions" and "Flow" is below 0.05, which indicates that there is sufficient evidence to reject the null hypothesis in this specific case.

Table 14.6 Values of the VIFs of model 1

MODEL 1	
Measured variable	VIF
Alkalinity	1.273
Pot. Tox. Cyanobacteria	1.119
EVI	3.003
NDVI	3.003
Occupation	1.000
Precipitation	1.000
Average temperature	1.000
WQI	1.273
TSI index (DS)	1.119

Table 14.7 Results of the bootstrapping analysis run on the model 1. All values higher than 0.05 are marked to reject the null hypothesis

Path	Original path coefficient (β)	Standard deviation (STDEV)	*P*-values
Water quality- eutrophication	0.219	0.299	**0.464**
Water quality- Eutrophication indexes	0.545	0.564	**0.334**
Flow-eutrophication	0.043	0.241	**0.858**
Weather conditions-flow	0.617	0.155	0.000
Weather conditions-eutrophication	0.119	0.316	**0.706**
Eutrophication indexes-eutrophication	0.784	0.691	**0.257**

Additionally, a look at the standard deviation of the coefficients shows that, for the most part, the figures obtained are not excessively high. However, there are some notable exceptions: when the link between "Water Quality" and "Eutrophication Indexes" and that between "Eutrophication Indexes" and "Eutrophication"are considered, for instance, higher standard deviations (0.564 and 0.691, respectively) are observed. These higher standard deviations could indicate greater variability in these particular relationships, suggesting that measurements may be more sensitive to different factors in these cases (Rousseau et al., 2018).

3.1.2 Model 2: Analysis of the Influence of the FAI Index

Model 2 is analyzed where the influence of the FAI index is studied. This second model is similar to the first, but there are some differences in the parameters and a further variable has been introduced over and above the maximum number previously considered in this model. In PLS-PM, a latent variable may be related to more measured variables than are used to estimate it. This is known as "over-identification" and can be useful in situations where the number of observations is limited or when the idea is to capture the variability of the latent variable more completely (McIntosh et al., 2014). In such cases, more measured variables representing different aspects or indicators of the latent variable can be included. The results obtained from model 2 are shown in Fig. 14.4.

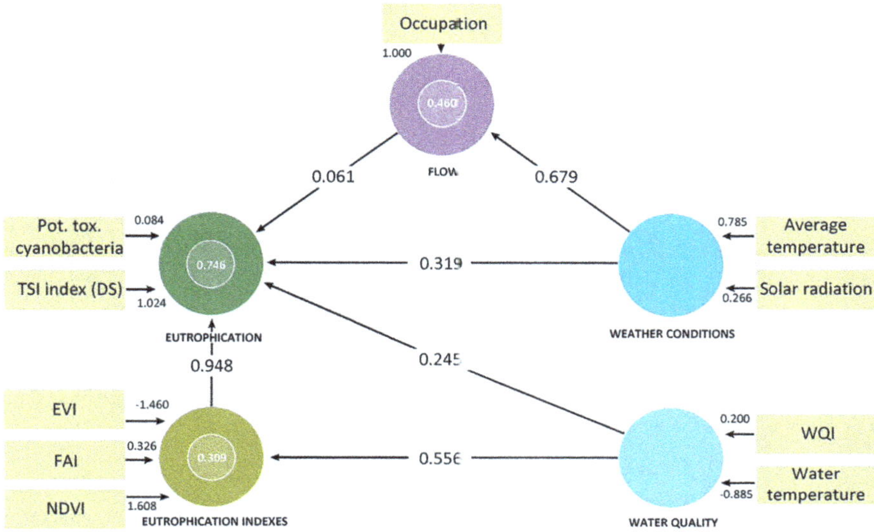

Fig. 14.4 PLS-PM results of model 2 with FAI index variable in the A Baxe reservoir

Starting with the results of the coefficient of determination R^2, model 2 shows similarities with the results obtained in model 1. Specifically, the R^2 value for the latent variable "Eutrophication"is 0.746, that for "Flow" is 0.460, and that for "Eutrophication Indexes" is 0.309. Within the context of this model, it is important to highlight the R^2 value associated with "Eutrophication," which reveals that in this model eutrophication is predicted at a level of 74.6%. In line with the variables investigated, this can justifiably be considered a high value. In this variable, the number of incoming arrows is observed to be 6, with a sample size of 23. According to the standards for a statistical significance of 5%, the R^2 value is expected to be around 0.75 for a dataset of this size. In this case, the estimated values are close to this figure, suggesting that the "Eutrophication"variable meets expectations in terms of statistical significance and ability to explain variability in the model. Regarding the weights assigned to the measured variables, it is observed that for "Water temperature" a negative weight of −0.885 is attributed with respect to the latent variable "Water Quality." Also worth highlighting is the positive weight of the "FAI"index within the "Eutrophication Indexes" ($W = 0.326$).

For the path coefficients, the results are very similar to those obtained with model 1. In the context of this model, we highlight the value of the coefficient obtained from the variable "Eutrophication Indexes" towards "Eutrophication"($\beta = 0.948$).

The following equations show the measured score (*MS*) (Eqs. (14.19), (14.20), (14.21), (14.22), and (14.23)) and predicted score (*PS*) (Eqs. (14.24), (14.25), and (14.26)) for each latent variable.

$$\begin{aligned} MS\,\text{Eutrophication} = {} & \text{Pot.tox.cyanobacteria} \times 0.084 \\ & + TSI\,\text{Index}\left(DS\right) \times 1.024 \end{aligned} \tag{14.19}$$

$$MS\,\text{Flow} = \text{Occupation} \times 1.000 \tag{14.20}$$

$$\begin{aligned} MS\,\text{Weather Conditions} = {} & \text{Solar radiation} \times 0.266 \\ & + \text{Average temperature} \times 0.785 \end{aligned} \tag{14.21}$$

$$MS\,\text{Water Quality} = \text{Water temperature} \times \left(-0.885\right) + WQI \times 0.200 \tag{14.22}$$

$$\begin{aligned} MS\,\text{Eutrophication Indexes} = {} & EVI \times \left(-1.460\right) + FAI \times 0.326 \\ & + NDVI \times 1.608 \end{aligned} \tag{14.23}$$

$$\begin{aligned} PS\,\text{Eutrophication} = {} & MS\,\text{Flow} \times 0.061 + MS\,\text{Weather Conditions} \\ & \times 0.319 + MS\,\text{Water Quality} \times 0.245 \\ & + MS\,\text{Eutrophication Indexes} \times 0.948 \end{aligned} \tag{14.24}$$

$$PS\,\text{Flow} = MS\,\text{Weather Conditions} \times 0.679 \tag{14.25}$$

$$PS\,\text{Eutrophication Indexes} = MS\,\text{Water Quality} \times 0.556 \tag{14.26}$$

As in model 1, in model 2 the VIFs are lower than 10 (Table 14.8). These results confirm the lack of collinearity between the MVs and the possibility of using the PLS-PM results in a robust way.

Finally, bootstrapping analysis was used to test the reliability of the model. The results are shown in Table 14.9.

Most of the paths assessed do not show sufficient statistical evidence for the null hypothesis to be rejected (Table 14.9). This suggests that, in the context of our model, many of the proposed links between variables are not statistically significant.

Table 14.8 Values of the VIFs of model 2

MODEL 2	
Measured variable	VIF
Pot. Tox. Cyanobacteria	1.119
EVI	3.297
FAI	1.564
NDVI	3.061
Occupation	1.000
Solar radiation	2.274
Water temperature	1.328
Average temperature	2.274
WQI	1.328
TSI index (DS)	1.119

Table 14.9 Results of the bootstrapping analysis run on model 2. All values higher than 0.05 are marked to reject the null hypothesis

Path	Original path coefficient (β)	Standard deviation (STDEV)	*P*-values
Water quality- Eutrophication	0.245	0.372	**0.511**
Water quality- Eutrophication indexes	0.556	0.570	**0.337**
Flow-Eutrophication	0.061	0.223	**0.758**
Weather conditions-flow	0.679	0.167	0.000
Weather conditions-Eutrophication	0.319	0.411	**0.437**
Eutrophication indexes-Eutrophication	0.984	0.734	**0.196**

Furthermore, the standard deviations of the path coefficients are observed to be high in some cases, although not excessively high. These standard deviations indicate the variability in the coefficient estimates and, ultimately, the uncertainty associated with the links identified in the model. Although the standard deviations are high in certain cases, it is important to note that they are not exceptionally high, which suggests that the estimates of the trajectories are relatively stable and reliable (Rousseau et al., 2018).

4 Discussion

4.1 *Model 1: Analysis of Influence of Weather Variables*

Starting with the weights established in model 1 (Fig. 14.3), the alkalinity value W = -0.900 is associated with a water acidification process. This could be influenced by several factors, including an increase in nitrogen levels, a finding supported by previous research (Camargo & Alonso, 2007). Alkalinity is a crucial indicator of a water body's ability to resist changes in pH, and its decrease could indicate an increased susceptibility to acidification. By contrast, the positive weight attributed to the WQI index ($W = 0.186$) reflects a direct, positive correlation between this index and water quality. In other words, as the WQI increases, water quality tends to improve. This is an expected result, since the WQI is used to assess overall water quality, taking into account various parameters such as nutrient concentration, turbidity, and the presence of contaminants. A higher WQI indicates higher water quality, which translates into healthier conditions and a lower presence of pollutants in the water body being analyzed. The weight ($W = 0.035$) assigned to the precipitation variable is supported by the consideration of an increase in the amount of rainfall, which leads to greater water availability. The temperature value of W = 0.999 is explained by Deng et al. (2014) as follows: The combination of warmer conditions, light availability, and enriched nutrients in less transparent waters creates an environment conducive to the proliferation of cyanobacteria, resulting in the formation of blooms. The weight of $W = -1.373$ associated with Eutrophication Indexes is explained by the fact that this a more complex metric that

considers atmospheric interactions, tends to reflect the influence of vegetation on near-infrared light absorption and scattering, and may be sensitive to other factors not directly related to eutrophication (Yan et al., 2022). The NDVI index ($W = 1.731$) is based on the difference between near-infrared and red reflectance, and tends to be more directly sensitive to the amount and vigor of vegetation (Pettorelli et al., 2005. A rise in the NDVI index could indicate a greater density of vegetation, which in turn could be related to areas of aquatic growth and, therefore, have some influence on eutrophication. However, a negative effect on the EVI index could suggest that other environmental factors or specific terrain characteristics (Matsushita et al., 2007) may be dominating its impact in relation to eutrophication. This divergence in effects between the EVI index and the NDVI index within the model may reflect the unique sensitivity of each index to different environmental aspects and their complex links to eutrophication in the reservoir studied.

Moving on to the values of the path coefficients, the figure recorded between the variable "Eutrophication Indexes" and "Eutrophication"is $\beta = 0.784$, with this phenomenon being explained via the interconnection between algal blooms and the increase in potentially toxic cyanobacteria. Likewise, the positive link established between "Water Quality" and "Eutrophication Indexes" ($\beta = 0.545$) is due to the fact that degradation of the physicochemical parameters of water (a characteristic of decreases in water quality) is conducive to the proliferation of algal blooms in the reservoir. In a feedback loop, declining water quality provides a more favorable environment for algal and cyanobacterial growth, which in turn increases the Eutrophication Indexes. Similarly, the positive link between "Flow" and "Eutrophication" ($\beta = 0.043$) can be explained in terms of nutrient transport. An increase in water flow may result in an increased flow of nutrients, such as nitrogen and phosphorus, from the surrounding areas into the study reservoir. These nutrients act as building blocks for the growth of algae and cyanobacteria. With increased nutrient input due to increased flow, an environment conducive to blooms of these organisms is created, which in turn can lead to further eutrophication of the water body (Tang et al., 2020). The link between flow and eutrophication is supported by the notion that higher water flow entails a higher nutrient load, thus stimulating the excessive growth of algae and cyanobacteria that characterizes eutrophication processes. Finally, the coefficient between "Weather Conditions" and "Flow" is $\beta = 0.617$, indicating that as precipitation increases the flow available in the reservoir tends to increase.

4.2 Model 2: Analysis of the Influence of the FAI Index

In model 2 (Fig. 14.4), the value of "Water temperature" ($W = -0.885$) stands out, indicating that increases in it favor the rapid growth of algae and cyanobacteria, leading to the formation of abundant blooms (Nazari Sharabian et al., 2018; Ho & Michalak, 2020). As temperatures rise, these microorganisms reproduce and multiply faster due to an increase in their metabolism (Lago Meijide et al., 2016). However,

this overgrowth can reduce oxygen levels in the water as algae die and decompose. In addition, certain cyanobacteria can produce dangerous toxins (Zanchett & Oliveira-Filho, 2013). This has negative implications for the quality of water and its safety for various uses. The positive weight of the "FAI" index ($W = 0.326$) is justified by its being a specific indicator of the presence of floating algae in water bodies, which are a common component of algal blooms associated with eutrophication. This means that a rising FAI index value indicates a greater presence of floating algae, which is related to eutrophication (Cao et al., 2020). The NDVI index also has a positive weight towards "Eutrophication indexes" because it reflects changes in the density of aquatic vegetation, which increases in response to eutrophication. However, the weight of the EVI index is negative, possibly due to its differences in sensitivity and focus with respect to vegetation growth in water bodies compared to FAI and NDVI.

As for the path coefficients, the value found between "Eutrophication Indexes" and "Eutrophication" ($\beta = 0.948$) is higher due to the incorporation of the variable measured from the FAI index. This is because FAI is a specific index for detecting floating algae directly associated with eutrophication (Zhang et al., 2021). The rationale behind the modification of this path coefficient is based on the idea that the FAI provides more accurate, more detailed information on the presence of floating algae, which play a crucial role in the eutrophication process. Eutrophication, as a phenomenon, is characterized by an excessive enrichment of nutrients in a water body, which leads to accelerated growth of algae and other organisms, generating adverse changes in the aquatic ecosystem (Akinnawo, 2023). By incorporating FAI as a measured variable in the model, its ability to more directly capture the presence of floating algae, an early and significant manifestation of the eutrophication process, is recognized. On the other hand, the coefficient established between "Water Quality" and "Eutrophication" is 0.245. This positive influence is due to the direct link between water temperature and the proliferation of cyanobacteria. The correlation between water temperature and cyanobacteria concentration is based on a number of fundamental factors. First, cyanobacteria are photosynthetic microorganisms, which means that their growth and metabolic activity are highly dependent on light availability and water temperature (Rahman et al., 2023). Increased water temperature in the summer months provides conditions more conducive to photosynthesis and reproduction of cyanobacteria, resulting in an increase in their biomass and, consequently, their concentration in the water. In addition, warmer temperatures often increase thermal stability in the water column, which can lead to thermal stratification (Coma et al., 2009). This stratification can trap nutrients in the deeper layers of the reservoir. Combined with the increased photosynthetic activity of cyanobacteria, this can provide an abundant supply of nutrients for their growth (Burch et al., 2021).

4.3 Pressure Change Scenario

In addition, an investigation was conducted into the effects of variations in pressure, trophic status, and projected changes in temperature and precipitation according to the scenario established by the administration. Under model 1 (Fig. 14.3), a decrease

of approximately 9% in eutrophication is anticipated by 2030, and of about 7% by 2050. These results lead to the conclusion that the assessment of the proposed scenario in relation to the trophic state of the reservoir and its impact on eutrophication shows a direct correlation between the degradation of the trophic state and increases in cyanobacterial blooms. More rigorous corrective measures directed towards an oligotrophic state are shown to result in a significant reduction in cyanobacteria concentration.

However, temperature increases have emerged as a key factor amplifying the dynamics of eutrophication. They not only promote the availability of nutrients, accelerating the growth of cyanobacteria, but also change the thermal stratification of the water, further contributing to the proliferation of these organisms (Augusto et al., 2007). It is evident that the complex interactions between trophic state, temperature, and the presence of cyanobacteria call for a comprehensive, adaptive approach in the management of these aquatic ecosystems. The implementation of early, effective corrective measures, combined with continuous monitoring of physicochemical parameters and temperature, is presented as an essential strategy for mitigating the adverse effects of eutrophication and cyanobacterial blooms.

4.4 General Discussion

After a full evaluation of the models developed, we conclude that model 2 stands out as the most robust and best explained option. Under this model, we obtain a coefficient of determination R^2 of 0.746 when considering the FAI index to define the variable "Eutrophication". According to Ketchen (2013), for a number of around 39 samples and six incoming arrows to the study LV, the coefficient of determination should be around 0.75. This highlights the high goodness of fit achieved (Romeo, 2020) and suggests that model 2 is able to explain a significant proportion of the variable "Eutrophication" and is therefore the correct choice as the best representation of this phenomenon. This means that 74.6% of the variable "Eutrophication"can be explained by the independent variables that we have analyzed. Several previous studies have used the PLS-PM methodology to investigate the factors affecting eutrophication and Cyano-HABs (Hu, 2009; Francy et al., 2016; Zhao et al., 2020; Gao et al., 2021; Acuña-Alonso et al., 2022; Yan et al., 2022; Shi et al., 2023). One example is the study by Gao et al. (2021), which focuses on the influence of weather conditions and particulate organic matter on the eutrophication of subtropical freshwaters. His results show a significant correlation with high R^2 values, ranging from 0.605 to 0.669, which underlines the importance of these factors in the eutrophication process. Francy et al. (2016) reports research addressing microcystin levels in several lakes. Her findings reveal close links, with R^2 values ranging from 0.8 to 0.9, highlighting the critical influence of microcystin levels on cyanobacterial proliferation and water quality. Additionally, Acuña-Alonso et al. (2022) focuses on the phenomenon of Cyano-HABs, considering variables such as weather conditions and the physicochemical parameters of water. His results show an R^2 value of 0.515, which reflects a significant correlation and provides

valuable information on the factors that drive the proliferation of cyanobacteria in water bodies. For its part, Shi et al. (2023) evaluated the effects of rainfall on algal growth and obtained an R^2 of 0.54 for the river zone and 0.55 for the lake zone, demonstrating the importance of rainfall on nutrient loading in the water. Likewise, our research here also highlights that "Eutrophication Indexes" (with a β coefficient of 0.948) emerge as the most influential variable in defining eutrophication in this specific context. Importantly, these results contrast with the ability of the FAI index to detect harmful algae, as indicated by Zhao et al. (2020). The combination of these indexes has been examined in other studies, such as those by Yan et al. (2022) and Hu (2009). In these cases, the FAI index has been noted to be more effective than the NDVI and EVI indexes in detecting floating algae, as it is not as sensitive to changes in environmental conditions. We also highlight the use of the WQI index in the model due to the limited number of samples because we obtain a representation of all the physicochemical parameters that make up this index.

However, it is important to qualify these results in the context of the models used. Despite the strength of the R^2 obtained in the model 2, the overall results of both interactions are considered to show a lack of robust statistical significance. This lack of significance is partly due to the small number of samples available for analysis in this specific model. Sample size plays a critical role in statistics as it influences the ability of statistical tests to identify significant relationships. In this case, the limited number of observations could have affected the ability of the analysis to reliably detect significant effects. It is important to keep in mind that with small samples, the inherent variability in the data can make it difficult to distinguish between signal and noise, which in turn can lead to results that lack statistical significance. Therefore, it is critical to interpret these results with caution and to consider the possibility that conclusions derived from them may be limited by lack of statistical power due to small sample size. This limitation may affect the generalizability of the findings and highlight the need for additional studies with a larger number of observations to confirm or refute the observed patterns. Ultimately, recognizing and addressing these limitations is essential to ensure the reliability and validity of the conclusions drawn from this analysis. We also highlight the importance of incorporating satellite images into our models. These images play an essential role in providing accurate geospatial data, which is fundamental for the assessment of eutrophication in the reservoir studied. In particular, they are crucial for the calculation of the NDVI, FAI, and EVI indexes. The importance of these indexes lies in their ability to detect and quantify the presence of algae and aquatic vegetation, which is directly related to eutrophication processes. One of the advantages of incorporating satellite images into the models developed is the ability to capture information in different spectral bands, which enables the detection of specific pigments associated with cyanobacteria. In addition, the availability of historical data facilitates long-term analysis of changes in cyanobacterial proliferation, allowing a better understanding of trends and patterns over time. However, one of the main limitations lies in the need to calibrate and validate the data with field measurements to ensure their accuracy and reliability in detecting cyanobacteria. In addition, it is important to note that Sentinel-2 images, while a valuable tool for monitoring water

bodies and detecting changes in their composition, are not specifically designed to identify cyanobacterial blooms. Although they can provide useful information on water quality and the distribution of aquatic vegetation, they are not capable of directly detecting the presence of cyanobacteria. Therefore, it is necessary to complement satellite image analysis with other monitoring methods, such as chemical or biological analyses, to obtain a complete and accurate assessment of the presence and proliferation of cyanobacteria in the water.

Furthermore, the WQI and TSI indexes provide a comprehensive approach by considering a wide range of physicochemical parameters in models that often have limited samples available. However, it is important to be aware of the limitations of these indexes. The WQI, although a valuable tool, was not designed specifically for reservoirs. This may limit its applicability in assessing water quality in these environments. Nonetheless, it is important to note that future adaptations could be made to make it more specific and sensitive to the detection of cyanobacterial blooms. This improvement would strengthen our monitoring and assessment capabilities. TSI, although useful in assessing the trophic state of water, fails to fully capture the dynamics of highly complex aquatic systems. To overcome these limitations, it is essential to supplement the TSI with other indicators and to perform an overall analysis of water quality.

Finally, numerous studies identify nitrogen and phosphorus as predominant factors in the eutrophication of water bodies (Huang et al., 2017; Savic et al., 2022). Comparisons of our results with those from previous research, such as the study by Acuña-Alonso (Acuña-Alonso et al., 2022) which reports weights of 0.690 for phosphorus and 0.010 for nitrogen, reveal similarities with the present study when both parameters are integrated in the WQI variable, where we obtain a weight of 0.200. To mitigate the presence of these pollutants, it is essential to address water management comprehensively in the framework of eutrophication, and in detecting cyanobacteria and floating algae. This involves implementing effective policies and regulations, promoting sustainable agricultural practices, and raising awareness of the importance of preserving water quality.

The development of these models is of crucial importance in the detection and monitoring of cyanobacterial blooms, and benefits both water security and environmental governance. Such models permit early, accurate identification of blooms, which facilitates the implementation of preventive and corrective measures and thus helps preserve water quality and protect public health. They also enrich environmental governance by providing a sound scientific basis for decision-making, which supports more informed and effective management of water resources and promotes the conservation of aquatic biodiversity. In this context, to improve governance it is essential to raise public awareness of the importance of preserving the health of rivers and reservoirs, promoting responsible practices that ensure optimum water quality. These practices include agricultural initiatives aimed at reducing the use of fertilizers, thus reducing runoff into water bodies. It is also crucial to regulate and control wastewater discharges and promote the conservation and restoration of riparian areas, enhancing the role of vegetation as a natural nutrient filter. The promotion of these measures by the competent administrations and the population would imply a considerable reduction in the cyanobacteria concentration, thus improving water quality and water safety.

5 Conclusions

This study focuses on developing realistic, predictive, explanatory models for anticipating eutrophication events in the A Baxe reservoir and assessing the probability of harmful algal blooms. Through data analysis, we seek to understand the underlying factors that contribute to eutrophication and to provide tools to support decision-making in water resource management and aquatic ecosystem preservation. Two models are drawn up which enable 67.4% and 74.6% of eutrophication to be predicted via the variables used to analyze this phenomenon. In addition, a future scenario is calculated in which the application of corrective measures in the reservoir improves eutrophication by 9%. Despite a lack of statistical significance due to the limited number of samples, this highlights the need for future research with more samples to identify hitherto undetected patterns and improve decision-making and governance.

6 Future Perspectives

As a result of the research carried out in the A Baxe reservoir, several predictive models have been developed that focus on analyzing several crucial variables related to the eutrophication process. These models represent a valuable tool to accurately anticipate the occurrence of this phenomenon and, equally significant, to quantify the specific influence of each of these variables. This, in turn, can provide a solid basis for the management and implementation of effective mitigation strategies. By virtue of the results and findings obtained in this study, interesting prospects open for future research. These future lines of research could focus on several key aspects, including:

- **Refinement of existing models**. Continuous improvement of existing prediction models could be carried out by incorporating more data and relevant variables, which would increase the accuracy and significance of the predictions and allow for more effective reservoir management.
- **Adaptation of water quality indices for the identification of Cyano-HABs**. Specific parameters that are indicative of the presence of Cyano-HABs in the water would be identified. This could include chlorophyll-a concentration, presence of cyanotoxins, water turbidity, temperature and other factors known to be related to cyanobacterial blooms.
- **Effects of land use changes**. Investigate how changes in land use, such as urbanization or intensive agriculture, affect cyanobacterial dynamics and how Sentinel-2 data can be used to track these relationships in different landscapes.
- **Development of adaptive management strategies.** Work on the formulation of adaptive management strategies that can be applied in response to changing reservoir conditions and the emergence of eutrophication events.

These are some of the possible areas of research that could be derived from the results presented in this research, with the aim of advancing our knowledge of eutrophication in A Baxe reservoir and contributing to informed decision-making and sustainable management of this water resource.

Acknowledgements The authors thank Augas de Galicia, which is answerable to the Xunta de Galicia regional government, for its collaboration. We would also like to thank Chris Pellow for proofreading the English and providing helpful comments.

Authors' Contributions Conceptualization, Carolina Acuña-Alonso and Xana Álvarez; Formal analysis, Carolina Acuña-Alonso, Elena Seoane-Martínez, Xana Álvarez and Fernando António Leal Pacheco; Funding acquisition, Xana Álvarez; Investigation, Carolina Acuña-Alonso, Elena Seoane-Martínez, Xana Álvarez, Fernando António Leal Pacheco and Luís Felipe Saches Fernandes; Methodology, Carolina Acuña-Alonso, Elena Seoane-Martínez and Xana Álvarez; Project administration, Xana Álvarez; Resources, Xana Álvarez; Software, Elena Seoane-Martínez; Supervision, Carolina Acuña-Alonso, Xana Álvarez, Fernando António Leal Pacheco and Luís Felipe Saches Fernandes; Visualization, Carolina Acuña-Alonso, Elena Seoane-Martínez and Xana Álvarez; Writing—original draft, Carolina Acuña-Alonso and Elena Seoane-Martínez; Writing—review & editing, Xana Álvarez, Fernando António Leal Pacheco and Luís Felipe Saches Fernandes.

Declarations Conflict of Interest

The authors have no competing interests to declare that are relevant to the content of this article.
Funding This research is part of the Grant TED**2021-130241**A-**I00** funded by MCIN/AEI/**10.13039/501100011033** and by the "European Union NextGenerationEU/PRTR" (X. Álvarez, E. Seoane and C. Acuña-Alonso). For the author who belongs to the CITAB research center, this work was further supported by National Funds from FCT—Fundação para a Ciência e Tecnologia under the project UIDB/**04033/2020**. The author who belongs to the CITAB research center also belongs to the Inov**4**Agro—Institute for Innovation, Capacity Building and Sustainability of Agri-food Production. The Inov**4**Agro is an Associate Laboratory comprising two R&D units (CITAB & GreenUPorto). For the author affiliated to the CQVR, the research received support from the FCT—Fundação para a Ciência e Tecnologia, via projects UIDB/**00616/2020** and UIDP/**00616/2020**.Data AvailabilityNot applicable.

References

Acuña-Alonso, C., Álvarez, X., et al. (2020a). Assessment of water quality in eutrophized water bodies through the application of indexes and toxicity. *Science of the Total Environment, 728*, 138775. https://doi.org/10.1016/j.scitotenv.2020.138775

Acuña-Alonso, C., Lorenzo, O., et al. (2020b). Influence of Microcystis sp. and freshwater algae on pH: Changes in their growth associated with sediment. *Environmental Pollution, 263*, 114435. https://doi.org/10.1016/j.envpol.2020.114435

Acuña-Alonso, C., et al. (2021). Water toxicity in reservoirs after freshwater algae harvest. *Journal of Cleaner Production, 283*, 124560. https://doi.org/10.1016/j.jclepro.2020.124560

Acuña-Alonso, C., et al. (2022). Modelling of threats that affect Cyano-HABs in an eutrophicated reservoir: First phase towards water security and environmental governance in watersheds. *Science of the Total Environment, 809*, 152155. https://doi.org/10.1016/j.scitotenv.2021.152155

Adaptecca (2023). Plataforma sobre Adaptación al Cambio Climático en España. Available at: https://adaptecca.es/.

Al-Tamimi, A. N. A., & Braak, M. M. (2019). Water quality, diatoms, pollution and Shannon diversity indices for the Euphrates river. CABI Agriculture and Bioscience, 2(1), Article 13. https://doi.org/10.5555/20193208356

Akinnawo, S. O. (2023). Eutrophication: Causes, consequences, physical, chemical and biological techniques for mitigation strategies. *Environmental Challenges, 12*, 100733. https://doi.org/10.1016/j.envc.2023.100733

Ambrose-Igho, G., et al. (2021). Spatiotemporal analysis of water quality indicators in Small Lakes using Sentinel-2 satellite data: Lake Bloomington and Evergreen Lake, Central Illinois, USA. *Environmental Processes, 8*(2), 637–660. https://doi.org/10.1007/s40710-021-00519-x

Augas de Galicia. (2023). Available at: https://augasdegalicia.xunta.gal/.

Augusto, C., Hoyos, B., & Toro, Á. W. (2007). Evaluación de la estratificación térmica y su relación con el oxígeno disuelto del agua en la ciénaga Cachimbero, municipio de Cimitarra, Santander Evaluation of thermal stratification and its relation with the water quality in the wetland Cachimbero. pp. 48–65.

Augusto Mendonça Asciutti, G., Noel Stanganini, F., & Augusto Melanda, E. (2019). Identificação Dos Diferentes Usos E Ocupação Do Solo Da Bacia Hidrográfica Do Rio Do Quilombo, São Carlos/Sp Utilizando Imagem Do Satélite Sentinel 2 E O Complemento Scp Do Qgis. *Anais do XIX Simpósio Brasileiro de Sensoriamento Remoto*, p. 4.

Benitez, J., Henseler, J., & Castillo, A. (2017). Development and update of guidelines to perform and report partial least squares path modeling in information systems research. *Proceedings ot the 21st Pacific Asia conference on information systems: '"Societal transformation through IS/IT"', PACIS 2017* [Preprint].

Borsuk, M. E., Stow, C. A., & Reckhow, K. H. (2004). A Bayesian network of eutrophication models for synthesis, prediction, and uncertainty analysis. *Ecological Modelling, 173*(2), 219–239. https://doi.org/10.1016/j.ecolmodel.2003.08.020

Burch, M., Brookes, J., & Chorus, I. (2021). Assessing and controlling the risk of cyanobacterial blooms. *Toxic Cyanobacteria in Water*, pp. 505–562. Available at: https://doi.org/10.1201/9781003081449-8.

Burns, N., McIntosh, J., & Scholes, P. (2005). Strategies for managing the lakes of the Rotorua district, New Zealand. *Lake and Reservoir Management, 21*(1), 61–72. https://doi.org/10.1080/07438140509354413

Camargo, J. A., & Alonso, A. (2007). Contaminación por nitrógeno inorgánico en los ecosistemas acuáticos: problemas medioambientales, criterios de calidad del agua, e implicaciones del cambio climático. *Revista Ecosistemas, 16*(2), 98–110. Available at: http://www.revistaecosistemas.net/index.php/ecosistemas/article/view/457

Cao, M., et al. (2020). Monitoring the spatial and temporal variations in the water surface and floating algal bloom areas in Dongting Lake using a long-term MODIS image time series. *Remote Sensing, 12*, 3622. https://doi.org/10.3390/rs12213622

Chavez, P. S. (1988). An improved dark-object subtraction technique for atmospheric scattering correction of multispectral data. *Remote Sensing of Environment, 24*(3), 459–479. https://doi.org/10.1016/0034-4257(88)90019-3

Coma, R., et al. (2009). Global warming-enhanced stratification and mass mortality events in the mediterranean. *Proceedings of the National Academy of Sciences of The United States of America, 106*(15), 6176–6181. https://doi.org/10.1073/pnas.0805801106

Deng, J., et al. (2014). Earlier and warmer springs increase cyanobacterial (Microcystis spp.) blooms in subtropical Lake Taihu, China. *Freshwater Biology, 59*, 1076. https://doi.org/10.1111/fwb.12330

Edeh, E., Lo, W.-J., & Khojasteh, J. (2023). Review of partial least squares structural equation modeling (PLS-SEM) using R: A workbook. *Structural Equation Modeling: A Multidisciplinary Journal, 30*, 165. https://doi.org/10.1080/10705511.2022.2108813

Fernandes, A. C., et al. (2019). The role of landscape configuration, season, and distance from contaminant sources on the degradation of stream water quality in urban catchments. *Water, 11*, 2025. https://doi.org/10.3390/w11102025

Fisher, D. R., et al. (2021). Environmental governance. In B. Schaefer Caniglia, A. Jorgenson, S. A. Malin, L. Peek, D. N. Pellow, & X. Huang (Eds.), *Handbook of environmental sociology. Handbooks of sociology and social research* (pp. 333–353). Springer. https://doi.org/10.1007/978-3-030-77712-8_16

Flood, M., Frabutt, D., Floyd, D., Powers, E. U., Devol, A., & Tiquia-Arashiro, S. M. (2015). Ammonia-oxidizing bacteria and archaea in sediments of the Gulf of Mexico. *Environmental Technology, 36*, 124–135. https://doi.org/10.1080/09593330.2014.942385

Francy, D. S., et al. (2016). Estimating microcystin levels at recreational sites in western Lake Erie and Ohio. *Harmful Algae, 58*, 23–34. https://doi.org/10.1016/j.hal.2016.07.003

GAIA (2023). Plataforma Galega de Información Ambiental. Available at: https://gaia.xunta.es/plataforma/.

Galicia, A. de. (2011). Plan Integral de Actuación sobre a Mycrocystis sp. no Encoro de Caldas de Reis no Río Umia.

Gao, X., et al. (2021). Particulate organic matter as causative factor to eutrophication of subtropical deep freshwater: Role of typhoon (tropical cyclone) in the nutrient cycling. *Water Research, 188*, 116470. https://doi.org/10.1016/j.watres.2020.116470

González Huelva, I. (2018). Modelos PLS-PM. (pp. 1–91). Available at: https://idus.us.es/bitstream/handle/11441/77637/Gonz%E1lezHuelvaIreneTFG.pdf?sequence=1.

Guo, L. (2007). Doing Battle with the green monster of Taihu Lake. *Science, 317*(5842), 1166. https://doi.org/10.1126/science.317.5842.1166

Ho, J. C., & Michalak, A. M. (2020). Exploring temperature and precipitation impacts on harmful algal blooms across continental U.S. lakes. *Limnology and Oceanography, 65*(5), 992–1009. https://doi.org/10.1002/lno.11365

Hu, C. (2009). A novel ocean color index to detect floating algae in the global oceans. *Remote Sensing of Environment, 113*(10), 2118–2129. https://doi.org/10.1016/j.rse.2009.05.012

Huang, J., et al. (2017). Nitrogen and phosphorus losses and eutrophication potential associated with fertilizer application to cropland in China. *Journal of Cleaner Production, 159*, 171. https://doi.org/10.1016/j.jclepro.2017.05.008

Huo, S., et al. (2013). Using artificial neural network models for eutrophication prediction. *Procedia Environmental Sciences, 18*, 310–316. https://doi.org/10.1016/j.proenv.2013.04.040

Igwaran, A., Kayode, A. J., Moloantoa, K. M., et al. (2024). Cyanobacteria harmful algae blooms: Causes, impacts, and risk management. *Water, Air, and Soil Pollution, 235*, 71. https://doi.org/10.1007/s11270-023-06782-y

'Instituto Nacional de Estadística'. (2023). Available at: https://www.ine.es/.

Jetoo, S., Grover, V. I., & Krantzberg, G. (2015). The Toledo drinking water advisory: Suggested application of the water safety planning approach. *Sustainability, 7*, 9787–9808. https://doi.org/10.3390/su7089787

Jiang, Z., et al. (2008). Development of a two-band enhanced vegetation index without a blue band. *Remote Sensing of Environment, 112*(10), 3833–3845. Available at:. https://doi.org/10.1016/j.rse.2008.06.006

Kassem, A., Abbas, L., Coutinho, O., Opara, S., Najaf, H., Kasperek, D., Li, X., & Tiquia-Arashiro, S. M. (2023). Applications of Fourier transform-infrared spectroscopy in microbial cell biology and environmental microbiology: Advances, challenges, and future perspectives. *Frontiers in Microbiology, 14*, 1304081. https://doi.org/10.3389/fmicb.2023.1304081

Ketchen, D. J. (2013). A primer on partial least squares structural equation modeling. *Long Range Planning, 46*, 184. https://doi.org/10.1016/j.lrp.2013.01.002

Khan, M. (2014). Eutrophication: Challenges and solutions. Available at: https://doi.org/10.13140/2.1.3673.8884.

Khatoon, Z., Huang, S., Bilal, A., Janjuhah, H. T., Kontakiotis, G., Antonarakou, A., Besiou, E., Wei, M., Gao, R., Zhang, T., et al. (2023). Current and previous green technologies, their efficiency, associated problems, and success rates to mitigate m. aeruginosa in aquatic environments. *Sustainability, 15*(10), 8048. https://doi.org/10.3390/su15108048

Kriegler, F. J., Malila, W. A., Nalepka, R. F., & Richardson, W. (1969). Preprocessing transformations and their effects on multispectral recognition.

Kutser, T., et al. (2016). Remote sensing of Black Lakes and using 810 nm reflectance peak for retrieving water quality parameters of optically complex waters. *Remote Sensing, 8*, 497. https://doi.org/10.3390/rs8060497

Lago Meijide, L., et al. (2016). Floraciones de Cianobacterias y Valores de Microcistina-lr sestónica y Disuelta En Embalses de la Cuenca Hidrográfica del Miño-Sil (NW-España). *MOL*, 113–125.

Lee, G., & Jones-Lee, A. (2005). Eutrophication (excessive fertilization). Available at: https://doi.org/10.1002/047147844X.sw1606.

Liang, S., et al. (2022). Accurate monitoring of submerged aquatic vegetation in a Macrophytic Lake using time-series Sentinel-2 images. *Remote Sensing, 14*, 640. https://doi.org/10.3390/rs14030640

Limon, P. J., & Webb, R. H. (1964). A magnetic resonance experiment for the undergraduate laboratory. *American Journal of Physics, 32*(5), 361–364. https://doi.org/10.1119/1.1970348

Ma, J., et al. (2021). Spatio-temporal variations and driving forces of harmful algal blooms in Chaohu Lake: A multi-source remote sensing approach. *Remote Sensing, 13*, 427. https://doi.org/10.3390/rs13030427

Martínez Ávila, M., & Fierro Moreno, E. (2018). *Aplicación de la técnica PLS-SEM en la gestión del conocimiento: Un enfoque técnico práctico/application of the PLS-SEM technique in knowledge management: A practical technical approach.* RIDE Revista Iberoamericana para la Investigación y el Desarrollo Educativo. https://doi.org/10.23913/ride.v8i16.336

Matsushita, B., et al. (2007). Sensitivity of the enhanced vegetation index (EVI) and normalized difference vegetation index (NDVI) to topographic effects: A case study in high-density cypress Forest. *Sensors, 7*, 2636–2651. https://doi.org/10.3390/s7112636

McKindles, K. M., & Tiquia-Arashiro, S. M. (2012). Functional gene arrays for analysis of microbial communities on ocean platform. In S. Tiquia-Arashiro (Ed.), *Molecular biological technologies for ocean sensing* (Springer Protocols Handbooks). Humana Press. https://doi.org/10.1007/978-1-61779-915-0_9

McIntosh, C. N., Edwards, J. R., & Antonakis, J. (2014). Reflections on partial least squares path modeling. *Organizational Research Methods, 17*(2), 210–251. https://doi.org/10.1177/1094428114529165

'Meteogalicia' (2023). Available at: https://www.meteogalicia.gal/web/inicio.action.

Moravec, D., et al. (2021). Effect of atmospheric corrections on NDVI: Intercomparability of Landsat 8, Sentinel-2, and UAV sensors. *Remote Sensing, 13*, 3550. https://doi.org/10.3390/rs13183550

Naeem, M., et al. (2014). Task of mineral nutrients in eutrophication BT – Eutrophication: Causes, consequences and control: Volume 2. In A. A. Ansari & S. S. Gill (Eds.), (pp. 223–237). Springer Netherlands. https://doi.org/10.1007/978-94-007-7814-6_16

Nassereddine, Z., Opara, S., Coutinho, O., Qyteti, F., Book, R., & Heinicke, M. P. (2024, December 20). Napieralski J and Tiquia-Arashiro SM (2024) critical perspectives on advancing antibiotic resistant gene (ARG) detection technologies in aquatic ecosystems. *Science of the Total Environment, 957*, 177775. https://doi.org/10.1016/j.scitotenv.2024.177775

Nazari Sharabian, M., Ahmad, S., & Karakouzian, M. (2018). Climate change and Eutrophication: A short review. *Engineering, Technology & Applied Science Research, 8*(6), 3668–3672. https://doi.org/10.48084/etasr.2392

Oest, A., Alsaffar, A., Fenner, M., Azzopardi, D., & Tiquia-Arashiro, S. M. (2018). Patterns of change in metabolic capabilities of sediment microbial communities in river and lake ecosystems. *International Journal of Microbiology, 2018*, 1. https://doi.org/10.1155/2018/6234931

Oliveira, C. F., et al. (2019). The modeling of pasture conservation and of its impact on stream water quality using partial least squares-path modeling. *Science of the Total Environment, 697*, 134081. https://doi.org/10.1016/j.scitotenv.2019.134081

Pagnucco, G., Overfield, D., Chamlee, Y., Fortuna, A., Sulaiman, F., Farinas, J., Schittenhelm, R., Shuler, C., Kassem, A., Opara, S., Najaf, F., Abbas, L., Coutinho, O., Catalfano, B., Li, X., & Tiquia-Arashiro, S. M. (2023). Metal tolerance and biosorption capacities of bacterial

strains isolated from an urban watershed. *Frontiers in Microbiology, 14*, 1278886. https://doi.org/10.3389/fmicb.2023.1278886

Parada, N. J. F., & Solano, F. O. (2007). Capítulo III: índices de calidad (ICAs) y de contaminación (ICOs) del agua de importancia mundial. *Índices De Calidad Y De Contaminación Del Agua*, 76. Available at: http://www.unipamplona.edu.co/unipamplona/portalIG/home_10/recursos/general/pag_contenido/libros/06082010/icatest_capitulo3.pdf

Paerl, H. W., & Otten, T. G. (2013). Harmful cyanobacterial blooms: Causes, consequences, and controls. *Microbial Ecology, 65*, 995–1010. https://doi.org/10.1007/s00248-012-0159-y

Patel, D., Gismondi, R., Alsaffar, A., & Tiquia-Arashiro, S. M. (2019). Applicability of API ZYM to capture seasonal and spatial variabilities in lake and river sediments. *Environmental Technology, 40*(24), 3227–3239. https://doi.org/10.1080/09593330.2018.1468492

Pettorelli, N., et al. (2005). Using the satellite-derived NDVI to assess ecological responses to environmental change. *Trends in Ecology and Evolution, 20*(9), 503–510. https://doi.org/10.1016/j.tree.2005.05.011

Pu, J., et al. (2022). Distinguishing algal blooms from aquatic vegetation in Chinese Lakes using sentinel 2 image. *Remote Sensing, 14*, 1988. https://doi.org/10.3390/rs14091988

'Q-GIS Development Team'. (2023). Available at: https://qgis.org/es/site/.

Rahman, M., et al. (2023). An assessment of the effects of light intensities and temperature changes on cyanobacteria's oxidative stress via the use of hydrogen peroxide as an indicator. *Water, 15*, 2429. https://doi.org/10.3390/w15132429

Rajamanickam, S., Sivarethinamohan, S., Gaddam, M. K. R., Velusamy, P., Gomathi, R., Ravindiran, G., Gurugubelli, T. R., & Muniasamy, S. K. (2023, December 1). Impact of climate change and anthropogenic activities on aquatic ecosystem – A review. *Environmental Research, 238*(Pt 2), 117233. https://doi.org/10.1016/j.envres.2023.117233

Romeo, G. (2020). *Chapter 13 – Data analysis for business and economics, elements of numerical mathematical economics with excel* (pp. 695–761). Academic. https://doi.org/10.1016/B978-0-12-817648-1.00013-X

Rosso, L., & Giannuzzi, L. (2017). Factores ambientales y antropogénicos que afectan la formación de floraciones de cianobacterias y cianotoxinas. *Cianobacterias como determinantes de la salud, 5*, 70–93. Available at: www.msal.gov.ar

Rousseau, R., Egghe, L., & Guns, R. (2018). *Statistics, becoming metric-wise*. Available at: https://www.sciencedirect.com/topics/economics-econometrics-and-finance/measure-of-dispersion.

Rousso, B. Z., et al. (2020). A systematic literature review of forecasting and predictive models for cyanobacteria blooms in freshwater lakes. *Water Research, 182*, 115959. https://doi.org/10.1016/j.watres.2020.115959

Savic, R., et al. (2022). Nitrogen and phosphorus concentrations and their ratios as indicators of water quality and eutrophication of the hydro-system Danube–Tisza–Danube. *Agriculture, 12*, 935. https://doi.org/10.3390/agriculture12070935

Shi, P., et al. (2023). Rainstorm events trigger algal blooms in a large oligotrophic reservoir. *Journal of Hydrology, 622*, 129711. https://doi.org/10.1016/j.jhydrol.2023.129711

Smith, V. H. (2009). Eutrophication. In: G.E.B.T.-E. of I.W. Likens (Ed.). (pp. 61–73). Academic. Available at: https://doi.org/10.1016/B978-012370626-3.00234-9.

'SNAP-ESA Sentinel Application Platform' (2023). Available at: https://step.esa.int/main/.

Srisuksomwong, P., & Pekkoh, J. (2020). Artificial neural network model to prediction of eutrophication and Microcystis Aeruginosa bloom. *Emerging Science Journal, 4*, 129–135. https://doi.org/10.28991/esj-2020-01217

Tang, X., et al. (2020). Response of eutrophication development to variations in nutrients and hydrological regime: A case study in the Changjiang River (Yangtze) basin. *Water, 12*, 1634. https://doi.org/10.3390/w12061634

Tiquia-Arashiro, S. M. (2012). *Molecular biological technologies for ocean sensing* (Springer protocols handbooks) (p. 307). Humana. https://doi.org/10.1007/978-1-61779-915-0

Tiquia-Arashiro, S. M., & Pant, D. (2020). *Microbial electrochemical technologies* (p. 518). CRC Press. https://doi.org/10.1201/9780429487118

Toming, K., et al. (2016). First experiences in mapping Lake water quality parameters with Sentinel-2 MSI imagery. *Remote Sensing, 8*, 640. https://doi.org/10.3390/rs8080640

Viso-Vázquez, M., et al. (2021). Remote detection of cyanobacterial blooms and chlorophyll-a analysis in a eutrophic reservoir using Sentinel-2. *Sustainability, 13*, 8570. https://doi.org/10.3390/su13158570

Wardlow, B. D., Egbert, S. L., & Kastens, J. H. (2007). Analysis of time-series MODIS 250 m vegetation index data for crop classification in the U.S. Central Great Plains. *Remote Sensing of Environment, 108*(3), 290–310. https://doi.org/10.1016/j.rse.2006.11.021

Wold, H. (1980). 'Model construction and evaluation when theoretical knowledge is scarce: Theory and application of partial least squares. In: J.A.N. KMENTA and J.B.B.T.-E. of E.M. RAMSEY (Eds.). (pp. 47–74). Academic. Available at: https://doi.org/10.1016/B978-0-12-416550-2.50007-8.

Xunta de Galicia-Augas de Galicia. (2013). Resolución do 29 de outubro de 2013 pola que se modifica a licitación do contrato de obras de mantemento e mellora do dominio público hidráulico das bacías de Galicia-Costa. Diario Oficial de Galicia, 214, 5 de novembro de 2013. https://www.xunta.gal/dog/Publicados/2013/20131105/AnuncioO143-291013-0001_gl.html

Yan, K., et al. (2022). Deep learning-based automatic extraction of cyanobacterial blooms from Sentinel-2 MSI satellite data. *Remote Sensing, 14*, 4763. https://doi.org/10.3390/rs14194763

Zanchett, G., & Oliveira-Filho, E. C. (2013). Cyanobacteria and cyanotoxins: From impacts on aquatic ecosystems and human health to anticarcinogenic effects. *Toxins, 5*(10), 1896–1917. https://doi.org/10.3390/toxins5101896

Zhang, Y., et al. (2021). Wind effects for floating algae dynamics in Eutrophic Lakes. *Remote Sensing, 13*, 800. https://doi.org/10.3390/rs13040800

Zhao, Y., et al. (2022). Relative contributions of different sources to DOM in Erhai Lake as revealed by PLS-PM. *Chemosphere, 299*, 134377. https://doi.org/10.1016/j.chemosphere.2022.134377

Zhao, Y., Liu, D., & Wei, X. (2020). Monitoring cyanobacterial harmful algal blooms at high spatiotemporal resolution by fusing Landsat and MODIS imagery. *Environmental Advances, 2*, 100008. https://doi.org/10.1016/j.envadv.2020.100008

Zheng, G., Gu, J., Zhao, W., Zhang, Y., Guan, Z., Lei, M., & He, C. (2024). Spatial, geographical, climatic, and edaphic influences on Moss Community structure: A case study from Qinhuangdao, China. *Forests, 15*(3), 424. https://doi.org/10.3390/f15030424

Chapter 15
Spatial and Temporal Analysis of the Water Quality in Brazilian Semi-Arid Reservoirs Using AlgaeMAp

Felipe de Lucia Lobo, Lino Augusto Sander de Carvalho, Daniel Andrade Maciel, Gustavo Willy Nagel, Danielle de Almeida Bressiani, Suzana Gico Montenegro, Gilberto Queiroz de Lima Filho, and Laura Martins Bueno

1 Introduction

1.1 Brazilian Northeastern Semi-Arid Region

The Brazilian Northeastern Semi-Arid region covers 11% of the country's territory and possesses distinct characteristics such as high spatial and temporal variability of precipitation, high temperatures, shallow and poorly structured soils, deciduous vegetation cover, drainage basins composed of intermittent rivers and streams, and a high potential for evapotranspiration (Monicelli & Diniz, 2023). With prolonged

F. de Lucia Lobo (✉) · D. de Almeida Bressiani · L. M. Bueno
Center for Technological Development (CDTec), Universidade Federal de Pelotas (UFPel), Pelotas, Brazil
e-mail: felipe.lobo@ufpel.edu.br

L. A. S. de Carvalho
Department of Meteorology, Federal University of Rio de Janeiro (UFRJ), Rio de Janeiro, Brazil

D. A. Maciel
Instrumentation Lab for Aquatic Systems (LabISA), Earth Observation Coordination of National Institute for Space Research (INPE), São José dos Campos, Brazil

G. W. Nagel
Ocean and Earth Science, University of Southampton, Southampton, UK

S. G. Montenegro
Centro de Tecnologia, Engenharia Civil, Universidade Federal de Pernambuco, Recife, Brazil

Agência Pernambucana de Águas e Clima (APAC), Recife, Brazil

G. Q. de Lima Filho
Agência Pernambucana de Águas e Clima (APAC), Recife, Brazil

S. M. Tiquia-Arashiro et al. (eds.), *Urban Watershed Microbiology, Volume 1*,
https://doi.org/10.1007/978-3-031-89071-0_15

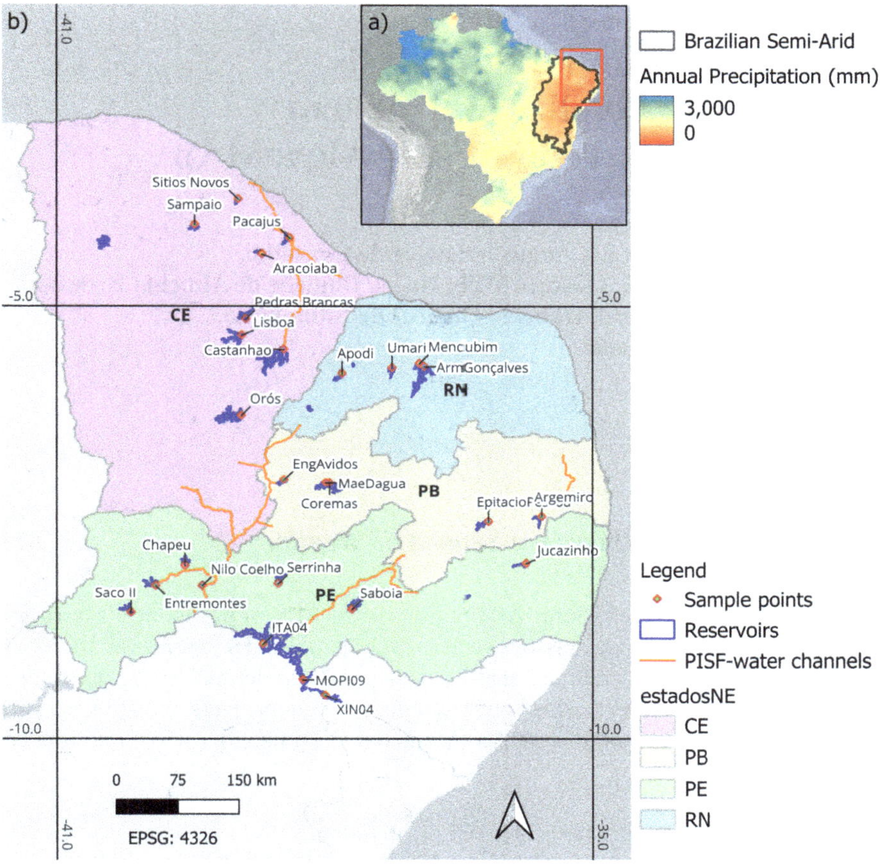

Fig. 15.1 Study area with indication of (**a**) annual precipitation in the region. (**b**) Sample points used—sample points in PB (Paraíba) and RN (Rio Grande do Norte) were not available for analysis, PISF water channels, water bodies ("açudes"), and Federal States.

dry periods, high temperatures, and limited vegetation cover, the region's climate poses significant challenges to water availability and management. Home to a diverse population, the Northeastern Semi-Arid region houses millions of residents who rely heavily on its limited water resources for various activities. These include agriculture, livestock, and industrial processes. The study area is defined as the semi-arid region within the Federal States of Pernambuco, Ceará, Paraíba, and Rio Grande do Norte (Fig. 15.1).

Water infrastructure plays a crucial role in the region's water management efforts. More than 500 dams have been constructed to combat water scarcity and support various functions such as irrigation, drinking water supply, and energy generation (BNDES, 2023). These reservoirs stand as a testament to the region's commitment to addressing water deficits, but the challenge of maintaining adequate

water availability persists due to the arid conditions. Despite the efforts to harness and manage water resources, recurring and extended droughts have had a profound impact on the region. These drought events disrupt agricultural productivity, threaten food security, and affect the livelihoods of the population. To address these challenges, the Brazilian National Water Agency (ANA) established a National Drought Monitoring and Alert System to provide timely information and aid in mitigation efforts. Water quality within the reservoirs has been a topic of concern due to its implications for both ecosystem health and potable water sources (Cho et al., 2012; Bowman et al., 2018; Pagnucco et al., 2023). Nutrient enrichment, often associated with agricultural runoff, can lead to eutrophication, adversely affecting aquatic life and the quality of available water.

1.2 PISF—São Francisco River Integration Program

The São Francisco River Integration Project (Projeto de Integração do Rio São Francisco [PISF]) stands as a monumental hydraulic engineering initiative devised to address the pressing issue of water scarcity in the arid and semi-arid regions of northeastern Brazil. This ambitious endeavor seeks to transfer water from the relatively water-abundant São Francisco River basin to the drought-prone regions of the Brazilian Northeast, specifically targeting the states of Pernambuco, Ceará, Paraíba, and Rio Grande do Norte.

The PISF project is divided into two primary axes: the Eastern Axis and the Northern Axis. Launched in April 2017, the Eastern Axis spans approximately 217 kilometers and serves to channel water from the São Francisco River through an elaborate system of canals and reservoirs to supply the states of Pernambuco and Paraíba. The axis also incorporates a network of pumping stations, strategically positioned to overcome the region's topographical challenges. The Northern Axis, which became operational in 2021, extends over approximately 402 kilometers and is responsible for providing water to the states of Ceará and Rio Grande do Norte. Like the Eastern Axis, this axis also involves an arrangement of channels, reservoirs, and pumping stations. The PISF project represents a substantial financial investment by the Brazilian government, with significant funds allocated to its planning and execution (BNDES, 2023). Despite efforts including the PISF and traditional water quality monitoring, comprehensive, real-time, and large-scale analysis of water quality remains limited.

1.3 Water Quality Conditions

The Northeastern Semi-Arid region of Brazil grapples with intricate water quality challenges stemming from its unique climatic and environmental conditions. Previous research works have shown that these reservoirs are generally prone to

eutrophication (Barbosa et al., 2012) in which the water quality may be affected by several factors combined, such as: precipitation (Ítalo et al., 2013), reservoir accumulated volume (Braga et al., 2015), external load of nutrients (Araújo et al., 2019), internal enrichment due to aquaculture (do Nascimento Moura et al., 2012), and sediment and phosphorus resuspension (de Farias Mesquita et al., 2020). In general, these authors found that during drought events, reservoir water levels decrease, causing changes in the water's physical and chemical properties. This can result in reduced water clarity, higher turbidity due to sediment disturbance, elevated nutrient levels, and increased phytoplankton biomass (Monicelli & Diniz, 2023). Following droughts, reservoirs experience reflooding as rainfall replenishes water levels. Reflooding may lead to improvements in water quality by diluting nutrient concentrations in reservoirs with increased water volume. Conversely, extreme rainfall events, including heavy rain, floods, and severe droughts, tend to deteriorate water quality by increasing suspended solids and nutrient levels, promoting eutrophication (Fourqurean & Robblee, 1999; Tiquia et al., 2006a; Tiquia et al., 2007; Dokulil & Teubner, 2010; Tiquia, 2011; Cooper et al., 2013; Flood et al., 2015; Oest et al., 2018; Patel et al., 2019; Fukushima et al., 2021; Moreira et al., 2023). The pressing concerns are eutrophic conditions, cyanobacteria blooms, and the potential presence of harmful cyanotoxins. In fact, do Nascimento Moura et al. (2018), based on literature review, indicate that the species with the highest occurrences were *Cylindrospermopsis raciborskii*, *Microcystis aeruginosa*, and *Planktothrix agardhii*—all of them can produce harmful toxins, and were found in all the Federal States considered in this study (do Nascimento Moura et al. (2018)).

These blooms not only disrupt the ecological balance of water bodies but also pose risks to human health. Cyanobacteria have the capacity to produce toxic compounds known as cyanotoxins, which can be harmful if ingested or come into contact with skin (Stewart et al., 2006; Funari & Testai, 2008; Tiquia-Arashiro & Mormile, 2013; Zanchett & Oliveira-Filho, 2013; Koreivienė et al., 2014; Testai et al., 2016). A pivotal event that underscored the gravity of water quality challenges in the region was the Caruaru incident in Pernambuco. In 1996, tragedy struck when cyanotoxins, sourced from a hypereutrophic dam, contaminated the water supply of a hemodialysis clinic. This catastrophe resulted in multiple fatalities and illuminated the critical importance of robust water quality management practices.

To address these challenges, the Brazilian National Water Agency (ANA) established the *Qualiagua* program (ANA, 2023), aiming to monitor and evaluate water quality nationwide. However, despite these efforts, comprehensive insights into water quality conditions across the Northeastern Semi-Arid region remain elusive. While certain states like Ceará and Pernambuco have made strides in enhancing their water quality monitoring programs, there remains a recognized need for broader coverage and more accessible data of water quality parameters. These parameters encompass chlorophyll-*a* concentration, nutrient levels (such as nitrogen and phosphorus), phytoplankton analysis, and turbidity measured in nephelometric turbidity units (NTU). Among others, these measures collectively furnish information about trophic conditions, potential algal blooms, and the likelihood of toxin production. Nevertheless, the challenge lies in efficiently collating and

integrating data from disparate sources, hindering the formulation of effective water quality management strategies.

Alternatively, water quality estimation by remote sensing has proven an effective approach for gathering crucial insights into water resource management (Tyler et al., 2006; Tiquia et al., 2006b; Chehbouni et al., 2008; McKindles & Tiquia-Arashiro, 2012; Tiquia-Arashiro, 2012; Wu et al., 2014; Gholizadeh et al., 2016; Topp et al., 2020; Nassereddine et al., 2024; Raghul & Porchelvan, 2024). With its capacity to deliver wide-ranging spatial coverage and frequent temporal resolution, remote sensing allows spatial and temporal analysis that is key to understanding the water resources dynamics (Thakur et al., 2017; Mashala et al., 2023). Through satellite imagery, remote sensing can estimate critical parameters such as chlorophyll-*a* concentration and turbidity, furnishing a comprehensive and recurring evaluation of water quality conditions.

1.4 Remote Sensing Approach

Remote sensing technology has emerged as a powerful tool for estimating water quality parameters, with success in assessing chlorophyll-*a* concentration—a key indicator of phytoplankton abundance and water trophic conditions (Giardino et al., 2019; El-Mezayen & Abd El-Hamid, 2023). The advantage of using Chl-*a* for water quality monitoring is that it can be estimated with satellite data and, therefore, be used to systematically monitor phytoplankton abundance at a relatively low cost (Tyler et al., 2006; Tiquia et al., 2008; Kampel et al., 2009; Tiquia, 2010a, b; Bowman et al., 2018; Moses et al., 2012; Dörnhöfer et al., 2018; Tiquia-Arashiro & Pant, 2020; Papenfus et al., 2020; Seegers et al., 2021). The interaction between light and photoactive pigments in phytoplankton cells, such as chlorophyll-*a* (Chl-*a*), enables the detection of algae bloom from space-borne imagery (Llodrà-Llabrés et al., 2023). The utilization of remote sensing data, particularly with the high spatial resolution (10–20 meters) and temporal resolution (5 days) offered by Sentinel-2/MSI (MultiSpectral Instrument) imagery, provides an unprecedented opportunity for in-depth analyses of the region's water quality dynamics (ESA, 2017). In fact, Sentinel-2/MSI data has been used in several applications for Chl-*a* (Llodrà-Llabrés et al., 2023) and algae bloom detection (Rolim et al., 2023), as indicated by these review articles.

One prominent approach is the use of the Normalized Difference Chlorophyll-*a* Index (NDCI) (Mishra & Mishra, 2012), which leverages the distinct absorption features of chlorophyll-*a* in the blue and green spectral bands of satellite imagery. By computing the NDCI, researchers can derive valuable insights into chlorophyll-*a* levels, shedding light on water quality variations for different water bodies around the world (Watanabe et al., 2017; Rodrigues et al., 2023; Makwinja et al., 2023; Aubriot et al., 2020; Buma & Lee, 2020; Lobo et al., 2021).

Notably, AlgaeMAp, an application that combines NDCI with Sentinel-2 imagery using the Google Earth Engine platform, has exemplified the potential of remote

sensing in this realm. Developed by Lobo et al. (2021), the AlgaeMAp project employs NDCI-based analyses to create an online application—the Algae Bloom Monitoring Application—dedicated to tracking algae blooms in dams and reservoirs across Latin America. The application uses Sentinel-2 images corrected for atmospheric and sun-glint effects to generate an image collection of the Normalized Difference Chlorophyll-*a* Index (NDCI) for the entire time-series (see Lobo et al., 2021 for image processing details). NDCI can be calculated as follows:

$$\text{NDCI} = \left(SR705 - SR665\right) / \left(SR705 + SR665\right) \tag{15.1}$$

For Sentinel-2, the central wavelength for Surface Reflectance at 705 nm (SR705) and at 665 nm (SR665) corresponds to Band 5 (red edge) and Band 4 (red), respectively.

NDCI is used to estimate both Chl-*a* concentration, based on a non-linear fitting model, and Trophic State Index (TSI), based on a tree-decision-model classification into five classes. The App benefits from the huge processing capability of GEE (Google Earth Engine) that allows any user with Internet access to easily extract detailed spatial (30 m) and long temporal Chl-*a* and TSI information (from August 2015 and with images every 5 days) throughout the most important reservoirs in Brazil. This innovative fusion of technology provides timely and accessible water quality information in graphs and images, enhancing monitoring efforts and bolstering management strategies (Free access to AlgaeMAp at felipellobo.users.earthengine.app/view/algaemapv10).

Therefore, the aim of this research is to comprehensively investigate water quality conditions, specifically Chl-*a* concentration, and TSI, within the primary dams of the Northeastern Semi-Arid region. By describing the relationship between satellite-derived data and hydrological information, the study seeks to provide an understanding of water quality dynamics in the region's critical water bodies. The specific objectives are as follows:

(i) To validate the estimations provided by the AlgaeMAp application by comparing them with in situ measurements. This validation process will serve as a foundational step to ensure the robustness of the following methodology in accurately estimating chlorophyll-*a* concentrations.
(ii) To conduct spatial and temporal analysis of water quality. This analysis will encompass the Chl-*a* time series and its trending overtime for the 30 reservoirs used for this research.
(iii) Further, the research aims to describe the relationships between water quality and hydrological conditions, including precipitation and dams' water levels, offering insights into the dynamics of trophic states.

2 Materials and Methods

The methodology of this research is divided into three main tasks. First, the validation of AlgaeMAp's Chl-*a* and TSI estimations involved accessing in situ measurements from official databases maintained by state agencies (Table 15.1).

The validation process of AlgaeMAp's Chl-*a* and TSI estimation was performed by following these steps:

(i) Access to in situ Chl-*a* measurements at the official database (Table 15.1) between 2017 and 2022. In order to gather enough data to perform a validation between AlgaeMAp's and in situ Chl-*a* values, three official databases were accessed online: FUNCEME (Water and Meteorological Foundation of Ceará State), APAC (Water and Climate Agency of Pernambuco State), and CHESF (Hydropower Company of São Francisco) (FUNCEME, 2023; APAC, 2023; CHESF, 2023). Although the data are provided by different sources, the agencies followed the same water quality parameter standards defined by QUALIAGUA (ANA, 2023), which allows us to integrate them into one database for further analysis. In the same database, we classify measured Chl-*a* into TSI classes (Table 15.2) resulting in a csv file as output.

(ii) Extract Chl-*a* and TSI estimated from AlgaeMAp between 2017 and 2022 at felipellobo.users.earthengine.app/view/algaemapv10 from 30 different water bodies (Fig. 15.1). Sample point's coordinates are usually located next to the dam's margin, which can cause an adjacency effect on Chl-*a* estimation from satellite images. In order to avoid this influence, Chl-*a* extraction plots were created with a buffer zone (120 m) within the main water body (Fig. 15.2a), as suggested by Paulino et al. (2022). The data extraction was carried out for each reservoir individually followed by a meticulous analysis to avoid cloud interference on Chl-*a* estimations. This was done at AlgaeMAp by looking into

Table 15.1 Type and source of data used in this research

Type	Unity	Description	Source
In situ chlorophyll-a	μg/L	Provided by several official sources	FUNCEME (2023); APAC (2023); CHESF (2023)
Satellite-estimated Chl-*a*	μg/L	Chlorophyll-a concentration derived from Sentinel-2 (considering one sample point)	AlgaeMAp
Satellite-estimated TSI	Trophic classes	Trophic State Index derived from Sentinel-2 (considering the whole water body)	AlgaeMAp
Precipitation	mm	CHIRPS Pentad(Climate Hazards Group InfraRed Precipitation With Station Data)	UCSB/CHG (Climate Hazards Center, UC Santa Barbara) , Funk et al. (2015)
Dam's water level	Volume percentage	Daily measurements	ANA/CHESF

Table 15.2 Chl-*a* concentration thresholds for Trophic State Levels according to Cetesb (2011)

Trophic state	Chlorophyll-*a* (ug/L)
Ultraoligotrophic[a]	Chl-*a* < 1.17
Oligotrophic	1.17 < Chl-*a* < 3.24
Mesotrophic	3.24 < Chl-*a* < 11.03
Eutrophic	11.03 < Chl-*a* < 30.55
Supereutrophic[b]	30.55 < Chl-*a* < 69.05
Hypereutrophic[b]	69.05 < Chl-*a*

[a]In this research, ultraoligotrophic class was combined with oligotrophic
[b] Considered Algae Bloom condition for this research

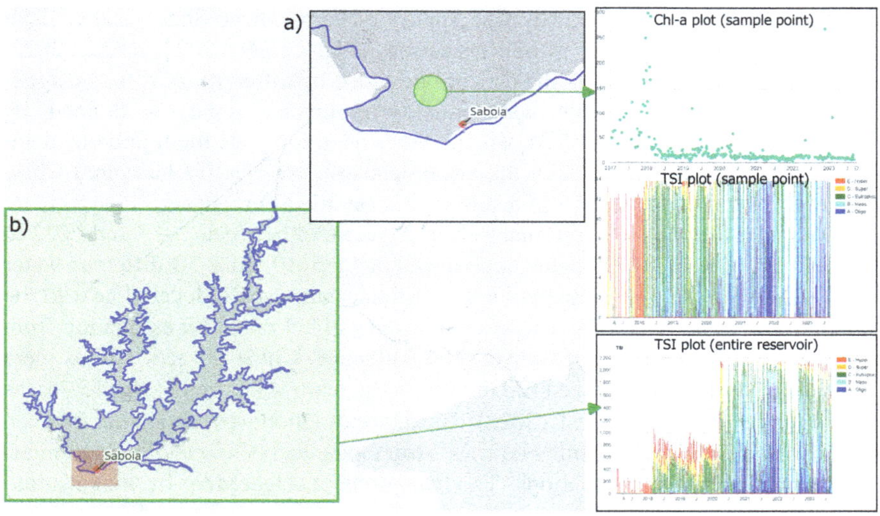

Fig. 15.2 Chl-*a* and TSI sampling method using AlgaeMAp. (**a**) Example of TSI extraction from the Sabóia Reservoir in Pernambuco considering the entire reservoir for temporal analysis. (**b**) For Chl-*a* and TSI validation, plots were created considering the sample point and compared with in situ Chl-*a* (and TSI) measurements

specific dates individually. The AlgaeMAp's estimations were combined into the database created in i).

(iii) Once the in situ and estimated databases were established, the next step was to define matchups between measured and estimated Chl-*a* coinciding within a 2-day window between Sentinel's date and sampling date. Here again, a meticulous analysis was carried out to avoid cloud interference on Chl-*a* estimations in some images in which cloud was not completely removed. As a result, a total of 188 pairs of measured Chl-*a* and estimated Chl-*a* samples were defined.

(iv) To gauge the precision of AlgaeMAp's Chl-*a* estimations, three key performance metrics were employed: the determination coefficient (R^2), the root mean square error (RMSE), and mean absolute percentage error (MAPE).

Table 15.3 Confusion matrix for TSI classes

Class	Precision	Recall	f1-score
Oligotrophic	0.43	0.71	0.54
Mesotrophic	0.65	0.63	0.64
Eutrophic	0.73	0.71	0.72
Supereutrophic	0.69	0.56	0.62
Hypereutrophic	0.71	0.45	0.56
Overall accuracy			0.65

Note: Precision, recall, f1-score, and overall accuracy are also displayed

(v) For TSI classification's assessment, measured Chl-*a* data were classified into five trophic classes (Table 15.3) and a confusion matrix was used to evaluate the classification performances in terms of accuracy and f1-score. All the calibration and validation processing were conducted in Python 3.6 (Colab).

The second task was to conduct a spatial and temporal analysis based on AlgaeMAp's estimation, encompassing the years 2017–2022. For every reservoir (Fig. 15.2a), a heat map was created showing monthly TSI conditions (Fig. 15.2b). The monthly TSI was extracted at AlgaeMAp using the *Monthly TSI* plot option. The heatmap is followed by Chl-*a* time series plots extracted from AlgaeMAp with indication of non-parametric Menn-Kendall (M-K) trend test and the Sen's slope. The M-K test computes a *p*-value, which measures the likelihood of observing a test statistic as extreme as the observed one under the assumption that the null hypothesis holds true. When the *p*-value is less than 0.05, it indicates statistical significance in the trend. Sen's slope determines whether the trend in the time series is positive or negative. A positive Sen's slope suggests that later observations tend to be larger than earlier ones, while a negative Sen's slope indicates that later observations are typically smaller than earlier ones.

The third task focused on a descriptive investigation that took into consideration other hydrological factors (Table 1). The time series data for Chl-*a* concentrations were plotted along with hydrological data (precipitation and dam's volume) and daily TSI (Fig. 15.2b). Here, we used TSI daily for the entire reservoir extracted at AlgaeMAp, which also indicates the water surface extension, along with precipitation and reservoir water level to describe the influence of hydrological effects onto Chl-*a* concentration. Monthly values of precipitation data were extracted via Google Engine accessing the CHIRPS (Climate Hazards Group InfraRed Precipitation With Station) data (Funk et al., 2015) by generating a csv file for the Jaguaribe River subbasins. The water level values were accessed at ANA database, which provides daily water level values. This descriptive analysis was performed for only one, as an example. The reservoir is the Castanhão (Ceará), one of the largest and most important in the region for public water supply and irrigation (Fig. 15.1). The description of hydrological effects in this reservoir illustrates an example of an application for Chl-*a* time series derived from satellite. Besides temporal analysis, AlgaeMAp allows spatial analysis for individual reservoirs. Here, a bloom frequency map was created for the Castanhão reservoir, illustrating the percentage of bloom occurrence

per pixel (representing supereutrophic condition or higher) between 2017 and 2022. For instance, if there are 10 satellite images available and a bloom condition is detected 7 times for a particular pixel within the reservoir, a value of 0.7 will be attributed to that specific pixel.

3 Results and Discussion

3.1 Validating AlgaeMAp

The assessment of Chl-*a* estimations derived from the AlgaeMAp model against in situ measurements revealed a high agreement between them (Fig. 15.3).

The determination coefficient (R^2) of 0.89 is a robust correlation, indicating a strong relationship between the estimated and measured Chl-*a* concentrations. The Root Mean Square Error (RMSE) of 10.64 µg/L further reinforces the model's

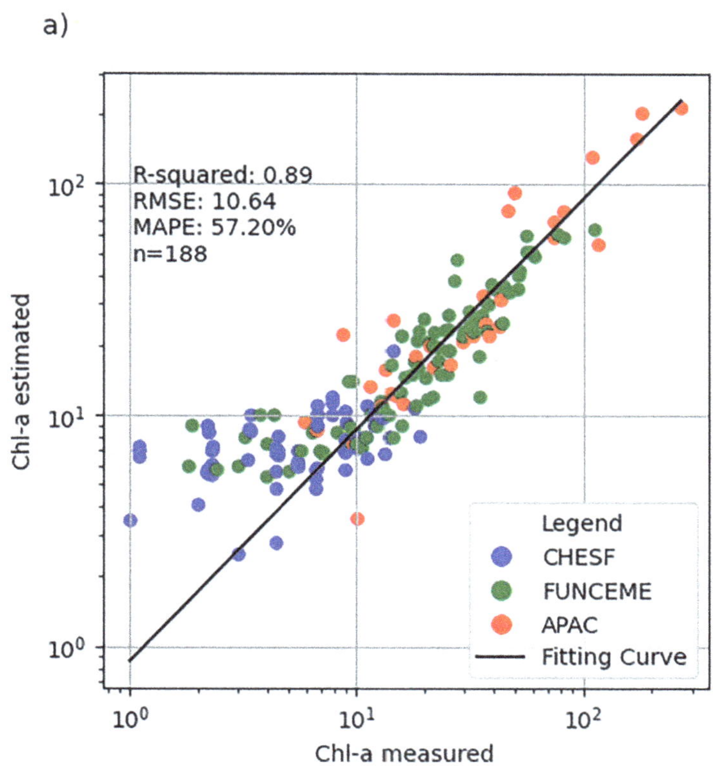

Fig. 15.3 (**a**) Scatterplot between measured (different data sources are indicated with different colors) and estimated Chl-*a* with fitting curve and statistics

accuracy, illustrating a minimal deviation from the measured values and is within the range found by other authors (Woo Kim et al., 2022; Molkov et al., 2019; Lobo et al., 2021). Notably, the Mean Absolute Percentage Error (MAPE) of 57.2% highlights that, on average, the model's estimations deviated by approximately 57.2% from the actual Chl-*a* concentrations, which is a reasonable error compared to literature (Watanabe et al., 2019; Cairo et al., 2020). For instance, Watanabe et al. (2016), investigating a eutrophic reservoir in Tietê River Basin (Barra Bonita), obtained an R^2 of 0.80 and a Mean Absolute Percentage Error (MAPE) of 39.44% for an NDCI algorithm, using in situ radiometric measurements.

In a more global application for a wider range of Chl-*a* concentrations (0–1000 μg/L), Pahlevan et al. (2020) reported that NDCI obtained from in situ radiometric data estimated Chl-*a* with errors of approximately 49%. Given that the in situ measurements acquisition plan was not originally intended for remote sensing applications like those referenced in the literature, we find the AlgaeMAp chlorophyll-*a* estimation to be suitable for spatial and temporal analysis.

When examining the three databases employed (APAC, FUNCEME, and CHESF), a noteworthy agreement emerged. Water bodies monitored by CHESF, predominantly characterized by large hydropower dams, exhibited consistently low Chl-*a* concentrations, indicating oligotrophic conditions. In contrast, APAC (Pernambuco State) and FUNCEME displayed higher Chl-*a* concentrations, with the majority exceeding 10 μg/L. This points toward eutrophic to supereutrophic conditions prevalent in these regions.

The validation of Trophic State Index (TSI) estimates derived from AlgaeMAp's classification against in situ TSI values based on Chl-*a* concentrations showed a satisfactory model's performance (Fig. 15.4).

The oligotrophic classification displayed a precision of 43%, indicating that only 0.43 (out of 21 samples) of water bodies identified as oligotrophic were accurately classified. The recall for oligotrophic conditions was relatively higher at 71%, suggesting that the model effectively identified water bodies with low nutrient levels. However, this discrepancy between precision and recall underscores a potential limitation in accurately categorizing water bodies into the oligotrophic class.

The mesotrophic classification exhibited a precision of 65% (out of 62 samples), and recall of 63%, suggesting a moderate level of accuracy in identifying water bodies with moderate nutrient levels.

The eutrophic classification demonstrated a commendable precision of 0.73, implying that 73% (out of 62 samples) of water bodies identified as eutrophic were correctly classified. The recall for eutrophic conditions was also 71%, indicating a consistent ability to identify water bodies with elevated nutrient levels. This balanced performance in both precision and recall metrics highlights the model's proficiency in categorizing water bodies as eutrophic.

The supereutrophic and hypereutrophic classifications displayed relatively lower precision scores of 69% (out of 32) and 71% (out of 11 samples), respectively. This indicates that a notable proportion of water bodies identified as supereutrophic or hypereutrophic was accurately classified. However, the recall for both classes was

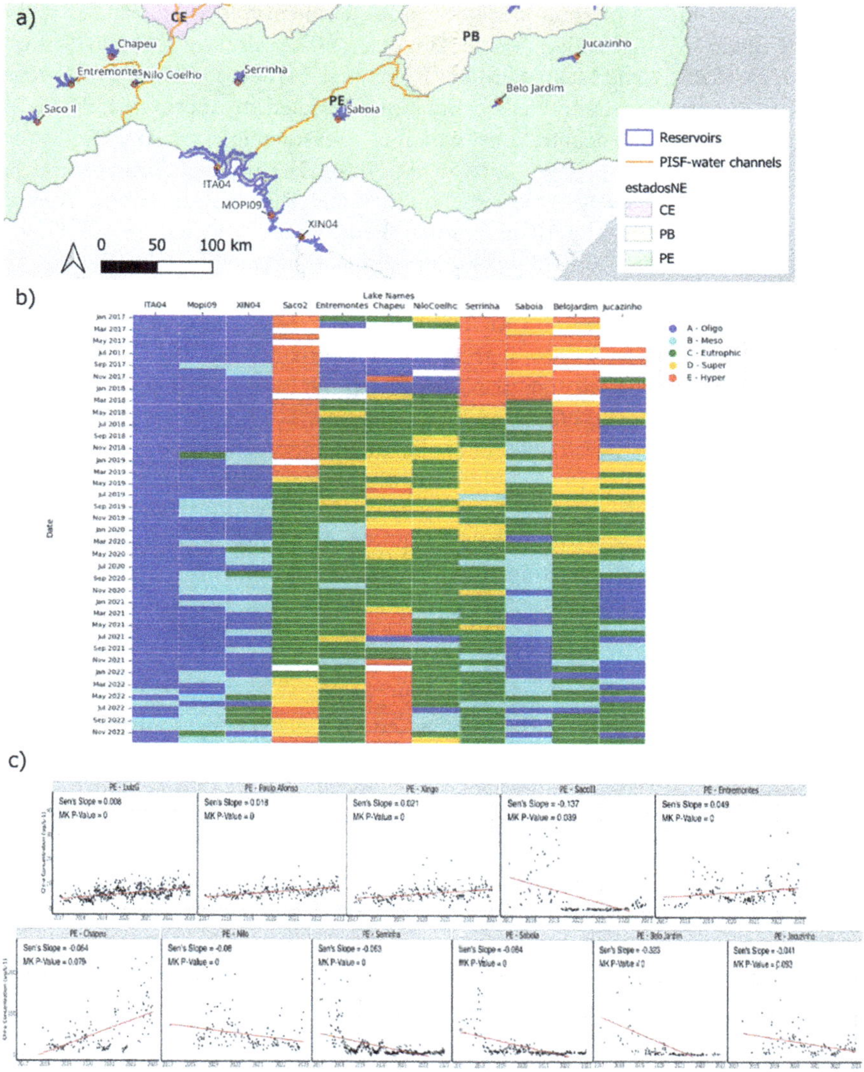

Fig. 15.4 (**a**) Location of the main reservoirs in the Pernambuco State considered for analysis. (**b**) Monthly TSI from 2017 to 2023. (**c**) Temporal scatterplots for chlorophyll-a concentration with linear regression to assess trends (Sen's value indicates if the trend is positive or negative; M-K lower than 0.05 indicates significant trending value, which indicates if the trend is positive or negative)

lower, indicating a sub-estimation in correctly identifying water bodies with extremely high nutrient levels.

Similar TSI classification performance was found in literature. Watanabe et al. (2020) investigated the inland water's trophic status classification based on machine

learning and remote sensing data and found global accuracies varying from 69 to 80% for different classifiers, such as Random Forest, and Support Vector Machine. Lobo et al. (2021), using the same methodology of this chapter, yielded an overall accuracy of 71% for TSI classification based on NDCI derived from Sentinel-2.

With both chlorophyll-*a* (Chl-*a*) and Trophic State Index (TSI) assessments yielding satisfactory results, despite AlgaeMap's initial development for a distinct region, a consistent foundation has been established for further analysis in the Northeastern Semi-Arid region of Brazil. This validation framework now allows for an in-depth exploration of spatial and temporal conditions within the Northeastern Brazilian context, enabling a comprehensive assessment of water quality dynamics.

3.2 *Spatial and Temporal Water Quality Analysis*

The following figures (Figs. 15.4, 15.5 and 15.6) provide an overview of the water quality conditions of the main reservoirs for each state showing their location, monthly Trophic State Index (TSI) condition, and temporal chlorophyll-*a* concentration with linear regressions to detect positive or negative trends.

3.2.1 Pernambuco State

For Pernambuco State (Fig. 15.4), the large reservoirs (Luiz Gonzaga—ITA04; Paulo Afonso—MOPI09; and Xingó—XIN04) in the São Francisco River show oligotrophic conditions most of the period analyzed with Chl-*a* concentration under 10 ug/L. These results are in accordance with those from literature (Lima & Severi, 2014). However, a positive linear trend is observed in all three reservoirs, meaning that Chl-*a* concentration has slightly increased over time from 2017 to 2022 (Fig. 15.4c). This suggests a gradual shift toward mesotrophic conditions and could be a potential concern for future reference because these reservoirs have been used as a source for water transportation to the other reservoirs in the region, as part of the PISF project.

It is worth mentioning that the Chl-*a* levels in the final reservoir of the cascade (Xingó—XIN04) are higher compared to those in the upstream reservoirs (Paulo Afonso—MOPI09; and Luiz Gonzaga—ITA04). Typically, in a series of consecutive reservoirs, each dam reduces nutrient levels, leading to a gradual decrease in water concentrations and lower trophic states (Watanabe et al., 2019). However, this pattern is not observed here, partly due to aquaculture activities and urban discharges occurring in the region (de Lima & Severi, 2014). Additionally, the Luiz Gonzaga and Moxotó reservoirs have intermittent tributaries—the Pajeú and Moxotó rivers, respectively—which exhibit limnological characteristics distinct from the portions of the reservoirs into which they discharge. These tributaries show higher values of electrical conductivity, suspended solids, turbidity, and nutrients

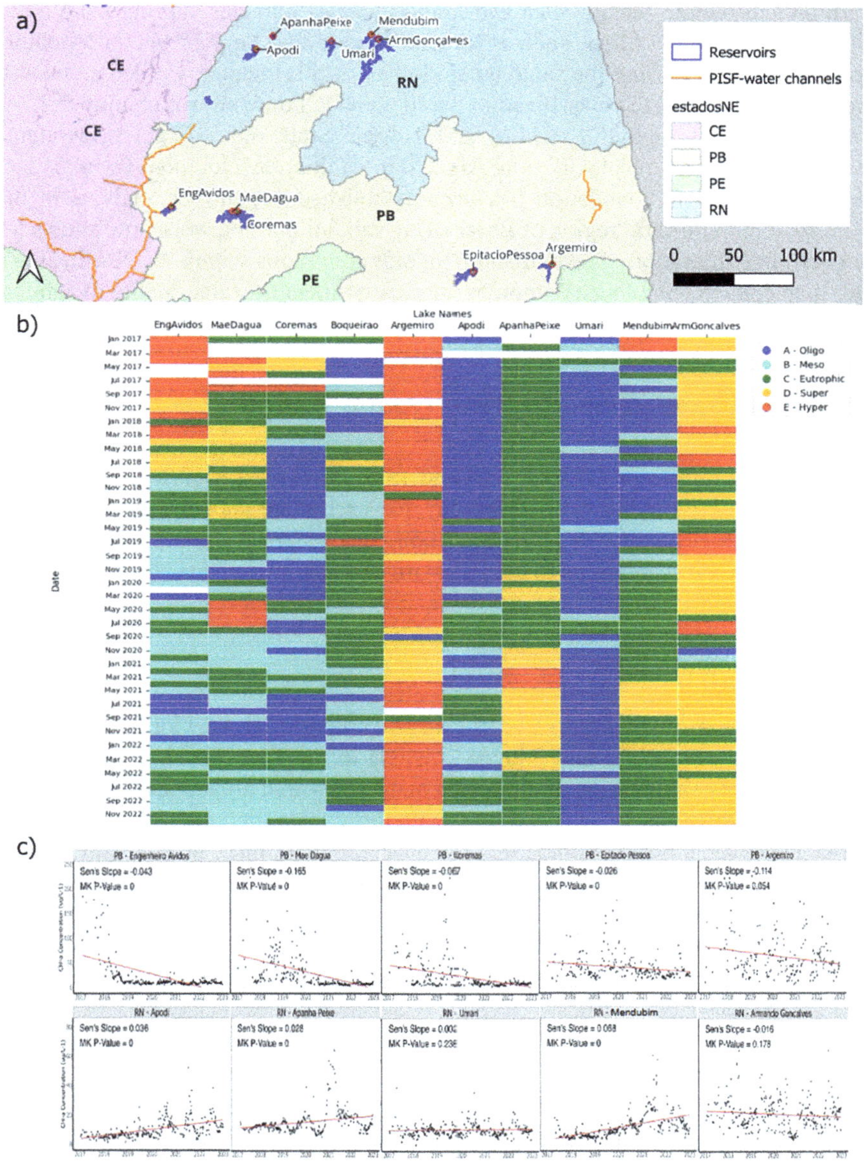

Fig. 15.5 (**a**) Location of the main reservoirs in the Paraíba and Rio Grande do Norte States considered for analysis. (**b**) Monthly TSI from 2017 to 2022. (**c**) Temporal scatterplots for chlorophyll-a concentration with linear regression to assess trends (Sen's value indicates if the trend is positive or negative; M-K lower than 0.05 indicates significant trending)

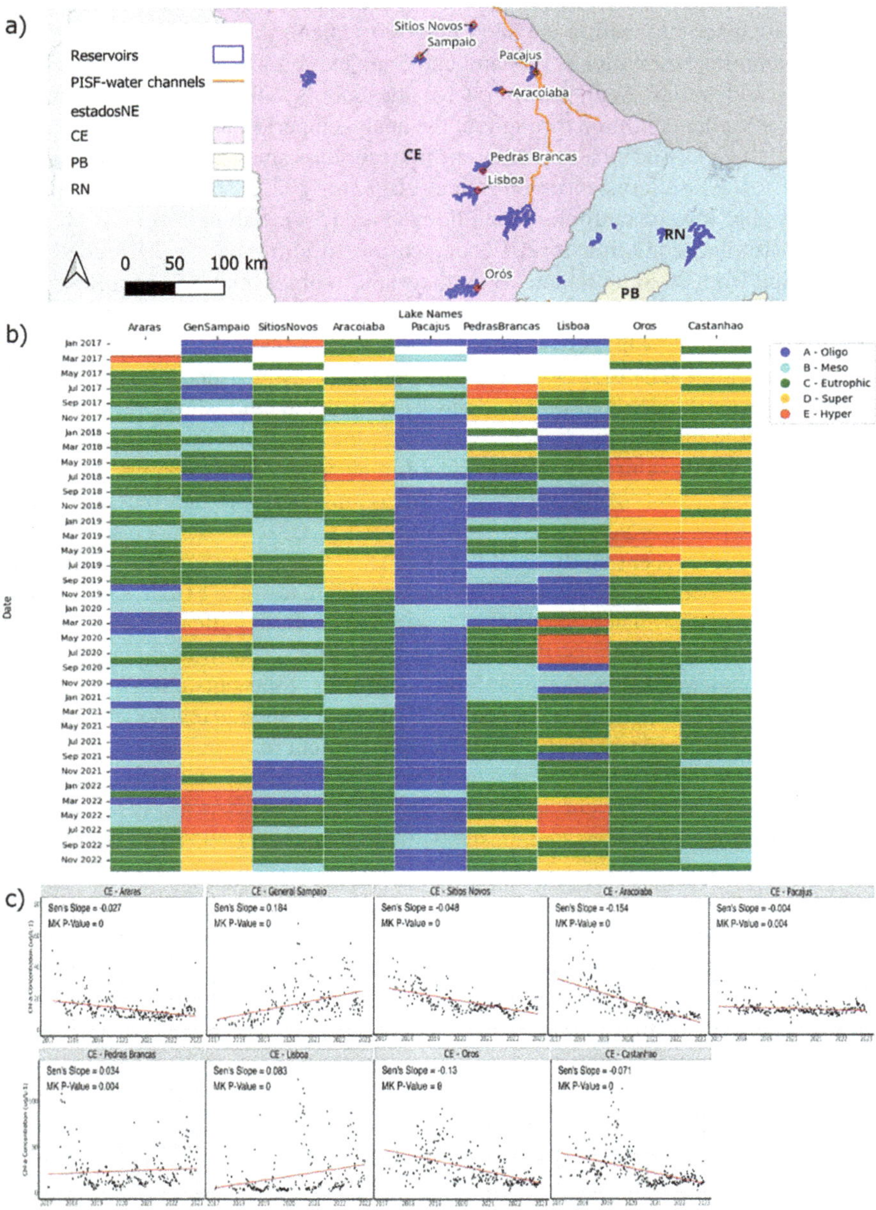

Fig. 15.6 (**a**) Location of the main reservoirs (so-called açudes) in the Ceará State considered for analysis. (**b**) Monthly TSI from 2017 to 2022. (**c**) Temporal scatterplots for chlorophyll-a concentration with linear regression to assess trends (Sen's value indicates if the trend is positive or negative; M-K lower than 0.05 indicates significant trending)

(FADURPE/CHESF, 2011), thereby elevating the trophic state of the water in these intermediate reservoirs within the São Francisco system.

The remaining reservoirs in Pernambuco State exhibit a higher range of trophic conditions and trends. With exception for the Saboia, all the reservoirs present eutrophic or higher condition throughout the analyzed period with Chl-*a* concentration above 10 ug/L. Serrinha and Belo Jardim reservoirs show a significant decreasing trend of Chl-*a* concentration from 2017 to 2022, shifting from super/hypereutrophic toward eutrophic conditions (Fig. 15.4). Saboia reservoir showed very high trophic condition in 2017, decreasing to eutrophic in 2018/2019 and mesotrophic from 2020 up to date. It is important to observe that Saboia Reservoir was one of the first reservoirs to receive water from PISF in March 2017 and in Belo Jardim Reservoir, the water arrived in February 2019.

The reservoir Chapéu showed a significant increasing trend from eutrophic toward hypereutrophic conditions. Saco II and Jucazinho show recent Chl-*a* increases not captured by linear regression due to their initial high levels. Entremontes remains eutrophic with unclear future trends. With exception to Saboia and Belo Jardim, none of the reservoirs in Pernambuco State had received water from the PISF program yet.

According to a recent study by Silveira et al. (2022), Nilo Coelho reservoir is found to be subject to some source of pollution, which, to some extent, explains the high Chl-*a* levels estimated for this reservoir. The authors indicate that while the overall water quality was classified as good according to the water quality index, specific parameters such as thermotolerant coliforms, total phosphorus (TP), and biochemical oxygen demand exceeded the legal limits, suggesting potential pollution sources in the reservoir. It was observed that nutrient (nitrogen and phosphorus) loads are higher during the rainy season. Analysis of land use and land cover indicates an expansion of agricultural areas surrounding the reservoir, resulting in a reduction of vegetation areas and, possibly, in an increase of nutrient load into the water system.

3.2.2 Rio Grande do Norte State

For the reservoirs in Rio Grande do Norte (Fig. 15.5), we can observe a wide range of trophic conditions. Umari reservoir showed oligotrophic state throughout the whole period of analysis and no significant trend was observed, indicating temporal stability. Santa Cruz do Apodí (shortly Apodi) and Mendubim reservoirs showed similar temporal variation shifting from oligotrophic state until 2019 toward eutrophic conditions in the last 3 years. The temporal Chl-*a* concentration plot confirms this observation with significant positive trends in these two reservoirs. Bezerra (2017) had already indicated that the high nutrient concentration found in the Apodi reservoir could trigger blooms in the reservoirs over the following years, complicating water treatment by conventional treatment plants and amplifying health risks

since cyanobacteria can produce toxins harmful to human health and animals that use these waters.

The Armando Gonçalves reservoir, the largest in the state and providing water for 40 municipalities, exhibited high trophic conditions based on our results from 2017 to 2023, with a predominant supertrophic state and no observable trend for chlorophyll-*a*, indicating temporal stability. The deterioration of water quality in this reservoir, characterized by its green color, garnered public attention, and was reported in national news (Globo, 2017). The water transfer from PISF program reached the reservoir in February 2022 and its effects cannot be observed yet. The Apanha Peixe reservoir showed predominantly eutrophic conditions with a slight positive trend for Chl-*a* concentration, although a high variation of Chl-*a* was observed over the years (Fig. 15.4).

3.2.3 Paraíba State

According to Fig. 15.5, we observed that the reservoirs located in the western part of the state (Eng. Ávidos, Coremas, and Mãe D'Água) showed a similar pattern shifting from high eutrophic levels until the beginning of 2020 to low eutrophication levels from 2020 onward. This pattern is clear in Fig. 15.5c, where linear regression for Chl-*a* concentration is significantly negative for these three reservoirs. Important to note that none of these reservoirs have received water from the PISF program until the current date.

One of the primary reservoirs in the state, Boqueirão or Eng. Epitácio Pessoa, responsible for supplying water to 24 municipalities, including Campina Grande, exhibited fluctuating trophic levels throughout the analyzed period, predominantly showing eutrophic conditions, with chlorophyll-*a* concentrations exceeding 10 μg/L most of the time. The trend line indicates a slight, yet statistically significant, negative trend, suggesting a minor improvement in water quality for this reservoir. Since 2017, this reservoir has benefited from water transportation via the PISF Eastern.

Caldas (2021) investigated water quality distribution from 2014 to 2019 along the Eastern channel from the Luiz Gonzaga reservoir—depicted in Fig. 1—toward the Boqueirão (Epitácio Pessoa) reservoir. They found that Boqueirão exhibited high values for water quality parameters, including total solids concentration, coliforms, nutrient load, and chlorophyll-*a* concentration (above 10 μg/L), confirming the eutrophic condition observed in our research.

Situated in the eastern part of the state, the Argemiro reservoir exhibited the poorest water quality condition among all the water bodies analyzed in this study—even though it has benefited from water transfer since March 2017. It consistently demonstrated hypereutrophic levels throughout the period between 2017 and 2022, with chlorophyll-*a* concentrations averaging around 100 μg/L (no significant trend was observed). In fact, signs of eutrophication in this reservoir have been evident since 2007. Studies conducted over a 2-year period (2007–2009) in this reservoir revealed frequent cyanobacterial blooms with higher biomass and toxin production

during dry periods. Of the 16 blooms observed during this period, 14 (87.5%) were found to contain microcystin (Lins et al., 2016).

3.2.4 Ceará State

In the Ceará state, nine reservoirs were analyzed (Fig. 15.6), with eight displaying high trophic conditions. The only exception was the Pacajus reservoir, which maintained oligotrophic state levels with chlorophyll-*a* concentrations consistently below 10 μg/L throughout the studied period, showing temporal stability. Among these reservoirs, three—Arrojado Lisboa, General Sampaio, and Pedras Brancas reservoirs—exhibited increasing trophic level trends from meso to supereutrophic levels, raising concerns about their water quality. This situation is particularly severe for General Sampaio, where the analysis of chlorophyll-*a* concentration trends showed a statistically significant positive slope of 0.18, indicating significant water degradation over recent years. Similar positive trends in chlorophyll-*a* concentration were observed for Arrojado Lisboa and Pedras Brancas, although not as pronounced as in General Sampaio. These findings are consistent with those of Wiegand and Costa (2021), who investigated water quality from 2008 to 2017 in 65 reservoirs in Ceará State.

The study revealed that the reduction in reservoir volume was accompanied by increases in phosphorus and chlorophyll-*a* concentrations and decreases in transparency. The Trophic State Index (TSI) values increased from the wet period (2008–2012) to the hydrological drought (2013–2017), with the trophic state transitioning from mesotrophic to eutrophic for 36 reservoirs (55.4%) and remaining unchanged for 28 reservoirs (43.1%), despite most of them showing an increase in TSI. These results confirm the substantial impact of droughts on the trophic state of semi-arid reservoirs and highlight the high levels of eutrophication estimated by AlgaeMAp.

Despite chlorophyll-*a* concentrations consistently exceeding 10 μg/L between 2017 and 2022, four eutrophic reservoirs—Aracoiaba, Araras, Orós, and Castanhão—showed some improvement in water quality. Chlorophyll-*a* estimations for Aracoiaba, Orós, and Castanhão decreased from approximately ~50 μg/L (supereutrophic) at the beginning of the series to ~10 μg/L (eutrophic) at the end of the temporal series, indicating a consistent improvement over the years. Araras followed a similar pattern of water quality improvement, transitioning from eutrophic to oligotrophic levels in 2021. However, in 2022, chlorophyll-*a* levels began to increase again in Araras, unlike in Orós and Castanhão (Fig. 15.6c). It is important to observe that only Castanhão e Pacajus has received water from PISF since March 2021—the remaining reservoirs are not benefited yet. In both cases, the water transfer, apparently, has contributed to water quality conditions observed within the last few years.

Using Sentinel-2 imagery to estimate Chl-*a* concentration in this region, Aranha et al. (2022) found very similar Chl-*a* estimations for the Pacajus, Castanhão, and Orós reservoirs. They show that the majority of reservoirs exhibit significant

eutrophication, with chlorophyll-*a* concentrations exceeding 40 μg/L and showing seasonal variations. Notably, the Pacajus reservoir shows comparatively lower chlorophyll-*a* concentrations, with values not exceeding 20 μg/L.

Literature reveals that in Orós, for instance, samples collected bimonthly from April 2008 to April 2010 showed the highest chlorophyll-*a* concentrations during the dry season, ranging from 3.8 to 26.5 μg/L, indicating high temporal and spatial variability. The water was classified as eutrophic based on the average TSI at all sampled points, indicating deteriorating water quality and the need for intervention to reduce waste release and improve the trophic state of Orós reservoir waters (dos Santos et al., 2014).

At the state level, another study evaluated correlations between total phosphorus (TP) and chlorophyll-*a* in 155 reservoirs in Ceará, identifying TP's ability to predict chlorophyll-*a* and the factors influencing correlation dynamics. The study found that average chlorophyll-*a* concentration increased in all basins during the dry period, similar to the trend observed for phosphorus concentrations in approximately 58.3% of the basins. Although the correlation models between TP and chlorophyll-*a* were not highly significant, they aligned with patterns described in classical literature models, suggesting that TP availability in the water column contributes to increased chlorophyll-*a* concentrations in most basins (Guimarães & Neto, 2023).

Figure 15.7 provides a comprehensive overview of the water quality trends over time for the primary reservoirs in each state. The size of each point corresponds to the respective reservoir's size. Blue points denote reservoirs where chlorophyll-*a* (Chl-*a*) concentration has decreased, indicating an improvement in water quality. Shades of dark blue represent varying degrees of improvement, with, for instance, reservoirs showing an average decrease in Chl-*a* concentration ranging from −0004 to −0.377 ug/L per year. Conversely, reservoirs exhibiting positive Chl-*a* values are depicted in a red palette, ranging from +0.002 ug/L per year (yellow) to +0.184 ug/L per year (dark red). Notably, most reservoirs exhibit a slight increase in water quality, with a negative Chl-*a* tendency ranging from 0.004 to −0.079 ug/L/year (light blue). However, the largest reservoir under analysis, the Luiz Gonzaga (ITA04) hydroelectric dam, and its downstream reservoirs experienced an increase in Chl-*a* concentration over time, ranging from +0.002 to +0.038 ug/L/year (yellow).

Figure 15.7 also depicts the prevailing TSI conditions for each reservoir spanning from 2017 to 2022. Among the 30 reservoirs analyzed, 6 exhibited an oligotrophic condition, 4 showed a mesotrophic condition, while 14 reservoirs predominantly displayed eutrophic states. Additionally, two reservoirs each indicated a supereutrophic and hypereutrophic dominance. These findings underscore a concerning scenario for the region. Conversely, the results indicate a more favorable trend, with 15 out of 30 reservoirs showing a negative trend compared to 9 showing positive trend. For five reservoirs (Umari, Armando Gonçalvez, Jucazinho, Chapéu, and Argemiro), the M-K test displayed values higher than 0.05 and, therefore, were not statistically significant.

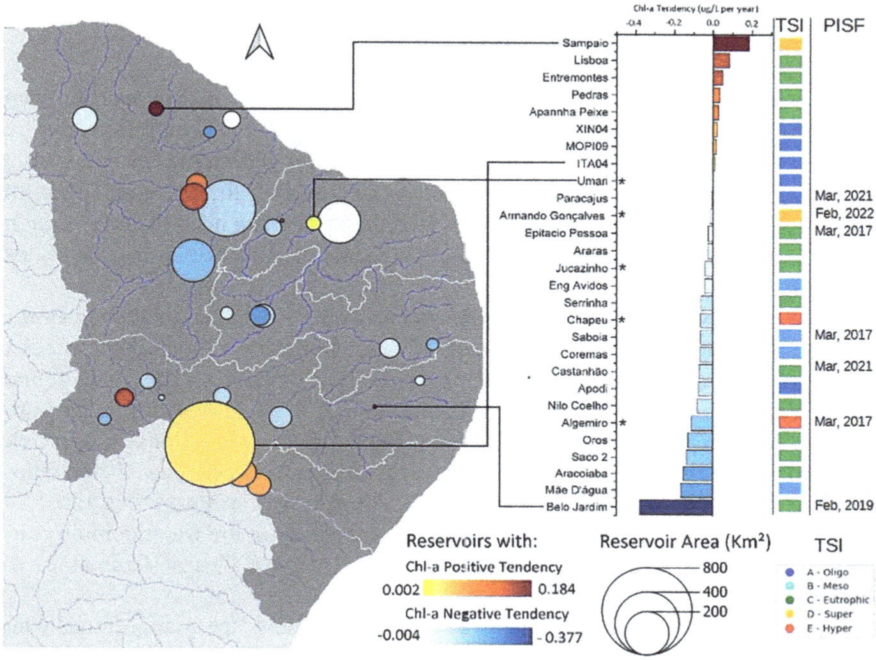

Fig. 15.7 Map of the study area, depicting reservoirs with negative trends (Sen's slope shown in blue tones) and positive trends (displayed in red tones) throughout the time series from 2017 to 2022. The trend lines with M-K values higher than 0.05 were not statistically significant and are indicated with a star symbol. The size of each icon corresponds to the area of the respective reservoir. The dominant TSI condition for each reservoir between 2017 and 2022 is also displayed. For those reservoirs that have received water from PISF program until 2023, a date indicating the initial month of receiving water from PISF is added on the right

Considering the water quantity challenges in the region, the variations in water quality appear intricately linked to the hydrological cycle, including precipitation events and water accumulation, as well as the contribution of water transfer from the PISF program. Figure 15.7 highlights the reservoirs that have received water from the PISF program, specifying the date when transferred water reached each specific reservoir. For instance, the Epitácio Pessoa reservoir in Paraíba State has been receiving water since March 2017. Interestingly, the seven reservoirs benefiting from the PISF program demonstrated a negative trend for Chl-*a* concentration—although three were not statistically significant, which prevents any conclusion. In fact, the precise magnitude of its direct impact remains unexplored presently, primarily due to insufficient data regarding the volume of water transferred to each reservoir. Investigating the direct influence of the PISF program should be prioritized for future research endeavors.

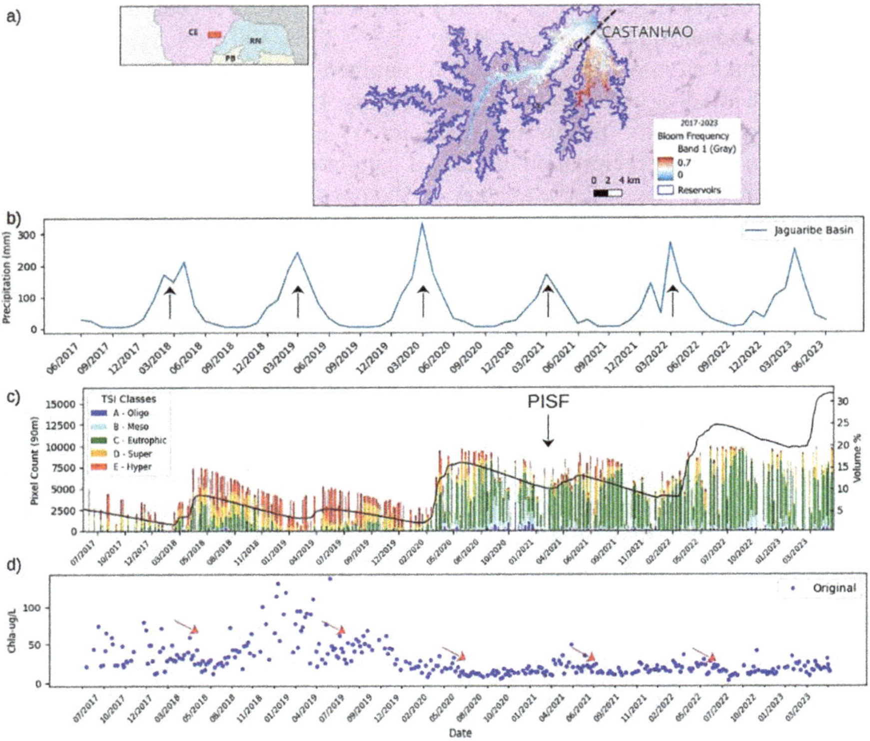

Fig. 15.8 Time series analysis for Castanhão/Ceará. (**a**) Location with a bloom frequency map from 2017 and 2023. (**b**) Precipitation in the Jaguaribe River Basin. (**c**) Reservoir volume and TSI for the entire reservoir. (**d**) Chl-*a* time series

3.3 Hydrological Effect on the Water Quality

An example of the effect of hydrological cycle on the water quality is the case of Castanhão. Figure 15.8 shows the location with a map indicating bloom frequency from 2017 and 2023 (Fig. 15.8a), precipitation (Fig. 15.8b), volume percentage with trophic state classification (Fig. 15.8c), and the Chl-*a* time series. Our findings regarding the effect of the hydrological cycle on water quality in Castanhão reservoir indicate a high trophic level over the analysis period spanning 2017–2022. The reservoir's volume remained predominantly below 15% from 2017 to 2020, with a significant increase to 20% following a precipitation event in March 2020. This increase positively impacted both the reservoir's volume and water quality, leading to a decline in trophic conditions from a dominant super/hyper level to a eutrophic state, accompanied by a decrease in chlorophyll-*a* concentration from 50 to 10 ug/L. Subsequently, the reservoir maintained a eutrophic level with slight seasonal variations, with chlorophyll-*a* concentration typically decreasing after each rainy

season. We observed that after every rainy season the Chl-*a* concentration seems to decrease as indicated by red arrows in Fig. 15.8.

Our findings align with the observations outlined by Laurent et al. regarding the impact of the hydrological cycle on water quality. Laurent et al. (2023) evaluated the water quality of 155 freshwater reservoirs in Ceará state. They noted that low annual accumulated precipitation correlated with inferior water quality, suggesting that reservoir water levels serve as a key predictor of bloom occurrence and overall water quality status. We observed that low reservoir volumes, particularly below 15%, were associated with higher trophic levels and elevated chlorophyll-*a* concentrations, indicating poorer water quality. In both studies, precipitation events were identified as significant factors influencing water quality dynamics. Our findings indicate that a substantial precipitation event in March 2020 positively impacted the reservoir's volume and water quality, leading to a decrease in trophic conditions and chlorophyll-*a* concentration. Laurent et al. also highlighted the importance of precipitation, noting that low rainfall contributes to reduced water quality and increased susceptibility to algal blooms. Both studies recognize the temporal variability in water quality parameters, particularly in response to seasonal changes and precipitation patterns. Our study observed seasonal variations in chlorophyll-*a* concentration, with levels typically decreasing after each rainy season. Similarly, Laurent et al. emphasized the need for enhanced precautions during dry periods when reservoirs are most prone to algal blooms, indicating a clear temporal link between precipitation patterns, water quality, and bloom occurrence.

These results suggest that the water transferred through the PISF program could potentially enhance water quality by introducing a greater volume of good-quality water into the system. However, Fig. 15.8 indicates that the direct impact of the PISF program was not immediately observable when water transfer commenced in March 2021. Once more, the exact magnitude of its direct influence remains unexplored at present, mainly due to inadequate data regarding the volume of water transferred to individual reservoirs.

From a spatial perspective, the analysis reveals that due to elevated trophic conditions, specific areas of the reservoir, particularly the southeastern section, consistently exhibited super/hypereutrophic conditions, indicative of algae blooms, in over 45% of the acquired images during the study period. In contrast, the main body of the reservoir experienced bloom conditions in approximately 30% of the images (refer to Fig. 15.8a). Additionally, the map illustrates the extent of low water levels, currently estimated at around 30% of the reservoir's surface area. These analyses underscore the capability of AlgaeMAp to not only offer temporal insights but also spatial information. Consequently, for effective water quality management in the Castanhão reservoir, prioritizing measures in the southeastern tributaries appears crucial based on these spatial patterns.

4 Conclusions

We have demonstrated that AlgaeMAp serves as a useful tool for analyzing water quality, specifically chlorophyll-*a* (Chl-*a*) and Trophic State Index (TSI), in Brazilian semi-arid reservoirs, offering an alternative to traditional monitoring methods. Despite the data initially not being intended for remote sensing applications, the Chl-*a* estimations demonstrated acceptable RMSE and MAPE values, supported by a high determination coefficient ($R^2 = 0.89$) when compared to literature. Given that the in situ measurements acquisition plan was not originally intended for remote sensing applications like those referenced in the literature, we find the AlgaeMAp chlorophyll-*a* estimation to be suitable for spatial and temporal analysis. Nonetheless, there is still potential for improvement, particularly in accurately estimating low Chl-*a* values. Enhanced precision could be achieved through dedicated field campaigns aligned with Sentinel-2 acquisition dates or by exploring alternative algorithms for Chl-*a* estimation. By utilizing AlgaeMAp products, users can save considerable time and effort as there is no need for image processing. With AlgaeMAp, users gain access to hundreds of pre-processed satellite images, eliminating the need to download and process individual images on their personal computers. Furthermore, no specific expertise in remote sensing or image processing is required to access water quality products or download spatial data such as Chl-*a* concentration maps, enabling easy access for users and professionals alike.

Regarding the spatial and temporal analysis of the water quality in Brazilian semi-arid reservoirs, we found that the reservoirs along the São Francisco River, which are the water source for the PISF program, exhibited oligo/mesotrophic conditions. However, a positive linear trend for Chl-*a* was observed in all three reservoirs, indicating a slight increase in Chl-*a* concentration over time. This suggests a gradual transition toward mesotrophic conditions, which could pose concerns for future water resources within the region, such as those under the PISF project.

Considering all reservoirs investigated, the trophic conditions of the analyzed reservoirs indicate that most of them (23 out of 30) exhibit low water quality (eutrophic or higher) throughout 2017 and 2022 indicating a scenario of degraded water condition. On the other hand, examining time-series data, we observed that 15 out of 30 reservoirs displayed a decreasing trend in Chl-*a* concentration, which could be explained by many environmental and hydrological factors. For example, the recovery of water storage in the reservoir plays a positive factor for water quality improvement as indicated by Laurent et al. (2023). In this chapter, we described the influence of hydrological effects on the water quality particularly for Castanhão reservoir; it is important to note that a severe drought persisted from 2012 until 2018 causing water degradation in Castanhão and most of the reservoirs.

With successive rainy periods this reservoir has recovered some water storage resulting in water quality improvements. However, it is evident that there is no uniform temporal pattern applicable to all reservoirs in the region and further research is suggested. It is worth mentioning that the PISF program may have played a role in the observed improvements in water quality in some reservoirs, particularly those

that have been receiving water since 2019 or earlier, such as Epitácio Pessoa (PB), Saboia (PE), and Belo Jardim (PE). However, for a more precise and robust analysis of the impact of the PISF program on water quality in the region, it is advisable to take into account the specific water inputs at each water body over a longer period of analysis, perhaps spanning 5 years or more after the implementation of PISF. This aspect will be the subject of further investigation.

Acknowledgments We acknowledge the partial funding provided by the Office of Naval Research (ONR), Proc. Number N629092112022 (*Felipe Lobo*). We are thankful for the partial funding provided by the FAPERJ E-26/201.406/2022 (*Lino Carvalho*). We also acknowledge the Brazilian National Council for Scientific and Technological Development (CNPq) for the support from the National Institute of Science and Technology (INCT): National Observatory for Water Security and Adaptative Management (Observatório Nacional de Segurança Hídrica e Gestão Adaptativa—ONSeadapta), Proc. Number 406919/2022-4 (*Danielle Bressiani*) and Proc. Number 406919/2022-4 (*Suzana Montenegro*), and for the Research Scholarship PQ Proc. Number 313392/2020-0 (*Suzana Montenegro*). We are thankful for the support extended by CAPES (National Coordination for the Improvement of Higher Education Personnel), which awarded Laura Bueno a master's scholarship.

Author Contributions Material preparation and data collection were performed by Felipe Lobo, Lino Carvalho, and Daniel Maciel. The first draft of the manuscript was written by Felipe Lobo and all authors commented on previous versions of the manuscript. All authors contributed to the data analysis and final version edition.

Conflict of interest The authors have no competing interests to declare that are relevant to the content of this article.

Data Availability All the data used in this article are freely available in their respective references. AlgaeMAp data are available at: felipellobo.users.earthengine.app/view/algaemapv10.

References

ANA. (2023). Portal Qualiágua/ANA. 2023. https://portalpnqa.ana.gov.br/Qualiagua.aspx.

APAC. (2023). Agência Pernambucana de Águas e Clima. 2023. https://www.apac.pe.gov.br/.

Aranha, T. R., Benevides, T., Martinez, J.-M., Souza, E. P., Barros, M. U. G., Eduardo Sávio, P. R., & Martins. (2022). Remote analysis of the Chlorophyll-a concentration using Sentinel-2 MSI images in a semiarid environment in Northeastern Brazil. *Water, 14*(3), 451. https://doi.org/10.3390/w14030451

Araújo, G. M., Lima Neto, I. E., & Becker, H. (2019). Phosphorus dynamics in a highly polluted urban drainage channel-shallow reservoir system in the Brazilian Semiarid. *Anais Da Academia Brasileira de Ciências. scielo, 91*(3), e20180441.

Aubriot, L., Zabaleta, B., Bordet, F., Sienra, D., Risso, J., Achkar, M., & Somma, A. (2020). Assessing the origin of a massive cyanobacterial bloom in the Río de La Plata (2019): Towards an early warning system. *Water Research, 181*, 115944. https://doi.org/10.1016/j.watres.2020.115944

Barbosa, J. E., de Lucena, E., Medeiros, S. F., Brasil, J., da Silva, R., Cordeiro, M. C., Crispim, B., Gonzaga, G. H., & da Silva. (2012). Aquatic systems in Semi-Arid Brazil: Limnology and management. *Acta Limnologica Brasiliensia, 24*(1), 103–118.

BEZERRA, Alana Jade De Lima et al.. (2017). Qualidade da água em reservatórios de abastecimento humano na bacia Apodi-Mossoró/RN. Anais II CONIDIS... Campina Grande: Realize Editora. Available at: https://editorarealize.com.br/artigo/visualizar/33805.

BNDES. (2023). PISF – Projeto de Integração Do Rio São Francisco Com as Bacias Do Nordeste Setentrional. 2023. https://bndes.gov.br/wps/portal/site/home/transparencia/desestatizacao/projetos-em-andamento/pisf.

Bowman, N., Patel, P., Xu, W., Alsaffar, A., & Tiquia-Arashiro, S. M. (2018). Enrichment and isolation of Pb-resistant bacteria from Saint Clair River sediments and their potential for Pb removal in aqueous solutions. *Applied Microbiology and Biotechnology, 102*, 2391–2398. https://doi.org/10.1007/s00253-018-8772-4

Braga, G. G., Becker, V., de Oliveira, J. N. P., de Mendonça Junior, J. R., de Medeiros Bezerra, A. F., Torres, L. M., Galvão, Â. M. F., & Mattos, A. (2015). Influence of extended drought on water quality in tropical reservoirs in a semiarid region. *Acta Limnologica Brasiliensia, 27*(1), 15–23.

Buma, W. G., & Lee, S.-I. (2020). Evaluation of Sentinel-2 and Landsat 8 Images for estimating Chlorophyll-a concentrations in Lake Chad, Africa. *Remote Sensing., 12*(15), 2437. https://doi.org/10.3390/rs12152437

Cairo, C. T., Barbosa, C., Lobo, F., Novo, E., Carlos, F., Maciel, D., Jnior, R. F., Silva, E., & Curtarelli, V. (2020). Hybrid Chlorophyll-a algorithm for assessing trophic states of a tropical Brazilian reservoir based on Msi/Sentinel-2 Data. *Remote Sensing, 12*(1), 40. https://doi.org/10.3390/RS12010040

Caldas, H. F. M. (2021). *Análise Da Evolução Espaço-Temporal Da Qualidade Da Água No Eixo Leste Do Projeto de Integração Do Rio São Francisco*. Universidade Federal de Pernambuco.

Cetesb. (2011). Guia Nacional de Coleta e Preservação de Amostras – Água, Sedimento, Comunidades Aquáticas e Efluentes Líquidos. *Companhia Ambiental Do Estado de São Paulo*, p. 326. https://doi.org/C737g.

Chehbouni, A., Escadafal, R., Duchemin, B., Boulet, G., Simonneaux, V., Dedieu, G., Sobrino, J. A., et al. (2008). An integrated modelling and remote sensing approach for hydrological study in arid and semi-arid regions: The SUDMED Programme. *International Journal of Remote Sensing, 29*(17–18), 5161–5181. https://doi.org/10.1080/01431160802036417

CHESF. (2023). Monitoramento Do Rio São Francisco. 2023. https://www.chesf.com.br/sustentabilidade/Pages/MeioAmbiente/Monitoramento-do-Rio-Sao-Francisco.aspx.

Cho, K., Zholi, A., Frabutt, D., Flood, M., Floyd, D., & Tiquia, S. M. (2012). Linking bacterial diversity and geochemistry of uranium-contaminated groundwater. *Environmental Technology, 33*, 1629–1640. https://doi.org/10.1080/09593330.2011.641036

Cooper, S. D., Lake, P. S., Sabater, S., et al. (2013). The effects of land use changes on streams and rivers in mediterranean climates. *Hydrobiologia, 719*, 383–425. https://doi.org/10.1007/s10750-012-1333-4

de Farias Mesquita, J. B., Neto, I. E. L., Raabe, A., & de Araújo, J. C. (2020). The influence of hydroclimatic conditions and water quality on evaporation rates of a tropical lake. *Journal of Hydrology, 590*, 125456.

do Nascimento Moura, A., do Nascimento, E. C., & Dantas, Ê. W. (2012). Temporal and spatial dynamics of phytoplankton near farm fish in eutrophic reservoir in Pernambuco, Brazil. *Revista de Biología Tropical, 60*(2), 581–597.

do Nascimento Moura, A., Aragao-Tavares, N. K. C., & Amorim, C. A. (2018). Cyanobacterial blooms in freshwater bodies from a semiarid region, Northeast Brazil: A review. *Journal of Limnology, 77*(2), 179–188.

Dokulil, M. T., & Teubner, K. (2010). Eutrophication and climate change: Present situation and future scenarios. In A. Ansari, S. Singh Gill, G. Lanza, & W. Rast (Eds.), *Eutrophication: Causes, consequences and control*. Springer. https://doi.org/10.1007/978-90-481-9625-8_1

Dörnhöfer, K., Klinger, P., Heege, T., & Oppelt, N. (2018, January 15). Multi-sensor satellite and in situ monitoring of phytoplankton development in a eutrophic-mesotrophic lake. *Science of the Total Environment, 612*, 1200–1214. https://doi.org/10.1016/j.scitotenv.2017.08.219

dos Santos, J. C. N., de Andrade, E. M., de Araújo Neto, J. R., Meireles, A. C. M., & de Queiroz Palácio, H. A. (2014). Land use and trophic state dynamics in a tropical Semi-Arid reservoir. *Revista Ciência Agronômica, 45*(1), 35–44.

El-Mezayen, M. M., & Abd El-Hamid, H. T. (2023). Assessment of water quality and modeling trophic level of Lake Manzala, Egypt using remotely sensed observations after recent enhancement project. *Journal of the Indian Society of Remote Sensing, 51*, 197–211. https://doi.org/10.1007/s12524-022-01635-3

ESA. (2017). Sentinel-2: MultiSpectral Instrument (MSI) overview. *Technical Guide*. 2017. https://sentinel.esa.int/web/sentinel/technical-guides/sentinel-2-msi/msi-instrument.

FADURPE/CHESF. (2011). Inventário Dos Ecossistemas Aquáticos Do Baixo São Francisco. Reservatórios Do Complexo Hidrelétrico de Paulo Afonso. Relatório Trianual – Dezembro/2007 a Novembro/2010. http://licenciamento.ibama.gov.br/Hidreletricas/Xingo/3º Relatório Anual (Reser. Xingó e Baixo São Francisco).pdf.

Flood, M., Frabutt, D., Floyd, D., Powers, E. U., Devol, A., & Tiquia-Arashiro, S. M. (2015). Ammonia-oxidizing bacteria and archaea in sediments of the Gulf of Mexico. *Environmental Technology, 36*, 124–135. https://doi.org/10.1080/09593330.2014.942385

Fourqurean, J. W., & Robblee, M. B. (1999). Florida Bay: A history of recent ecological changes. *Estuaries, 22*, 345–357. https://doi.org/10.2307/1353203

Fukushima, T., Kitamura, T., & Matsushita, B. (2021). Lake water quality observed after extreme rainfall events: Implications for water quality affected by stormy runoff. *SN Applied Sciences, 3*, 841. https://doi.org/10.1007/s42452-021-04823-x

Funari, E., & Testai, E. (2008). Human health risk assessment related to cyanotoxins exposure. *Critical Reviews in Toxicology, 38*(2), 97–125. https://doi.org/10.1080/10408440701749454

FUNCEME. (2023). Portal Hidrológico Do Ceará. 2023. http://www.hidro.ce.gov.br/hidro-ce-zend/mi/download.

Funk, C., Peterson, P., Landsfeld, M., Pedreros, D., Verdin, J., Shukla, S., Husak, G., et al. (2015). The climate hazards infrared precipitation with stations—A new environmental record for monitoring extremes. *Scientific Data, 2*(1), 1–21.

Gholizadeh, M. H., Melesse, A. M., & Reddi, L. (2016). A comprehensive review on water quality parameters estimation using remote sensing techniques. *Sensors, 16*(8), 1298. https://doi.org/10.3390/s16081298

Giardino, C., Brando, V. E., Gege, P., et al. (2019). Imaging spectrometry of inland and coastal waters: State of the art. *Achievements and Perspectives. Surv Geophys, 40*, 401–429. https://doi.org/10.1007/s10712-018-9476-0

Globo. (2017). Esverdeada, Água Da Maior Barragem Do RN Chama Atenção; Veja Vídeo. Anderson Barbosa. 2017. https://g1.globo.com/rn/rio-grande-do-norte/noticia/esverdeada-agua-da-maior-barragem-do-rn-chama-atencao-veja-video.ghtml.

Guimarães, B. M. D. M., & Neto, I. E. L. (2023) Analysis of total nitrogen and chlorophyll a correlations in reservoirs in Ceará/Brazil. *Engenharia Sanitaria e Ambiental, 28*, e20230015.

Ítalo, F., Chaves, B., de Freitas Lima, P., Leitão, R. C., Paulino, W. D., & Santaella, S. T. (2013). Influence of rainfall on the trophic status of a Brazilian semiarid reservoir. *Acta Scientiarum Biological Sciences, 35*(4), 505–511. https://www.redalyc.org/articulo.oa?id=187129138007

Kampel, M., Lorenzzetti, J. A., Bentz, C. M., Nunes, R. A., Paranhos, R., Rudorff, F. M., & Politano, A. T. (2009). Simultaneous measurements of chlorophyll concentration by lidar, fluorometry, above-water radiometry, and ocean color MODIS images in the Southwestern Atlantic. *Sensors (Basel)., 9*(1), 528–541. https://doi.org/10.3390/s90100528

Kim, W., Young, T. H. K., Shin, J., Lee, D.-S., Park, Y.-S., Kim, Y., & Cha, Y. K. (2022). Validity Evaluation of a Machine-Learning Model for Chlorophyll a Retrieval Using Sentinel-2 from Inland and Coastal Waters. *Ecological Indicators, 137*, 108737. https://doi.org/10.1016/j.ecolind.2022.108737

Koreivienė, J., Anne, O., Kasperovičienė, J., et al. (2014). Cyanotoxin management and human health risk mitigation in recreational waters. *Environmental Monitoring and Assessment, 186*, 4443–4459. https://doi.org/10.1007/s10661-014-3710-0

Laurent, A., Godoy, R. F. B., Elias, M. C., Franco, G., Carril, A. F., Trevisan, E., & Flombaum, P. (2023). Extremes in water quality parameters modulated by weather and climate conditions in Ceara State, Brazil. *Ecohydrology & Hydrobiology, 23*(4), 650–661.

Lima, A. E., & Severi, W. (2014). Trophic state of reservoirs in the cascade of a river in the Brazilian Semiarid.

Lins, R. P. M., Barbosa, L. G., Minillo, A., & De Ceballos, B. S. O. (2016). Cyanobacteria in a eutrophicated reservoir in a Semi-Arid Region in Brazil: Dominance and microcystin events of blooms. *Brazilian Journal of Botany, 39*, 583–591.

Llodrà-Llabrés, J., Martínez-López, J., Postma, T., Pérez-Martínez, C., & Alcaraz-Segura, D. (2023). Retrieving water chlorophyll-a concentration in inland waters from Sentinel-2 imagery: Review of operability, performance and ways forward. *International Journal of Applied Earth Observation and Geoinformation, 125*, 103605. https://doi.org/10.1016/j.jag.2023.103605

Lobo, F. D., Nagel, G. W., Maciel, D. A., Carvalho, L. A., Martins, V. S., Barbosa, C. C., & Novo, E. M. (2021). AlgaeMAp: Algae Bloom monitoring application for inland waters in Latin America. *Remote Sensing., 13*(15), 2874. https://doi.org/10.3390/rs13152874

Makwinja, R., Inagaki, Y., Sagawa, T., Obubu, J. P., Habineza, E., & Haaziyu, W. (2023). Monitoring trophic status using in situ data and Sentinel-2 MSI algorithm: Lesson from Lake Malombe, Malawi. *Environmental Science and Pollution Research, 30*(11), 29755–29772. https://doi.org/10.1007/s11356-022-24288-8

Mashala, M. J., Dube, T., Mudereri, B. T., Ayisi, K. K., & Ramudzuli, M. R. (2023). A systematic review on advancements in remote sensing for assessing and monitoring land use and land cover changes impacts on surface water resources in Semi-Arid tropical environments. *Remote Sensing., 15*(16), 3926. https://doi.org/10.3390/rs15163926

McKindles, K. M., & Tiquia-Arashiro, S. M. (2012). Functional Gene Arrays for analysis of microbial communities on ocean platform. In S. Tiquia-Arashiro (Ed.), *Molecular biological technologies for Ocean Sensing*. Springer Protocols Handbooks, Humana Press. https://doi.org/10.1007/978-1-61779-915-0_9

Mishra, S., & Mishra, D. R. (2012). Normalized difference chlorophyll index: A novel model for remote estimation of chlorophyll-a concentration in turbid productive waters. *Remote Sensing of Environment, 117*, 394–406. https://doi.org/10.1016/j.rse.2011.10.016

Molkov, A. A., Fedorov, S. V., Pelevin, V. V., & Korchemkina, E. N. (2019). Regional models for high-resolution retrieval of chlorophyll a and TSM concentrations in the gorky reservoir by Sentinel-2 imagery. *Remote Sensing., 11*(10), 1215. https://doi.org/10.3390/rs11101215

Monicelli, F., da Rocha Junior, C. A. N., Lima, S., Diniz, I. S. M., & Becker, V. (2023). Trophic state in Brazilian Semiarid reservoirs after reflooding. *RBRH, 28*, e37.

Moreira, C. V., Matos, M. R., da Costa, A., & Becker, V. (2023). Impacts of extreme precipitation events in water quality: A scientometric analysis in global scale. *Acta Limnologica Brasiliensia, 35*, e17.

Moses, W. J., Gitelson, A. A., Perk, R. L., Gurlin, D., Rundquist, D. C., Leavitt, B. C., Barrow, T. M., & Brakhage, P. (2012 March 15). Estimation of chlorophyll-a concentration in turbid productive waters using airborne hyperspectral data. *Water Research, 46*(4), 993–1004. https://doi.org/10.1016/j.watres.2011.11.068

Nassereddine, Z., Opara, S., Coutinho, O., Qyteti, F., Book, R., Heinicke, M. P., Napieralski, J., & Tiquia-Arashiro, S. M. (2024). Critical perspectives on advancing antibiotic resistant gene (ARG) detection technologies in aquatic ecosystems. *Science of the Total Environment, 957*(20), 177775. https://doi.org/10.1016/j.scitotenv.2024.177775

Oest, A., Alsaffar, A., Fenner, M., Azzopardi, D., & Tiquia-Arashiro, S. M. (2018 May). Patterns of change in metabolic capabilities of sediment microbial communities in river and lake ecosystems. *International Journal of Microbiology, 27*(2018), 6234931. https://doi.org/10.1155/2018/6234931

Pagnucco, G., Overfield, D., Chamlee, Y., Fortuna, A., Sulaiman, F., Farinas, J., Schittenhelm, R., Shuler, C., Kassem, A., Opara, S., Najaf, H., Abbas, L., Coutinho, O., Catalfano, B., Li, X., & Tiquia-Arashiro, S. M. (2023). Metal tolerance and biosorption capacities of bacterial strains isolated from an urban watershed. *Frontiers in Microbiology, 14*, 1278886. https://doi.org/10.3389/fmicb.2023.1278886

Pahlevan, N., Smith, B., Schalles, J., Binding, C., Cao, Z., Ma, R., Alikas, K., et al. (2020, February). Seamless retrievals of chlorophyll-a from Sentinel-2 (MSI) and Sentinel-3 (OLCI) in inland and coastal waters: A machine-learning approach. *Remote Sensing of Environment, 240*, 111604. https://doi.org/10.1016/J.RSE.2019.111604

Papenfus, M., Schaeffer, B., Pollard, A. I., et al. (2020). Exploring the potential value of satellite remote sensing to monitor chlorophyll-a for US lakes and reservoirs. *Environmental Monitoring and Assessment, 192*, 808. https://doi.org/10.1007/s10661-020-08631-5

Patel, D., Gismondi, R., Alsaffar, A., & Tiquia-Arashiro, S. M. (2019). Applicability of API ZYM to capture seasonal and spatial variabilities in lake and river sediments. *Environmental Technology, 40*(24), 3227–3239. https://doi.org/10.1080/09593330.2018.1468492

Paulino, R. S., Martins, V. S., Novo, E. M. L. M., Barbosa, C. C. F., de Carvalho, L. A. S., & Begliomini, F. N. (2022). Assessment of adjacency correction over inland waters using Sentinel-2 MSI images. *Remote Sensing., 14*(8), 1829. https://doi.org/10.3390/rs14081829

Raghul, M., & Porchelvan, P. (2024). A critical review of remote sensing methods for inland water quality monitoring: Progress, limitations, and future perspectives. *Water, Air, and Soil Pollution, 235*, 159. https://doi.org/10.1007/s11270-024-06957-1

Rodrigues, V. F., Wiesenhutter, R. K., Shimabukuro, K., & Machado, R. E. (2023). *Chlorophyll-a concentration in the salto grande reservoir (Americana, Brazil) estimated by satellite images.* Research Square Platform LLC.

Rolim, S. B. A., Veettil, B. K., Vieiro, A. P., Kessler, A. B., & Gonzatti, C. (2023). Remote sensing for mapping algal blooms in freshwater lakes: A review. *Environmental Science and Pollution Research, 30*(8), 19602–19616. https://doi.org/10.1007/s11356-023-25230-2

Seegers, B. N., Werdell, P. J., Vandermeulen, R. A., Salls, W., Stumpf, R. P., Schaeffer, B. A., Owens, T. J., Bailey, S. W., Scott, J. P., & Loftin, K. A. (2021 December 1). Satellites for long-term monitoring of inland U.S. lakes: The MERIS time series and application for chlorophyll-a. *Remote Sensing of Environment, 266*, 1–14. https://doi.org/10.1016/j.rse.2021.112685

Silveira, N. T., Silveira, H. T., Tibúrcio, I. M., & Galvíncio, J. D. (2022). Avaliação Da Qualidade Da Água Do Reservatório Nilo Coelho, Terra Nova, Pernambuco, Brasil. *Revista Brasileira de Geografia Física, 15*(06), 2866–2877.

Stewart, I., Schluter, P. J., & Shaw, G. R. (2006). Cyanobacterial lipopolysaccharides and human health—A review. *Environmental Health, 5*, 7. https://doi.org/10.1186/1476-069X-5-7

Testai, E., Scardala, S., Vichi, S., Buratti, F. M., & Funari, E. (2016). Risk to human health associated with the environmental occurrence of cyanobacterial neurotoxic alkaloids anatoxins and saxitoxins. *Critical Reviews in Toxicology, 46*(5), 385–419. https://doi.org/10.3109/10408444.2015.1137865

Thakur, J. K., Singh, S. K., & Ekanthalu, V. S. (2017). Integrating remote sensing, geographic information systems and global positioning system techniques with hydrological modeling. *Applied Water Science, 7*, 1595–1608. https://doi.org/10.1007/s13201-016-0384-5

Tiquia, S. M. (2010a). Salt-adapted bacteria isolated from the Rouge River and potential for degradation of contaminants and biotechnological applications. *Environmental Technology, 31*, 967–978. https://doi.org/10.1080/09593331003706226

Tiquia, S. M. (2010b). Metabolic diversity of the heterotrophic microorganisms and potential link to pollution of the Rouge River. *Environmental Pollution, 158*, 1435–1443. https://doi.org/10.1016/j.envpol.2009.12.035

Tiquia, S. M. (2011). Extracellular hydrolytic enzyme activities of the heterotrophic microbial communities of the Rouge River: An approach to evaluate ecosystem response to urbanization. *Microbial Ecology, 62*, 679–689. https://doi.org/10.1007/s00248-011-9871-2

Tiquia, S. M., Gurczynski, A., Zholi, A., & Devol, A. (2006a). Diversity of biogeochemical cycling genes from puget sound sediments using DNA microarrays. *Environmental Technology, 27*(12), 1377–1389. https://doi.org/10.1080/09593332708618756

Tiquia, S. M., Masson, S. A., & Devol, A. (2006b) Vertical distribution of nitrite reductase genes (nirS) in continental margin sediments of the Gulf of Mexico. *FEMS Microbiology Ecology, 58*(3), 464–475. https://doi.org/10.1111/j.1574-6941.2006.00173.x

Tiquia, S. M., Davis, D., Hadid, H., Kasparian, S., Ismail, M., & Murray, K. S. (2007). Halophilic and halotolerant bacteria from river waters and shallow groundwater along the Rouge River of Southeastern Michigan. *Environmental Technology, 28*, 297–307. https://doi.org/10.1080/09593332808618789

Tiquia, S. M., Schleibak, M., Schlaff, J., Floyd, J. C., Benipal, B., Zakhem, E., & Murray, K. S. (2008). Microbial community profiling and characterization of some heterotrophic bacterial isolates from river waters and shallow groundwater wells along the Rouge River, Southeast Michigan. *Environmental Technology, 29*, 651–663. https://doi.org/10.1080/09593330801986998

Tiquia-Arashiro, S. M. (2012). *Molecular biological technologies for ocean sensing* (Springer protocols handbooks) (p. 307). Humana. https://doi.org/10.1007/978-1-61779-915-0

Tiquia-Arashiro, S. M., & Mormile, M. (2013). Sustainable technologies: Bioenergy and biofuel from biowaste and biomass. *Environmental Technology, 34*(13-14), 1637–1638. https://doi.org/10.1080/09593330.2013.834162

Tiquia-Arashiro, S. M., & Pant, D. (2020). *Microbial electrochemical technologies* (p. 518). CRC Press. https://doi.org/10.1201/9780429487118

Topp, S. N., Pavelsky, T. M., Jensen, D., Simard, M., & Ross, M. R. V. (2020). Research trends in the use of remote sensing for inland water quality science: Moving towards multidisciplinary applications. *Water., 12*(1), 169. https://doi.org/10.3390/w12010169

Tyler, A. N., Svab, E., Preston, T., Présing, M., & Kovács, W. A. (2006). Remote sensing of the water quality of shallow lakes: A mixture modelling approach to quantifying phytoplankton in water characterized by high-suspended sediment. *International Journal of Remote Sensing, 27*(8), 1521–1537. https://doi.org/10.1080/01431160500419311

Watanabe, F., Mishra, D. R., Astuti, I., Rodrigues, T. W. P., Alcântara, E., Imai, N. N., & Barbosa, C. C. F. (2016). Parametrization and calibration of a quasi-analytical algorithm for tropical eutrophic waters. *ISPRS Journal of Photogrammetry and Remote Sensing, 121*, 28–47. https://doi.org/10.1016/j.isprsjprs.2016.08.009

Watanabe, F., Alcântara, E., Rodrigues, T., Rotta, L., Bernardo, N., Imai, N., Sayuri, F., & Watanabe, Y. (2017). Remote sensing of the chlorophyll-a based on OLI/Landsat-8 and MSI/Sentinel-2A (Barra Bonita Reservoir, Brazil). *Annals of the Brazilian Academy of Sciences, 90*(2 suppl 1), 1987–2000. https://doi.org/10.1590/0001-3765201720170125

Watanabe, F., Alcântara, E., Bernardo, N., de Andrade, C., Gomes, A. C., do Carmo, A., Rodrigues, T., & Rotta, L. H. (2019). Mapping the chlorophyll-a horizontal gradient in a cascading reservoirs system using MSI sentinel-2A images. *Advances in Space Research, 64*(3), 581–590. https://doi.org/10.1016/j.asr.2019.04.035

Watanabe, F. S. Y., Miyoshi, G. T., Rodrigues, T. W. P., Bernardo, N. M. R., Rotta, L. H. S., Alcântara, E., & Imai, N. N. (2020). Inland water's trophic status classification based on machine learning and remote sensing data. *Remote Sensing Applications: Society and Environment, 19*, 100326. https://doi.org/10.1016/j.rsase.2020.100326

Wiegand, M. C., do Nascimento, A. T. P., Costa, A. C., & Neto, I. E. L. (2021). Trophic state changes of Semi-Arid reservoirs as a function of the hydro-climatic variability. *Journal of Arid Environments, 184*, 104321. https://doi.org/10.1016/j.jaridenv.2020.104321

Wu, J. L., Ho, C. R., Huang, C. C., Srivastav, A. L., Tzeng, J. H., & Lin, Y. T. (2014 November 28). Hyperspectral sensing for turbid water quality monitoring in freshwater rivers: Empirical relationship between reflectance and turbidity and total solids. *Sensors (Basel), 14*(12), 22670–22688. https://doi.org/10.3390/s141222670

Zanchett, G., & Oliveira-Filho, E. C. (2013). Cyanobacteria and cyanotoxins: From impacts on aquatic ecosystems and human health to anticarcinogenic effects. *Toxins., 5*(10), 1896–1917. https://doi.org/10.3390/toxins5101896

Chapter 16
Satellite for Probing Aquatic Microbiological Processes: A Review

Rejane S. Paulino, Evlyn M. L. M. Novo, and Vitor S. Martins

1 Introduction

This chapter summarizes how satellite remote sensing allows the study of aquatic microscopic organisms that live in aquatic environments. However, it will concentrate on phytoplankton, which comprises photosynthetic cyanobacteria and the algae that drift in the euphotic layers of inland, coastal, and ocean aquatic systems. As phytoplankton absorb specific wavelengths according to their different species, they cause changes in the surface water's color as they carry out photosynthesis. By registering the light scattered back from within the water surface, satellite images allow the retrieval of fundamental information on microorganisms responsible for life on Earth as primary and O_2 producers and CO_2 absorbers. Phytoplankton's importance as the primary producer that supports the aquatic food web is well known, but its role as the principal factor responsible for most of the ocean's biologically driven carbon sequestration from the atmosphere and land runoff to the ocean interior and seafloor is underrated sometimes.

The emphasis on phytoplankton also reflects the increasing concern regarding the threats to global water security (Oki & Quiocho, 2020; Vörösmarty et al., 2021; Mishra et al., 2021; He et al., 2021) and the continuous spread and resilience of vector-borne infectious diseases (Nichols et al., 2018; Mora et al., 2022) with the intensification of the climate change in the twenty-first century. Inland and coastal waters are widely used for human activities, such as recreation and consumption

R. S. Paulino (✉) · V. S. Martins
Department of Agricultural & Biological Engineering, Mississippi State University (MSU), Starkville, MS, USA
e-mail: rdp282@msstate.edu

E. M. L. M. Novo
Earth Observation and Geoinformatics Division (DIOTG), National Institute for Space Research (INPE), São José dos Campos, SP, Brazil

S. M. Tiquia-Arashiro et al. (eds.), *Urban Watershed Microbiology, Volume 1*,
https://doi.org/10.1007/978-3-031-89071-0_16

("blue water"), and for sustaining ecosystems ("green water"). The degradation of blue water threatens human health and compromises ecosystem functions and structure, thereby limiting the provision of aquatic ecosystem services (Costanza et al., 2017). The agriculture food systems currently respond to a third of the world's greenhouse gas emissions and are a primary driver of biodiversity loss and water pollution. Additionally, by 2050, projections suggest that water demand will increase by more than 20%, and billions of people across the globe will suffer from water scarcity (UNESCO, 2023), while the demand for food is expected to increase by more than 50% by 2040.

Much evidence indicates that potentially harmful phytoplankton blooms have risen worldwide. This increase in blooms has negatively impacted human and ecosystem health, as they have become more abundant and spread to more expansive geographical areas. Several factors are connected to this rise, including higher water temperatures, increased nutrient input from agriculture, sludge coming from industrial processes, wastewater treatment, lack of sanitation system, increased aquaculture production, and overfishing (O'Neil et al., 2012; Brooks et al., 2016).

The widespread water quality degradation has demanded new monitoring tools to provide information on aquatic systems at several spatial and temporal scales for a long time. Historically, National Aeronautics and Space Administration (NASA) has led the efforts on satellite imaging systems and launched the first Earth Observation Satellites of the Landsat program in 1972. The Landsat-1, former Earth Research Satellite (ERTS-1), had a polar orbit satellite carrying on board a new sensor, the Multispectral Scanner Subsystem, with four spectral bands in the visible and near-infrared wavelengths and nominal 80 × 80 m spatial resolution. However, technological constraints prevented its application in optically complex water monitoring, such as inland and coastal waters. In response, perhaps, to the 1972 United Nations Conference on the Human Environment (Smith, 1974), in 1978, NASA launched the Coastal Zone Color Scanner (CZCS) on the Nimbus-7 spacecraft, making it the first satellite mission aiming to probe the phytoplankton, which, since then, had become an increasing concern given its importance for climate regulation (Brewin et al., 2021; Wang et al., 2023).

The first satellites used for monitoring microorganisms were the so-called ocean color sensors. These sensors typically had low spatial resolution (around 800 m to 1.1 km at the nadir) and were only applicable to ocean studies. The 1-year mission proved that CZCS images provided helpful information on the spatial extent and temporal development of a phytoplankton bloom for 28 days (April through May 1982) using image time series (Brown et al., 1985; Tyler & Stumpf, 1989). Not long after, the health microbiology community started to search for ways of applying satellite images to infer the presence of the bacterium *Vibrio cholerae* based on interdisciplinary cross-cut of oceanography, ecology, microbiology, marine biology, epidemiology, medicine, and satellite imagery (space science) for the understanding and prevention of global pandemics of this disease. Lobitz et al. (2000) summarize the results of over 25 years of research to develop a cholera prediction model based on remote sensing data to provide early warning of conditions associated with cholera outbreaks. Although satellite data

cannot directly detect the bacterium *Vibrio cholerae*, the authors aimed to determine the relationship between remote sensed coastal water parameters and cholera case data for Bangladesh, where frequent disease outbreaks occur. The authors linked the sequence of cholera cases to ocean remotely sensed parameters (e.g., phytoplankton blooms feeding zooplankton bloom), sea surface temperature, and sea surface height.

Since then, new Earth and Ocean observation sensors (e.g., MODIS [Moderate Resolution Imaging Spectroradiometer], VIIRS [Visible Infrared Imaging Radiometer Suite], MSI [Multispectral Instrument]/Sentinel-2, OLCI [Ocean and Land Colour Instrument]/Sentinel-3, PRISMA, OLI [Operational Land Imager]/ Landsat, OCI/PACE [Plankton, Aerosol, Cloud, ocean Ecosystem], and EnMAP) with improved spectral and radiometric characteristics have successfully addressed some of the previous challenges, thereby inspiring significant progress in the remote sensing of water through continuous observation of water quality for both inland and coastal systems. Initially, the accurate retrieval of optically active phytoplankton pigments from satellite had significant challenges due to the (1) small water reflectance compared to the atmospheric contribution to the signal detected by the orbital sensors, (2) stratification of organisms in the water column, (3) limited number of spectral bands in the orbital sensors, (4) inaccuracies on the image processing, and others. However, to address those issues, considerable progress has been made in atmospheric methods and adjacent correction methods (Gordon, 2021; Pahlevan et al., 2021; Paulino et al., 2022). The intricate nature of the interaction between light and microbial particles within water also experimented an extensive effort in algorithm development from empirical models (O'Reilly et al., 1998; Matthews et al., 2012), semi-analytical models (Gitelson et al., 2008; Campbell et al., 2011; Le et al., 2009; Hunter et al., 2010; Lee et al., 2016; Li et al., 2022), and more recently machine learning models (Cao et al., 2020; Li et al., 2023).

This way, in this review, we discuss fundamental concepts and key applications of aquatic remote sensing focused on microbiology, offering new perspectives on exploring microbiological organisms, especially phytoplankton. To address the fundamental aspects, we explore issues related to the applicability of aquatic remote sensing to microbiology through a global view that encompasses the recorded literature spanning the past 44 years on this subject. We describe the basic concepts of remote sensing and water optical properties and explain how the microorganisms interact within the water column in complex inland and coastal waters where many other particles may affect their estimations from satellite images. Then, we explore the scientific aspects of remote sensing applied to phytoplankton, the core of water remote sensing, and discuss the future directions for investigating these organisms through the lens of remote sensing.

2 Insights into Water Microbiology via Remote Sensing

Water microbiology, a field that explores the complex interactions of various microorganisms within aquatic systems, plays a crucial role across multiple scientific disciplines, including but not limited to ecology, epidemiology, and environmental protection and conservation. In this comprehensive review, we used Elsevier's Scopus database (1980–April 2024) to identify studies correlated to the microbiology of inland and coastal waters remote sensing, and 1,063 studies were identified on this topic (Fig. 16.1). Notably, studies in coastal waters are more extensively investigated in comparison to studies on inland waters, accounting for 78% of the identified literature. However, a trend observed in the past decade (2014–April 2024) suggests a surge in published studies dedicated to inland waters, indicating a twofold increase compared to previous years related to improving satellite sensors and computer processing capabilities, as reported in the previous topic. Across the globe, countries in the Northern Hemisphere, particularly the United States and China, lead the volume of publications on this topic, collectively contributing to over 50% of the identified literature. Differently, regions in South America, such as Brazil, have a relatively low contribution, representing less than 3% of the studies. At the same time, just a few countries on the African continent show a low percentage of publications (Fig. 16.1b). The predominant keywords from the analyzed studies relate remote sensing to phytoplankton and water quality (Fig. 16.1c), highlighting the primary water remote sensing's focus within microbiology.

Aquatic microbiology is related to phytoplankton, microscopic organisms that live in aquatic environments, salty or fresh, freely floating and drifting with water currents and waves. The phytoplankton is composed of bacteria and algae. The bacteria are unicellular microorganisms with various shapes (i.e., spheres, spirals, filaments) and sizes ranging from diameters from 0.5 to 5 μm. The algae can be either unicellular or multicellular and differ from the bacteria due to the presence of a nuclear membrane encompassing its DNA. Both microorganisms have in common to interact with the incoming electromagnetic energy (Pagnucco et al., 2023). This interaction will depend on the phytoplankton species. The energy that interacts with the constituents in the water column and is captured by the orbital sensor will then be used in different ways depending on the study goals and spectral information. Phytoplankton, the core of water remote sensing, are organisms of global importance. They are the subject of extensive research due to their significance as indicators of the health and equilibrium of aquatic systems (Cupertino et al., 2019; Chandel et al., 2024). The use of remote sensing in studying phytoplankton relies on changes in the watercolor resulting from the optically active constituents (OACs) concentration, type, strength of their covariation, and optical properties, and a comprehensive overview of this relationship is provided in Sect. 4.

Investigations into remote sensing applications involving phytoplankton rely on bio-optical characterization and modeling (Mandal et al., 2015; Nair et al., 2008; Kiefer et al., 2015; Kudela et al., 2015; Salgado-Hernanz et al., 2019; Kramer et al., 2022; Albin et al., 2023) and analysis of long-term trends (Dai et al., 2023; Feng

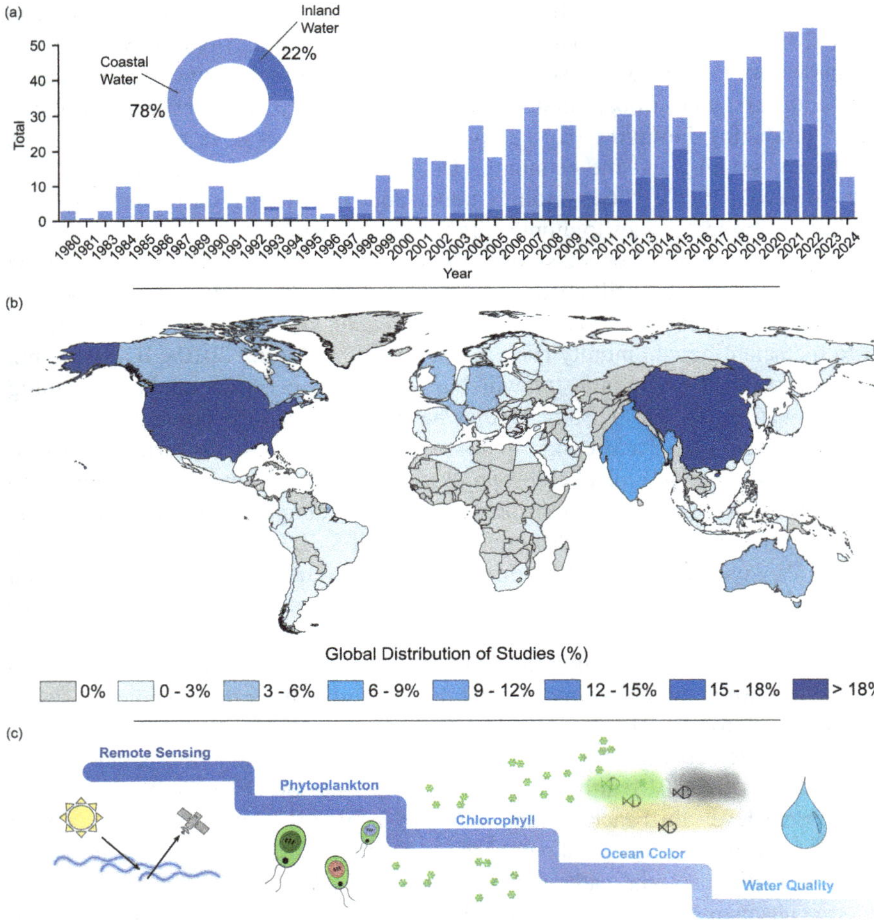

Fig. 16.1 Overview of historical literature in microbiology using remote sensing. The following keywords were used in the review in Elsevier's Scopus database: "microbiology + remote sensing + inland/coastal waters" and "phytoplankton + remote sensing + inland/coastal waters." A total of 1,063 studies were identified. In (**a**) the number/proportion of studies from 1980 to April 2024 for both inland and coastal waters is shown; (**b**) is the cartogram of the study distribution across the globe considering the corresponding author information; and (**c**) refers to five keywords used more often in the studies (note that generic keywords, e.g., coastal/inland waters and sensor names, were not considered).

et al., 2021; Ho et al., 2019; Löptien & Dietze, 2022), producing critical information for the development and assessment of management strategies aimed at ensuring good-quality water for both environment and human safety. Remarkably, Dai et al. (2023) documented alterations in global patterns of phytoplankton blooms in coastal waters (2003–2020) through the utilization of daily orbital sensors, demonstrating a considerable rise in the frequency of such events over around 59%. Similarly, using

a couple of satellite images from Landsat at a spatial resolution of 30 meters (1984–2012), Ho et al. (2019) also identified rises in the intensity of phytoplankton blooms within inland water systems on a global scale. The dynamic of phytoplankton blooms assumes critical importance due to potentially harmful organisms capable of producing biotoxins (e.g., cyanobacteria) that can pose threats to public health, ecosystems, and the economy (Dodds et al., 2009; Ibelings et al., 2014; Pearson et al., 2016). These findings underscore the motivation behind applying remote sensing techniques to monitoring those organisms.

Aquatic remote sensing is also used to investigate bacterial groups, although its importance in the topic still needs to be explored. The bacterial groups associated with microbiological pollution (e.g., *Escherichia coli* and *Enterococci*) are of interest to the scientific community (Fogarty et al., 2003; Tiquia, 2010a, b, 2011) and, especially, to the tourism industry, which is affected by the water quality of bathing areas (Tiwari et al., 2021). Some studies have emphasized the potential of remote sensing-based data for addressing this issue, using variables like surface temperature, precipitation, turbidity, chlorophyll-a, trophic state, and nutrient levels to establish relationships with these organisms. For example, Bruschi et al. (2021) developed two indexes to simulate the dynamics of fecal bacteria in coastal bathing waters. They used hydrodynamic and biogeochemical models, primarily relying on in situ and remote sensing observations, such as temperature and ocean color images. In a study by Laureano-Rosario et al. (2017), the environmental factors linked to the variability of culturable *Enterococci* in recreation areas were investigated. The authors observed that 20% of the variability in *Enterococci* levels could be accounted for by environmental parameters, some of which were obtained through remote sensing observations (e.g., precipitation, temperature, and turbidity). Zhang et al. (2015) utilized in situ and satellite observations to predict the *Enterococci* levels in beach waters, and they achieved a prediction accuracy of 82.63% by training an artificial neural network algorithm.

Additionally, bacterial pathogens for humans like *Vibrio cholerae*, the microorganism causative of cholera, have also been correlated to remote sensing data and concluded that watercolor and climate patterns might help predict these organisms in aquatic systems (Jutla et al., 2012; Magny et al., 2008; Xu et al., 2016; Nassereddine et al., 2024). Similarly, methylotrophic bacteria groups in coastal waters have been investigated through remote sensing data (Dixon et al., 2013). These organisms assimilate methanol and play a significant role in marine carbon cycling (Chistoserdova et al., 2009; McKindles & Tiquia-Arashiro, 2012; Peres et al., 2023). Although water remote sensing exhibits versatility in its applications, ranging from addressing issues related to water pollution and human health to exploring the ecological benefits associated with microorganisms, some challenges limit its wide application in bacterial study. Firstly, bacteria do not exhibit significant optical activity that can be observed directly by orbital optical sensors. Secondly, the numerous variables included in modeling these water-dwelling organisms prevent the transferability and generalization of models to other regions. And third, bacteria have a short life cycle of approximately 3 days, making it challenging to match the remote sensing data with in situ observations.

3 Remote Sensing of the Watercolor

Inland and coastal waters encompass a wide range of aquatic systems, including rivers, lakes, lagoons, reservoirs, estuaries, deltas, and intertidal zones. These systems are often negatively impacted by anthropogenic activities that cause water quality degradation (Khan et al., 2023; Oest et al., 2018; Patel et al., 2019). The intensification of pollution within these aquatic systems, coupled with the effects of global climate change (Downing et al., 2021; Khelifa et al., 2021), has raised considerable concern within the scientific community regarding water depletion and the ecological imbalance that has persisted for the past few decades (Tiquia et al., 2007; Delpla et al., 2009; Bowman et al., 2018; Lin et al., 2022; Bashir et al., 2020; Cavicchioli et al., 2019; Reid et al., 2019; Beaulieu et al., 2019; Woolway & Maberly, 2020; Li et al., 2023; Woolway et al., 2021; Priya et al., 2023). Water quality is critical to many of the United Nations Sustainable Development Goals (SDGs), including SDG 3 (good health and well-being), SDG 6 (clean water and sanitation for all), and SDG 14 (life below the water) (UNESCO, 2023). These goals, coupled with water's threats and importance, highlight the urgent need for continuous monitoring and efficient management of aquatic systems. In this context, remote sensing through both in situ and orbital observations has been crucial to preserving the health and productivity of inland and coastal waters (Cavender-Bares et al., 2022; Hestir et al., 2015). Orbital remote sensing data covers large regions and provides frequent information on various scales. This data complements traditional water quality monitoring methods and offers crucial insights into aquatic systems affected by land use, land cover changes, and climate change.

Remote sensing of water bodies' color relies on how various constituents in the water interact with the light. These optically active constituents, or OACs, include colored dissolved organic matter (CDOM), phytoplankton, non-algal particles, and the water itself (Giardino et al., 2019; Kirk, 2011). Each OAC affects how light is absorbed and scattered, which in turn changes the color of the water (Mobley, 1999; Gordon & Wang, 1994; Gordon et al., 1983). The most influential OACs are illustrated in Fig. 16.2. It is essential to highlight that OACs' spectral variability in each water body is affected in time and space due to their dependence on environmental factors such as seasonal dynamics, hydrodynamics, and nutrient load (Ward, 1989). For instance, Paulino et al. (2023), through an analysis based on spectral similarity metrics for estimating the hydrological connectivity in floodplain areas, proved that decreasing hydrological connectivity is associated with differences in water spectra features between rivers and lakes, attributed to the seasonal dynamic of sediment flux in white-water rivers. The relationship between the optical constituents and the spectral composition of the water bodies supports the development of bio-optical models, which encompass a series of empirical or analytical equations designed to quantify various water quality parameters (e.g., CDOM, sediment loads, Secchi deep, chlorophyll-a, phycocyanin, and turbidity) (Gitelson et al., 2008; Mishra & Mishra, 2012; Nechad et al., 2010; Tiquia-Arashiro, 2012; Maciel et al., 2023a). Currently, the remote sensing community of water quality also applies empirical

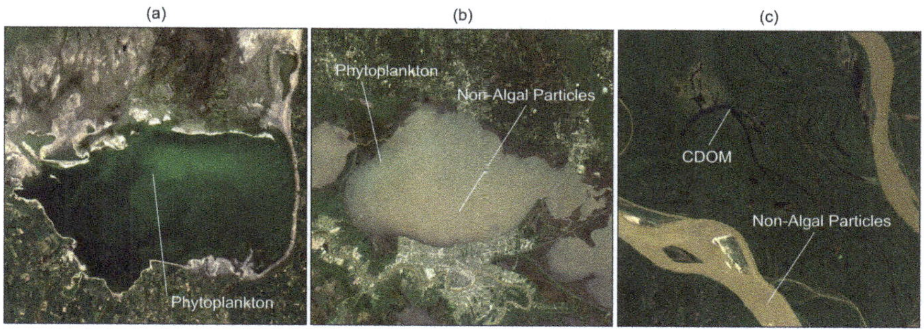

Fig. 16.2 Watercolor feature RGB (composite in true color) from Landsat 8/9 to (**a**) Mar Chiquita (Argentina) on February 2, 2024, (**b**) Lake Pontchartrain (United States) on April 7, 2022, and (c) Amazon's floodplain lakes and rivers (Brazil) on August 21, 2023

methodologies based on machine and deep learning to solve optical complexity problems and address many other watercolor tasks (Cao et al., 2020; Maciel et al., 2023; O'Shea et al., 2021).

Satellite images taken over water bodies can be affected by atmospheric effects, such as scattering and absorption of gasses and aerosols, and environmental factors, such as adjacent targets, which reduce OACs' retrieval accuracy. Figure 16.3 shows how the energy from the aquatic systems reaches the orbital sensors and the main factors affecting it. Note that energy travels a long way from the source to the water and back to the sensor, suffering from attenuation processes caused by atmospheric constituents (e.g., aerosol and gaseous), targets adjacent to the water body, specular reflection of light from the water surface, and bottom effects in optically shallow waters. We must correct the atmospheric/environmental effects before retrieving OACs from images. This process involves complex steps with different approaches and assumptions (Harmel et al., 2018; Vanhellemont, 2019; Paulino et al., 2022), but significant progress has been achieved in the correction of adjacency (Paulino et al., 2022) and atmosphere effects (Vanhellemont, 2019) benefiting remote sensing watercolor applications.

Moreover, the limitations of optical sensor characteristics, in terms of the balance between spectral bandwidth, spatial and radiometric quality, and revisit frequency, present challenges for studying small and narrow water bodies and temporary-spatially non-homogeneous regions. The water remote sensing community has tested new advancements to overcome challenges, including hyperspectral data and innovative methodologies. Additional information regarding these new capabilities is available in Sect. 4.

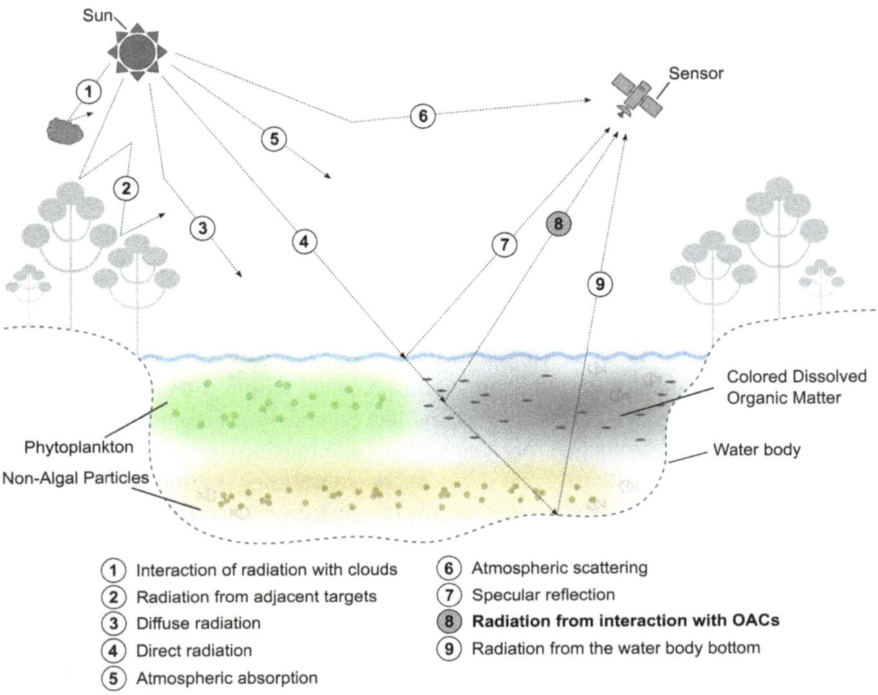

Fig. 16.3 Scheme of the energy path from sun-target and target-sensor describing the main processes that affect this transference for inland and coastal aquatic systems

4 Remote Sensing of the Phytoplankton and New Capabilities

Phytoplankton, the focus of this review, are photosynthetic microorganisms highly abundant in aquatic systems. They play a crucial ecological role by significantly contributing to oxygen production in the Earth's atmosphere (Falkowski, 2012). The growth of phytoplankton relies on elements such as light, carbon dioxide, and inorganic nutrients, with their pigment composition (e.g., Chlorophyll-a, Chlorophyll-b, Chlorophyll-c, and Phycobiliproteins) and cellular architecture and chemical composition being essential to their ability to utilize light efficiently (Edwards et al., 2016; Schulze et al., 2014; Kassem et al., 2023). Photosynthetic pigments play a crucial role in remote sensing by altering the light spectral composition, enabling the detection and differentiation of these microorganisms in water. Among the various pigments found in phytoplankton, chlorophyll-a stands out as the primary green pigment more often in these organisms, with its absorption peaks centered around the blue (λ: 440 nm) and red (λ: 665 nm) wavelengths (Matthews, 2017; Keeling, 2013). However, certain phytoplankton groups, including cyanobacteria and red algae, use accessory pigments like phycoerythrin (λ: 565 nm) and

phycocyanin (λ: 620 nm) to increase their capacity to capture light across a broader spectrum (Hohmann-Marriott & Blankenship, 2011). The light absorbed at smaller wavelengths (<670 nm) is reemitted as fluorescence, that is, the mechanism through which the cells dissipate the excess energy not used in photosynthesis (Huot & Babin, 2010). The fluorescence process is characterized by emission at longer wavelengths, particularly within the red and red-edge spectral regions, depending on chlorophyll-a configuration within phytoplankton's cell photosystems. For instance, cyanobacteria have fluorescence peaks at wavelengths exceeding 700 nm, while other algal groups display fluorescence around 681 nm and beyond 700 nm (Simis et al., 2012; Huot & Babin, 2010).

Although microscopic, phytoplankton significantly change the observed light field through absorption, scattering, and fluorescence processes, producing a distinctive spectral signature that indicates their presence within the water's column. Figure 16.4 shows two water spectra with distinct compositions regarding the dominant phytoplankton group and pigments. The relationship between absorption-scattering/fluorescence intensity and pigment concentration in bio-optical modeling indicates chlorophyll-a or phycocyanin levels in aquatic systems. Both pigments are standard measures used as proxies of algal biomass and cyanobacteria in water quality studies based on water reflectance (Begliomini et al., 2023; Gitelson et al., 2010; Matthews et al., 2012; Wynne et al., 2008). The effectiveness of the modeling process, which uses the spectral response of phytoplankton, relies on factors such as the sensibility of optical sensors, intracellular pigment concentration, and spectral overlap among multiple OACs. Due to the shape of the absorption feature, satellite data exhibit higher sensitivity to chlorophyll-a than phycocyanin. Chlorophyll-a demonstrates a maximal specific absorption of 0.027 $m^2 mg^{-1}$ (at 665–681 nm); in contrast, phycocyanin shows 0.007–0.0095 $m^2 mg^{-1}$ (at 620 nm), indicating that higher phycocyanin concentration is required to produce an equivalent response to chlorophyll-a in reflectance bands (Johnsen & Sakshaug, 2007; Simis et al., 2005).

Additionally, the sensor loses sensitivity to changes in pigment concentration in environments with very high concentrations. As the pigment concentration increases, the reflectance decreases at the absorption bands, reducing the sensor's ability to distinguish tiny reflectance differences at the absorption diagnostic features. For instance, the poor sensor signal-to-noise ratio of the first sensors of the Landsat family sensors aggravated this problem (Gitelson et al., 1999). The spatial–temporal variability of intracellular pigment concentration also challenges modeling efforts. Phytoplankton can reduce or increase the pigment concentration within cells according to light availability to optimize growth potential (Schulze et al., 2014), impacting their absorption and scattering properties (Morel & Bricaud, 1986). Another factor involves the CDOM's absorption features at 440 nm, which overlap with the chlorophyll-a absorption. Modeling becomes more uncertain in inland and coastal waters with multiple OACs, given that a combination of OACs can lead to a unique reflectance spectrum (Defoin-Platel & Chami, 2007).

The estimates of phytoplankton pigments, for example, chlorophyll-a and phycocyanin, to assess eutrophication levels and detect harmful algal blooms have

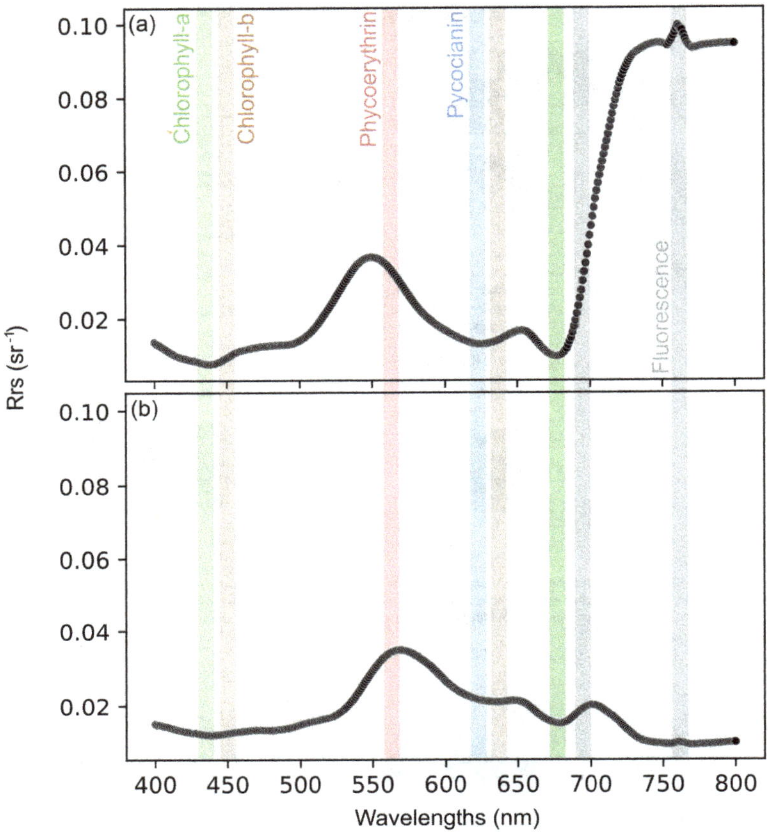

Fig. 16.4 In situ spectra collected over eutrophic inland waters in Brazil. (**a**) Water spectrum acquired over an algal bloom with a higher cyanobacteria concentration of the Microcystis aeruginosa type (162,210 cells/mL). (**b**) Water spectrum with a lower concentration of these organisms (358 cells/mL). The colored shadow areas indicate the absorption from cyanobacteria and other algal pigments and their fluorescence peaks

made significant advancements in the last few years. New capabilities related to hyperspectral orbital sensors (e.g., PRISMA and EnMap) (Cetinić et al., 2024; Giardino et al., 2019; Saberioon et al., 2023) highlight the effectiveness of remote sensing techniques in monitoring water quality. Hyperspectral sensors, like OCI aboard PACE (Plankton, Aerosol, Cloud, ocean Ecosystem) (Petro et al., 2020), feature hundreds of spectral bands across the visible, near- and shortwave-infrared domains, allowing for accurate identification of pigment's absorption features and the simultaneous mapping of large areas due to its wide swath width of 2500 km. Additionally, future directions about the remote sensing of phytoplankton involve deriving synthetic spectral bands, forecasting algal blooms, and extracting algal genera from satellite images. The main new capabilities of remote sensing applied

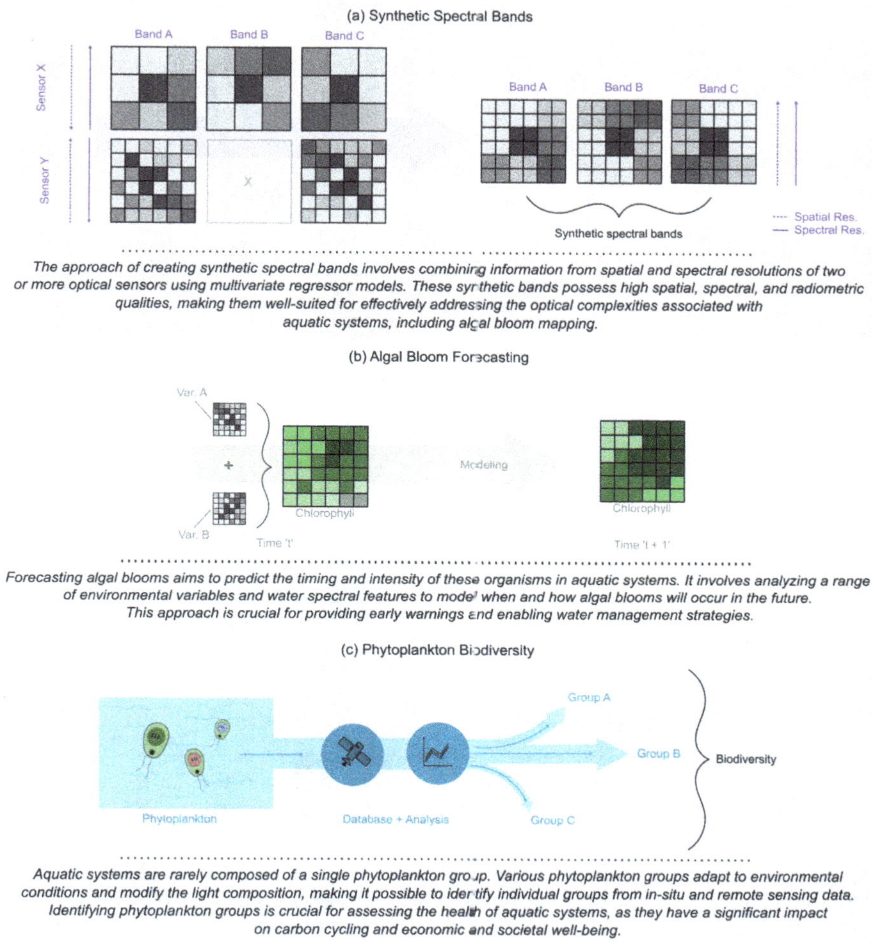

Fig. 16.5 Description of main new capabilities of water remote sensing concerning phytoplankton. In (**a**), (**b**), and (**c**), definitions from synthetic spectral bands, algal bloom forecasting, and phytoplankton biodiversity are illustrated, respectively

to phytoplankton study are illustrated in Fig. 16.5, and their detailed description can be found below:

- *Synthetic spectral bands* result from a data fusion method that combines two or more optical sensors to create synthetic spectral multi-bands (Paulino et al., 2025). In this approach, the unique characteristics of each sensor (i.e., spatial, spectral, and radiometric resolutions) are combined to produce a new set of spectral bands that are more suitable for addressing aquatic tasks. Synthetic spectral bands are a novel and timely topic, as they allow for algal bloom mapping in high

spatial resolution (around 10 meters) with similar accuracy provided by ocean color sensors (more than 250 meters of spatial resolution).

- *Algal bloom forecasting* involves using current environmental characteristics to predict future occurrences and severity of algal blooms in aquatic systems (Caballero et al., 2025). The spectral and physical data, along with climate variables (Tewari et al., 2022; Silva et al., 2023), are combined to forecast algal bloom dynamics at a future time, ranging from 1 day to several months. Algal bloom forecasting is valuable for detecting early signs of bloom formation, which can help mitigate the effects of harmful algal blooms on ecosystems and human health.
- *Phytoplankton biodiversity* refers to identifying the algal genera from satellite images. Previous studies, as in Legleiter et al. (2022), employed spectral mixture analysis with a standard spectral library of 12 cyanobacteria genera for mapping, including planktonic cyanobacteria (e.g., *Microcystis, Aphanizomenon,* and *Gloeotrichia*), promising outcomes (root mean square error of less than 0.03) via hyperspectral images. Similarly, Kraus et al. (2021) noted a relationship between light field distribution in the water's column and the presence of specific algae genera using multispectral sensors. Despite the complexity and challenges associated with this subject in inland and coastal waters, its benefits to water quality and aquatic system management are considerable and multifaceted.

5 Summary and Conclusion

This chapter evaluated how water remote sensing contributes to aquatic microbiology from a space view, focusing on a detailed review of the relationship, challenges, and benefits of this technology applied to aquatic systems such as inland and coastal waters. The primary outcomes of this review can be summarized in four points:

- Inland and coastal waters are necessary social and ecological resources threatened by anthropogenic pollution and climate change. Observing aquatic systems from space is a challenge because many aspects relative to optical water and environmental factors are complexly related. Although current orbital sensors still have limitations in meeting these challenges, significant improvements have been made by the remote sensing community of water. These improvements include enhanced atmospheric processors specialized for complex aquatic environments, robust machine- and deep-learning models, and new methods to derive synthetic spectral bands with improved spatial and spectral resolutions.
- For over 44 years, remote sensing has supported microbiology investigations in aquatic ecosystems. In particular, inland waters have experienced a significant increase in research on this topic in the past decade (2014–2024), motivated by increased spatial, temporal, spectral, and radiometric resolutions of orbital sensors.

- Aquatic remote sensing mainly focuses on optically active constituents, and phytoplankton is one the most extensively studied organisms. Pigments from these microorganisms, such as chlorophyll-a or phycocyanin, are combined with $\boldsymbol{R}_{rs}$ to model and derive information on phytoplankton concentration, distribution, and frequency within aquatic systems. Even though some of these organisms produce biotoxins (e.g., cyanobacteria), they play a vital role in aquatic ecosystems as they are primary producers, providing food for organisms ranging from microscopic (e.g., zooplankton) to macroscopic organisms (e.g., whales). The study of phytoplankton from a remote sensing perspective is crucial for addressing essential public and ecosystem safety issues.
- New capabilities for hyperspectral and multispectral orbital sensors allow new methodologies to study phytoplankton. Future directions involve investigating phytoplankton group biodiversity, forecasting algal blooms, and combining multiple sensors to create accurate synthetic spectral bands in a high spatial resolution suitable for phytoplankton study in different aquatic system types and contexts. These new capabilities in remote sensing enable more accurate predictions and detections of these organisms, but challenges relating to fine-scale and real-time phytoplankton mapping still exist and need to be addressed.

Author Contributions Conceptualization: Rejane S. Paulino, Evlyn M.L.M. Novo; Literature search: Rejane S. Paulino, Evlyn M.L.M. Novo; Formal analysis: Rejane S. Paulino; Writing—original draft preparation: Rejane S. Paulino, Evlyn M.L.M. Novo, Vitor S. Martins; Writing—review and editing: Rejane S. Paulino, Evlyn M.L.M. Novo, Vitor S. Martins.

Conflict of Interest The authors declare no conflict of interest.

Data Availability The literature database is freely available from Elsevier's Scopus website (https://www.scopus.com/home.uri [accessed on April 25, 2024]).

References

Albin, K. J., Jyothibabu, R., Santhi Krishnan, S., Alok, K. T., Sherin, C. K., & Gupta, G. V. M. (2023). Winter phytoplankton size classes in the Northeastern Arabian Sea based on in-situ and remote sensing methods. *Marine Environmental Research, 187*. https://doi.org/10.1016/j.marenvres.2023.105972

Bashir, I., Lone, F. A., Bhat, R. A., Mir, S. A., Dar, Z. A., & Dar, S. A. (2020). Concerns and threats of contamination on aquatic ecosystems. In *Bioremediation and biotechnology: Sustainable approaches to pollution degradation* (pp. 1–26). Springer. https://doi.org/10.1007/978-3-030-35691-0_1

Beaulieu, J. J., DelSontro, T., & Downing, J. A. (2019). Eutrophication will increase methane emissions from lakes and impoundments during the 21st century. *Nature Communications, 10*(1). https://doi.org/10.1038/s41467-019-09100-5

Begliomini, F. N., Barbosa, C. C. F., Martins, V. S., Novo, E. M. L. M., Paulino, R. S., Maciel, D. A., Lima, T. M. A., O'Shea, R. E., Pahlevan, N., & Lamparelli, M. C. (2023). Machine learning for cyanobacteria mapping on tropical urban reservoirs using PRISMA hyperspectral data. *ISPRS Journal of Photogrammetry and Remote Sensing, 204*, 378–396. https://doi.org/10.1016/j.isprsjprs.2023.09.019

Bowman, N., Patel, P., Xu, W., Alsaffar, A., & Tiquia-Arashiro, S. M. (2018). Enrichment and isolation of Pb-resistant bacteria from Saint Clair River sediments and their potential for Pb removal in aqueous solutions. *Applied Microbiology and Biotechnology, 102*, 2391–2398. https://doi.org/10.1007/s00253-018-8772-4

Brewin, R. J. W., Sathyendranath, S., Platt, T., Bouman, H., Ciavatta, S., Dall'Olmo, G., Dingle, J., Groom, S., Jönsson, B., Kostadinov, T. S., Kulk, G., Laine, M., Martínez-Vicente, V., Psarra, S., Raitsos, D. E., Richardson, K., Rio, M. H., Rousseaux, C. S., Salisbury, J., et al. (2021). Sensing the ocean biological carbon pump from space: A review of capabilities, concepts, research gaps and future developments. *Earth-Science Reviews, 217*, 103604. https://doi.org/10.1016/j.earscirev.2021.103604

Brooks, B. W., Lazorchak, J. M., Howard, M. D. A., Johnson, M. V. V., Morton, S. L., Perkins, D. A. K., Reavie, E. D., Scott, G. I., Smith, S. A., & Steevens, J. A. (2016). Are harmful algal blooms becoming the greatest inland water quality threat to public health and aquatic ecosystems? *Environmental Toxicology and Chemistry, 35*(1), 6–13. https://doi.org/10.1002/etc.3220

Brown, O. B., Evans, R. H., Brown, J. W., Gordon, H. R., Smith, R. C., & Baker, K. S. (1985). Phytoplankton blooming off the U.S. East coast: A satellite description. *Science, 229*(4709), 163–167. https://doi.org/10.1126/science.229.4709.163

Bruschi, A., Lisi, I., De Angelis, R., Querin, S., Cossarini, G., Di Biagio, V., Salon, S., Solidoro, C., Fassina, D., Ancona, S., & Silvestri, C. (2021). Indexes for the assessment of bacterial pollution in bathing waters from point sources: The northern Adriatic Sea CADEAU service. *Journal of Environmental Management, 293*, 112878. https://doi.org/10.1016/j.jenvman.2021.112878

Caballero, C.B., Martins, V.S., Paulino, R.S., Butler, E., Sparkes, E., Lima, T.M., Novo, E.M.L.M. (2025). Understanding algal bloom forecasting: a review of remote sensing and modeling approaches. *Ecological Indicators, 172*, 113244. https://doi.org/10.1016/j.ecolind.2025.113244

Campbell, G., Phinn, S. R., Dekker, A. G., & Brando, V. E. (2011). Remote sensing of water quality in an Australian tropical freshwater impoundment using matrix inversion and MERIS images. *Remote Sensing of Environment, 115*(9), 2402–2414. https://doi.org/10.1016/j.rse.2011.05.003

Cao, Z., Ma, R., Duan, H., Pahlevan, N., Melack, J., Shen, M., & Xue, K. (2020). A machine learning approach to estimate chlorophyll-a from Landsat-8 measurements in inland lakes. *Remote Sensing of Environment, 248*, 111974. https://doi.org/10.1016/j.rse.2020.111974

Cavender-Bares, J., Schneider, F. D., Santos, M. J., Armstrong, A., Carnaval, A., Dahlin, K. M., Fatoyinbo, L., Hurtt, G. C., Schimel, D., Townsend, P. A., Ustin, S. L., Wang, Z., & Wilson, A. M. (2022). Integrating remote sensing with ecology and evolution to advance biodiversity conservation. *Nature Ecology and Evolution, 6*(5), 506–519. https://doi.org/10.1038/s41559-022-01702-5

Cavicchioli, R., Ripple, W. J., Timmis, K. N., Azam, F., Bakken, L. R., Baylis, M., Behrenfeld, M. J., Boetius, A., Boyd, P. W., Classen, A. T., Crowther, T. W., Danovaro, R., Foreman, C. M., Huisman, J., Hutchins, D. A., Jansson, J. K., Karl, D. M., Koskella, B., Mark Welch, D. B., et al. (2019). Scientists' warning to humanity: Microorganisms and climate change. *Nature Reviews Microbiology, 17*(9), 569–586. https://doi.org/10.1038/s41579-019-0222-5

Cetinić, I., Rousseaux, C. S., Carroll, I. T., Chase, A. P., Kramer, S. J., Werdell, P. J., Siegel, D. A., Dierssen, H. M., Catlett, D., Neeley, A., Soto Ramos, I. M., Wolny, J. L., Sadoff, N., Urquhart, E., Westberry, T. K., Stramski, D., Pahlevan, N., Seegers, B. N., Sirk, E., et al. (2024). Phytoplankton composition from sPACE: Requirements, opportunities, and challenges. *Remote Sensing of Environment, 302*, 113964. https://doi.org/10.1016/j.rse.2023.113964

Chandel, P., Mahajan, D., Thakur, K., Kumar, R., Kumar, S., Brar, B., Sharma, D., & Sharma, A. K. (2024). A review on plankton as a bioindicator: A promising tool for monitoring water quality. *World Water Policy, 10*(1), 213–232. https://doi.org/10.1002/wwp2.12137

Chistoserdova, L., Kalyuzhnaya, M. G., & Lidstrom, M. E. (2009). The expanding world of methylotrophic metabolism. *Annual Review of Microbiology, 63*, 477–499. https://doi.org/10.1146/annurev.micro.091208.073600

Costanza, R., de Groot, R., Braat, L., Kubiszewski, I., Fioramonti, L., Sutton, P., Farber, S., & Grasso, M. (2017). Twenty years of ecosystem services: How far have we come and how far do we still need to go? *Ecosystem Services, 28*, 1–16. https://doi.org/10.1016/j.ecoser.2017.09.008

Cupertino, A., Gücker, B., Von Rückert, G., & Figueredo, C. C. (2019). Phytoplankton assemblage composition as an environmental indicator in routine lentic monitoring: Taxonomic versus functional groups. *Ecological Indicators, 101*, 522–532. https://doi.org/10.1016/j.ecolind.2019.01.054

Dai, Y., Yang, S., Zhao, D., Hu, C., Xu, W., Anderson, D. M., Li, Y., Song, X. P., Boyce, D. G., Gibson, L., Zheng, C., & Feng, L. (2023). Coastal phytoplankton blooms expand and intensify in the 21st century. *Nature, 615*(7951), 280–284. https://doi.org/10.1038/s41586-023-05760-y

Defoin-Platel, M., & Chami, M. (2007). How ambiguous is the inverse problem of ocean color in coastal waters? *Journal of Geophysical Research: Oceans, 112*(3). https://doi.org/10.1029/2006JC003847

Delpla, I., Jung, A. V., Baures, E., Clement, M., & Thomas, O. (2009). Impacts of climate change on surface water quality in relation to drinking water production. *Environment International, 35*(8), 1225–1233. https://doi.org/10.1016/j.envint.2009.07.001

Dixon, J. L., Sargeant, S., Nightingale, P. D., & Colin Murrell, J. (2013). Gradients in microbial methanol uptake: Productive coastal upwelling waters to oligotrophic gyres in the Atlantic Ocean. *ISME Journal, 7*(3), 568–580. https://doi.org/10.1038/ismej.2012.130

Dodds, W. K., Bouska, W. W., Eitzmann, J. L., Pilger, T. J., Pitts, K. L., Riley, A. J., Schloesser, J. T., & Thornbrugh, D. J. (2009). Eutrophication of U. S. freshwaters: Analysis of potential economic damages. *Environmental Science and Technology, 43*(1), 12–19. https://doi.org/10.1021/es801217q

Downing, J. A., Polasky, S., Olmstead, S. M., & Newbold, S. C. (2021). Protecting local water quality has global benefits. *Nature Communications, 12*(1). https://doi.org/10.1038/s41467-021-22836-3

Edwards, K. F., Thomas, M. K., Klausmeier, C. A., & Litchman, E. (2016). Phytoplankton growth and the interaction of light and temperature: A synthesis at the species and community level. *Limnology and Oceanography, 61*(4), 1232–1244. https://doi.org/10.1002/lno.10282

Falkowski, P. (2012). The power of plankton. *Nature, 483*, 17–20. https://doi.org/10.1038/483S17a

Feng, L., Dai, Y., Hou, X., Xu, Y., Liu, J., & Zheng, C. (2021). Concerns about phytoplankton bloom trends in global lakes. *Nature, 590*(7846), E35–E47. https://doi.org/10.1038/s41586-021-03254-3

Fogarty, L. R., Haack, S. K., Wolcott, M. J., & Whitman, R. L. (2003). Abundance and characteristics of the recreational water quality indicator bacteria Escherichia coli and enterococci in gull faeces. *Journal of Applied Microbiology, 94*(5), 865–878. https://doi.org/10.1046/j.1365-2672.2003.01910.x

Giardino, C., et al. (2019). Imaging spectrometry of inland and coastal waters: State of the art, achievements and perspectives. *Surveys in Geophysics, 40*, 401–429. https://doi.org/10.1007/s10712-018-9476-0

Gitelson, A. A., Schalles, J. F., Rundquist, D. C., Schiebe, F. R., & Yacobi, Y. Z. (1999). Comparative reflectance properties of algal cultures with manipulated densities. *Journal of Applied Phycology, 11*, 345–354. https://doi.org/10.1023/A:1008143902418

Gitelson, A. A., Dall'Olmo, G., Moses, W., Rundquist, D. C., Barrow, T., Fisher, T. R., Gurlin, D., & Holz, J. (2008). A simple semi-analytical model for remote estimation of chlorophyll-a in turbid waters: Validation. *Remote Sensing of Environment, 112*(9), 3582–3593. https://doi.org/10.1016/j.rse.2008.04.015

Gitelson, A. A., Gitelson, A. A., Zhou, J., Gurlin, D., Moses, W., Ioannou, I., & Ahmed, S. A. (2010). Algorithms for remote estimation of chlorophyll-a in coastal and inland waters using red and near infrared bands. *Optics Express, 18*(23), 24109. https://doi.org/10.1364/OE.18.024109

Gordon, H. R. (2021). Evolution of ocean color atmospheric correction: 1970–2005. *Remote Sensing, 13*(24), 1970–2005. https://doi.org/10.3390/rs13245051

Gordon, H. R., & Wang, M. (1994). Retrieval of water-leaving radiance and aerosol optical thickness over the oceans with SeaWiFS: A preliminary algorithm. *Applied Optics, 33*(3), 443. https://doi.org/10.1364/ao.33.000443

Gordon, H. R., Clark, D. K., Brown, J. W., Brown, O. B., Evans, R. H., & Broenkow, W. W. (1983). Phytoplankton pigment concentrations in the Middle Atlantic Bight: Comparison of ship determinations and CZCS estimates. *Applied Optics, 22*(1), 20–36. https://doi.org/10.1364/ao.22.000020

Harmel, T., Chami, M., Tormos, T., Reynaud, N., & Danis, P. A. (2018). Sunglint correction of the multispectral instrument (MSI)-SENTINEL-2 imagery over inland and sea waters from SWIR bands. *Remote Sensing of Environment, 204*, 308–321. https://doi.org/10.1016/j.rse.2017.10.022

He, C., Liu, Z., Wu, J., Pan, X., Fang, Z., Li, J., & Bryan, B. A. (2021). Future global urban water scarcity and potential solutions. *Nature Communications, 12*(1). https://doi.org/10.1038/s41467-021-25026-3

Hestir, E. L., Brando, V. E., Bresciani, M., Giardino, C., Matta, E., Villa, P., & Dekker, A. G. (2015). Measuring freshwater aquatic ecosystems: The need for a hyperspectral global mapping satellite mission. *Remote Sensing of Environment, 167*, 181–195. https://doi.org/10.1016/j.rse.2015.05.023

Ho, J. C., Michalak, A. M., & Pahlevan, N. (2019). Widespread global increase in intense lake phytoplankton blooms since the 1980s. *Nature, 574*(7780), 667–670. https://doi.org/10.1038/s41586-019-1648-7

Hohmann-Marriott, M. F., & Blankenship, R. E. (2011). Evolution of photosynthesis. *Annual Review of Plant Biology, 62*, 515–548. https://doi.org/10.1146/annurev-arplant-042110-103811

Hunter, P. D., Tyler, A. N., Carvalho, L., Codd, G. A., & Maberly, S. C. (2010). Hyperspectral remote sensing of cyanobacterial pigments as indicators for cell populations and toxins in eutrophic lakes. *Remote Sensing of Environment, 114*(11), 2705–2718. https://doi.org/10.1016/j.rse.2010.06.006

Huot, Y., & Babin, M. (2010). In D. J. Suggestt, O. Prášil, & M. A. Borowitzka (Eds.), *Chlorophyll a fluorescence in aquatic sciences: Methods and applications*. Springer.

Ibelings, B. W., Backer, L. C., Kardinaal, W. E. A., & Chorus, I. (2014). Current approaches to cyanotoxin risk assessment and risk management around the globe. *Harmful Algae, 40*, 63–74. https://doi.org/10.1016/j.hal.2014.10.002

Johnsen, G., & Sakshaug, E. (2007). Biooptical characteristics of PSII and PSI in 33 species (13 pigment groups) of marine phytoplankton, and the relevance for pulseamplitude-modulated and fast-repetition-rate fluorometry. *Journal of Phycology, 43*(6), 1236–1251. https://doi.org/10.1111/j.1529-8817.2007.00422.x

Jutla, A. S., Akanda, A. S., & Islam, S. (2012). Satellite remote sensing of space-time plankton variability in the Bay of Bengal: Connections to cholera outbreaks. *Remote Sensing of Environment, 123*, 196–206. https://doi.org/10.1016/j.rse.2012.03.005

Kassem, A., Abbas, L., Coutinho, O., Opara, S., Najaf, H., Kasperek, D., Li, X., & Tiquia-Arashiro, S. M. (2023). Applications of Fourier Transform-Infrared spectroscopy in microbial cell biology and environmental microbiology: Advances, challenges, and future perspectives. *Frontiers in Microbiology, 14*, 1304081. https://doi.org/10.3389/fmicb.2023.1304081

Keeling, P. J. (2013). The number, speed, and impact of plastid endosymbioses in eukaryotic evolution. *Annual Review of Plant Biology, 64*, 583–607. https://doi.org/10.1146/annurev-arplant-050312-120144

Khan, M. H. R. B., Ahsan, A., Imteaz, M., Shafiquzzaman, M., & Al-Ansari, N. (2023). Evaluation of the surface water quality using global water quality index (WQI) models: Perspective of river water pollution. *Scientific Reports, 13*(1), 20454. https://doi.org/10.1038/s41598-023-47137-1

Khelifa, R., Mahdjoub, H., Baaloudj, A., Cannings, R. A., & Samways, M. J. (2021). Effects of both climate change and human water demand on a highly threatened damselfly. *Scientific Reports, 11*(1), 7725. https://doi.org/10.1038/s41598-021-86383-z

Kiefer, I., Odermatt, D., Anneville, O., Wüest, A., & Bouffard, D. (2015). Application of remote sensing for the optimization of in-situ sampling for monitoring of phytoplankton abundance in a large lake. *Science of the Total Environment, 527–528*, 493–506. https://doi.org/10.1016/j.scitotenv.2015.05.011

Kirk, J. T. O. (2011). *Light and photosynthesis in aquatic ecosystems* (3rd ed.). Cambridge University Press.

Kramer, S. J., Siegel, D. A., Maritorena, S., & Catlett, D. (2022). Modeling surface ocean phytoplankton pigments from hyperspectral remote sensing reflectance on global scales. *Remote Sensing of Environment, 270*, 112879. https://doi org/10.1016/j.rse.2021.112879

Kraus, C. N., Maciel, D. A., Bonnet, M. P., & Novo, E. M. L. D. M. (2021). Phytoplankton genera structure revealed from the multispectral vertical diffuse attenuation coefficient. *Remote Sensing, 13*(20). https://doi.org/10.3390/rs13204114

Kudela, R. M., Palacios, S. L., Austerberry, D. C., Accorsi, E. K., Guild, L. S., & Torres-Perez, J. (2015). Application of hyperspectral remote sensing to cyanobacterial blooms in inland waters. *Remote Sensing of Environment, 167*, 196–205. https://doi.org/10.1016/j.rse.2015.01.025

Laureano-Rosario, A. E., Symonds, E. M., Rueda-Roa, D., Otis, D., & Muller-Karger, F. E. (2017). Environmental factors correlated with culturable enterococci concentrations in tropical recreational waters: A case study in escambron beach, San Juan, Puerto Rico. *International Journal of Environmental Research and Public Health, 14*(12). https://doi.org/10.3390/ijerph14121602

Le, C., Li, Y., Zha, Y., Sun, D., Huang, C., & Lu, H. (2009). A four-band semi-analytical model for estimating chlorophyll a in highly turbid lakes: The case of Taihu Lake, China. *Remote Sensing of Environment, 113*(6), 1175–1182. https://doi.org/10.1016/j.rse.2009.02.005

Lee, Z., Shang, S., Qi, L., Yan, J., & Lin, G. (2016). A semi-analytical scheme to estimate Secchi-disk depth from Landsat-8 measurements. *Remote Sensing of Environment, 177*, 101–106. https://doi.org/10.1016/j.rse.2016.02.033

Legleiter, C. J., King, T. V., Carpenter, K. D., Hall, N. C., Mumford, A. C., Slonecker, T., Graham, J. L., Stengel, V. G., Simon, N., & Rosen, B. H. (2022). Spectral mixture analysis for surveillance of harmful algal blooms (SMASH): A field-, laboratory-, and satellite-based approach to identifying cyanobacteria genera from remotely sensed data. *Remote Sensing of Environment, 279*, 113089. https://doi.org/10.1016/j.rse.2022.113089

Li, Z., Yang, W., Matsushita, B., & Kondoh, A. (2022). Remote estimation of phytoplankton primary production in clear to turbid waters by integrating a semi-analytical model with a machine learning algorithm. *Remote Sensing of Environment, 275*, 113027. https://doi.org/10.1016/j.rse.2022.113027

Li, X., Yang, Y., Ishizaka, J., & Li, X. (2023). Global estimation of phytoplankton pigment concentrations from satellite data using a deep-learning-based model. *Remote Sensing of Environment, 294*, 113628. https://doi.org/10.1016/j.rse.2023.113628

Lin, L., Yang, H., & Xu, X. (2022). Effects of water pollution on human health and disease heterogeneity: A review. *Frontiers in Environmental Science, 10*. Frontiers Media S.A. https://doi.org/10.3389/fenvs.2022.880246

Lobitz, B., Beck, L., Huq, A., Wood, B., Fuchs, G., Faruque, A. S. G., & Colwell, R. (2000). Climate and infectious disease: Use of remote sensing for detection of *Vibrio cholerae* by indirect measurement. *PNAS, 97*(4), 1438. https://doi.org/10.1073/pnas.97.4.1438

Löptien, U., & Dietze, H. (2022). Retracing cyanobacteria blooms in the Baltic Sea. *Scientific Reports, 12*(1), 10873. https://doi.org/10.1038/s41598-022-14880-w

Maciel, D. A., Pahlevan, N., Barbosa, C. C. F., Martins, V. S., Smith, B., O'Shea, R. E., Balasubramanian, S. V., Saranathan, A. M., & Novo, E. M. L. M. (2023). Towards global long-term water transparency products from the Landsat archive. *Remote Sensing of Environment, 299*, 113889. https://doi.org/10.1016/j.rse.2023.113889

Magny, C. G., Murtugudde, R., Sapiano, M. R. P., Nizam, A., Brown, C. W., Busalacchi, A. J., Yunus, M., Nair, G. B., Gil, A. I., Lanata, C. F., Calkins, J., Manna, B., Rajendran, K., Kumar Bhattacharya, M., Huq, A., Sack, R. B., & Colwell, R. R. (2008). Environmental signatures

associated with cholera epidemics. *Proceedings of the National Academy of Sciences of the United States of America, 105*(46), 17676–17681. https://doi.org/10.1073/pnas.0809654105

Mandal, S. K., Patel, V. R., Temkar, G., George, B. M., & Raman, M. (2015). Bio-optic characterization of Discosphaera tubifer bloom occurs in an overcrowded fishing harbour at Veraval, India. *Environmental Monitoring and Assessment, 187*(9). https://doi.org/10.1007/s10661-015-4798-6

Matthews, M. W. (2017). Bio-optical modeling of phytoplankton chlorophyll-a. In *Bio-optical modeling and remote sensing of inland waters* (pp. 157–188). Elsevier. https://doi.org/10.1016/B978-0-12-804644-9.00006-9

Matthews, M. W., Bernard, S., & Robertson, L. (2012). An algorithm for detecting trophic status (chlorophyll-a), cyanobacterial-dominance, surface scums and floating vegetation in inland and coastal waters. *Remote Sensing of Environment, 124*, 637–652. https://doi.org/10.1016/j.rse.2012.05.032

McKindles, K. M., & Tiquia-Arashiro, S. M. (2012). Functional gene arrays for analysis of microbial communities on ocean platform. In S. Tiquia-Arashiro (Ed.), *Molecular biological technologies for ocean sensing* (Springer protocols handbooks). Humana Press. https://doi.org/10.1007/978-1-61779-915-0_9

Mishra, S., & Mishra, D. R. (2012). Normalized difference chlorophyll index: A novel model for remote estimation of chlorophyll-a concentration in turbid productive waters. *Remote Sensing of Environment, 117*, 394–406. https://doi.org/10.1016/j.rse.2011.10.016

Mishra, B. K., Kumar, P., Saraswat, C., Chakraborty, S., & Gautam, A. (2021). Water security in a changing environment: Concept, challenges and solutions. *Water, 13*(4). https://doi.org/10.3390/w13040490

Mobley, C. D. (1999). Estimation of the remote-sensing reflectance from above-surface measurements. *Applied Optics, 38*(36), 7442. https://doi.org/10.1364/ao.38.007442

Mora, C., McKenzie, T., Gaw, I. M., Dean, J. M., von Hammerstein, H., Knudson, T. A., Setter, R. O., Smith, C. Z., Webster, K. M., Patz, J. A., & Franklin, E. C. (2022). Over half of known human pathogenic diseases can be aggravated by climate change. *Nature Climate Change, 12*, 869. https://doi.org/10.1038/s41558-022-01426-1

Morel, A., & Bricaud, A. (1986). Inherent optical properties of algal cells including picoplankton: Theoretical and experimental results. In T. Piatt & W. K. W. Li (Eds.), *Photosynthetic picoplankton* (Vol. 214, pp. 521–559) Canadian Bulletin of Fisheries and Aquatic Sciences.

Nair, A., Sathyendranath, S., Platt, T., Morales, J., Stuart, V., Forget, M. H., Devred, E., & Bouman, H. (2008). Remote sensing of phytoplankton functional types. *Remote Sensing of Environment, 112*(8), 3366–3375. https://doi.org/10.1016/j.rse.2008.01.021

Nassereddine, Z., Opara, S., Coutinho, O., Qyteti, F., Book, R., Heinicke, M. P., Napieralski, J., & Tiquia-Arashiro, S. M. (2024). Critical perspectives on advancing antibiotic resistant gene (ARG) detection technologies in aquatic ecosystems. *Science of the Total Environment, 957*(20), 177775. https://doi.org/10.1016/j.scitotenv.2024.177775

Nechad, B., Ruddick, K. G., & Park, Y. (2010). Calibration and validation of a generic multi-sensor algorithm for mapping of total suspended matter in turbid waters. *Remote Sensing of Environment, 114*(4), 854–866. https://doi.org/10.1016/j.rse.2009.11.022

Nichols, G., Lake, I., & Heaviside, C. (2018). Climate change and water-related infectious diseases. *Atmosphere, 9*(10). https://doi.org/10.3390/atmos9100385

O'Shea, R. E., Pahlevan, N., Smith, B., Bresciani, M., Egerton, T., Giardino, C., Li, L., Moore, T., Ruiz-Verdu, A., Ruberg, S., Simis, S. G. H., Stumpf, R., & Vaičiūtė, D. (2021). Advancing cyanobacteria biomass estimation from hyperspectral observations: Demonstrations with HICO and PRISMA imagery. *Remote Sensing of Environment, 266*, 112693. https://doi.org/10.1016/j.rse.2021.112693

Oest, A., Alsaffar, A., Fenner, M., Azzopardi, D., & Tiquia-Arashiro, S. M. (2018). Patterns of change in metabolic capabilities of sediment microbial communities in river and lake ecosystems. *International Journal of Microbiology, 2018*, 6234931. https://doi.org/10.1155/2018/6234931

Oki, T., & Quiocho, R. E. (2020). Economically challenged and water scarce: Identification of global populations most vulnerable to water crises. *International Journal of Water Resources Development, 36*(2–3), 416–428. https://doi.org/10.1080/07900627.2019.1698413

O'Neil, J. M., Davis, T. W., Burford, M. A., & Gobler, C. J. (2012). The rise of harmful cyanobacteria blooms: The potential roles of eutrophication and climate change. *Harmful Algae, 14*, 313–334. https://doi.org/10.1016/j.hal.2011.10.027

O'Reilly, J. E., Maritorena, S., Mitchell, B. G., Siegel, D. A., Carder, K. L., Garver, S. A., Kahru, M., & McClain, C. (1998). Ocean color chlorophyll algorithms for SeaWiFS. *Journal of Geophysical Research: Oceans, 103*(C11), 24937–24953. https://doi.org/10.1029/98JC02160

Pagnucco, G., Overfield, D., Chamlee, Y., Fortuna, A., Sulaiman, F., Farinas, J., Schittenhelm, R., Shuler, C., Kassem, A., Opara, S., Najaf, H., Abbas, L., Coutinho, O., Catalfano, B., Li, X., & Tiquia-Arashiro, S. M. (2023). Metal tolerance and biosorption capacities of bacterial strains isolated from an urban watershed. *Frontiers in Microbiology, 14*, 1278886. https://doi.org/10.3389/fmicb.2023.1278886

Pahlevan, N., Mangin, A., Balasubramanian, S. V., Smith, B., Alikas, K., Arai, K., Barbosa, C., Bélanger, S., Binding, C., Bresciani, M., Giardino, C., Gurlin, D., Fan, Y., Harmel, T., Hunter, P., Ishikaza, J., Kratzer, S., Lehmann, M. K., Ligi, M., et al. (2021). ACIX-aqua: A global assessment of atmospheric correction methods for Landsat-8 and Sentinel-2 over lakes, rivers, and coastal waters. *Remote Sensing of Environment, 258*, 112366. https://doi.org/10.1016/j.rse.2021.112366

Patel, D., Gismondi, R., Alsaffar, A., & Tiquia-Arashiro, S. M. (2019). Applicability of API ZYM to capture seasonal and spatial variabilities in lake and river sediments. *Environmental Technology, 40*(24), 3227–3239. https://doi.org/10.1080/09593330.2018.1468492

Paulino, R. S., Martins, V. S., Novo, E. M. L. M., Barbosa, C. C. F., Carvalho, L. A. S. D., & Begliomini, F. N. (2022). Assessment of adjacency correction over inland waters using Sentinel-2 MSI images. *Remote Sensing, 14*(8), 1829. https://doi.org/10.3390/rs14081829

Paulino, R. S., Martins, V. S., Novo, E. M. L. M., Maciel, D. A., Correia-Lima, D. L., Barbosa, C. C. F., Bonnet, M. P., & Uhde, A. (2023). A framework based on spectral similarity to estimate hydrological connectivity in Juruá River floodplain lakes using 3-m PlanetScope data. *Journal of Hydrology, 625*, 130156. https://doi.org/10.1016/j.jhydrol.2023.130156

Paulino, R.S., Martins, V.S., Novo, E.M.L.M., Barbosa, C.C.F., Maciel, D.M., Wanderley, R.L.N., Portela, C.I., Caballero, C.B., Lima, T. (2025). Generation of robust 10-m Sentinel-2/3 Synthetic Aquatic Reflectance Bands over Inland Waters. *Remote Sensing of Environmental, 318*, 114593. https://doi.org/10.1016/j.rse.2024.114593

Pearson, L. A., Dittmann, E., Mazmouz, R., Ongley, S. E., D'Agostino, P. M., & Neilan, B. A. (2016). The genetics, biosynthesis and regulation of toxic specialized metabolites of cyanobacteria. *Harmful Algae, 54*, 98–111. https://doi.org/10.1016/j.hal.2015.11.002

Peres, F. V., Paula, F. S., Bendia, A. G., et al. (2023). Assessment of prokaryotic communities in Southwestern Atlantic deep-sea sediments reveals prevalent methanol-oxidising Methylomirabilales. *Scientific Reports, 13*, 12782. https://doi.org/10.1038/s41598-023-39415-9

Petro, S., Pham, K., & Hilton, G. (2020). Plankton, aerosol, cloud, ocean ecosystem (PACE) mission integration and testing. In *IEEE Aerospace Conference* (pp. 1–20). Big Sky. https://doi.org/10.1109/AERO47225.2020.9172326

Priya, A. M., Muruganandam, M., Rajamanickam, S., Sivarethinamohan, S., Gaddam, M. K. R., Velusamy, P. R. G., Ravindiran, G., Gurugubelli, T. R., & Muniasamy, S. K. (2023). Impact of climate change and anthropogenic activities on aquatic ecosystem—A review. *Environmental Research, 238*, 117233. https://doi.org/10.1016/j.envres.2023.117233

Reid, A. J., Carlson, A. K., Creed, I. F., Eliason, E. J., Gell, P. A., Johnson, P. T. J., Kidd, K. A., MacCormack, T. J., Olden, J. D., Ormerod, S. J., Smol, J. P., Taylor, W. W., Tockner, K., Vermaire, J. C., Dudgeon, D., & Cooke, S. J. (2019). Emerging threats and persistent conservation challenges for freshwater biodiversity. *Biological Reviews, 94*(3), 849–873. https://doi.org/10.1111/brv.12480

Saberioon, M., Khosravi, V., Brom, J., Gholizadeh, A., & Segl, K. (2023). Examining the sensitivity of simulated EnMAP data for estimating chlorophyll-a and total suspended solids in inland waters. *Ecological Informatics, 75*, 102058. https://doi.org/10.1016/j.ecoinf.2023.102058

Salehi, M. (2022). Global water shortage and potable water safety; Today's concern and tomorrow's crisis. *Environment International, 158*. https://doi.org/10.1016/j.envint.2021.106936

Salgado-Hernanz, P. M., Racault, M. F., Font-Muñoz, J. S., & Basterretxea, G. (2019). Trends in phytoplankton phenology in the Mediterranean Sea based on ocean-colour remote sensing. *Remote Sensing of Environment, 221*, 50–64. https://doi.org/10.1016/j.rse.2018.10.036

Schulze, P. S. C., Barreira, L. A., Pereira, H. G. C., Perales, J. A., & Varela, J. C. S. (2014). Light emitting diodes (LEDs) applied to microalgal production. *Trends in Biotechnology, 32*(8), 422–430. https://doi.org/10.1016/j.tibtech.2014.06.001

Silva, E., Counillon, F., Brajard, J., Pettersson, L. H., & Naustvoll, L. (2023). Forecasting harmful algae blooms: Application to Dinophysis acuminata in northern Norway. *Harmful Algae, 126*, 102442. https://doi.org/10.1016/j.hal.2023.102442

Simis, S. G. H., Peters, S. W. M., & Gons, H. J. (2005). Remote sensing of the cyanobacterial pigment phycocyanin in turbid inland water. *Limnology and Oceanography, 50*(1), 237–245. https://doi.org/10.4319/lo.2005.50.1.0237

Simis, S. G. H., Huot, Y., Babin, M., Seppälä, J., & Metsamaa, L. (2012). Optimization of variable fluorescence measurements of phytoplankton communities with cyanobacteria. *Photosynthesis Research, 112*(1), 13–30. https://doi.org/10.1007/s11120-012-9729-6

Smith, G. P. (1974). Toward an international standard of environment. *Pepperdine Law Review, 2*, 28.

Tewari, M., Kishtawal, C. M., Moriarty, V. W., Ray, P., Singh, T., Zhang, L., Treinish, L., & Tewari, K. (2022). Improved seasonal prediction of harmful algal blooms in Lake Erie using large-scale climate indices. *Communications Earth & Environment, 3*(1). https://doi.org/10.1038/s43247-022-00510-w

Tiquia, S. M. (2010a). Salt-adapted bacteria isolated from the Rouge River and potential for degradation of contaminants and biotechnological applications. *Environmental Technology, 31*, 967–978. https://doi.org/10.1080/09593331003706226

Tiquia, S. M. (2010b). Metabolic diversity of the heterotrophic microorganisms and potential link to pollution of the Rouge River. *Environmental Pollution, 158*, 1435–1443. https://doi.org/10.1016/j.envpol.2009.12.035

Tiquia, S. M. (2011). Extracellular hydrolytic enzyme activities of the heterotrophic microbial communities of the Rouge River: An approach to evaluate ecosystem response to urbanization. *Microbial Ecology, 62*, 679–689. https://doi.org/10.1007/s00248-011-9871-2

Tiquia, S. M., Davis, D., Hadid, H., Kasparian, S., Ismail, M., & Murray, K. S. (2007). Halophilic and halotolerant bacteria from river waters and shallow groundwater along the Rouge River of Southeastern Michigan. *Environmental Technology, 28*, 297–307. https://doi.org/10.1080/09593332808618789

Tiquia-Arashiro, S. M. (2012). *Molecular biological technologies for ocean sensing* (Springer protocols handbooks) (p. 307). Humana Press. https://doi.org/10.1007/978-1-61779-915-0

Tiwari, A., Oliver, D. M., Bivins, A., Sherchan, S. P., & Pitkänen, T. (2021). Bathing water quality monitoring practices in Europe and the United States. *International Journal of Environmental Research and Public Health, 18*(11). https://doi.org/10.3390/ijerph18115513

Tyler, M. A., & Stumpf, R. P. (1989). Feasibility of using satellites for detection of kinetics of small phytoplankton blooms in estuaries: Tidal and migrational effects. *Remote Sensing of Environment, 27*(1), 233–249. https://doi.org/10.1016/0034-4257(89)90085-0

UNESCO. (2023). *The United Nations world water development report 2023: Partnerships and cooperation for water*. UNESCO. https://www.unesco.org/reports/wwdr/2023/en

Vanhellemont, Q. (2019). Adaptation of the dark spectrum fitting atmospheric correction for aquatic applications of the Landsat and Sentinel-2 archives. *Remote Sensing of Environment, 225*, 175–192. https://doi.org/10.1016/j.rse.2019.03.010

Vörösmarty, C. J., Stewart-Koster, B., Green, P. A., Boone, E. L., Flörke, M., Fischer, G., Wiberg, D. A., Bunn, S. E., Bhaduri, A., McIntyre, P. B., Sadoff, C., Liu, H., & Stifel, D. (2021). A

green-gray path to global water security and sustainable infrastructure. *Global Environmental Change, 70*, 102344. https://doi.org/10.1016/j.gloenvcha.2021.102344

Wang, W. L., Fu, W., Le Moigne, F. A. C., Letscher, R. T., Liu, Y., Tang, J. M., & Primeau, F. W. (2023). Biological carbon pump estimate based on multidecadal hydrographic data. *Nature, 624*(7992), 579–585. https://doi.org/10.1038/s41586-023-06772-4

Ward, J. V. (1989). The four-dimensional nature of lotic ecosystems. *Journal of the North American Benthological Society, 8*(1), 2–8. https://doi.org/10.2307/1467397

Woolway, R. I., & Maberly, S. C. (2020). Climate velocity in inland standing waters. *Nature Climate Change, 10*(12), 1124–1129. https://doi.org/10.1038/s41558-020-0889-7

Woolway, R. I., Sharma, S., Weyhenmeyer, G. A., Debolskiy, A., Golub, M., Mercado-Bettín, D., Perroud, M., Stepanenko, V., Tan, Z., Grant, L., Ladwig, R., Mesman, J., Moore, T. N., Shatwell, T., Vanderkelen, I., Austin, J. A., DeGasperi, C. L., Dokulil, M., La Fuente, S., et al. (2021). Phenological shifts in lake stratification under climate change. *Nature Communications, 12*(1). https://doi.org/10.1038/s41467-021-22657-4

Wynne, T. T., Stumpf, R. P., Tomlinson, M. C., Warner, R. A., Tester, P. A., Dyble, J., & Fahnenstiel, G. L. (2008). Relating spectral shape to cyanobacterial blooms in the Laurentian Great Lakes. *International Journal of Remote Sensing, 29*(12), 3665–3672. https://doi.org/10.1080/01431160802007640

Xu, M., Cao, C. X., Wang, D. C., Kan, B., Xu, Y. F., Ni, X. L., & Zhu, Z. C. (2016). Environmental factor analysis of cholera in China using remote sensing and geographical information systems. *Epidemiology and Infection, 144*(5), 940–951. https://doi.org/10.1017/S095026881500223X

Zhang, Z., Deng, Z., Rusch, K. A., & Walker, N. D. (2015). Modeling system for predicting enterococci levels at Holly Beach. *Marine Environmental Research, 109*, 140–147. https://doi.org/10.1016/j.marenvres.2015.07.003

Chapter 17
Remote Sensing for Freshwater Quality Assessment: Current Developments and Future Trends

Widana Gamage Shirani Manel Kumari, Keerthi Sri Senarathna Atapaththu, Kanaji Masakorala, Kelum Sanjaya, and Yapa Mudiyanselage Ajith Lalith Weerasinghe Yapa

1 Introduction

Freshwater environments are among the most sensitive and vulnerable ecosystems on Earth. The global consumption of freshwater for drinking, irrigation, hydroelectric power generation, recreational activities, transportation, and fisheries and aquaculture has increased dramatically during the past few decades with the increasing global population and industrialization. The United Nations (UN) World Water Development Report (2024) indicates 1% annual increase of global freshwater demand. This 1% annual increase is mainly demanded by agriculture accounting for 70% of freshwater resources, while the remaining is for industries (20%) and domestic uses (10%). In parallel, freshwater resources and connected ecosystems are severely threatened. Expected expansion in agriculture not only demands more water but also utilizes more fertilizer, and other agrochemicals, and consequently, more agrochemical residues accumulate in water resources. Besides agrochemical residues, certain other chemical substances including pharmaceuticals, antibiotics, heavy metals, micro and nano plastics, oil and grease, polychlorinated biphenyls (PCBs), and e-waste are added to aquatic environments by various anthropogenic activities through waste disposal (Nassereddine et al., 2024). The annual industrial waste disposal to aquatic ecosystems has been estimated at approximately 300–400 megatons (du Plessis, 2022). Altogether, these pollutants in the shadow of climate

W. G. S. M. Kumari (✉) · K. Masakorala
Department of Botany, Faculty of Science, University of Ruhuna, Matara, Sri Lanka
e-mail: shirani@bot.ruh.ac.lk

K. S. S. Atapaththu · K. Sanjaya
Department of Limnology and Water Technology, Faculty of Fisheries and Marine Sciences & Technology, University of Ruhuna, Matara, Sri Lanka

Y. M. A. L. W. Yapa
Department of Chemistry, Faculty of Science, University of Ruhuna, Matara, Sri Lanka

S. M. Tiquia-Arashiro et al. (eds.), *Urban Watershed Microbiology, Volume 1*,
https://doi.org/10.1007/978-3-031-89071-0_17

change have been identified as the principal driving forces behind eutrophication, salinization, microbial pollution, and toxic element accumulation in aquatic environments (Naderian et al., 2024). Further, degradation of water quality in rivers due to nitrogen pollution will triple the water scarcity in the next couple of decades (Wang et al., 2024). However, there are region-specific drivers of aquatic ecosystem pollution and their consequences. For instance, the river Ganges (India), and the river Citarum (Indonesia) are the two most polluted rivers in the world. Specifically, millions of impoverished residents of Jakarta are at risk of frequent outbreaks of waterborne disease. According to Pink (2016), contaminated water and unhealthy sanitation jointly account for 10,000 deaths per day, including 5000 children under the age of five.

Climate change directly and indirectly makes significant qualitative and quantitative changes in aquatic ecosystems. Changes in rainfall patterns cause drought leading to water scarcity while flooding may enhance accumulation of pollutants, pathogens risking human and animal health. Further, changing rainfall patterns cause long-term variations in flow dynamics, and thereby habitat modifications can happen. Such changes would probably impair the local biodiversity (Panella et al., 2020). Hence, wise management of freshwater resources is vitally important to maintain sustainability and ecological balance. Frequent monitoring of water resources to monitor changes in water quality, detection of sources of pollution, and identifying emerging threats are essential in making appropriate decisions at the right time for water quality management.- In-situ monitoring and laboratory analyses of water quality parameters are the most widely used global conventional practices in freshwater quality management programs. Although the conventional methods are reliable and provide direct measurements, these methods are expensive, time-consuming, and practically difficult in long-term, regional, and national monitoring programs (Dong et al., 2016). In terms of water quality degradation, the conventional methods fail to provide information on real-time detection of water quality deterioration (Ahmed et al., 2020). On the other hand, there are many databases freely available to extract data to predict water quality. For instance, chlorophyll data can be freely downloaded by using satellite images. Further, sea surface temperature, wind velocity, and precipitation data are also available to extract without expenses, and labor. Therefore, the emerging technologies are being widely applied in water quality assessments substituting conventional monitoring techniques. Under this context, remote sensing of important water quality parameters and estimation of water quality is becoming a global trending alternative approach to monitor freshwater quality. Specifically, remote sensing techniques are capable of detecting and identifying water quality status in wide geographic areas and large-scale water bodies more effectively and efficiently. Further, in-situ data collections are snapshots of water quality conditions in time and space and practically impossible to employ to obtain temporal and spatial data for large water bodies. Furthermore, the data format of remotely sensed data is easily readable in computer processing and artificial intelligence (AI) tools (Gholizadeh et al., 2016; Ritchie, 2003). Therefore, countries like Australia, Canada, and United States have implemented regional monitoring programs already. This chapter focuses on reviewing

the rationale of using remote sensing for water quality monitoring, current and future remote sensing platforms, and associated limitations.

1.1 Water: The Most Important Natural Resource on Earth

Water is the most important natural resource next to air essential for the continuation of life on Earth. Unique physico-chemical properties make the water fundamental for life and it is the major constituent of the human body which accounts for 60–75% body weight. Loss of 4% of total body water causes dehydration and loss of 15% can be fatal. The molecular structure of water makes water an excellent solvent and plays an important role in biochemical reactions, transportation of nutrients, gases, and other biomolecules throughout the body, maintaining body temperature and cellular structure. Besides its biological role, water plays an important ecological role by linking and maintaining all ecosystems on Earth.

About 71% of the Earth's surface is covered with water which is approximately 1.4 billion km^3. Freshwater represents only 2.5% of global water resources and the most are inaccessible due to frozen glaciers and ice caps while the rest is groundwater (Fig. 17.1). That leaves only 1.2% surface water and other forms of freshwater resources that are readily available for consumption (Holden, 2014). Thus, freshwater is the most important, limited and essential water resource for sustaining life on Earth.

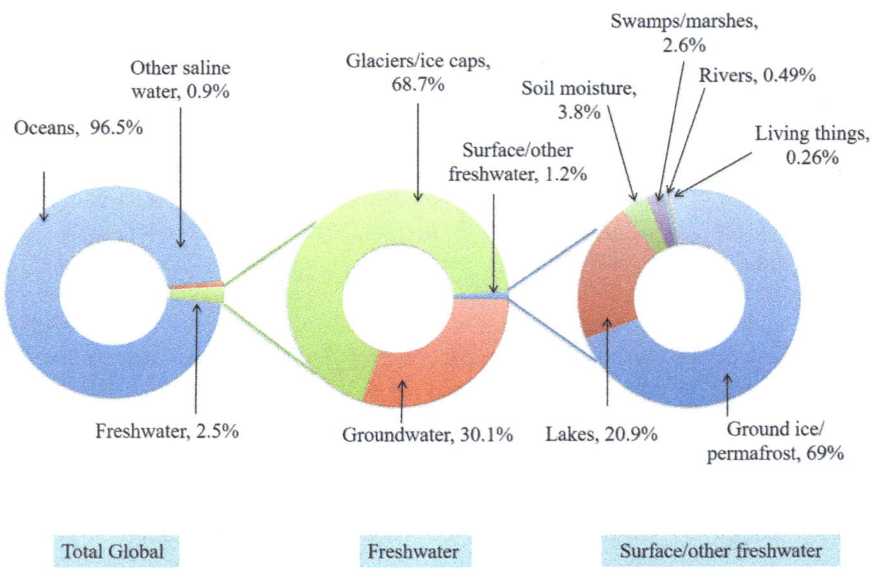

Fig. 17.1 Global distribution of water resources (Shiklomanov, 1993)

Access to water is recognized by the UN as a human right fundamental for everyone's health, dignity, and prosperity. It entitles everyone to have access to sufficient, safe, acceptable, physically accessible, and affordable water for personal and domestic use. Because freshwater resources are unequally distributed throughout the world, some parts of the world do not have access to clean, safe drinking water. Global freshwater resources are mostly concentrated in Brazil, Canada, China, Colombia, Indonesia, Russia, and the United States. According to the statistics of World Health Organization (WHO, 2023), 1.7 billion people across the world lacked good quality drinking water in 2022. The heterogeneity in the distribution of freshwater resources at local or global scale exists because of a number of factors including climate, physical topography, size of the catchment, accessibility of water resources, and the status of socioeconomic development.

Access to freshwater is also important for economic development in both local and global contexts for activities such as agriculture and fisheries. According to the World Bank resources (2017), in many parts of the world, more than 70% of freshwater is used for agriculture and estimates 15% increase in the future to meet 50% increase in agriculture production to feed the 9 billion population by 2050. This will put more pressure on limited freshwater resources on Earth. Apart from that, freshwater fisheries extend significant contributions for food security and employment in tropical Asia in countries like Bangladesh, Cambodia, China, India, and Myanmar and tropical Africa in countries like Congo, Kenya, Mozambique, Tanzania, and Uganda. The UN's Food and Agriculture Organization (FAO) and United States Geological Survey (USGS) highlight that changes in water quality and water pollution are among the major threats to the freshwater fisheries globally.

2 Freshwater Quality

Water quality is one of the main global challenges faced during the twenty-first century by many countries of the world at various magnitudes. Poor quality and degraded water risks human health, limits food production, hinders economic growth, and threatens ecosystems and causes social, economic, and environmental problems. Freshwater quality is closely linked to the expansion of agriculture, industry, and energy production and population and economic growth that produce more solid waste and wastewater. The present global annual wastewater production of 380 billion m^3 is expected to increase by 24 and 51% by 2030 and 2050, respectively (Qadir et al., 2020). According to Jones (2021), a considerable portion (48%) of global wastewater is being released to the natural environment as untreated status. Water is primarily contaminated with chemicals and microorganisms from either point or non-point sources of pollution. Nutrient loading, which is often associated with pathogens, is one of the most prevalent water quality problems (Tiquia et al., 2007, 2008). The single-largest contributor of non-point source pollution is the nutrient loading from agricultural runoff. Industrial effluent containing toxic chemicals, heavy metals, dyes, and many other pollutants has also become a major

contributor to water pollution (Tiquia, 2010a, b; Cho et al., 2015; Bowman et al., 2018). In addition, hundreds of other chemicals such as pharmaceuticals, hormones, detergents, newly discovered industrial chemicals, cyanotoxins, and nanomaterials are considered as emerging contaminants in water environments (Sauvé & Desrosiers, 2014; Kassem et al., 2023; Pagnucco et al., 2023). Water pollution has significant health impacts on humans and the environment. Approximately, 4000 children die every day from diseases caused by polluted water and poor sanitation, while some 1.8 million people die every year by diarrhea alone (WHO, 2023). Discharge of untreated sewage to water triggers pathogen growth and provides breeding grounds for vectors of water-borne diseases.

Eutrophication is the major cause of pollution in freshwaters. Eutrophication is defined as the acceleration in primary production in water bodies as a result of the enrichment of nutrients such as nitrate (N) and phosphate (P) in association with hydrological factors and nutrient retention time (Tiquia, 2011; Chapman, 2021). Natural eutrophication is a slow process and is accelerated due to external loading of nutrients. Extensive use of chemical fertilizers brings nutrients such as N and P with agricultural runoff to the water bodies. In addition, N, P, and ammonium (NH_4) nutrients released from various point-sources such as industrial effluents and domestic wastewater also discharge nutrients to water. At eutrophication, growth of three major groups of phytoplankton, green algae, cyanobacteria, and diatoms, is accelerated leading to the formation of excessive biomass termed algal bloom. Algal blooms form thin, greenish surface film or thick scum on the water. Algal blooms impede light penetration and water circulation. Consequently, aquatic life at bottom layers of the water column is affected. Algal biomass increases biological oxygen demand (BOD) and deplete availability of oxygen for other aquatic organisms. Prolonged algal blooms can even from "dead zones" in water bodies. Finally, the entire ecosystem may collapse. Algal blooms reduce drinking water quality by adding foul odor and taste to the water. Further, if the bloom is caused by cyanotoxin-producing cyanobacterial species, water can be toxic to human, fish, and other organisms. A high nitrate concentration (>10 mg/L) in drinking water reacts with hemoglobin and reduces oxygen level in blood in infants leading to blue baby syndrome (APHA, 2005). Thus, ensuring water quality is one of the biggest global challenges. Under this context, temporal and spatial monitoring of contaminants has paramount validity in water security, particularly targeting the SDG Goal 6. However, timely detection of these contaminants poses a significant challenge due to their inherent dynamic nature. Further, the impact assessments of these contaminants are challenging because such impacts vary with the ecology of freshwaters. For instance, limits of tolerance and susceptibility of flora and fauna vary with pollutants, while the same species shows different responses for different pollutants. As a result, formulation of regulations and management strategies to overcome such issues is a complicated task. Thus, it should be a collective effort of not only scientists, researchers, water professionals, and politicians but also people of all ages, gender, and ethnic groups. The United Nations' Environment Program (UNEP) has identified the necessity in having reliable and accessible water quality data for translating the information into realistic and tangible actions to safeguard water resources.

GEMStat (Global Freshwater Quality Database) is a program of UNEP that collects a variety of freshwater quality data from countries worldwide with good geographical and temporal coverage with a free access data portal (https://gemstat.org/).

2.1 *Parameters of Freshwater Quality*

Water quality is determined by testing or monitoring physical, chemical, and biological properties of water. These properties can be indicated by using a certain set of quality parameters (Table 17.1). Among the many different water quality parameters, several are considered essential parameters in water quality testing and monitoring programs.

2.1.1 Dissolved Oxygen (DO)

Dissolved oxygen (DO) is one of the most important water quality parameters of freshwaters. DO is available for aquatic organisms for respiration and other biochemical reactions. In addition, DO participates in various chemical reactions in water. The source of oxygen in the DO is from gaseous molecular oxygen (O_2) originating from the atmosphere or from the oxygenic photosynthesis. In the dissolution of oxygen in water, no chemical reaction takes place between water molecules and oxygen and oxygen just fills the spaces between water molecules. The solubility of oxygen in water is quite low and the concentration is constantly affected by diffusion, aeration, photosynthesis, respiration, and decomposition. The DO

Table 17.1 Parameters of water quality

	Water quality parameters	
Physical parameters	Chemical parameters	Biological parameters
Color	Ammonia/ammonium	Algae
Electrical conductivity (EC)	Biological oxygen demand (BOD)	Bacteria
Secchi depth (SD)	Chemical oxygen demand (COD)	Cyanobacteria
Temperature	Dissolved oxygen (DO)	Protozoa
Total suspended solids (TSS)	Hardness	
Total dissolved solids (TDS)	Nitrate	
Turbidity	Oxidation-reduction potential (ORP)	
Dissolved organic matter (DOM)	pH	
Colored dissolved organic matter (CDOM)	Phosphorus	
	Salinity	

concentration in water is inversely related to the temperature and salinity while directly related to the pressure. Further, DO concentration in surface water is affected by diurnal and seasonal variations in temperature. In late spring and warm summer, DO is often low due to high consumption of oxygen for aerobic decomposition of organic matter and limited water flow. The DO level of surface waters of normal salinity in the summers is typically more than 8 mg/L. At 20 °C, saturation concentration of DO is about 9 mg/L (Tchobanoglus et al., 2003). However, in summer, it can fall below 2 mg/L when there is excessive phytoplankton growth and accelerated respiration rate. In such a situation, water is defined as hypoxic. Hypoxia kills many aquatic life forms. Cold water can hold more DO than warm water. Thus, in winter and early spring, DO concentration is higher. At 0 °C, saturation concentration of DO is about 14.6 mg/L (Tchobanoglus et al., 2003). During the rainy season, DO concentration usually goes up due to water mixing and strong wind action. The desired DO concentration in water ranges from 5 to 15 mg/L (Boyd, 1998). Higher the concentration of DO better the water quality. Although DO has no direct influence on public health, it reflects biological and physical conditions that exist in a waterbody that support aquatic life (Oest et al., 2018; Patel et al., 2019). Therefore, DO is considered as an important water quality parameter. In water quality testing and monitoring, DO is measured with titrimetric, colorimetric, and sensors of either optical or electrochemical.

2.1.2 Electrical Conductivity (EC)

Electrical conductivity (EC) measures the ability of water to conduct an electrical current (Tchobanoglus et al., 2003). Since ions in a solution conduct an electrical current, EC varies according to the concentration of ions, their mobility, and temperature. The temperature affects the viscosity of a solution. When the temperature increases, viscosity decreases and the mobility of ions increases and hence temperature positively affects the EC of a solution. Salinity and total dissolved solid measurements of water are derived from EC measurements. Since the EC of water indicates the concentration of ions, it is one of the main parameters which determine the use of freshwater for different purposes such as drinking and irrigation. The EC of freshwater is <600 μS/cm. Generally, each water body tends to have a relatively constant range of EC. Thus, a significant change in EC may then may be an indicator that a discharge or some other pollutant has entered the waterbody. In water quality testing and monitoring, EC is measured with a conductivity meter with a probe.

2.1.3 Nitrogen/Nitrate

The most abundant element in Earth's atmosphere (78%) is nitrogen. Gaseous N does not react with water but dissolves in water with very low solubility which is about 20 mg/L at 20 °C and 1 atm. However, other nitrogenous ions such as

ammonia (NH_3), ammonium (NH_4^+), nitrite (NO_2^-), and nitrate (NO_3^-) and organic N are readily soluble in freshwater (APHA, 2005). Nitrogen may enter into freshwater through different types of point sources such as sewage inlets, wastewater treatment plants, leaky septic systems, industrial effluents, and different types of non-point sources such as agricultural runoff and atmospheric depositions. Organic N represents the N present in living, dead, or decomposing animals, plants, and other aquatic organisms. In sewage-contaminated water, organic N and ammonia are the major forms of nitrogen and they are transformed to nitrites and nitrates by the action of aerobic nitrifying bacteria. Therefore, a higher amount of organic N and ammonia in freshwater indicates sewage pollution. Some nitrogen-containing compounds that originate from automobiles and other sources may enter freshwater from the atmosphere. Typically, nitrate is the most common form of inorganic nitrogen in freshwater which can enter into surface and groundwater as a pollutant from chemical fertilizers applied in the agricultural lands through surface runoff or leaching. Ammonium is the most reactive form of N in aquatic systems. Being positively charged, it tends to adhere to sediments and soil (Tiquia et al., 2006a, b; Flood et al., 2015) . Under anaerobic condition, nitrate in the aquatic environment is converted into gaseous N (N_2 or N_2O) by denitrifying bacteria. Since nitrate nitrogen is an important basic nutrient for growth, presence of excessive amounts of N in freshwater may stimulate phytoplankton and algal growth leading toward eutrophication (Ansari et al., 2010; Domingues et al., 2011; Tiquia-Arashiro & Pant, 2020). Eutrophication provides favorable conditions for the growth of aquatic organisms including harmful cyanobacteria that produce an array of cyanotoxins that directly influence aquatic life. This rapid proliferation of toxin-producing phytoplankton is termed as harmful algal blooms (HABs). HABs are formed by cyanobacteria, microalgae, diatoms, dinoflagellates, and raphidophytes and may severely impact on people, animals, and the local ecology. Specifically, the HABs are important in food security as certain health issues appear with the presence of HABs via the consumption of shellfish. Under this context, mollusks, and crustaceans are particularly important. Amnesic shellfish poisoning, Azaspiracid shellfish poisoning, Ciguatera fish poisoning, diarrhetic shellfish poisoning, neurotoxic shellfish poisoning, and paralytic shellfish poisoning are some of the health issues caused by cyanotoxins of HABs. In water quality monitoring, total nitrogen and Kjeldahl-nitrogen are measured. Total nitrogen in freshwater refers to the sum of both organic and inorganic nitrogen compounds while Total Kjeldahl Nitrogen (TKN) indicates total nitrogen concentration, which is composed of the sum of both organic nitrogen compounds and ammonium nitrogen.

2.1.4 Phosphorus/Phosphate

Phosphorus naturally exists in rocks as rock phosphates and other different types of mineral deposits. During the natural weathering process of rocks, phosphorus element is gradually released as phosphate ions and solubilize in water. Phosphates

may exist in three different forms including orthophosphate, organically bound phosphate, and metaphosphate (or polyphosphate). In freshwater systems, phosphates are contained in living organisms, decaying organic matter, free ions, absorbed to soils and sediments and mineralized compounds in rocks, soil, and sediments. Organic phosphate in living organisms, organic matter, and waste solids is converted to orthophosphate following decomposition. This orthophosphate is readily bioavailable but concentration in unpolluted freshwater is typically very low. However, in polluted water orthophosphate concentration is fairly high. A larger quantity of phosphate compounds is brought to water bodies by untreated laundering and commercial cleaning discharge and agriculture runoff. In unpolluted freshwaters, mostly orthophosphate is the limiting nutrient for plant and phytoplankton growth. The desired concentration of phosphate in water ranges from 0.005 to 0.2 mg/L (Boyd, 1998). The WHO (2011) recommended level for drinking water is 2.5 mg/L phosphate. In water quality testing and monitoring, phosphate concentration is measured using colorimetric methods.

2.1.5 pH

pH is one of the most important parameters of freshwater quality. pH stands for negative logarithm of hydrogen ion concentration [$-\log(H^+)$] (Spellman, 2017). The numerical value of pH indicates acidity and alkalinity of water. pH effects on other physico-chemical and biological parameters of water and it determines the use of freshwater for different purposes. The water at extreme ends of pH can be detrimental for aquatic organisms and not suitable for many purposes. The sensitivity of organisms for pH is species specific and also dependent on their stage of life. Thus, a slight change in pH might be detrimental for life. At high pH, DO concentration is high, while at low pH, bioavailability of toxic metals is high. The safe ranges of pH for drinking, domestic use, and for living organisms range from 6.5 to 8.5 (WHO, 2011). Both electrometric and colorimetric methods are used in water quality testing and monitoring for the determination of pH of freshwater.

3 Monitoring and Assessment of Freshwater Quality: Conventional Approach

The quality of freshwater is naturally determined by hydrological, topographical, and lithological properties of the waterbody and the atmospheric and climatic factors of the environment. Apart from that, nature and magnitude of anthropogenic activities also play a major role. Their influence on water quality is reflected through physical, chemical, and biological properties of the waterbody. Deterioration of quality of freshwater ecosystems is becoming one of the serious environmental issues in many countries. As a result, in many countries water quality monitoring is

among the top priorities in aquatic ecosystem management. Presently, a range of water quality assessment methods are in practice. These methods include assessment of physical, chemical, and biological aspects of the waterbody either as a single or combined approach. The most widely practiced conventional water quality assessment methods utilize a combination of physical, chemical, and biological quality parameters of water to determine the overall quality by employing different matrices. These combined approaches are said to be more accurate in predicting water quality since the overall quality of water is determined by interactive effects of physical, chemical, and biological properties of that water.

Water quality assessment provides the baseline information on water safety and suitability of water for specific purposes. Since water quality changes with time and other factors, continuous monitoring and assessment of water is essential. Well-designed monitoring programs facilitate identification of temporal changes or trends in water quality and emerging threats to the water quality. Thereby it permits taking proactive measures to improve water quality, responding to emerging pollution situations, designing future plans for water quality management, implementation of pollution control actions, and development of guidelines or regulations to ensure water safety, etc. There are many ways to conduct water quality monitoring. It can be monitored on continuous, seasonal, or emergency basis at regular sites, selected sites, random sites, or at sites of emergency and measure or assess physical, chemical, and biological properties of water. Investments on water quality assessment are expensive. Hence, countries adopt national policies and programs for water quality monitoring. Our literature survey showed that many of the regional, national and state monitoring programs are designed mostly based on WHO guidelines but with slight modifications to meet regional or national requirements. A well-planned water quality monitoring program contains several stepwise activities which include setting objectives, ensuring resource availability, reconnaissance survey, sampling, laboratory analyses, data management, and interpretation.

Objectives of Water Quality Monitoring

Identification of the requirement and setting the objectives of the monitoring program is the first step. Usually with large-scale regional or national investments, monitoring programs are designed on multiple objectives. There can be an exhaustive list of objectives, while some of them are identification of baseline conditions in water bodies, detection of any signs of water quality deterioration, identification of contaminated areas, evaluation of the effectiveness of a water quality management practice, development of guidelines for water quality, and pollution control, etc. In general, the standard methods of water quality monitoring are done by comparing the experimental data with reference values. These standards are available for single parameters, as well as for a group of water quality parameters. There are different indices to assess water quality by using a group of water quality parameters, and such an index is known as water quality index (WQI). Among them, the National Sanitation Foundation WQI (NSFWQI), Oregon WQI (OWQI), British Columbia WQI (BCWQI), Fuzzy WQI (FWQI), and Canadian Council of Ministers of Environment (CCME) are popular in water quality assessments (Kumar et al.,

2024). Electrical conductivity, pH, DO, TDS, BOD, turbidity, and most probable number (MPN) are some of the common water quality parameters considered for WQI calculation. Although the fecal coliform count has been used in conventional WQI to evaluate the drinking water quality, the newly developed WQI models use *E. coli* instead of coliforms (Uddin et al., 2021). Based on overall assessment, water bodies are rated and categorized. For instance, drinking water sources are being rated as excellent, good, and poor water quality, but these categories are index specific. Although these WQI are being widely employed in water quality assessments, there are many index-specific limitations. For instance, NSFWQI has the eclipsing effect, while the OWQI is unable to assess all health risks such as metals, toxins, and germs.

Resource Availability
Once objectives are set, consideration on the resource availability to continue the program such as laboratory facilities, transport and storage facilities, and human resources is important.

Reconnaissance Survey
This survey includes a collection of background information such as a map of the area of concern, weather and hydrological data, potential pollution sources, extent of anthropogenic activities and determination of sampling procedure. The effectiveness of a water quality monitoring program is largely dependent on adopting an accurate sampling procedure which in turn depends on the objectives of the program. This includes determination of sampling stations, number of samples, parameters to be measured, and frequency of sampling followed by sampling method, sample preservation, and transportation. The status of water quality can be described by using a single parameter or a combination of more than 100 parameters in complex monitoring programs. However, for most purposes, water quality can be sufficiently described by using less than 20 physical, chemical, and biological parameters (UNEP/WHO, 1996). Selection of parameters depends on the program objectives and use of water such as drinking and irrigation. Sampling frequency varies with anticipated changes in water quality. If considerable changes in water quality is expected, more frequent sampling should be conducted than the water bodies where relatively constant water quality is expected. For example, for lakes and reservoirs, monthly sampling frequency per year with twice a month in summer months is generally recommended for monitoring for eutrophication.

Sampling
A substantial proportion of total water quality monitoring cost is spent on sampling. Apart from the cost, sampling is a labor and time-consuming task. Therefore, well-planned field work for sample collection, preservation, and transportation needs to be practiced in coordination with the laboratory. Samples are obtained to represent the entire waterbody from representative sites. The choice of representative sites and use of appropriate sampling techniques requires trained personnels. Nowadays, many of the important water quality parameters can be measured in the field with instruments that are specifically designed for field applications. Obtaining field

(*in-situ*) measurements are essential for some of the quality parameters such as temperature, Secchi depth, and pH while parameters such as DO, conductivity, phytoplankton require fixation in the field until analyses are completed in the laboratory. Transportation of samples to the laboratory also should be carried out with care to prevent sample degradation. Therefore, field work is the most challenging task in a water quality monitoring program. Erroneous sampling can lead to false conclusions. However, field analysis of samples can be advantageous than laboratory analyses. One of the significant advantages is that fresh samples can be analyzed without changing characteristic features. Other advantages include that it saves time and labor for sample collection and analysis and minimizes likelihood of losing or mislabeling of samples. However, there are certain limitations in field analysis of water quality parameters. Typically, accuracy of field measured data is less than the data obtained from standard laboratory analyses although instruments are specifically designed for field applications. Usually, detection range and reproducibility of field instruments have limitations. Further, it is difficult to control the quality of field generated data.

Laboratory Analyses
Samples that are brought to the laboratory are analyzed with standard laboratory procedures and analytical methods. There is an exhaustive list of standard laboratory tests to measure water quality parameters. However, those are not included in this chapter.

3.1 Data Analysis and Interpretation

Field and laboratory analyses generate huge sets of numerical data which are computerized along with location data obtained from Geographical Information System (GIS). These data must be processed and translated to meet objectives of the water quality monitoring program. For instance, this translation should be able to describe spatial and temporal patterns in water quality, impact of anthropogenic activities on water quality, consequences of management practices, identify sources of pollution, and status of water quality. Important conclusions and decisions based on water quality monitoring activities can be drawn only after the assessment of water quality. Often, statistical analysis of data is essential, such as for long-term water quality management programs. However, good and statistically analyzed data do not give meaningful translation or interpretation of water quality. A fair judgment can always be made with comparison of data with standard or ideal values. This can be done with the aid of water quality standards and indices.

3.2 Water Quality Standards

The suitability of water for a particular purpose such as drinking, irrigation or recreation depends on the quality of water which is determined by many different physical, chemical, and biological parameters. Light intensity, temperature, and water level are some of the physical properties, while dissolved oxygen concentration, pH, conductivity, nutrient concentrations (NO_3^- and PO_4^{3-}), and biological oxygen demand are the common chemical properties used in WQI calculations. Besides, primary production, *E. coli* count, total coliform count, and the presence of indicator species are some of the biological parameters used in water quality detections. Presently, there are no universally applicable standards of water quality parameters for quality determination of water. Therefore, different national and organizational level water quality standards are adopted by decision makers, researchers, laboratories, water quality monitoring programs, and organizations to recommend the utilization of freshwater for specific purposes. The main objective of setting up quality standards is to protect the health of stakeholders. Recognizing the significance of maintaining and improving water quality, WHO has developed a series of normative "guidelines" that present an authoritative assessment of health risks associated with exposure to health hazards through water (WHO, 2022). The WHO further describes that guidelines and guideline values do not necessarily mean that the water is degraded if the value deviates from the recommended level. The guidelines are there to make continuous efforts to maintain water quality at the highest possible level.

To date, the WHO guidelines for various types of water such as drinking-water, irrigation, aquaculture, and recreation have been developed. Many countries have adopted or developed water quality standards at regional and national scale based on the WHO guidelines. However, it is important that each country should review their requirements and capacity prior to adopting quality standards from another source. Because, the nature and the form of available water may vary from country to country. Also, prior to drawing the acceptable or safety level, the resulting benefit and the cost of maintaining such water quality level and implementation of programs for further improvement should also be considered.

3.3 Water Quality Indices

Assessment of water quality requires collection and analysis of large datasets of water quality parameters (Uddin et al., 2021). Certain individual and combinations of measurements of quality parameters are used to classify quality of water bodies. For example, water bodies are classified into four categories based on their hardness level as soft (0–60 ppm), moderate (61–120 ppm), hard (121–180 ppm), and very hard (>181 ppm) based on calcium carbonate concentration, while to classify water bodies based on eutrophication level, combination of nitrate, phosphate, and

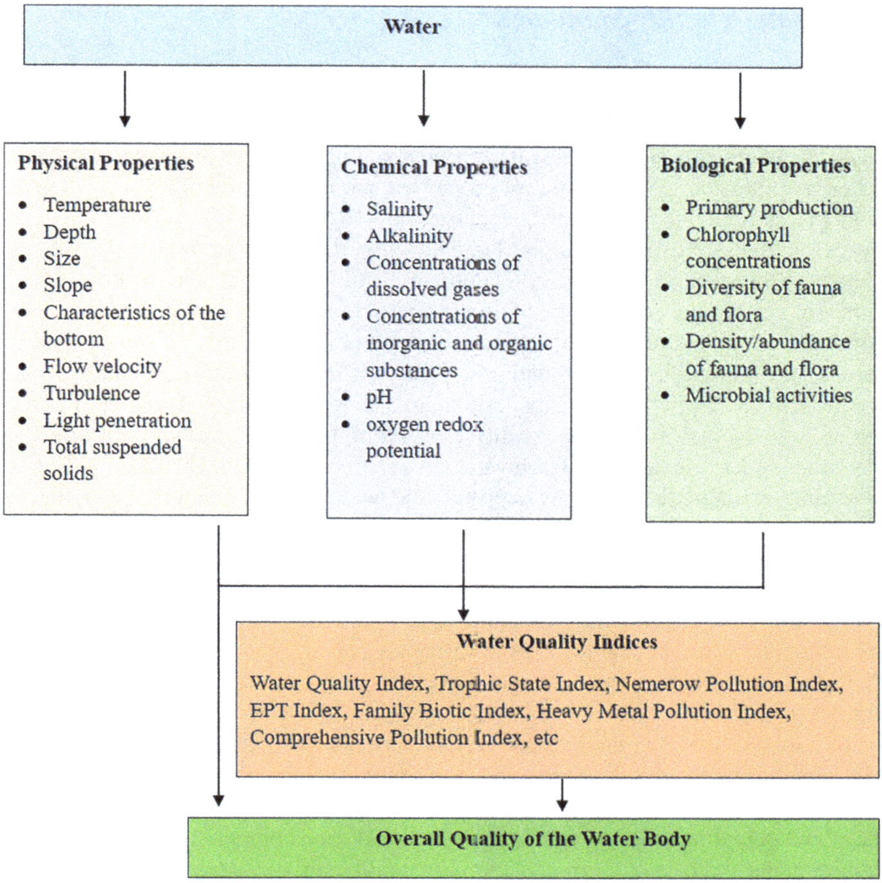

Fig. 17.2 A diagram illustrating factors affecting water quality and widely used water quality indices that determine the overall quality of the water body

chlorophyll measurements is used. However, comparable and meaningful fair judgement of water quality can be made with the help of quality standards and indices (Fig. 17.2). Among these indices, the Water Quality Index (WQI) proposed by Horton (1965) has been extensively applied for decades with different modifications (Abbasi & Abbasi, 2012). The WQI is a popular index as it is relatively easy to use, and it converts complex water quality datasets into a single measure of water quality that is easy to understand (Uddin et al., 2021). There are four steps in the development of an index for water quality assessments (i) selection of water quality parameters, (ii) generation of parameter sub-indices, (iii) assignment of parameter-weight values, and (iv) computation of the water quality index using an aggregation function. Scientists applied WQI with their own modifications while adhering to these four basic steps (Dinius, 1987; Hasan et al., 2015; Horton, 1965; House, 1980; Liou et al., 2004; Naubi et al., 2016; Sutadian et al., 2017). Similar approaches have also

been applied to predict pollution status of water bodies. For example, the heavy metal pollution index (HPI) has been employed to determine the level of heavy metal pollution in aquatic ecosystems (Asim & Nageswara Rao, 2021; Chettri et al., 2022; Kumar et al., 2019).

In some water quality indices, physical properties of water have also been coupled with chemical properties to evaluate the quality of water bodies. For example, Secchi depth (SD) transparency has been employed to evaluate the trophic status of water bodies. According to Carlson (1977), the Trophic State Index (TSI) of lakes is a function of the Secchi depth as given below.

$$TSI = 10(6 - \log SD)$$

Similarly, biological properties of water have been also used as proxies of water quality (Table 17.2).

Although water quality indices provide assessment of water to a satisfactory level, requirement of large set of numerical data of several parameters is not often convenient when regular monitoring of water quality has to be conducted, because acquisition and laboratory analysis of a large set of data is not only time and

Table 17.2 Bioindices used in water quality assessment.

Index	Model	Classification of water body	References
Carlson Tropic State Index	TSI = 9.81 ln (Chl − *a*) + 30.6 $CTSI = \frac{TSI(SD)+TSI(chl)+TSI(TP)}{3}$	Oligotrophic(0 < CTSI≤ 30) Mesotrophic (30 < CTSI≤ 50) Eutrophic(50 < CTS≤ 70) Hypereutrophic(70 < CTSI ≤ 100)	Carlson (1977)
EPT Index	$EPT\ \text{index} = \frac{\text{Number of } EPT \text{ Taxa}}{\text{Total number of Taxa}}$	In general, the more EPT taxa, the better the water quality	Mandaville (2002)
Family Biotic Index	$FBI = \frac{\sum_{i=1}^{s} n_i x t_i}{\sum_{i=1}^{s} t_i}$	0.00–3.75: Excellent 3.76–4.25: Very Good 4.26–5.00: Good 5.01–5.75: Fair 5.76–6.50: Fairly Poor 6.51–7.25: Poor 7.26–10.00: Very Poor	Carter et al. (2017)
Biological Monitoring Working Party score system and Average Score per Taxon	BMWP = sum of score of organisms found $ASPT = \frac{\text{BMWP score}}{\text{Number of groups found}}$	>5 excellent 4–4.5 good 3–3.5 moderate 2–2.5 poor 1–1.5 very poor	(Hawkes, 1998)

Chl-a chlorophyll-a, *TSI* Trophic State Index, FBI richness of aquatic insect families, *EPT* Taxa richness of Ephemeroptera, Plecopteran and Trichopteran families

labor-consuming but also costly (Czyczula Rudjord et al., 2022; Gunawardana et al., 2022). Hence, routing monitoring of water bodies for quality assessments is not generally practiced in many countries. However, due to the absence of a unique variable that can sufficiently describe water quality, assessments based on a series of physico-chemical and biological parameters thus far the available option for water quality monitoring where WQIs play a pivotal role (Kachroud et al., 2019).

Since the first introduction of WQI by Horton (1965), many different water quality indices have been adopted by different countries and organizations (Kachroud et al., 2019). Thus, comparison of water quality in published literature based on WQI is difficult mainly due to three reasons. First, due to the differences in limits of classes used in the classification of quality levels although the same WQI was used. Second, use of different WQI but with same parameters leading to a different classification system. Third, use of different WQI with different parameters which also result in different classification systems. Generally, selection of variables for WQI is done following an expert's opinion. Presently, from 5 to 78 variables are used in WQIs (Kachroud et al., 2019) leading to the comparison more difficult. Once variables are selected, weights for selected variables are determined considering the impact of those variables on water quality and water quality guidelines in practice. Hence, the choice of variables and weighting can change WQI significantly when the expert panel and guidelines are changed. Introduction of statistics and multivariate analytical methods such as factor analysis, cluster analysis, multiple regression analysis, and discriminant analysis during the last decade allowed unbiased selection of variables and more accurate water quality prediction (Boyacioglu & Boyacioglu, 2007; Ji et al., 2016; Koklu et al., 2010). In addition, Bhargava (1983) and Icaga (2007) proposed a more innovative approach for an accurate WQI using "fuzzy logic" in which the selection of variables is made according to the water management objectives and evaluates the impact of each variable in the final WQI. However, neither Bhargava nor Icaga was able to define weighting factors for specific water use. Thus, it is still practically impossible to define a universal WQI.

Further, detection of specific water qualities, for instance, toxins and residual chemicals of pesticides, dyes, and pharmaceuticals needs high-tech instrumentation such as HPLC, GC-MS, and LC-MS. Moreover, estimation of bio-indices such as diversity indices, FBI index, and Taxa richness of Ephemeroptera, Plecopteran and Trichopteran families (EPT) index needs an extensive survey of flora and fauna to collect abundance of individual taxa. Thus, these sampling techniques are exhaustive and destructive.

In summary, practical difficulties in data collection and analysis and disagreements between water quality indices stimulate scientists to discover simple, and low-cost water quality detection techniques and assessment methods for efficient management of aquatic ecosystems. Under this context, application of artificial intelligence (Czyczula Rudjord et al., 2022; Nasir et al., 2022; Nova, 2023) and remote sensing techniques (Gunawardana et al., 2022; Taloor et al., 2022; Yang et al., 2022b) have been identified as promising tools to address the challenges and problems caused by the conventional water quality assessment methods.

3.4 Problems Associated with Conventional Water Quality Monitoring and Assessment

There are practical difficulties in conducting conventional-type water quality programs. However, it is still the most common practice of water quality monitoring in many countries mainly because of high precision of obtained data, lack of modern technology, and reluctance in changing the existing policies.

Apart from high time and labor requirement, high operational cost on large-scale monitoring is a major concern in water quality monitoring programs because data have to be collected simultaneously from many different locations (Chawla et al., 2020; Yang et al., 2022a). Besides, obtaining a fair number of sampling locations to represent the entire region is challenging because the selected locations should be able to adequately represent the region in concern. Sometimes, conclusions or decisions taken based on data obtained locations to represent the region may not always be highly accurate as water quality varies with time, it relies on several environmental factors that are highly dynamic from a temporal perspective. The dynamic nature of water bodies complicates monitoring efforts because water quality parameters are constantly changing due to a variety of natural and human-influenced factors. These dynamics can make it difficult to obtain representative samples and create accurate assessments of the overall health of the water body. Under this context, the temporal variability such as seasonal and diurnal changes in the shadow of climate change is particularly important. Further, spatial variability of flow pattern and stratifications makes significant influences on sampling. Furthermore, hydrological variabilities of river discharge and tidal changes are also significant. Hence, continuous in- situ hydrometric stations are essential but expensive to maintain for many developing countries (Fig. 17.3). On the other hand, certain locations are difficult to access due to its geographical location and safety. Further, a single water quality monitoring round may not cover the entire water body due to its larger size. However, selecting representative sampling sites for water quality assessment is crucial for obtaining accurate and meaningful data. Researchers employ several strategies and statistical methods to ensure that selected sites capture the spatial variability of the water body. Among these sampling strategies, targeted sampling, random sampling, systematic grid sampling, stratified sampling, and judgmental sampling are being employed to minimize sampling errors. The spatial variability of water quality can be addressed by employing several statistical approaches such as spatial autocorrelation, geostatistical interpolation, cluster analysis, and principal component analysis. Therefore, monitoring programs are tending to operate intermittently. Other than insufficient investment in infrastructure, these countries face lack of institutional framework, government interest, and absence of long-term or time-series data flatforms (Chawla et al., 2020). Therefore, regional scale or global scale assessments, forecasting, and modeling of water-related issues are practically challenging.

With these limitations allied with ground-based observations, remote sensing provides a worthwhile solution for monitoring large-scale water bodies at global and regional scales with reasonable accuracy together with appropriate temporal

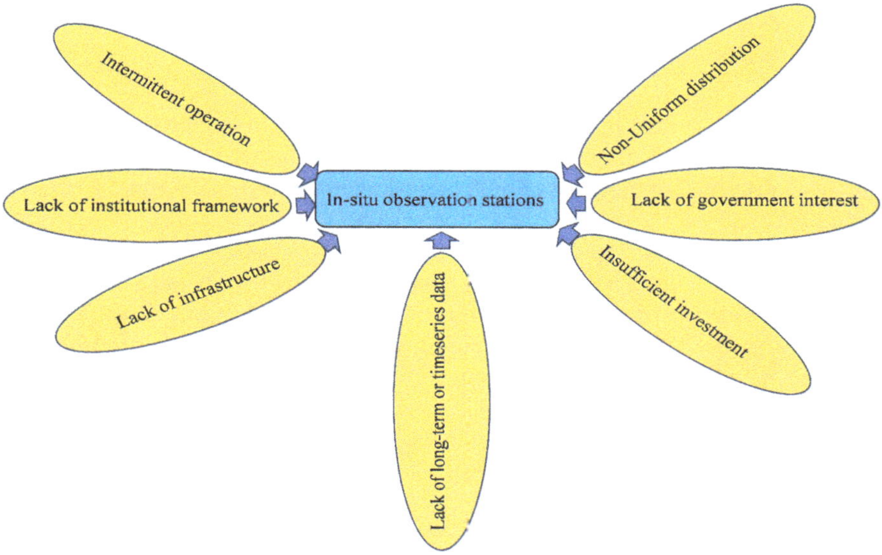

Fig. 17.3 A diagram illustrating challenges faced by in-situ observation stations for water quality monitoring at regional or global scale

coverage (Gholizadeh et al., 2016; Ritchie et al., 2003; Yang et al., 2022a). Large-scale viewing with reasonable temporal coverage helps accurate understanding of the progression of changes in water quality which may be leading to a pollution or nutrient depletion status and thus enables water resource managers and policymakers to take proactive measures. Therefore, provisioning of comprehensive, real-time, and spatially extensive data on various aspects of water bodies is vital for water quality assessments, and thus, traditional water quality sampling, which is limited by logistics and costs, is no longer supportive in this regard. Therefore, the promising modern tools of remote sensing offer the ability to monitor large areas, even in remote or inaccessible locations, at frequent intervals.

4 Rationale of Using Remote Sensing for Water Quality Monitoring

Although many physical, chemical, and biological components in water contribute to the quality of that water, practically, not all contributing parameters can be sensed remotely. However, the ability to measure several water quality parameters with optical and thermal sensors enables remote sensing of water quality a reality. This chapter focuses on remote sensing of water quality by optical properties of water quality determinants. Optical property of a material can be defined as how it

interacts with light. Optical properties include absorption, refraction, and transmission of light. So that optical properties of water are given by how different constituents in water interact with light. Solar radiation is the sole source of natural light that reaches water. Solar radiation is measured in wavelengths or frequency. As light travels in a wave, a wavelength is defined as the distance from one peak to the next peak and is measured in nanometers (nm), whereas frequency is defined as wavelength cycles per second and is measured in hertz (Hz). Although solar radiation consists of all types of radiation, only a part of the spectrum between 100 and 1 mm reaches the Earth. This band is divided into three ranges, ultraviolet (UV), visible (photosynthetically active radiation, PAR), and infrared (IR). UV falls wavelengths between 100 and 400 nm, while visible and IR fall within the ranges of 400–700 nm and 700 nm to over 1 mm, respectively. Of the total solar radiation that reaches Earth's surface, the majority of 49.4% is IR, while 42.3% is visible and 8% is UV, each having a different impact on the environment and its life. A large fraction of solar radiation that strikes on the water surface is absorbed and penetrates through the water column (Wetzel, 2001). The rest is scattered back or reflected toward the atmosphere at angles less than 90° (Ma et al., 2009). Penetration of absorbed solar radiation in water depends on water clarity. In waters with high turbidity and high concentrations of suspended solid particles, light penetration is restricted into deeper levels.

Only a small fraction (5–10%) of radiation is backscattered or reflected in clear, infinitely deep water (Bélanger et al., 2013) and this is the signal measured by remote sensors (Fig. 17.4). Water color and turbidity are the main optical properties that affect remote sensing signals. Watercolor increases light absorption in the water resulting in reduction in the remote sensing signal. Changes in water color can be detected by a camera or multispectral scanning sensor. Turbidity in water affects scattering and absorption of light resulting in interference in light being transmitted. Increase in turbidity results in increase in remote sensing signal as more light is backscattered or reflected by the particles that produce turbidity. Both water color and turbidity could not be directly and quantitatively related to water quality as many other variables such as sun elevation angle, season of the year, atmospheric concentration of water vapor and condition in the water surface such as presence of film, foam, debris, floating and submerged vegetation, depth of water, and reflectance of bottom sediments influence on both. At higher elevation and when the sun is directly overhead, more radiation reaches the water body resulting in more absorption and reflection whereas, sun angle results in less absorption and reflection from water bodies. Generally, the angle of the sun is dependent on latitude, time of the year, and the day. Therefore, remotely sensed signals from the same water body are different at different times of the year. Further, presence of light scattering elements in the atmosphere such as clouds and water vapor reduces solar radiation that hits the water surface. On the other hand, the presence of scattering elements on the water surface and in the water column such as foam, film, vegetation, suspended particles causes light to scatter in different directions.

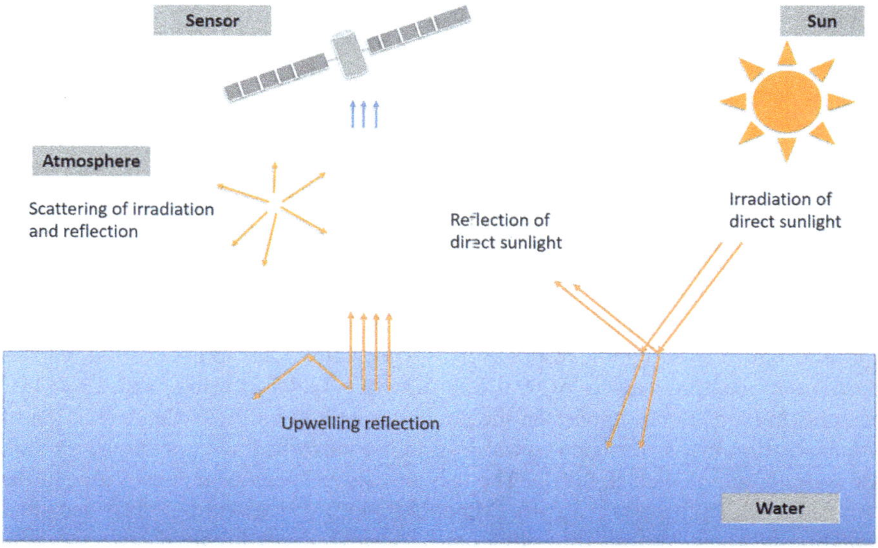

Fig. 17.4 A diagram illustrating the interaction of sunlight with the atmosphere and water. A large fraction of solar radiation that strikes on the water surface is absorbed and penetrated through the water column while the rest is scattered back or reflected toward the atmosphere and this is the signal captured by satellite sensors. (Adopted from Yang et al. (2022a) and revised to show absorption scattering)

Optical Properties of Water

When solar radiation or any other light source hits the surface of a water body, a large portion of the light is absorbed by water in the red and near-infrared regions (NIR) of the electromagnetic spectrum and interacts with suspended and dissolved matter in water (Zhang et al., 2014), while the other part is reflected off the surface. Therefore, no detectable reflectance signals can be obtained from a clear water body at wavelengths at NIR or longer (Zhang et al., 2014), while very low reflectance is given at the blue region of the spectrum. Thus, the key rationale of retrieving water quality data with remote sensing relies on inherent optical properties (IOPs). The IOPs of water depend only upon the medium and independent of the ambient light field (Mobely, 1995). The IOP of water is given by optically active components (OAC) in water which correspond to the fraction of suspended and dissolved matter. These components interact with electromagnetic radiation by absorbing, refracting, and scattering the radiation that hits on the surface of the water body. For each OAC, the measured absorbance, reflection and light scattering are specific and thus referred to as IOP (IOCCG, 2006). Reflection of light from a surface of water is determined by the refractive index of water which, in turn, is a function of salinity, temperature, and wavelength. The direction of light that passes through water is changed in the presence of suspended or dissolved matter in water resulting in scattering of light in different directions. Scattering depends on the size, shape, and color of particles. Taken together, radiation that passes through water is absorbed by

attenuation of light due to absorbance, reflection, scattering, and other processes. In addition, substances that give fluorescence are also important OAC in water of which photosynthetic pigments are the main contributors. Autofluorescence of chlorophyll and phycobili-protein pigments enable taking in-field and laboratory measurements of these pigments. Their excitation and emission wavelengths are specific to the fluorescent component or a group component (Skoog et al., 2019). Therefore, absorption and scattering properties of surface waters equals the sums of the constituents of IOP.

4.1 Optically-Active Water Quality Determinants

Physical, chemical, and biological constituents and their interactions in water contribute to the OAC in water and are classified based on their spectral water-leaving radiance which is the upwelling radiance from the water surface. Those OAC classes are pure water, chlorophyll [(Chal)-a (which also include other light-harvesting pigments)], non-algal particles, and colored dissolved organic matter (CDOM) (IOCCG, 2018).

Chlorophyll-a

Pure water typically stands as an “optical standard” and used as a calibration standard in in- situ measurements of IOPs. Chl-a represents phytoplankton and macrophyte constituents in water. Chlorophyll is the major light-harvesting pigment in phytoplankton and plants involved in photosynthesis. In addition, accessory pigments such as carotenoids and phycobilin proteins of phytoplankton also play a significant role in light harvesting and contribute to IOP in water. Chlorophyll absorb energy from sunlight at specific region in the blue and red region of the spectrum and reflect light at wavelengths in the green region while other accessory pigments also reflect light at specific regions of the spectrum such as blue region by phycocyanin and red region by phycoerythrin. However, in many laboratory and remote sensing applications, Chl-a in surface water is measured. In remote sensing, the light reflected from surface water is compared relative to the absorbance spectrum of the Chl-a.

Chlorophyll-a in Remote Sensing of Water Quality

Spectral signature of Chl-a has been widely used in remote sensing of algal blooms in freshwaters and marine waters in monitoring programs worldwide. CyAN is a multi-agency collaborative project of US-EPA which was initiated in 2015 with the aim of developing a uniform and systematic approach for identifying cyanobacterial blooms in the United States using ocean satellites (US-EPA, 2023). The ultimate goal was to develop an early warning indicator system to detect algal blooms in US freshwaters to ensure water quality and protect aquatic and human health. The project has successfully developed a public health advisory system connected with cyanobacterial incidences with health, economic, and environmental conditions using multiple public interfaces such as mobile and web-based applications, EviroAtlas,

Report on the Environment and NASA's Ocean Color Web with daily and weekly coverage. CyAN utilizes remote sensing data obtained from Sentinel-3 OLCI sensors.

More recently, a group of researchers developed a Bayesian spatial-temporal model for monitoring algal blooms with 89% overall prediction accuracy, 86% sensitivity, and 90% specificity (Schaeffer et al., 2024). This model provides seven-day advanced notice of algal blooms which may exceed WHO Alert Level 1 threshold. Thus, this prediction enables taking proactive management of algal blooms in order to ensure public health. Further to that, Environment and Climate Change Canada (ECCC) developed EOLakeWatch monitoring program to fulfil a long-felt requirement to monitor algal blooms in Canadian freshwaters using European Space Agency's OLCI (Ocean and Land Color Instrument) sensor (Binding et al., 2021).

Current research with innovative designs and cutting-edge technologies combined with new hyperspectral sensors provide more spectral, temporal, and spatial resolution and information on water quality in coming years that we never thought possible getting from observing the space. One example is NASA's PACE (Plankton, Aerosol, Cloud Ocean Ecosystem) mission which was launched in February 2024 with the aim of understanding how climate change is affecting blooms of ocean phytoplankton (https://science.nasa.gov/mission/pace/). It has spatial resolution of 1 km, one to two days global repeats, spanning 340–890 nm with a spectral resolution of 5 nm and seven short-wave infrared bands at wider spectral resolution. Data produced by PACE aims to monitor water quality, air quality, as well as respond to natural disasters such as flood, volcanic, and wildfire incidences. Another recent technical advancement in water quality monitoring is the development of Geostationary Littoral Imaging and Monitoring Radiometer (GLIMR) which is a geostationary satellite designed to monitor coastal algal blooms and oil spills (Schaeffer et al., 2023). GLIMR will be integrated on a NASA platform and launched in 2026 into a geosynchronous orbit above the Gulf of Mexico and southeastern United States and will be able to monitor for up to 15 hours a day (https://eospso.nasa.gov/missions/geosynchronous-littoral-imaging-and-monitoring-radiometer-evi-5). The instrument will measure reflectance of solar radiation at 300 nm spatial resolution, 350–890 nm spectral resolution at less than 10 nm and one near-infrared band at 1020 nm. Recently, Australia also has taken the initiative to monitor inland and coastal water quality through CSIRO's AquaWatch Australia program using satellite data and artificial intelligence to develop an advanced new system (https://www.awa.asn.au/resources/latest-news/monitoring-water-quality-from-space).

Apart from those applications, many research showed that remote sensing combined with IOPs of water can be a promising alternative to the conventional water quality monitoring which would resolve spatial-temporal issues adhered to conventional approaches (Adjovu et al., 2023; Elhag et al., 2019; Lioumbas et al., 2023; Ross et al., 2019; Topp et al., 2020).

Colored Dissolved Organic Matter (CDOM)

Colored dissolved organic matter (CDOM) includes dissolved organic carbon (DOC) and total organic carbon (TOC) in water. Decomposing organic matter that originate from aquatic plants and phytoplankton and that originate outside the water body such as terrestrial soil and plants are the major contributors to the DOC (Kritzberg et al., 2004). Total organic carbon in water represents a wide range of organic compounds such as carbohydrates, amino acids, fatty acids, and high molecular weight humic acids and fulvic acids. These humic substances contribute significantly to the color of natural water due to the presence of unsaturated and aromatic bonds. They absorb solar radiation in the UV and visible spectrum. In addition, there is fluorescent dissolved organic matter in water (fDOM). Both absorption and fluorescence are considered in quantification of dissolved organic matter in water. In remote sensing applications, CDOM is measured as surface reflectance while fDOM is measured at specific wavelengths. Mostly in commercial sensors, fDOM is measured to estimate organic matter content in water because of the higher sensitivity of the measurement than measuring absorbance (Tiquia, 2008; Huguet et al., 2009).

Colored Dissolved Organic Matter in Remote Sensing of Water Quality

The second most widely used OAC in water for remote sensing of water quality is CDOM. CDOM absorbs solar radiation at UV and visible wavelengths and a part of the absorbed radiation is emitted as blue color fluorescence. CDOM affects the productivity of the water body by blocking UV radiation in the upper part of the water column while blocking visible radiation at depth. Although CDOM is a promising OAC in water, remote sensing of CDOM for quality determination is still challenging due to segregation of spectral signatures of Chl-a and suspended particulate matter with CDOM. However, remote sensing of CDOM is a choice in large-scale water quality monitoring programs because it is more practicable than intensive laboratory analyses. Initially, Moderate Resolution Imaging Spectroradiometer (MODIS) on Terra and Aqua satellites was widely used to monitor CDOM in water because of the given daily monitoring, signature wavelengths and adequate spatiotemporal resolution. More recently, Landsat-8 and 9 and Sentinel- 2 data are trending to use in remote sensing applications due to increased spectral, spatial, and temporal resolutions. Presently, in Minnesota in the United States, a statewide water quality database is in operation including the measurements of water clarity, chlorophyll, and CDOM. It can be accessed through Minnesota LakeBrowser (https://lakes.rs.umn.edu/).

Non-algal Particles

Non-algal fraction of OAC constitute suspended particulate matter in the range of 0.2–07 mm. The major contributors include, organic debris, phytoplankton (non-pigmented components), other aquatic microflora such as bacteria and zooplankton and inorganic particles that come from the terrestrial surrounding. These suspended particulate matters absorb or scatter downwelling light and lead to increased reflectance across the visible and NIR spectrum. As a result, transparency through the water column is reduced (Dörnhöfer & Oppelt, 2016).

Thus, OAC in water determines the IOP in that water body in a particular time frame and influences the attributes of water leaving radiance which is then detected by a sensor mounted on a remote sensing platform (Fig. 17.4). However, although OACs provide promising directives in water quality monitoring programs with remote sensing, applications of remote sensing in water quality assessments have certain limitations since several important water quality determinants are non-OACs.

4.2 Non-optically Active Water Quality Determinants

These are the components that do not absorb light and thus could not be directly measured by spectral methods. Such important non-OAC water quality determinants include major nutrients; nitrogen and phosphorus and some other parameters of water such as conductivity, DO, total suspended solids (TSS), Secchi disc depth (SDD), and pH. Initial efforts of remote sensing of these water quality parameters were based on obtaining surface reflectance data and establishment of regression models with field data to derive a relationship between remotely sensed and field or laboratory measured data. In this context, single bands, band combinations or their ratios of remotely obtained spectrum were regressed with measured data to obtain a statistically best fitting linear or non-linear regression model to estimate a specific water quality parameter with significant prediction accuracy. There are numerous attempts that used remote sensing to develop models to estimate water quality. One such early effort estimated several water quality parameters such as Chl-a, total P, SDD, TSS, salinity, and temperature in the Norfolk Broads in north-east London using Landsat TM data (Baban, 1993). Later, total N and P concentrations in Taihu Lake in China was estimated also utilizing Landsat TM imagery (Chen & Quan, 2011) and recently total P in Yangtze-Huaihe Lake in China is estimated using the satellite imagery from MODIS (Wang et al., 2023).

Of the non-optically active water quality parameters, N and P are the most remotely sensed quality parameters because they are the two major nutrients that drive phytoplankton productivity in aquatic environments (McKindles & Tiquia-Arashiro, 2012). Waters enriched with N and P cause eutrophication and form algal blooms. Eutrophication can be considered as a state of nutrient pollution in aquatic environments. On the other hand, algal blooms downgrade water quality in many ways such as adding unpleasant taste, odor and color to the water and also deplete DO. Hence, N and P levels in water are essential quality parameters and essential components in water quality assessment. Although both N and P do not directly impact the visible spectrum of the water body, indirectly both affect the optically active Chl-a, CDOM, and TSS. This is an indirect way of establishing the relationship between the data obtained by remote sensing and field observation. First, an algorithm is developed to estimate selected OAC concentration based on remote sensing of surface reflectance. Then, an empirical relationship between OAC and non-OAC quality parameters such as N or P concentration is established. Previous studies were able to establish relationships between total N or P and OAC in water

primarily using individual bands or band ratios or their natural logarithm and statistical regressions based on field N or P data and spectral reflectance data (Mortula et al., 2020; Wang et al., 2023; Xiong et al., 2019; Yao & Hui, 2013; Zhang et al., 2022). Although this approach works well with small-scale water bodies, when the water body is large, those simple correlations could not accurately capture spatial variability of the water quality. There can be concentration gradients of N and P in different locations of the water body. Such natural variations can be minimized by the application of multivariate statistical modeling-based segmentation method to segment larger water bodies to represent spatial heterogeneity of nutrient concentrations. In a study by Gao et al. (2015), spectral data of water surface from Sentinel-2 and 3 to were first integrated to improve spatial coverage and time window. Then, spectral and in- situ measured data were fit into an inversion model obtained through a machine learning algorithm for remote sensing. Finally, N and P were estimated with considerable accuracy. Recently, Wang et al. (2023) estimated total P concentration in large water bodies with reasonable accuracy ($R^2 = 0.53$). For the model development they have used a combination of field data and spectral data retrieved from Sentinel-3A powered by advanced machine learning algorithms. More recently, several non-OAC of water including DO, total-N, and total-P were estimated using the machine learning methods (Gao et al., 2024). However, there are still limitations in remote sensing of non-OAC for water quality determination such as requirement of large sets of measured data, high spatial resolution, and poor estimation associated with machine learning.

5 Remote Sensing Platforms for Freshwater Quality Monitoring

Remote sensing technology offers effective ways to observe surface water dynamics such as water extent and water quality. Remote sensing is much more efficient compared to traditional in- situ measurements because of its ability to continuously monitor at multiple scales. Remote sensing can provide data to monitor water quality at regular and frequent time scales at regional and global scale with sufficient spatial and temporal coverage. Presently, satellite-based observation data are the most widely used remote sensing data in water monitoring compared to non-satellite-based data which is the focus of this chapter.

5.1 Satellite-Based Remote Sensing Data

Remote sensing of water quality offers uniquely advantageous options over traditional in- situ monitoring such as extensive spatial and temporal coverage, cost effectiveness, and less labor requirement. Sensors that are onboard on various Earth

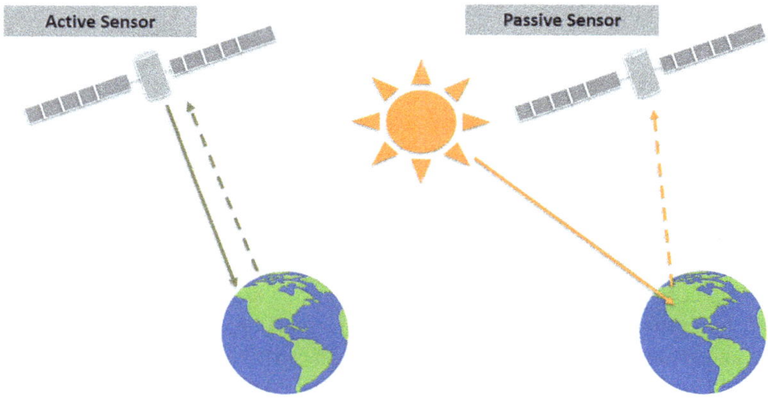

Fig. 17.5 A diagram comparing active and passive sensors used in surface water monitoring. On the left, an active sensor satellite emits a signal towards Earth and receives the reflected signal. On the right, a passive sensor satellite detects sunlight reflected from Earth. The active sensor operates independently of sunlight, while the passive sensor relies on sunlight

observation (EO) satellites measure Earth surface reflectance and capture a large volume of data from the space at regular and frequent time scale. These are the remotely sensed data used in various water and land monitoring applications. Basically, there are two types of sensors that can provide data for surface water monitoring; active sensors and passive sensors (Fig. 17.5). Active remote sensors provide their own energy source for illumination. Most active sensors operate in the microwave portion of the electromagnetic spectrum. Passive sensors measure radiant energy that is reflected or emitted by the Earth atmosphere or from the Earth.

Microwave sensors operate at long wavelength regions of the electromagnetic spectrum range from approximately 1–1 m. Compared to visible and infrared radiation, microwaves have special properties that are important for remote sensing. They have the ability to penetrate cloud cover, haze, dust, and vegetation cover which is not possible with shorter optical wavelengths. Further, microwave sensors operate independent of solar radiation. Thus, they can work day and night under any weather condition. Active microwave sensors transmit microwave signals toward the target on the Earth surface and record the back scattered or reflected portion of the signal which depends on the angle of the signal that strikes on the target and properties of the target. Most widely utilized sensors include Synthetic Aperture Radar (SAR) and Light Detection and Ranging (LIDAR). Active microwave sensors are further categorized as imaging and non-imaging. The most widely used imaging active microwave sensor is radio detection and ranging (RADAR). The sensor transmits a microwave signal towards the target and detects the backscattered fraction of the signal from which 2-D images are created. The strength of the signal differs between different targets and distance to the target. Non-imaging microwave sensors include altimeters and scatterometers that take measurements in one linear dimension.

Passive microwave sensors are on the other hand optical sensors that depend on the energy of solar radiation and detect naturally emitted energy within the field view. These sensors are referred to as microwave radiometers. Depending on the object's physical properties, such as atomic composition and crystalline structure, emitted microwave energy changes allow discrimination between objects. Although passive microwave remote sensing is advantageous over optical sensors as microwave energy can penetrate through atmospheric interferences like cloud cover and rain and also can operate at day and night, a major disadvantage is that the emitted energy level by the Earth is low. Also, spatial resolution is generally low. Thus, not suitable to track small-scale changes.

Microwave remote sensing data are mainly used in the determination of atmospheric and soil water content, sea surface temperature, salinity, and wind speed. Recently, microwave remote sensing is starting to gain more applications in water quality monitoring by estimating water temperature, salinity, surface roughness as indicators of water quality (Shareef et al., 2016; Schroeder et al., 2015; Tiquia-Arashiro, 2012). In addition, microwave remote sensing has been shown to detect pollutants such as oil spills and algal blooms with high resolution at any weather (Wang et al., 2010; Saleem et al., 2024). Researchers have shown direct estimation of water quality parameters like salinity, electrical conductivity and TDS with microwave remote sensing (Shareef et al., 2016) which otherwise could not be directly measured with optical remote sensors since these parameters are non-optically active. In this study, they have used thermal images as passive microwave remote sensing to extract temperature data for specific locations in the ground. Then in an inversion process, they obtained the reflectivity from backscattering coefficients which were calculated from active microwave sensing of SAR. This method allows calculation of water quality parameters directly from different remote sensing data with high accuracy. Some studies have shown that integration of microwave data with optical and in situ measurements improve the accuracy of water quality assessments, providing a holistic view of aquatic ecosystems and their responses to environmental changes (Zhang et al., 2022; Mohammadi et al., 2021; Saleem et al., 2024). Such an integrated approach was shown to be successfully used in the estimation of chlorophyll-a concentration in turbid water of Lake Taihu, China which was highly challenging with optical remote sensors alone due to complex optical conditions in turbid waters (Zhang et al., 2018).

Satellite-based optical sensors are the most widely utilized sensors for monitoring surface water. According to the Union of Concerned Scientists (UCS), there were 7560 active EO satellites that collected imagery and measurements of the Earth surface at the beginning of 2024. The number of EO satellites is continuing to increase, and yet the majority are optical imaging satellites. There were 480 optical imaging satellites in operation in 2023. Onboard sensors have different spatial, spectral, and temporal resolutions (Table 17.3). Landsat-8 and 9, Sentinel-2A and 2B, Sentinel-3A and 3B, Terra & Aqua, EnviSAT, and Worldview are the top 5 satellites used in remote sensing since the past to date. Each sensor has its own spectral, spatial, and temporal resolution levels which altogether determines overall sensor sensitivity of remote sensing.

Table 17.3 Commonly used satellite-based sensors for surface water quality monitoring.

Satellite	Sensor	Overall sensor resolution	Number of spectral bands	Spatial resolution (m)	Temporal resolution (days)	Data availability
NOAA	AVHRR (Advanced Very High-Resolution Radiometer)	Coarse	5	1100	0.5	Since 1978
Terra & Aqua	MODIS (Moderate resolution Imaging Spectroradiometer)	Coarse	36	250–1000	0.5	Since 1999
EnviSAT	MERIS (Medium-Resolution Imaging Spectrometer)	Coarse	15	300	3	2002–2012
Sentinel-3	OLIC (Ocean and Land Color Instrument), SLSTR (Sea and Land Surface Temperature Radiometer), SRAL (Synthetic Aperture Radar Altimeter), DORIS (Doppler Orbitography and Radiopositioning Integrated by Satellite), MWR (MicroWave Radiometer), LRR (Laser retroreflector), GNSS (Global Navigation Satellite System)	Coarse	21	300	2	Since 2016
Landsat-7 and 8	OLI (Operational Land Imager)	Medium	4–11	15–80	16	Since 1972
Sentinel-2	MSI (Multi Spectral Imager)	Medium	13	10–60	5	Since 2015
Worldview		High	4–17	0.31–2.4	1–4	Since 2007

Spectral resolution refers to the ability cf a satellite sensor to measure specific wavelengths of the electromagnetic spectrum. The finer the spectral resolution, the narrower the wavelength range for a particular channel or band. High spectral resolution facilitates fine discrimination between different targets based on their spectral response in each of the narrow bands. Based on the spectral resolution, two types of image data can be obtained from satellites: multispectral and hyperspectral imagery. Multispectral imagery is the most common and widely used image data in remote sensing. Multispectral imagery typically contains 3–10 spectral bands with wide bandwidth and descriptive band names whereas hyperspectral imagery contains hundreds or thousands of bands with much narrow bandwidths of 10–20 nm spectral resolution. Therefore, the main difference between the two is the number and

the spectra of electromagnetic radiation that each band contains. For example, Landsat-8 contains a multispectral sensor with 11 bands which produces 11 images from a single observation. The majority of these bands have a spatial resolution of 30 m. The spatial resolution defines the surface covered by a pixel or a cell. The other multispectral sensors used in water quality monitoring include Sentinel, EnviSAT, National Oceanic, and Atmospheric Administration Satellite (NOAA), Mobile Satellite Service (MSS), Thematic Mapper (TM), and Operational Land Imager (OLI). Due to low spectral resolution of multispectral imagery, empirical methods are mostly used when developing water quality prediction models for specific periods and water areas (Matthews, 2011). On the other hand, hyperspectral systems have very high spectral resolution because of their narrow bandwidths. Since the reflectance is measured over several small wavelength ranges, images are built with a continuous spectrum of the radiation detected for each pixel. This facilitates fine differentiation between targets. Therefore, the most reliable bands that could accurately predict water quality can be selected more precisely. Some of the present hyperspectral satellite missions include NASA's PACE (Plankton, Aerosol, Cloud and ocean Ecosystem) and Hyperion, Germany's Environmental Mapping and Analysis Program (EnMap), Italy's PRecursore IperSpettrale della Missione Applicativa (PRISMA), Japan's HISUI, and China's HJ-1. However, this increased sensitivity comes from a significant increase in the volume of data collected making storage and manipulation of data difficult.

Spatial resolution is another parameter that determines the overall sensitivity of remote sensing. The spatial resolution can be defined as the surface covered by a pixel which is by simply the size of one pixel, the smallest spot visible to the sensor. Spatial resolution of 10 m means that one pixel represents an area of 10 m × 10 m on the ground. Resolution depends upon satellite orbit configuration and sensor design. Thus, different sensors have different resolutions. Generally, the higher the spatial resolution, the less the area covered by a single image. Therefore, an appropriate sensor with required spatial resolution should be selected according to the remote sensing objectives. Since the remote sensing technology is constantly developing, categorization of sensors into low, medium, and high spatial resolution is constantly changing. For example, in the 1980s, a resolution of 60 m pixel of Landsat was considered as high, but today it is considered very low.

Temporal resolution is the time it takes for a satellite to complete one orbit cycle which is also called "revisit time". It depends on satellite/sensor capabilities, swath (the area imaged) overlap, and latitude. Some satellites have greater temporal resolution because they can maneuver their sensors. Revisit time of the satellite is crucial when the monitoring program is planned to collect corresponding in- situ data, as sampling should be done near to the time of the satellite overpass. On the other hand, for monitoring programs, frequent revisiting satellites could provide frequent data that are useful in identifying recent changes in water quality. In reality, no single sensor is available having high spectral, spatial, and temporal resolution. Therefore, the choice depends on remote sensing objectives.

5.2 *Non-satellite Remote Sensing Data*

Water quality can also be monitored remotely without satellite data but with other types of remote sensing data such as that obtained from unmanned aerial vehicle (UAV) systems and ground measured spectral data (Yang et al., 2022a). The UAVs equipped with multi-spectral cameras, spectroradiometers, and sensors have been used since the past few decades in water monitoring programs. NASA's Airborne Visible InfraRed Imaging Spectrometer (AVIRIS) is a unique optical sensor which takes images of upwelling spectral radiance in 224 spectral bands. AVIRIS is mounted on four aircraft platforms and flies across the US, Canada and Europe capturing images for various purposes including coastal and inland waters. Compact Airborne Spectrographic Imager (CASI) of NASA is a hyperspectral imager which has been mounted on light aircrafts. It has 48 bands and is designed for a variety of applications including water quality monitoring. Ground-based remote sensing is another way of obtaining real-time and continuous data for water quality programs over the last few years (Liew et al., 2019; Sun et al., 2022). Sensors are fixed on pre-selected sampling points above the ground with a high-resolution camera to acquire images of the water surface. Compared to satellite-based water quality monitoring, these methods have the limitations of high operational cost for flying flights and maintaining ground-based sampling points and limited observation area. However, possibility in obtaining high spectral and spatial resolution and less atmospheric effect on data are the advantageous features of non-satellite based remote sensing data. Due to the complementarity of satellite and UAV data, the concept of synergy between the two data are in discussion as an emerging trend to overcome existing limitations of both remote sensing methods (Alvarez-Vanhard et al., 2021). For instance, UAV data can be useful in analyzing and calibrating satellite data.

6 Water Quality Retrieval from Remote Sensing Data

Generally, water quality information is retrieved by relating water quality parameters to remotely sensed fraction of incoming solar radiation that is reflected from Earth's surface (surface reflectance) by empirical, semi-empirical, analytical or artificial intelligence-based approaches. What is measured by remote sensing is a function of the changes in the spectral signature of the water surface reflectance. This spectral signature is related to water quality by measuring changes in empirical or analytical models.

Empirical Method
Empirical method is purely a statistical approach where the relationship between ground-measured water quality parameters and surface reflectance are established by correlation and regression analyses without having prior knowledge. These models are simple and easy to operate and more precise estimations can be made through refining the most optimal band or band combination that is better regressed with

ground data. However, since IOP of water is affected by many components in water, accurate estimation of a single or a few water quality parameters by simple models is challenging. On the other hand, empirical models are hard to generalize due to regional and temporal variations, a model developed for one region or season may not be applicable in another region or season. Further, empirical modelling requires a large amount of ground data for more accurate estimation.

Analytical Method

In this method, physical relationships between water quality parameters, underwater light field, and the surface reflectance are derived. First, bio-optical and transmission models are derived to simulate the propagation of light in the water and then the relationship between water quality parameter and reflection is established.

Semi-empirical Method

This method is a combination of empirical and analytical methods for retrieving water quality. In this method, the spectral characters of water quality parameters are known. Therefore, water quality can be estimated with selected quality parameters with only a small set of ground and water surface reflectance data. This is the most widely used method in monitoring water quality (Wang & Yang, 2019).

Artificial Intelligence (AI) in Water Quality Modeling

Satellite remote sensing technologies generate large volumes and a variety of geospatial data that are difficult to handle manually during data analysis and interpretation in many fields including water quality monitoring. Therefore, cluster-based high-performance computing systems are necessary for handling such data. As a result, AI techniques are nowadays becoming a popular tool in analyzing satellite image data in water quality monitoring because AI models are capable of processing large volumes of data quickly and accurately. AI allows identification of trends and patterns in water quality data that may be difficult to track with other data analyzing techniques and thereby improve water quality monitoring (Kadam et al., 2019; Tinelli & Juran, 2019; Rana et al., 2023; Isık & Akkan, 2024; Palabıyık & Akkan, 2024).

Scientists have used many different AI models to estimate water quality parameters such as, Convolutional neural network (CNN), Artificial Neural Network (ANN), Dense Neural Network (DNN), M5 Model Tree (MT), Multivariate Adaptive Regression Spline (MARS), Multilayer Perceptrons (MLP) Gene Expression Programming (GEP), Evolutionary Polynomial Regression (EPR), Cubist Regression Model (CRM), Support Vector Machine (SVR), and Long Short-Term Memory (LSTM) model. Yang et al. (2022b) nicely summarized chronology of used remote sensing data types and AI models used in water quality monitoring. Each AI model has different challenges and limitations in water quality monitoring from satellite imagery (Yang et al., 2022b). Over the past few years, LMST, ANN, MLP, and DNN are the most widely used AI models in water quality monitoring. LMST has been shown to give a more accurate prediction of algal blooms over the other models (Lee & Lee, 2018).

Landsat, MODIS, and Sentinel are the mostly used satellite image data with these models. AI models are used to develop the most reliable and accurate relationships between satellite-based remote sensing data and ground measured water quality data. Recently, AI was used to establish conceptual relationships between properties of spectral bands of Landsat-8 OLI-TIRS images and water quality index with AI without using optical or non-optically active water quality parameters (Najafzadeh & Basirian, 2023). As with the conventional approach, they first developed multiple linear regression models between water quality parameters and spectral bands of Landsat-8 OLI-TIRS images. Then the most correlated spectral bands were selected as input variables of AI models. Finally, AI estimated water quality index and actual water quality index calculated from measured data were compared. They evaluated the efficiency of AI models in the training and testing phases by using index of agreement (IOA), root mean square error (RMSE), mean absolute error (MAE), and scatter index (SI). IOA is a standardized measure of the degree of numerical model estimation error which varies between 0 and 1 (Willmot, 1981). Here, +1 is considered as the highest IOA and that indicates the best performance of the AI model while 0 is the lowest level. *RMSE*, MAE, and SI were the error function values, ranging from 0 to $+\infty$. Among the tested models, MARS model performed best with high accuracy (IOA = 0.99) in estimating water quality index. Although accurate, in our opinion AI may have limited applicability in resolving real-time problems since the process of algorithm development is highly mathematical. However, continued advances in technology and the development of AI techniques are likely to improve the accuracy and accessibility of remote sensing data for water quality monitoring in the future.

7 Workflow for Monitoring and Assessment of Water Quality Using Remote Sensing

A generalized work flow for monitoring and assessment of water quality using remote sensing applications is presented in Fig. 17.6. Identification of objectives is the most important step in designing a water quality monitoring program which is then followed by method development. The method going to apply in the monitoring program depends on available resources such as funds, infrastructure, laboratory facilities, accessibility to remote sensing data, and human resources. Then proper planning for in- situ and remote sensing data collection is required. Identification of most appropriate parameters to determine water quality has a significant influence on final estimation of water quality. Optically active parameters such as chlorophyll, CDOM and TSS have direct relationship with sensor detected surface reflectance while optically inactive parameters like N and P levels, pH, chemical oxygen demand (COD) have no direct relationship with surface reflectance. However, their relationship with optically active parameters allows their applications in water quality monitoring. These relationships are discussed in Sect.

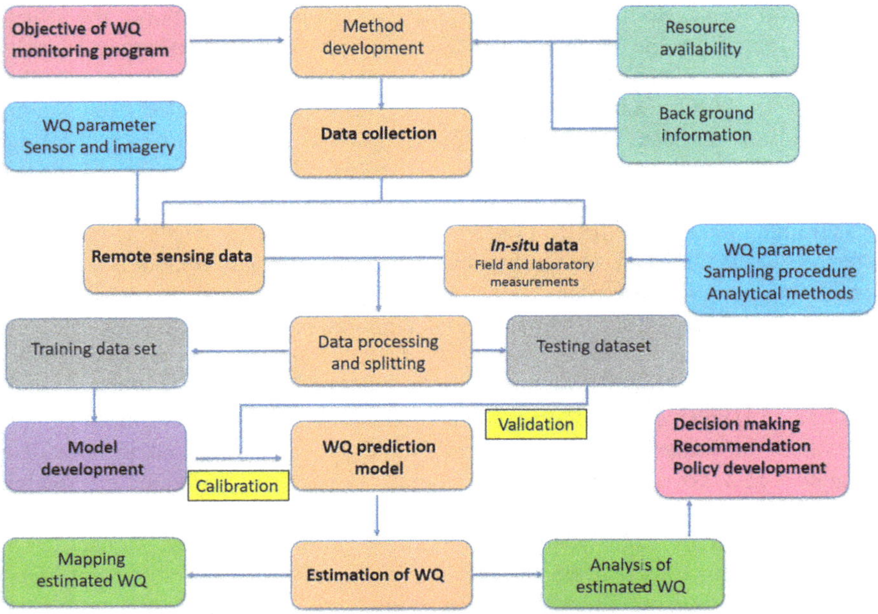

Fig. 17.6 A flowchart illustrating the process of water quality (WQ) monitoring with remote sensing

4.2 in this chapter. Parameters like chlorophyll, N, and P are more sensitive in estimating algal blooms and trophic status while CDOM, TSS, N and P levels, and COD can better reflect water pollution. Further, CDOM, TSS, and turbidity can be used to indicate the level of surface run-off. Thus, the choice of parameter/s depends on the objective/s of the monitoring program.

In- situ data of selected quality parameters can be obtained from field measurements and laboratory analyses while remote sensing data can be obtained from satellites or airborne vehicles. Required sensor and imagery type can be selected based on required spatial and temporal resolutions. With current satellite sensors, one has to make a tradeoff between high spatial and temporal resolution as both do not perfectly align. For example, the MSI sensor of Sentinel-2 has medium spatial resolution while it has fairly good temporal resolution (Table 17.3). Raw image data is received with pixel values or in other words digital numbers (DNs) which correlate with the energy that is measured by the sensor. The DN is affected by the light source used to illuminate the Earth surface, sensor, atmosphere, and surface material. During data preprocessing DNs are converted to radiance values which still have the effects of light source, atmosphere, and surface material. Therefore, for quantitative analyses it is essential to remove the effects of light source and atmosphere. When light passes through the atmosphere, part of it would not reach the Earth surface due to the scattering of light when hit on the particles in the atmosphere. In parallel, part of the reflected light from the Earth surface also scatters in the atmosphere. In addition, radiance of the light source is also affected by materials

in the atmosphere. Both of these atmosphere effects may superimpose the reflectance coming from the Earth surface. Thus, preprocessing of raw image data is essential to remove atmosphere effects which is done though atmosphere algorithms such as Dark Object Subtraction (DOS). Then atmosphere corrected data is combined with in- situ data which is then split into two. One set of data is used for the model development while the other part is used to validate the developed model. The set of in- situ and remotely sensed data used in model development is referred to as training data while the data used in the validation of the developed model is referred to as testing data. During model development regression models are developed between in- situ water quality parameters and remotely sensed surface reflectance values of various spectral bands and combinations of bands to select the best performing model.

The models developed by remote sense data need to be properly calibrated using in- situ data to enhance the accuracy of the model. Then the model estimated numerical values of water quality parameters should be compared with in- situ data through a regression analysis. This comparison gives validation to the model. During calibration and validation, the model performance is always evaluated for its prediction accuracy using statistical matrices such as Root Mean Square Error (RMSE), Percent Bias (PBIAS), Mean Absolute Error (MAE), and Percent Mean Absolute Relative Error (PMARE). Then, based on the estimated quality parameter, decisions, and recommendations can be made on the water quality of a particular water body by comparing that with established country or region specific WQIs or guidelines. Thereby, much insightful information can be derived such as status of trophic level that may lead to the development of harmful algal blooms, time of issuing harmful algal bloom alert, status of water pollution, and identifying sources and hotspots of contaminations. In addition, trends in water quality changes can also be identified. Then appropriate short- and long-term water quality management strategies can be designed, policies can be made, and research priorities can be identified.

8 Limitations of Remote Sensing in Water Quality Monitoring

Like with any other technologies, remote sensing also has certain limitations when applied in water quality monitoring programs. One of the main problems is that some water quality parameters cannot be directly measured with optical sensors as they are optically inactive. Thus, important quality parameters like total N, NH_3-N, total P, TDS, and pH cannot be directly estimated by remote sensing. Therefore, most of the direct remote sensing measurements are limited to few optically active water quality parameters such as Chl-a, CDOM, and TSS while the other parameters are estimated by establishing correlation with one or more optically active quality parameters. Optically active parameters can be sensed directly from satellite

observations since they absorb light in the solar electromagnetic spectrum and influence optical properties of water. Optically inactive parameters like pH, TDS, N, and P have a low signal-to noise ratio. The relationship of those parameters with optically active parameters are not always linear for instance TDS. Another limitation with remote sensing is high dependency on the resolution of the sensor of choice which sometimes may not fulfill the requirement of the monitoring program. For instance, the revisit time of the selected sensor may be too long to detect trends in temporal variations of water quality in some programs.

Lack of sufficient volume of ground data due to various reasons such as absence of past records, high operational cost for collection and lack of technical expertise for field and laboratory measurements is another problem with remote sensing of water quality. Since ground data is required for the calibration and validation of the remote sensing model, sufficient ground data may affect accuracy of the model and the water quality estimation made from the model.

Errors that can happen during satellite image acquisition is another limitation of remote sensing applications. The signals that reach the sensor can be affected by the atmospheric conditions such as cloud cover, fog, and haze. These interferences may restrict the optical signals sensed from the water surface and quantitatively affect the incoming radiance. Due to the images acquired in the visible and near-infrared regions of the electromagnetic spectrum of the target waterbody may blur or unclear (McCarthy et al., 2017; Yang et al., 2022a). Both image-based such as Top of Atmosphere (ToA), Dark Object Subtraction (DOS) and Cosine of the Sun Zenith Angle (COST) and physical models such as Atmospheric Correction for Flat Terrin (ATCOR), Fast Line-of-sight Atmospheric Analysis of Spectral Hypercubes (FLAASH) and ACOLITE are applied for atmosphere corrections. A comparison between the two models showed image-based models outperformed the physical models in terms of accuracy (El Alem et al., 2021). However, erroneous image processing for atmospheric corrections and pan-sharpening can lead to false estimation of water quality. Satellite images acquired by sensors with coarse spatial resolution may overlap with adjacent pixels in the neighboring land or vegetation surfaces leading to the detection of spectral signatures outside the waterbody. Sometimes, during image processing, actual intensity of the reflectance is not measured by the sensor when there is high reflectance in the near-infrared region of the spectrum by ocean color algorithms in situations like thick cyanobacterial blooms due to high reflectance in the near-infrared range of the electromagnetic spectrum which is not anticipated by ocean color algorithms. Thus, the surface reflectance is underestimated by the sensor. Pan-sharpening is a method of obtaining high-resolution images by enhancing spatial resolution artificially (Li et al., 2017). However, this alters radiometric and spectral characters of the satellite image. Therefore, although pan-sharpening gives better visualization of the image, it is not recommended to use such artifact image data for analytical purposes such as calibration and validation of a model (USDA, 2021). Hence, satellite image processing can lead to false estimation of water quality.

9 Conclusion and Future Perspectives

Access to water is one of the basic human rights. Usable form of water is mostly limited to freshwater. Thus, freshwater environments are among the most vulnerable ecosystems on Earth due to high exploitation. Demand for freshwater resources is expected to increase in the future to meet the demand of increasing population. In parallel, freshwater resources and connected ecosystems are at high risk. In addition, climate change directly and indirectly makes significant quantitative and qualitative changes in freshwater ecosystems leading to water scarcity, increased salinity, deterioration of water quality, and pollution. Degraded water quality risks human health, limits food production and economic growth, and threatens biodiversity. Water quality is determined by monitoring a set of physical, chemical, and biological parameters. Water quality assessment provides baseline information for water safety. Thus, its management is among the top priorities in aquatic ecosystem management in many countries. Since water quality changes with time and other factors, continuous monitoring with well-designed programs is required to identify trends in water quality changes and emerging threats. Accurate monitoring provides opportunities for taking proactive measures to improve water quality, responding to emerging pollution sources, designing future plans for water quality management, implementation of pollution control actions and development of guidelines, regulations, or policies to ensure water safety, etc. Suitability of water for a particular purpose is decided according to the local or global water quality standards such as those of WHO and water quality indices.

In- situ data collection is the major problem allied with these conventional water quality programs due to high operational costs for instruments, labor, time, and laboratory analyses. In addition, accurate representation of the waterbody with sufficient number of samples with sufficient temporal and spatial coverage adds more to the cost. Therefore, many countries do not have continuous monitoring programs that collect data on a daily basis. With these limitations connected to in- situ ground-based observations, remote sensing provides a worthwhile solution for large-scale monitoring of water bodies with high spatial and temporal coverage. Presently, both satellite-based and non-satellite-based platforms such as unmanned aerial vehicles are used in water quality monitoring. Although many different physical, chemical, and biological parameters collectively determine the quality of water, not all such parameters can be sensed directly by remote sensing. However, parameters that can be sensed remotely can provide optimal estimation of the status of water to assist quality determination such as chlorophyll-a, CDOM, TSS, and turbidity. Other important non-optically active parameters like pH, total N and P, COD, and TDS can be correlated with optically active parameters in water quality determination. Free accessibility to satellite image data given by the majority of satellite missions reduces the cost for monitoring programs compared to conventional operational costs. New satellite missions added each year with technologically advanced sensors onboard provide more spatial resolution for fine mapping and frequent revisit time for retrieving data at frequent time intervals. Remote sensing also allows retrieving water quality data from physically inaccessible locations.

There are several water quality monitoring programs that successfully use remote sensing in monitoring programs at global, national, or provincial levels such as CyAN of US-EPA, EOLakeWatch of Canada and Minnesota LakeBrowser for identifying cyanobacterial blooms in freshwaters and NASA's PACE for understanding the effect of climate change on ocean algal blooms. Although remote sensing is a promising tool for monitoring water quality, accuracy of data compared to ground data is questionable at some instances. However, reasonably accurate and comparable results to ground data could be obtained when the underlying limitations are properly addressed such as with appropriate atmospheric corrections and image processing methods. Obtaining fine spatial and temporal resolution is critical in water quality monitoring. Recent satellites such as Sentinel-3 provide significant temporal resolution to data by having two-day revisit time. However, obtaining fine spatial resolution together with high temporal resolution is not possible with Sentinel-3 and many other current sensors. Therefore, integration of satellite data with data obtained by UAV airborne sensors may add fine spatial resolution and permit better estimation of water quality. Data manipulation and interpretation in remote sensing is one of the issues due to the generation of enormous satellite image data. In here cloud-based data manipulation and storage and AI come as promising problem solvers. AI not only facilitates analysis and interpretation of big data but also allows real-time monitoring. CSIRO's AquaWatch is such an initiative taken by Australia to monitor inland and coastal water quality through integration of satellite imaging and AI. Likewise, recent technological advances are trying to solve the issues allied with remote sensing of water quality. The role of remote sensing should not be limited to water quality monitoring but it has a much broader role in environmental management to ensure public health and safety and maintaining biodiversity by replacing expensive and time consuming conventional strategic methodologies. However, continuous advancement in remote sensing and associated technologies such as UAVs and AI is essential in this context to solving real-world problems in real-time. Thus, we highlight the importance of continued investments in remote sensing technology development and training in data analysis techniques to broaden the applications of remote sensing.

Authors' Contributions Conceptualization: Widana Gamage Shirani Manel Kumari; Original draft preparation: Widana Gamage Shirani Manel Kumari, Keerthi Sri Senarathna Atapaththu, Kanaji Masakorala, Kelum Sanjaya and Yapa Mudiyanselage Ajith Lalith Weerasinghe Yapa; Reviewing and editing: Widana Gamage Shirani Manel Kumari; preparation of revised manuscript: Widana Gamage Shirani Manel Kumari and Keerthi Sri Senarathna Atapaththu.

Conflict of Interest Authors declare no conflict of interest.

References

Abbasi, T., & Abbasi, S. A. (2012). Water-quality indices: Looking back, looking ahead. In T. Abbasi & S. A. Abbasi (Eds.), *Water quality indices* (pp. 353–356). Elsevier. https://doi.org/10.1016/B978-0-444-54304-2.00016-6

Adjovu, G. E., Stephen, H., James, D., & Ahmad, S. (2023). Overview of the application of remote sensing in effective monitoring of water quality parameters. *Remote Sensing, 15*(7), 1938.
Ahmed, U., Mumtaz, R., Anwar, H., Mumtaz, S., & Qamar, A. M. (2020). Water quality monitoring: From conventional to emerging technologies. *Water Supply, 20*(1), 28–45. https://doi.org/10.2166/ws.2019.144
Alvarez-Vanhard, E., Corpetti, T., & Thomas, H. (2021). UAV & Satellite synergies for optical remote sensing applications: A literature review. *Science of Remote Sensing., 3*, 100019. https://doi.org/10.1016/j.srs.2021.100019
Ansari, A. A., Gill, S. S., & Khan, F. A. (2010). Eutrophication: Threat to aquatic ecosystems. In A. Ansari, S. Singh Gill, G. Lanza, & W. Rast (Eds.), *Eutrophication: Causes, consequences and control*. Springer. https://doi.org/10.1007/978-90-481-9625-8_7
APHA. (2005). *Standard methods for the examination of water and wastewater*. A. P. H. Association.
Asim, M., & Nageswara Rao, K. (2021). Assessment of heavy metal pollution in Yamuna River, Delhi-NCR, using heavy metal pollution index and GIS. *Environmental Monitoring and Assessment, 193*(2), 103.
Baban, S. M. (1993). Detecting water quality parameters in the Norfolk Broads, UK, using Landsat imagery. *International Journal of Remote Sensing, 14*(7), 1247–1267.
Bélanger, S., Cizmeli, S. A., Ehn, J., Matsuoka, A., Doxaran, D., Hooker, S., & Babin, M. (2013). Light absorption and partitioning in Arctic Ocean surface waters: Impact of multiyear ice melting. *Biogeosciences, 10*, 6433–6452. https://doi.org/10.5194/bg-10-6433-2013
Bhargava, D. S. (1983). Use of water quality index for river classification and zoning of Ganga River. Environmental Pollution Series B, *Chemical and Physical, 6*(1), 51–67.
Binding, C., Pizzolato, L., & Zeng, C. (2021). EOLakeWatch; delivering a comprehensive suite of remote sensing algal bloom indices for enhanced monitoring of Canadian eutrophic lakes. *Ecological Indicators, 121*, 106999.
Bowman, N., Patel, P., Xu, W., Alsaffar, A., & Tiquia-Arashiro, S. M. (2018). Enrichment and isolation of Pb-resistant bacteria from Saint Clair River sediments and their potential for Pb removal in aqueous solutions. *Applied Microbiology and Biotechnology, 102*, 2391–2398. https://doi.org/10.1007/s00253-018-8772-4
Boyacioglu, H., & Boyacioglu, H. (2007). Surface water quality assessment by environmetric methods. *Environmental Monitoring and Assessment, 131*, 371–376.
Boyd, C. E. (1998). Pond water aeration systems. *Aquacultural Engineering, 18*(1), 9–40.
Carlson, R. E. (1977). A trophic state index for lakes. *Limnology and Oceanography, 22*(2), 361–369. https://doi.org/10.4319/lo.1977.22.2.0361
Carter, J. L., Resh, V. H., & Hannaford, M. J. (2017). Macroinvertebrates as biotic indicators of environmental quality. In F. R. Hauer & G. A. Lamberti (Eds.), *Methods in stream ecology* (pp. 293–318). Elsevier.
Chapman, D. (2021). *Water quality assessments: A guide to the use of biota, sediments and water in environmental monitoring*. CRC Press.
Chawla, I., Karthikeyan, L., & Mishra, A. K. (2020) A review of remote sensing applications for water security: Quantity, quality, and extremes. *Journal of Hydrology, 585*, 124826.
Chen, J., & Quan, W. (2011). Using Landsat/TM imagery to estimate nitrogen and phosphorus concentration in Taihu Lake, China. *IEEE Journal of Selected Topics in Applied Earth Observations and Remote Sensing, 5*(1), 273–280.
Chettri, U., Chakrabarty, T., & Joshi, S. (2022). Pollution index assessment of surface water and sediment quality with reference to heavy metals in Teesta River in Eastern Himalayan range, India. *Environmental Nanotechnology, Monitoring & Management, 18*, 100742.
Cho, J., Hyun, S., Han, J. H., Kim, S., & Shin, D. H. (2015). Historical trend in heavy metal pollution in core sediments from the Masan Bay, Korea. *Marine pollution bulletin, 95*(1), 427–432.
Dinius, S. (1987). Design of an index of water quality 1. *Journal of the American Water Resources Association, 23*(5), 833–843.

Domingues, R. B., Barbosa, A. B., Sommer, U., et al. (2011). Ammonium, nitrate and phytoplankton interactions in a freshwater tidal estuarine zone: Potential effects of cultural eutrophication. *Aquatic Sciences, 73*, 331–343. https://doi.org/10.1007/s00027-011-0180-0

Dong, Z., Li, F., Beheshti, B., Mickelson, A., Panero, M., & Anid, N. (2016). Autonomous real-time water quality sensing as an alternative to conventional monitoring to improve the detection of food, energy, and water indicators. *Journal of Environmental Studies and Sciences, 6*, 200–207. https://doi.org/10.1007/s13412-016-0383-8

Dörnhöfer, K., & Oppelt, N. (2016). Remote sensing for lake research and monitoring–Recent advances. *Ecological Indicators, 64*, 105–122.

du Plessis, A. (2022). Persistent degradation: Global water quality challenges and required actions. *One Earth, 5*(2), 129–131.

El Alem, A., Lhissou, R., Chokmani, K., & Oubennaceur, K. (2021). Remote retrieval of suspended particulate matter in inland waters: Image-based or physical atmospheric correction models? *Water, 13*(16), 2149.

Elhag, M., Gitas, I., Othman, A., Bahrawi, J., & Gikas, P. (2019). Assessment of water quality parameters using temporal remote sensing spectral reflectance in arid environments, Saudi Arabia. *Water, 11*(3), 556.

Flood, M., Frabutt, D., Floyd, D., Powers, E. U., Devol, A., & Tiquia-Arashiro, S. M. (2015). Ammonia-oxidizing bacteria and archaea in sediments of the Gulf of Mexico. *Environmental Technology, 36*, 124–135. https://doi.org/10.1080/09593330.2014.942385

Gao, Y., Gao, J., Yin, H., Liu, C., Xia, T., Wang, J., & Huang, Q. (2015). Remote sensing estimation of the total phosphorus concentration in a large lake using band combinations and regional multivariate statistical modeling techniques. *Journal of Environmental Management, 151*, 33–43.

Gao, L., Shangguan, Y., Sun, Z., Shen, Q., & Shi, Z. (2024). Estimation of non-optically active water quality parameters in Zhejiang Province Based on machine learning. *Remote Sensing, 16*(3), 514.

Gholizadeh, M. H., Melesse, A. M., & Reddi, L. (2016). A comprehensive review on water quality parameters estimation using remote sensing techniques. *Sensors, 16*(8), 1298. https://doi.org/10.3390/s16081298

Gunawardana, M. H. M. A. S. V., Sanjaya, K., Atapaththu, K. S., Yapa Mudiyanselage, A. L., Masakorala, K., & Widana Gamage, S. M. K. (2022). Quantitative prediction of toxin-producing Aphanizomenon cyanobacteria in freshwaters using Sentinel-2 satellite imagery. *Journal of Water and Health, 20*(9), 1364–1379.

Hasan, H. H., Jamil, N. R., & Aini, N. (2015). Water quality index and sediment loading analysis in Pelus River, Perak, Malaysia. *Procedia Environmental Sciences, 30*, 133–138.

Hawkes, H. A. (1998). Origin and development of the biological monitoring working party score system. *Water Research, 32*(3), 964–968.

Holden, J. (2014). Water fundamentals. In J. Holden (Ed.), *Water resources: An integrated approach* (1st ed., p. 371). Routledge Taylor and Francis Group.

Horton, R. K. (1965). An index number system for rating water quality. *Journal of Water Pollution Control Fed, 37*(3), 300–306.

House, M. (1980). A water quality index for rivers. *Water International, 5*, 16–21.

https://gemstat.org/. *Global Freshwater Quality Database*. UNEP.

Huguet, A., Vacher, L., Relexans, S., Saubusse, S., Froidefond, J.-M., & Parlanti, E. (2009). Properties of fluorescent dissolved organic matter in the Gironde Estuary. *Organic Geochemistry, 40*(6), 706–719.

Icaga, Y. (2007). Fuzzy evaluation of water quality classification. *Ecological Indicators, 7*(3), 710–718.

IOCCG. (2006). *Remote sensing of inherent optical properties: Fundamentals, tests of algorithms,and applications*.

IOCCG. (2018). *Earth observations in support of global water quality monitoring.*

Isık, H., & Akkan, T. (2024). Water quality assessment with artificial neural network models: Performance comparison between SMN, MLP and PS-ANN methodologies. *Arabian Journal for Science and Engineering, 50*, 369. https://doi.org/10.1007/s13369-024-09238-5

Ji, X., Dahlgren, R. A., & Zhang, M. (2016). Comparison of seven water quality assessment methods for the characterization and management of highly impaired river systems. *Environmental Monitoring and Assessment, 188*, 1–16.

Jones, E. R., van Vliet, M. T. H., Qadir, M., & Bierkens, M. F. P. (2021). Country-level and gridded estimates of wastewater production, collection, treatment and reuse. *Earth System Science Data., 13*, 237–254. https://doi.org/10.5194/essd-13-237-2021

Kachroud, M., Trolard, F., Kefi, M., Jebari, S., & Bourrié, G. (2019). Water quality indices: Challenges and application limits in the literature. *Water, 11*(2), 361.

Kadam, A. K., Wagh, V. M., Muley, A. A., Umrikar, B. N., & Sankhua, R. N. (2019). Prediction of water quality index using artifcial neural network and multiple linear regression modelling approach in Shivganga River basin, India. *Modeling Earth Systems and Environment, 5*(3), 951–962. https://doi.org/10.1007/s40808-019-00581-3

Kassem, A., Abbas, L., Coutinho, O., Opara, S., Najaf, H., Kasperek, D., Li, X., & Tiquia-Arashiro, S. M. (2023). Applications of Fourier Transform-Infrared spectroscopy in microbial cell biology and environmental microbiology: Advances, challenges, and future perspectives. *Frontiers in Microbiology, 14*, 1304081. https://doi.org/10.3389/fmicb.2023.1304081

Koklu, R., Sengorur, B., & Topal, B. (2010). Water quality assessment using multivariate statistical methods—a case study: Melen River System (Turkey). *Water Resources Management, 24*, 959–978.

Kritzberg, E. S., Cole, J. J., Pace, M. L., Granéli, W., & Bade, D. L. (2004). Autochthonous versus allochthonous carbon sources of bacteria: Results from whole-lake 13C addition experiments. *Limnology and Oceanography, 49*(2), 588–596.

Kumar, V., Parihar, R. D., Sharma, A., Bakshi, P., Sidhu, G. P. S., Bali, A. S., et al. (2019). Global evaluation of heavy metal content in surface water bodies: A meta-analysis using heavy metal pollution indices and multivariate statistical analyses. *Chemosphere, 236*, 124364.

Kumar, D., Kumar, R., Sharma, M., Awasthi, A., & Kumar, M. (2024). Global water quality indices: Development, implications, and limitations. *Total Environment Advances, 9*, 200095. https://doi.org/10.1016/j.teadva.2023.200095

Lee, S., & Lee, D. (2018). Improved prediction of harmful Algal Blooms in four major South Korea's rivers using deep learning models. *International Journal of Environmental Research and Public Health, 15*, 1322.

Li, Z., Zhang, H. K., Roy, D. P., Yan, L., Huang, H., & Li, J. (2017). Landsat 15-m panchromatic-assisted downscaling (LPAD) of the 30-m reflective wavelength bands to Sentinel-2 20-m resolution. *Remote Sensing, 9*(7), 755.

Liew, S., Choo, C., Lau, J., Chan, W., & Dang, T. (2019). Monitoring water quality in Singapore reservoirs with hyperspectral remote sensing technology. *Water Practice & Technology, 14*(1), 118–125.

Liou, S.-M., Lo, S.-L., & Wang, S.-H. (2004). A generalized water quality index for Taiwan. *Environmental Monitoring and Assessment, 96*, 35–52.

Lioumbas, J., Christodoulou, A., Katsiapi, M., Xanthopoulou, N., Stournara, P., Spahos, T., et al. (2023). Satellite remote sensing to improve source water quality monitoring: A water utility's perspective. *Remote Sensing Applications: Society and Environment, 32*, 101042.

Ma, R., Pan, D., Duan, H., & Song, Q. (2009). Absorption and scattering properties of water body in Taihu Lake, China: Backscattering. *International Journal of Remote Sensing, 30*(9), 2321–2335. https://doi.org/10.1080/01431160802549385

Mandaville, S. (2002). *Benthic macroinvertebrates in freshwaters: Taxa tolerance values, metrics, and protocols* (Vol. 128). Soil & Water Conservation Society of Metro Halifax Nova Scotia.

Matthews, M. W. (2011). A current review of empirical procedures of remote sensing in inland and near-coastal transitional waters. *International Journal of Remote Sensing, 32*(21), 6855–6899.

McCarthy, M. J., Colna, K. E., El-Mezayen, M. M., Laureano-Rosario, A. E., Méndez-Lázaro, P., Otis, D. B., et al. (2017). Satellite remote sensing for coastal management: A review of successful applications. *Environmental Management, 60*, 323–339.

McKindles, K. M., & Tiquia-Arashiro, S. M. (2012). Functional gene arrays for analysis of microbial communities on ocean platform. In S. Tiquia-Arashiro (Ed.), *Molecular biological technologies for ocean sensing. Springer protocols handbooks*. Humana Press. https://doi.org/10.1007/978-1-61779-915-0_9

Mobley, C. D. (1995). The optical properties of water. *Handbook of Optics, 1*(43), 43.

Mohammadi, M., Sharifi, A., Hosseingholizadeh, M., & Tariq, A. (2021). Detection of oil pollution using SAR and optical remote sensing imagery: A case study of the Persian Gulf. *Journal of the Indian Society of Remote Sensing, 49*(10), 2377–2385.

Mortula, M., Ali, T., Bachir, A., Elaksher, A., & Abouleish, M. (2020). Towards monitoring of nutrient pollution in coastal lake using remote sensing and regression analysis. *Water, 12*(7), 1954.

Naderian, D., Noori, R., Heggy, E., Bateni, S. M., Bhattarai, R., Nohegar, A., & Sharma, S. (2024). A water quality database for global lakes. *Resources, Conservation and Recycling, 202*, 107401. https://doi.org/10.1016/j.resconrec.2023.107401

Najafzadeh, M., & Basirian, S. (2023). Evaluation of river water quality index using remote sensing and artificial intelligence models. *Remote Sensing, 15*(9), 2359.

Nasir, N., Kansal, A., Alshaltone, O., Barneih, F., Sameer, M., Shanableh, A., & Al-Shamma'a, A. (2022). Water quality classification using machine learning algorithms. *Journal of Water Process Engineering, 48*, 102920. https://doi.org/10.1016/j.jwpe.2022.102920

Nassereddine, Z., Opara, S., Coutinho, O., Qyteti, F., Book, R., Heinicke, M. P., Napieralski, J., & Tiquia-Arashiro, S. M. (2024). Critical perspectives on advancing antibiotic resistant gene (ARG) detection technologies in aquatic ecosystems. *Science of the Total Environment, 957*, 177775. https://doi.org/10.1016/j.scitotenv.2024.177775

Naubi, I., Zardari, N. H., Shirazi, S. M., Ibrahim, N. F. B., & Baloo, L. (2016). Effectiveness of water quality index for monitoring Malaysian river water quality. *Polish Journal of Environmental Studies, 25*(1), 231.

Nova, K. (2023). AI-enabled water management systems: An analysis of system components and interdependencies for water conservation. *Eigenpub Review of Science and Technology, 7*(1), 105–124.

Oest, A., Alsaffar, A., Fenner, M., Azzopardi, D., & Tiquia-Arashiro, S. M. (2018). Patterns of change in metabolic capabilities of sediment microbial communities in river and lake ecosystems. *International Journal of Microbiology, 2018*, 6234931. https://doi.org/10.1155/2018/6234931

Pagnucco, G., Overfield, D., Chamlee, Y., Fortuna, A., Sulaiman, F., Farinas, J., Schittenhelm, R., Shuler, C., Kassem, A., Opara, S., Najaf, H., Abbas, L., Coutinho, O., Catalfano, B., Li, X., & Tiquia-Arashiro, S. M. (2023). Metal tolerance and biosorption capacities of bacterial strains isolated from an urban watershed. *Frontiers in Microbiology, 14*, 1278886. https://doi.org/10.3389/fmicb.2023.1278886

Palabıyık, S., & Akkan, T. (2024). Evaluation of water quality based on artificial intelligence: Performance of multilayer perceptron neural networks and multiple linear regression versus water quality indexes. *Environment, Developemt and Sustainability*. https://doi.org/10.1007/s10668-024-05075-6

Panella, T., Fernandez Illescas, C., Cardascia, S., van Beek, E., Guthrie, L., Hearne, D., et al. (2020). *Asian water development outlook 2020: Advancing water security across Asia and the Pacific*. Asian Development Bank.

Patel, D., Gismondi, R., Alsaffar, A., & Tiquia-Arashiro, S. M. (2019). Applicability of API ZYM to capture seasonal and spatial variabilities in lake and river sediments. *Environmental Technology, 40*(24), 3227–3239. https://doi.org/10.1080/09593330.2018.1468492

Pink, R. M. (2016). Indonesia: One of the most polluted countries in the world. In *Water rights in Southeast Asia and India*. Palgrave Macmillan. https://doi.org/10.1057/9781137504234_5

Qadir, M., Drechsel, P., Jiménez Cisneros, B., et al. (2020). Global and regional potential of wastewater as a water, nutrient and energy source. *Natural Resoucer Forum, 44*, 40–51. https://doi.org/10.1111/1477-8947.12187

Rana, R., Kalia, A., Boora, A., Alfaisal, F. M., Alharbi, R. S., Berwal, P., Alam, S., Khan, M. A., & Qamar, O. (2023). Artificial intelligence for surface water quality evaluation, monitoring and assessment. *Water, 15*(22), 3919. https://doi.org/10.3390/w15223919

Ritchie, J. C., Zimba, P. V., & Everitt, J. H. (2003). Remote sensing techniques to assess water quality. *Photogrammetric Engineering & Remote Sensing, 69*(6), 695–704.

Ross, M. R., Topp, S. N., Appling, A. P., Yang, X., Kuhn, C., Butman, D., et al. (2019). AquaSat: A data set to enable remote sensing of water quality for inland waters. *Water Resources Research, 55*(11), 10012–10025.

Rudjord, Z. C., Reid, M., Schwermer, C., & Lin, Y. (2022). *AI system for real time monitoring of water quality*. Available at SSRN: https://ssrn.com/abstract=4103356

Saleem, U., Haghighi, A. K., Klöve, B., & Oussalah, M. (2024). Citizen science applications for water quality monitoring. A review. *EGUsphere, 2024*, 1–18.

Sauvé, S., & Desrosiers, M. (2014). A review of what is an emerging contaminant. *Chemistry Central Journal, 8*, 1–7.

Schaeffer, B. A., Whitman, P., Vandermeulen, R., Hu, C., Mannino, A., Salisbury, J., et al. (2023). Assessing potential of the Geostationary Littoral Imaging and Monitoring Radiometer (GLIMR) for water quality monitoring across the coastal United States. *Marine Pollution Bulletin, 196*, 115558.

Schaeffer, B. A., Reynolds, N., Ferriby, H., Salls, W., Smith, D., Johnston, J. M., & Myer, M. (2024). Forecasting freshwater cyanobacterial harmful algal blooms for Sentinel-3 satellite resolved US lakes and reservoirs. *Journal of Environmental Management, 349*, 119518.

Schroeder, R., McDonald, K. C., Chapman, B. D., Jensen, K., Podest, E., Tessler, Z. D., et al. (2015). Development and evaluation of a multi-year fractional surface water data set derived from active/passive microwave remote sensing data. *Remote Sensing, 7*(12), 16688–16732.

Shareef, M. A., Khenchaf, A., & Toumi, A. (2016). Integration of passive and active microwave remote sensing to estimate water quality parameters. In *2016 IEEE radar conference (RadarConf)* (pp. 1–4). IEEE.

Shiklomanov, I. A. (1993). World freshwater resource. In P. H. Gleick (Ed.), *Water in crisis*. Oxford University Press.

Skoog, D. A., Holler, F. J., & Crouch, S. R. (2019) *Principles of instrumental analysis*. Cengage learning.

Spellman, F. R. (2017). *The drinking water handbook*. CRC Press.

Sun, X., Zhang, Y., Shi, K., Zhang, Y., Li, N., Wang, W., et al. (2022). Monitoring water quality using proximal remote sensing technology. *Science of the Total Environment, 803*, 149805.

Sutadian, A. D., Muttil, N., Yilmaz, A. G., & Perera, B. (2017). Using the Analytic Hierarchy Process to identify parameter weights for developing a water quality index. *Ecological Indicators, 75*, 220–233.

Taloor, A. K., Thakur, P. K., & Jakariya, M. (2022). Remote sensing and GIS applications in water science. *Groundwater for Sustainable Development, 19*, 100817. https://doi.org/10.1016/j.gsd.2022.100817

Tchobanoglus, G., Burton, F., & Stensel, H. D. (2003). Wastewater engineering: Treatment and reuse. *American Water Works Association. Journal, 95*(5), 201.

The United Nations World Water Development Report (2024). Water for Prosperity and Peace. UNESCO, Paris.

Tinelli, S., & Ilan Juran, I. (2019). Artificial intelligence-based monitoring system of water quality parameters for early detection of non-specific bio-contamination in water distribution systems. *Water Supply, 19*(6), 1785–1792. https://doi.org/10.2166/ws.2019.057

Tiquia, S. M. (2008). Diversity of sulfate-reducing genes (dsrAB) in sediments from Puget Sound. *Environmental Technology, 29*(10), 1095–1108. https://doi.org/10.1080/09593330802190608

Tiquia, S. M. (2010a). Salt-adapted bacteria isolated from the Rouge River and potential for degradation of contaminants and biotechnological applications. *Environmental Technology, 31*, 967–978. https://doi.org/10.1080/09593331003706226

Tiquia, S. M. (2010b). Metabolic diversity of the heterotrophic microorganisms and potential link to pollution of the Rouge River. *Environmental Pollution, 158*, 1435–1443. https://doi.org/10.1016/j.envpol.2009.12.035

Tiquia, S. M. (2011). Extracellular hydrolytic enzyme activities of the heterotrophic microbial communities of the Rouge River: An approach to evaluate ecosystem response to urbanization. *Microbial Ecology, 62*, 679–689. https://doi.org/10.1007/s00248-011-9871-2

Tiquia, S. M., Gurczynski, S., Zholi, A., & Devol, A. (2006a). Diversity of biogeochemical cycling genes from Puget Sound sediments using DNA microarrays. *Environmental Technology, 27*(12), 1377–1389. https://doi.org/10.1080/09593332708618756

Tiquia, S. M., Masson, S. A., & Devol, A. (2006b). Vertical distribution of nitrite reductase genes (nirS) in continental margin sediments of the Gulf of Mexico. *FEMS Microbiology Ecology, 58*(3), 464–475. https://doi.org/10.1111/j.1574-6941.2006.00173.x

Tiquia, S. M., Davis, D., Hadid, H., Kasparian, S., Ismail, M., & Murray, K. S. (2007). Halophilic and halotolerant bacteria from river waters and shallow groundwater along the Rouge River of Southeastern Michigan. *Environmental Technology, 28*, 297–307. https://doi.org/10.1080/09593332808618789

Tiquia, S. M., Schleibak, M., Schlaff, J., Floyd, J. C., Benipal, B., Zakhem, E., & Murray, K. S. (2008). Microbial community profiling and characterization of some heterotrophic bacterial isolates from river waters and shallow groundwater wells along the Rouge River, Southeast Michigan. *Environmental Technology, 29*, 651–663. https://doi.org/10.1080/09593330801986998

Tiquia-Arashiro, S. M. (2012). *Molecular biological technologies for ocean sensing. Springer protocols handbooks* (p. 307). Humana. https://doi.org/10.1007/978-1-61779-915-0

Tiquia-Arashiro, S. M., & Pant, D. (2020). *Microbial electrochemical technologies* (p. 518). CRC Press. https://doi.org/10.1201/9780429487118

Topp, S. N., Pavelsky, T. M., Jensen, D., Simard, M., & Ross, M. R. (2020). Research trends in the use of remote sensing for inland water quality science: Moving towards multidisciplinary applications. *Water, 12*(1), 169.

Uddin, M. G., Nash, S., & Olbert, A. I. (2021). A review of water quality index models and their use for assessing surface water quality. *Ecological Indicators, 122*, 107218. https://doi.org/10.1016/j.ecolind.2020.107218

UNEP/WHO. (1996). *Water quality monitoring: A practical guide to the design and implementation of freshwater quality studies and monitoring programs*. E & FN Spon.

USDA. (2021). *Resampling and pansharpening using Raster functions in ArcPro; Forest Service*. USDA.

US-EPA. (2023). *Cyanobacteria Assessment Network (CyAN)*. https://www.epa.gov/water-research/cyanobacteria-assessment-network-cyan

Wang, X., & Yang, W. (2019). Water quality monitoring and evaluation using remote sensing techniques in China: A systematic review. *Ecosystem Health and Sustainability, 5*(1), 47–56.

Wang, S., Zhang, F., Shao, Y., Tian, W., & Gong, H. (2010). Microwave remote sensing for marine monitoring: An example of Enteromorpha prolifera bloom monitoring. In *2010 IEEE international geoscience and remote sensing symposium* (pp. 4530–4533).

Wang, X., Jiang, Y., Jiang, M., Cao, Z., Li, X., Ma, R., et al. (2023). Estimation of total phosphorus concentration in lakes in the Yangtze-Huaihe region based on Sentinel-3/OLCI images. *Remote Sensing, 15*(18), 4487.

Wang, M., Bodirsky, B. L., Rijneveld, R., et al. (2024). A triple increase in global river basins with water scarcity due to future pollution. *Nature Communications, 15*, 880. https://doi.org/10.1038/s41467-024-44947-3

Wetzel, R. G. (2001). *Limnology: Lake and river ecosystems* (3rd ed.). Academic Press.

WHO. (2011). *World Health Organization guidelines for drinking-water quality*. World Health Organization.

WHO. (2022). *Guidelines for drinking-water quality*. World Health Organization.
WHO. (2023). *Drinking water*. WHO. https://www.who.int/news-room/fact-sheets/detail/drinking-water
Willmott, C. J. (1981). On the validation of models. *Physical Geography, 2*, 184–194.
World Bank Blogs. https://blogs.worldbank.org/en/opendata/chart-globally-70-freshwater-used--agriculture#:~:text=In%20most%20regions%20of%20the,percent%20increase%20in%20water%20withdrawals.
Xiong, J., Lin, C., Ma, R., & Cao, Z. (2019). Remote sensing estimation of lake total phosphorus concentration based on MODIS: A case study of Lake Hongze. *Remote Sensing, 11*(17), 2068.
Yang, H., Kong, J., Hu, H., Du, Y., Gao, M., & Chen, F. (2022a). A review of remote sensing for water quality retrieval: Progress and challenges. *Remote Sensing, 14*(8), 1770.
Yang, L., Driscol, J., Sarigai, S., Wu, Q., Lippitt, C. D., & Morgan, M. (2022b). Towards synoptic water monitoring systems: A review of AI methods for automating water body detection and water quality monitoring using remote sensing. *Sensors, 22*(6), 2416. https://www.mdpi.com/1424-8220/22/6/2416
Yao, L., & Hui, J. (2013). Retrieval of total phosphorus concentration in the surface waters of Poyang lake based on remote sensing and analysis of its spatial-temporal characteristics. *Journal of Natural Resources, 28*(12), 2169–2177.
Zhang, Y., Ma, R., Duan, H., Loiselle, S. A., Xu, J., & Ma, M. (2014). A novel algorithm to estimate algal bloom coverage to subpixel resolution in Lake Taihu. *IEEE Journal of Selected Topics in Applied Earth Observations and Remote Sensing, 7*(7), 3060–3068.
Zhang, Y., Hallikainen, M., Zhang, H., Duan, H., Li, Y., & San Liang, X. (2018). Chlorophyll-a estimation in turbid waters using combined SAR Data with hyperspectral reflectance Data: A case study in Lake Taihu, China. *IEEE Journal of Selected Topics in Applied Earth Observations and Remote Sensing, 11*(4), 1325–1336.
Zhang, L., Zhang, L., Cen, Y., Wang, S., Zhang, Y., Huang, Y., et al. (2022). Prediction of total phosphorus concentration in macrophytic lakes using chlorophyll-sensitive bands: A case study of Lake Baiyangdian. *Remote Sensing, 14*(13), 3077.

Chapter 18
Recent Advances in Watershed-Scale Modeling of Microbial Fate and Transport

Victor Ella

1 Introduction

The threat to public health posed by microbial contamination of drinking water sources due to the deposition and movement of pathogenic microorganisms in watersheds has long been recognized. Microbiologically contaminated drinking water transmits diseases like diarrhea, cholera, dysentery, typhoid, and polio (WHO, 2024). Worldwide, the World Health Organization (2024) reported an average of 505,000 deaths due to diarrhea alone every year. Recently, it was estimated that nearly 2 billion people around the world use fecally contaminated water for drinking purposes, although the extent of microbial contamination varies substantially between urban and rural settings and between countries (WHO, UNICEF, World Bank, 2022).

Microbial contamination of water supplied by watersheds is mostly caused by microbiological loading within these hydrologic systems from point sources like direct deposition of wastewater effluents, non-point sources like land application of livestock wastes, and other sources like wildlife wastes, residential wastes, leakages from septic tanks, and other wastewater treatment systems. Microbiological contaminants like bacteria that come from various sources have been well documented and reported in published literature (e.g., Benham et al., 2006; ASAE, 1998). However, the identification and quantification of these various sources of microbial contamination, particularly the non-point sources, continue to remain a challenge.

Monitoring microbiological water quality over space and time is generally costly and time-consuming and may even be infeasible particularly if performed over long

V. Ella (✉)
Land and Water Resources Engineering Division, Institute of Agricultural and Biosystems Engineering, College of Engineering and Agro-industrial Technology, University of the Philippines Los Baños, College, Laguna, Philippines
e-mail: vbella@up.edu.ph

S. M. Tiquia-Arashiro et al. (eds.), *Urban Watershed Microbiology, Volume 1*,
https://doi.org/10.1007/978-3-031-89071-0_18

periods of time for large watersheds. Actual monitoring requires rigorous water sampling at various points in the watershed, proper handling and storage of collected samples, accurate laboratory analysis among other intermediate activities, which are all resource-intensive. The cost of monitoring microbiological contaminants such as bacteria and protozoa may reach up to three times as much as that for monitoring inorganic and organic contaminants as reported by Baffaut and Sadeghi (2010) and Pachepsky et al. (2006). In many areas in the world, particularly in developing countries, long-term direct microbiological water quality monitoring in watersheds would be practically infeasible.

As an alternative, watershed-scale simulation models may be employed for predicting the fate and transport of pathogenic microorganisms in watersheds. Like hydrologic models, these microbial fate and transport models are essentially mathematical representations of the various processes governing the movement, transformation, and eventual deposition of these microorganisms in the various portions of the watershed and its waterways (Cho et al., 2016a; Benedini & Tsakiris, 2013; Fetter, 1993). With the use of these models, the resulting outputs in terms of microbial concentrations of water along the waterways of the watershed may be simulated and predicted for any given model inputs, biophysical conditions, and microbial loading scenarios. Moreover, the complexities of hydrologic systems in the presence of external drivers such as land use change and climate change can more conveniently be handled by these mathematical models.

While mathematical modeling would still require observed data for proper calibration and validation, properly calibrated models would prove to be highly useful as a management tool and may obviate the need for extensive and costly microbiological water quality monitoring. More importantly, model simulations and predictions under various scenarios could be used as basis for formulating preventive and mitigative measures and effective policies to avoid compromising the water quality of water supply sources. With models, these mitigating measures and policies can even be refined or fine-tuned because of the inherent capability of the models not just to predict but to quantify these predictions. The inherent power of watershed-scale microbial fate and transport models, therefore, makes them potentially useful for sustainable water resources management. Nevertheless, it should be noted that the applicability of these models may be constrained by data availability and adequacy and the model assumptions used may pose some uncertainties in model prediction.

Modeling has been a part of management decision support tools in developed countries like the United States. Microbial fate and transport models could be used for setting total maximum daily loads (TMDLs) of potential microbiological sources in watersheds (Benham et al., 2006). These models could also be used for assessing the effectiveness of potential mitigating strategies (Baffaut & Sadeghi, 2010). Moreover, the use of models for legal and regulation purposes has even been encouraged as reported by Cho et al. (2016a). They also pointed out the potential of watershed-scale microbial fate and transport models for assessing whether certain regulatory standards set by agencies, commodity groups and private companies

usually in terms of concentration of fecal coliforms, *Escherichia coli (E. coli)* or *Enterococci* are complied with.

For many years, numerous watershed-scale models have been developed primarily for simulating and predicting hydrologic processes (Singh, 2018; Todini, 2007; Anderson & Burt, 1985). Over the past two decades, however, additional model components or submodels to handle microbial fate and transport have been either incorporated in or coupled with these hydrologic models. Other types of models for microbial fate and transport on a watershed-scale have also evolved in recent years. Some reviews of the various watershed models for microbial fate and transport have been made (e.g., Cho et al., 2016a, Behnam et al., 2006). Since then, however, some other watershed-scale models for microbial fate and transport have evolved and further improvements have been made on some of the existing models. While a recent comprehensive review of the various water quality models was published by Yuan et al. (2020), they focused only on models developed in the United States. This chapter aims to present the recent advances in microbial fate and transport modeling on a watershed-scale and provide an updated review of existing and newly developed watershed-scale models based on published peer-reviewed literature worldwide focusing on recent model application for microbial fate and transport and recent model improvements. Consequently, this chapter attempts to identify and formulate recommendations for further research to maximize the potential of watershed-scale microbial fate and transport models as a management tool for preventing adverse water quality impacts of microbiological contaminants on water resources supplied by watersheds so as to minimize the potential public health and environmental risks.

2 Processes, Pathways, and Mechanics of Microbial Fate and Transport in Watersheds

Numerous biophysical processes govern the fate and transport of pathogenic microorganisms emanating from various point and non-point pollution sources in the watershed and the mechanics of microbial fate and transport is highly dynamic and complex. Depending on the nature of the pathogenic microorganisms and the location of their entry into the watershed, these microbiological contaminants may be carried by water through the processes of advection, dispersion, and diffusion along the surface and subsurface flow regimes and by eroded soil particles through adsorption and sediment transport along the streams. The fundamental principles of these hydrodynamic processes are described in several published literature (e.g., Bennett, 2013; Benedini & Tsakiris, 2013; Weiskerger & Phanikumar, 2020; Fetter, 1993). While in transport, the microorganisms would also undergo transformations such as growth, decay, or die-off. The fundamental principles of microbial kinetics are described in published literature (e.g., Panikov, 1995). Along the streams, settlement and resuspension of microbial-laden sediments may also occur (Jamieson

et al., 2005; East et al., 2018; Pachepsky & Shelton, 2011; Patel et al., 2019; Bradshaw et al., 2021) and this further complicates the fate and transport of microorganisms. In some cases, microbial transport in soils may also be enhanced by preferential flow (Wang et al., 2013) and this may accelerate the entry of pathogens to the groundwater systems, which are also often used for domestic water supply purposes. The hydraulically connected nature of watersheds makes the interplay of all the processes all the more complex.

Compounding the microbial fate and transport processes even more are the possible enhancing or inhibitory effects of changes in environmental conditions like temperature and pH along with the presence of other contaminants (metals and organic compounds) which could all potentially affect the survival and movement of pathogenic microorganisms (Bowman et al., 2018; Cho et al., 2012a; Flood et al., 2015; Kassem et al., 2023; Pagnucco et al., 2023). Various pathogenic microorganisms such as bacteria, viruses, and protozoa would also behave differently under changing environmental conditions. Furthermore, the hydrologic processes which influence the microbial fate and transport processes could also be affected by land use change and climate change. Watershed denudation could lead to higher surface runoff and lower groundwater recharge thereby making the transport of microorganisms over land surface more pronounced than through the subsurface. Increasing temperatures due to climate change would lead to higher evapotranspiration, while extreme rainfall events also due to climate change would result in greater surface runoff. Combined together, these factors would affect the entire hydrologic cycle which would consequently affect the fate and transport of pathogenic microorganisms.

Figure 18.1 shows a conceptual diagram of the various processes that govern the fate and transport of microorganisms emanating from point sources, non-point sources, wildlife wastes, and leaking septic tanks and other wastewater treatment systems in a watershed. Excessive precipitation would lead to the occurrence of surface runoff, which serves as the primary driver of the advective, diffusive, and dispersive transport of microorganisms released from non-point sources and wildlife wastes. Other processes such as soil adsorption, partitioning, soil erosion, and microbial transformation also occur during this transport within the watershed (Kassem et al., 2023; Pagnucco et al., 2023). The microbial-laden runoff and eroded soil particles will then reach the tributaries and the mainstream of the watershed contributing to microbial loading together with the discharge from point sources like wastewater effluents. On the other hand, infiltration and percolation will drive the downward transport of microorganisms from non-point sources, wildlife wastes, and leaking septic tanks through the subsurface by advection, diffusion, and dispersion. Soil adsorption, partitioning, and microbial transformations will also occur during this transport. In some cases, the vertical subsurface transport may be accelerated by preferential flow through the soil macropores. Microorganisms reaching the groundwater or aquifer system are then transported laterally by groundwater movement through the processes of advection, diffusion, and dispersion until they reach the streams and join those fed by surface runoff and erosion (Gentile et al., 2006; Wu et al., 2005). Along the stream, such processes as sediment deposition,

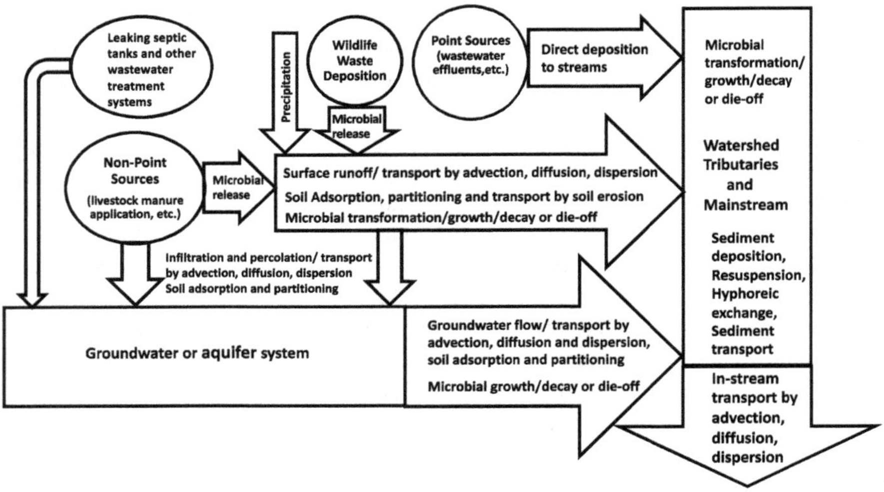

Fig. 18.1 Microbial fate and transport processes in a watershed

sediment resuspension, hyporheic exchange, sediment transport, microbial transformations may occur as the microorganisms are transported downstream by advection, diffusion, and dispersion. The apparent complexities of the interrelated processes in the watershed governing the fate and transport of microbiological contaminants are best handled by watershed-scale mathematical models. These models are capable of not only in simulating and predicting microbial concentrations over space and time but also in quantifying the fate and transport of pathogenic microorganisms under changing environmental conditions.

3 Modeling Approaches for Watershed-Scale Microbial Fate and Transport

Modeling the fate and transport of microbiological contaminants on a watershed scale may be done in a number of ways. The most widely accepted approach is to make the models process-based or physically based instead of using simpler empirical or conceptual approaches. Process-based or physically based models employ fundamental hydrologic and environmental engineering principles such as water balance, material balance, reaction kinetics, adsorption principles along with contaminant transport principles (Cho et al., 2016a, Behnam et al., 2006; Yuan et al., 2020; Fetter, 1993; Ella, 2001, 2009a). The governing mathematical equations are then used in the watershed model code to represent these processes.

Process-based or physically based watershed-scale models are developed based on the assumption of either homogeneity of model parameters over space or heterogeneity, the latter of which considers the spatial variability of the various model

parameters and factors governing the various processes in the watershed. Process-based hydrologic watershed-scale models that consider the spatial variability of various factors influencing the various hydrologic processes are generally called distributed-parameter watershed models. The same term could be applied to watershed-scale, process-based modeling of microbial fate and transport if the spatial variabilities are taken into account. Other process-based watershed-scale models use large discretization in handling spatial variability of various parameters, some of which subdivide the entire watershed into sub-basins or hydrologic units or grids within which the model parameters are assumed to be uniform. These models are often treated as semi-distributed models. Most of the microbial fate and transport watershed-scale models are of the semi-distributed parameter type. On the other hand, watershed-scale models that assume uniform model parameters for the whole catchment are lumped-parameter models. Obviously, distributed-parameter and semi-distributed models generally provide better prediction accuracy than lumped-parameter models as the former types are able to capture the heterogeneity of watersheds.

Physically based models are also generally called deterministic models. Other models may be stochastic, which is essentially based on the assumption that model parameters are governed by certain probability distributions and are therefore not deterministic (Anderson & Burt, 1985; Overton & Meadows, 1976). The simpler models such as the conceptual models or empirical, which are not necessarily process-based, are generally inadequate for microbial fate and transport modeling because of their inability to capture the inherent complexity and dynamic nature of these biological contaminants. There are other models that employ special computational techniques such as inverse methods and models that are integrative in nature. Watershed-scale models that are of practical value for water quality management are those that are continuous, that is, model simulation can run over long periods of time using reasonable time steps or intervals, rather than based on single storm event or event-based. Continuous watershed-scale microbial fate and transport models are capable of simulating or predicting the concentration of pathogenic microorganisms over space and time in the watershed (Nassereddine et al., 2024). This chapter is focused on these types of microbial fate and transport models.

4 Process-Based Watershed-Scale Models of Microbial Fate and Transport

Among the most widely used process-based watershed-scale models that have proven to be applicable for microbial fate and transport include the Soil and Water Assessment Tool (SWAT), Hydrologic Simulation Program Fortran (HSPF), and the Agricultural Policy/Environmental eXtender (APEX) models. A more extensive review of these models is therefore presented in this section. Other process-based watershed-scale hydrologic models that have been coupled with microbial fate and

transport models are also presented. In all cases, some evidence of their applicability for microbial fate and transport modeling and recent model modifications or improvements, if there are any, are presented and discussed.

4.1 The SWAT Model

4.1.1 Model Description

The SWAT model is a physically based, watershed-scale, continuous-event hydrologic model developed to quantify the impact of land management practices on water, sediment, and agricultural chemical yields in large, complex watersheds with varying soils, land use, and management conditions over long periods of time (Arnold et al., 1998). It was developed by the United States Department of Agriculture (USDA)- Agricultural Research Service (ARS) in collaboration with various agencies and academic institutions and agencies including the United States Environmental Protection Agency (USEPA), Natural Resources Conservation Service, National Oceanic and Atmospheric Administration, and Bureau of Indian Affairs among others (Neitsch et al., 2005).

SWAT can model the various physical processes associated with water movement, sediment transport, crop growth, nutrient cycling among others using such input data as climate, soil, topography, vegetation, and land management occurring in each watershed (Arnold et al., 2009; Ella, 2010). A group of hydrologic response units (HRUs) constitutes the watershed model for predicting runoff and sediment yield. SWAT has proven its applicability in various countries around the world as evidenced by numerous peer-reviewed publications (Gassman et al., 2007; Ella, 2009b; Arnold et al., 2009; SWAT, 2024).

In the early to mid-2000s, the modeling capability of SWAT was expanded with the incorporation of a bacterial submodel (Arnold & Fohrer, 2005). The equations were developed by Sadeghi and Arnold (2002) based on field studies on fecal coliform from cattle manure application on pasture and corn plots in Virginia (Baffaut & Sadeghi, 2010). The bacterial fate and transport equations were consequently introduced into SWAT and made available in the public domain through the updated SWAT model documentation published by Neitsch et al. (2005).

The bacterial fate and transport modeling component in SWAT are fundamentally based on material balance principles accounting for various bacterial inputs in soil and stream in terms of loading rates and incorporating the effects of bacterial transformation in the form of decay or die-off, adsorption onto soil particles and sediments, soil redistribution, and advective transport by surface runoff, percolation, and streamflow. Detailed description of the various equations used in the bacterial submodel in SWAT was presented by Sadeghi and Arnold (2002), Neitsch et al. (2005) and Baffaut and Sadeghi (2010).

4.1.2 Recent Applications for Microbial Fate and Transport Modeling

Among the most recent SWAT applications for microbial fate and transport modeling was done by Sowah et al. (2020) in the Clouds Creek watershed in Georgia, United States to understand the sources and drivers of microbial water quality. Results of model application showed satisfactory performance with NSE of 0.81 and 0.55 for model calibration and validation, respectively, and NSE of 0.32–0.34 when predicting *E. coli* and fecal coliforms, respectively. Their modeling study also showed that in-stream cattle manure deposition contributes the most to bacterial contamination followed by cattle manure application, poultry manure application, septic systems discharges, and wildlife manure contribution.

Iqbal and Hofstra (2019) applied the unmodified SWAT model to evaluate the contribution of various bacterial sources to *E. coli* contamination in Kabul River basin with an area of 22,000 sq. km. Using monthly time steps and three years of data, the calibrated and validated SWAT model performed generally well in simulating bacterial fate and transport with R^2 of 0.70 to 0.72, NSE of 0.66 to 0.69, and Percentage Bias (PBIAS) of 1.9 to 3.7. Their modeling results showed that *E. coli* concentrations peak during the wet season and that human feces and livestock manure contribute 44% to the *E. coli* contamination. The study also identified that the model parameters BACTKDQ, BACTMIX, and THBACT in SWAT are the most sensitive. Comparison with observed data also showed that both the high and low *E. coli* concentrations were slightly underestimated by the SWAT model.

Jeong et al. (2019) also applied the SWAT model to determine critical sources of *E. coli* and the effectiveness of conservation practices for mitigating *E. coli* contamination in the Arroyo Colorado watershed in Texas, United States by linking a bacterial source tracking (BST) assessment scheme to the model. Their modeling study demonstrated better prediction of *E. coli* source distribution with the incorporation of BST in SWAT ($R^2 = 0.90$) than without BST ($R^2 = 0.59$). Their model simulations under various scenarios also showed that wildlife contribution to microbial contamination was still significant regardless of land use change.

Application of the unmodified SWAT for bacterial fate and transport modeling was also done by Kim et al. (2018) in the Houay Pano watershed located in northern Lao People's Democratic Republic to assess the impacts of land use change on the transfer of *E. coli* from soils to streams. They found out that the fate and transport of *E. coli* is more sensitive to the initial *E. coli* concentration than surface runoff and suspended solids. They also reported that leaf area index significantly influences the determination of the initial concentration of *E. coli* on the soil surface.

Thilakarathne et al. (2018) applied the unmodified SWAT model to simulate *E. coli* concentrations in Upper Stroubles Creek, Virginia, United States, and used the Spatially Explicit Load Enrichment Calculation Tool (SELECT) to account for spatial variability of bacterial load. The model performance for in-stream simulation of *E. coli* concentration was, however, relatively poor without sediment bacteria resuspension and deposition (NSE of −0 41 to 0.34) and even with resuspension and deposition (−0.19 to 0.36). Their modeling study also showed that overestimation and underestimation of in-stream bacterial concentration occurred at midrange flows and low flows, respectively.

Kim et al. (2017) also applied the unmodified SWAT model to simulate the fate and transport of *E. coli* in a 60-ha watershed in Northern Lao and assessed the influences of bacterial deposition and resuspension, bacterial regrowth, and hyporheic exchange on bacterial concentration. Their results showed a relatively poor performance of the unmodified model during the dry season and recommended to consider the effects of deposition and resuspension, regrowth, and hyporheic exchange to improve the SWAT model performance.

A modified version of SWAT was applied by Cho et al. (2016b) to simulate the seasonal variabilities of fecal coliform and *E. coli* concentrations in four watersheds, namely, the Stillwater River watershed in Massachusetts, the Jumping Run Creek watershed in North Carolina, Little Cove Creek watershed in Pennsylvania, and the Komacwon Creek watershed in Korea. In-stream model simulation showed better performance of the modified SWAT model in capturing the temporal variation in fecal coliform and *E. coli* concentration compared to the original SWAT model. However, their reported coefficient of determination only ranged from 0.10 to 0.28. Other measures of goodness of model fit to observed data were not reported.

Niazi et al. (2015) applied the unmodified SWAT to simultaneously simulate *E. coli* and fecal coliform in a 37 sq. km. agricultural watershed in New Jersey. Modeling results showed a poor-to-good agreement between monthly simulated and observed values of *E. coli* and fecal coliform concentrations. During model calibration, the NSE ranged from 0.03 to 0.39 for *E. coli* and from 0.07 to 0.47 for fecal coliform. Validation results showed an NSE of −0.81 to 0.31 for *E. coli* and −0.94 to 0.33 for fecal coliform.

Jayakody et al. (2014) applied the SWAT model to examine the spatial and seasonal variability of fecal coliform bacteria in 527 sq. km Pelahatchie watershed in Mississippi, United States. The model enabled them to identify the major microbial sources as livestock manure, poultry litter, and effluent from septic tanks. Model calibration and validation showed good agreement between simulated and observed values with NSE of 0.67 to 0.75.

Coffey et al. (2013) applied SWAT to predict the daily concentrations of *E. coli* in Kilshanvey catchment, a 6 sq.km small agricultural watershed in Ireland. Using only a year-long observed data, the SWAT model performance proved to range from unsatisfactory to fair with coefficient of determination R^2 of 0.03 to 0.35 and NSE of −0.42 to 0.29. Their sensitivity analysis showed that parameters associated with stream deposition and die-off rates were the most sensitive. Their study suggests that SWAT is adequate for assessing the magnitude of microbial sources but inadequate for simulating daily *E. coli* concentrations.

Earlier applications of the SWAT model for microbial fate and transport modeling were reported by Frey et al. (2013), Cho et al. (2010, 2012b), Coffey et al. (2010, 2009), Baffaut and Sadeghi (2010), Chin et al. (2009), Parajuli et al. (2009), Benham et al. (2006) among others.

4.1.3 Recent Model Improvements

Among the most recent improvements in the SWAT model for microbial fate and transport modeling was done by Meshesha et al. (2020). They modified the SWAT-bacteria module to account for the effect of pH on bacterial growth and die-off. They tested this modification by simulating *E. coli* concentration in the Athabasca River Basin in the province of Alberta in Canada. Results of model calibration based on three years of data and model validation using two years of data showed generally good agreement between simulated and observed values with coefficient of determination (R^2)of 0.70 to 0.80, Nash-Sutcliffe Efficiency (NSE) of 0.59 to 0.68, and Percentage Bias (PBIAS) of 7.94% to 17.85% for model calibration and R^2 of 0.59 to 0.72, NSE of 0.55 to 0.55, and PBIAS of 10% to 22% for model validation. Their results also demonstrated the significant effect of pH on *E. coli* simulation using SWAT.

Jeon et al. (2019) also modified the SWAT source code for the soil and in-stream bacteria modules by introducing additional mathematical representation for bacterial growth and hyporheic exchange to account for seasonal temperature variations for better calculation of bacterial re-growth and die-off and to enable better simulation of bacterial concentration during the dry season or low-flow conditions. They tested the modified SWAT model in the Taehwa River watershed in Korea to evaluate climate change impacts on fecal coliform bacteria. The modified SWAT model showed better performance than the unmodified version for dealing with seasonal variability of fecal coliform bacteria. Their modeling study also showed that the in-stream bacterial load was most dominantly affected by resuspension from the streambed.

An earlier improvement in the SWAT model for microbial fate and transport was done by Cho et al. (2016b) to address its limitation in simulating seasonal variability of fecal coliform due to the exclusion of alternating bacterial growth. They developed a new bacteria subroutine for SWAT that considers critical temperatures as a parameter for simulating the onset of bacterial growth or die-off so as to more accurately model the seasonal variability of bacterial concentrations. The critical temperatures introduced represent threshold values for determining whether re-growth or die-off would become dominant.

In previous years, the SWAT modeling capability for microbial fate and transport was also improved with the inclusion of an in-stream module to handle and account for in-stream processing of fecally derived microorganisms (e.g., Cho et al., 2012b; Pandey et al., 2013; Kim et al., 2010). All the model improvements would certainly enhance the capability of the SWAT model as a tool for water quality management.

4.2 The HSPF Model

4.2.1 Model Description

The Hydrologic Simulation Program Fortran (HSPF) is a process-based continuous, lumped-parameter watershed model. It was originally developed as a hydrologic model called the Stanford watershed model in the 1960s, was expanded in the 1970s with the addition of water quality processes, was further developed by the United States Geological Survey (USGS) and the United States Environmental Protection Agency (USEPA) in the 1980s in terms of algorithm enhancements data processing (USGS, 2024). The current release version 11 was developed and documented by Bicknell et al. (1997). While HSPF is treated to be a lumped-parameter type of model, it accounts for spatial variability of model parameters by subdividing the entire watershed into hydrologic response units (HRUs), which are relatively homogeneous areas in terms of physical characteristics, similar to what is done in SWAT.

The bacterial fate and transport are modeled in HSPF in almost the same manner as in SWAT. Simulation of bacterial concentrations is based on fundamental biophysical principles including material balance, reaction kinetics, adsorption, sediment transport, advective transport among others. However, in HSPF the release of microorganisms from sources like livestock manure is taken to be exponential in nature unlike in SWAT which treats the release as a linear function. Based on reaction kinetics, the HSPF model would seem to provide a more realistic representation of microbial release than SWAT. However, the release of microorganisms from manure is complex and is affected by numerous factors like rainfall intensity, depth and duration, manure age, method of application, among other factors (Alegbeyele & SantAna, 2020; Benham et al., 2006).

HSPF is a complex model and involves the use of numerous mathematical relationships to capture the hydrologic and microbial transport processes. The various equations used in HSPF for bacterial fate and transport were reported by Benham et al. (2006). The hydrologic equations used in HSPF were reported by Bicknell et al. (1997).

4.2.2 Recent Model Applications

Among the most recent application of the HSPF model for watershed-scale microbial fate and transport modeling was done by Abbas et al. (2021) to simulate the transport of *E. coli* in a 0.6 sq. km small watershed in Lao PDR. They used HSPF and a deep learning model employing the Long-Short-Term Memory (LSTM) technique for comparison purposes. They calibrated both models for surface and subsurface flow and consequently performed simulation of *E. coli* transport with a time step of six minutes using both models. Results showed a relatively poor simulation using HSPF with an NSE of −3.01, whereas the deep learning model performed better with an NSE of 0.35. They attributed the poor performance of HSPF on its limitation to capture the dynamics of *E. coli* due to land use change.

In another recent application, HSPF was integrated with the Tidal Prism Model (TPM) by Sobel et al. (2017) to simulate bacteria in coastal watersheds in Texas, United States. The Tidal Prism Model is a one-dimensional mass balance model that considers loading from tidal exchange, runoff, point sources, and bacterial die-off. They calibrated the integrated model using discharge and *E. coli* concentration data for HSPF and salinity *enterococci* data for the TPM using hourly time steps. They then used the linked model to simulate and evaluate management scenarios and assess the impact of reducing runoff load and in determining areas in the watershed needing such reduction. They also assessed the impact of improving wastewater treatment effluent water quality using the combined HSPF and TPM.

Coffey et al. (2015) used the HSPF model to evaluate the impacts of climate change and land management on microbial water quality in the Pigg River watershed, located in southwest Virginia, the United States. Their model simulation results for years 2040 to 2069 showed that future trends in seasonal and annual precipitation would lead to increased microbial loading in the watershed. However, they highlighted a high level of model uncertainty during high and low flow scenarios.

Kim et al. (2014) used the HSPF model to develop the Total Maximum Daily Loads (TMDLs) for bacteria in the Hardware River watershed in Virginia, the United States, in order to meet water quality standards for bacteria in Virginia, which at the time of their study was set to not more than 126 cfu/100 mL of *E. coli* based on the calendar-month geometric mean concentration with no sample exceeding a concentration of 235 cfu/100 mL. They calibrated and validated the hydrologic and water quality components of the HSPF model and eventually used the model for identifying and estimating the bacterial contributions from various sources. Model simulations under various scenarios were also performed.

Earlier applications of the HSPF model were published by Rolle et al. (2012), Benham et al. (2006), Chin et al. (2009) among many others.

4.2.3 Recent Model Improvements

Liao et al. (2015) modified the HSPF model by accounting for the interaction between bacteria and sediments during in-stream microbial transport. They calculated the total in-stream microbial concentration in water based on the free-phase microbial concentration and the microbial concentration attached to the suspended clay and silt particles, which were obtained by applying the principles of advection, adsorption, and desorption using linear reversible isotherm, accounting for sediment deposition and resuspension and assuming first order decay. The modified HSPF model showed better simulation performance compared to previous studies.

While Benham et al. (2006) made numerous other research recommendations that would improve the performance of the HSPF model, no other model improvement or modification on HSPF has been reported in published peer-reviewed literature. The absence of model improvements in HSPF may be due to several reasons like the complexity of the model, data requirements, the availability of other more widely used modeling options like SWAT among other reasons.

4.3 The APEX Model

4.3.1 Model Description

The Agricultural Policy/Environmental eXtender (APEX) model is a process-based distributed-parameter, continuous watershed-scale hydrologic, and water quality model (Williams & Izaurralde, 2006; Hong et al., 2017). The APEX model has 12 major components, namely, climate, hydrology, crop growth, pesticide fate, nutrient cycling, erosion-sedimentation, carbon cycling, management practices, soil temperature, plant environment control, economic budgets, and subarea/channel routing. Agricultural management practices such as irrigation, fertilizer application, manure management, crop rotation and selection, cover crops, biomass removal, grazing, and tillage can be handled by the APEX model (Hong et al., 2017; Williams & Izaurralde, 2006). The watershed-scale version of the APEX model subdivides the entire catchment into several subbasins, each of which is assumed to be homogeneous in terms of crop, land use, soil, climate, topography, and management practice and the subbasins interact hydrologically with each other for hydrologic routing (Wang et al., 2011).

About a decade after APEX was introduced, Hong et al. (2018) developed a bacterial fate and transport subroutine for APEX. Their submodel handles either a single strain or a group of fecally derived bacteria from livestock manure. Such processes as manure erosion, microbial survival and die-off, manure transport by runoff, and in-stream processes are considered in their microbial fate and transport model. A two-stage thermal time-dependent model is used to simulate bacterial survival, which is done separately for each manure application occurrence. In-stream processes for microbial fate and transport are simulated by APEX by considering "the reach-scale release of bacteria with resuspended streambed sediment during high flow events, the release of bacteria from streambed sediment due to the hypothetic exchange, the deposition with settling sediment, and the thermal time-dependent survival" (Hong et al., 2018). The microbial transport processes are tied up with the hydrologic processes such as runoff, infiltration, sediment discharge, and streamflow, which are provided by the basic APEX hydrologic model.

The model equations for the microbial fate and transport subroutine in APEX along with the algorithms employed were presented by Hong et al. (2018).

4.3.2 Recent Model Applications

The APEX model was tested for microbial fate and transport modeling by Hong et al. (2018) in Toenepi watershed, with an area of 15.38 sq. km, in New Zealand using about 5.5 years of hydrologic and water quality data as basis for model calibration and validation. Model calibration and validation showed an RMSE value of 0.627 and 0.820, respectively, and an overall NSE of 0.33. The pattern of temporal variability of *E. coli* concentration at the watershed outlet was replicated by the

model. During model calibration, 89% of simulated values deviated by one order of magnitude from observed values and the rest by one to two orders of magnitude. During model validation, a model discrepancy of one order of magnitude was observed in 70% of the paired values and one to two orders of magnitude for the rest. Comparison between simulated and observed values was, however, made based on modeled daily average *E. coli* concentration and instantaneous *E. coli* concentration from grab samples collected from the stream. Overall, the APEX model proved to be generally satisfactory in modeling bacterial fate and transport in the test watershed.

Aside from the work of Hong et al. (2018), no other applications of the APEX model for microbial fate and transport exist in published peer-reviewed literature. The absence of other model applications may be due to several reasons like the complexity of the model, data requirements, the availability of other more widely used modeling options like SWAT among other reasons.

4.3.3 Recent Model Improvements

Since the incorporation of a microbial fate and transport submodel in APEX in 2018, no recent improvements have been published in peer-reviewed literature to the best knowledge of the author. The absence of model improvement is obviously due to the limited published applications of this model, which in turn may be due to the availability of more widely applied models like SWAT among other reasons. The results of the study by Hong et al. (2018) could serve as a take off point for further model application and improvement.

4.4 The WAM Model

4.4.1 Model Description

The Watershed Assessment Model (WAM) is a process-based watershed-scale hydrologic and water quality model developed by Bottcher et al. (2012). First developed in the 1980s, the model was transformed into a Geographic Information System (GIS)-based model and became a fully integrated ArcMap application in the 2010s (Bottcher et al., 2014). In this model, the entire watershed is subdivided into scalable grids within which mass balance is applied for various constituents such as water, total suspended solids, biological oxygen demand, nitrogen, and phosphorus. Outflows from each cell are then routed to stream reaches via surface runoff and subsurface runoff considering the assimilative processes along the flow paths and are subsequently routed to the watershed outlet (Bottcher, 2012).

Microbial fate and transport modeling in WAM basically involves the application of mass balance to each grid or cell in the watershed considering microbial mass loading, attenuation, and diffusion assuming constant die-off rate and time

dependent diffusion rate (Tian et al., 2002). The microbiological contaminant is then routed to the stream from surface runoff and subsurface flows and simulation of in-stream processes of deposition and entrainment and simulation of bacterial die-off in land and water as a function of temperature and solar radiation (Collins & Reutherford, 2004). These various processes are represented in the WAM model by mathematical relationships including differential equations.

4.4.2 Recent Model Applications

Application of the WAM model for bacterial fate and transport is limited based on published peer-reviewed literature. Among the rare WAM applications done even before its full-blown GIS-based version include simulation of fecal indicator organism concentrations in streams to improve land management practices in terms of reduction of fecal contamination of waterways in a 2.6-sq km watershed in New Zealand by Collins and Rutherford (2004). In their study, *E. coli* concentrations estimated from daily records of grazing livestock were used to simulate bacterial transport with both surface and subsurface flows incorporating any *E. coli* deposited directly into the streams. Their results showed that the WAM model was able to broadly reproduce the observed *E. coli* concentrations, but they reported some uncertainties in some of the processes represented. They also found that the distance over which surface runoff delivers bacteria to a stream and the amount of excretion direct to streams and onto seepage zones are the most sensitive model parameters in WAM. They also performed scenario analysis using the model and suggested that buffer strips could minimize bacterial contamination of water in the stream.

The other early application of WAM for microbial fate and transport modeling was done by Tian et al. (2002) in a 1.4 sq. km watershed in the United States to examine fecal coliform contamination in the stream posed by animal-grazed farmlands. Their results showed a good agreement between simulated and observed *E. coli* concentration. They, however, suggested that further experiments should be done to expand the model's functionality to cover more options for mitigation.

The limited application of the WAM model may be due to several reasons like model complexity, numerous data requirements, lack of user-friendliness, the availability of better modeling alternatives like SWAT among other reasons.

4.4.3 Recent Model Improvements

No reported model improvement on the WAM model can be found in recent peer-reviewed published literature. Obviously, this is tied up with the limited application of the WAM model. The results of the previous model applications could serve as the basis for further model improvement especially in view of advances in computing technology to enhance its capability for water quality management.

4.5 The SWMM Model

4.5.1 Model Description

The Storm Water Management Model (SWMM) is a physically based, event or continuous watershed-scale rainfall-runoff model used for simulating runoff quantity and quality primarily from urban areas (Rossman, 2015). The watershed is treated as a collection of subcatchments receiving precipitation and generating runoff, which is then routed through a system of conveyance channels. Both the quantity and quality of runoff generated within each subcatchment are tracked by this model, along with the flow characteristics and water quality in the conveyance channels during simulation. This model was developed by the US Environmental Protection Agency.

The microbial fate and transport model in SWMM is based on mass balance and advective transport principles. Mass balance is applied to each subcatchment, which is treated as a linear reservoir. Modeling of the build-up of contaminants on surfaces is set by the user using a power, exponential, or Langmuir saturation function. The rate constant for the mobilization process representing contaminant wash off follows a power function of overland flow and assumed to be proportional to the remaining mass of the contaminant. The decay of pollutants like bacteria is assumed to follow a first-order reaction with a user-supplied reaction rate constant. (Niazi et al., 2017).

4.5.2 Recent Model Applications

The most recent application of the SWMM model for microbial fate and transport modeling was done by Hong et al. (2021) to simulate the impact of *E. coli* from stormwater drainage from a 0.95 sq. km urban watershed in France on water quality of a recreational lake. In their study, SWMM was linked to two lake models, namely, the Delft3D-Flow model and the Delft3D-WAQ model to form an integrated modeling system. The stormwater sewer flow and the flux of fecal contaminants were generated by the SWMM model and served as inputs to the Delft3D-Flow model and Delft3D-WAQ model, respectively. Both event-based and long-term simulations with a time step of five seconds reported every two minutes were performed. Results showed generally good agreement between observed and simulated *E. coli* concentrations. However, they reported some model uncertainties associated with build-up and wash-off parameters which caused simulation discrepancies of up to two orders of magnitude. Nevertheless, this integration of the SWMM model with widely accepted Delft3D flow and water quality models represents an innovative approach for water quality impact assessment. Such an innovative approach could enhance SWMM model's functionality for dealing with complex urban water watersheds and recreational water systems to prevent microbial contamination of water supply sources and for better water quality management in general.

The SWMM model was also applied previously by Bai et al. (2006) to model the fate and transport of *enterococci* in a tidal river in the United States. They used the SWMM model to generate the combined sewer outflows from an urban catchment in Delaware in combination with other hydrologic models and a modified fluid dynamics model. Their results showed that the tidal river is mainly affected by the *enterococci* coming from the upstream creeks because of manure washed off from agricultural land during rainfall events and direct deposition of livestock wastes into the streams.

Other than the above, no other applications of SWMM for microbial fate and transport on a watershed-scale exist in published peer-reviewed literature. The limited application of SWMM in published literature may be due to the data limitations and the availability of other more widely used alternatives like the SWAT model.

4.5.3 Recent Model Improvements

To date, no reported recent model improvement on the SWMM model can be found in peer-reviewed published literature. The absence of model improvement in SWMM for microbial fate and transport is obviously due to its limited number of users as evidenced by the scarcity of published literature on its applications. More model applications would lead to more recommendations on how the model can be further improved particularly on how the fate and transport of pathogenic microorganisms can be better simulated and predicted.

4.6 The WATFLOOD Model

4.6.1 Model Description

The WATFLOOD model is a distributed parameter hydrologic model that can be used for both flood forecasting and streamflow modeling developed by Kouwen and his colleagues at the University of Waterloo in Canada for the Ontario Ministry of Natural Resources and Forestry (Kouwen & Mousavi, 2002; Kouwen, 2022). The WATFLOOD model has been recently called the WATFLOOD/CHARM model, with WATFLOOD constituting a set of DOS-based executable programs including pre- and post-processing while CHARM or the Canadian Hydrological and Routing Model is the model itself (Kouwen, 2022). In this model, the whole watershed is subdivided into grid cells which are grouped by land use which then generates hydrologic response. This model can then be coupled with a microbial fate and transport model (Cho et al., 2016b).

4.6.2 Recent Model Applications

The WATFLOOD model was applied by Wu et al. (2009) to predict the temporal and spatial variation in *E. coli* in water and sediments in Blackstone River in Massachusetts with a catchment area of 186.7 sq. km. The study was intended to demonstrate the role of sediments in the fate and transport of potential pathogens such as *Cryptosporidium, Giardia, Campylobacter, E. coli O157:H7* using *E. coli* as tracer. Both the hydrologic model and microbial fate and transport model were calibrated and validated using observed data. Model results showed high correlation between observed and simulated streamflow values, with R^2 ranging from 0.64 to 0.92, with the low value attributed by the authors to the inadequacy of the model to capture diurnal variations related to the wastewater input into the river. On the other hand, the microbial fate and transport model underpredicted the observed *E. coli* concentration values. The authors found improved results if sediment resuspension is included during model calibration indicating the significant contribution of sediments on microbial fate and transport.

Dorner et al. (2006) applied and augmented the WATFLOOD model to predict the fate and transport of bacteria and pathogens in a watershed in Southwestern Ontario, Canada, with a catchment area of 130 sq. km. The WATFLOOD model was augmented by considering the microbial transport to be due to overland flow, subsurface flow to tile drainage systems, and in-stream routing. Their study was intended to determine the primary sources of pathogenic contamination and the factors influencing their survival and transport. Their modeling results showed that most of the pathogens in the stream come from point sources like tile drainage systems rather than from overland flow and microbial transport. They also found the significant role of sediment resuspension on the transport of these pathogens.

4.6.3 Recent Model Improvements

To date, no reported recent model improvement on the mathematical representation of microbial fate and transport processes in the WATFLOOD model can be found in peer-reviewed published literature since Dorner et al. (2006) did its augmentation. However, the model itself is subjected to continuous software updates for users (Kouwen, 2022).

4.7 *The GWLF Model*

4.7.1 Model Description

The Generalized Watershed Loading Function (GWLF) model is a continuous lumped-parameter watershed-scale hydrologic and water quality model developed by Haith and Shoemaker (1987) at Cornell University for predicting streamflow

discharge and nutrients like nitrogen and phosphorus in streamflow. In this model, streamflow consists of both surface runoff and groundwater flow calculated from water balance while nutrients are calculated based on transported amounts through runoff and eroded sediments coming from various sources. Runoff is calculated in the model using SC Curve Number, while soil erosion is estimated using the Universal Soil Loss Equation. Like the other watershed-scale models, subdivision of the entire watershed into subbasins serving as hydrologic response units is also done in GWLF model based on land use. Detailed model description and equations used are presented by Haith et al. (1992).

Owing to the presence of water quality components, the GWLF model can be extended for microbial fate and transport modeling on a watershed-scale. The GWLF model considers the contribution of both rural and urban runoff to pollutant loading. It also uses a mass balance approach for both the unsaturated zone and the shallow saturated zone in the subsurface. Unique in this model is the presence of a septic or onsite wastewater disposal system component, which accounts for four possible pollutant sources such as effluents from properly sited septic systems, short-circuited effluents from septic systems built near surface water bodies, effluents due to hydraulic failure of septic tanks and direct septic discharge, each of which is represented by mathematical relationships reported by Haith et al. (1992). However, other processes that affect the growth, behavior, and die-off of microorganisms need to be considered for proper microbial fate and transport modeling.

4.7.2 Recent Model Applications

To date, no recent application of GWLF exists in published peer-reviewed literature. However, this model has been proven to be applicable for microbial fate and transport modeling as demonstrated by Walker and Stedinger (1999). In their study, they developed this model to simulate the fate and transport of *Cryptosporidium parvum* in the Catskill-Delaware watershed based on the assumption that the microorganisms move in response to runoff events and that the microbial degradation follows the first-order reaction kinetics incorporating a first-order decay function for the stream routing and reservoir modeling. This demonstration of GWLF's capability for microbial fate and transport modeling could have paved the way for more model applications and further research geared towards model improvement.

The absence of other recent applications of GWLF may be attributable to several reasons including data availability for the unique model components, shifts in model preferences by users leaning more towards widely used models like SWAT, other advances in microbial fate and transport modeling superseding GWLF's capabilities among other reasons.

4.7.3 Recent Model Improvements

While no improvement on GWLF for microbial fate and transport modeling has been documented in published peer-reviewed literature, Yuan et al. (2020) reported that the model itself has been integrated into the Mapshed GIS-based model developed and documented by Evans (2022) and maintained by Pennsylvania State University. This integration of the GWLF model into a GIS-based MapShed model could lead to a more refined spatial prediction of the fate and transport of pathogenic microorganisms. The use of this GIS-platform could also greatly facilitate the handling, storage, management, and visual representation of the various parameters associated with microbial fate and transport.

4.8 *The PCB Model*

4.8.1 Model Description

The Pathogen Catchment Budget (PCB) model is a watershed-scale model that was developed to quantify microbial contamination from primary sources such as livestock manure, wildlife wastes, and point sources like wastewater treatment plants and septic tanks and to predict pathogen and fecal indicator outputs under various weather conditions such as "dry" condition (5 mm of rainfall in 24 hr), "intermediate wet weather" (30 mm of rainfall in 24 h) and "wet weather flood events" (100 mm of rainfall in 24 h), and other weather conditions which can be specified in the model (Ferguson et al., 2007). The model has five model components each representing the hydrologic, land budget, on-site systems, sewage treatment plant, and in-stream transport.

In the PCB model, the entire watershed is subdivided into subcatchments, which form the basic mass balance units defined by land slope, soil type, elevation, stream density, and land use and that different governing equations are employed for dominant weather conditions (Cho et al., 2016b).

The various model assumptions and governing equations for representing the mass transport of microorganisms from various sources and through various hydrologic pathways including survival rates of microorganisms under each of the predominant weather conditions are described fully in Ferguson et al. (2007).

4.8.2 Recent Model Applications

The PCB model was applied by Ferguson et al. (2007) in combination with a hydrologic model to estimate pathogen budgets in a 9050 sq. km watershed in Sydney, Australia as a case study when they developed and tested the model. The large watershed has mixed land use having unsewered and sewered urban areas, rural residential, pastures, horticulture, native vegetation, grazing areas, and a protected

upland swamp. Subdivision of the watershed into 52 subcatchments from which water quality sampling was performed and other model parameters such as animal and human population were estimated. Their study proved that the PCB model could generate estimates of pathogenic loading or inputs for each microorganism such as *Cryptosporidium, Giardia, and E. coli* and the pathogenic outputs routed downstream for each subcatchment under various predominant weather conditions."

To date, no other publications demonstrating the application of the PCB model for microbial fate and transport exist in published peer-reviewed literature. This may be due to model complexity, data availability, and higher preference by modelers on the use of more widely accepted models like SWAT over other models among other reasons.

4.8.3 Recent Model Improvements

No reported model improvement on the PCB model can be found in recent peer-reviewed published literature The absence of model improvements is mainly due to the absence of more model applications in published peer-reviewed literature, which could generate potential recommendations for improving model performance. The modeling results published by Ferguson et al. (2007) could serve as takeoff point for further research on the improvement of the various modeling components particularly those directly associated with watershed-scale microbial fate and transport processes. While the PCB model is considerably complex and requires numerous datasets, its unique capability to account for the effects of rainfall conditions on the fate and transport of pathogenic microorganisms and in dealing with both agricultural, wildlife and domestic pathogen sources in urban watersheds with mixed land use could pave the way for the evolution of a new landscape for microbial transport modeling.

4.9 The RIVERSTRAHLER Model

4.9.1 Model Description

The RIVERSTRAHLER model is a continuous watershed-scale model that simulates the biogeochemical processes in river systems for a given set of morphological characteristics of the drainage network, hydrometerological conditions and sources of pollution, both point and non-point sources. It was developed by Billen et al. (1994) and Garnier et al. (2002) and was further developed by Ruelland et al. (2007) with the introduction of a GIS-interface that led to the evolution of SENEQUE/RIVERSTRAHLER model, applicable to watersheds with a catchment area of 100 to 100,000 km^2 or more. The drainage network consists of upstream basins, branches, and ponds or reservoirs. Depending on the spatial resolution required, the

drainage network is treated as either a single basin or as a collection of connected basins and branches. The routed water fluxes are coupled by the model with the biological, microbiological, and physico-chemical processes.

The microbiological component of the SENEQUE/RIVERSTRAHLER model was developed by Servais et al. (2007). They included both point and non-point or diffused microbial sources. The microbial release of microorganisms from effluents of wastewater treatment plants is determined based on the treatment plant capacity and the specific microbial load per capita per day. On the other hand, the microbial release from non-point sources is calculated based on land use in each subcatchment accounting for both surface runoff and soil leaching. Modeling of the microbial fate and transport assumes first order kinetics with a temperature-dependent reaction rate constant.

4.9.2 Recent Model Applications

The RIVERSTRAHLER model was applied by Ouattara et al. (2013) to model fecal contamination in the Sheldt drainage network with a catchment area of 20,000 sq km. in northern continental Europe which covers portions of France, Belgium and the Netherlands. They considered both point source and non-point sources of microbial contamination and used SENEQUE-STRAHLER to facilitate the handling and management of spatial data. For each stretch of the river in the large drainage network, they considered three stocks of *E. coli* namely, free *E. coli* in water, *E. coli* adsorbed onto suspended sediments, and *E. coli* attached to deposited sediments. The three stocks are governed by a different mortality rate and were assumed to follow first-order kinetics with a temperature-dependent reaction constant. They also allowed settlement and resuspension of the *E. coli* attached to sediments and resuspension. The model was calibrated using model parameter values reported in published literature and consequently validated by comparing the predicted *E. coli* concentration values with observed data from the various river stretches. Results showed the capability of the SENEQUE/RIVERSTRAHLER model to simulate average or seasonal *E. coli* concentrations with an acceptable NSE of 0.57 and RMSE of 0.59. However, the model proved to be incapable of simulating short time variations in microbial concentration like those due to tidal occurrences or extreme conditions because of its low temporal resolution of 10 days. Nevertheless, this study proved the capability of the RIVERSTRAHLE model for predicting seasonal microbial water quality in large and complex transnational river systems like the Scheldt drainage network.

Since then, no other recent peer-reviewed publications exist demonstrating further application of the RIVERSTRAHLER/EC model for microbial fate and transport modeling. This may be due to model complexity, accessibility, data availability and preference to other more widely accepted models. In previous years, however, the RIVERSTRAHLER/EC model was applied for modeling microbial contamination in a 75,000-sq. km. watershed in France by Servais et al. (2007) when they developed the microbiological component of the model as described in the

preceding section. They were able to demonstrate that the model, after validation with observed data, can handle various predictive scenarios through prospective and retrospective modeling, which would prove useful for water quality management.

4.9.3 Recent Model Improvements

No reported model improvement on the RIVERSTRAHLER/EC model can be found in recent peer-reviewed published literature. The absence of model improvements is due to the absence of recent model applications. The modeling research contributions of Servais et al. (2007) and Ouattara et al. (2013) could serve as basis for future research to further enhance the capability of the RIVERSTRAHLER model for microbial fate and transport modeling on a watershed scale.

4.10 Other Process-Based Models That Can Be Used for Predicting Microbial Fate and Transport

Apart from the most popular and widely used watershed-scale models presented in the preceding sections, there are other process-based models that are capable of simulating and predicting microbial fate and transport. Among them is the commercially available MIKE SHE models developed in Europe. It is one of the most credible distributed and physically-based hydrologic models extensively used in practice and is fundamentally based on widely accepted hydrologic principles. Its merits, superior features, applications and challenges were sufficiently reported by Ma et al. (2016). Among the challenges often encountered in its application include the need for multiple scientific disciplines other than hydrology, adequate computational hardware requirements and large amounts of data.

To date, no peer-reviewed publications are available to show evidence of MIKE SHE model application for microbial fate and transport as pointed out by Cho et al. (2016b) and even from recent literature searches made by the author. The seeming underrepresentation of the MIKE SHE models for watershed-scale microbial fate and transport modeling in peer-reviewed literature may be due to the higher preference of modelers to use free and equally credible models like SWAT over commercial modeling software and the complexity and enormous data requirements of the MIKE SHE model. Moreover, the use of the MIKE SHE models by practitioners may be intended more for actual and practical water resources planning, development and management purposes than for scientific publications.

There are other process-based based models such as the Watershed Analysis Risk Management Framework (WARMF) and the Source Loading and Management Model (SLAMM) models, both developed in the United States, that can handle microbial fate and transport modeling as reported by Yuan et al. (2020). WARMF is a comprehensive distributed parameter and physically based watershed-scale

hydrologic and water quality model that can simulate 13 hydrologic parameters and 43 water quality constituents including coliform bacteria (Herr & Chen, 2012). The model essentially accounts for advective transport processes, adsorption onto soil particles and suspended sediments, reaction rates and temperature. On the other hand, the SLAMM model is a model intended primarily for determining sources of both particulate and dissolved pollutants and the effective management of stormwater to control the pollutants in a watershed such as detention ponds, percolation ponds, infiltration devices, porous pavements, grass swales, catch basin cleaning, street cleaning, and other management alternatives. All its unique features are described by Pitt (1997). This model incorporates a unique description of processes instead of using common assumptions and simplifications employed in drainage design models for a more accurate prediction of sources of pollutants. The model essentially uses a mass balance approach to simulate or predict water quality constituents, which may include pathogenic microorganisms. However, no recent peer-reviewed publications exist that the author is aware of to show their applications for watershed-scale microbial fate and transport modeling based on searchable literature databases. This may be due to their less popularity compared to more widely accepted models like SWAT and HSPF and other possible reasons like model accessibility issues, data requirements among others.

Other researchers developed their own process-based model for microbial fate and transport modeling, among the most notable of which include the models developed by Niu and Phanikumar (2015) and by Wilkinson et al. (2011). Niu and Phanikumar (2015) developed a process-based watershed-scale model based on an operator splitting strategy and a combination of Lagrangian particle transport for advection and Eulerian approach for dispersion and reactions. They employed differential equations for overland flow and transport in both the vadose and groundwater zones and other equations for the interaction among flow domains. The model was tested in the Red Cedar River watershed in Michigan, the United States, to simulate *E. coli* transport and results showed good agreement between simulated and observed values with log-RMSE ranging from 0.59 to1.00. On the other hand, Wilkinson et al. (2011) developed a process-based model using equations based on first-order kinetics for bacterial die-off and dispersion due to mixing for in-stream microbial transport allowing for microbial release from bedload and an assumption that the number of bacteria released from animal waste manure is directly proportional to the runoff depth and that delivery to streams occurs at the same time. They tested the model to simulate the transport of *E. coli* in a 2074-sq.km watershed in New Zealand under dry and wet weather conditions. Since the development of these models, however, no other applications exist in published peer-reviewed literature. Its wider application may have been hampered by the fact that they were developed for site-specific purposes. Besides, modelers have plenty of options and tend to adopt more widely used models like SWAT.

5 Non-process-Based Watershed-Scale Models of Microbial Fate and Transport

Watershed-scale models that are not physically based or process-based are generally of the conceptual type. These models employ equations with adjustable model parameters to estimate hydrologic processes. Some of these simple hydrologic models have been coupled with microbial fate and transport models. The most commonly used conceptual models that have shown applicability for microbial fate and transport modeling on a watershed scale are presented in this section.

5.1 The SIMHYD-ED Model

5.1.1 Model Description

The SIMHYD model is a simple conceptual rainfall-runoff model consisting of reservoirs for interception storage, soil or vadose zone storage and groundwater storage operated using daily time steps to simulate streamflow from watersheds (Chiew et al., 2002; Cho et al., 2016b). Daily precipitation is routed through the conceptual storage reservoirs to generate the various hydrologic processes including surface runoff and baseflow, which are summed up to simulate daily total streamflow. This model is basically similar to the conceptual water yield simulation model developed by the author (Ella, 1986 and Ella, 2016). The SIMHYD was developed and tested in numerous Australian watersheds. Like other hydrologic models, a microbial fate and transport model that account for advection, diffusion, die-off and other processes can be combined with the SIMHYD model to simulate and predict bacterial concentrations in watersheds over time.

5.1.2 Recent Model Applications

The SIMHYD model was applied by Haydon and Deletic (2006) in combination with a microbial fate and transport model to simulate *E. coli* concentrations generated by animal and human wastes in large Australian watersheds ranging in area from 780 ha to 11,900 ha. They used SIMHYD to predict streamflow and coupled a separate conceptual pathogen transport model they developed based on assumed pathways for inflows like land deposition and septic tanks and outflows or wash offs through evapotranspiration, infiltration, overflow from conceptual soil moisture storage, interflow, surface runoff, and subsurface flows. They then developed the corresponding parametric equations with adjustable model parameters to calculate the microbial concentration stored or transported through the assumed pathways. The model was tested in three Australian watersheds and results showed a generally good agreement between simulated and observed *E. coli* concentration values with

R^2 of 0.93 for all watersheds. However, the authors did not include uncertainty analysis and suggested further model improvement to minimize timing errors in the simulation.

To date, no other application of SIMHYD exists in published peer-reviewed literature for microbial fate and transport modeling on a watershed scale. This may be due to higher reliability of process-based models over non-process-based models.

5.1.3 Recent Model Improvements

No reported model improvement on the SIMHYD model can be found in recent peer-reviewed published literature. While Haydon and Deletic recommended further model improvements, the absence of follow up work is most likely due to greater preference by researchers and modelers on the use and improvement of more widely applied process-based models over less widely used non-processed-based models.

5.2 *The WALRUS Model*

5.2.1 Model Description

The Wageningen Lowland Runoff Simulator (WALRUS) model is a parametric conceptual rainfall-runoff model consisting of reservoirs for soil, quick flow and surface water used for simulating total runoff from precipitation inputs. It was developed by Brauer et al. (2014) and is basically similar in simulation approach as the SIMHYD model and the conceptual model developed by the author (Ella, 1986 and Ella, 2016). The WALRUS model is suitable for lowlands where both surface water and shallow groundwater contribute to runoff (Cho et al., 2016b). Like other hydrologic models, the WALRUS model can be linked to a microbial fate and transport model that account for advection, diffusion, die-off and other processes to simulate and predict the bacterial concentrations in watersheds over time.

5.2.2 Recent Model Applications

Sterk et al. (2016) applied the WALRUS model supplemented with a microbial fate and transport module to evaluate the effect of climate change on the runoff of such microorganisms as *Campylobacter* and *Cryptosporidium* from land to surface water in a watershed in the Netherlands. They used WALRUS to predict runoff and calculated microbial concentration by dividing the microbial concentration over the flow routes proportional to the water flow and assuming instant microbial mixing for all fluxes. They then developed the corresponding parametric equations accounting for advection processes and decay rates. The coupled model was then used to predict

and evaluate the concentration of Cryptosporidium and Campylobacter in runoff under different climate change scenarios and from fecal pollution generated by dairy cows and geese and manure fertilization. Results showed minimal overall impact of climate change on the microbial concentration in runoff. Nevertheless, this study demonstrated the capability of non-process-based models to perform scenario analysis and evaluate the impact of contemporary issues like climate change on hydrologic processes in the watershed that may affect public health.

To date, no other recent peer-reviewed publication exists showing further applications of the WALRUS model for microbial fate and transport on a watershed scale. Again, this may be due to higher reliability of process-based models over non-process-based models.

5.2.3 Recent Model Improvements

No reported model improvement on the WALRUS model can be found in recent peer-reviewed published literature. The absence of model improvement is mainly due to the absence of follow up work to Sterk et al. (2016), which in turn may be due to greater shift by modelers on the use of more widely applied process-based models over non-processed based models.

6 Integrative Models or Modeling Systems

A framework or system that integrates various types of models is generally called a modeling system or an integrative model. This system of models is often used for making risk assessment on a watershed scale. While they are not generally used for describing the fate and transport of microbiological contaminants the way process-based models do, these modeling systems have become one of the useful tools for making microbiological assessments on a watershed scale and hence, they are included in this chapter.

The most widely used modeling systems include the Framework for Risk Analysis in Multimedia Environmental Systems (FRAMES) and the Quantitative Microbial Risk Assessment (QMRA). FRAMES is a component-based framework that integrates various model types such as watershed models, groundwater model, surface water model, exposure risk model and other models together with environmental databases (Cho et al., 2016b). QMRA is a modeling system that also integrates microbial fate and transport and other data that characterize the distribution of pathogenic microorganisms within the watershed (Whelan et al., 2018).

The most recent application of QMRA for waterborne disease risk assessment and mapping was done by Nakamura et al. (2023). They combined QMRA with a hydraulic simulation model to evaluate the spatio-temporal changes in risks associated with waterborne disease in a village in Cambodia. Simulation of *E. coli* transport was done using a 3D hydraulic model. Only a limited number of observed

E. coli concentration values were, however, available for proper assessment of the goodness of model fit. The QMRA was then used for generating disease risk maps.

Another application of QMRA was made by Whelan et al. (2018) for determining the distribution of microbial sources in a mixed-use watershed in the United States. They used the Bacterial Indicator Tool (BIT) developed by the USEPA (2000) for the microbial modeling part as input to QMRA. They were able to demonstrate the capability of QMRA to organize, capture, and execute microbial data within a modeling workflow and emphasized that source-term data are critical to QMRA development. Consequently, this may pose a big challenge in data collection and management for the widespread adoption of QMRA and FRAMES for the risk management of waterborne diseases emanating from the fate and transport of pathogenic microorganisms.

Earlier application of FRAMES and QMRA was done by Roser et al. (2006) for assessing the microbial exposure of an urban recreational lake in the United States in which they applied new risk-based guidelines. While this earlier application basically made use of general conceptual frameworks mainly intended for managing risks associated with microbial contamination in drinking and recreational waters, it served as an important initial step towards the full development of the integrative modeling approaches.

7 Other Modeling Approaches for Watershed-Scale Microbial Fate and Transport

7.1 Deep Learning Models

Deep learning models are among the new models that can be used for microbial fate and transport modeling. Essentially of the data-driven type, deep learning models are based on extensive data used to develop relationships for microbial fate and transport prediction. Deep learning models are, however, not process-based and cannot properly represent physical processes and hence cannot fully describe the spatial and temporal dynamics of bacterial processes. However, with adequate historical data, properly calibrated or trained deep learning models can handle complex and nonlinear relationships and could offer a sound alternative for predicting the fate and transport of pathogenic microorganisms particularly in large and complex watersheds. The availability of an enormous amount of spatial and temporal hydrologic and microbiological data could, however, limit the use of this type of models for watershed-scale microbial fate and transport modeling. On the other hand, in watersheds where data is not a constraint, such problems as overfitting and interpretability may pose additional challenges on the use of deep learning models.

Among the most recent applications of deep learning models for microbial fate and transport include in-stream modeling of *E. coli* performed by Abbas et al. (2021) in a 0.6 sq. km. watershed in Lao PDR. They applied the deep learning model based on long short-term memory technique and found a better model fit to

observed values in comparison with a process-based model, the HSPF. Results of their modeling study showed an NSE value of 0.35 for deep learning model simulation and a low NSE of −3.01 for HSPF based on observed and simulated *E. coli* concentrations.

7.2 Inverse Modeling Approach

Inverse modeling, as the name implies, essentially involves backward instead of forward calculation or backtracking of inputs or loads based on observed output or concentration values. For microbial fate and transport this means tracing and calculating the pathogenic sources and loading rates in upstream areas and other possible zones of microbial entry based on observed pathogenic concentrations downstream of the watershed. Its outputs can then be used as inputs to watershed-scale hydrologic models for the simulation or prediction of microbial fate and transport. The approach is basically like the inverse groundwater modeling performed by Ella et al. (2002a, b) and Ella (1999).

Inverse modeling approach was performed by Sisson et al. (2016) to estimate bacterial loading and transport in a tidal embayment. They used an inverse approach to estimate loading from unknown bacterial sources based on observations in the embayment and consequently combined them with loadings from human, wildlife, agriculture, and other known sources. They then used these loadings as inputs to the watershed model. Their study suggests that inverse approaches combined with watershed models and bacterial transport models improves model simulation of bacterial transport generated from unknown sources not solely driven by precipitation.

8 Comparison of Existing Microbial Fate and Transport Models

Table 18.1 shows a summary of the various watershed-scale models that have proven applicability for modeling microbial fate and transport, microbial parameters simulated, sample watershed applications, and recent model improvements. The applicability of these models varied across the board in terms of watershed size, soil, climate and other biophysical factors. Overall, the SWAT and HSPF models have shown wider applicability in terms of number of peer-reviewed publications. However, while both models were developed in the United States, the SWAT model has been applied in six other countries outside the United States for microbial fate and transport modeling while HSPF was applied mostly in the United States for the same purpose and only one recent application in another country. The US-developed APEX model was also applied in two other countries outside the United States but the number of peer-reviewed publications demonstrating its application for

Table 18.1 Watershed-scale models for microbial fate and transport, evidences of application and recent major improvements

Model	Microbial parameter or microorganism simulated	Peer-reviewed evidences of model application	Country location of test watersheds	Recent model improvements
SWAT	Fecal coliform concentration, *E.coli* concentration, bacterial loading, pathogen source	Meshesha et al. (2020), Sowah et al. (2020), Jeong et al. (2019), Jeon et al. (2019), Thilakarathne et al. (2018), Iqbal and Hofstra (2019), Kim et al. (2017), Cho et al. (2016a, b), Jayakody et al. (2014), Coffey et al. (2013), Frey et al. (2013), Baffaut and Sadeghi (2010), Chin et al. (2009), Coffey et al. (2009), Parajuli et al. (2009), Benham et al. (2006)	USA, Canada, France, Ireland, Korea, Afghanistan, Lao People's Democratic Republic	Meshesha et al. (2020) modified the SWAT-bacteria module to account for the effect of pH on the fate and transport of *E. coli*. Jeon et al. (2019) modified the SWAT source code and introduced improved mathematical representation of bacterial growth and hyporheic exchange to account for seasonal temperature variations for improved calculation of bacterial re-growth and die-off. Cho et al. (2016b) developed a new bacteria subroutine for SWAT that takes into account critical temperatures as a parameter for simulating the onset of bacterial growth or die-off
HSPF	Fecal coliform concentration, *E. coli* concentration, bacterial loading	Abbas et al. (2021), Sobel et al. (2017), Coffey et al. (2015), Iudicello and Chin (2013), Seong et al. (2013), Rolle et al. (2012), Chin (2011), Desai et al. (2011), Petersen et al. (2011), Chin et al. (2009), Hall et al. (2008), Benham et al. (2006), LaWare and Rifai (2006), Im et al. (2004), Paul et al. (2004), Moyer and Hyer (2003), Brannan et al. (2002)	USA, Lao PDR	Liao et al. (2015) modified the HSPF model by accounting for the interaction between bacteria and sediments during in-stream microbial transport, calculating the total in-stream microbial concentration in water based on the free-phase microbial concentration and the microbial concentration attached to the suspended clay and silt particles. No other recent model improvement exists in published peer-reviewed literature although the HSPF model has been linked to other models like the Tidal Prism Model and some Deep Learning models to improve the accuracy of model performance

APEX	*E. coli* concentration	Hong et al. (2018)	USA, Korea, New Zealand	None over the last 10 years
WAM	*E. coli*	Bottcher et al. (2012), Collins and Rutherford (2004), Tian et al. (2002)	USA, New Zealand	None over the last 10 years
SWMM	*E. coli*, *Enterococci*	Hong et al. (2021), Bai et al. (2006)	USA, France	None over the last 10 years
WATFLOOD	*E. coli*	Wu et al. (2009), Dorner et al. (2006)	Canada, USA	None over the last 10 years
GWLF	*Cryptosporidium parvum*	Walker and Stedinger (1999)	USA	None over the last 10 years
PCB	*Cryptosporidium*, *Giardia*, and *E. coli*	Ferguson et al. (2007)	Australia	None over the last 10 years
RIVERSTRAHLER	*E. coli*	de Brauwere et al. (2014), Ouattara et al. (2013), Holz (2010), Servais et al. (2007)	France, Belgium, The Netherlands	None over the last 10 years
SYMHYD-ED	*E. coli*	Haydon and Deletic (2006)	Australia	None over the last 10 years
WALRUS	*Campylobacter*, *Cryptosporidium*	Sterk et al. (2016)	The Netherlands	None over the last 10 years
FRAMES QMRA	*Enterococci* density, Rotaviruses, Enteroviruses	Nakamura et al. (2023), Whelan et al. (2018), Roser et al. (2006)	USA, Cambodia	None over the last 10 years

microbial fate and transport is rather limited. The WAM and SWMM models, both developed in the United States, have been applied in another country aside from local application. The WATFLOOD model developed in Canada proved to be also applicable in the United States. The RIVERSTRAHLER/EC model developed in Europe has been applied in three countries. All other models that are originally hydrologic models linked to microbial fate and transport models have only been applied in countries where they were developed. Of all the models reviewed, the SWAT, HSPF, SWMM, PCB, FRAMES, and WALRUS models were able to demonstrate versatility in terms of variety of pathogenic microorganisms they can simulate while the other models only proved to be able to simulate a single form of microorganism, notably *E. coli*.

In terms of accuracy in model simulation, no standardized comparison can be made for all the models presented in the preceding sections owing to varying nature of modeling objectives, watershed area, watershed complexity, temporal and spatial discretization, microbiological contaminants dealt with, varying length of record for model calibration and validation, overall watershed modeling conditions, varying measures used for assessing the goodness of model fit to observed data among other variables. Based on peer-reviewed publications on the application of these various models, the reported model simulation accuracy varied widely for each model. For instance, in SWAT, the Nash-Sutcliffe model efficiency (NSE), one of the best measures of goodness of fit, varied from poor at −0.42 (Coffey et al., 2013) with only one year of data used to satisfactory at +0.69 (Iqbal & Hofstra, 2019) who used three years of observed data. Other recent SWAT applications also showed varying degrees of accuracy in terms of NSE which ranged from as low as −0.94 (Niazi et al., 2015) to as high as 0.81 (Sowah et al., 2020). In HSPF, a relatively low NSE of −3.01 was obtained during model calibration and validation for a small watershed based on about seven years of data (Abbas et al., 2021). The APEX model application showed an overall NSE of 0.33 using about 5.5 years of hydrologic and water quality for model calibration and validation (Hong et al., 2018). Nevertheless, because of the wide application of these models and the soundness of their modeling approach, these models remain to be potentially useful for microbial fate and transport modeling on a watershed scale for assessing microbiological contamination of water supply sources from watersheds among other practical environmental, water resources management and public health applications.

9 Future Directions and Research Needs in Watershed-Scale Modeling of Microbial Fate and Transport

The fate and transport of pathogenic microorganisms in watersheds are highly complex and an accurate prediction of their spatial and temporal distribution in river systems remains a continuing challenge for modelers. Model uncertainties reported in a number of peer-reviewed publications on the application of watershed-scale microbial fate and transport models of various forms would certainly necessitate

improvements in the mathematical representation of the various biophysical processes governing microbial fate and transport, more accurate estimation of highly sensitive model parameters, and collection of sufficiently long records of observed microbiological data for better model calibration and validation.

Improvement in mathematical equations governing microbial fate and transport in watersheds would require further extensive experimental and modeling research. Most of the existing models for fate and transport of microorganisms account for advective transport by runoff and streamflow, adsorption onto eroded soil particles and sediments, and bacterial growth and decay. The transport of contaminants in soils is usually governed not only by advection and adsorption but also by dispersion. Inclusion of dispersion components in microbial transport equations may have to be explored. While the effect of sediment resuspension from bedload and hyporheic exchange has already been considered in some models, further research on these processes still needs to be pursued for various soil materials forming the channel bed. Similarly, studies on the effect of vegetative growth in the channel and other physical and environmental conditions on these processes should be undertaken. Bacterial growth and decay for most models are assumed to follow first-order kinetics. Higher orders of reaction kinetics may have to be explored if model accuracy is to be improved. The effects of environmental conditions such as temperature and pH on bacterial growth and decay should also be further examined considering the temporal and spatial variability of these parameters over the watershed. Research on the fate and transport of other pathogenic microorganisms in watersheds is also needed. Most of the model applications focused on fecal coliform and *E. coli*. Other forms of microorganisms including viruses, protozoa among others need to be included in modeling studies. The various forms of pathogenic microorganisms behave differently, and their transport dynamics would change under varying environmental conditions. Consequently, this would require the necessary adjustments in modeling. Advances in microbiological research may have to be tapped for improving the representation of microbial behavior in soil and water under varying watershed physical and environmental conditions. Advances in understanding microbial life cycles and interactions could lead to significantly improved rate constants or reaction kinetics in models.

The most sensitive model parameters that significantly influence the accuracy of model simulation as reported in various model applications should be further examined. Research on the various factors that influence the values of these model parameters needs to be undertaken to serve as additional basis for model calibration. The use of finer spatial and temporal resolution, with the application of modern tools, instruments, and techniques, should also be explored for the proper identification and refinement of the sensitive model parameter values especially in the most widely used models to further improve their predictive capacities and enhance their applicability for water quality management. Other sensitive model parameters associated with physical and microbiological processes would require further laboratory or field experimental research to further improve parameter estimation and optimization.

Model uncertainties due to insufficiently short calibration data could obviously be addressed with the collection of sufficiently long observed microbiological and hydrologic data in watersheds, particularly in those with potential sources for microbiological contamination. Technological advances in instrumentation for data collection methods including real-time monitoring using wireless sensors, automation, and Internet of Things (IoT) could be tapped for this purpose. Other model uncertainties due to inadequate representation of microbial fate and transport processes, simplifying assumption on homogeneity for heterogeneous spatial scales, use of coarse spatial and temporal resolutions among others should also be addressed in future model developments and improvements of models that have shown some applicability.

The adequacy of microbial fate and transport modeling could also be improved using coupled surface and groundwater modeling approaches. While the theories of contaminant transport in porous media like soils in both the vadose and unsaturated zones have long been developed (e.g., Fetter, 1993; Yates et al., 1988; Simunek et al., 2009) and a number of field and modeling studies have been done (e.g., Banasiak et al., 2023; Ella et al., 2001, 2002a, b; Gannon et al., 1991; Corapcioglu & Haridas, 1985), studies on the coupling of subsurface models with surface models for microbial transport remain scarce. In fact, Wang and Chen (2021) reported that this model coupling approach like the coupled SWAT-MODFLOW model has been applied mostly for hydrologic and nutrient transport and no study on its application for microbial fate and transport exists in published peer-reviewed literature that the author is aware of. Also related to subsurface and groundwater flow is the need to consider the possible occurrence of preferential flow in the coupled surface and groundwater models. Preferential flow in soil macropores and in cracking soils would certainly accelerate the transport of pathogens into the groundwater systems. Because groundwater systems are also widely used as drinking water supplies and that groundwater flow also contributes to streamflow, more field and modeling research needs to be conducted on the groundwater–surface water interactions and contributions to microbiological contamination of both groundwater and surface drinking water supply sources.

With the generation of enormous amounts of microbiological data over time and space on a watershed-scale, modern modeling tools such as machine learning and other Artificial Intelligence (AI) models could provide an alternative to process-based modeling in the future. The use of these new modeling techniques for assessing the combined impacts of climate change, land use/land cover change, and other factors on microbiological water quality may also be explored.

While the availability of enormous amount of climatic, hydrologic, microbiological, and other relevant modeling data with finer spatial and temporal resolutions would pose a constraint in the speedy takeoff of machine learning models and AI application for microbial fate and transport modeling, several emerging hardware and software technologies could pave the way to address these issues. For instance, the use of Climate Hazards Group Infrared Precipitation with Stations (CHIRPS) to generate rainfall inputs for SWAT modeling was demonstrated by Alejo et al. (2021). On the other hand, some sensor-based hydrologic data collection approaches

were discussed by Llaban and Ella (2022) and highlighted that these emerging methods are most effective for real-time monitoring of streamflow and applicable to almost any terrain and river size. State-of-the-art hardware and sensors and Internet of Things (IoT) application was also presented by Ramirez et al. (2022) for real-time monitoring of field data relevant for modeling purposes such as soil moisture, water level, and weather conditions. This technology may be extended to other hydrologic applications to generate data with finer spatial and temporal resolutions, which are needed by machine learning and AI models. Given these technologies among others which may still be under development, it is not a remote possibility that AI may take the place of traditional modeling approaches in the future including microbial fate and transport modeling. However, caution must be observed on the use of machine learning models being essentially a data-driven and black-box approach and hence, their reliability rests largely on the quality of observed data. Moreover, overdependence on AI tools must be avoided so as not to suppress the further scientific development of physical theories through human mind-led empirical and theoretical research which are essential for the full understanding of the various hydrologic and microbiological fate and transport processes under various prevailing physical and environmental conditions.

10 Conclusion

The review of the recent advances in microbial fate and transport on a watershed scale presented in this chapter provides a fresh perspective on the state of the art in modeling pathogenic microorganisms in hydrologic systems. Numerous process-based, non-process-based watershed-scale models, integrative, and other types have evolved and have been demonstrated to be applicable for microbial fate and transport modeling for various specific purposes and watershed conditions. Over the years, model applications have differed in terms of purpose, watershed size, watershed complexity, spatial and temporal discretization, length of observed data for model calibration and validation, extent of application among other variables. Results of this review showed that while other process-based, non-processed, and emerging types of models have been developed through the years, the SWAT and HSPF models remain as the most widely used and versatile for microbial fate and transport modeling in the world. This may be due to higher preference by modelers to use more widely accepted and more accessible models that continue to undergo model improvements. Other models were mainly developed for certain country-specific purposes, most of which require numerous data, and some are complex to use, which add up to their low rate of adoption. On the other hand, newer models like machine learning and deep learning models are still at their infancy stage and are just beginning to take off in microbial fate and transport modeling, although the availability of enormous amounts of data to enable their application would remain a constraint if their wider application is to be desired.

While model application for various purposes elicited recommendations to enhance model performance, no significant model improvements have been made in most of the existing models in recent years based on available peer-reviewed literature. Based on the results of this review, it is recommended that further research be conducted on improving the mathematical representation of the various biophysical processes governing the fate and transport of the various forms of microbiological contaminants in watersheds under varying environmental conditions. Moreover, improvement in data collection using IoT and wireless sensor technologies among other modern technologies needs to be pursued to generate sufficiently long water quality data for model calibration and validation not only to improve model performance but also to broaden the extent of applicability of these models. With the generation of enormous microbiological water quality data, the use of modern modeling tools such as machine learning and other AI models may then be explored for various purposes. Furthermore, research on the use of coupled surface water and groundwater models and consideration of possible preferential flow in the subsurface of watersheds should be explored for watershed-scale microbial fate and transport modeling especially that groundwater not only contributes to streamflow but also serves as a domestic water supply source in many areas around the world. Ultimately, microbial fate and transport modeling type of research and all other related experimental research that could contribute to the further improvement of these models in predicting the spatial and temporal distribution of these pathogenic microorganisms in watersheds and for identifying potential contamination sources should be continuously supported and sustained to enhance the usefulness of these models for water quality management.

Overall, the information presented in this chapter on the recent advances in watershed-scale microbial fate and transport modeling and the research recommendations offered could serve as a basis for future scientific work geared towards the enhancement of watershed-scale microbial fate and transport models as management tools to make them even more powerful for preventing the adverse water quality impacts of microbiological contaminants emanating from various sources in agricultural and urban watersheds and ultimately contribute better and to a larger extent to address public health concerns, environmental risks, and water resources sustainability.

Acknowledgements The author would like to thank his co-editor, Prof. Sonia M. Tiquia-Arashiro for the opportunity to work on this review article and contribute to this book.

Authors' Contributions The conceptualization, literature search, organization, writing, preparation of figure and table, revision, and preparation of the entire manuscript were done solely by the author.

Conflict of Interest While the author is among the editors of this book, he has no competing interests to declare that are relevant to the content of this article.

Funding No funds, grants or other support was received by the author in the preparation of this manuscript.Data AvailabilityNot applicable.

References

Abbas, A., Baek, S., Silvera, N., Soulileuth, B., Pachepsky, Y., Ribolzi, O., Boithias, L., & Cho, K. H. (2021). In-stream *Escherichia coli* modeling Using high temporal-resolution data with deep learning and process-based models. *Hydrology and Earth System Sciences*. https://doi.org/10.5194/hess-2021-98

Alegbeleye, O. O., & Sant'Ana, A. S. (2020). Manure-borne pathogens as an important source of water contamination: An update on the dynamics of pathogen survival/transport as well as practical risk mitigation strategies. *International Journal of Hygiene and Environmental Health, 227*, 113524. https://doi.org/10.1016/j.ijheh.2020.113524

Alejo, L. A., Ella, V. B., & Saludes, R. B. (2021). Applicability of the Climate Hazards Group precipitation with stations as rainfall input for SWAT watershed modeling. *Water Resources, 48*(6), 925–935.

Anderson, M. G., & Burt, T. P. (Eds.). (1985). *Hydrological forecasting*. Wiley.

Arnold, J. G., & Fohrer, N. (2005). SWAT2000: Current capabilities and research opportunities in applied watershed modeling. *Hydrological Processes, 19*(3), 563–572. https://doi.org/10.1002/hyp.5611

Arnold, J. G., Srinivasan, R., Muttiah, R. S., & Williams, J. R. (1998). Large-area hydrologic modeling and assessment: Part I. Model development. *Journal of American Water Resources Association, 34*(1), 73–89.

Arnold, J., Srinivasan, R., Neitsch, S., George, C., Abbaspour, K., Hao, F. H., van Griensven, A., Gosain, A., Debels, P., Kim, N. W., Somura, H., Ella, V. B., Leon, L., Jintrawet, A., Reyes, M., & Sombatpanit, S. (Eds.). (2009). *Soil and Water Assessment Tool (SWAT) global applications. World Association of Soil and Water Conservation (WASWAC)* (p. 397). Special Publication No. 4. ISBN: 978–974–613-722-5.

ASAE. (1998). *ASAE standards*. American Society of Agricultural and Biosystems Engineers.

Baffaut, C., & Sadeghi, A. (2010). Bacteria modeling with SWAT for assessment and remediation studies: A review. *Transactions of the ASABE, 53*(5), 1585–1594.

Bai, S., Morton, M., & Parker, A. (2006). Modeling *enterococci* in the tidal Christina River. Estuary and Coastal modeling. In M. L. Spaulding et al. (Eds.), *Proceedings of the ninth international conference Estuarine and coastal modeling* (pp. 305–318). ASCE.

Banasiak, L. J., Weaver, L., Humphries, B., Dann, R., Blaurock, M., Gosses, M., Pang, L., & Close, M. (2023). Microbial and solute transport through intact vadose zone cores of heterogeneous alluvial gravel under variably saturated conditions. *Vadose Zone Journal, 22*, e20250. https://doi.org/10.1002/vzj2.20250

Benedini, M., & Tsakiris, G. (2013). *Water quality modeling for rivers and streams*. Springer.

Benham, B. L., Baffaut, C., Zeckoski, R. W., Mankin, K. R., Pachepsky, Y. A., Sadeghi, A. M., Brannan, K. M., Soupir, M. L., & Habersack, M. J. (2006). Modeling bacteria fate and transport in watersheds to support TMDLS. *Transactions of the ASABE, 49*(4), 987–1002. https://doi.org/10.13031/2013.21739

Bennett, T. (2013). *Transport by advection and diffusion: Momentum, heat and mass transfer*. Wiley.

Bicknell, B., Imhoff, J., Kittle, J., Jr., Donigan, A., & Johanson, R. (1997). *Hydrological simulation program FORTRAN. User's manual for version 11. EPA/600/R–97/080*. USEPA National Exposure Research Laboratory.

Billen, G., Garnier, J., & Hanset, P. (1994). Modeling phytoplankton development in whole drainage networks: The RIVERSTRAHLER model applied to the Seine River system. *Hydrobiologia, 289*, 119–137.

Bottcher, A. B., Whiteley, B. J., James, A. I., & Hiscock, J. G. (2012). Watershed assessment model (WAM): Model use, calibration, and validation. *Transactions of ASABE, 55*(4), 1367–1383.

Bowman, N., Patel, P., Xu, W., Alsaffar, A., & Tiquia-Arashiro, S. M. (2018). Enrichment and isolation of Pb-resistant bacteria from Saint Clair River sediments and their potential for Pb

removal in aqueous solutions. *Applied Microbiology and Biotechnology, 102*, 2391–2398. https://doi.org/10.1007/s00253-018-8772-4

Bradshaw, J. K., Snyder, B., Spidle, D., Sidle, R. C., Sullivan, K., & Molina, M. (2021). Sediment and fecal indicator bacteria loading in a mixed land use watershed: Contributions from suspended sediment and bedload transport. *Journal of Environmental Quality, 50*, 598. https://doi.org/10.1002/jeq2.20166

Brannan, K. M., Mostaghimi, S., Dillaha, T. A., Heatwole, C. D., & Wolfe, M. L. (2002). Fecal Coliform TMDL for Big Otter River, Virginia: A case study. In A. Saleh (Ed.), *Total Maximum Daily Load (TMDL) environmental regulations: Proceedings of March 11–13, 2002, Conference* (pp. 367–376). ASAE Publication Number 701P0102.

Brauer, C., Teuling, A., Torfs, P., & Uijlenhoet, R. (2014). The Wageningen Lowland runoff simulator (WALRUS): A lumped rainfall runoff model for catchments with shallow groundwater. *Geosciences Model Development., 7*(5), 2313–2332.

Bottcher, A.B., James, A.I. (2014). Watershed Assessment Model (WAM): Florida Applications, 21st Century Watershed Technology Conference and Workshop Improving Water Quality and the Environment Conference Proceedings, 3–6, University of Waikato, New Zealand.https://doi.org/10.13031/wtcw.2014-006

Chiew, F. H. S., Peel, M. C., & Western, A. W. (2002). Application and testing of the simple rainfall-runoff model SIMHYD. In V. P. Singh & D. K. Frevert (Eds.), *Mathematical models of small watershed hydrology and applications* (pp. 335–367). Water Resources Publications.

Chin, D. A., Sakura-Lemessy, D., Bosch, D. D., & Gay, P. A. (2009). Watershed-scale fate and transport of bacteria. *Transactions of the ASABE, 52*(1), 145–154. https://doi.org/10.13031/2013.25955

Chin, D.A. (2011). Quantifying pathogen sources in streams by hydrograph separation Journal of Environmental Engineering, 137 (9): 770–781

Cho, K. H., Pachepsky, Y., Kim, J. H., Guber, A., Shelton, D., & Rowland, R. (2010). Release of *Escherichia coli* from the bottom sediment in a first-order creek: experiment and reach-specific modeling. *Journal of Hydrology, 391*(3), 322–332.

Cho, K. H., Pachepsky, Y. A., Kim, J. H., Kim, J. W., & Park, M. H. (2012a). The modified SWAT model for predicting fecal coliforms in the Wachusett Reservoir Watershed, USA. *Water Research, 46*(15), 4750–4760.

Cho, K., Zholi, A., Frabutt, D., Flood, M., Floyd, D., & Tiquia, S. M. (2012b). Linking bacterial diversity and geochemistry of uranium-contaminated groundwater. *Environmental Technology, 33*, 1629–1640. https://doi.org/10.1080/09593330.2011.641036

Cho, K. H., Pachepsky, Y. A., Oliver, D. M., Muirhead, R. W., Park, Y., Quilliam, R. S., & Shelton, D. R. (2016a). Modeling fate and transport of fecally-derived microorganisms at the watershed scale: State of the science and future opportunities. *Water Research, 100*(2016), 38–56.

Cho, K. H., Pachepsky, Y. A., Kim, M., Pyo, J., Park, M., Kim, Y. M., Kim, J. W., & Kim, J. H. (2016b). Modeling seasonal variability of fecal coliform in natural surface waters using the modified SWAT. *Journal of Hydrology, 535*, 377–385.

Coffey, R., Cummins, E., Bhreathnach, N., O'Flaherty, V., & Cormican, M. (2009). Development of a pathogen transport model for Irish catchments using SWAT. *Agricultural Water Management, 97*(2010), 101–111.

Coffey, R., Cummins, E., O'Flaherty, V., & Cormican, M. (2010). Pathogen sources estimation and scenario analysis using the Soil and Water Assessment Tool (SWAT). *Human and Ecological Risk Assessment: An International Journal, 16*(4), 913–933. https://doi.org/10.1080/10807039.2010.502051

Coffey, R., Dorai-Raj, S., O'Flaherty, V., Cormican, M., & Cummins, E. (2013). Modeling of pathogen indicator organisms in a small-scale agricultural catchment using SWAT. *Human and Ecological Risk Assessment: An International Journal, 19*(1), 232–253. https://doi.org/10.1080/10807039.2012.701983

Coffey, R., Benham, B., Kline, K., Wolfe, M. L., & Cummins, E. (2015). Modeling the impacts of climate change and future land use variation on microbial transport. *Journal of Water and Climate Change, 6*(3), 449–471. https://doi.org/10.2166/wcc.2015.049

Collins, R., & Rutherford, K. (2004). Modeling bacterial water quality in streams draining pastoral land. *Water Resources, 38*(3), 700–712.

Corapcioglu, M. Y., & Haridas, A. (1985). Microbial transport in soils and groundwater: A numerical model. *Advances in Water Resources, 8*(4), 188–200. https://doi.org/10.1016/0309-1708(85)90063-6

Desai, A.M., Rifai, H.S., Petersen, T.M., Stein, R. (2011). Mass Balance and Water Quality Modeling for Load Allocation of Escherichia coli in an Urban Watershed. Journal of Water Resources Planning and Management, 137: 412–427

de Brauwere, A., Gourgue, O., de Brye, B., Servais, P., Ouattara, N. K., & Deleersnijder, E. (2014). Integrated modeling of faecal contamination in a densely populated river-sea continuum (Scheldt River and Estuary). *Science of the Total Environment, 468*, 31–45.

Dorner, S. M., Anderson, W. B., Slawson, R. M., Kouwen, N., & Huck, P. M. (2006). Hydrologic modeling of pathogen fate and transport. *Environmental Science and Technology, 40*(15), 4746–4753.

Ella, V. B. (1986). *A conceptual water yield simulation model for small Philippine and Belgian watersheds*. M.S. Thesis, Catholic University of Leuven, Belgium.

Ella, V. B. (1999). *Spatial analysis and simulation of selected hydrogeologic and groundwater quality parameters in a glacial till aquitard*. Ph.D. Dissertation, Iowa State University, Ames, Iowa, USA.

Ella, V. B. (2001). *Potential applicability of hydrologic models for sustainable water resources management in the Philippines*. Paper presented and published in the proceedings of the 2001 international conference on water resources management, Philippine Water Works Association, Davao City. September 5–7, 2001.

Ella, V. B. (2009a). *Contaminant transport in soils: Measurement, modeling and potential applications for groundwater quality protection in The Philippines* (p. 60). SEARCA Professorial Chair Lecture, SEARCA, College, Laguna.

Ella, V. B. (2009b). *Potential applications of the SWAT model for sustainable soil and water resources management in Southeast Asia*. Keynote paper presented at the 1st International SWAT Conference in Southeast Asia, Chiang Mai, Thailand, January 7–8, 2009.

Ella, V. B. (2010). Soil *erosion and sediment yield prediction technologies for sustainable watershed management in the Philippines: Implications for the conservation of Mount Makiling and Laguna Lake*. UPLB Centennial Professorial Chair Lecture, University of the Philippines Los Banos, College, Laguna. 51 p.

Ella, V. B. (2016). Simple hydrologic model for predicting streamflow in small watersheds for irrigation system planning. *International Agricultural Engineering Journal, 25*(4), 1–13.

Ella, V. B., Lee, J. H., & Horton, R. (2001). *Swine manure transport measurements in cracking soils using time domain reflectometry*. ASAE Meeting Paper No. 01-2111. St. Joseph, Michigan. Paper presented at the 2001 Annual International ASAE Conference, Sacramento, California, USA, July 29–August 2, 2001.

Ella, V. B., Melvin, S. W., Kanwar, R. S., Jones, L. C., & Horton, R. (2002a). Inverse three- dimensional groundwater modeling using the finite difference method for recharge estimation in a glacial till aquitard. *Transactions of the ASAE, 45*(3), 703–715.

Ella, V. B., Lee, J. H., & Horton, R. (2002b). *Simulating swine manure transport in soils using the mobile-immobile contaminant transport model*. ASAE-CIGR Meeting Paper No. 02-2129. St. Joseph, Michigan. Paper presented at the 2002 annual international ASAE meeting-CIGR World Congress, Chicago, Illinois, USA, July 28–31, 2002.

Evans, B. M. (2022). *MapShed GIS-BasedWatershed Modeling Tool*. Available online: https://wikiwatershed.org/help/model-help/mapshed/. Accessed on 30 March 2024.

Ferguson, C. M., Croke, B. F. W., Beatson, P. J., Ashbolt, N. J., & Deere, D. A. (2007). Development of a process-based model to predict pathogen budgets for the Sydney drinking water catchment. *Journal of Water Health, 5*(2), 187–208.

Fetter, C. W. (1993). *Contaminant hydrogeology*. McMillan Publishing.

Flood, M., Frabutt, D., Floyd, D., Powers, E. U., Devol, A., & Tiquia-Arashiro, S. M. (2015). Ammonia-oxidizing bacteria and archaea in sediments of the Gulf of Mexico. *Environmental Technology, 36*, 124–135. https://doi.org/10.1080/09593330.2014.942385

Frey, S. K., Topp, E., Edge, T., Fall, C., Gannon, V., Jokinen, C., Marti, R., Neumann, N., Ruecker, N., Wilkes, G., & Lapen, D. R. (2013). Using SWAT, Bacteroidales microbial source tracking markers, and fecal indicator bacteria to predict waterborne pathogen occurrence in an agricultural watershed. *Water Research, 47*(16), 6326–6337. https://doi.org/10.1016/j.watres.2013.08.010

Gannon, J. T., Mixgelgrin, M., Alexander, M., & Wagenet, R. J. (1991). Bacterial transport through homogeneous soil. *Soil Biology and Biochemistry, 23*(12), 1155–1160. https://doi.org/10.1016/0038-0717(91)90028-I

Garnier, J., Billen, G., Hannon, E., Fonbonne, S., Videnina, Y., & Soulie, M. (2002). Modeling transfer and retention of nutrients in the drainage network of the Danube River. *Estuarine Coastal and Shelf Science, 54*, 285–308.

Gassman, P. W., Reyes, M. R., Green, C. H., & Arnold, J. G. (2007). The soil and water assessment tool: Historical development, applications, and future research directions. *Transactions of the ASABE, 50*(4), 1211–1250.

Gentile, M., Yan, T., Tiquia, S. M., Fields, M. W., Nyman, J., Zhou, J., & Criddle, C. S. (2006). Stability and resilience in a denitrifying fluidized bed reactor. *Microbial Ecology, 52*(2), 311–321. https://doi.org/10.1007/s00248-006-9024-1

Haith, D. A., & Shoemaker, L. L. (1987). Generalized watershed loading functions for stream flow nutrients. *Journal of American Water Resources Association, 23*, 471–478.

Haith, D.A., Mandel, R., Wu, R.S. (1992). GWLF Generalized Watershed Loading Functions Version 2.0 User's Manual,https://owl.cwp.org/mdocs-posts/generalized-watershed-loading-functions-version-2-0-users-manual

Hall, K.M., Zeckoski, R.W., Brannan, K.M. (2008). FTABLE generation method effects on instream fecal bacteria concentrations simulated with HSPF. Journal of American Water Resources Association, 44 (2) :2008 489-495

Haydon, S., & Deletic, A. (2006). Development of a coupled pathogen-hydrologic catchment model. *Journal of Hydrology, 328*(3–4), 467–480.

Herr, J. W., & Chen, C. W. (2012). WARMF: Model use, calibration, and validation. *Transactions of the ASABE, 55*, 1385–1394.

Hong, E., Park, Y., Muirhead, R., Jeong, J., & Pachepsky, Y. A. (2018). Development and evaluation of the bacterial fate and transport module for the Agricultural Policy/Environmental eXtender (APEX) model. *Science of the Total Environment, 615*(2018), 47–58.

Hong, E., Park, Y., Muirhead, R., Jeong, J., Pachepsky, Y.A. (2017). Development and evaluation of the bacterial fate and transport module for the Agricultural Policy/Environmental eXtender (APEX) model, Science of the Total Environment 615 :2018, 47–58

Hong, Y., Soulignac, F., Roguet, A., Li, C., Lemaire, B. J., Martins, R. S., Lucas, F., & Vinçon-Leite, B. (2021). Impact of Escherichia coli from stormwater drainage on recreational water quality: An integrated monitoring and modeling of urban catchment, pipes and lake. *Environmental Science and Pollution Research, 28*(2), 2245–2259. https://doi.org/10.1007/s11356-020-10629-y

Holz (2010). Sources and processes of contaminant loss from an intensively grazed catchment inferred from patterns in discharge and concentration of thirteen analytes using high intensity sampling. Journal of Hydrology, 383: 2010, 194–208

Im, S., Brannan, K.M., Mostaghimi, S., Cho, J. (2004). Simulating fecal coliform bacteria loading from an urbanizing watershed. Journal of Environmental Science and Health Part a-Toxic/Hazardous Substances and Environmental Engineering, 39: 663–679

Iqbal, M. S., & Hofstra, N. (2019). Modeling Escherichia coli fate and transport in the Kabul River Basin using SWAT. *Human and Ecological Risk Assessment: An International Journal, 25*(5), 1279–1297. https://doi.org/10.1080/10807039.2018.1487276

Iudicello, J.J., Chin, D.A. (2013). Multimodel, multiple watershed examination of in-stream bacteria modeling. Journal of Environmental Engineering, 139 (5): 719-727

Jamieson, R. C., Joy, D. M., Lee, H., Kostaschuk, R., & Gordon, R. J. (2005). Transport and deposition of sediment-associated Escherichia coli in natural streams. *Water Research, 39*(12), 2665–2675.

Jayakody, P., Parajuli, P. B., & Brooks, J. P. (2014). Evaluating spatial and temporal variability of Fecal Coliform bacteria loads at the Pelahatchie watershed in Mississippi. *Human and Ecological Risk Assessment: An International Journal, 20*(4), 1023–1041. https://doi.org/10.1080/10807039.2013.784155

Jeon, D. J., Ligaray, M., Kim, M., Kim, G., Lee, G., Pachepsky, Y. A., Cha, D. H., Kyung, & Cho, H. (2019). Evaluating the influence of climate change on the fate and transport of fecal coliform bacteria using the modified SWAT model. *Science of the Total Environment, 658*, 753–762. https://doi.org/10.1016/j.scitotenv.2018.12.213

Jeong, J., Wagner, K., Flores, J. L., Cawthon, T., Her, Y., Osorio, J., & Yen, H. (2019). Linking watershed modeling and bacterial source tracking to better assess E. coli sources. *Science of the Total Environment, 648*, 164–175. https://doi.org/10.1016/j.scitotenv.2018.08.097

Kassem, A., Abbas, L., Coutinho, O., Opara, S., Najaf, H., Kasperek, D., Li, X., & Tiquia-Arashiro, S. M. (2023). Applications of Fourier Transform-Infrared spectroscopy in microbial cell biology and environmental microbiology: Advances, challenges, and future perspectives. *Frontiers in Microbiology, 14*, 1304081. https://doi.org/10.3389/fmicb.2023.1304081

Kim, J. W., Pachepsky, Y. A., Shelton, D. R., & Coppock, C. (2010). Effect of streambed bacteria release on *E. coli* concentrations: Monitoring and modeling with the modified SWAT. *Ecological Modeling, 221*(12), 1592–1604.

Kim, S. M., Brannan, K. M., Zeckoski, R. W., & Benham, B. L. (2014). Development of total maximum daily loads for bacteria impaired watershed using the comprehensive hydrology and water quality simulation model. *Journal of Environmental Science and Health Part A Toxic/Hazardous Substances & Environmental Engineering, 49*(9), 1077–1089. https://doi.org/10.1080/10934529.2014.895567

Kim, M., Boithias, L., Cho, K. H., Silvera, N., Thammahacksa, C., Latsachack, K., Rochelle-Newall, E., Sengtaheuanghoung, O., Pierret, A., Pachepsky, Y. A., & Ribolzi, O. (2017). Hydrological modeling of Fecal Indicator Bacteria in a tropical mountain catchment. *Water Research, 119*, 102–113. https://doi.org/10.1016/j.watres.2017.04.038

Kim, M., Boithias, L., Cho, K. H., Sengtaheuanghoung, O., & Ribolzi, O. (2018). Modeling the impact of land use change on Basin-scale transfer of fecal indicator bacteria: SWAT model performance. *Journal of Environmental Quality, 47*, 1115. https://doi.org/10.2134/jeq2017.11.0456

Kouwen, N. (2022). *Flow forecasting manual for WATFLOOD/CHARM*. University of Waterloo.

Kouwen, N., & Mousavi, S. F. (2002). WATFLOOD/SPL9 hydrological model and flood forecasting system. In V. P. Singh & D. K. Frevert (Eds.), *Mathematical models of large watershed hydrology* (pp. 649–685). Water Resource Publications.

LaWare, P., Rifai, H.S. (2006). Modeling fecal coliform contamination in the Rio Grande. Journal of the American Water Resources Association, https://doi.org/10.1111/j.1752-1688.2006.tb03842.x

Liao, H., Krometis, L. H., Kline, K., & Hession, W. C. (2015). Long-term impacts of bacteria–sediment interactions in watershed-scale microbial fate and transport modeling. *Journal of Environmental Quality, 44*(5), 1483–1490. https://doi.org/10.2134/jeq2015.03.0169

Llaban, A. B., & Ella, V. B. (2022). Conventional and sensor-based streamflow data acquisition system for sustainable water resources management and agricultural applications: An extensive review of literature. *IOP Conference Series: Earth and Environmental Science, 1038*(2022), 012040. https://doi.org/10.1088/1755-1315/1038/1/012040

Ma, L., He, C., Bian, H., & Sheng, L. (2016). MIKE SHE modeling of ecohydrological processes: Merits, applications and challenges. *Ecological Engineering, 96*, 137. https://doi.org/10.1016/j.ecoleng.2016.01.008

Meshesha, T. W., Wang, J., & Melaku, N. D. (2020). A modified hydrological model for assessing effect of pH on fate and transport of Escherichia coli in the Athabasca River basin. *Journal of Hydrology, 582*, 124513. https://doi.org/10.1016/j.jhydrol.2019.124513

Moyer, D. L., Hyer, K.E. (2003). Use of the Hydrological Simulation Program–FORTRAN and bacteria source tracking for development of the fecal coliform Total Maximum Daily Load (TMDL) for Christians Creek, Augusta County, Virginia. USGS Water–Resources Investigations Report 03–4162. Washington, D.C.: U.S. Geological Survey Available at: www.deq.virginia.gov/tmdl/apptmdls/shenrvr/chrstnfc.pdf.

Nakamura, T., Fujii, H., Watanabe, T., Ly, S., Lun, S., Fujihara, Y., Hoshikawa, K., Miyanaga, K., & Yoshimura, C. (2023). Waterborne disease risk assessment and mapping for a floating village by combining 3D hydraulic simulation and quantitative microbial risk assessment. *Water, 15*, 4199. https://doi.org/10.3390/w15234199

Nassereddine, Z., Opara, S., Coutinho, O., Qyteti, F., Book, R., Heinicke, M. P., Napieralski, J., & Tiquia-Arashiro, S. M. (2024). Critical perspectives on Advancing Antibiotic Resistant Gene (ARG) detection technologies in aquatic ecosystems. *Science of the Total Environment, 957*, 177775. https://doi.org/10.1016/j.scitotenv.2024.177775

Neitsch, S. L., Arnold, J. G., Kiniry, J. R., & Williams, J. R. (2005). *Soil and Water Assessment Tool: Theoretical documentation, version 2005*. Grassland, Soil and Water Research Laboratory.

Niazi, M., Obropta, C., & Miskewitz, R. (2015). Pathogen transport and fate modeling in the Upper Salem River Watershed using SWAT model. *Journal of Environmental Management, 151*, 167–177. https://doi.org/10.1016/j.jenvman.2014.12.042. CorpusID: 28457251.

Niazi, M., Nietch, C., Maghrebi, M., Jackson, N., Bennett, B. R., Tryby, M., & Massoudieh. (2017). A storm water management model: Performance review and gap analysis. *Journal of Sustainable Water Built Environment, 3*(2), 04017002.

Niu, J., & Phanikumar, M. S. (2015). Modeling watershed-scale solute transport using an integrated, process-based hydrologic model with applications to bacterial fate and transport. *Journal of Hydrology, 529*, Part 1, 35–48. https://doi.org/10.1016/j.jhydrol.2015.07.013

Ouattara, N. K., de Brauwere, A., Billen, G., & Servais, P. (2013). Modeling faecal contamination in the Scheldt drainage network. *Journal of Marine Systems, 128*, 77–88.

Overton, D. E., & Meadows, M. E. (1976). *Storwater modeling*. Academic Press.

Pachepsky, Y.A., Sadeghi, A.M., Bradford, S.A., Shelton, D.R., Guber, A.K., Dao, T.,(2006). Transport and fate of manure-borne pathogens: Modeling perspective. Agricultural Water Management, 86:1-2, 81–92

Pachepsky, Y. A., & Shelton, D. R. (2011). Escherichia coli and fecal coliforms in freshwater and estuarine sediments. *Critical Reviews in Environmental Science and Technology, 41*, 1067–1110.

Pagnucco, G., Overfield, D., Chamlee, Y., Fortuna, A., Sulaiman, F., Farinas, J., Schittenhelm, R., Shuler, C., Kassem, A., Opara, S., Najaf, H., Abbas, L., Coutinho, O., Catalfano, B., Li, X., & Tiquia-Arashiro, S. M. (2023). Metal tolerance and biosorption capacities of bacterial strains isolated from an urban watershed. *Frontiers in Microbiology, 14*, 1278886. https://doi.org/10.3389/fmicb.2023.1278886

Pandey, P. K., & Soupir, M. L. (2013). Assessing the impacts of *E. coli* laden streambed sediment on *E. coli* loads over a range of flows and sediment characteristics. *Journal of American Water Resources Association, 49*(6), 1261–1269.

Panikov, N. S. (1995). *Microbial growth kinetics*. Springer Science and Business Media.

Parajuli, P., Mankin, K. R., & Barnes, P. L. (2009). Source specific fecal bacteria modeling using soil and water assessment tool model. *Bioresource Technology, 100*(2), 953–963. https://doi.org/10.1016/j.biortech.2008.06.045

Petersen, T.M., Rifai, H.S., Villarreal, G.C., Stein, R., (2011). Modeling Escherichia Coli and Its Sources in an Urban Bayou with Hydrologic Simulation Program-FORTRAN. Journal of Environmental Engineering, 137: 487–503

Patel, D., Gismondi, R., Alsaffar, A., & Tiquia-Arashiro, S. M. (2019). Applicability of API ZYM to capture seasonal and spatial variabilities in lake and river sediments. *Environmental Technology, 40*(24), 3227–3239. https://doi.org/10.1080/09593330.2018.1468492

Paul, S., Haan, P.K., Matlock, M., Mukhtar, S., Pillai, S. (2004). Analysis of the HSPF water quality parameter uncertainty in predicting peak in-stream fecal coliform concentrations. Transactions of the ASAE, 47(1): 69

Pitt, R. (1997). Unique features of the source loading and management model (SLAMM). *Journal of Water Resources Planning and Management, 6*, 13–37.

Ramirez, R. C., Agulto, E. S., Glaser, S. D., Zhang, Z., Hermocilla, J. C., & Ella, V. B. (2022). Development of a real-time wireless sensor network-based information system for efficient irrigation of upland and lowland crop production systems. *IOP Conference Series: Earth and Environmental Science, 1038*, 012028. https://doi.org/10.1088/1755-1315/1038/1/012028

Rolle, K., Gitau, M. W., Chen, G., & Chauhan, A. (2012). Assessing fecal coliform fate and transport in a coastal watershed using HSPF. *Water Science and Technology, 66*(5), 1096–1102.

Roser, D. J., Davies, C. M., Ashbolt, N. J., & Morison, P. (2006). Microbial exposure assessment of an urban recreational lake: A case study of the application of new risk-based guidelines. *Water Science and Technology, 54*(3), 245–252.

Rossman, L. A. (2015). *Storm water management model user's manual, Version 5.1, EPA-600/R-14/413b*. United States Environment Protection Agency.

Ruelland, D., Billen, G., Brunstein, D., & Garnier, J. (2007). SENEQUE: A multi-scaling GIS interface to the Riverstrahler model of the biogeochemical functioning of river systems. *Science of the Total Environment, 375*(1), 257–273.

Sadeghi, A. M., & Arnold, J. G. (2002). *A SWAT microbial sub-model for predicting pathogen loadings in surface and groundwater at watershed and basin scales*. ASAE Paper No. 062196. St Joseph, Michigan: ASAE.

Servais, P., Billen, G., Goncalves, A., & Garcia-Armisen, T. (2007). Modeling microbiological water quality in the Seine River drainage network: Past, present and future situations. *Hydrology and Earth System Sciences, 11*(5), HESS, 11, 1581–1592. https://doi.org/10.5194/hess-11-1581-2007

Seong, C. H., Benham, B. L., Hall, K. M., Kline, K. (2013). Comparison of alternative methods to simulate bacteria concentrations with HSPF under low-flow conditions. Applied Engineering in Agriculture, 29(6):917–931

Šimůnek, J., Jacques, D., Twarakavi, N. K. C., et al. (2009). Selected HYDRUS modules for modeling subsurface flow and contaminant transport as influenced by biological processes at various scales. *Biologia, 64*, 465–469. https://doi.org/10.2478/s11756-009-0106-7

Singh, V. P. (2018). Hydrologic modeling: Progress and future directions. *Geoscience Letters, 5*, Article number: 15. https://doi.org/10.1186/s40562-018-0113-z

Sisson, M., Shen, J., & Schlegel, A. (2016). Combining inverse and transport modeling to estimate bacterial loading and transport in a tidal embayment. *Journal of Marine Science Engineering, 4*, 69. https://doi.org/10.3390/jmse4040069

Sobel, R. S., Rifai, H. S., & Petersen, C. M. (2017). Integration of Tidal Prism Model and HSPF for simulating indicator bacteria in Coastal Watersheds. *Estuarine, Coastal and Shelf Science, 196*, 248–257.

Sowah, R. A., Bradshaw, K., Snyder, B., Spidle, D., & Molina, M. (2020). Evaluation of the soil and water assessment tool (SWAT) for simulating E. coli concentrations at the watershed-scale. *Science of the Total Environment, 746*, 140669. https://doi.org/10.1016/j.scitotenv.2020.140669

Sterk, A., Schijven, J., de Roda Husman, A. M., & de Nijs, T. (2016). Effect of climate change on runoff of *Campylobacter* and *Cryptosporidium* from land to surface water. *Water Resources, 95*, 90–102.

SWAT. (2024). SWAT Literature Database. https://swat.tamu.edu/publications/. Date accessed 15 February 2024.

Thilakarathne, M., Sridhar, V., & Karthikeyan, R. (2018). Spatially explicit pollutant load-integrated in-stream E. coli concentration modeling in a mixed land-use catchment. *Water Research, 144*, 87–103. https://doi.org/10.1016/j.watres.2018.07.021

Tian, Y. Q., Gong, P., Radke, J. D., & Scarborough, J. (2002). Spatial and temporal modeling of microbial contaminants on grazing farmlands. *Journal of Environmental Quality, 31*(3), 860–869.

Todini, E. (2007). Hydrological catchment modeling: Past, present and future. *Hydrology and Earth System Sciences, 11*(1), 468–482.

USGS. (2024). *Summary of HSPF*. Water Resources Applications Software. https://water.usgs.gov/cgi-bin/man_wrdapp?hspf. Accessed on 30 March 2024.

Walker, F. R., & Stedinger, J. R. (1999). Fate and transport model of *Cryptosporidium*. *Journal of Environmental Engineering, 125*(4), 325–333.

Wang, Y., & Chen, N. (2021). Recent progress in coupled surface–ground water models and their potential in watershed hydro-biogeochemical studies: A review. *Watershed Ecology and the Environment, 3*(2021), 17–29.

Wang, X., Williams, J., Atwood, J. D., Lee Norfleet, M., King, A. D. (2011). *APEX model upgrades, data inputs, and parameter settings for use in CEAP cropland modeling*. http://www.nrcs.usda.gov/wps/portal/nrcs/detail/national/technical/?cid=nrcs143_014165.

Wang, Y., Bradford, S. A., & Simunek, J. (2013). Transport and fate of microorganisms in soils with preferential flow under different solution chemistry conditions. *Water Resources Research, 49*(5), 2424. https://doi.org/10.1002/wrcr.20174

Weiskerger, C. J., & Phanikumar, M. S. (2020). Numerical modeling of microbial fate and transport in natural waters: Review and implications for normal and extreme storm events. *Water, 12*, 1876. https://doi.org/10.3390/w12071876

Whelan, G., Kim, K., Parmar, R., Laniak, G. F., Wolfe, K., Galvin, M., Molina, M., Pachepsky, Y. A., Duda, P., Zepp, R., Prieto, L., Kinzelmane, J. L., Kleinheinz, G. T., & Borchardt, M. A. (2018). Capturing microbial sources distributed in a mixed-use watershed within an integrated environmental modeling workflow. *Environmental Modeling Software, 99*, 126–146. https://doi.org/10.1016/j.envsoft.2017.08.002

WHO, UNICEF, World Bank. (2022). *State of the world's drinking water: an urgent call to action to accelerate progress on ensuring safe drinking water for all*. World Health Organization. License: CC BY-NC-SA 3.0 IGO.

Wilkinson, R. J., McKergow, L. A., Davies-Colley, R. J., Ballantine, D. J., & Young, R. G. (2011). Modeling storm-event *E. coli* pulses from the Motueka and Sherry Rivers in the South Island, New Zealand. *New Zealand Journal of Marine and Freshwater Research, 45*(3), 369–393.

Williams, J. R., & Izaurralde, R. C. (2006). The APEX model. In V. P. Singh & D. K. Frevert (Eds.), *Watershed models* (pp. 437–482). CRC Press.

World Health Organization (WHO). (2024). https //www.who.int/news-room/fact-sheets/detail/drinking-water. Date accessed 11 September 2024.

Wu, W., Gu, B., Fields, M. W., Gentile, M., Ku, Y.-K., Yan, H., Tiquia, S. M., Yan, T., Nyman, J., Zhou, J., Jardine, P. M., & Criddle, C. S. (2005) Uranium (VI) reduction by denitrifying biomass. *Bioremediation Journal, 9*, 1–13. https://doi.org/10.1080/10889860590929628

Wu, J., Rees, P., Storrer, S., Alderisio, K., & Dorner, S. (2009). Fate and transport modeling of potential pathogens: The contribution from sediments. *Journal of the American Water Resources Association, 45*(1), 35–44. https://doi.org/10.1111/j.1752-1688.2008.00287.x

Yates, M. V., Yates, S. R., & Gerba, C. P. (1988). Modeling microbial fate in the subsurface environment. *Critical Reviews in Environmental Science and Technology, 17*(4), 307–344. https://doi.org/10.1080/10643388809388339

Yuan, L., Sinshaw, T., & Forshay, K. J. (2020). Review of watershed-scale water quality and nonpoint source pollution models. *Geosciences, 10*(1), 25. https://doi.org/10.3390/geosciences10010025

USEPA. (2000). Bacterial Indicator Tool: User's Guide. Office of Water, U.S. Environmental Protection Agency, Washington, DC: 2000EPA-823-B-01-003

Chapter 19
Citizen Science Monitoring of Freshwater Mussels (Unionidae)

Amani Tolin

1 Introduction

Freshwater mussels, belonging to the family Unionidae, play critical roles in the health and stability of Michigan's aquatic ecosystems. Their activities benefit the environment in multiple ways. Primarily, they are filter feeders; mussels siphon water, trapping and ingesting bacteria, thereby cleaning the water (Green et al., 1989; Shevchuk et al., 2021). This function not only purifies water but also mitigates the effects of pollution and eutrophication. According to Mulcrone and Rathbun (2018), these bivalves also contribute to the food web by serving as a food source for a variety of aquatic predators. Furthermore, they enhance biodiversity by providing living spaces for macroinvertebrates, often acting as micro-ecosystems themselves (Vaughn & Spooner, 2006). Importantly, the presence and health of mussel populations are indicators of water quality, signaling the overall condition of the water bodies they inhabit. Despite these essential functions, freshwater mussel populations in Michigan have been experiencing significant declines due to the impact of human activities, including but not limited to pollution, habitat destruction, climate change, and the introduction of invasive species. Such pervasive threats have led to dire consequences for the sustainability of these mussel communities.

Currently identified in Michigan are 43 species of freshwater mussels, but their existence is at risk with 18 of these species designated as state threatened, endangered, or presumed extirpated under Part 365, Endangered Species Protection, of the Michigan Natural Resources and Environmental Protection Act (Michigan PA 451 of 1994). Going further, six of these species are recognized at the federal level for additional protections under the Endangered Species Act (1983), which means their survival is of national concern (Legislative Service Bureau, 1994; Mulcrone &

A. Tolin (✉)
School for Environment and Sustainability, University of Michigan, Ann Arbor, MI, USA
e-mail: atolin@umich.edu

S. M. Tiquia-Arashiro et al. (eds.), *Urban Watershed Microbiology, Volume 1*,
https://doi.org/10.1007/978-3-031-89071-0_19

Rathbun, 2018). The precarious status of these mussels signals a need for immediate and proactive conservation measures. Rigorous monitoring programs are essential to track the health and numbers of these populations, facilitating swift action when necessary. Conservation efforts not only aim to safeguard the existing mussel populations from further decline but are also a step toward possible revitalization of species that have suffered great losses. These efforts include habitat restoration, research on species' life histories and ecological requirements, and the development and implementation of management strategies that can mitigate the impact of harmful human activities. In summary, the Unionidae are invaluable constituents of Michigan's freshwater systems, and the alarming decline of these mussels is a call to action. To restore and maintain the ecological balance, as well as to ensure water quality and biodiversity, Michigan needs to invest substantially in the protection and recovery of its threatened and endangered mussel species. This comprehensive approach promises not only the survival of the mussels themselves but also the health of the ecosystems they support.

1.1 Michigan Unionidae

Michigan is home to 43 known species of mussels within the order Unionidae, including species found widespread throughout the state such as cylindrical papershell (*Anodontoides ferussacianus*), giant floater (*Pyganodon grandis*), creeper (*Strophitus undulatus*), plain pocketbook (*Lampsilis cardium*), fat mucket (*Lampsilis siliquoidea*), and spike (*Eurynia dilata*) (Mulcrone & Rathbun, 2018). The distribution of mussels is related to habitat and fish host preferences that vary between mussel species, both of which are made more complex by the limited amount of high-quality habitat that exists in Michigan (Maio & Corkum, 1995; Mulcrone, 2006a, b; Dickinson & Sietman, 2008; Jones, 2015; Pandolfo et al., 2016; Mulcrone & Rathbun, 2018; Michigan Natural Features Inventory, 2022).

Of the 34 unionid species observed in the Huron River Watershed (Fig. 19.1), 12 are listed as state threatened or endangered and 4 are listed as federally threatened or endangered, including Northern riffleshell (*Eplioblasma torulosa rangiana*), snuffbox (*Epioblasma triquetra*), rayed bean (*Villosa fabalis*), and round hickorynut (*Obovaria subrotunda*) (Badra, 2010; Mulcrone & Rathbun, 2018; U.S. Fish and Wildlife Service, 2020; Williams, 2023). Past surveys have examined parts of southeastern Michigan rivers and lakes to assess mussel abundance and diversity. Van der Schalie's 1938 report of the Huron River has extensive comparisons of species distributions throughout the watershed. Van der Schalie (1938) observed and recorded accounts of 25 species of unionids along 37 sites in the Huron River and described their abundance across different habitat types. Peter Badra (2010) conducted a survey of mussels in the Huron-Clinton Metroparks for the Michigan Natural Features Inventory (MNFI) and found 16 mussel species in the Huron River Watershed across 21 sites, including live specimens from 12 species and dead shells from 4 species. Of these Metroparks' surveys, the Huron Meadows was the most

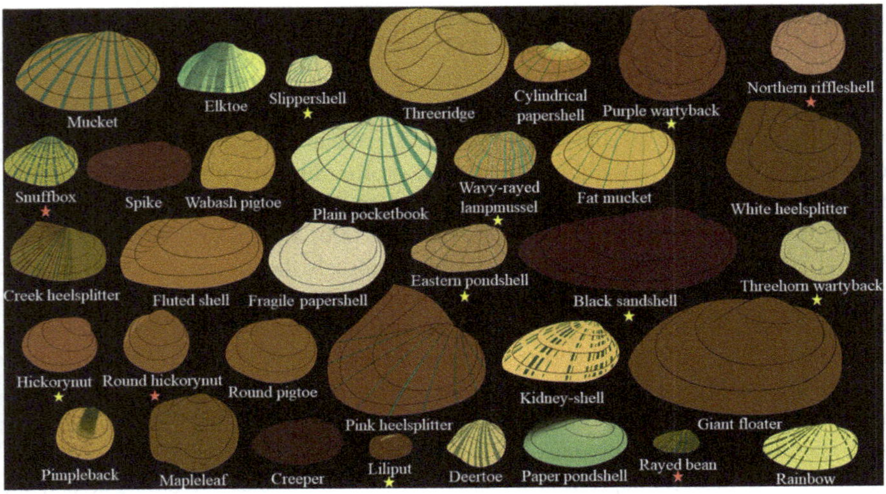

Fig. 19.1 Illustrations of the 34 unionid species that are found in the Huron River Watershed. All the starred species are state listed. The four species with red stars are also federally listed, including the federally endangered Northern riffleshell (*Epioblasma torulosa*), the federally endangered snuffbox (*Epioblasma triquetra*), the federally threatened round hickorynut (*Obovaria subrotunda*), and the federally endangered rayed bean (*Villosa fabalis*)

speciose with 14 species observed (Badra, 2010). These past surveys provided a basis to aid in site selection and to compare with Huron River Watershed Council's (HRWC's) mussel survey data.

1.2 Threats to Unionidae

Overall, the decline of freshwater mussels can be attributed to a multitude of threats. Habitat alteration, pollution, climate change, invasive species, and other environmental threats reduce unionid reproductive capacity, decrease food availability, and cause mussel mortality (Sousa et al., 2021; Aldridge et al., 2023). Since mussels are sedentary animals, habitat alteration can be devastating to mussel populations. River impoundment limits the population growth of unionids because dams act as barriers to the movement of the fish hosts that mussels require for the dispersal of their parasitic larvae (Tiemann et al., 2007; Pilger & Gido, 2012; Mulcrone & Rathbun, 2018; Badra, 2020). By restricting the movement of these fish hosts, mussel populations can be isolated from each other causing declines in genetic diversity, which is necessary for the long-term sustainability of mussel populations (Badra, 2020). Dams can also alter mussel assemblages by changing river habitats from lotic to lentic environments, which can be detrimental for species that are dependent on systems with flowing water (Tiemann et al., 2007). Similarly, the sedimentation that occurs from increasing siltation from dams can also be harmful for mussel populations

because of the ensuing reduction in dissolved oxygen availability and substrate quality (Tiemann et al., 2007; Staton et al., 2000; Maceda-Veiga, 2013).

Exotic species are another major threat to native mussels. Zebra mussels (*Dreissena polymorpha*) were introduced to the Great Lakes in the 1980s through the ballast water of cargo ships and have devastated aquatic ecosystems in the region ever since (Mulcrone & Rathbun, 2018). The proliferation of zebra mussels is detrimental to unionids because they attach to the shell making it hard for unionids to move, feed, and reproduce (Baker & Hornbach, 2000; Mulcrone & Rathbun, 2018; Badra, 2020). Abundant populations of zebra mussels then filter out nutrients and food resources from the water that native mussels need to survive. Exotic Asian clams (*Corbicula* spp.) and quagga mussels (*Dreissena bugensis*) can also reach high densities and compete with native unionids (Badra, 2020).

The filter feeding nature of mussels makes them especially susceptible to various forms of pollution. Organic pollution from herbicides, wastewater, and other industrial chemicals can cause toxicity in mussels, inducing oxidative stress, higher susceptibility to disease, reproductive dysfunction, genomic DNA damage, and an overall reduction in metabolism (Hassel & Farris, 2007; Viarengo et al., 2007; Tiquia, 2011; Hellou et al., 2012; Tiquia-Arashiro, 2012; Zhang et al., 2021). Nanoplastics ingested by mussels can cause damage in the functional role of their gills as "brooding chambers" for larval glochidia, which thus threatens their reproductive capacity (Abdelsaleheen, 2023). Unionids can even suffer unintentional impacts from conservation strategies such as from poison applications for the control of other organisms like invasive sea lampreys (Gruber et al., 2012).

Their filter feeding behaviors also make unionids vulnerable to infestation from parasites such as trematodes, which can hinder mussel growth, energy production and storage, and reproductive capacity (Zieritz & Aldridge, 2011; Abdelsaleheen, 2023). A wide variety of parasites including oligochaetes, nematodes, trematodes, watermites, and bitterlings can infest unionids and prevent their populations from recovering (Gillis & Mackie, 1994; McElwain et al., 2019; Taskinen et al., 2020).

Climate change and global warming pose long-term threats to unionids because of their sensitivity to temperature changes (Woodward et al., 2010). Since unionids must allocate their energy between regulating their valves for feeding and breathing, burrowing into the sediment, and occasionally moving, their survival is dependent on their metabolism (Abdelsaleheen, 2023). Mussels can vary in their tolerance to temperature change; however, increases in average temperatures due to global warming can present risks to mussels in the form of thermal stress reducing mussel growth, metabolic rate, and oxygen uptake (Payton et al., 2016; Abdelsaleheen, 2023). At temperatures above 25 °C, mussels may also cease the display of their reproductive lures or even expel their glochidia to relieve thermal and respiratory stress, which may prevent mussels from attracting fish hosts and reproducing successfully (Gascho Landis et al., 2012; Gibson, 2019).

1.3 Mussel Conservation

While mussel populations have heavily declined, conservation measures can be effective at protecting and restoring remnant populations. Species are reviewed to be listed as threatened or endangered based on information including "population, distribution, habitat needs, limiting factors, and other biological and ecological data" (Legislative Service Bureau, 1994). Since 31 of Michigan's 43 species are currently listed as state or federally threatened, endangered, or special concern, there is substantial progress to be made.

The removal of old dams can be a significant tool to increase the access of mussels to host fishes. Removing dams can restore a lotic flow regime, connectivity with the rest of the watershed, the passage of fish, and dissolved oxygen flow, all of which can be beneficial for mussel populations (McCombs, 2014). However, it should also be noted that any harmful impacts of dam removal should be assessed to mitigate potential negative effects of desiccation and exposure, to relocate mussels from impact zones where reservoir deposits would be transported, and to evaluate the potential of the dam removal to increase the spread of invasive species like zebra mussels and sea lamprey (Sethi et al., 2004). The colonization and recovery of mussel populations in habitats where dams have been removed can also be slow, so continuous monitoring and management of those habitats is important to ensure the viability of these populations (Doyle et al., 2005). Thus, dam removal can be a powerful restoration measure with many different long-term and short-term impacts to consider.

Relocation can be an effective tool to use alongside dam removals or construction projects. Relocating mussels before dams are removed can prevent them from desiccation in the newly exposed habitat or from being buried by the deposited sediments. Impact avoidance measures, including relocation, are also currently required in Michigan for construction projects and dredging (Mulcrone & Rathbun, 2018). These relocation projects provide a unique opportunity for mussel surveys to be applied and to monitor how mussel populations are faring post-relocation.

Sedimentation and pollution reduction strategies can help to alleviate the impacts of excessive siltation and runoff (Tiquia, 2010a, b; Bowman et al., 2018; Oest et al., 2018; Patel et al., 2019; Nassereddine et al., 2024). Riparian buffer zones can be effective at protecting streams from sediment inputs and pollutants (U.S. Fish and Wildlife Service, 2006; Tiquia et al., 2008). Rain gardens can also be implemented to filter out runoff for streams in more urban areas (Chaffin et al., 2016). These strategies paired with education on the application of harmful industrial products like herbicides can improve water quality conditions for mussels in streams near areas of human use (Reid et al., 2019).

Controlling the impact of non-native mussels is a challenging yet important element of mussel conservation. Since zebra mussels have already contributed to the decline of many of the Great Lakes' native unionid populations, the protection of remaining refuges from zebra mussels is vital for the preservation of native unionids

(Zanatta et al., 2002). Simple strategies to prevent the spread of zebra mussels include inspecting and cleaning off any equipment that has been in contact with water including boats, waders, fishing gear, buckets, and other equipment.

The successful conservation of freshwater mussels is also dependent on the preservation of their fish host species. Fish species that are commonly used as hosts in the reproductive cycle of mussels in Michigan include large-mouth bass (*Micropterus salmoides*), small-mouth bass (*Micropterus dolomieu*), bluegill (*Lepomis macrochirus*), green sunfish (*Lepomis cyanellus*), rainbow darter (*Etheostoma caeruleum*), white crappie (*Pomoxis annularis*), long-nosed gar (*Lepisosteus osseus*), Orange spotted sunfish (*Lepomis humilis*), yellow perch (*Perca flavescens*), walleye (*Sander vitreus*), spotfin shiner (*Cyprinella spiloptera*), and state threatened lake sturgeon (*Acipenser fulvescens*) (Keller & Ruessler, 1997; Zanatta et al., 2007; Zanatta & Woolnough, 2011; Hnytka et al., 2022). Dam removal and the creation of fish passageways can help to improve the connectivity and movement of fish hosts throughout watersheds (Ferreira-Rodríguez et al., 2019). Protecting populations of these fish species so that they can move throughout the Huron River Watershed carrying mussel glochidia can help to facilitate the growth and dispersal of mussel populations.

Lab propagation is a larger-scale conservation measure that can also be used to reintroduce mussel populations into watersheds. Lab propagation can produce genetically diverse and viable juvenile unionids (VanTassel et al., 2021), and hatcheries have cultured and reintroduced juvenile mussels in watersheds in Missouri, Wisconsin, Minnesota, Arkansas, Tennessee, Georgia, Virginia, North Carolina, and Ohio (Neves, 2004; Eads et al., 2007). Central Michigan University and the Michigan Department of Natural Resources (MDNR) have collaborated on the first mussel propagation facility in Michigan, with research focusing on the federally endangered snuffbox mussel (*Epioblasma triquetra*) (Woolnough, n.d.).

Along with these different conservation measures, the monitoring of mussel distribution is essential to know where populations exist and to effectively implement these restoration strategies. The Huron River Watershed Council (HRWC) located in Ann Arbor, Michigan, has many programs and projects that protect and restore other components of the watershed, such as macroinvertebrate water quality monitoring, water chemistry sampling, green infrastructure implementation, dam and impoundment assessment, river cleanups, and habitat restoration (Huron River Watershed Council, 2023). By piloting a citizen science mussel monitoring program for HRWC, these other programs can benefit from the data obtained through mussel surveys and from the connections made with volunteers and other organizations working with unionids. Volunteer mussel monitoring surveys pose an apt opportunity to engage local communities in meaningful science to aid in the conservation of unionids.

2 Program Design

2.1 Mussel Survey Training

HRWC's mussel monitoring protocol was developed for the purposes of leading monitoring surveys with local volunteers to assess mussel abundance and species distribution in the Huron River Watershed. Standard survey instructions were assembled from the "Michigan Freshwater Mussel Survey Protocols and Relocation Procedures for Rivers and Streams" developed by the Michigan Natural Feature Inventory (MNFI) (Hanshue et al., 2021). This document of protocols and procedures details instructions for how to conduct surveys, definitions of stream groups, and state and federal permit requirements.

Initial mussel survey training was obtained from Dr. Renee Mulcrone, a malacologist and aquatic biologist with over 25 years of experience in mussel research. Two training mussel surveys were conducted under Dr. Mulcrone. The first outing took place on September 17, 2022, at Riverside Park in Ypsilanti, Michigan (Fig. 19.2). During this survey, Dr. Mulcrone guided a one-hour timed survey in the stretch of the Huron River located at 42.24475°N, 83.61110°W. Through this survey, strategies to search for, collect, and handle mussels were demonstrated as live specimens of three species were collected and identified.

Fig. 19.2 Mussels collected during the first training survey at Riverside Park in Ypsilanti, Michigan. Live specimens of three species were collected, including nine wavy-rayed lampmussels (*Lampsilis fasciola*), ten plain pocketbooks (*Lampsilis cardium*), and one black sandshell (*Ligumia recta*). Dead shells were also collected for elktoe, Wabash pigtoe, and spike mussels

The second mussel survey training took place on September 30, 2022, in the stretch of the Huron River under Whitmore Lake Road and the US-23 Bridge at −42.471198°N, −83.756501°W. This survey was part of a relocation contract job for the Michigan Department of Transportation (MDOT) in preparation for a construction project on the US-23 Bridge above the Huron River. For this project, a transect survey was conducted to search for federally endangered snuffbox mussels that were to be relocated as an impact avoidance measure for the construction project.

Guidance for designing the program's survey protocol was also received from experts from the Michigan Department of Natural Resources (MDNR). Cleyo Harris, a fisheries technician for the Lake Erie Management Unit of the MDNR, conferred the requirements and potential opportunities for running a volunteer mussel monitoring program regarding required permits and the level of involvement that volunteers could have. Additionally, the 2022 Annual MiCorps Conference hosted a virtual presentation by Joe Rathbun, a retired mussel biologist from the Michigan Department of Environment, Great Lakes, and Energy (EGLE), that discussed mussel survey protocol and species identification information (Rathbun, 2022). Rathbun was also able to provide mussel species posters and brochures for HRWC volunteers to use.

2.2 *Mussel Survey Protocol*

The Michigan surveying season for mussel monitoring should take place from summer to early fall, around June 1 to October 15, when water temperatures are greater than 10 °C. Conditions should be such that it is warm enough for more mussels to be near the surface and the water level is low enough for surveyors to wade safely (Rathbun, 2022). Recommendations for suitable monitoring days include a depth of water transparency of about 1 ft., and water flow speeds that are safe for wading.

At the start of each survey, site habitat details including the extent of the stream being surveyed (i.e., acres, sq. m.), water flow speed, water depth, and habitat descriptions (i.e., substrate, topography, nearby dominating plant species) should be recorded. Volunteers can be separated into two teams based on interest and ability to enter the water. The field team consists of volunteers wading in the stream to search for and gather mussel specimens, while the dry team includes those who cannot or would not like to enter the water, instead staying on the stream bank to help sort mussel species. A few more experienced or willing members of the dry team can also be appointed as liaisons to answer questions from any passersby.

2.2.1 Survey Format

Two survey types are recommended based on the purpose of the survey: timed surveys and transect surveys.

Timed Surveys: For most general volunteer events, a timed survey is recommended. In the timed survey, a section of the stream is identified and marked (i.e., from a bridge to a tree). The group spends a set amount of time (typically around 60 minutes) to sample throughout that section of the stream collecting mussels. The collected specimens are to be identified and counted by number of individuals per species, noting distinctions for adult, juvenile, or dead shell material. The total number of specimens and number per species are to be recorded. The team will also photograph each species before returning it to the water to hold as evidence of observation and to submit with permit reports (Fig. 19.3). Photographs should be labeled with the name of the stream each mussel specimen was collected from and

Fig. 19.3 When taking photographs for species identification during mussel monitoring surveys, capture a shot of the profile view (**a**) and a shot of the beak sculpture (**b**). Pictured are an alive specimen of plain pocketbook (*Lampsilis cardium*) and a dead shell of creeper (*Strophitus undulatus*); specimens of both dead and alive observed species should be photographed. Ruler measurements are not required to include in photos but may help to validate the identification of species after a survey if necessary

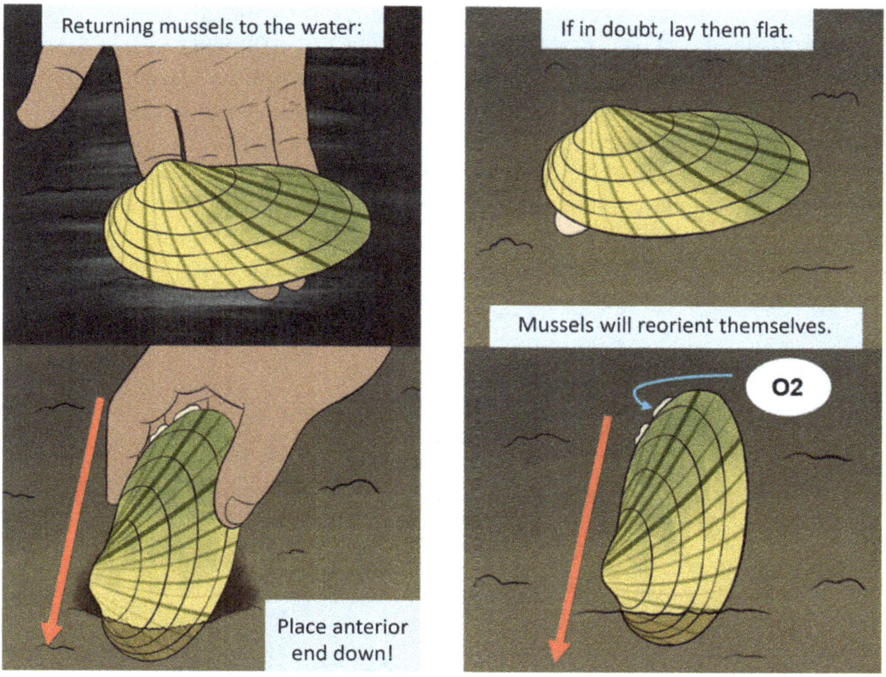

Fig. 19.4 Diagram of the orientation in which mussels should be returned to the river. This allows the mussels to reposition themselves, so they do not suffocate

the date of the survey. Afterward, the mussels are returned to the stream and equipment is cleaned. As the mussels are placed back into the stream, it is essential to ensure that they are positioned correctly to not suffocate them; when in doubt, place mussels flat against the bottom of the stream so they will be able to reposition themselves (Fig. 19.4).

Transect Survey Format: The transect method is recommended for surveys that are being conducted for specific purposes, such as in conjunction with other restoration projects or for relocation before small-scale construction projects for structures like bridges and pipelines. While this method is more time and energy consuming, it is preferable over the timed method for these specific projects as it is more effective at finding every species and individual mussels present in the stream. Most instances where the transect method is used will be for projects where volunteers help personnel from the MDNR, MDOT, etc , and so in those contexts their instructions supersede the recommendations provided here.

For this method, the area of concern in the stream is divided up into transects of ten meters placed perpendicular to stream flow. As per MNFI recommendations: "Transect spacing in small and medium rivers (drainage area less than 300 square miles) should not exceed 10 m and in large rivers transects will be spaced 25 m apart" (Hanshue et al., 2021). Accordingly, in surveys in the main branch of the

Huron River, transects should be spaced 25 m apart, and in surveys in the smaller tributaries of the Huron River, such as Portage Creek, Mill Creek, Honey Creek, and Fleming Creek, transects should be spaced less than 10 m apart. Each transect is searched thoroughly multiple times with a minimum of 30 seconds/m^2, or 2 minutes for streams that support mussels sized ≤4 cm, to find all mussels present in each transect (Hanshue et al., 2021). In instances where the transect method is being used to survey for mussels before construction projects, additional instructions and requirements may also be given on a case-by-case basis according to the MDNR and USFWS (U.S. Fish and Wildlife Service).

2.2.2 Survey Procedures

1. (a) Record site habitat information: this includes water flow speed, water depth, substrate type, and GPS location. (b) Take photos of the site (Appendix D).
2. Record the name(s) of the survey leader(s) and the date of the survey.
3. For transect surveys: Set up 10 m survey transects in stream along area of concern.
4. Start a timer to record how long the survey lasted to gauge general surveying effort.
5. Collect live mussels and dead shell material and place in mesh bags or buckets to carry under water. Mussels should always be kept cool and wet, minimizing the time they are out of water.
6. Spread volunteers out in the stream to get samples across the site and prevent any rocks or sediment from getting kicked downstream into another volunteer. Surveyors should stay in shallow areas to ensure safety. Snorkel/scuba may be necessary for deeper waterbodies.
7. Keep mussels wet after collection by laying them out over a surface (such as mesh bags) in shallow water at the bank of the stream, or in large trays filled with stream water (Fig. 19.5).
8. Take photos of live and dead specimens and identify, count, and record species. Dead shells may only be collected according to the survey leader's scientific collector's permit.
9. When returning mussels to the stream after collection and identification, place them anterior end down (or if in doubt lay flat to allow them to reorient themselves) to not suffocate the mussels.
10. Clean all equipment after the survey has concluded to avoid spreading pathogens or invasive species between streams.

2.2.3 List of Recommended Equipment

- Glassbottom buckets or snorkeling masks to search underwater for mussels.
- Mesh collection bags to hold mussels while collecting.
- Buckets or trays to hold mussels in water during species identification.

Fig. 19.5 Example of how collected mussels can be kept in trays/buckets, or above a mesh bag in shallow water, to keep mussels in water with suitable temperature. Mussels should be identified to species and photographed as quickly as possible to minimize stress

- Waders or swimming clothing.
- Digging tools to scoop mussels that are burrowed in sediment.
- Waterproof gloves for pulling mussels from the water.

- Pack of stake markers to outline transects with rope.
- Depth gauge or stick to record depth of stream.
- Water velocity meter/thermometer to measure water flow speed and temperature.
- GPS unit (or smartphones) to track and record site location coordinates.
- Field guides with species identification information.

2.3 Site Selection

Potential sites can be identified using the Michigan Mussels Web App from the Michigan Natural Features Inventory to find stream groups and determine which sites can be surveyed based on survey leader permit level. Past mussel studies in the Huron River conducted by Van der Schalie (1938) and Badra (2010) were also used to identify potential sites for HRWC's mussel program.

Sites were ranked for volunteer mussel surveys based on factors of accessibility for parking, walking, and sampling access; substrate type; site ownership; and stream conditions. Qualitative substrate measurements were collected and noted through visual observations based on characteristics of rock, sand, gravel, and mud.

The most accessible and higher-quality sites were recommended as primary sites, which consisted of streams with high accessibility for volunteers to park, walk to, and access streams; higher stream quality; stable substrate for wading; and public ownership.

Secondary potential sites were ranked based on lower parking availability, longer or more difficult walks through forested areas to the stream, and less accessible ways to enter a stream. Secondary sites also may have lower water quality with a less probable chance of observing live mussels, with more mucky streambeds that make wading more difficult.

2.3.1 Site Recommendations

Out of 46 sites ranked for recommendations for HRWC's volunteer monitoring program, 31 were recommended as primary sites and 15 were recommended as secondary sites. These sites include locations in four Michigan counties in the Huron River Watershed (Washtenaw, Wayne, Oakland, and Livingston) that span across the main branch of the Huron River and seven branching tributaries (Pettibone Creek, Portage Creek, Fleming Creek, Mill Creek, Livingston Country Honey Creek, Washtenaw County Honey Creek, and Woodruff Creek) and represent a variety of habitats that range from urban to rural and vary between substrate types (i.e., rock, gravel, sand, and mud). Nine of the visited sites were not recommended for the survey program because of difficult access to the site or stream, poor habitat quality, and/or low probability of mussel presence.

2.4 Program Implementation

The mussel monitoring program was piloted during the summer of 2023 under Dr. Paul Steen and Amani Tolin as mussel survey leaders. Seven survey outings were conducted using volunteers from a range of experience levels and ages including local citizens, HWRC staff, ecologists, HWRC summer interns, and children (Fig. 19.6).

The seven visited sites comprised of four tributaries (Huron River, Portage Creek, Mill Creek, and Fleming Creek) from three counties in the watershed (Washtenaw, Oakland, and Livingston). Eight live species were observed across the seven surveys, including creeper, fat mucket, fluted-shell, kidneyshell, plain pocketbook, rainbow, spike, and the state threatened wavy-rayed lampmussel. Through all seven sites, a total of 129 live mussels were observed with an average of 2.4 species observed per site (Table 19.1). Over 200 intact shells and shell fragments from approximately 12 species were also observed across the seven sites.

General trends were discernable for mussel distribution from this limited sample size of surveys. Sites located toward the upper part of the Huron River Watershed, consisting of the two main branch Huron River sites and the Portage Creek site,

Fig. 19.6 Mussel monitoring survey coordinator Amani Tolin (right) leading Huron River Watershed Council summer interns in the identification of collected freshwater mussels from the Huron River at Proud Lake Recreation Area

Table 19.1 Total number of mussels and species observed at each site visited during the 2023 pilot for HRWC's mussel monitoring program

Stream	Total no. of specimens	Total no. of species
Huron River: Proud Lake Recreation Area	35	6
Huron River: Millrace Park	49	5
Portage Creek: Hiland Lake Dam	41	4
Mill Creek: Jackson Rd	3	1
Mill Creek: Sloan Preserve	1	1
Fleming Creek: Parker Mill County Park	0	0
Fleming Creek: University of Michigan Botanical Gardens	0	0

hosted richer mussel assemblages than the sites located lower in the watershed, consisting of the Mill Creek and Fleming Creek sites. At the upper watershed sites, species richness and the overall number of live individuals observed were no less than 35 total individuals and 4 total species found at each site, which meets the MNFI's benchmark of diverse mussel communities supporting at least 4 species (Hanshue et al., 2021). For the lower watershed sites, two sites in Mill Creek had a total of four spike (*Eurynia dilatata*) mussels observed and the two sites in Fleming Creek had no live mussels observed.

These trends support relationships that have been found between mussel density and habitat parameters. Streams that tend to support larger mussel communities often share habitat characteristics that include lower hydrological flashiness, a mix of flow velocities and depth present throughout the stream, and substrate compositions that consist of more boulder and cobble and less fine sediment (McRae et al., 2004; Nicklin & Balas, 2007). While surveys for other taxa often have more quantitative assessments of habitat quality, such as the biotic scoring indices used for aquatic macroinvertebrates, the population metrics used here for these freshwater mussel surveys and any corresponding implications for habitat quality are more qualitative due the complexity of the determining factors affecting mussel presence (Rathbun, 2022).

3 Educational Materials

A set of informational materials were produced and acquired that can be used in species identification, environmental education, and spreading awareness about mussel conservation. These materials include species identification keys, brochures, posters, a set of three-dimensional (3D) models, and an illustrated mussel life cycle mural.

3.1 Species Identification Field Guide

The Field Guide to the freshwater mussels of Michigan (Mulcrone & Rathbun, 2018) and the Pocket field guide to the freshwater mussels of Michigan (2nd ed.) (Mulcrone & Rathbun, 2020) were used to inform the development of this program and the production of species identification materials. A field guide was also developed for species exclusively from the Huron River Watershed that provides illustrations of shell characters such as shape, color, and other features to aid volunteers with identifying species (Figs. 19.7 and 19.8).

Volunteers looking to become survey leaders after gaining enough experience handling and identifying mussels can take the Michigan Chapter of the American Fisheries Society and MDNR Freshwater Mussels Identification Test to help with their certification training. This test was intended for for-profit businesses and consulting firms looking to do projects and is a recommendation, not a requirement, for citizen and scientific surveyors (Rathbun, 2022). Applicants can take this test at no cost. The test covers species identification for adult specimens of 41 native Michigan

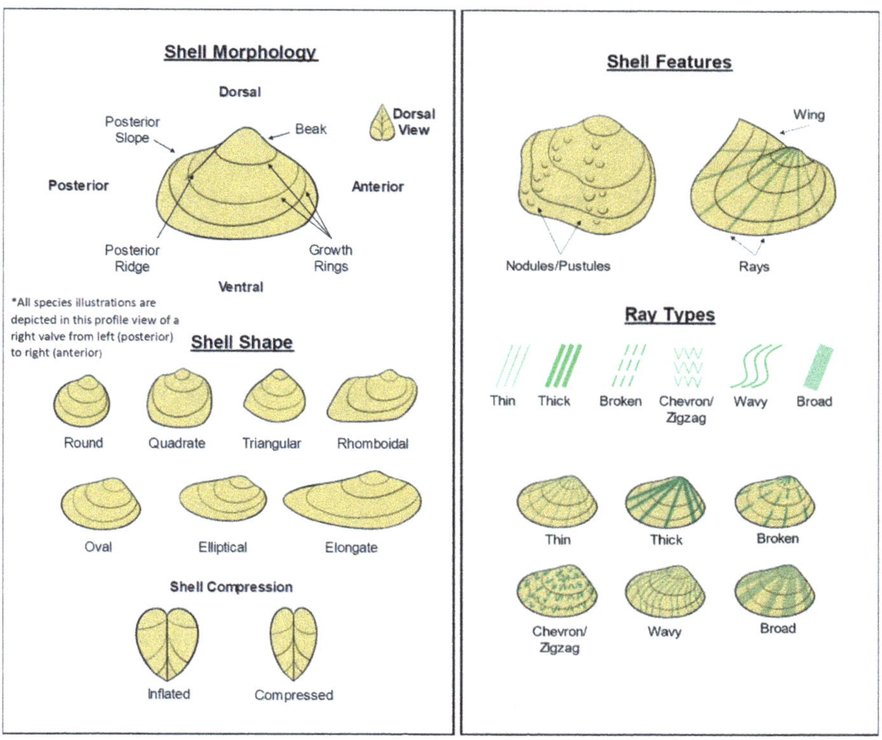

Fig. 19.7 Unionid morphology illustrations from the developed "Field Guide to the Freshwater Mussels of the Huron River Watershed." The full field guide can be viewed online at: https://www.artbyaskari.com/science#h.ix7cbxsxx2mq

Snuffbox, *Epioblasma triquetra* (Federally Endangered) ★

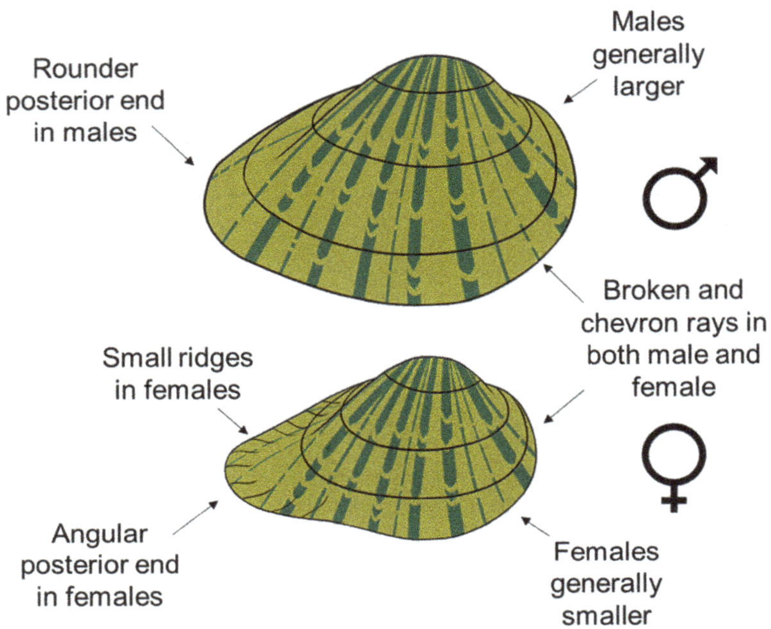

Shell Features

Size: up to 7.5 cm (3 in)

Shape: triangular or elongate; inflated shell; sexually dimorphic*

***Males:** shorter, round posterior end; larger than females

***Females:** longer, angular posterior end; smaller than males; ridges on posterior slope

Color: yellow to yellowish green; white nacre

Rays: green rays and arrow/chevron shaped markings

Fig. 19.8 Example of a species description for the snuffbox mussel (*Epioblasma triquetra*) from the illustrated "Field Guide to the Freshwater Mussels of the Huron River Watershed." Species descriptions list size, shape, color, presence of rays, and any other helpful additional shell characteristics to aid in species identification

species. Interested applicants can register for this test at https://michigan.fisheries.org/annual-meetings/upcoming-meeting/ and David Strayer can be contacted for more information (strayerd@caryinstitute.org).

3.2 Huron River Mussels iNaturalist Page

An iNaturalist page was set up to record past and future observations of mussels in the Huron River Watershed, found at: https://www.inaturalist.org/projects/freshwater-mussels-of-the-huron-river-watershed. The parameters for this iNaturalist project page are set to include all observations of freshwater mussels in the Michigan counties of Washtenaw, Oakland, Livingston, Ingham, Jackson, Wayne, and Monroe. This tool allows HRWC survey leaders as well as certified volunteers in the watershed to post mussel observations and receive quick feedback on species identification when needed.

3.3 Mussel 3D Models

Since mussel shells are difficult to acquire for use as references in species identification due to permitting restrictions, a set of 3D models for species found in the Huron River was produced. Shell specimens were loaned from the University of Michigan Museum of Zoology Mussel Mollusk Collection (University of Michigan LSA Museum of Zoology, 2023) and then scanned through collaboration with Syrena Kapsa and other personnel at CultureVerse, an Ann Arbor based art and technology non-profit gallery. Twenty-five species prints were donated to HRWC by CultureVerse, including: mucket (*Actinonaias ligamentina*), elktoe (*Alasmidonta marginata*), threeridge (*Amblema plicata*), cylindrical papershell (*Anodontoides ferussacianus*), rainbow (*Cambarunio iris*), pimpleback (*Cyclonaias pustulosa*), spike (*Eurynia diltata*), Wabash pigtoe (*Fusconaia flava*), plain pocketbook (*Lampsilis cardium*), white heelsplitter (*Lasmigona complanata*), creek heelsplitter (*Lasmigona compressa*), fluted-shell (*Lasmigona costata*), male and female fat mucket, fragile papershell (*Leptodea fragilis*), threehorn wartyback (*Obliquaria reflexa*), round pigtoe (*Pleurobema sintoxia*), pink heelsplitter (*Potamilus alatus*), kidneyshell (*Ptychobranchus fasciolaris*), giant floater (*Pyganodon grandis*), hickorynut (*Obovaria olivaria*), mapleleaf (*Quadrula quadrula*), deertoe (*Truncillla truncata*), creeper (*Strophitus undulatus*), and paper pondshell (*Utterbackia imbecillis*) (Fig. 19.9). This set of models will aid in species identification and volunteer training, since permit restrictions limit the collection of live and dead specimens from the field. These 3D models are also available to view online in a Sketchfab collection at: https://sketchfab.com/atolin/collections/freshwater-mussels-of-the-huron-river-mi-e9de9db859c64b3297d3348461bcf9fd (Fig. 19.10). The museum specimens were scanned with a Shining 3D EinScan-SP scanner using the software EXScan S. The models were then printed using a Prusa i3 MK3 3D printer, taking around ten hours per model to print.

Fig. 19.9 Photograph of painted 3D models of Huron River mussel species. Species depicted include plain pocketbook, threeridge, male and female fat mucket, pink heelsplitter, kidneyshell, giant floater, mucket, spike, round pigtoe, hickorynut, cylindrical papershell, fragile papershell, mapleleaf, pimpleback, threehorn wartyback, creek heelsplitter, rainbow, and elktoe

3.4 *Mural of Unionidae Life Cycle*

A scientific illustration was created to depict the life history and reproduction strategies of unionids (Fig. 19.11). In the mural, there is a plain pocketbook (*Lampsilis cardium*) displaying its lure to attract a bluegill host. The image shows the reproductive cycle from left to right as the bluegill swims upstream. At the very left, a male pocketbook is shown releasing its sperm as it is carried downstream to a nearby

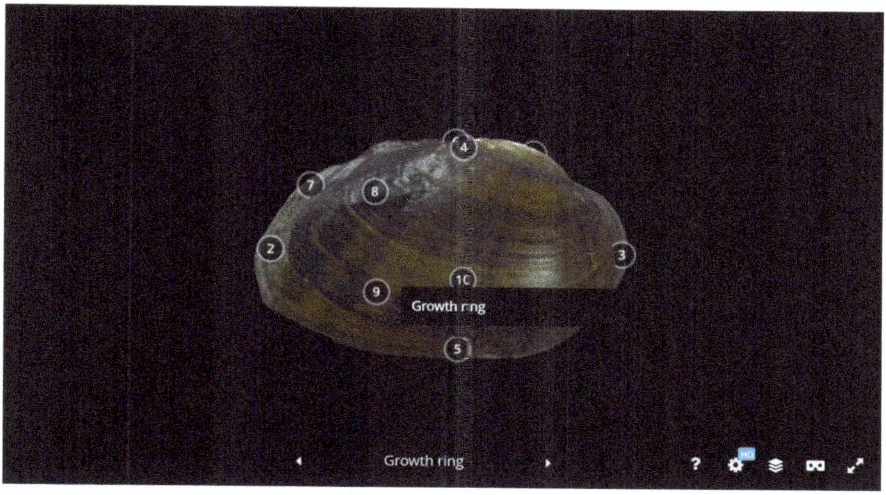

Fig. 19.10 Example of a mussel scan with annotated shell characteristics in Sketchfab to aid in species identification. A scan of a mucket (*Actinonaias ligamentina*) is depicted in this image. The full collection of 3D models can be viewed at: https://www.artbyaskari.com/science#h.az2blcow6zod

Fig. 19.11 Mural of unionid life history and reproductive strategy that requires a fish host. The mussel species depicted is the plain pocketbook (*Lampsilis cardium*) with a bluegill (*Lepomis macrochirus*) as its fish host. The diagram will be used to help draw viewers in to learn about unionids and the volunteer program. The mural can also be viewed at: https://www.artbyaskari.com/science#h.r74ki5lay2ph

female. This sperm enters the female pocketbook and fertilizes its eggs. Continuing to the right, the female mussel is shown displaying its lure in the shape of a blue fish to attract the bluegill fish host. The bluegill is attracted to the lure and the female is depicted releasing its glochidia larvae to attach to the gills of the bluegill. The bluegill continues swimming upstream to the right until a larval pocketbook is shown dropping off its bluegill fish host to burrow into the sediment and develop into an adult.

4 Discussion

The main purposes of the program's volunteer mussel surveys are (a) to assess the status of unionid mussels in the Huron River Watershed, (b) to provide data to inform other HRWC projects, and (c) to educate and engage volunteers with freshwater mussels to promote awareness for a lesser appreciated and declining group of animals.

4.1 Role of Mussel Surveys

HRWC's mussel monitoring program is one of the first organized mussel programs in Michigan for volunteer mussel surveys. Most mussel surveys in Michigan have been conducted by mussel experts from research, government, and consulting groups such as Central Michigan University, MDOT, ASTI Environmental - PEA Group, EGLE, MDNR, and MNFI. This is principally due to Michigan's strict permitting restrictions and the lack of public awareness about unionids. This monitoring program makes it possible for volunteers from around the Huron River Watershed to participate in mussel surveys and contribute to meaningful conservation efforts.

The species diversity and abundance results from the mussel surveys help to evaluate the status of mussels in the Huron River Watershed. Since mussels are especially susceptible to habitat quality, the presence of mussels can indicate that a stream has sufficient dissolved oxygen levels, low levels of pollutants and turbidity, stable substrates and stream hydrology, and/or healthy populations of host fishes (Badra, 2020). Generally, streams that have at least four species of unionids present are considered to host diverse mussel communities (Hanshue et al., 2021). Conversely, small/absent mussel communities observed during surveys do not necessarily indicate low habitat quality. There could be a variety of probable limiting factors unrelated to the suitability of a stream, including an insufficient population of host fish, high patchiness leading to surveyors missing spots with higher mussel density, or even that mussels were not observed just due to environmental stochasticity affecting mussel and host fish populations (Rathbun, 2022). Therefore, the presence of mussels can indicate if a stream has healthy, favorable conditions; however, the absence of mussels does not entirely exclude a stream as suitable habitat (Rathbun, 2022).

Information from the surveys on where mussels are located throughout the Huron River Watershed can help guide HRWC as to which sites might be higher priority for protection and restoration measures. High-quality sites with healthy, diverse communities of mussels can be marked for protection, especially sites with state and federally listed species. These high-quality sites should also be monitored to protect them from the spread of zebra mussels, quagga mussels, and Asian clams. In sites where new occurrences of state or federally listed species are recorded, stronger protections can be obtained for those streams. In these instances, after finding a

federally listed species in Group 1 and 2 streams where they are not expected to be found, the survey should be stopped to contact USFWS for guidance at the Michigan Field Office in Lansing at (517) 351–2555. Additionally, high-quality sites that are identified through this survey program could potentially have their mussel populations augmented or reintroduced, for example following work done by the Michigan mussel hatchery project from Central Michigan University and the MDNR (Central Michigan University Communications, 2020).

Surveys will also help to find mussels affected by construction activities to relocate them to protected sites. This type of relocation is currently required for state and federally listed species and recommended for non-listed species (Hanshue et al., 2021). These activities also provide opportunities for volunteer monitoring to assess how populations fare after relocation.

Sites that have mussels observed in habitats facing any threats can be prioritized for restoration. Sites with sparse mussel communities in streams with turbid water could be optimal locations to implement rain gardens, riparian buffers, and other forms or green infrastructure to lessen excessive sediment inputs. Streams that have mussels present but hold low populations of host fish or have low connectivity to the rest of the watershed may require measures to protect or restore host fish populations or to reconnect pathways to the watershed. Improving the access of mussels to host fish may contribute to rationale that supports the removal of old dams and the replacement or improvement of ineffective culverts.

4.2 Social Value of Mussel Citizen Science Program

The mussel surveys provide a valuable opportunity for scientific public engagement and education. Volunteerism has proven to be a valuable tool in restoration because it can increase knowledge and awareness about ecology and environmental issues, improve community relationships with the environment, promote sustainable behaviors and lifestyles, and encourage social interaction between community members (Dresner et al., 2015; Stepenuck & Green, 2015). Freshwater mussel conservation depends heavily on the promotion of sustainable activities such as cleaning equipment to prevent the spread of zebra mussels, spreading awareness about the use of herbicides, advocating for the implementation of rain gardens and green infrastructure, and the overall practice of safe recreational behavior in Michigan waterways as to not disturb or harm native mussels. The program also can bring young volunteers to the field and introduce them to mussels, nature, and conservation. HRWC's surveys were valuable in this way with their inclusion of volunteers across backgrounds, ages, and levels of experience.

4.3 Similar Programs

Several organizations across the United States have volunteer programs for mussel monitoring. These programs vary in levels of volunteer involvement.

The Anacostia Watershed Society (AWS) has a freshwater mussel survey program for the Anacostia River Watershed in Washington D.C. and Maryland. This program has involved around 120 volunteers that work under AWS staff to collect data during field surveys and to monitor floating baskets of mussel cultures propagated at the Harrison Lake National Fish Hatchery (J. Bogantes, personal communication, August 1, 2022). These surveys have helped AWS gauge the health of mussel communities over time by producing data about the distribution, growth, and survival of propagated and released mussels (J. Bogantes, personal communication, August 1, 2022).

Wisconsin has a mussel monitoring program that was developed by the Wisconsin Aquatic and Terrestrial Resources Inventory and sponsored by the Wisconsin Department of Natural Resources. This program conducts monitoring training for volunteers and has a project page on the online network iNaturalist for their volunteers to submit photos of mussels (Weingzinger & Kitchel, 2018). Unlike in Michigan, the collection of live mussels in Wisconsin without a permit like the Collector's Permit is legal, so volunteers are free to conduct their own surveys (Weingzinger & Kitchel, 2018).

The Texas Mussel Watch of the Wildlife Diversity Program is similar to Wisconsin's program with the less supervised participation of volunteers. Fishing licenses and endorsements are required in Texas for residents to collect freshwater mussels and can be purchased with a yearly resident freshwater package for $30, a non-resident freshwater package for $58, or a senior freshwater package for residents 65 and older for $12 (Texas Parks and Wildlife Department, n.d.). The Texas Mussel Watch program encourages volunteers to search for mussels and submit photos to their iNaturalist page. By April 2023, over 1300 observations had been submitted to the Texas Mussel Watch iNaturalist page (Freshwater Mussels of Texas, 2012).

5 Limitations

Mussels are heavily regulated in the state of Michigan; thus, surveys were limited to sites permitted by the acquired Common and State level permits that cover Group 1 and 2 streams. Since the purposes of this program are volunteer based, many suitable sites were also excluded based on accessibility with permits and considering volunteer abilities with walking, climbing, wading, etc. However, future iterations of the program would be able to visit more sites protected under higher permits due to the connections that were made with mussel experts over the course of this study.

There were also limitations to producing the set of 3D mussel models. Since the set of 3D models was produced by CultureVerse at no cost using specimens from the University of Michigan Museum of Zoology, not every species present in the Huron River was scanned and printed. The scanning and printing process took considerable time, and the museum mussel specimens were loaned on a limited time basis, so only 25 species of the Huron River were produced. Since a mussel expert would already be required to identify most protected species, mostly unlisted species were chosen to produce along with a few protected species.

6 Future Perspectives

Through the development of HRWC's mussel monitoring program, a foundation has been built to help other watershed groups in Michigan, such as the Clinton River Watershed Council, the Flint River Watershed Coalition, Friends of the Rouge, and Friends of the St. Clair River, to share program materials for the development of mussel monitoring programs for their own watersheds. By expanding volunteer mussel monitoring efforts beyond the Huron River Watershed, mussel populations across the state can be assessed and protected further than solely by government, contract, and university efforts. While other states vary in restrictions for interacting with mussels, HRWC's program design can also serve as reference for other emerging programs throughout the Great Lakes region and beyond.

The program can present opportunities to introduce a broader audience to the plight of mussels through species identification workshops with mussel experts, volunteer survey outings, and other educational activities. Increasing public involvement could in turn produce greater public investment in the conservation status of threatened and endangered species.

With the summer of 2023 serving as a pilot for the HRWC mussel program, more applications of the mussel surveys are also possible due to the experience gained by HRWC staff and volunteers. With plans for the Peninsular Paper Dam in Ypsilanti, Michigan, to be removed soon, HRWC aims to use the mussel program under the guidance of mussel expert Dr. Renee Mulcrone to have volunteers help to survey and relocate mussels before the dam removal. Talks with the MDNR have also proved productive for the potential of HRWC volunteers conducting long-term return surveys for projects such as mussel relocations, which have often been lacking in Michigan.

7 Conclusion

Innovative methods that expand beyond what academic and governmental approaches can handle are especially critical to the success of conservation in the face of anthropogenic threats that have increasingly degraded natural habitats.

Volunteerism and citizen science are posed as valuable opportunities to involve local communities in meaningful conservation. This project extended across various fields of ecology, the arts, and social contexts to create an initiative that helps protect Michigan's native freshwater mussels and contributes to improving the overall well-being of the Huron River Watershed. Future improvements as the program continues may provide an example to other watershed coalitions in Michigan and abroad to adopt their own mussel survey programs. This program indicates the potential opportunities that are available through collaborations across fields to improve the impact of conservation in watershed ecology.

Acknowledgments The content of this chapter was started and submitted in a thesis by Amani Tolin in fulfillment of the University of Michigan School for Environment and Sustainability's Master of Science degree (Tolin, 2023).

Special thanks to Dr. Sara Adlerstein-Gonzalez from the University of Michigan and Dr. Paul Steen and Jason Frenzel from the Huron River Watershed Council for their guidance throughout this study. Additional thanks to Dr. Renee Mulcrone for providing her expertise with freshwater mussels, as well as Syrena Kapsa from the Ann Arbor art gallery CultureVerse and Dr. Taewhan Lee from the University of Michigan Museum of Zoology for their help in producing the 3D models.

Funding Financial support to acquire field equipment was received from the University of Michigan Rackham Graduate School.

References

Abdelsaleheen, O. (2023). *Reproduction biology and environmental threats for freshwater mussels (Bivalvia: Unionidae)*. Dissertation, University of Eastern Finland.

Aldridge, D. C., Ollard, I. S., Bespalaya, Y. V., Bolotov, I. N., Douda, K., Geist, J., Haag, W. R., Klunzinger, M. W., Lopes-Lima, M., Mlambo, M. C., Riccardi, N., Sousa, R., Strayer, D. L., Torres, S. H., Vaughn, C. C., Zając, T., & Zieritz, A. (2023). Freshwater mussel conservation: A global horizon scan of emerging threats and opportunities. *Global Change Biology, 29*(3), 575–589. https://doi.org/10.1111/gcb.16510

Badra, P. (2010). *Unionid mussel surveys in the Huron-Clinton Metroparks*. Michigan Natural Features Inventory. https://mnfi.anr.msu.edu/reports/MNFI-Report-2010-26.pdf.

Badra, P. (2020). *Freshwater mussels of Michigan*. [Brochure]. Michigan Natural Features Inventory, Michigan State University Extension.

Baker, S., & Hornbach, D. (2000). Physiological status and biochemical composition of a natural population of unionid mussels (*Amblema plicata*) infested by zebra mussels (*Dreissena polymorpha*). *The American Midland Naturalist, 143*(2), 443–452. https://doi.org/10.1674/0003-0031(2000)143[0443:PSABCO]2.0.CO;2

Bowman, N., Patel, P., Xu, W., Alsaffar, A., & Tiquia-Arashiro, S. M. (2018). Enrichment and isolation of Pb-resistant bacteria from Saint Clair River sediments and their potential for Pb removal in aqueous solutions. *Applied Microbiology and Biotechnology, 102*, 2391–2398. https://doi.org/10.1007/s00253-018-8772-4

Central Michigan University Communications. (2020). CMU helps an endangered river cleaner. Retrieved May 26, 2023 from https://www.cmich.edu/news/details/cmu-helps-an-endangered-river-cleaner

Chaffin, B. C., Shuster, W. D., Garmestani, A. S., Furio, B., Albro, S. L., Gardiner, M., Spring, M., & Green, O. O. (2016). A tale of two rain gardens: Barriers and bridges to adaptive manage-

ment of urban stormwater in Cleveland, Ohio. *Journal of Environmental Management, 183*(Pt 2), 431–441. https://doi.org/10.1016/j.jenvman.2016.06.025

Dickinson, B., & Sietman, B. (2008). Recent observation of metamorphosis without parasitism in *Utterbackia imbecillis*. *Ellipsaria, 10*(1), 7–8.

Doyle, M., Stanley, E., Orr, C., et al. (2005). Stream ecosystem response to small dam removal: Lessons from the Heartland. *Geomorphology, 71*(1–2), 227–244. https://doi.org/10.1016/j.geomorph.2004.04.011

Dresner, M., Handelman, C., Braun, S., & Rollwagen-Bollens, G. (2015). Environmental identity, pro-environmental behaviors, and civic engagement of volunteer stewards in Portland area parks. *Environmental Education Research, 21*(7), 991–1010. https://doi.org/10.1080/13504622.2014.964188

Eads, C., Raley, M., Schubert, E., Bogan, A., & Levine, J. (2007). *Propagation and culture of freshwater mussels for release into North Carolina waters* (No. FHWA/NC/2006–37).

Endangered Species Act, 16 U.S.C.§ 1531–1544 (1983).

Ferreira-Rodríguez, N., Akiyama, Y., Aksenova, O., et al. (2019). Research priorities for freshwater mussel conservation assessment. *Biological Conservation, 231*, 77–87. https://doi.org/10.1016/j.biocon.2019.01.002

Freshwater Mussels of Texas. (2012). *iNaturalist*. Retrieved April 11, 2023, from https://www.inaturalist.org/projects/freshwater-mussels-of-texas.

Gascho Landis, A., Mosley, T., Haag, W., & Stoeckel, J. (2012). Effects of temperature and photoperiod on lure display and glochidial release in a freshwater mussel. *Freshwater Science, 31*(3), 775–786. https://doi.org/10.1899/11-082.1

Gibson, R. (2019). *A metabolic approach to assessing interactions between climate change, agrochemicals, and freshwater mussel reproduction*. Master's thesis, Auburn University.

Gillis, P., & Mackie, G. (1994). Impact of the zebra mussel, *Dreissena polymorpha*, on populations of Unionidae (Bivalvia) in Lake St. Clair. *Canadian Journal of Zoology, 72*, 1260–1271.

Green, R., Bailey, R., Hinch, S., Metcalfe, J., & Young, V. (1989). Use of freshwater mussels (Bivalvia: Unionidae) to monitor the nearshore environment of lakes. *Journal of Great Lakes Research, 15*(4), 635–644. https://doi.org/10.1016/S0380-1330(89)71517-3

Gruber, B., Gonzalez, S., Badra, P., & Herbert, M. (2012). Distribution and habitat of unionid mussels and invasive sea lamprey larvae in the Paw Paw River, a tributary of Lake Michigan. *Freshwater Biology, 57*(6), 1293–1305. https://doi.org/10.1111/j.1365-2427.2012.02777.x

Hanshue, S., Rathbun, J., Badra, P, et al. (2021). *Michigan freshwater mussel survey protocols and relocation procedures for rivers and streams*. Michigan Natural Features Inventory. https://mnfi.anr.msu.edu/pdfs/MI_Freshwater_Mussel_Survey_Protocol_May_2021_V.pdf.

Hassel, J., & Farris, J. (2007). A review of the use of unionid mussels as biological indicators of ecosystem health. In *Freshwater bivalve ecotoxicology* (pp. 19–49). Routledge.

Hellou, J., Ross, N. W., & Moon, T. W. (2012). Glutathione, glutathione S-transferase, and glutathione conjugates, complementary markers of oxidative stress in aquatic biota. *Environmental Science and Pollution Research, 19*, 2007–2023. https://doi.org/10.1007/s11356-012-0909-x

Hnytka, S., Watkinson, D., Enders, E., Friesen, C., Murray, C., & Church, C. (2022). *The ecology, distribution, and identification of native freshwater mussels (Unionidae) of Manitoba* (Vol. 3435). Canadian Technical Report of Fisheries and Aquatic Sciences.

Huron River Watershed Council. (2023). *Programs*. Retrieved April 11, 2023, from https://www.hrwc.org/what-we-do/programs.

Jones, J. (2015). Freshwater Mussels of Virginia (Bivalvia: Unionidae): An introduction to their life history, status and conservation. *Virginia Journal of Science, 66*(3), 6. https://doi.org/10.25778/3baj-mq15

Keller, A., & Ruessler, D. (1997). Determination or verification of host fish for nine species of unionid mussels. *American Midland Naturalist*, 402–407. https://doi.org/10.2307/2426831

Legislative Service Bureau. (1994). Natural Resources and Environmental Protection Act: (act 451 of 1994). http://www.legislature.mi.gov/documents/mcl/pdf/mcl-451-1994-iii-1- endangered-species-365.pdf

Maceda-Veiga, A. (2013). Towards the conservation of freshwater fish: Iberian Rivers as an example of threats and management practices. *Reviews in Fish Biology and Fisheries, 23*, 1–22. https://doi.org/10.1007/s11160-012-9275-5

Maio, J., & Corkum, L. (1995). Relationship between the spatial distribution of freshwater mussels (Bivalvia: Unionidae) and the hydrological variability of rivers. *Canadian Journal of Zoology, 73*(4), 663–671. https://doi.org/10.1139/z95-078

McCombs, E. (2014). *Dam removal and freshwater mussels: effective restoration and prioritization through case studies*. Presentation at the International Conference on Engineering and Ecohydrology for Fish Passage, University of Massachusetts Amherst, Amherst.

McElwain, A., Warren, M., Pereira, F., et al. (2019). Pathobiology and first report of larval nematodes (*Ascaridomorpha* sp.) infecting freshwater mussels (*Villosa nebulosa*, Unionidae), including an inventory of nematode infections in freshwater and marine bivalves. *International Journal for Parasitology Parasites and Wildlife, 10*, 41–58. https://doi.org/10.1016/j.ijppaw.2019.05.006

McRae, S., Allan, J., & Burch, J. (2004). Reach-and catchment-scale determinants of the distribution of freshwater mussels (Bivalvia: Unionidae) in south-eastern Michigan, USA. *Freshwater Biology, 49*(2), 127–142. https://doi.org/10.1046/j.1365-2426.2003.01165.x

Michigan Natural Features Inventory. (2022). Michigan Mussels Web App. Retrieved August 21, 2023 from https://mnfi.maps.arcgis.com/apps/webappviewer/index.html?id=3860be5d7f28471396d44e0b384abb12

Mulcrone, R. (2006a). *Lasmigona costata*. Animal Diversity Web. Retrieved May 26, 2023, from https://animaldiversity.org/accounts/Lasmigona_costata/#ecosystem_roles

Mulcrone, R. (2006b). *Ligumia recta*. Animal Diversity Web. Retrieved May 26, 2023, from https://animaldiversity.org/accounts/Ligumia_recta/

Mulcrone, R., & Rathbun, J. (2018). *Field guide to the freshwater mussels of Michigan*. Michigan Department of Natural Resources.

Mulcrone, R., & Rathbun, J. (2020). *Pocket field guide to the freshwater mussels of Michigan* (2nd ed.). Michigan Department of Natural Resources.

Nassereddine, Z., Opara, S., Coutinho, O., Qyteti, F., Book, R., Heinicke, M. P., Napieralski, J., & Tiquia-Arashiro, S. M. (2024). Critical perspectives on advancing antibiotic resistant gene (ARG) detection technologies in aquatic ecosystems. *Science of the Total Environment, 957*, 177775. https://doi.org/10.1016/j.scitotenv.2024.177775

Neves, R. (2004). Propagation of endangered freshwater mussels in North America. *Journal of Conchology Special Publication, 3*, 69–80.

Nicklin, L., & Balas, M. (2007). Correlation between unionid mussel density and EPA habitat-assessment parameters. *Northeastern Naturalist, 14*(2), 225–234.

Oest, A., Alsaffar, A., Fenner, M., Azzopardi, D., & Tiquia-Arashiro, S. M. (2018). Patterns of change in metabolic capabilities of sediment microbial communities in river and lake ecosystems. *International Journal of Microbiology, 2018*, 6234931. https://doi.org/10.1155/2018/6234931

Pandolfo, T., Kwak, T., & Cope, W. (2016). Microhabitat suitability and niche breadth of common and imperiled Atlantic slope freshwater mussels. *Freshwater Mollusk Biology and Conservation, 19*(2), 27–50. https://doi.org/10.31931/fmbc.v19i2.2016.27-50

Patel, D., Gismondi, R., Alsaffar, A., & Tiquia-Arashiro, S. M. (2019). Applicability of API ZYM to capture seasonal and spatial variabilities in lake and river sediments. *Environmental Technology, 40*(24), 3227–3239. https://doi.org/10.1080/09593330.2018.1468492

Payton, S., Johnson, P., & Jenny, M. (2016). Comparative physiological, biochemical and molecular thermal stress response profiles for two unionid freshwater mussel species. *Journal of Experimental Biology, 219*(22), 3562–3574. https://doi.org/10.1242/jeb.140129

Pilger, T., & Gido, K. (2012). Variation in Unionid assemblages between streams and a reservoir within the Kansas River Basin. *The American Midland Naturalist, 167*(2), 356–365. https://doi.org/10.1674/0003-0031-167.2.356

Rathbun, J. (2022). *An introduction to freshwater mussels*. Presented at Annual MiCorps Conference. Retrieved November 18, 2022, from https://www.youtube.com/watch?v=6y0Z8UkjiEM&feature=youtu.be

Reid, A. J., Carlson, A. K., Creed, I. F., Eliason, E. J., Gell, P. A., Johnson, P. T. J., Kidd, K. A., MacCormack, T. J., Olden, J. D., Ormerod, S. J., Smol, J. P., Taylor, W. W., Tockner, K., Vermaire, J. C., Dudgeon, D., & Cooke, S. J. (2019). Emerging threats and persistent conservation challenges for freshwater biodiversity. *Biological Reviews of the Cambridge Philosophical Society, 94*(3), 849–873. https://doi.org/10.1111/brv.12480

Sethi, S., Selle, A., Doyle, M., Stanley, E., & Kitchel, H. (2004). Response of unionid mussels to dam removal in Koshkonong Creek, Wisconsin (USA). *Hydrobiologia, 525*, 157–165.

Shevchuk, L., Vasilieva, L., Romaniuk, R., & Pavliuchenko, O. (2021). Species diversity of unionid mussels (Mollusca: Bivalvia: Unionidae) as a bioindicator of the state water environment of river basins of Ukraine. *IOP Conference Series: Earth and Environmental Science, 915*(1), 012006. https://doi.org/10.1088/1755-1315/915/1/01200

Sousa, R., Halabowski, D., Labecka, A. M., Douda, K., Aksenova, O., Bespalaya, Y., Bolotov, I., Geist, J., Jones, H. A., Konopleva, E., Klunzinger, M. W., Lasso, C. A., Lewin, I., Liu, X., Lopes-Lima, M., Mageroy, J., Mlambo, M., Nakamura, K., Nakano, M., Österling, M., Pfeiffer, J., Prié, V., Paschoal, L. R. P., Riccardi, N., Santos, R., Shumka, S., Smith, A. K., Son, M. O., Teixeira, A., Thielen, F., Torres, S., Varandas, S., Vikhrev, I. V., Wu, X., Zieritz, A., & Nogueira, J. G. (2021). The role of anthropogenic habitats in freshwater mussel conservation. *Global Change Biology, 27*(11), 2298–2314. https://doi.org/10.1111/gcb.15549

Staton, S., Metcalfe-Smith, J., & West, E. (2000). Status of the Northern riffleshell, *Epioblasma torulosa rangiana* (Bivalvia: Unionidae), in Ontario and Canada. *Canadian Field-Naturalist, 114*(2), 224–235.

Stepenuck, K., & Green, L. (2015). Individual-and community-level impacts of volunteer environmental monitoring: A synthesis of peer-reviewed literature. *Ecology and Society, 20*(3). https://doi.org/10.5751/ES-07329-200319

Taskinen, J., Urbańska, M., Ercoli, F., et al. (2020). Parasites in sympatric populations of native and invasive freshwater bivalves. *Hydrobiologia, 848*(12), 3167–3178. https://doi.org/10.1007/s10750-020-04284-0

Texas Parks and Wildlife Department. (n.d.). *Fishing licenses and packages*. Retrieved May 26, 2023, from https://tpwd.texas.gov/regulations/outdoor-annual/licenses/fishing-licenses-stamps-tags-packages/fishing-licenses-and-packages.

Tiemann, J., Dodd, H., Owens, N., & Wahl, D. (2007). Effects of lowhead dams on unionids in Fox River, Illinois. *Northeastern Naturalist, 14*(1), 125–138. https://doi.org/10.1656/1092-6194(2007)14[125:EOLDOU]2.0.CO;2

Tiquia, S. M. (2010a). Salt-adapted bacteria isolated from the Rouge River and potential for degradation of contaminants and biotechnological applications. *Environmental Technology, 31*, 967–978. https://doi.org/10.1080/09593331003706226

Tiquia, S. M. (2010b). Metabolic diversity of the heterotrophic microorganisms and potential link to pollution of the Rouge River. *Environmental Pollution, 158*, 1435–1443. https://doi.org/10.1016/j.envpol.2009.12.035

Tiquia, S. M. (2011). Extracellular hydrolytic enzyme activities of the heterotrophic microbial communities of the Rouge River: An approach to evaluate ecosystem response to urbanization. *Microbial Ecology, 62*, 679–689. https://doi.org/10.1007/s00248-011-9871-2

Tiquia, S. M., Schleibak, M., Schlaff, J., Floyd, C., Benipal, B., Zakhem, E., & Murray, K. S. (2008). Microbial community profiling and characterization of some heterotrophic bacterial isolates from river waters and shallow groundwater wells along the Rouge River, southeast Michigan. *Environmental Technology, 29*(6), 651–663. https://doi.org/10.1080/09593330801986998

Tiquia-Arashiro, S. M. (2012). *Molecular biological technologies for ocean sensing. Springer protocols handbooks* (p. 307). Humana. https://doi.org/10.1007/978-1-61779-915-0

Tolin, A. (2023). *Citizen science monitoring of freshwater mussels (Unionidae) in the Huron River Watershed, Michigan. Master's thesis*, University of Michigan.

U.S. Fish and Wildlife Service. (2006). *Riparian buffers – Management recommendations for site-specific water quality protection and restoration planning in waters supporting federally-listed aquatic species*. Retrieved May 26, 2023, from https://downloads.regulations.gov/FWS-R4-ES-2017-0017-0013/content.pdf

U.S. Fish and Wildlife Service. (2020). Species profile for round hickorynut (*Obovaria subrotunda*). Environmental Conservation Online System. Retrieved May 26, 2023, from https://ecos.fws.gov/ecp/species/9879.

University of Michigan LSA Museum of Zoology. (2023). *Mollusk collection*. Retrieved May 26, 2023, from https://lsa.umich.edu/ummz/mollusks/collections.html

Van der Schalie, H. (1938). *The naiad fauna of the Huron River, in southeastern Michigan*. University of Michigan Museum of Zoology. https://deepblue.lib.umich.edu/bitstream/handle/2027.42/56285/MP040.pdf?sequence=1

VanTassel, N., Morris, T., Wilson, C., & Zanatta, D. (2021). Genetic diversity maintained in comparison of captive-propagated and wild populations of *Lampsilis fasciola* and *Ptychobranchus fasciolaris* (Bivalvia: Unionidae). *Canadian Journal of Fisheries and Aquatic Sciences, 78*(9), 1312–1320. https://doi.org/10.1139/cjfas-2020-0373

Vaughn, C., & Spooner, D. (2006). Unionid mussels influence macroinvertebrate assemblage structure in streams. *Journal of the North American Benthological Society, 25*(3), 691–700. https://doi.org/10.1899/0887-3593(2006)25[691:UMIMAS]2.0.CO;2

Viarengo, A., Lowe, D., Bolognesi, C., Fabbri, E., & Koehler, A. (2007). The use of biomarkers in biomonitoring: A 2-tier approach assessing the level of pollutant-induced stress syndrome in sentinel organisms. *Comparative Biochemistry and Physiology Part C: Toxicology & Pharmacology, 146*(3), 281–300. https://doi.org/10.1016/j.cbpc.2007.04.011

Weingzinger, J., & Kitchel, L. (2018). *Wisconsin Mussel Monitoring Program training manual*. Wisconsin Aquatic and Terrestrial Resources Inventory. Retrieved May 26, 2023, from https://wiatri.net/inventory/mussels/Monitoring/pdf/MusselTrainingManual.pdf

Williams, M. (2023). *Endangered and threatened wildlife and plants; Threatened species status with section 4(d) rule for longsolid and round hickorynut and designation of critical habitat*. U.S. Fish and Wildlife Service. Retrieved May 26, 2023, from https://www.federalregister.gov/documents/2023/03/09/2023-03998/endangered-and-threatened-wildlife-and-plants-threatened-species-status-with-section-4d-rule-for

Woodward, G., Perkins, D. M., & Brown, L. E. (2010). Climate change and freshwater ecosystems: Impacts across multiple levels of organization. *Philosophical Transactions of the Royal Society of London. Series B, Biological Sciences, 365*(1549), 2093–2106. https://doi.org/10.1098/rstb.2010.0055

Woolnough, D. (n.d.). *Woolnough lab research*. Woolnough Lab. Retrieved May 26, 2023, from https://daelynwoolnough.weebly.com/research.html

Zanatta, D., & Woolnough, D. (2011). Confirmation of *Obovaria olivaria*, Hickorynut Mussel (Bivalvia: Unionidae), in the Mississagi River, Ontario, Canada. *Northeastern Naturalist, 18*(1), 1–6. https://doi.org/10.1656/045.018.0101

Zanatta, D., Mackie, G., Metcalfe-Smith, J., & Woolnough, D. (2002). A refuge for native freshwater mussels (Bivalvia: Unionidae) from impacts of the exotic zebra mussel (*Dreissena polymorpha*) in Lake St. Clair. *Journal of Great Lakes Research, 28*(3), 479–489.

Zanatta, D., Fraley, S., & Murphy, R. (2007). Population structure and mantle display polymorphisms in the wavy-rayed lampmussel, *Lampsilis fasciola* (Bivalvia: Unionidae). *Canadian Journal of Zoology, 85*(11), 1169–1181. https://doi.org/10.1139/Z07-089

Zhang, L. J., Qian, L., Ding, L. Y., Wang, L., Wong, M. H., & Tao, H. C. (2021). Ecological and toxicological assessments of anthropogenic contaminants based on environmental metabolomics. *Environmental Science and Ecotechnology, 5*, 100081. https://doi.org/10.1016/j.ese.2021.100081

Zieritz, A., & Aldridge, D. (2011). Sexual, habitat-constrained and parasite-induced dimorphism in the shell of a freshwater mussel (*Anodonta anatina*, Unionidae). *Journal of Morphology, 272*(11), 1365–1375. https://doi.org/10.1002/jmor.10990

Index

A

B

C

S. M. Tiquia-Arashiro et al. (eds.), *Urban Watershed Microbiology, Volume 1*, https://doi.org/10.1007/978-3-031-89071-0

M

N

O

P

R

S

T

U

V

www.ingramcontent.com/pod-product-compliance
Lightning Source LLC
Chambersburg PA
CBHW071113211025
34252CB00002B/11
* 9 7 8 3 0 3 1 8 9 0 7 0 3 *